Introduction to Organic and Biological Chemistry

An Electron Micrograph of Collagen

This photograph is a very enlarged view of some fibers of the protein collagen, which serves as a tough, flexible structural material in such tissues as tendons and skin. You could almost guess its function from its ropelike appearance. Now look more closely. Each strand is banded, or "striated." The bands are the result of the lining up of the individual overlapping collagen molecules. Each molecule is four bands long.

In this one photo we see both chemical features and biological function. It is my hope in this text to continue toward that goal—to bring chemistry and biological function together.

Photo by Glenn K. Walker

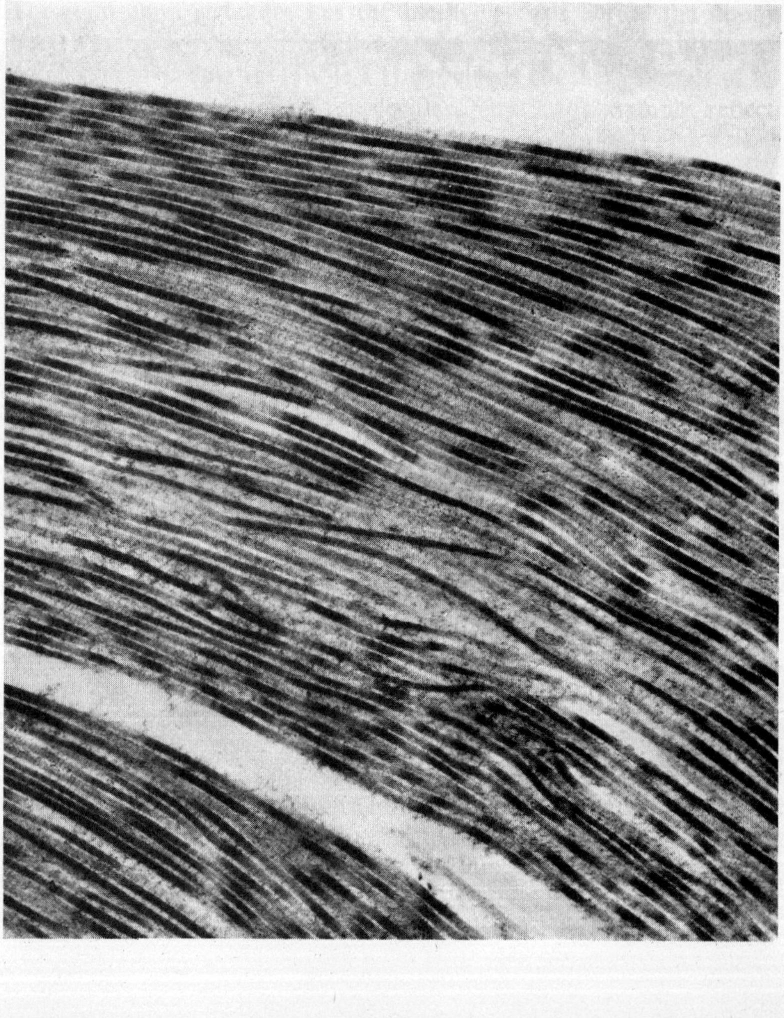

Introduction to Organic and Biological Chemistry

Ronald M. Scott

Eastern Michigan University

Harper & Row, Publishers
San Francisco

Cambridge
Hagerstown
New York
Philadelphia

London
Mexico City
São Paulo
Sydney

1817

Our cover photograph is a photomicrograph of cyclohexanone oxime under polarized light with a magnification of 40X. (Photo by Manfred P. Kage, Peter Arnold, Inc.)

Sponsoring Editor: Malvina Wasserman
Project Editor: Molly Scully
Copy Editor: Karen Judd
Production Manager: Marian Hartsough
Designer: Marjorie Spiegelman
Illustrator: Syntax; Larry Jansen
Photo Researcher: Myra Schachne
Cover Artist: Dare Porter
Compositor: Syntax International
Printer & Binder: R. R. Donnelley & Sons

INTRODUCTION TO ORGANIC AND BIOLOGICAL CHEMISTRY

Library of Congress Cataloging in Publication Data

Scott, Ronald McLean, 1933–
Introduction to organic and biological chemistry.

1. Chemistry, Organic. 2. Biological chemistry.
I. Title. [DNLM: 1. Chemistry, Organic.
2. Biochemistry. QD251.2 S428i]
QD251.2.S38 547 79-21590
ISBN 0-06-388450-X

Dedication

To my wife, for her support and help, and to my professional colleagues who have made many contributions to this project.

Contents

Chapter 3 Unsaturated Hydrocarbons 42

Focus: Double and triple bonds, terpenes, polymers. To the inert alkane structures we add the reactivity of double and triple carbon-to-carbon bonds, producing alkenes and alkynes, respectively. Reactivity is due to the addition of pi electrons to the sigma electrons of single bonds. Geometric isomers can result from the simple rearrangement of the groups attached to a double bond. Terpenes are biologically important molecules that are often alkenes. Specific examples of terpene functions include the perfume of flowers, the chemistry of vision, pheromones (compounds by which insects communicate), and insect juvenile hormone. The addition of small molecules to double bonds is illustrated. Some biochemical examples of addition reactions are presented, and enzymes, the protein catalysts of these reactions, are introduced. The formation of long chains called polymers by addition reactions results in many of the materials we commonly call plastics.

Chapter 4 Conjugated Double Bonds 68

Focus: Special chemistry of conjugated double bonds. When a structure contains a conjugated system of double and single bonds, addition reactions occur across the entire system. Natural and synthetic rubbers are examples of polymers formed by this type of addition. Benzene, a six-member ring containing three conjugated double bonds, is the simplest aromatic structure. Such structures are named and some of their substitution reactions are described. The relationship between chemical structure, especially conjugated double bonds, and color is discussed.

Chapter 5 Alcohols, Ethers, and Thiols 86

Focus: Oxygen and sulfur single-bonded functional groups. Compounds containing single-bonded oxygen and sulfur functional groups are named, their physical properties are assessed, and their substitution, elimination, and oxidation reactions are presented. We consider medicinal and biochemical examples. Stereoisomerism is explained and its importance in biochemistry is discussed.

*Optional sections.

Chapter 6 Aldehydes, Ketones, and Carbohydrates *116*

Focus: Aldehydes, ketones, and carbohydrates. The nomenclature and reactions of the carbonyl functional groups, aldehydes and ketones, are introduced. Carbohydrates, much of whose chemistry depends on carbonyl groups, are discussed in detail from the chemical, biological, and commercial points of view.

Chapter 7 Carboxylic Acids *154*

Focus: Carboxylic acids and lipids. The carboxyl functional group introduces us to the acid-base chemistry of organic compounds. We name carboxylic acids, then prepare and react esters, amides, acid chlorides, and acid anhydrides. Condensation polymers are prepared and discussed. We introduce the chemistry and functions of simple and complex lipids, waxes, soaps and detergents, and prostaglandins. Cell membranes are revealed as complex lipid structures. Special sections on flavor-creating compounds and medicinal derivatives of carboxylic acids are included.

Chapter 8 Steroids *200*

Focus: Steroids and hormone function. We look at steroid structure in general, then focus on such examples as cholesterol, vitamin D, bile salts, and hormones. Hormone function is presented in detail, illustrating how hormones provide regulation.

Chapter 9 Amines and Amino Acids 218

Focus: Amines and amino acids. Amines, their acid-base chemistry, and their reactions are presented. We consider the structure of naturally occurring amino acids and their roles in nutrition and protein structure. The relationship between mutation and protein structure is explained. The electrical charges on proteins as a result of acid-base reactions and the utilization of this charge for purification by electrophoresis are discussed. We are introduced to quaternary amines, exemplified by detergents and catecholamine hormones.

Chapter 10 Heterocycles 242

Focus: Important heterocyclic compounds. Some compounds include a ring structure containing atoms other than carbon. The variety of examples includes alkaloids, drugs of abuse, medicinals, and dyes. The biologically important nucleosides are discussed.

Chapter 11 Protein Shape and Function 262

Focus: The structure and function of proteins. The structure and function of proteins, the most important biological molecules, are surveyed. We learn about protein folding patterns, the forces maintaining these shapes, and the character of protein binding sites. Antibodies and the transport protein hemoglobin provide examples of functional proteins.

Chapter 12 Enzymes 284

Focus: Enzyme function. Enzymes catalyze the chemical changes in living systems. The enzyme active site selectively binds molecules that are then specifically altered. The chemical alteration involves functional groups of the enzyme possibly with the help of metal ions or the organic cofactor structures. Carbohydrate and protein digestion and blood clotting exemplify enzymes at work. The control provided by allosteric enzymes and isozymes is presented as part of an overview of metabolism. We consider the future of enzymes used as industrial catalysts.

Chapter 13 Nucleic Acids: Genes and Protein Synthesis 324

Focus: Genetic information and its translation into protein structure. We learn the structure and genetic function of DNA. The genetic message is preserved when synthesizing new DNA or RNA. The message finally is translated into the amino acid sequences of proteins. Virus structure is introduced as the minimum expression of this system.

Chapter 14 The Cell, Cellular Metabolism, and Muscle Contraction 348

Focus: The cell and biological energy. Now we look at the entire cell and consider the relationship between all chemical changes and the generation, storage, and use of biochemical energy. Muscle contraction is presented as an energy utilizing system.

Chapter 15 Carbohydrate Metabolism 362

Focus: The use of hexoses by cells. The chemistry of hexose metabolism is discussed in detail. We learn how blood glucose levels are stabilized and what the disease diabetes mellitus involves.

Chapter 16 Fatty Acid Oxidation 386

Focus: Fats as fuels. We consider the digestion, transportation, storage, and metabolism of fats.

Chapter 17 The Krebs Cycle 400

Focus: The formation of CO_2 in the oxidation of foods. The portion of metabolism between carbohydrate and fat breakdown to acetyl groups and the formation of CO_2 are presented.

Chapter 18 Oxidative Phosphorylation 408

Focus: Cofactor oxidation and ATP formation. The final process in biological energy capture is the reoxidation by oxygen of all the cofactors reduced in metabolism. This reoxidation is accompanied by the addition of phosphate to ADP, producing ATP.

Chapter 19 Amino Acid Metabolism 418

Focus: Amino acids as fuels. The use of amino acids, the third major fuel in foods, requires a pathway to eliminate nitrogen. We study a few of the several individual pathways by which amino acids are completely oxidized. Examples of genetic diseases are presented.

Chapter 20 Biosynthesis and Photosynthesis 432

Focus: Pathways that build structures. We observe the pathways that synthesize hexoses and fatty acids. The ultimate biosynthetic process, photosynthesis, is outlined.

Chapter 21 Membranes and Transport 448

Focus: Transport across membranes. We learn about the movement of substances across membranes. The transmission of nerve impulses along nerve fibers and across synapses and the initiation of muscle contraction provide important examples of this process.

Chapter 22 Body Fluids 464

Focus: The composition of blood and urine. We consider water as the biological solvent and look in detail at the composition of blood and urine. Urine formation and the importance to this process of transport in the kidney is stressed. Normal and pathological states are discussed.

Preface

Every scientist is enthralled with his or her special field and seldom is heard to say, "My field is of only average interest and importance." A good case can be built, however, for stating that the place where chemistry and biology overlap has produced more exciting new knowledge and has answered more "Why-is-the-grass-green?"-types of questions—the questions that led people to study science in the first place—than any other field in recent decades. In this book I hope to probe deeply enough into the biology–chemistry interface to allow the student to sense and perhaps share some of the excitement.

To use this text a student should have already been introduced to the basics of chemistry. One semester or one term of a general chemistry survey should be sufficient.

Covalent bonding and organic structures are the opening topics, and by the end of Chapter 9 I have introduced the common functional groups. Discussions of biochemically important families of compounds, carbohydrates, lipids, steroids, proteins, and terpenes, are integrated into the presentation of basic organic chemistry. These, after all, are organic compounds undergoing predictable chemical changes. Often students feel that a carbohydrate is magically more than a polyhydroxyaldehyde or ketone because it isn't discussed until biochemistry is being taught. Such integration also serves to compact the information a bit, perhaps allowing topics not ordinarily discussed to be reached in that always inadequate amount of time allotted to a class. Students for whom the objective of the class is to learn some biochemistry should be more inclined to learn the rudiments of organic chemistry when early in the course they encounter biochemical examples.

By carefully selecting the organic chemistry to be included, and by covering the families of biochemicals early, room has been cleared to include a range of topics normally encountered only in pure biochemistry courses beyond the introductory level. I have taken care to lay the necessary theoretical foundation in a simplified fashion in hopes that students can understand rather than just memorize biochemical processes such as metabolism.

The conscientious instructor tries to build the course so as to stimulate the interests and serve the needs of the specific group of students. To build some potential for tailoring the class to particular interests, and to allow shrinking or enlarging the material to match the time available, sections of several chapters have been designated as optional topics. These may be omitted without leaving out background necessary for later chapters. Obviously, instructors have been editing out material since the first class was held, but such designation reassures the student that the instructor's omission is not going to haunt the student later in the course.

Great care has been taken with the preparation of exercises. On the premise that students learn material in a text more efficiently if they can challenge their understanding of a topic immediately after they have read it, straightforward, relatively easy exercises are inserted throughout each chapter. All of these exercises have answers in the back of the book, allowing the student to self-test his or her understanding. Self-teaching is also possible for students interested in pursuing a topic in more depth than the text provides. Sections called "Challenge Problems" introduce totally new material. Explanation accompanies the introduction of the new idea, and questions test the student's understanding in a fashion not unlike programmed learning.

Several study aids are incorporated into the book. Each chapter has an introductory statement of goals, a summary, and a range of problems. In the back of the book are a glossary of terms and a glossary of reactions. Additional material is available in a study guide.

Acknowledgments

I wish to express appreciation to the reviewers who contributed important ideas and suggested needed revisions. These include Professors Nancy Adin, California State University at Los Angeles; W. Robert Coley, Montgomery College; Duane W. Fish, West Valley College; Lester Gorelic, Wayne State University; Joe J. Lehman, Colorado State University; Robert N. Lindquist, San Francisco State University; and David Newton, Salem State College.

The chemistry editor of Harper & Row, Malvina Wasserman, along with the Harper & Row staff, provided strong support and encouragement.

I wish also to recognize my academic colleagues who generously contributed illustrations. Glenn Walker, of the Biology

Department of Eastern Michigan University, supplied and interpreted the electron micrographs and provided important suggestions, and Richard Oltmann, of the Media Services Department of the EMU library, and Catherine Feryok, a student assistant in that department, provided photographs.

Finally, my thanks go to all who worked on the manuscript preparation, including Madeen Pinnock and Judy Neinas of the Chemistry Department secretarial staff, and Linda Watkins and Janice McWilliams, chemistry students. I wish most of all to thank my wife, who not only had to put up with the disruptions caused by a husband writing a book, but also did most of the typing of the manuscript, provided endless corrections and suggestions, and did much painstaking proofreading.

R. M. S.

Introduction to Organic and Biological Chemistry

1
Chemical Bonding

Objectives

In this chapter we:

1. Define organic chemistry and relate it to living systems.
2. Define covalent single, double, and triple bonds.
3. Draw Lewis structures of organic compounds.
4. Observe the angles between groups attached to carbon.
5. Consider the character of polar covalent bonds and their effect on the properties of a compound.
6. Learn about hydrogen bonds and their effect on the properties of a compound.
7. Discover what functional groups are.

What is organic chemistry? As with any complex field, definitions are a bit tricky, but basically organic chemistry is the study of the covalent compounds of carbon. There are some exceptions to this generalization. For example, carbon dioxide (CO_2), carbon monoxide (CO), cyanide ion (CN^-), and carbonate ion (CO_3^{2-}) are structures we would generally term inorganic. However, more than 3 million carbon compounds are clearly organic; in fact, more than 90 percent of all the compounds chemists have made and identified are classed as organic compounds.[1]

1.1 How Does Organic Chemistry Relate to the Study of Living Systems?

As students, you are trying to understand how living systems function, particularly the living systems we call human beings. We can ask a series of questions:

1. ***What is the composition of a living system?*** Living systems are composed almost entirely of organic compounds. Some of these compounds belong to classes or groups of organic compounds that are as familiar as the evening paper—sugars or carbohydrates, proteins, fats or lipids. The sugar in our coffee is really an example of a carbohydrate; meat is protein; the bulge round our middles is a collection of fat molecules. Others seem more mysterious. What is the DNA that has made so many headlines recently? Knowing the structures of organic compounds is essential to understanding life processes.
2. ***How are these compounds produced?*** Living organisms are incredibly clever organic chemists. In the wink of an eye they

[1] See Appendix 1.

can convert one organic compound to another using basically the same reactions we would in the laboratory. In the laboratory, however, the same process might take hours or days. To understand life we need to understand the reactions of organic compounds. ***3. How does all this chemistry result in a living being?*** As living beings, we are a bit like factories. Factories need raw materials that are readily available. Humans need food, and the substances we use for food are produced by other living organisms around us. We eat carbohydrates, such as wheat or potato starch; protein, such as animal, fish, or poultry muscle; fats, such as lard or corn oil.

Like a factory we need machinery, and our individual machines are a series of protein structures called *enzymes*. Enzymes cause the chemical changes that convert the raw materials into body parts. The food is not converted to the product in a single step; instead it moves along an assembly line of enzymes called a *metabolic pathway*, being changed a small amount by each enzyme. Because the raw material and the body parts are often chemically similar, we sometimes don't need to change the raw materials very much.

Like a factory we need energy, and we obtain this energy by burning a fuel, just as most factories do. This fuel in our case is the same as the raw material we make into new body parts. It is the carbohydrates, proteins, and fats in our diet. Again enzymes are the machinery for obtaining energy from the fuel. Just as burning coal or oil requires oxygen, so does the burning of fats, carbohydrates, and proteins. We stockpile fuel as body fat and as the carbohydrate glycogen in muscle and liver.

Our living factory can store energy, just as a car stores energy in its battery. Our energy storage takes the form of "high-energy" organic chemicals.

Our factory needs to maintain itself in good running order. Living organisms are capable of an impressive amount of self-repair, because we keep all the machinery for making parts in good working order even after we have made a complete set of parts.

No factory can operate long without a transportation system to move the products to storage, shipping, or other assembly lines. In humans the most important transportation system is the blood circulatory system. Many of the carriers in that system are proteins.

A factory needs a communication system. In humans the nervous system handles much of the communication function.

Our living factory needs management. There must be systems that sense how much is being produced by each assembly line and what the market looks like for each product. The factory needs to adapt to changes in the supply and kind of raw materials.

This management is very modern; it even has a computer. Learning about the machinery for this "coordination" or management is the object of much of the present research in biochemistry.

We must end the factory analogy at one point. Our living system can consult its blueprints, the DNA in the nucleus of the cells, and from those plans create a new organism. I have yet to see a factory that builds copies of itself—and I hope I won't.

In summary, it is not incorrect to view life as a collection of directed and controlled organic chemical reactions, with much of the direction and control also being provided by organic chemicals.

1.2 Covalent Compounds

Our definition of organic chemistry specified "covalent compounds." Recall that there are two extremes of chemical interaction between atoms, *ionic* and *covalent*. In ionic interactions, electrons are transferred completely from one atom to another; in covalent interactions, electrons are shared between two atoms.

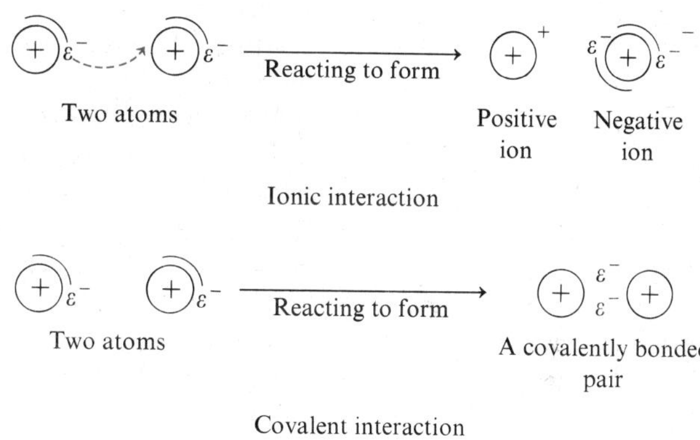

Once formed, the ions can separate; that is, they are not bound together as a molecule. In fact, in solution they do move independently. In the solid state, however, the ions are fixed into a lattice arrangement. In covalent bonding, by contrast, the atoms must remain next to one another at all times to allow each atom to claim its share of the electrons. In fact, the distances between atoms (bond lengths) and the angles formed by these bonds are fairly rigidly fixed under usual circumstances.

In both ionic and covalent interaction the atoms are manipulating electrons to obtain a more desirable electron configuration, and if they succeed, a stable compound results. Saying "a stable compound results" is another way of saying "energy is released." How stable the compound is can be determined by measuring how much energy is released (Figure 1.1) or, conversely, by measuring how much energy must be added to break (reverse the formation of) the bond (Figure 1.2).

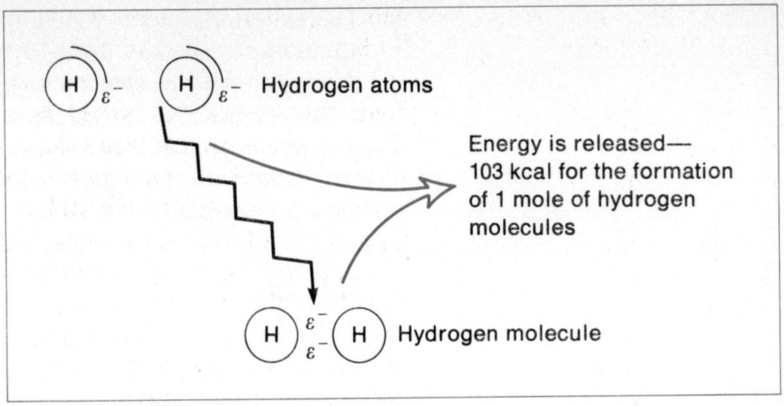

Figure 1.1
Energy and covalent bond formation
The formation of a covalent bond creates a more stable environment for the atoms. The increase in stability is measured by the amount of energy released.

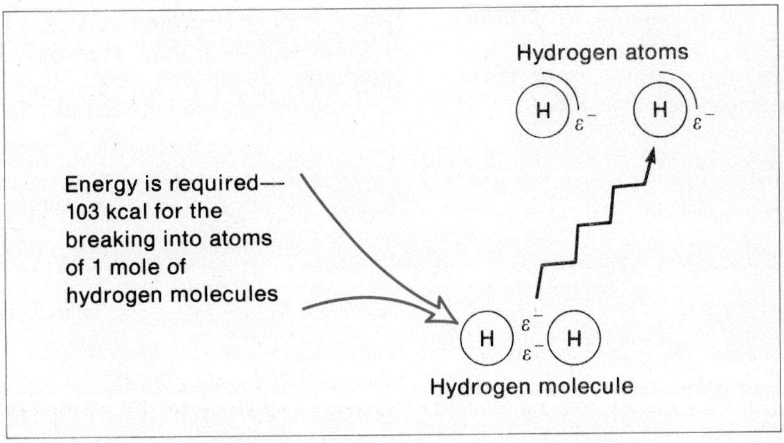

Figure 1.2
Breaking the covalent bond
The exact amount of energy released during bond formation must be added to the molecule to break it apart.

Exercise 1.1

Consider a variety of reactions where 2 moles of atoms combine to form 1 mole of covalently bonded molecules with the release of the indicated amounts of energy.

Atoms	*Molecule*	*Energy released*
2H	H_2	103 kcal
2F	F_2	38 kcal
2O	O_2	37 kcal
2N	N_2	118 kcal

a) How much energy must be added to 1 mole of fluorine molecules to convert them to individual atoms?
b) Of the molecules listed above, which requires the greatest energy to break the covalent bonds?
c) Which of the molecules listed above has the weakest covalent bond? Which has the strongest?

Examine a periodic chart and notice that carbon is in group IVA, meaning that its outer electron shell has four electrons. As a fair generalization, atoms would like to gain, lose, or share electrons to have an outer shell of eight.[2] An important exception to this rule is hydrogen, which is content with two electrons. To interact in an ionic fashion, carbon would have to gain or lose four electrons. Gaining or losing that many electrons would require that a very large charge be carried by a very small atom. This is unlikely. Therefore, carbon is most likely to offer its four outer shell electrons to be matched by and shared with other atoms. This covalent sharing results in eight electrons (an octet) surrounding carbon: four carbon electrons and four from the atoms with which the carbon is sharing. We represent this on paper by *Lewis structures*, named for G. N. Lewis, an early student of bonding. In these structures, only outer-shell electrons are shown. These are indicated as dots around the symbol for the atom. Shared electrons are drawn between the bonded atoms. For example, consider the compound methane, which consists of one carbon and four hydrogens (CH_4). Hydrogen (H) has one outer shell electron (H·), and carbon (C) has four outer shell electrons (·Ċ·, with one dot on each side).

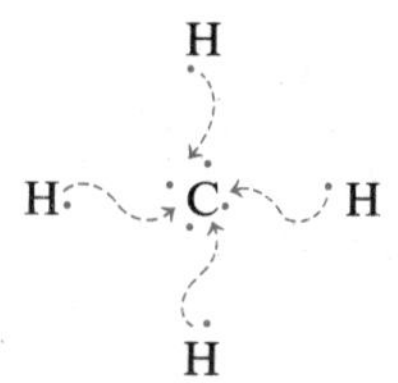

becomes

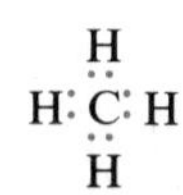

methane (the chief component of natural gas)

Notice that by sharing in this fashion carbon has an octet and each hydrogen has a pair. Don't be upset that the electrons are counted both for the carbon and for the hydrogen. That is what we mean by sharing the electrons. The pair of electrons shared between the carbon and a hydrogen constitute a covalent single bond between the two atoms. In representing the structure, the pair of electrons is often replaced by a short straight line called a *bond*.

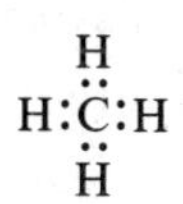

is the same as

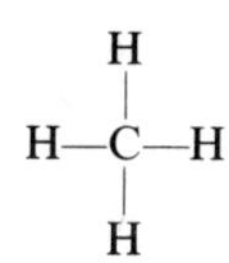

A single bond is a pair of electrons.

---

[2] Chemists commonly say that the atoms are creating the electron pattern of the inert gases at the right of the periodic chart (argon, neon, helium, etc.). However, this does not explain why those atoms are chemically inert (stable). There is a large body of theory and of experimental results that attempts to explain this stability. It is complex to understand, however, and a student can deal successfully with a surprising amount of organic chemistry just by creating outer shells of eight electrons.

Exercise 1.2

How many electrons does nitrogen have in its outer shell? How many additional electrons would it require by way of covalent sharing to be stable? Predict the structure that nitrogen forms with hydrogen by drawing the Lewis structure.

1.3 The Variety of Carbon Compounds

Why are there so many organic compounds? The answer lies in the ability of one carbon to share electrons with other carbons. Consider ethane (C_2H_6):

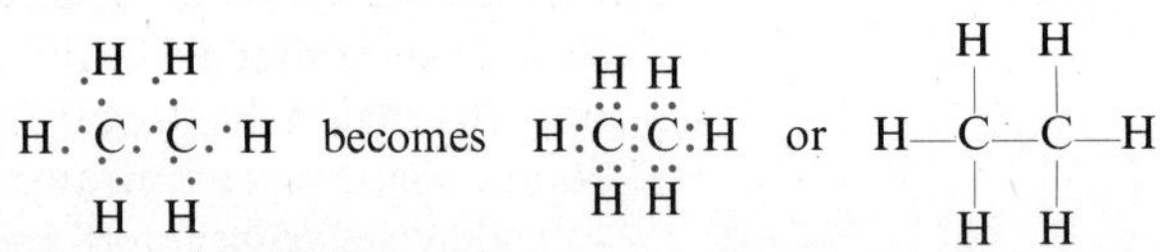

Similarly, we have propane (C_3H_8) and butane (C_4H_{10}):

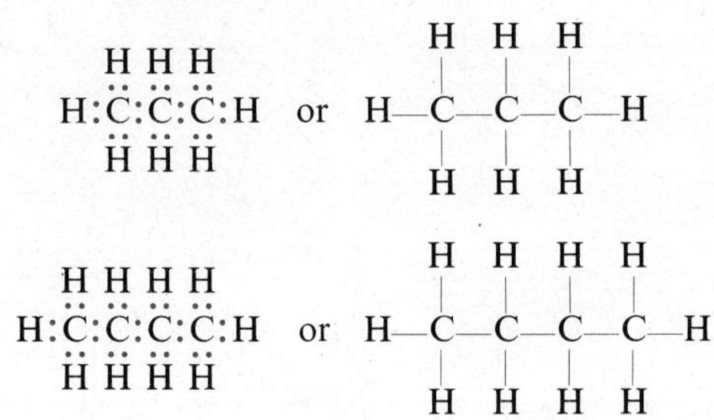

Obviously, we can generate a whole family of compounds simply by inserting additional methylene 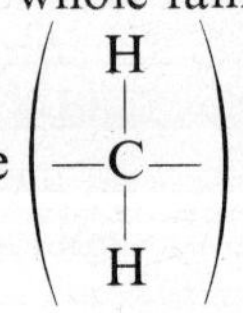 units into the chain.

Exercise 1.3

Draw a Lewis structure for pentane (C_5H_{12}).

1.4 The Shape of Organic Molecules

At first thought it may seem that the shape of something as tiny as a molecule is unimportant. However, in biological systems, molecules very often interact in useful and predictable ways because they fit together like pieces in a jigsaw puzzle. Similarly, interactions that are not desirable don't occur if the shapes of two molecules are wrong for joining together. Our machinery for immunity from disease, the immunity provided by previous exposure to the disease or by vaccination, depends on our body's producing molecules that have the correct shape to fit onto the potentially harmful agent. With this in mind, let us develop a simple understanding of the shapes of organic molecules. We are not representing organic molecules properly by writing them flat on a piece of paper. First, molecules are three-dimensional, not flat. In the compounds considered so far, the atoms are actually attached to each carbon as if that carbon were at the center of a pyramid and the attached atoms were at the corners of the pyramid. Carbon surrounded by four groups in this fashion is referred to as *tetrahedral* (Figure 1.3). The angles between the atoms attached to the carbon are not 90°, as they seem to be on the flat drawing, but somewhat larger, 109.5°. Writing flat structures of organic compounds usually distorts the true shape of the structure, but it is much easier than trying to draw and interpret pictures that represent the three-dimensional shape. Therefore we will continue drawing them flat, but at the same time we will watch out for any misunderstandings.

A second type of confusion arises by representing structures with flat drawings. It is possible to arrange a structure several ways and become convinced that each is a different compound. For example, suppose we draw C_3H_8 several different ways:

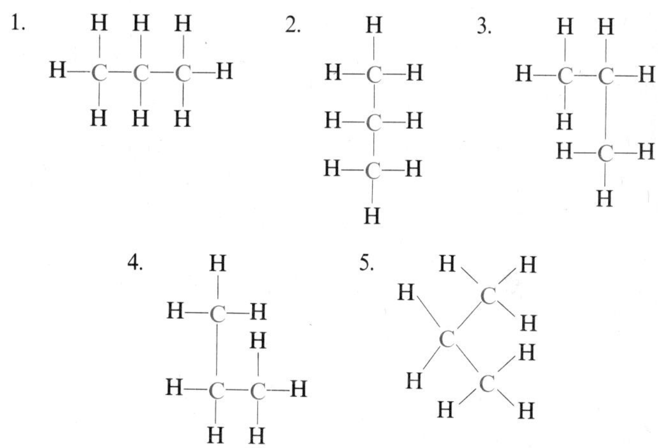

Are these all the same? Look at each carbon to see what is attached to it. In all five structures the carbons on the end each have 3 H and 1 C no matter which way we orient them on the paper, while the middle carbon has 2 H and 2 C attachments. At this point we will say that this test is sufficient to indicate that all the drawings represent the same

Figure 1.3
The tetrahedral carbon
Carbon with four covalent attachments is not flat with bond angles of 90° as shown on the left, but rather a pyramid, or tetrahedron, with bond angles of 109.5°.

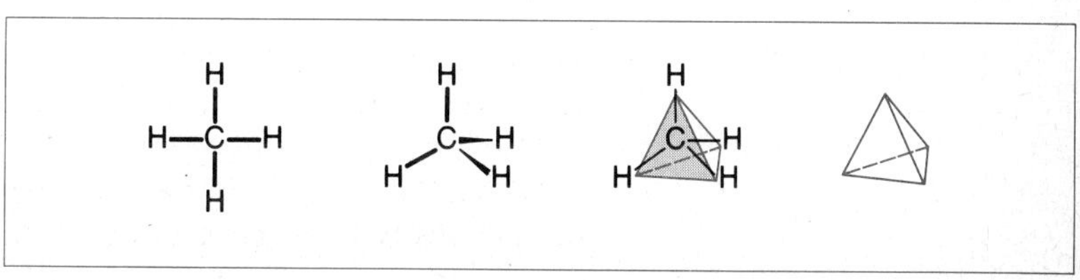

structure. Now let us try the same thing with C_4H_{10}:

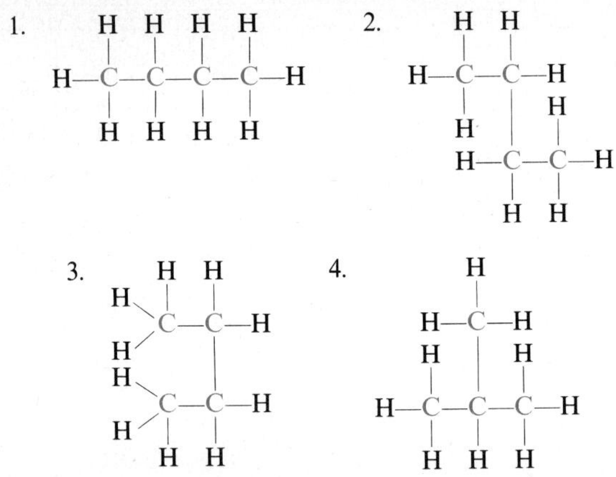

Notice that 1, 2, and 3 all meet the requirement for being the same; that is, the end carbons each attach 3 H and 1 C while the two middle carbons attach 2 H and 2 C. However, structure 4 is different; it has three carbons each bonding 3 H and 1 C, and one bonding 1 H and 3 C. Structure 4 is, in fact, a different organic compound from the other three, with different chemical and physical properties.

Exercise 1.4

You can put C_5H_{12} together in three ways. Can you find them?

1.5 *Are All Covalent Bonds Single Bonds?*

It is possible for more than two electrons to be shared. For example, if we are told that the compound C_2H_4, ethene or ethylene, is a stable covalent molecule, we could try to draw a Lewis structure of it. First connect the two carbons by a shared pair:

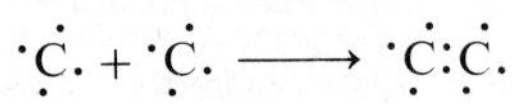

Then have each carbon share electrons with two of the four hydrogens:

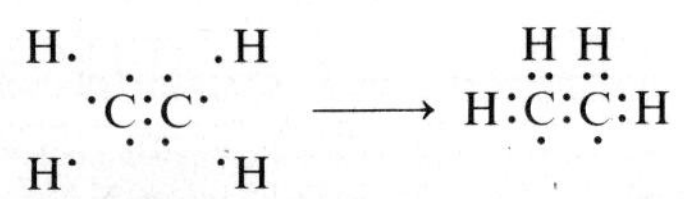

We see that the hydrogens have satisfied their need for a pair of electrons, but each carbon lacks one electron for an octet. If the carbons each donate one more electron to the bond between them, they will each have an octet:

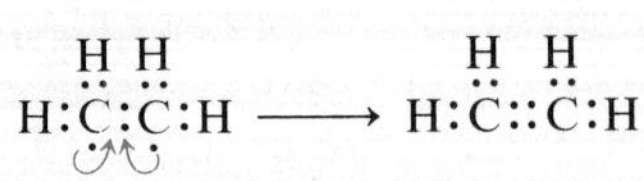

The sharing of four electrons between two atoms is called a *covalent double bond*, and such a bond can be represented by two parallel lines between the symbols for the atoms.

This time our drawing of the molecule on paper is not a bad representation of its shape. A carbon with three groups attached to it (here two hydrogens and one carbon) *is* flat, and the angles between the atoms are not 109.5° but rather 120°.

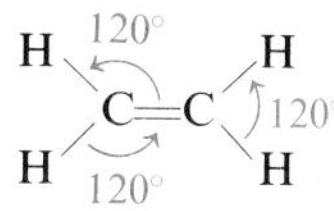

Exercise 1.5

Can carbon be connected to hydrogen by a covalent double bond?

Another stable compound that contains two carbons with hydrogen is C_2H_2, ethyne or acetylene. Using the same reasoning as was applied to the ethene example, we have the following:

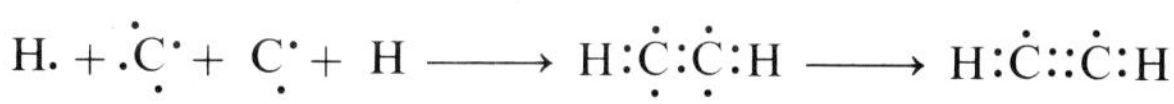

Once again we have carbons lacking one electron for an octet. The same solution as before can work—share one more pair of electrons.

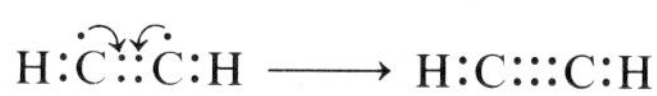

This is called a *covalent triple bond* and can be represented by three parallel straight lines:

H:C:::C:H becomes H—C≡C—H

In a triple bond the atoms are arranged in a straight line; in other words, the angle between the atoms attached to the carbon is 180°.

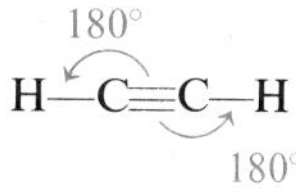

This is as far as we can go with multiple bonds. It is impossible for each of two carbons to move all four electrons to the same side of the atoms to create a quadruple bond between them.

It is possible for one compound to contain more than one kind of covalent bond; for example,

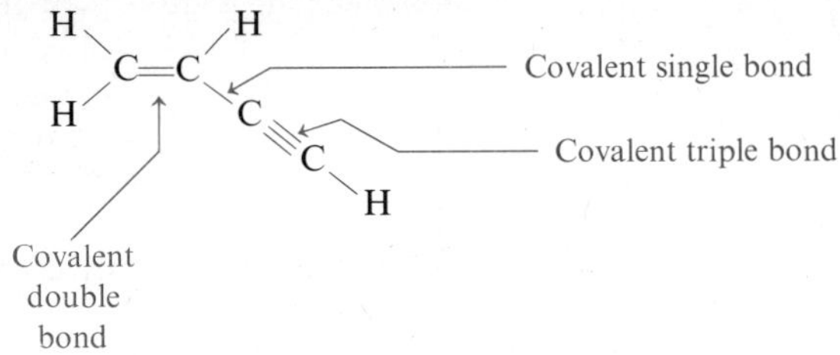

> ### *Exercise 1.6*
> Draw the Lewis structures of two compounds with four carbons in a row. In one structure join the carbons by a single bond, a triple bond, and a single bond. In the other join them by a double bond, a single bond, and a double bond. Be careful to add the correct numbers of hydrogens to each. Look at the lines that represent the shared electrons. Each carbon always has how many lines leading to it? Does it make any difference whether the bonds are single, double, or triple? How does the number of bonds relate to the number of electrons in the outer shell of carbon?

1.6 Covalent Bonds with Unequal Electron Sharing

The examples we have used thus far involve atoms that share electrons equally. When two carbons share electrons, their ability to attract the electrons is equal, so there is no reason to imagine that one of them could pull the electrons extra close, leaving the other carbon with less than its share. Usually the bond between carbon and hydrogen involves approximately equal sharing of the electrons too. Bonds in which electrons are equally shared are called *nonpolar*.

Water is a molecule in whose covalent bonds the electrons are not equally shared. With the exception of hydrogen, atoms can attract electrons more strongly (are more electronegative) the farther to the right or the nearer to the top row of the periodic chart they are placed. Oxygen is in the same row as carbon and is to the right of it; therefore we conclude that it is more electronegative.

> ### *Exercise 1.7*
> Which of the atoms in each of the following pairs is the more electronegative? C or F, Br or F, C or B

A more electronegative atom draws the electrons toward it more strongly, resulting in unequal sharing. We call this bond *polar*. At the end of the bond with the more electronegative atom, the higher electron concentration produces a weak negative charge, while the other end has an equal but positive weak charge.[3]

---

[3] It might help to think of polar bonds as intermediate cases between the totally covalent and totally ionic extremes presented on page 5. They are joined together tightly as covalent bonds are but have some of the ionic charge.

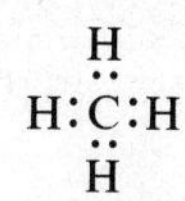

Approximately
equal sharing,
no charge

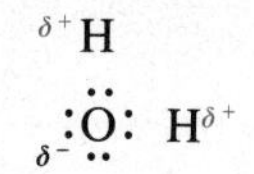

Unequal sharing,
small + charge on H,
small − charge on O

These charges are much smaller in magnitude than the charges on ions.

Organic molecules may contain polar bonds. For example, the —O—H bond shown above in water is also found attached to carbon:

methanol (wood alcohol)

A polar double bond is also found:

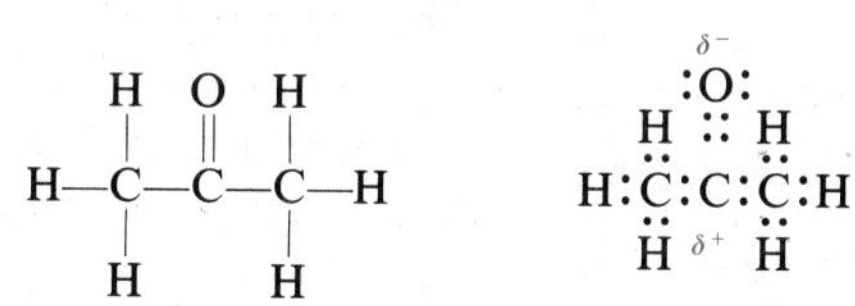

propanone (acetone)

In compounds of carbon and hydrogen, all the outer-shell electrons become part of the pattern of covalent bonds. Notice in the case of oxygen, which starts out with six electrons in the outer shell, that an octet is obtained by sharing only two of the six electrons. The other two pairs of electrons remain unshared. Although not involved in the covalent bonding, these unshared electrons contribute to the chemical properties of the molecule.

> ### *Exercise 1.8*
> Consider a covalent compound between hydrogen and chlorine. Draw a Lewis structure. How many unshared electron pairs are on the chlorine?

1.7 Hydrogen Bonds

When the very small hydrogen atom is bonded to a very electronegative atom, such as oxygen, nitrogen, or fluorine, it is possible for a second electronegative atom with unshared electrons to approach the hydrogen and attract it strongly enough that the hydrogen actually becomes

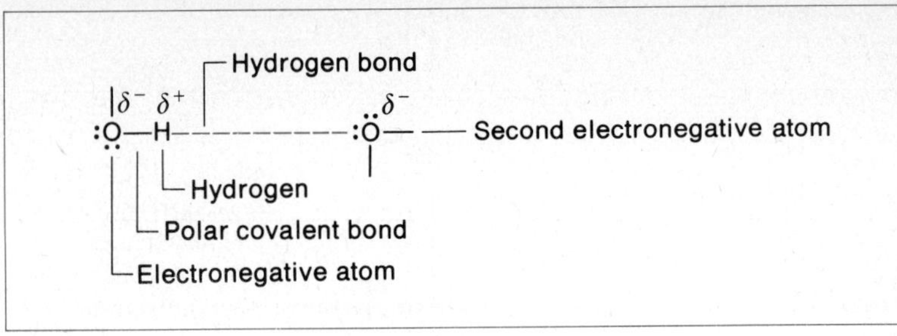

Figure 1.4
The hydrogen bond
The hydrogen bond is partly electrostatic and partly covalent. The electrostatic component is the attraction of the slightly positive hydrogen to a nearby somewhat negative O, N, or F. The covalent contribution represents the hydrogen sharing electrons most of the time with its assigned partner and the rest of the time with a convenient passerby. Hydrogen, we see, can be fickle.

weakly bonded to the second atom. The hydrogen becomes a sort of bridge between the two electronegative atoms. This bridge is called a *hydrogen bond*, and it is about one-twentieth as strong as a regular covalent bond (Figure 1.4).

1.8 Properties of Organic Molecules

First consider melting point and boiling point. A crystalline solid melts when the amount of heat energy added to the solid becomes greater than the forces holding the molecules next to each other in the crystal lattice. Similarly, a liquid boils when the heat energy added to the liquid is greater than the forces between the closely packed molecules of the liquid. All atoms and molecules have small attractive forces between them if they can approach each other closely enough. These are called *van der Waals forces.* The more closely the atoms and molecules can pack together, the greater the size of the van der Waals attractive forces. Small molecules do not have as many atoms as larger molecules to place very close to a neighboring molecule in either a crystal or a liquid, and as a result the total force for each molecule is smaller. Look at the melting points and boiling points of a small series of familiar carbon–hydrogen molecules (Table 1.1) to verify this fact. The shape of a molecule has an effect on the number of identical molecules that can approach it closely. For example, having a methyl[4] group protruding from an otherwise straight carbon chain makes it harder to pack molecules together tightly, just as it is harder to bundle branched sticks as compactly as straight sticks. This looser packing of the branched molecules reduces the van der Waals interactions, and the smaller attraction between the molecules means lower melting and boiling points (Table 1.2).

The presence of a polar covalent bond in a covalent compound has a dramatic effect on the melting and boiling points. The negative

---

[4] A methyl group is 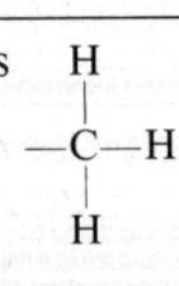

Table 1.1
Properties of Organic Molecules: Effect of Chain Length

Structure	*Name*	*Melting Point (°C)*	*Boiling Point (°C)*
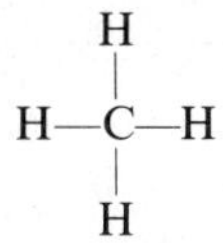	Methane	−183	−162
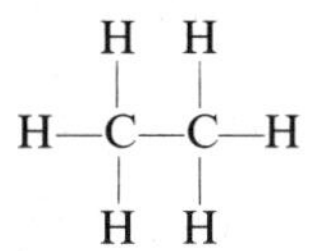	Ethane	−172	−89
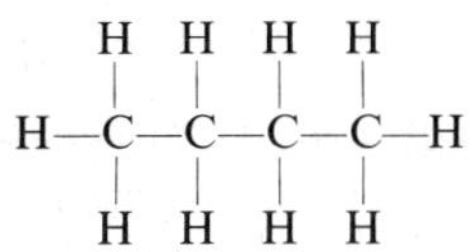	Butane	−138	0
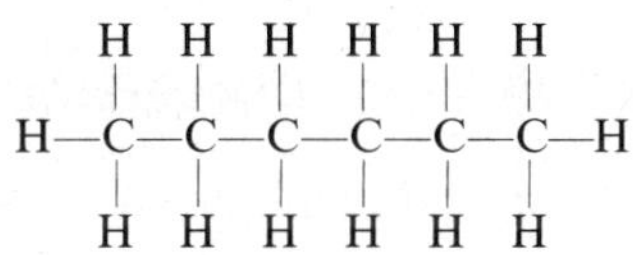	Hexane	−95	69
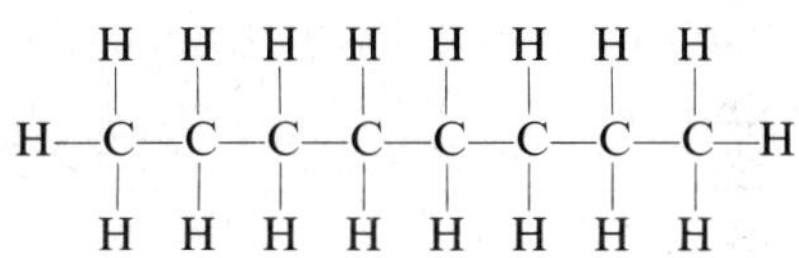	Octane	−57	126
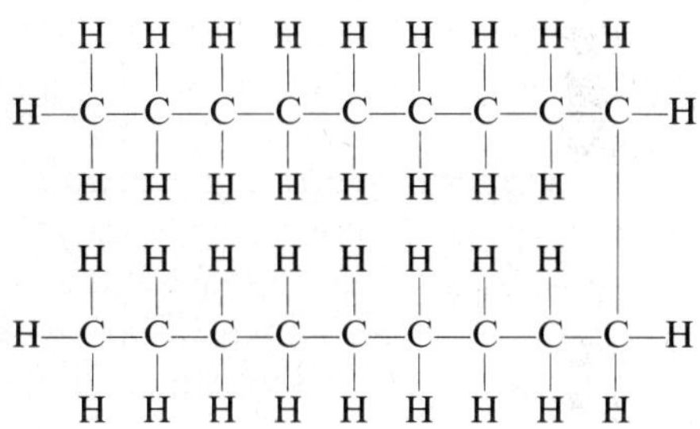	Octadecane	28	316

Table 1.2
Properties of Organic Molecules: Effect of Branching

Structure	*Name*	*Melting Point (°C)*	*Boiling Point (°C)*
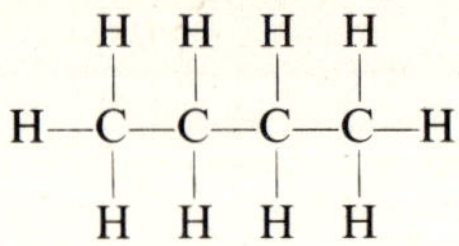	Butane	−138	0
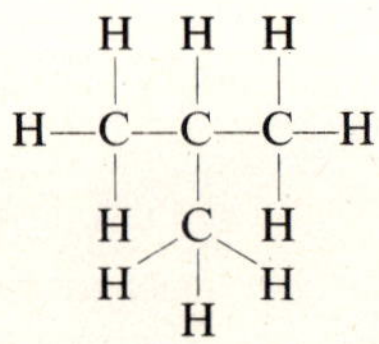	Isobutane	−160	−12

Table 1.3
Properties of Organic Molecules: Effect of Polar Bonds

	Structure	*Name*	*Melting Point (°C)*	*Boiling Point (°C)*	*Solubility in Water*
Nonpolar Bond	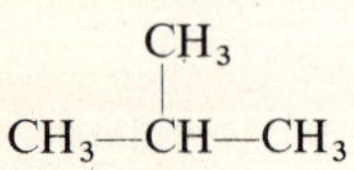	Isobutane	−160	−12	Insoluble
Polar Bond	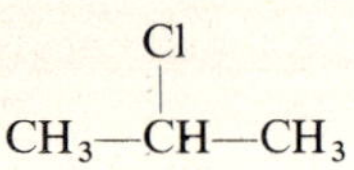	2-Chloropropane	−117	37	0.31 g per 100 g
Very Polar Bond	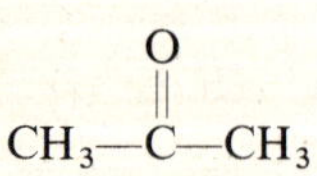	Acetone	−95	57	Completely soluble
Very Polar Bond and Hydrogen Bond	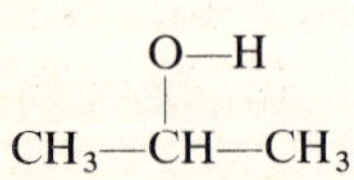	2-Propanol	−86	83	Completely soluble

Table 1.4
Properties of Organic Molecules: Effect of Hydrogen Bonding on Properties

Compound	*Molecular Weight*	*Melting Point (°C)*	*Boiling Point (°C)*
CH_4	16	−182.5	−161.5
NH_3	17	−77.7	−33.4
H_2O	18	0	100.0
HF	20	−83.1	19.5

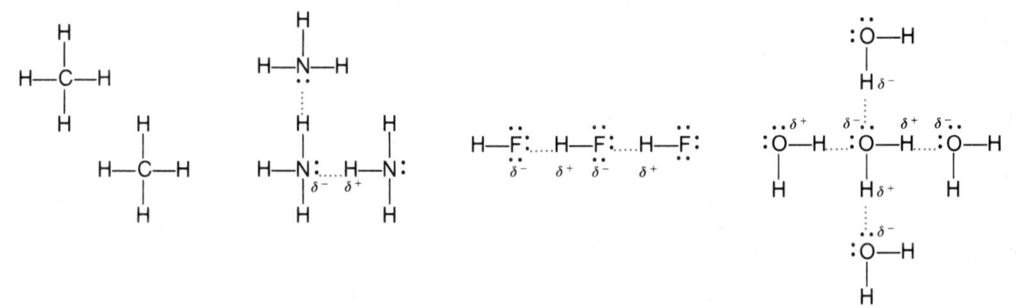

Figure 1.5
The hydrogen bonding patterns of some simple molecules
The possibilities range from CH_4, which does not hydrogen bond because there are no unshared electrons and carbon is not sufficiently electronegative, through NH_3 and HF, which have either one hydrogen and three unshared pairs of electrons or vice versa, to the optimum case of water, with two hydrogens and two unshared pairs of electrons per molecule.

end of the polar bond of one molecule and the positive end of the bond on a neighboring molecule attract one another, adding greatly to the heat energy required to separate them and raising the melting and boiling point sharply (Table 1.3).

The presence of hydrogen bonds provides even stronger forces between molecules. Compare the properties of CH_4, which has only nonpolar bonds, with those of NH_3 and HF. In HF and NH_3, the electronegative atoms have unshared electrons and are bonded to hydrogen, so they can form hydrogen bonds. Now consider water, which has an electronegative atom with two hydrogens and two unshared electron pairs, an optimum situation for forming the most hydrogen bonds (Table 1.4 and Figure 1.5). Forces between molecules are summarized in Figure 1.6.

If molecules associate only by van der Waals forces, there is no reason for them not to mix freely with one another. However, if we attempt to mix polar with nonpolar molecules, the strong attractive

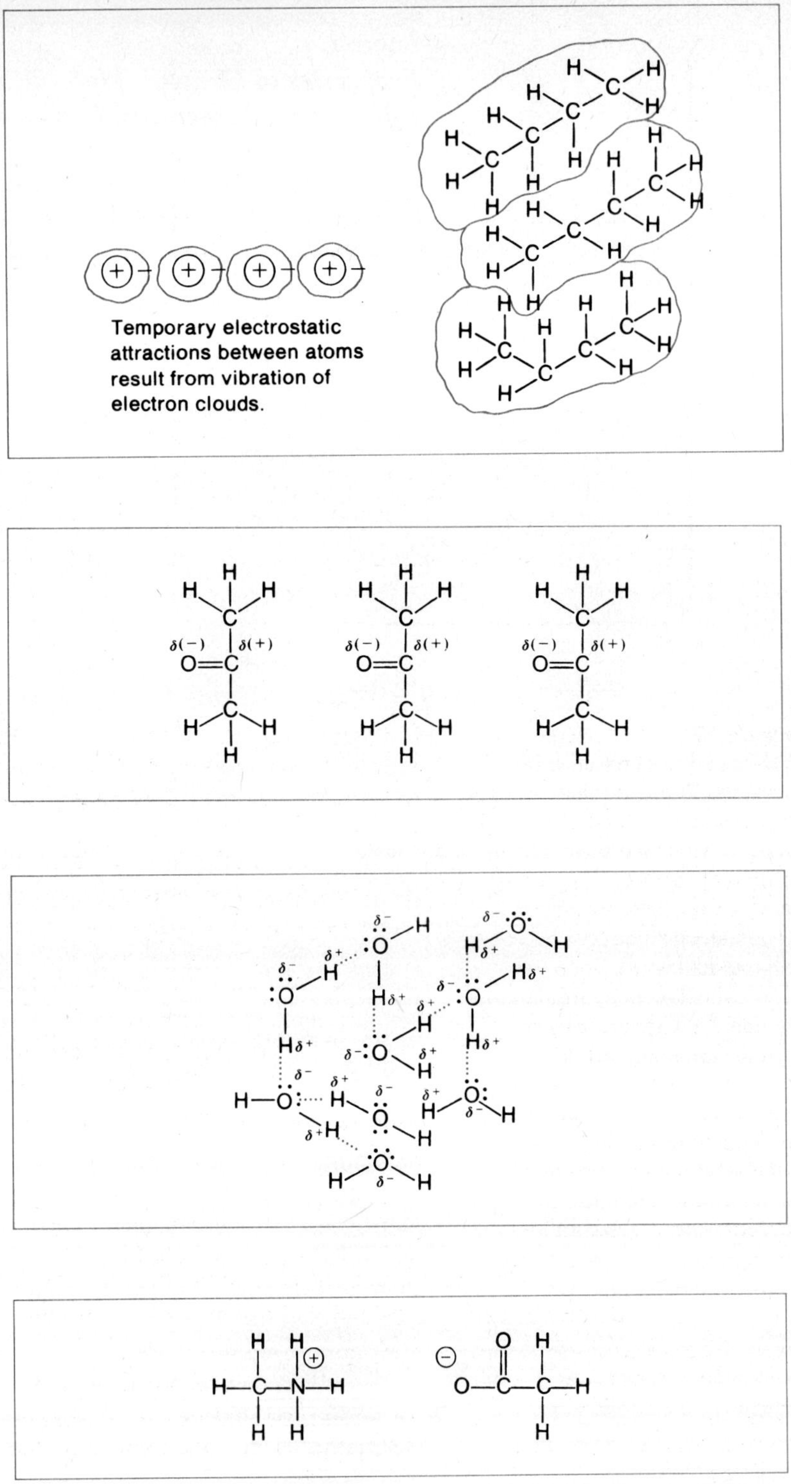

Figure 1.6
Types of forces between molecules
(a) van der Waals: very weak, because the molecules must fit closely together. (b) Charge attractions of polar covalent bonds: charges much less than full ionic charge, stronger than van der Waals attractions. (c) Hydrogen bonds: a hydrogen is shared between two electronegative atoms. It is a relatively strong interaction with the character partly of a strong polar interaction and partly of a covalent bridge.
(d) Electrostatic attraction: fully charged ions leave strong attractive forces. Some organic structures can carry a full charge.

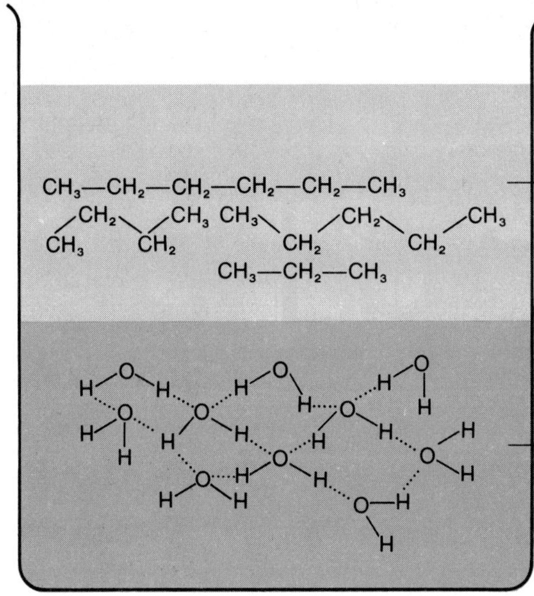

Figure 1.7
The layering of oil on water
The less dense oil floats on top of water instead of mixing with it. Water molecules have strong attractive forces for one another and are drawn together by these forces. The oil molecules are not polar and cannot form hydrogen bonds. As a result, they do not attract water. Given a choice between associating with other water molecules or with oil molecules, the waters associate with one another and exclude the oil, which forms a separate layer.

force of the polar molecules for one another causes them to aggregate only with one another. The well-known layering of oil (very nonpolar) on top of water (very polar) is a good example of this situation (Figure 1.7).

The more polar the organic molecule is, the more likely it is to dissolve in water, a very polar solvent. This is of great importance in living organisms. Because the solvent in a living cell is water, it is necessary for the compounds that need to be dissolved to be polar. For example, proteins that must be suspended in water to function are arranged so as to have polar groups where the water touches the protein. At the same time the cell must have structural parts that are not water soluble. Often this is achieved by using largely nonpolar compounds in these insoluble structures.

1.9 Bonding and Chemical Properties

Chemical change in organic molecules involves the breaking and forming of covalent bonds. Two important simple generalizations can be made:

1. In a covalent bond, two shared electrons form the bond. In double or triple bonds, the additional pairs of electrons strengthen the bond, but they are more free to move around and are easier for other atoms or ions to reach.

2. Polar bonds can attract other polar or charged molecules and so are more likely to be involved in chemical change than nonpolar bonds are.

Reactions of covalent molecules occur most readily at bonds that are double, triple, or polar. Furthermore, a specific grouping of atoms tends to react the same way every time, a fact that greatly simplifies the study of organic chemistry. The commonplace arrangements are called *functional groups*. So much importance is attached to functional groups that names of compounds are built around them. For example, an organic molecule that includes an —OH group is called an *alcohol*.

Table 1.5
Functional Groups

Group	*Example*	*Group Name*
R—OH	CH_3—OH	Alcohol
R—O—R	CH_3—O—CH_3	Ether
R—SH	CH_3—SH	Thiol, mercaptan
R—C(R)=C(R)—R	CH_2=CH_2	Alkene
R—C≡C—R	H—C≡C—H	Alkyne
R—C(=O)—H	CH_3—C(=O)—H	Aldehyde
R—C(=O)—R	CH_3—C(=O)—CH_3	Ketone
R—C(=O)—O—H	CH_3—C(=O)—O—H	Carboxylic acid
R—C(=O)—O—R′	CH_3—C(=O)—O—CH_3	Ester
R—C(=O)—NH_2	CH_3—C(=O)—NH_2	Amide
R—NH_2	CH_3—NH_2	Amine

Note: R and R′ represent "any" covalent carbon structure.

The following are different compounds but they are all alcohols:

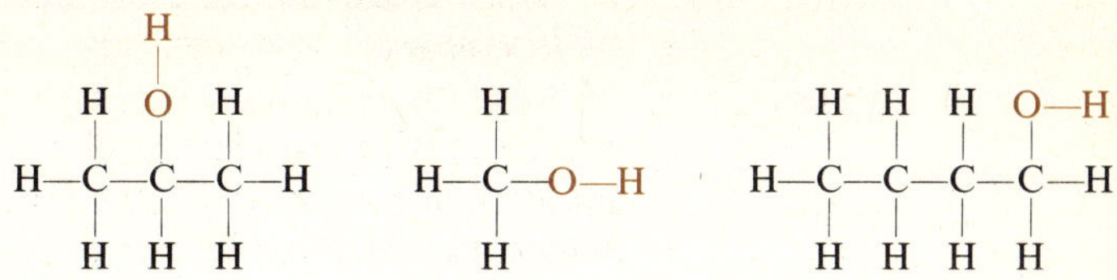

Once you identify a compound as an alcohol, you can predict much of its chemistry. For example, you can predict that it will react with a carboxylic acid if the two are mixed, and that the product will be an ester.

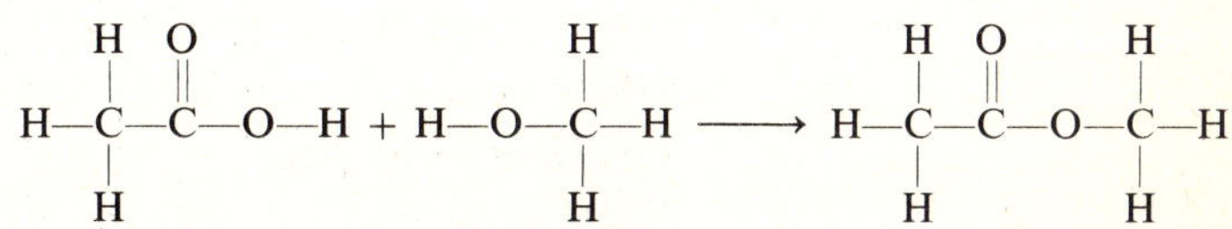

A carboxylic acid — An alcohol — An ester

Table 1.5 lists common functional groups. The next chapters will introduce the chemistry of functional groups.

Summary

An understanding of the structure and reactions of organic compounds is essential to an understanding of life processes. Most of the chemicals we recognize are organic compounds, that is, covalent compounds of carbon. Covalent and ionic are the two extremes in chemical bonding. When two atoms are covalently bonded, each atom moves toward a more stable electron pattern by sharing electrons with the other atom. To share they must stay together. In an ionic bond, one atom gives electrons to the other outright. They may then move apart.

Carbon has four outer-shell electrons and forms covalent bonds. Carbons can bond to each other to form large and complex molecules. It is necessary to describe the exact pattern of bonding because the same set of atoms may form several different arrangements of covalent bonds. The exact shape of these arrangements is known and is important to chemical and biochemical behavior. In a given bond, when two electrons are shared we have a single bond, four electrons a double bond, and six electrons a triple bond. Two carbons or a carbon and a hydrogen share the electron relatively equally, that is, form a nonpolar covalent bond. Two atoms of different electronegativities do not share equally. The more electronegative atom claims a greater share of the electrons and becomes somewhat negative while the other atom becomes positive to the same degree. This is a polar covalent bond. When hydrogen is attached to the very electronegative nitrogen, oxygen, or fluorine, and when another electronegative atom with unshared electrons approaches, the hydrogen becomes shared between the two atoms in a hydrogen bond.

Melting and boiling temperatures are higher when forces between molecules are greater. All atoms interact by van der Waals forces, and large molecules have more atoms, so they melt and boil at higher temperatures. Polar bonds and hydrogen bonds further increase the attractive forces between molecules and so increase the melting and boiling point. Solubility depends on polarity. Nonpolar molecules mix with other nonpolar molecules, whereas polar molecules mix with other polar molecules.

Chemical changes involve moving electrons, breaking bonds, and forming bonds. Double or triple bonds and polar bonds are more reactive than single bonds. Special arrangements of atoms, including double or triple bonds, polar bonds, or both, have predictable chemical behavior. They are called functional groups.

Problems

1. How many electrons does chlorine have in its outer shell? How many must it obtain to complete its outer shell? In the following cases, is the need of the chlorine for an octet satisfied by ionic or covalent means?

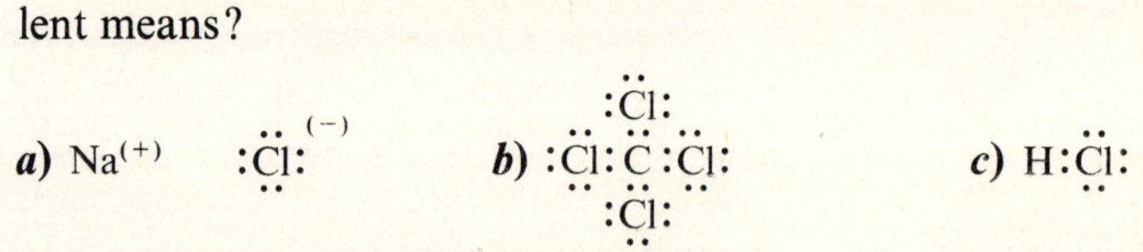

2. Add the correct numbers of hydrogens in the correct locations to the following incomplete structures:

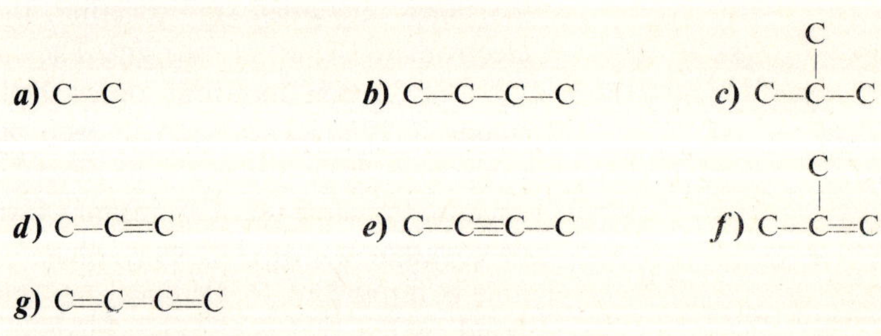

3. In the following structure, what are the true bond angles between the bonds indicated by a, b, c, and d?

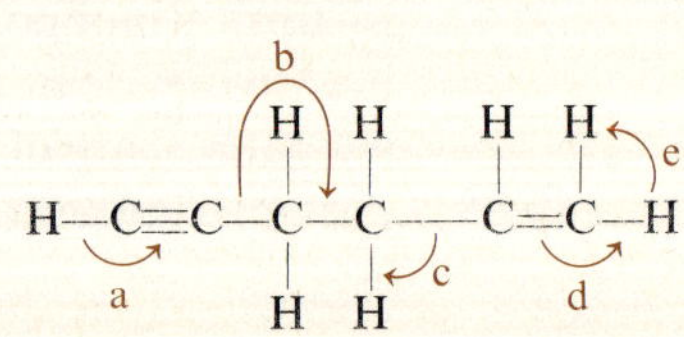

4. 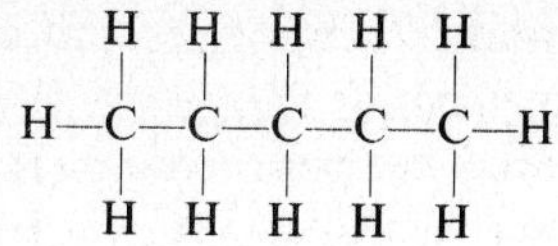 is the same as which of the following:

a) 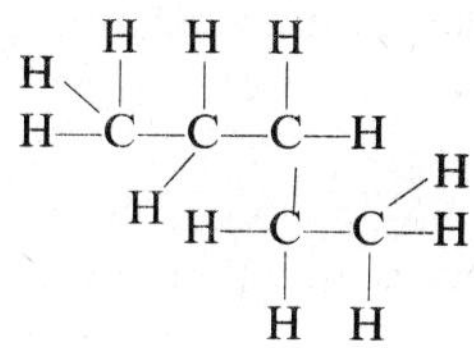

b) 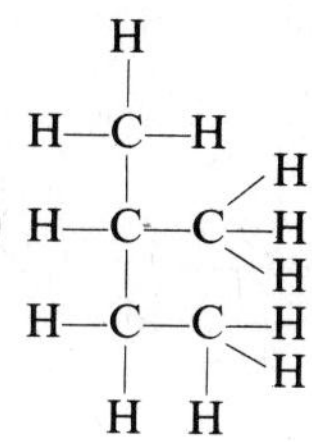

c) 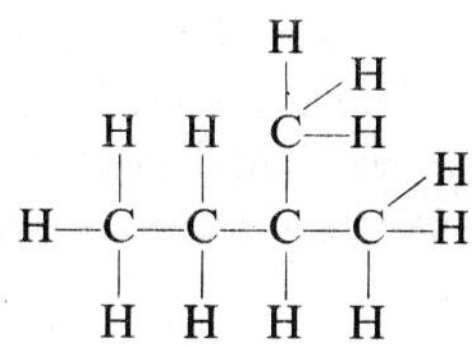

d) 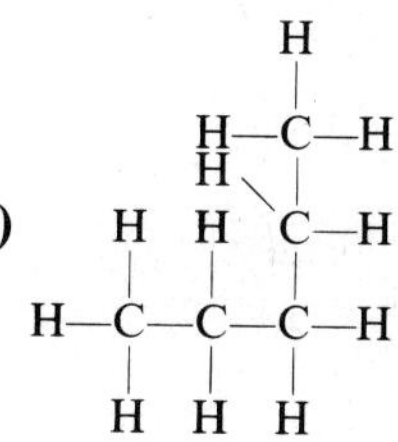

e) 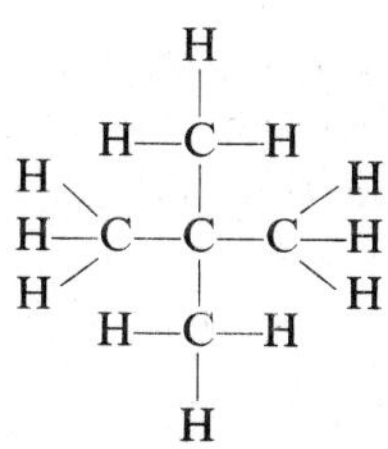

f) 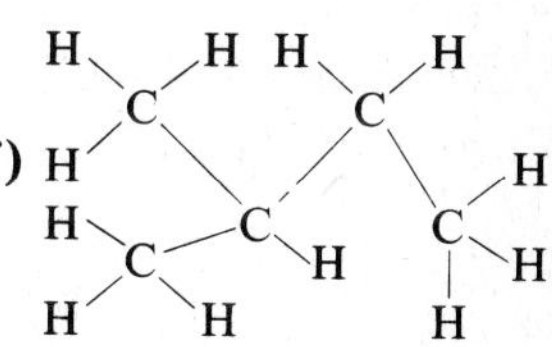

g) 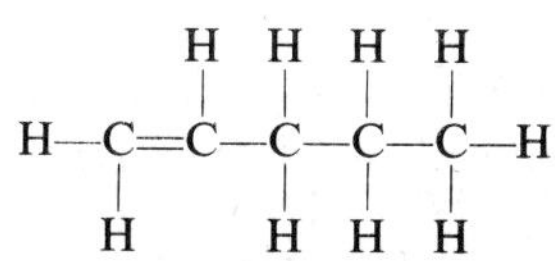

5. Which element would you expect to be the *most* electronegative: Br, As, F, B, P?

6. Arrange the following in order of probable boiling point:

a) 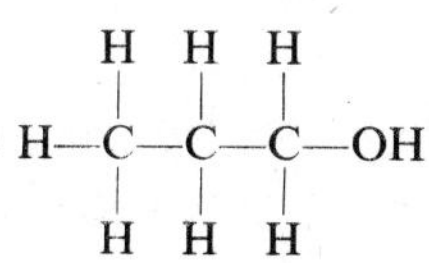 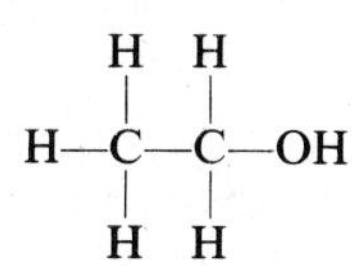

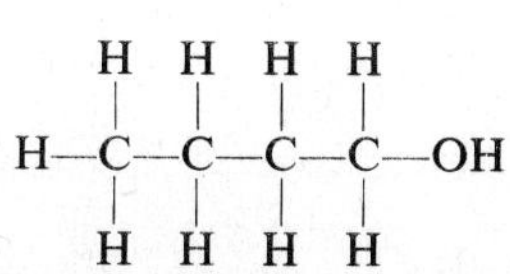 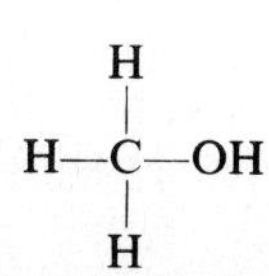

b) 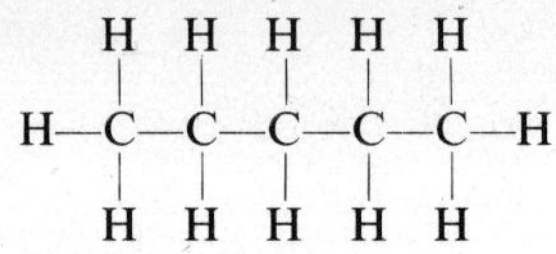

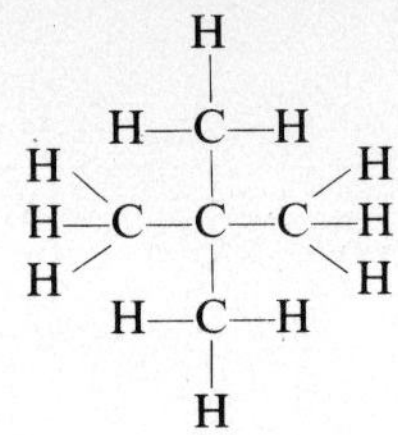

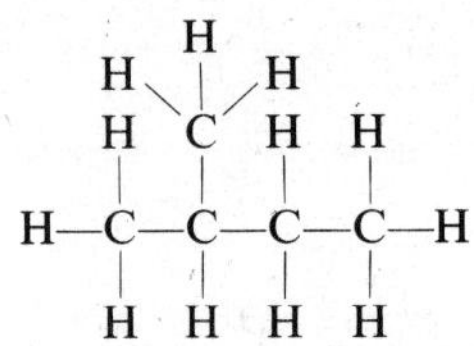

7. Draw all the different structures that have the formula C_6H_{14}.

8. Water is a very polar solvent and tends to dissolve only the most polar or charged solutes. Rank the compounds below by what you would expect their solubility in water to be according to their structure.

a) 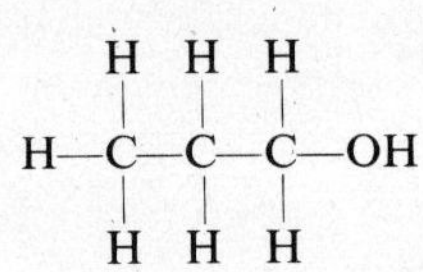

b) 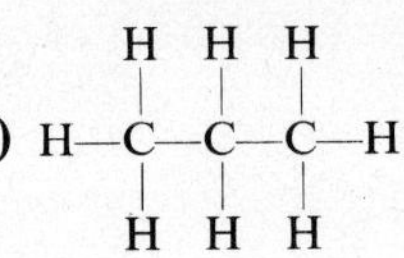

c) 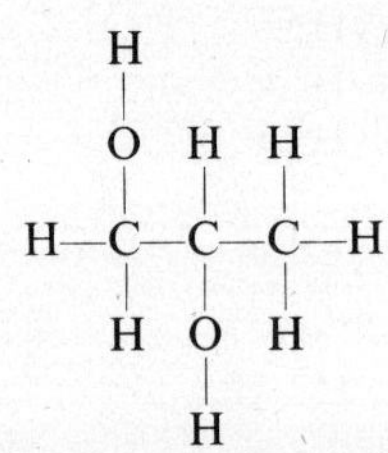

d) 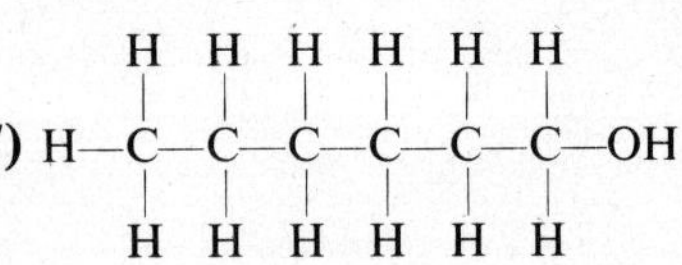

Now rank them as you would expect their solubility to be in a very nonpolar solvent, such as

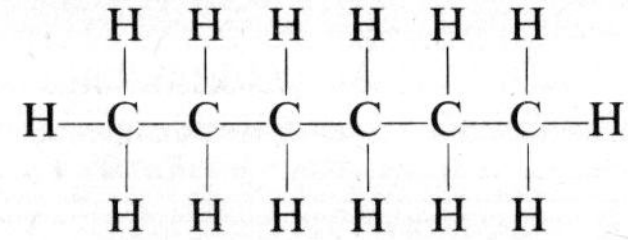

9. Classify the following by functional group:

a) 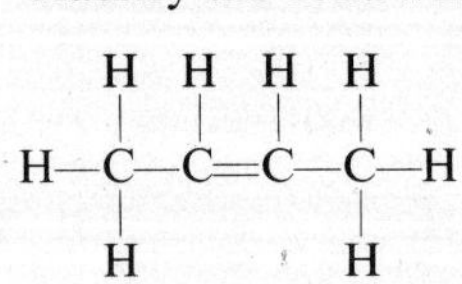

b) 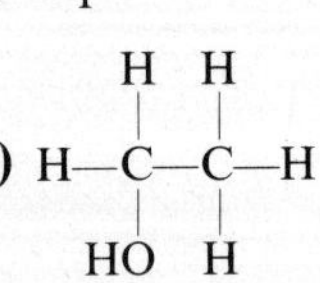

c) 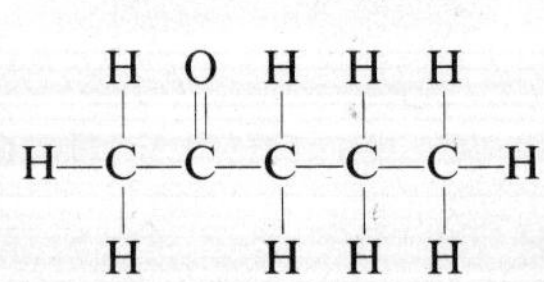

d) 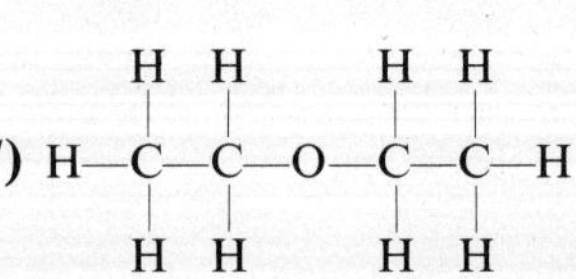

e) 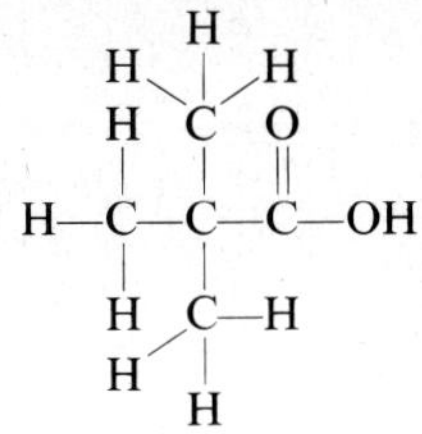

f) 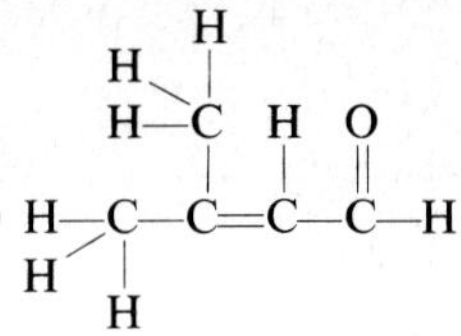

g) 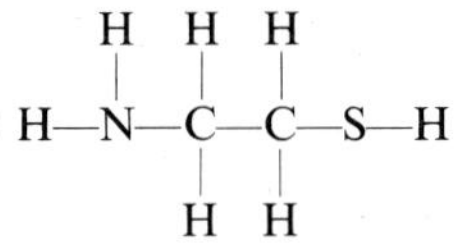

h) 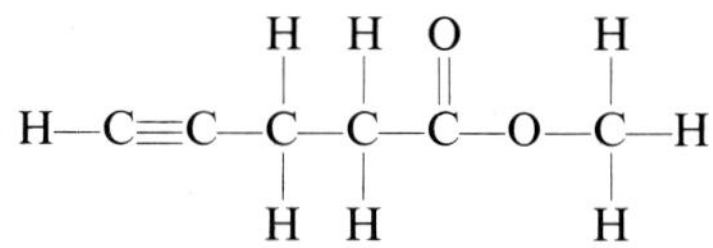

i) 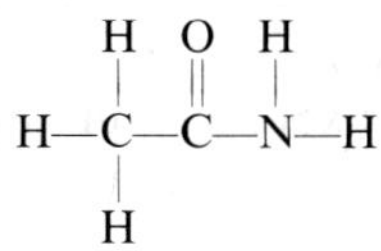

2
Saturated Hydrocarbons

Objectives

In this chapter we:

1. *Name and draw abbreviated structures of hydrocarbons and simple hydrocarbon derivatives.*
2. *Discuss the fractionation of petroleum.*
3. *Consider chemical oxidation as a source of energy.*
4. *Survey uses of hydrocarbons and their halogen derivatives.*

The simplest of the organic molecules are the saturated hydrocarbons, which contain only carbon, hydrogen, and single covalent bonds. Because of this structural simplicity, hydrocarbons form the foundation for the system by which organic molecules are named. In the previous chapter, such structures were introduced and were represented by Lewis structures. These structures are cumbersome and are often abbreviated by listing after each carbon the noncarbon substituents attached to it.

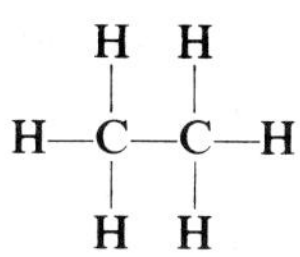
becomes CH_3—CH_3

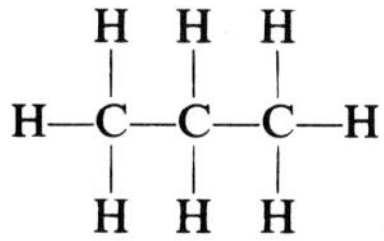
becomes CH_3—CH_2—CH_3

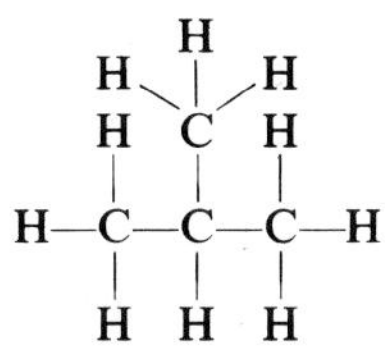
becomes CH_3—$CH(CH_3)$—CH_3

Exercise 2.1

a) Draw the Lewis structure of

$$CH_3—CH(CH_3)—CH(CH_3)—C(CH_3)_2—CH_3$$

Exercise 2.1 continued

b) Convert the following Lewis structure to an abbreviated form:

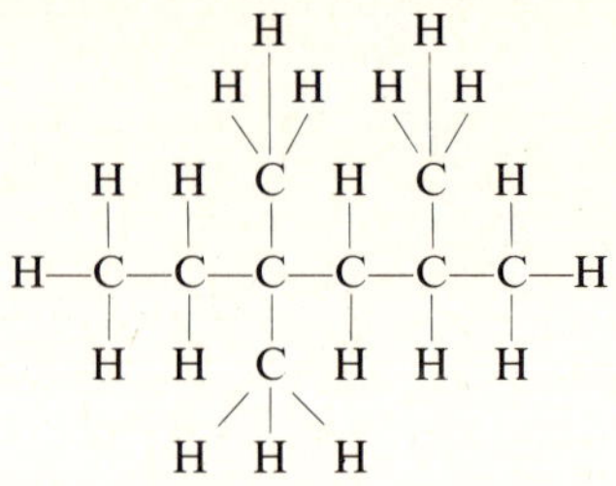

2.1 Naming Hydrocarbons

The simplest of the hydrocarbons to name are those in which the carbons are arranged in a straight line. The names we give these (Table 2.1) become the basis for many of the names used in organic chemistry.[1]

The saturated hydrocarbons are sometimes called the alkane family, and the members are identified by the -ane ending. Where the

Table 2.1
Names of Linear Alkanes

Structure	*Name*
CH_4	Methane
$CH_3—CH_3$	Ethane
$CH_3—CH_2—CH_3$	Propane
$CH_3—CH_2—CH_2—CH_3$	Butane
$CH_3—CH_2—CH_2—CH_2—CH_3$	Pentane
$CH_3—CH_2—CH_2—CH_2—CH_2—CH_3$	Hexane
$CH_3—CH_2—CH_2—CH_2—CH_2—CH_2—CH_3$	Heptane
$CH_3—CH_2—CH_2—CH_2—CH_2—CH_2—CH_2—CH_3$	Octane
$CH_3—CH_2—CH_2—CH_2—CH_2—CH_2—CH_2—CH_2—CH_3$	Nonane
$CH_3—CH_2—CH_2—CH_2—CH_2—CH_2—CH_2—CH_2—CH_2—CH_3$	Decane

[1] For chemists to all "speak the same language," a group called the International Union of Pure and Applied Chemistry (IUPAC) sets the values for physical constants and atomic weights and selects the system by which compounds are named. In this text we present the IUPAC rules for naming organic molecules. Many compounds also have recognized nonsystematic names. Biochemists are notorious for using nonsystematic names, largely because the IUPAC names of the most common biochemical compounds are very long and complex.

carbons forming the framework of the molecule are not in a straight line, we locate and name the longest continuous chain (the "main" chain).

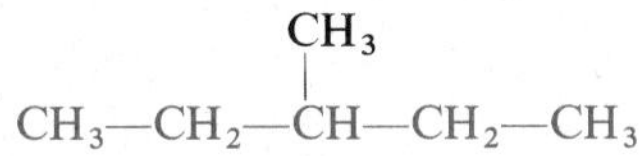

The main chain (in color) is 5 carbons long: *pentane.*

The rest of the carbons are named as "side groups" on the main chain. These are designated by derivatives of the alkane name, which end in -yl rather than -ane (Table 2.2). The side group or groups are listed before the alkane name.

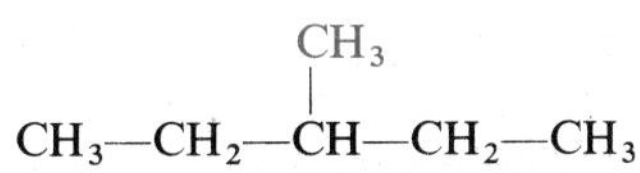

The side group is one carbon: *methyl*pentane.

We soon realize that more than one structure can be accurately described by the name methylpentane:

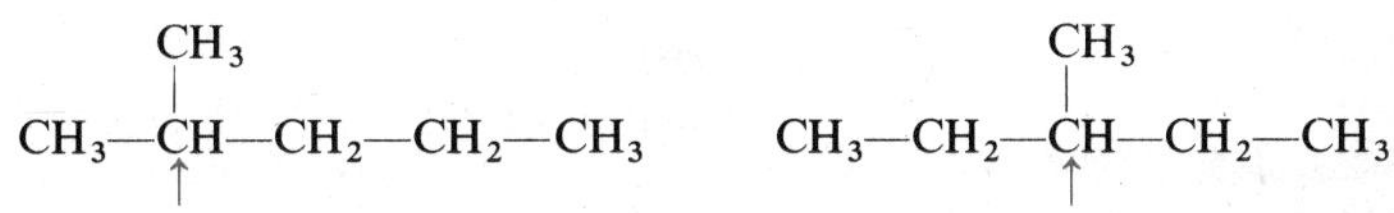

Table 2.2
Alkyl Side Chains

Group	*Prefix*
CH_3-	Methyl
CH_3-CH_2-	Ethyl
$CH_3-CH_2-CH_2-$	*n*-Propyl (*n* = normal)
$CH_3-\underset{\vert}{CH}-CH_3$	Isopropyl
$CH_3-CH_2-CH_2-CH_2-$	*n*-Butyl
$CH_3-\underset{\vert}{CH}-CH_2-CH_3$	*sec*-Butyl (*sec* = secondary)
$CH_3-\underset{CH_3}{\underset{\vert}{CH}}-CH_2-$	Isobutyl
$CH_3-\underset{CH_3}{\underset{\vert}{\overset{CH_3}{\overset{\vert}{C}}}}-$	*t*-Butyl (*t* = tertiary)

The difference between them is the point of attachment of the side group to the main chain. To indicate this point of attachment, the main chain is numbered, starting at the end closest to the side group.

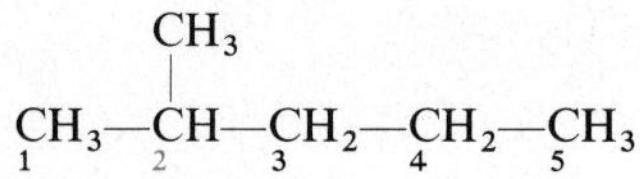

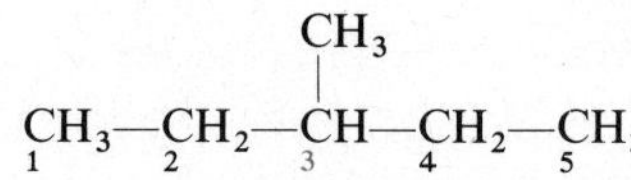

The number of the carbon to which the side group is attached is now written in front of the name of the side group, separated from it by a hyphen.

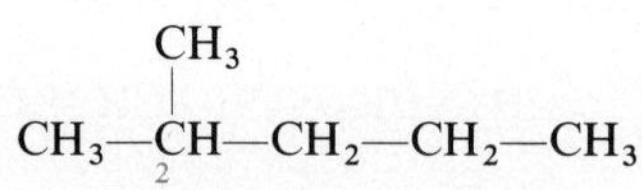

2-methylpentane

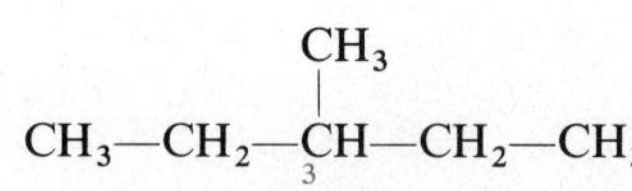

3-methylpentane

If the same group is present more than once, it is not necessary to repeat its name once for each time it is found in the molecule. Instead it is given the correct prefix from Table 2.3, and all the numbers are listed, separated by commas, in front of the modified side group name. There should be as many numbers as the number of side groups the prefix indicates, even if that means using the same number twice. Sometimes numbers are omitted where there is no possibility of confusion. For example, a propane with a methyl group substituted on it can have that group only in the 2 position. It can therefore be called simply methylpropane. See the table at the top of page 31.

Table 2.3
Prefixes

Prefix	***Number of Times a Group Is Present***
Mono[a]	One
Di	Two
Tri	Three
Tetra	Four
Penta	Five
Hexa	Six

[a] Usually omitted.

Exercise 2.2

Indicate the longest chain in the following:

a)

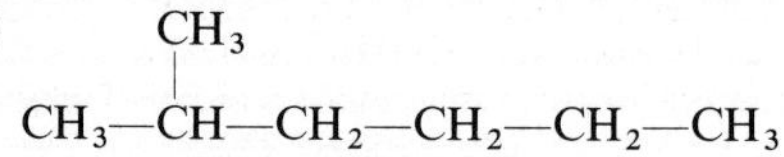

b)

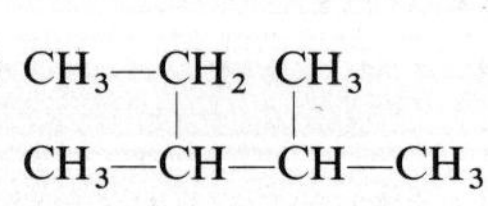

c)

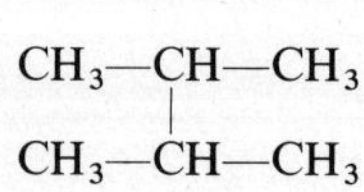

d)

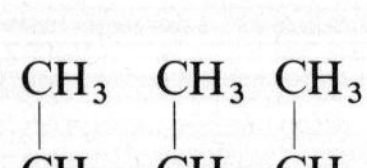

Structure	***Progressive Steps in Arriving at a Correct Name***	
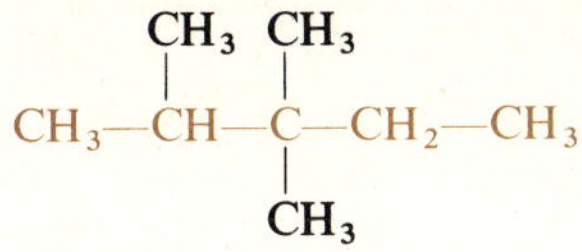	*pentane*	(Select the longest chain.)
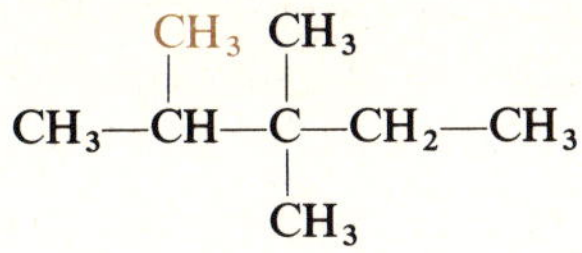	*methyl*pentane	(Identify and name the side groups attached to the chain.)
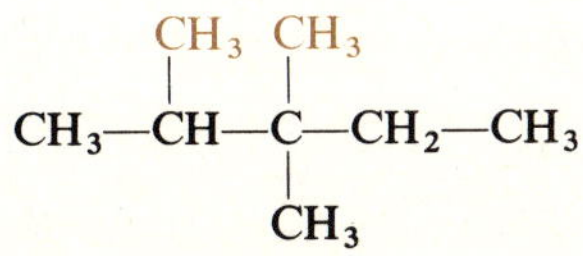	*di*methylpentane	
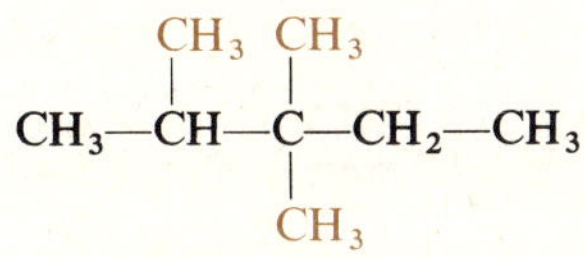	*tri*methylpentane	
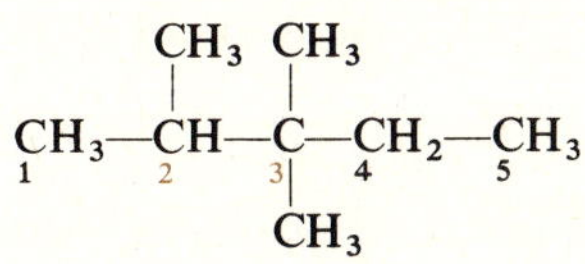	*2,3,3*-trimethylpentane	(Number the chain to produce the lowest total.)

Exercise 2.3

Complete the following names:

a)

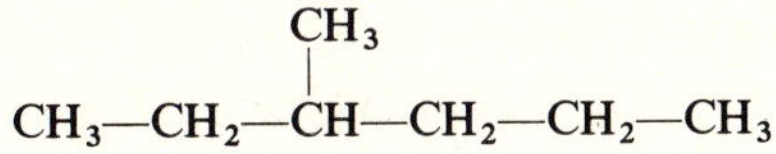

3-methyl ______________

b)

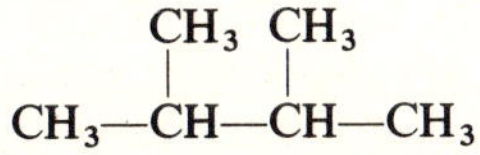

2,3- ______________ butane

c)

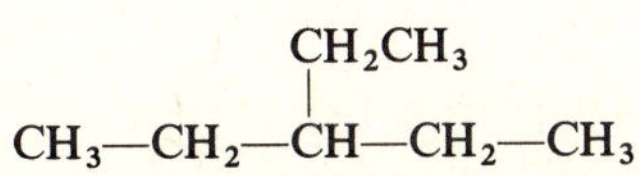

__ - ______________ pentane

Exercise 2.3 continued

d) 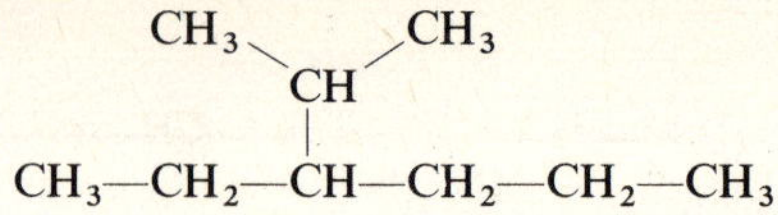

___ - ______________ ______________

or

___ - ______________ - ___ - ______________ ______________

2.2 Substituted Alkanes

The next level of complexity is reached when we replace the hydrogens of alkane structures with other groups. Some of the atoms or groups of atoms replacing the hydrogens were designated as functional groups in the previous chapter. Functional groups are chemically reactive and modify the character of the molecule so much that we segregate molecules containing them from the alkanes both in discussing the chemical properties and in naming them. The more inert substitutions, however, do not result in this sort of segregation. Some of these groups are illustrated and named in Table 2.4. As in the case of the carbon side groups, prefixes are used to indicate the presence of more than one of a particular group, and the locations are designated by numbering the carbons of the chain.[2]

Table 2.4
Nonalkyl Side Groups

Group	*Prefix*
F—	Fluoro
Cl—	Chloro
Br—	Bromo
I—	Iodo
NO_2—	Nitro
N≡C—	Cyano

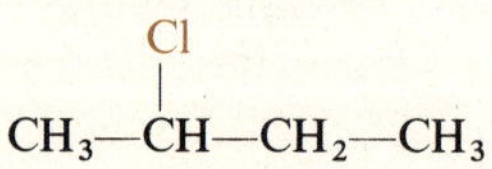

2-chlorobutane

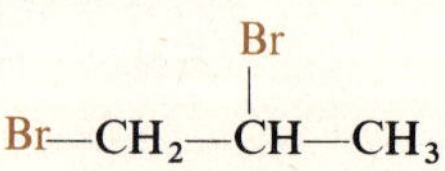

1,2-dibromopropane

$CH_3—CF_3$

1,1,1-trifluoroethane

CH_3NO_2

nitromethane

$CHCl_3$

trichloromethane (chloroform, an anesthetic)

CHI_3

triiodomethane (iodoform, an antiseptic)

Exercise 2.4

Name the following:

a) 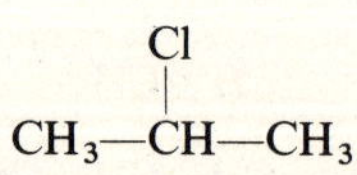

b) $CH_2Cl—CH_2—CH_3$

c) $CH_3—CH_2—CH_2Cl$

d) $CH_2Cl—CH_2—CH_2Cl$

e) 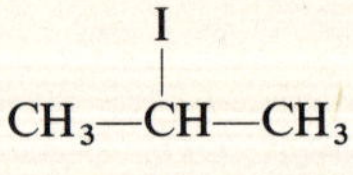

f) $CH_3—CH_2—CH_2—NO_2$

[2] It is general practice to list the side groups in alphabetical order.

g) $CH_3—CH_2—CH_2—Br$

h) $CH_2F—CH_2—CH_3$

i) $CH_2Br—CH_2—CH_2Br$

j)
I
|
$CH_3—CH—CH_2—Cl$

k) $CH_3—CH_2—CH_2NO_2$

l)
Cl
|
$CH_2Cl—CH—CH_2Cl$

2.3 Cycloalkanes

A variation of the alkane structure not yet discussed is the ring compound. A carbon chain can "grab its own tail" to form a stable molecule. Rings are named by prefixing by cyclo- the alkane name for the number of carbons in the ring. All attachments to the ring are named as side groups. Notice that with a ring there is no end, so we select as carbon 1 the position that will generate the lowest set of numbers.

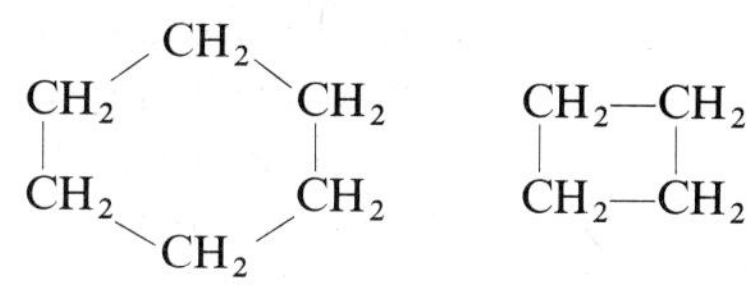

cyclohexane cyclobutane

It is common to abbreviate the writing of these structures by replacing the carbons and hydrogens with a simple geometric shape, cyclohexane by a hexagon, cyclobutane by a square, and so forth.

cyclohexane cyclobutane

It is easy to forget that there are two hydrogens at each of those corners. When something other than hydrogen is substituted onto the structure, it is drawn attached to the corner of the geometric figure. Thus a ring with a methyl group and a hydrogen on it will be drawn showing the methyl group only.

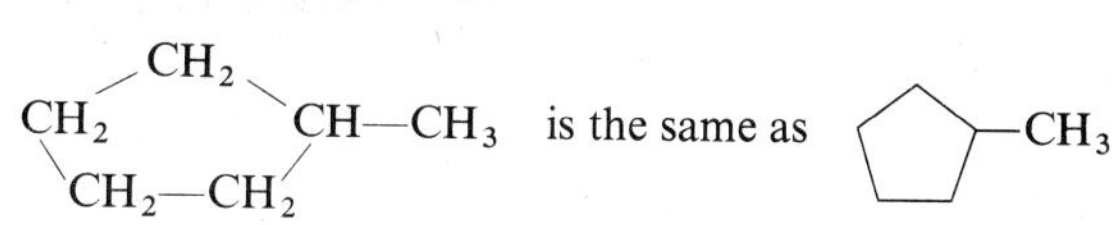

methylcyclopentane

Here are some sample structures and names:

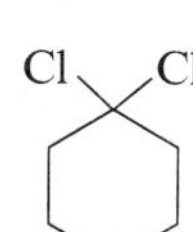

1,1-dichlorocyclohexane

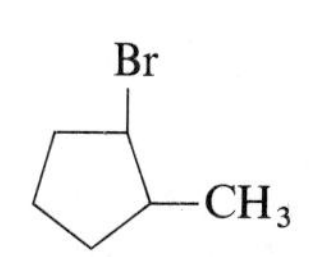

1-bromo-2-methylcyclopentane

Exercise 2.5

How many hydrogens are there in 1,1-dichlorocyclohexane? in 1-bromo-2-methylcyclopentane?

2.4 Strained and Unstrained Rings

Recall that a carbon substituted by four groups normally has bond angles of 109.5° between those groups (Chapter 1). Cycloalkanes present special problems. Cyclohexane as represented on paper seems to have bond angles of 120° between the carbons of a ring. However, this assumption is based on the same mistaken reasoning that would have us list the bond angles in methane as 90°, namely that the molecule is flat as we represent it on the page. In fact, cyclohexane is not flat. The ring puckers into two general shapes called the *chair form* and the *boat form* (Figure 2.1).

In each case we are looking at the ring as if it were sitting on a low table in front of us. In practice, cyclohexane spends part of its time in each of these forms, flipping back and forth between them. More time is spent in the chair form than in the boat form. In these puckered rings the bond angles are 109.5°. The same sort of puckering occurs in rings with more than six atoms. Smaller rings are different. The five-membered ring is almost as flat as its representation on paper, and three-membered rings *are* flat. Four-membered rings were once considered flat but now are seen as bent. The bond angles of cyclopentane are close to 109.5°, so there is little problem. However, consider the angles of the ring carbons in cyclobutane and cyclopropane. Clearly they are less than 109.5°. These rings have been called "strained" because of the distortion of the carbon bond angles. As a result, they are less likely to be formed. In living systems there are a few examples of the cyclopropane and cyclobutane ring, in the terpenes (Chapter 3), for example, but only a few. Rings with five or six atoms are the overwhelming preference whether in conventional organic chemistry or in biological systems.

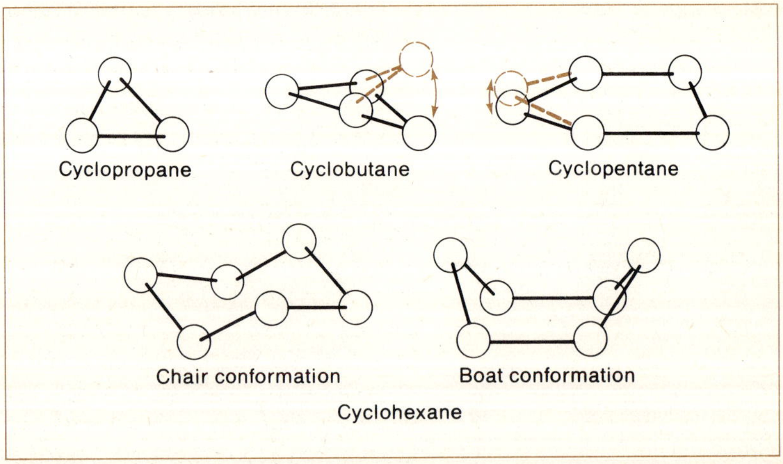

Figure 2.1
Shapes of cycloalkanes
Cycloalkanes have shapes that result from their desire to establish bond angles of 109.5°, as do all carbons bonded to four other atoms. For the smaller rings it is impossible to bend the structure such that every bond angle is ideal. Cyclopentane comes close to achieving the desired angles as a flat molecule. In rings larger than cyclohexane, the 109.5° bond angle can be formed, but the resulting ring must pucker.

Figure 2.2
A refinery
A modern refinery has a dramatic skyline. Many of these tall structures are fractionating towers, used to separate petroleum into fractions with relatively narrow boiling point ranges. (Photo by Design Photographers International, Inc.)

2.5 *The Source of Hydrocarbons*

Petroleum (crude oil) is largely a mixture of hydrocarbons. It represents the decomposition products of plant and animal carbon compounds from an earlier period in the history of the earth.

As a complex mixture, crude oil is less useful to us than when it has been sorted out into somewhat more homogeneous fractions. Primarily this splitting of crude oil into fractions is performed at a refinery in tall structures called *fractionating towers* (Figure 2.2). Recall that for hydrocarbons, higher molecular weight results in higher boiling points. In a fractionating tower, conditions are such that the temperature varies up and down the tower. The crude oil is boiled, and individual hydrocarbons condense in a region of the tower that is just cooler than the boiling point of the particular hydrocarbon. Fractions with lower boiling points may be tapped off to become gasoline, higher boiling points to be diesel oil, and so on (Figure 2.3). Natural gas, which is primarily methane, also arises from once-living material and is usually found in conjunction with petroleum. Methane is also produced by bacterial action on sewage. In some water-treatment plants this methane is then burned to raise the temperature of the sewage to speed the treatment process. This is a practical example of speeding a reaction by raising the temperature.

2.6 *Uses of Hydrocarbons*

Our largest use of hydrocarbons is to burn them as a source of energy. About 85 percent of refinery output is destined for use as fuel, slightly more than half of this as gasoline. Hydrocarbons represent a relatively clean, convenient, and portable fuel. Replacing coal with petroleum as a fuel for generating electricity is responsible for improving the air quality in a number of cities. The supply of petroleum is limited, however. Our use of hydrocarbons in other ways, for instance as lubricants and as the starting material for the manufacture of plastics,

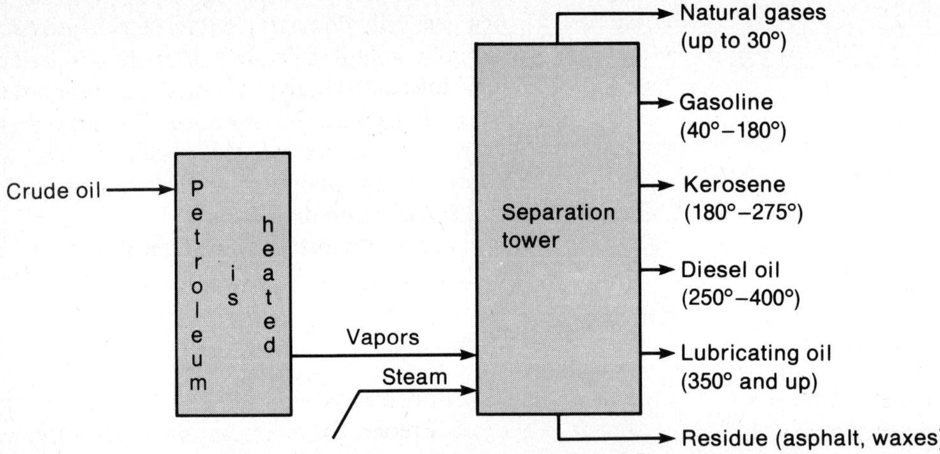

Figure 2.3
Petroleum fractionation
In a refinery the crude oil is heated, then transferred to a tower in which the temperature varies with height. Vapors condense when they reach a region below their boiling point. In this fashion hydrocarbons of different boiling point (molecular weight) are separated for various uses. (From James O. Schreck, *Organic Chemistry*, St. Louis, Mo., Mosby, 1975.)

raises questions about the wisdom of continued wholesale combustion of this resource. In fact, our dependence on hydrocarbon fuels is at the heart of our energy problem. Either we must generate our energy differently (by solar or wind power), or we must find a new source of hydrocarbon (the liquification of coal). Unfortunately, we don't have a practical alternative to the hydrocarbon fueled engine for autos.

The complete combustion of the hydrocarbon is its reaction with abundant oxygen to yield carbon dioxide and water.

$$CH_4 + 2O_2 \longrightarrow CO_2 + 2H_2O + \text{heat}$$

This reaction is very exothermic (heat releasing). The heat so produced warms our homes, propels our cars, planes, and trains, and generates our electricity. As a generalization, energy is released by oxidation. In fact, oxidation is the source of the energy in nonphotosynthetic living organisms. Rather than beginning with hydrocarbons, however, organisms use carbohydrates, lipids, or proteins. We can represent this process with a simple carbohydrate.

$$\underset{\text{carbohydrate}}{C_6H_{12}O_6} + 6O_2 \longrightarrow 6CO_2 + 6H_2O$$

In the organism the oxidation of the carbohydrate is not done in a single step as this equation suggests.

Although alkanes are capable of releasing sizable amounts of energy on oxidation, they are otherwise quite inert chemically. Thus alkanes serve as solvents in which reactions between other nonpolar molecules (therefore hydrocarbon-soluble molecules) can be run without interference by the solvent.

Alkanes have a minimal effect on humans. Contact with these compounds is not generally damaging to the skin, although repeated removal of skin oils by volatile alkanes can cause skin irritation. The nonvolatile (high-molecular-weight) alkanes are often deliberately rubbed on the skin either pure (for example, Vaseline) or as a component of a lotion. The alkanes act in the same fashion as natural skin oils to soften the skin and to provide a waterproof film, for instance to protect babies from diaper rash.

Inhaling volatile alkanes can eventually lead to damage to the lungs by dissolving nonpolar lung cell structures.

2.7 *Uses of Substituted Hydrocarbons*

Hydrocarbons substituted with chlorine and fluorine are used for a variety of functions. Freons, for example, are used as the refrigerant in household appliances. Formerly NH_3 or SO_2 were used for this purpose, but they were irritating if they leaked into the home. Freons, however, are odorless, nontoxic, and inert.

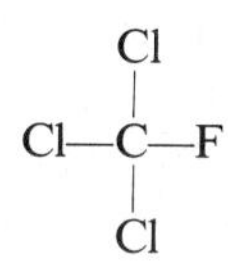

Freon 11

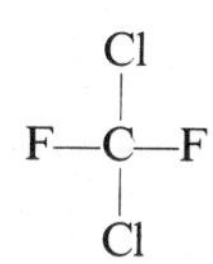

Freon 12

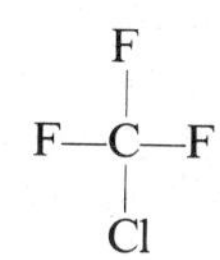

Freon 13

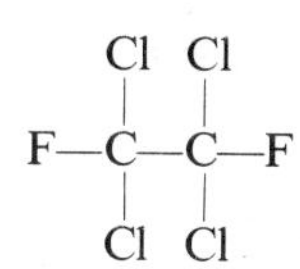

Freon 112

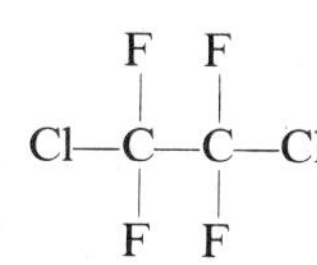

Freon 114

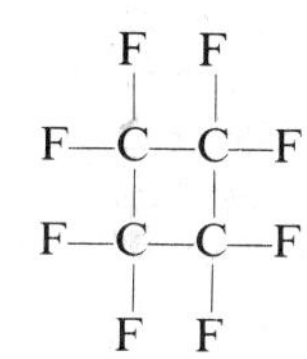

Freon C–318

Exercise 2.6

Can you give each of the above Freons an IUPAC name?

Figure 2.4
An aerosol spray container
Freons have been widely used as the propelling gas in aerosol cans. Recent concern about the effect of large quantities of chlorine Freons on the upper atmosphere has prompted a shift to other gases as propellants and to hand-pumped spray containers. (Photo by Catherine Feryok, Eastern Michigan University.)

Various Freons were approved by the Food and Drug Administration for use as propellants in aerosol cans, and such uses of Freons expanded very rapidly (Figure 2.4). Recently scientists have become concerned about the widespread use of Freons containing chlorine. When these Freons rise to very high altitudes in the atmosphere, they may be degraded by intense solar radiation. The fragments may react to eliminate the ozone (O_3) layer that shields us from a portion of the sun's ultraviolet radiation. Exposure to excess ultraviolet can cause skin cancer. This concern has led to the decision to phase out Freons as aerosol propellants.

2.8 Free Radical Reactions

Chlorine-substituted hydrocarbons are used as grease solvents in dry-cleaning plants and in baths for stripping lubricants from machinery under repair. Alkanes are converted directly to these chlorinated derivatives by free radical reactions. In these reactions, covalent bonds break such that one electron of the shared pair remains on each of the previously bonded atoms. This bond breaking can be stimulated by special chemical catalysts that readily form free radicals themselves, or by high-energy light such as ultraviolet light. For example, light converts a chlorine molecule into two chlorine free radicals.

$$:\ddot{\underset{\cdot\cdot}{Cl}}:\ddot{\underset{\cdot\cdot}{Cl}}: \xrightarrow[\text{light}]{\text{ultraviolet}} :\ddot{\underset{\cdot\cdot}{Cl}}\cdot + \cdot\ddot{\underset{\cdot\cdot}{Cl}}:$$

Lacking outer-shell octets, these free radicals are very reactive species and tend to extract electrons from whatever molecules they encounter, often to create a bond with this other atom or molecule.

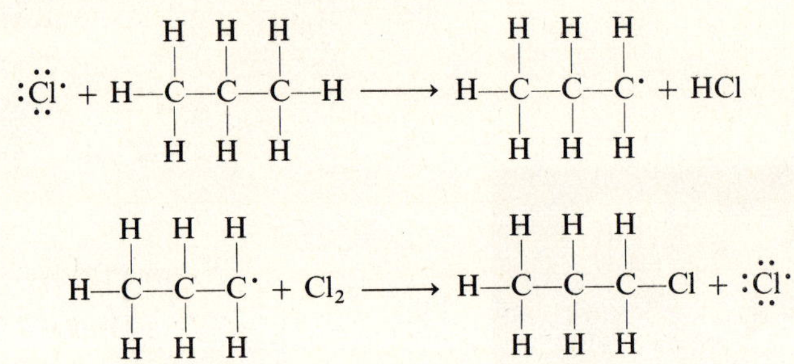

The process is rather random and can create a complex mixture of products.

Free radical changes are often responsible for the destructive effects of sunlight on dyes, fabrics, plastics, and paints. Bonds are broken due to the light energy, and the free radicals produced cause other chemical changes as they collide with surrounding molecules.

Summary

Saturated hydrocarbons are simple combinations of carbon and hydrogen containing only single covalent bonds. The system for naming these hydrocarbons is fundamental to the naming of most organic compounds. Petroleum is the primary source of hydrocarbons on a large scale. Most hydrocarbons so produced are burned to generate energy. Chemically, hydrocarbons and their chlorine or fluorine derivatives are very inert. For this reason they are often used as solvents of one sort or another. Randomly chlorinated hydrocarbons are produced by stimulating the formation of free radicals in a mixture of hydrocarbon and chlorine.

Problems

1. Name the following compounds:

a) $CH_3—CH_2—CH_2—CH_3$

b) 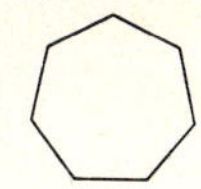

c) 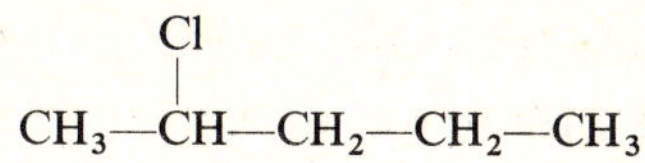

d) 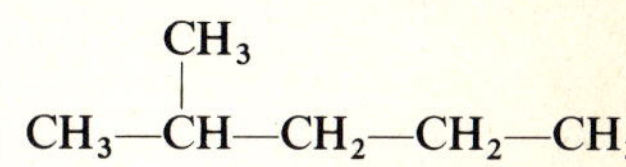

e) 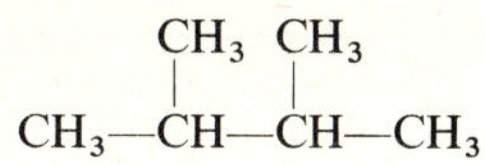

f) $CH_2Cl—CHCl_2$

g) CHF_3

h) 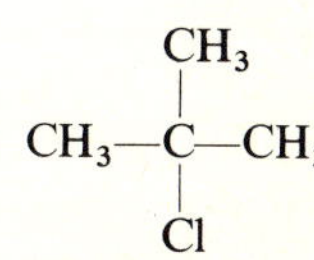

i) 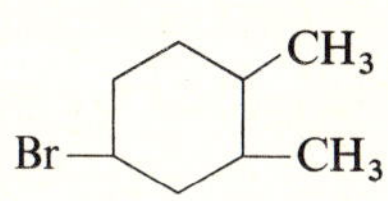

j) 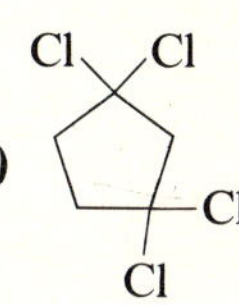

k) 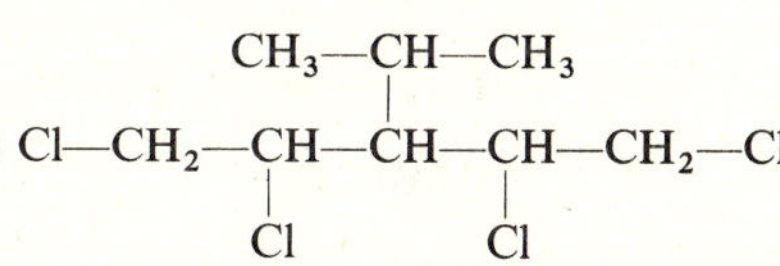

2. *a)* A compound called DBCP by newscasters was used as a soil fumigant and was found to cause sterility in male workers in the plant where it was produced. Its correct name is 1,2-dibromo-3-chloropropane. Draw the correct structure.

b) A compound called halothane has replaced chloroform in operating rooms as a less toxic, nonflammable anesthetic. Now the safety of halothane is being questioned. Given the structure, provide an IUPAC name:

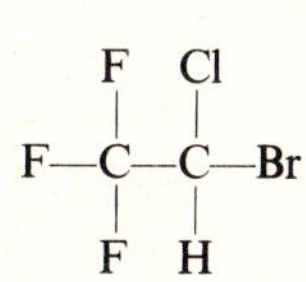

3. Draw the structures that correspond to these names:

a) diiodomethane

b) 1,2-dibromoethane

c) 1,1,3-trichlorocyclohexane

d) 2-chloro-5,5-dibromo-1-fluoroheptane

e) 2,2-dimethylpropane

f) isopropylcyclopentane

g) 3-bromo-3-ethylpentane

h) *t*-butylcyclohexane

4. Something is wrong with these names. Tell what it is, and correct them.

a) 1-methylpropane

b) 2-ethylcyclopentane

c) 3,3,3-tribromopropane

d) 2-isopropylpropane

5. Write a balanced equation for the complete oxidation of hexane (C_6H_{14}).

6. What is wrong with these structures?

a) 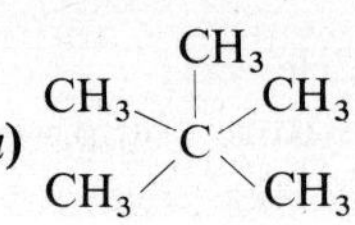

b) $CH_3—CH—CH_3$

c) $CH_3—CCl_3—CH_3$

d) 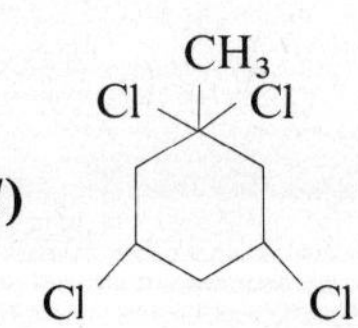

e) 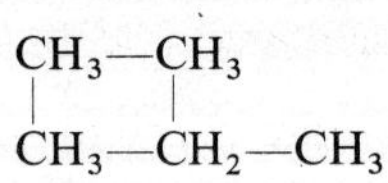

7. You are a research chemist assigned to create a new Freon. You produce a substance based on propane that you find has four chlorines and four fluorines. Based on that much information, what are the possible structures for your new Freon?

8. What are the ring bond angles in each of the following?

a) 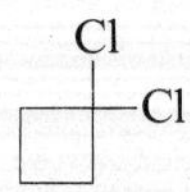

b) 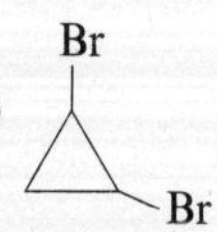

c) 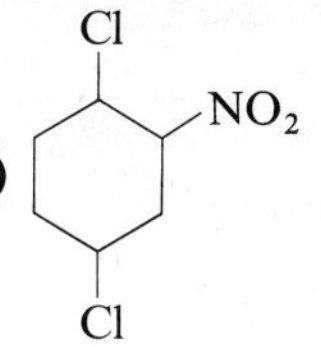

d) 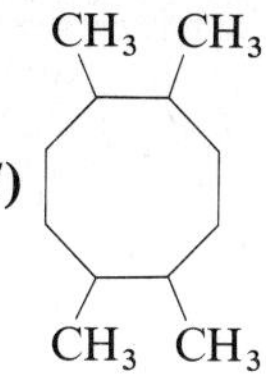

9. Air pollution due to combustion is being combated in a variety of ways. Automobiles emit nitric oxide (NO), the result of bringing nitrogen and oxygen together at high temperatures. Unburned hydrocarbons and carbon monoxide (CO) are also produced. What do you imagine leads to the production of carbon monoxide? A catalytic muffler solves the problem of CO and unburned hydrocarbons by catalyzing their conversion to CO_2. Why does this muffler get so hot?

10. Imagine that you are working in an industrial chemical plant that is chlorinating propane using light as a catalyst.

a) What are the products of the light-catalyzed breakdown of chlorine (Cl_2)?

b) If you consider the chlorination of the propane to occur randomly, each hydrogen on the propane can be replaced by chlorine, and a propane with a chloro group can be substituted again. Write structures for all the possible mono-, di-, and trichloropropanes.

c) For a reaction mixture composed of a particular combination of chlorine and propane, the product mix includes mostly monochloropropanes with smaller amounts of dichloropropanes and unreacted propane. Only traces of propane substituted with more than two chloro groups are found. Predict the effect on the product mix of increasing the concentration of chlorine and decreasing the concentration of propane present.

d) Consider the effect on the product mix in part c if the propane concentration is increased and the chlorine concentration decreased.

e) What would be the effect of increasing the intensity of the light catalyst?

3 Unsaturated Hydrocarbons

Objectives

In this chapter we:

1. *Learn to name alkenes and alkynes.*
2. *Consider the character of sigma and pi bonds.*
3. *Learn about geometric isomerism.*
4. *Introduce terpenes, the biochemicals based on the five-carbon structure isoprene.*
5. *Discuss pheromones, animal communication compounds.*
6. *Introduce addition reactions and consider specific examples for both alkenes and alkynes.*
7. *Learn about addition polymers, their synthesis and properties.*

The discussion of bonding (Chapter 1) introduced covalent bonds with four or six shared electrons, the double and triple bond. Including such bonds between the carbons of a hydrocarbon makes the compound more likely to undergo chemical change. Furthermore, the change produced by a particular reagent is always the same, so double and triple bonds fit our concept of a functional group (Chapter 1).

How does it affect the properties of a hydrocarbon to replace a carbon-to-carbon single bond by a double bond or triple bond? Early chemists saw reagents being "absorbed" by hydrocarbons with double or triple bonds a little like a sponge absorbs water. Based on this faulty analogy, the compounds were described as "unsaturated." When the reactions were complete, the double and triple bonds had become single bonds, the ability to absorb a reagent had disappeared, and the compounds had become "saturated." A simple test for unsaturation is based on the addition of bromine (Br_2) to a sample. If the sample is unsaturated, the bright orange-red color of the bromine disappears, providing a simple visual test. The more unsaturation present, the more bromine absorbed.

Exercise 3.1

Arrange the following compounds in a list with the structure that would require the most bromine to saturate it at the top of the list and the one requiring little or no bromine at the bottom:

a) $CH_3{-}CH{=}CH_2$ ***b)*** $HC{\equiv}C{-}CH_3$

c) $CH_3{-}CH_2{-}CH_3$ ***d)*** $CH_2{=}CH{-}CH{=}CH{-}CH{=}CH_2$

In nature the triple bond is rare, but the double bond is common and important. Extensive use of the reactions of double bonds is made by the chemical industry to produce such commonplace household items as plastics.

3.1 Naming the Compounds

Carbon-to-carbon double bonds in hydrocarbons characterize a family of compounds called *alkenes*, and the presence of triple bonds creates the family called *alkynes*. The rules for naming alkenes are exactly as for alkanes, but the -ane ending of the alkane is changed to -ene.

$CH_3{-}CH_3$ — eth*ane*

$CH_2{=}CH_2$ — eth*ene* (common name: ethylene)

To describe the position of the double bonds, number the carbon chain beginning at the end closer to the double bond. Then place the smaller of the two numbers associated with the double bond directly in front of the alkene name.

$\underset{1}{CH_3}{-}\underset{2}{CH}{=}\underset{3}{CH}{-}\underset{4}{CH_3}$ — 2-butene

The double bond is between carbons 2 and 3. Use the smaller number (2) to indicate the location.

$\underset{1}{CH_2}{=}\underset{2}{CH}{-}\underset{3}{CH_2}{-}\underset{4}{CH_3}$ — 1-butene

Alkynes are handled in exactly the same way, except that the -ane ending becomes -yne.

$CH_3{-}C{\equiv}CH$ — prop*yne*[1]

$CH_3{-}C{\equiv}C{-}CH_2{-}CH_3$ — 2-pent*yne*

When other groups replace the hydrogens of the alkene or alkyne, they are listed and located by numbers, just as was done with alkanes. The components of the name are put together in this order: number of substituted group, name of substituted group, number indicating location of double bond, alkene name.

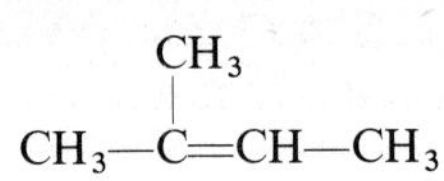

2-methyl-2-butene

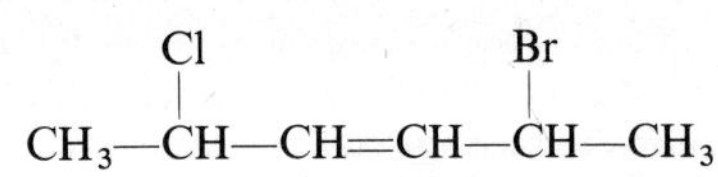

2-bromo-5-chloro-3-hexene

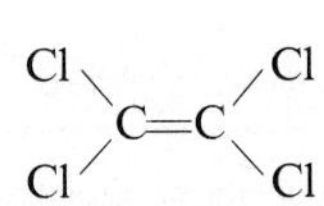

tetrachloroethene (an antihelminthic agent[2])

$Cl{-}CH_2{-}C{\equiv}C{-}CH_2{-}CH_3$

1-chloro-2-pentyne

---

[1] As discussed in Chapter 2, the numbers can be omitted when there is no possibility of confusion.

[2] An agent for combating worms, especially intestinal worms.

Exercise 3.2

Name the following:

a) $CH_3—CH═CH—CH_2—CH_3$ ***b)*** $Br—CH═CH—CH_3$

c)
$$\begin{array}{c} \quad\ \ Br \\ \quad\ \ | \\ CH_2═C—CH_3 \end{array}$$

d)
$$\begin{array}{c} \qquad\qquad\quad CH_3 \\ \qquad\qquad\quad | \\ CH_3—CH_2—CH—CH═CH_2 \end{array}$$

e)
$$\begin{array}{c} \quad Cl\quad Cl \\ \quad \backslash \quad / \\ CH_3—C═C—CH_3 \end{array}$$

f)
$$\begin{array}{c} \qquad\qquad\quad CH_3 \\ \qquad\qquad\quad | \\ CH_3—CH═CH—CH—CH_2—CH_3 \end{array}$$

Ring compounds are named by the same rules.

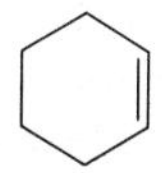

cyclohexene

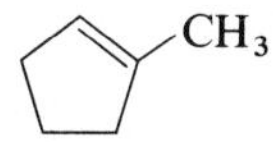

1-methyl-1-cyclopentene

Confusion can occur in naming ring compounds. Consider this compound:

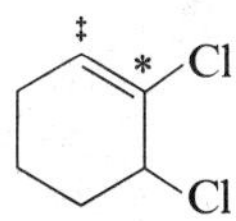

Suppose we call the * position carbon carbon 1 and name the structure 1,2,-dichloro-1-cyclohexene. But, then, what is this?

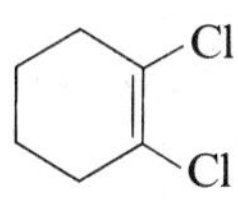

We need a rule to resolve this problem. In our example, difficulty arises when we try to number the ring so that the double bond is 1—6. Therefore we make that pattern illegal. If one side of the double bond is numbered 1, the other must be 2. Bonds *must* be between *consecutively numbered* carbons. Now if we reconsider the first structure, we call the ‡ position carbon 1 and name the compound 2,3-dichloro-1-cyclohexene.

Exercise 3.3

Test your understanding of the rules described. Which of the following choices represents the best numbering of the substituted cyclopentene ring?

a) 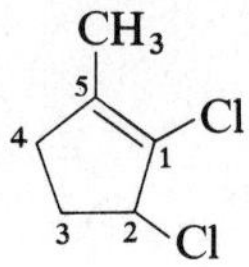

b) 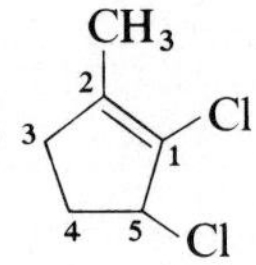

c) 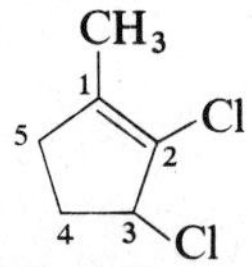

d) 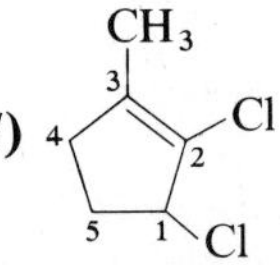

3.2 *Electron Arrangement and Shape*

A feeling for the shape and rigidity of a structure, and the location of electrons in it, is important for understanding its chemistry and biochemistry. Frequently compounds are selected by a living organism as acceptable for a particular function because they have (or can bend to have) the correct shape. Sometimes drugs function as they do because they have a shape similar to that of an important naturally occurring compound. Therefore they are accepted by the organism in place of the natural compound. Once accepted, their different chemistry results in a change in biochemical function. Several antibiotics function in this fashion.

In Chapter 1 we described double bonds as flat (planar) structures with bond angles of 120° between groups attached to the carbon. We described the second pair of electrons in the bond as being less tightly held. Looking at the double bond more closely, we see that the second pair of electrons are in a different place from the first pair. The first pair of electrons, or sigma electrons, are found between the two carbon atoms; the second pair, or pi, electrons are found above and below the sigma electrons (Figure 3.1).

Figure 3.1
Electrons: single and double bonds

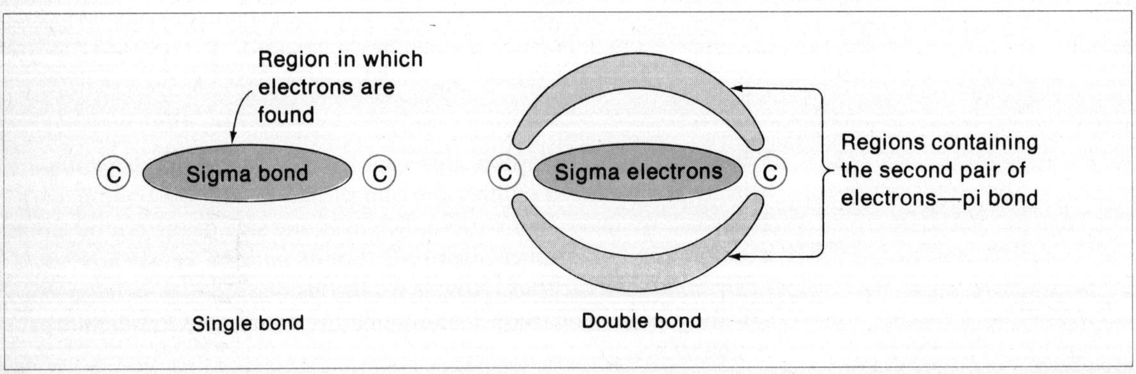

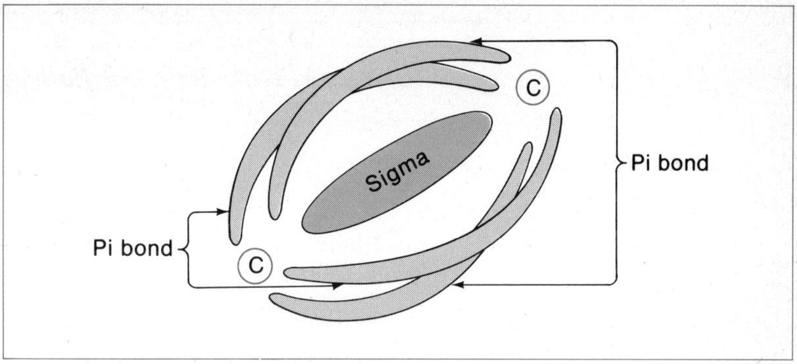

Figure 3.2
Electrons: triple bonds

The sigma electrons generate the sigma bond, and every pair of atoms joined covalently have a sigma bond between. The pi electrons form the pi bond. Pi bonds occur only in double or triple bonds. In a covalent triple bond there are one sigma and two pi bonds (Figure 3.2).

3.3 Geometric Isomerism

Atoms joined by a sigma bond only (by a single covalent bond) can rotate around that bond. For example, a methyl group can spin like a windmill.[3]

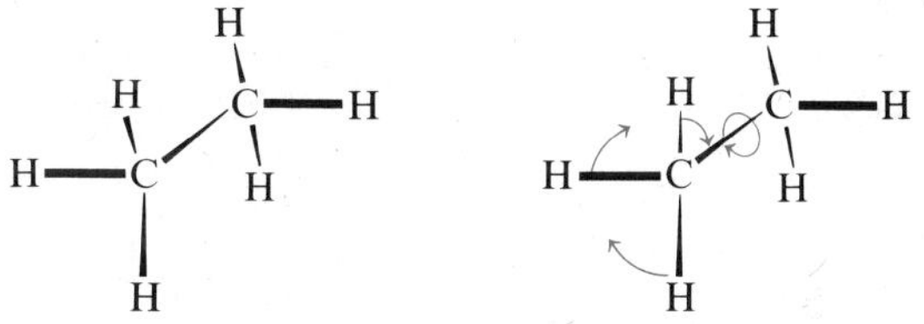

ethane

We call this "free rotation" at that bond. Molecules composed of such bonds are relatively flexible and can change shape easily. Once pi electrons exists, however, the bond no longer has free rotation. The presence of double or triple covalent bonds makes the molecule much less flexible. It also makes it possible to have two molecules that are different even though they possess exactly the same atoms. An example is provided by 2-butene.

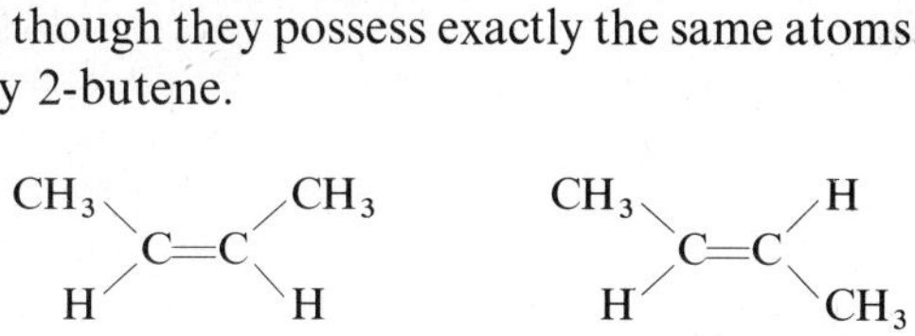

2-butene

[3] It not only *can*, but *does*. In fact, the higher the temperature, the faster such movements are. Generating such movement by absorption of light energy happens too. This is discussed in Chapter 4.

Table 3.1
Physical Properties of a Pair of Geometric Isomers

	Melting Point (°C)	*Boiling Point (°C)*	*Density at 20°C (g/ml)*
***cis*-2-butene**	−138.9	3.7	0.62
***trans*-2-butene**	−105.6	0.9	0.60

Because the double bond prevents free rotation, there is no way the right-hand structure can twist around to look like the left-hand structure. We say that the two forms are *geometric isomers* of one another. Isomers are compounds that have the same atoms arranged differently. The left-hand structure of 2-butene has both the methyl groups on the same side of the double bond (here above) and is called the *cis* isomer. The right-hand structure has the methyl groups across the double bond from each other and is called the *trans* isomer. Geometric isomers differ in their properties (Table 3.1), providing another example of the importance of the shape of a molecule. Density, for example, reflects how tightly the molecules are packed together; melting and boiling points reflect the size of the attractive forces between the molecules. These two characteristics are related. The more closely the molecules can pack together, the greater the attractive forces between them.

In 2-butene, each of the two carbons involved in the double bond has two other groups attached. The two groups are different from one another. Now consider 1-butene:

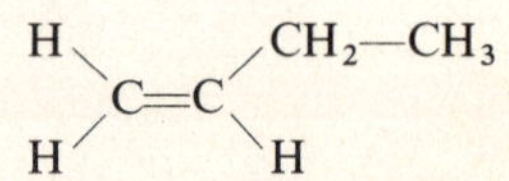

1-butene

The right-hand carbon of the double bond has an ethyl group and a hydrogen, so we can distinguish one side from the other. The left carbon, however, has two identical substitutions, two hydrogens. If we exchanged the upper hydrogen with the lower hydrogen, the molecule would not be changed. This molecule cannot form cis and trans isomers. In order to have geometric isomerism, each carbon of the double bond must have two different substituents.

Exercise 3.4

Which of these structures can form geometric isomers?

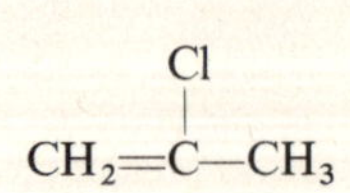

2-chloro-1-propene

$Cl{-}CH{=}CH{-}CH_3$

1-chloro-1-propene

Table 3.2
Some Naturally Occurring Terpenes with Alkene Structures

1. 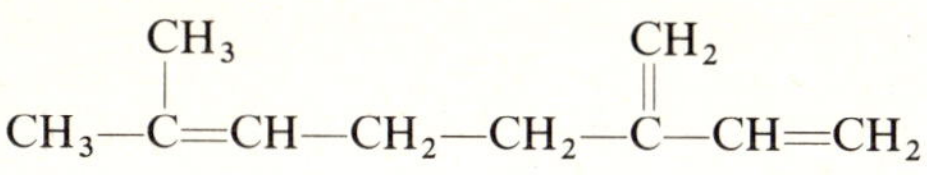

myrcene
(oils of bay, hops, verbena)

2. 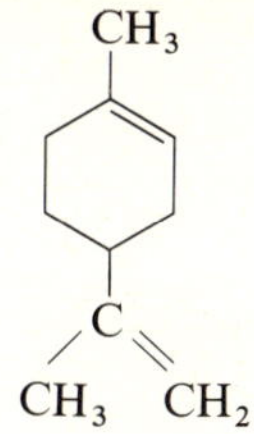

limonene
(oils of orange,
lemon, caraway,
and bergamot)

3. 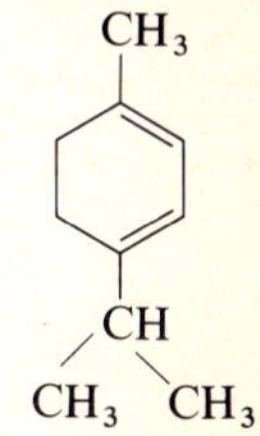

α-terpinene
(oil of marjoram)

4. 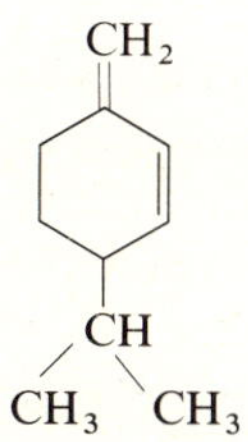

β-phellandrene
(oil of eucalyptus)

5. 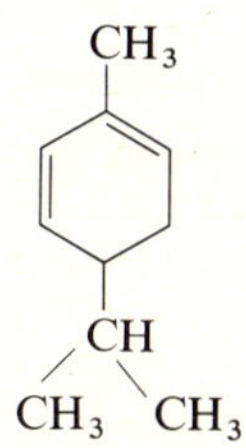

α-phellandrene
(oil of eucalyptus)

6. 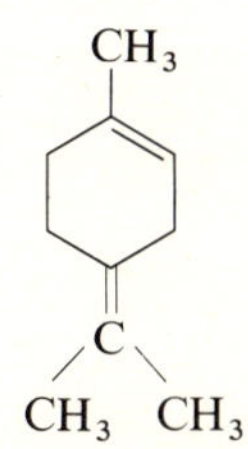

terpinolene
(oil of cupressus,
macrocarpa)

7. 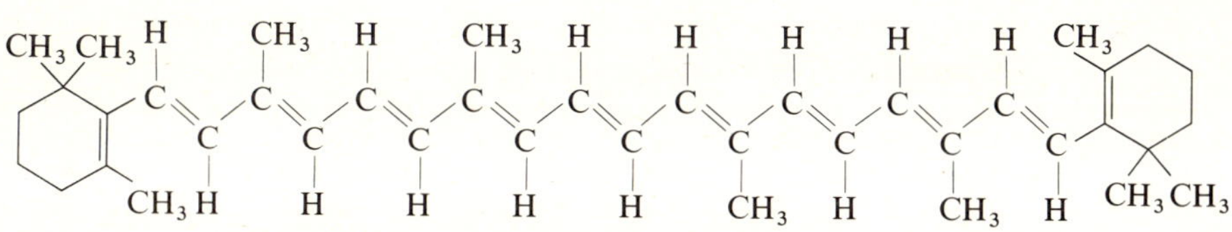

β-carotene (principal carrot pigment)

8. 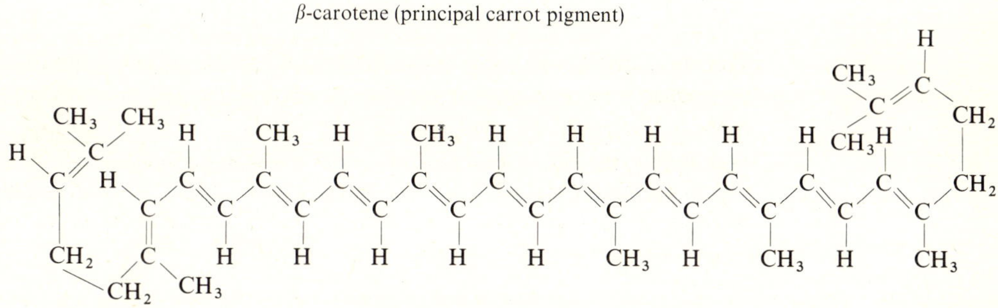

lycopene (principal tomato pigment)

3.4 Terpenes

A family of compounds most often termed *terpenes* is assembled from a basic unit whose five carbons are arranged as they are in isoprene.

$$CH_2{=}\overset{\overset{\displaystyle CH_3}{|}}{C}{-}CH{=}CH_2$$

isoprene

For this reason the compounds are sometimes called *isoprenoid.* Terpenes are among the compounds that have been isolated from plants since the Middle Ages to be used as medicines, poisons, or perfumes. Such plant extracts, which often contain a variety of terpenes, have been called "essential oils." Once we know the structure of a single component of a plant extract, we can decide whether or not it is likely to be a terpene by seeing if it has a structure containing some multiple of five carbons, some carbons of which are present as methyl side groups. Terpenes can have long carbon chains like myrcene (Table 3.2), which is found in oil of bay, oil of hops, and verbena oil. They might include one or more rings, particularly six-membered rings, formed by the carbon chains folding around and bonding to themselves.

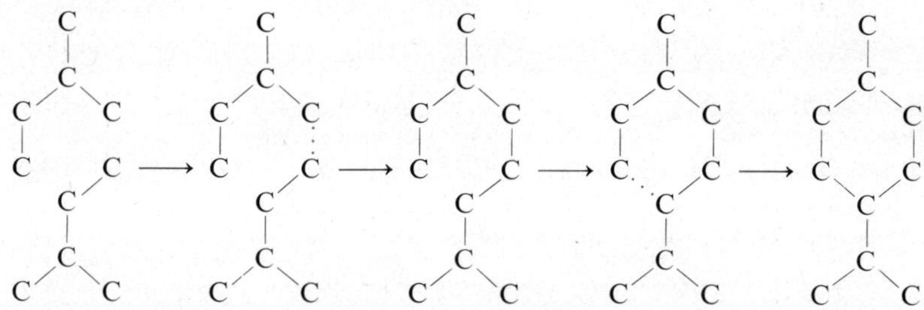

Notice in Table 3.2 that limonene, α-terpinene, α- and β-phellandrene, and terpinolene differ only in the location of the two double bonds, yet they have significantly different properties. In the case of β-carotene, all of the double bonds are trans. This is an example of an important generalization: Living organisms produce one specific isomer where more than one is possible. This kind of exacting chemical synthesis is often impossible in the laboratory.

A great many terpenes have oxygen functional groups (Table 3.3). Notice that the only difference between geraniol and nerol is whether one double bond is cis or trans. Of special interest are the three vitamins that are entirely or partly isoprenoid in structure. In vitamins A and K, geometric isomers of the structures are possible. Only the specific isomer drawn is biologically functional.

3.5 The Chemistry of Vision

Vitamin A is involved in the detection of light in the retina of the eye. The alcohol group of vitamin A is converted to an aldehyde group, producing the compound retinal. Retinal combines with

Table 3.3
Some Naturally Occurring Terpenes with Oxygen Functional Groups

1. 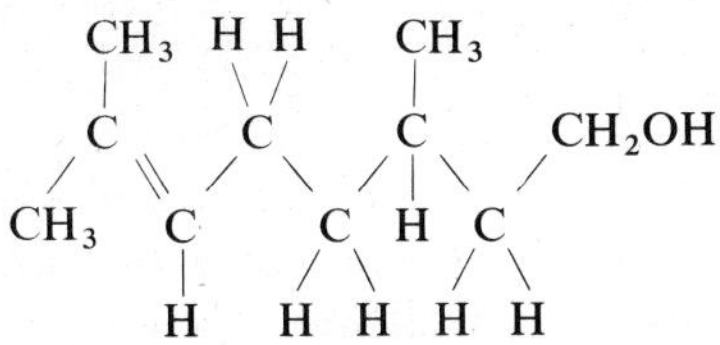

citronellol (oil of rose, geranium, citronella)

2. 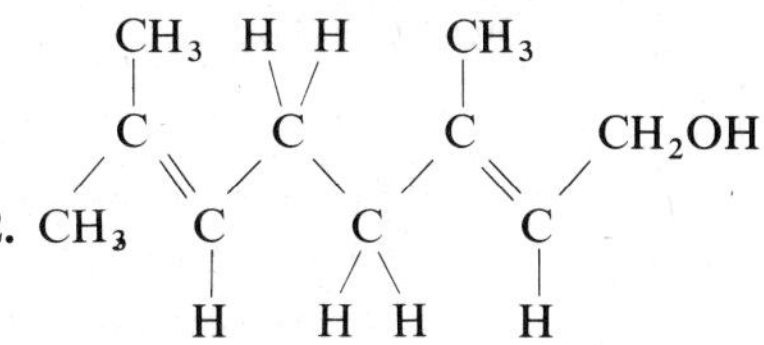

geraniol (oil of geranium)

3. 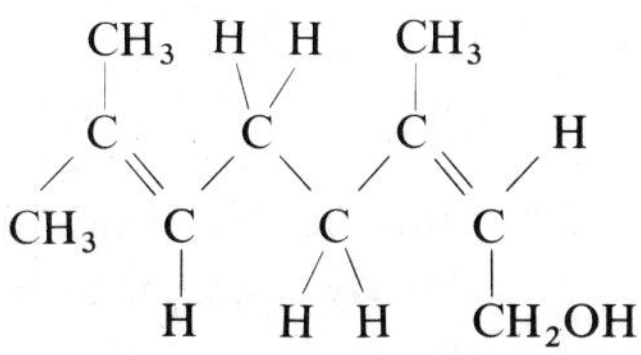

nerol (oil of neroli, bergamot)

4. 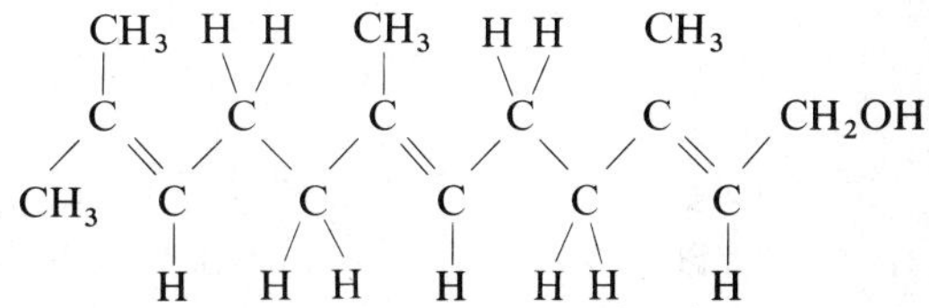

farnesol (lily-of-the-valley perfume)

5. 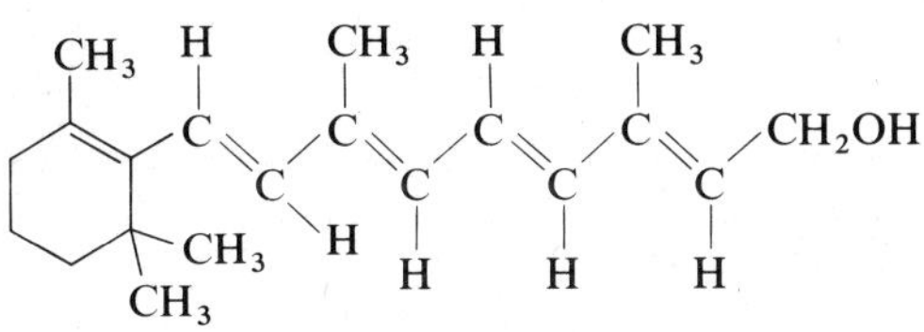

vitamin A

6. 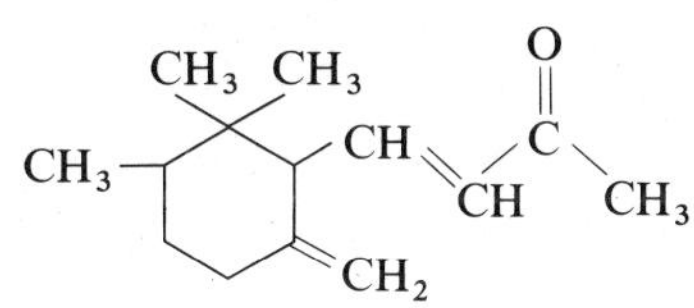

irone (violet perfume, oil of orris root)

7. 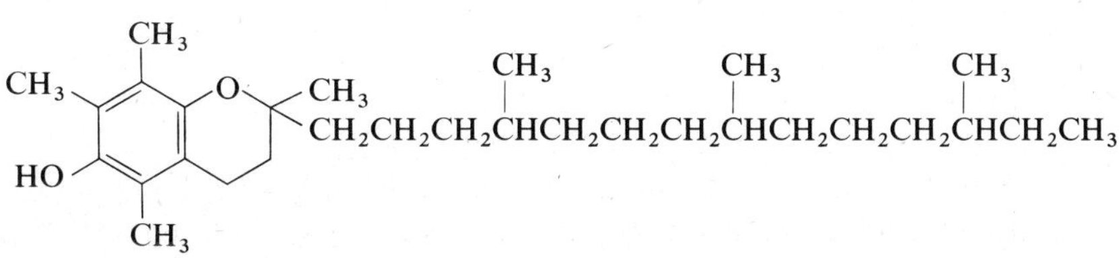

vitamin E

8. 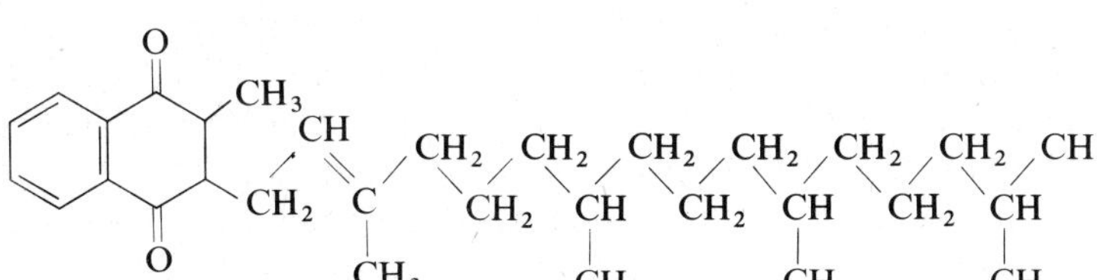

vitamin K

9. 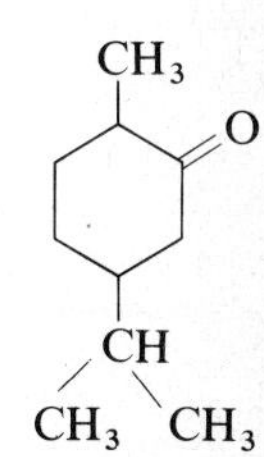

menthol (oil of peppermint)

the protein opsin in the retina, the combination being called *rhodopsin*. Light causes the retinal and opsin to separate. Before they can recombine, the retinal, which has all trans double bonds, must change shape. The double bond at carbon 11 must be changed to cis. Once the 11-*cis*-retinal and opsin are recombined, they are ready to respond to light again. In the course of these chemical changes, the nearby nerve ending of the retina is stimulated. We must obtain vitamin A in our diet to maintain the supply of retinal in the retina. A vitamin A deficiency results in night blindness, a weakening of vision at low light levels.

Exercise 3.5

Locate the structure of vitamin A in Table 3.3. Find carbon 11 by numbering the ring 1–6, then continuing down the side arm beginning with 7. Redraw the structure of the vitamin with two changes:

a) Change the alcohol functional group to an aldehyde.
b) Convert the carbon 11 double bond to cis. You have thus drawn 11-*cis*-retinal.

3.6 Pheromones

Pheromones are compounds produced by animals for communication. Those produced by insects are particularly interesting. For example, when honey bees find food, they excrete the terpenes citral and geraniol to attract other bees (Figure 3.3). Sex attractants are excreted by insects to help the male and female of the species to locate one another. These sex attractants have a high degree of species specificity. Chemically, sex attractants are often alkenes with simple oxygen functional groups. Sample structures are presented in Table 3.4. Notice that they are often terpenes.

Pheromones have enormous potential value because at exceedingly low concentration they affect the behavior of particular groups of insects. Pheromones may help scientists destroy insect pests without using environmentally dangerous insecticides, and they may help us avoid the degree of disruption to nature that would be produced by the elimination of all insects, some of which help to pollinate plants or serve as a food supply for birds, fish, and small animals.

3.7 Reactions of Alkenes

Double bonds react in a predictable fashion by the *addition reaction*. In this reaction, an agent composed of two parts held together by a single covalent bond breaks into its two components, one of which is positively charged, the other negatively charged. The positive ion attaches to one carbon using the pi electrons of the double bond. The

Figure 3.3
Phermones
Insects communicate by releasing chemicals that other insects detect with their antennae. Bees signal the location of food by releasing citral and geraniol. (Photo by Catherine Feryok, Eastern Michigan University.)

other carbon of the double bond, now lacking its share of the pi electrons, has a positive charge. Such a structure with a positive charge is called a *carbonium ion.* The carbonium ion lasts only as long as is necessary for the negative half of the reagent to reach it. A bond forms to that carbon using the electrons of the negative half of the agent. Notice that the ability of the pi electrons to move toward one of the carbons of the bond or toward the positively charged portion of the incoming addition reagent is important in the addition reaction.

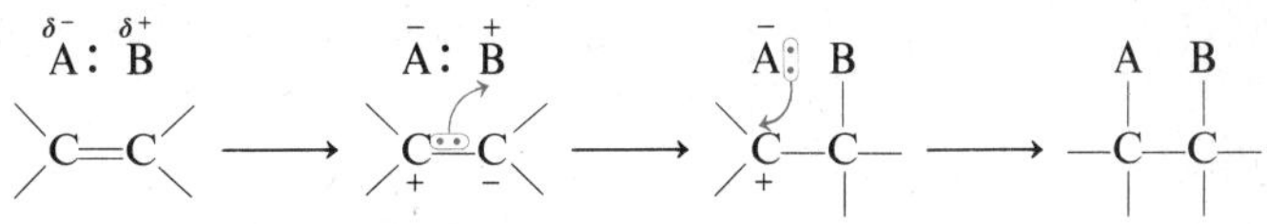

A variety of agents can add to a carbon-to-carbon double bond. Examples include H_2, HX, H_2O, and X_2 (where X is Cl, Br, or I). In the addition of hydrogen or water the reaction goes very slowly by itself, so a catalyst is added to accelerate it.

Alkene	*Agent*	*Catalyst*	*Product*	
$CH_2{=}CH_2$	H_2	Pt or Ni[4]	$H-CH_2-CH_2-H$	(CH_3-CH_3)
$CH_2{=}CH_2$	HCl		$H-CH_2-CH_2-Cl$	(CH_3-CH_2Cl)
$CH_2{=}CH_2$	Br_2		$Br-CH_2-CH_2-Br$	
$CH_2{=}CH_2$	H_2O	H^+	$H-CH_2-CH_2-OH$	(CH_3-CH_2OH)

Exercise 3.6
Complete these reactions:

a) $CH_3-CH{=}CH-CH_3 + HBr \longrightarrow$

b)  $+ H_2O \xrightarrow{H^+}$

(*Continued on page 56*)

[4] These metals cause the reaction to proceed faster by concentrating the hydrogen on their surface. Whether one metal or another is used is unimportant in understanding the principle of addition reactions. For this reason, the word *catalyst*, rather than the symbol of a specific metal, will be used hereafter.

Table 3.4
Insect Pheromones

Released by Bees to Signal the Location of Food

1. 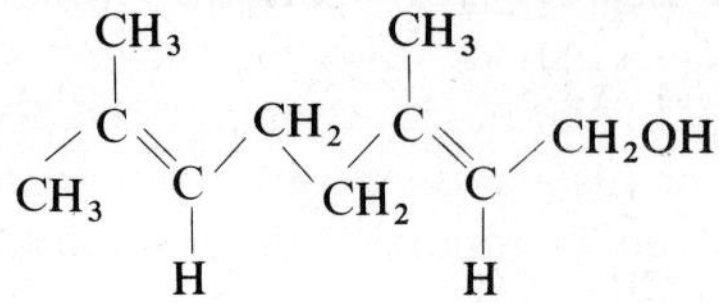

2. 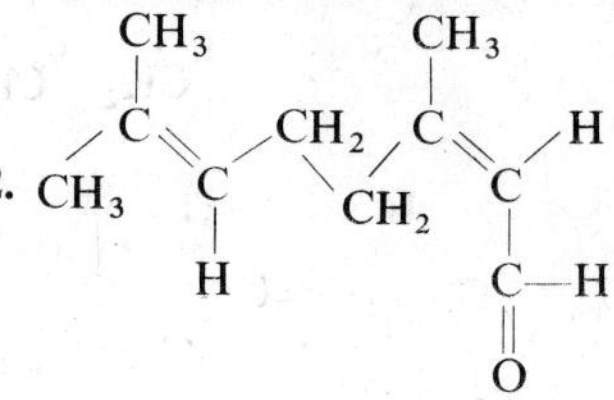

Beetle Sex Attractants

3. 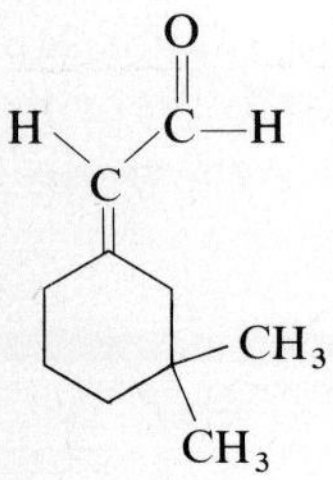

4. 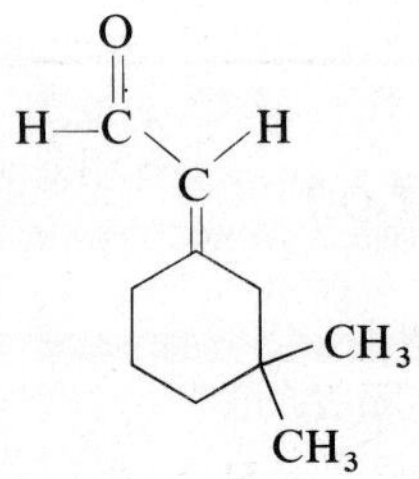

5. 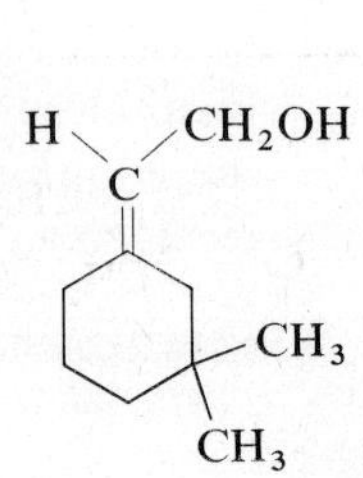

6. 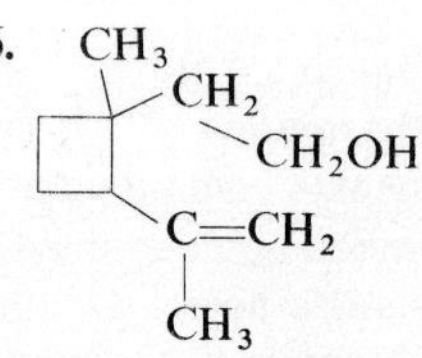

7. 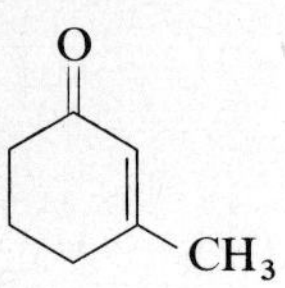

8. 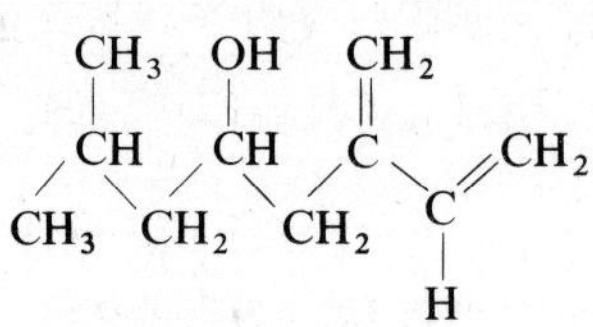

9. 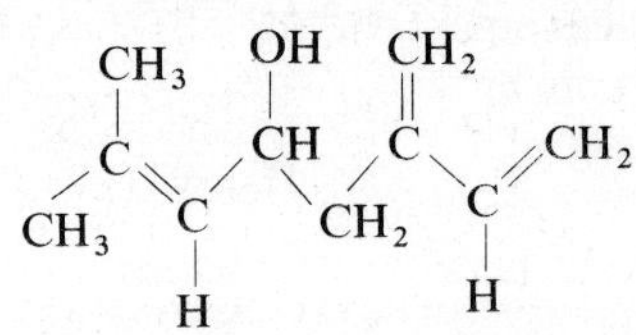

10. 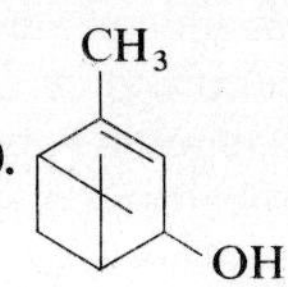

11. 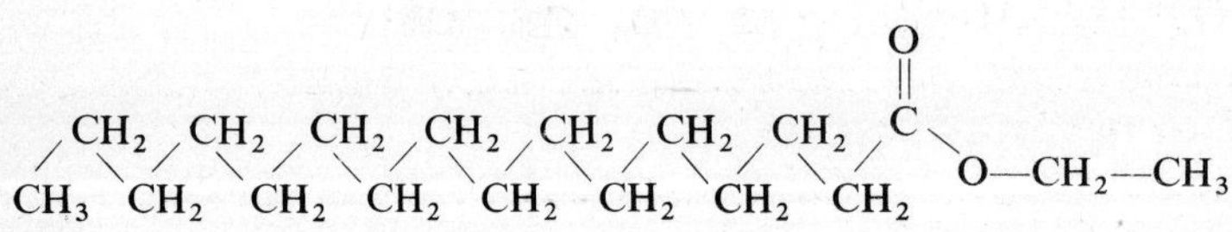

12. 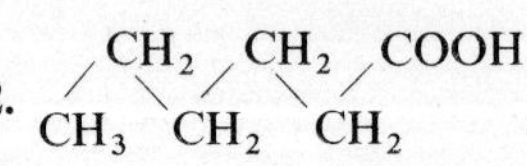

Moth Sex Attractants

13. 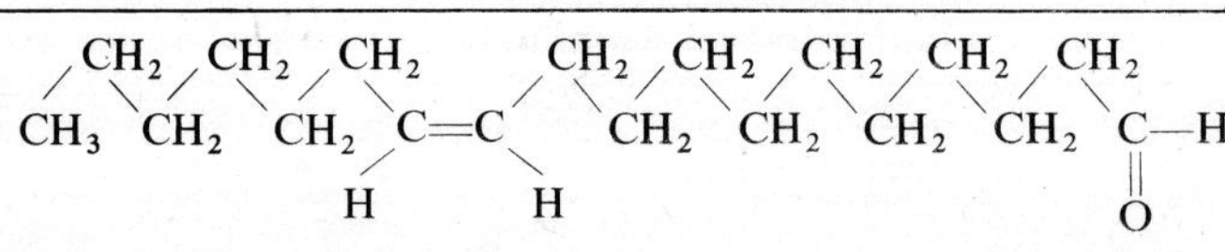

14. 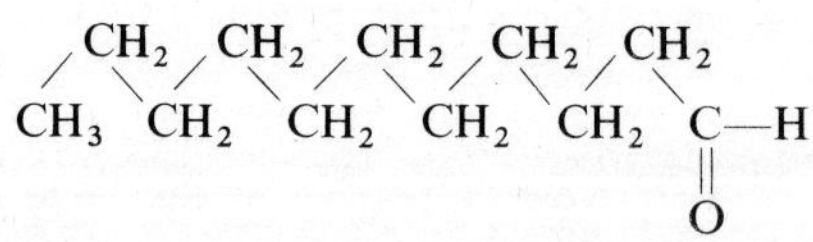

15. 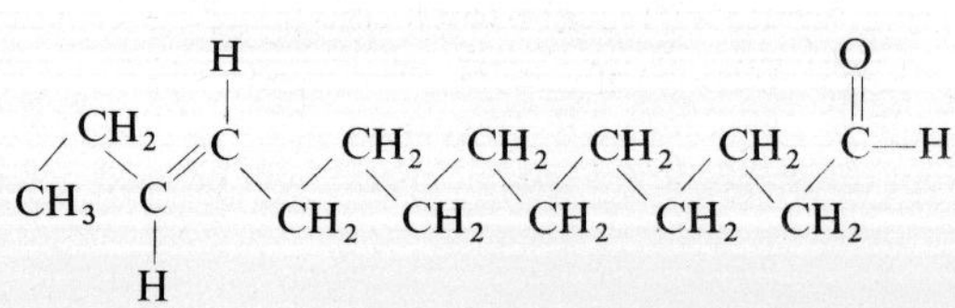

Table 3.4 (Continued)

Moth Sex Attractants (Continued)

16. 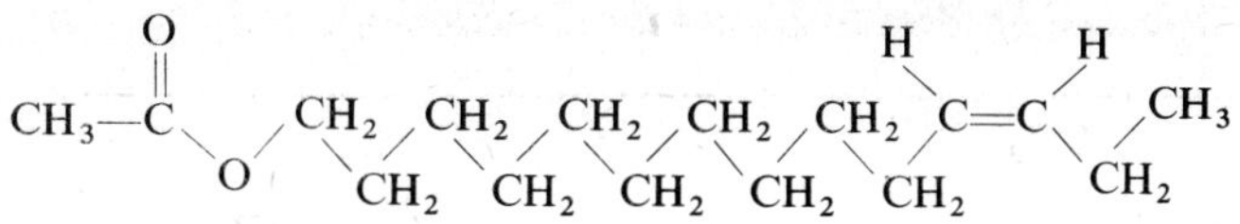

17. 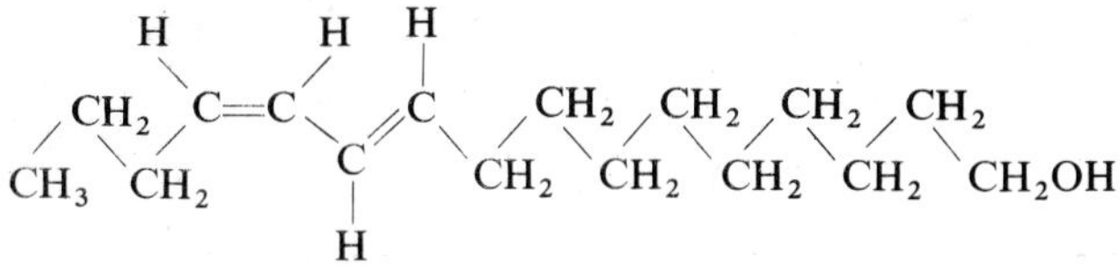

Fly Sex Attractants

18. 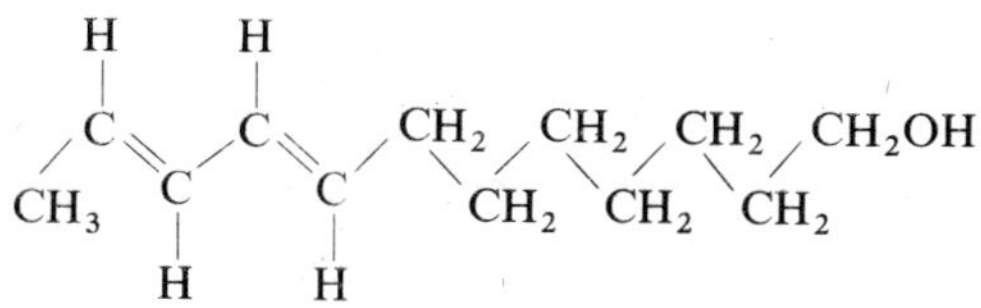

19. 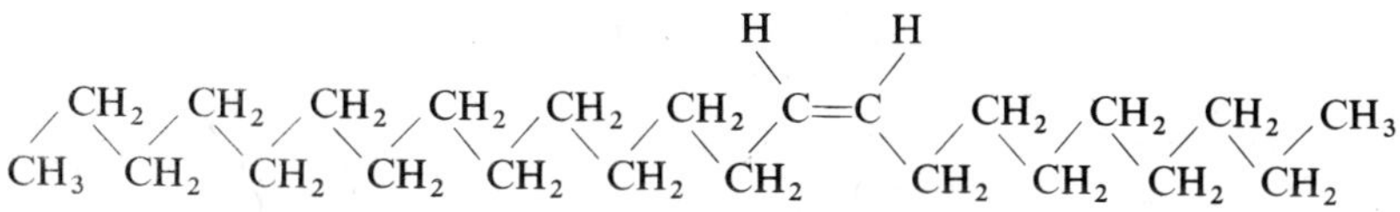

20. 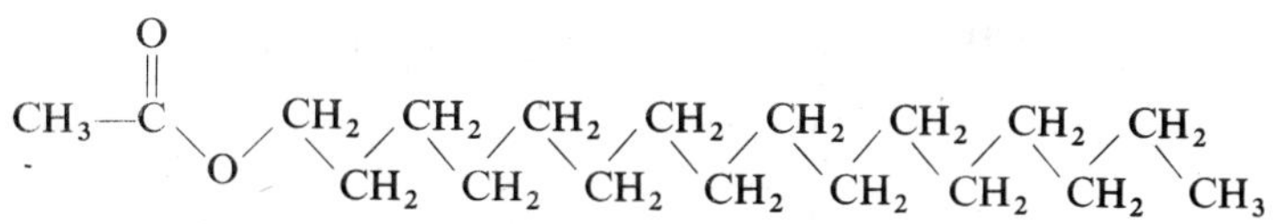

Bee Sex Attractants

21. 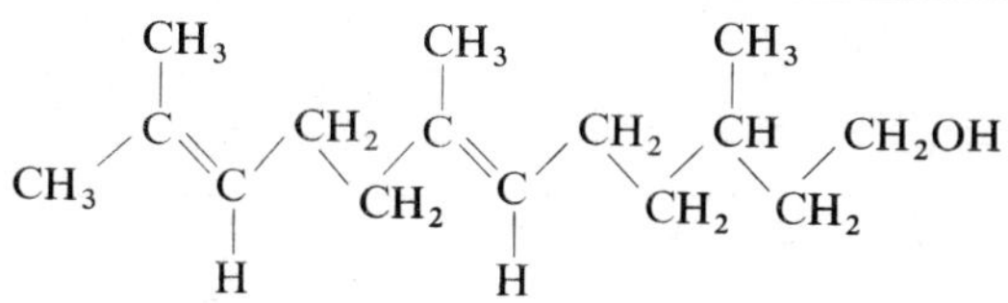

22. 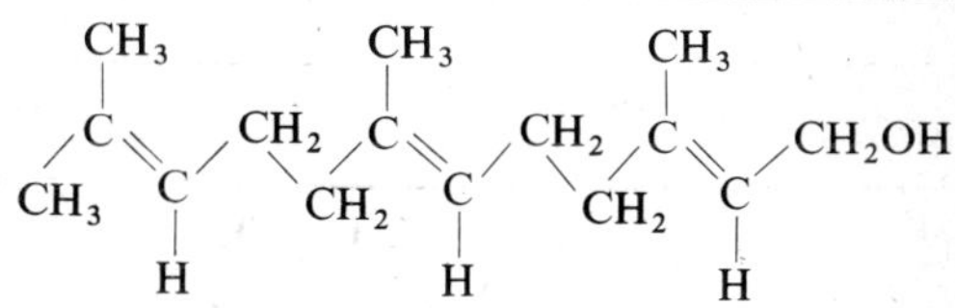

23. 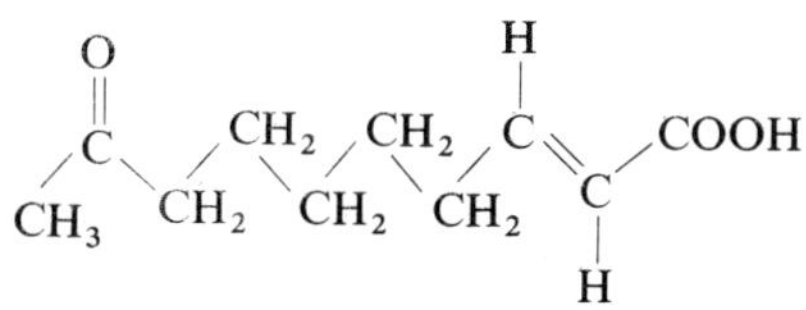

24. 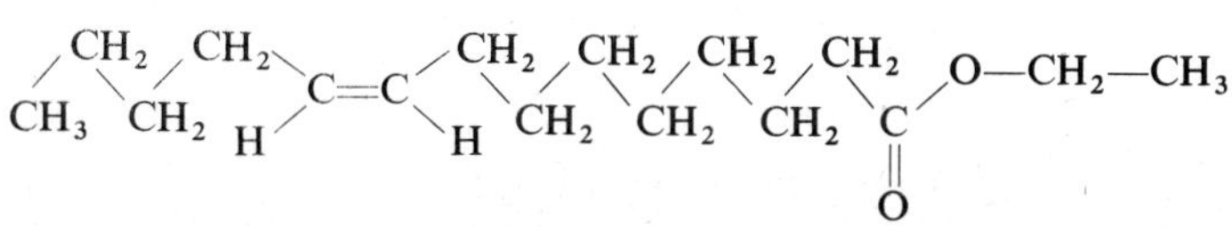

Exercise 3.6 (Continued)

c) $CH_3{-}CH{=}CH_2 + H_2 \xrightarrow{\text{catalyst}}$

d) $CH_2{=}CH{-}CH_2{-}CH_3 + Cl_2 \longrightarrow$

Unsymmetrical agents (HX, H_2O) can sometimes form more than one product when added to a double bond.

$$\underset{\text{propene}}{CH_3{-}CH{=}CH_2} + HBr \longrightarrow \underset{\text{2-bromopropane}}{CH_3{-}\overset{Br}{\overset{|}{C}}H{-}\overset{H}{\overset{|}{C}}H_2} + \underset{\text{1-bromopropane}}{CH_3{-}\overset{H}{\overset{|}{C}}H{-}\overset{Br}{\overset{|}{C}}H_2}$$

In such cases, one of the two products is made in greater yield than the other. A simple rule predicts that *the hydrogen adds to the carbon that already has the most hydrogens.* This is sometimes called Markovnikov's rule. It happens because in the double bond the pi electrons move most readily toward the carbon with the most hydrogens. The hydrogen of the unsymmetrical addition reagent is the positive end of that polar bond and is attracted to the pi electrons. The addition of the hydrogen, as described earlier, produces a carbonium ion. The negative half of the agent adds at the positive charge, completing the reaction.

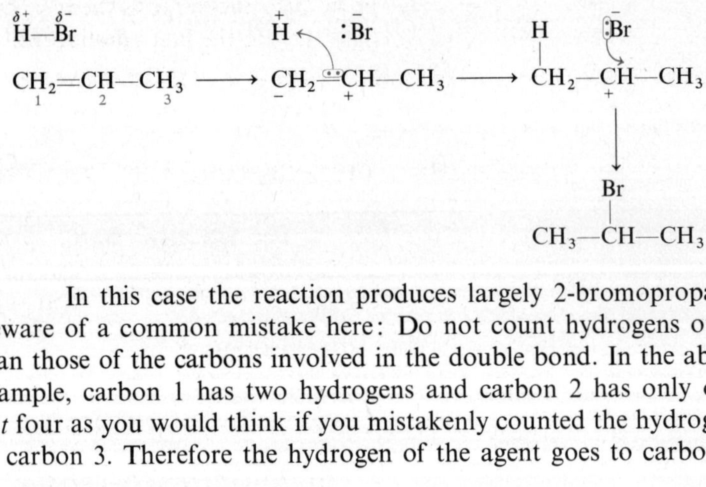

In this case the reaction produces largely 2-bromopropane. Beware of a common mistake here: Do not count hydrogens other than those of the carbons involved in the double bond. In the above example, carbon 1 has two hydrogens and carbon 2 has only one, *not* four as you would think if you mistakenly counted the hydrogens of carbon 3. Therefore the hydrogen of the agent goes to carbon 1.

Exercise 3.7

Complete these reactions:

a) $CH_3{-}CH{=}CH_2 + H_2O \xrightarrow{H^+}$

b) $CH_3{-}CH_2{-}CH{=}CH_2 + HCl \longrightarrow$

c) 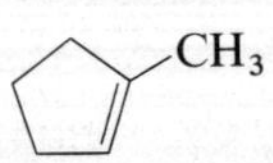 $+ H_2O \xrightarrow{H^+}$

3.8 Reactions of Alkynes

Alkynes perform addition reactions as do the alkenes. However, because there are two pairs of pi electrons in the bond rather than one, it is possible to add the agent twice. Compare the addition of Cl_2 to 1-butene and 1-butyne.

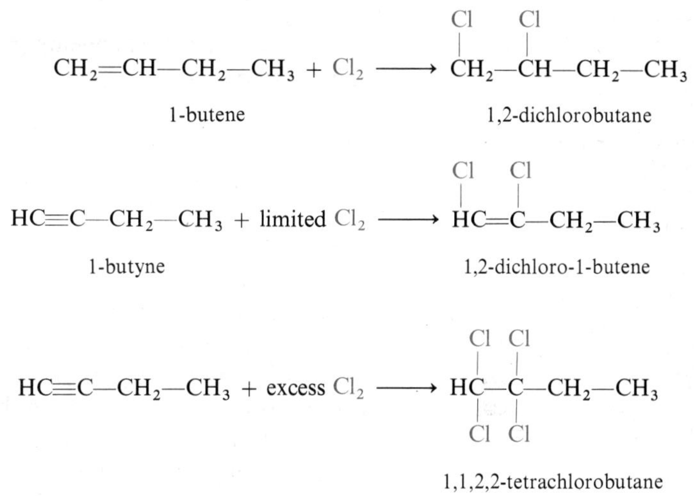

Notice that when we add HX to an alkyne, the second HX to add will be in the same order as the first, because the carbon that had the most hydrogen at the first addition will now have one more.

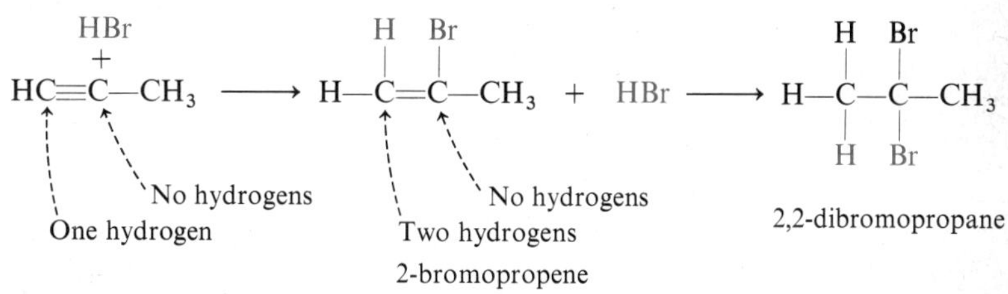

Exercise 3.8

Complete these reactions:

a) $HC{\equiv}CH$ + limited Br_2 $\longrightarrow$

b) $HC{\equiv}CH$ + excess Br_2 $\longrightarrow$

c) $HC{\equiv}CH$ + limited HCl $\longrightarrow$

d) $HC{\equiv}CH$ + excess HCl $\longrightarrow$

e) $CH_3CH_2C{\equiv}CH$ + excess HBr $\longrightarrow$

3.9 Addition Reactions in Biochemistry

Fats are very large molecules that frequently contain double bonds (Chapter 7).

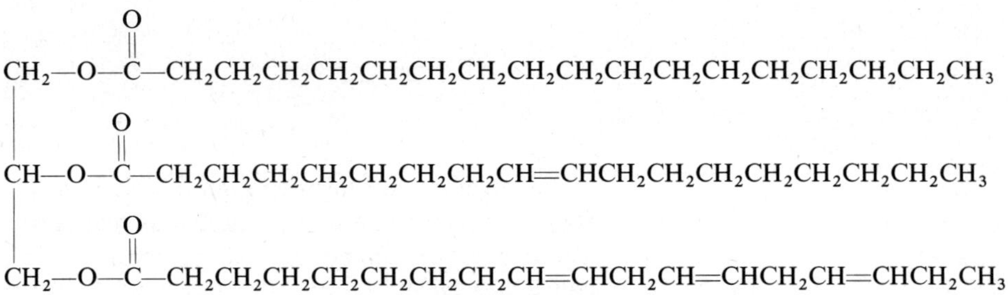

A representative fat molecule

The double bonds are almost always cis. Notice that this structure has several double bonds. Nutritionists therefore say that it is polyunsaturated.

The more double bonds a fat has, the lower its melting point (Table 3.5). Edible oils from plant sources, such as peanut, corn, or

Table 3.5
Fats and Oils
The Effect of Double Bonds on the Melting Point

Fat or Oil[a]	*Approximate Number of Double Bonds*	*Melting Point (°C)*
Cocoa butter	1–1.5	48–53
Beef tallow	1.5	43–44
Mutton tallow	2	34–49
Goose fat	2–2.5	37–40
Neat's foot oil	2–3	29–41
Lard	2–3	33–38
Peanut oil	3.5–4	26–36
Cottonseed oil	3.5–4	35
Corn oil	4–5	17–20
Pumpkinseed oil	5	26
Sunflower oil	5	22–24
Soybean oil	5	26–27
Poppyseed oil	5–5.5	21
Hempseed oil	6	17–21
Candlenut oil	6.5	20–21

[a] These fats and oils have approximately the same molecular weight: 900.

soybean oil, contain relatively large numbers of double bonds and are liquids at room temperature. Liquid plant oils can be converted to solids by adding hydrogen to some of the double bonds in the presence of a catalyst.

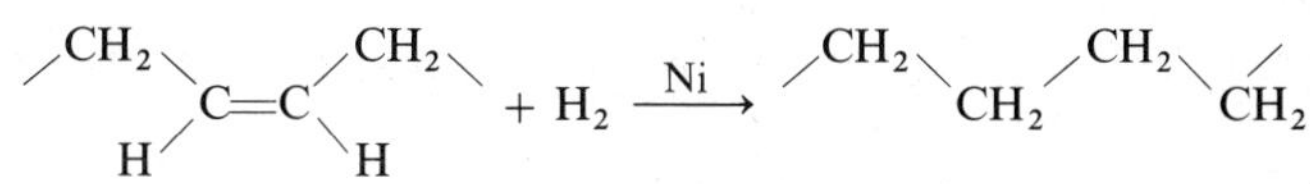

If the vegetable oil loses enough double bonds, it becomes a solid at room temperature and is marketed as margarine or solid shortening. The process of raising the melting point in this fashion is sometimes called *hardening* the oil. The labels of food products read "*hydrogenated.*" It is possible by controlling the amount of hydrogen added to produce a margarine that melts higher (more hydrogen) in hot weather and lower (less hydrogen) in cold weather.

The utilization of a fat such as margarine in the body cells involves a series of chemical changes. Some of these changes involve the same alkene reactions we have been discussing. A trans double bond is formed in the fat, and then water is added across this double bond to create an alcohol. How these reactions contribute to the use of fats by cells is discussed in detail in Chapter 16.

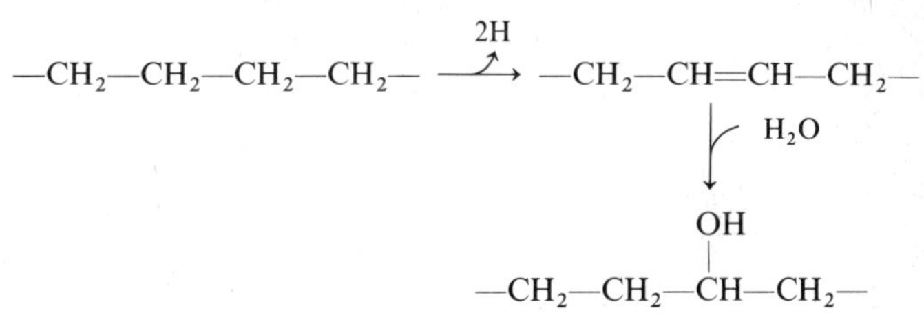

3.10 *Enzymes, the Catalysts of Living Organisms*

The reactions shown above require catalysts. So far we have used metals and acid as examples of catalysts (page 53). These catalysts speed the rate of all examples of a type of reaction. For example, acid makes the addition of water to any double bond occur more quickly. Inside the body each and every chemical reaction is catalyzed, and there is a separate catalyst for each specific reaction. In the case of adding water to a double bond, there is an individual catalyst for each compound to which we add water. These catalysts as a group are called *enzymes*. Enzymes are not simple metals or acids but rather large and complex molecules (Chapter 12). Thousands of different enzymes are produced to catalyze all the reactions necessary to life. Enzymes are superior catalysts, often capable of stimulating a reaction that is very difficult or even impossible to perform in the laboratory. The creation of a trans double bond at a specific place on a hydrocarbon chain, or the conversion of a trans double bond to the cis form as in the chemistry of vision (page 50), are examples of such reactions.

3.11 Juvenile Hormone

Another compound with the potential for controlling insect pests is the terpene structure called *juvenile hormone*. Immature insects produce the hormone to allow adequate growth to occur before developing to sexual maturity. If artificial supplies of the hormone are provided, the insects never mature but become giant larvae instead. Lacking sexual maturity, they cannot produce a new generation. Attempts have been made to produce compounds in the laboratory that are similar enough in reactivity and shape to the natural compound to have juvenile hormone activity. One such effort involves the modification of a known and readily obtained terpene with no hormone activity by the addition of HCl to some of the double bonds. The product does act like juvenile hormone.

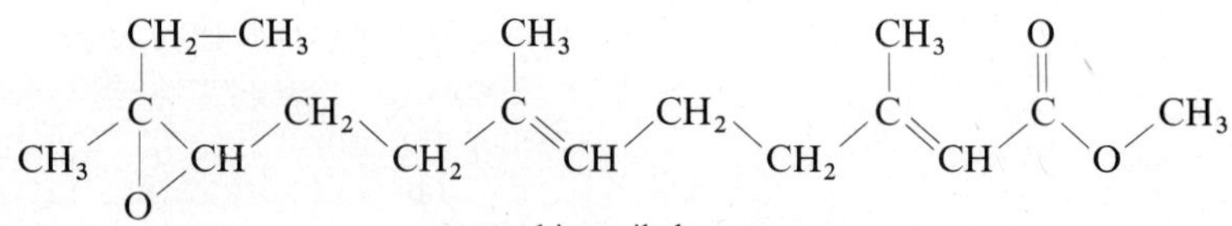

natural juvenile hormone

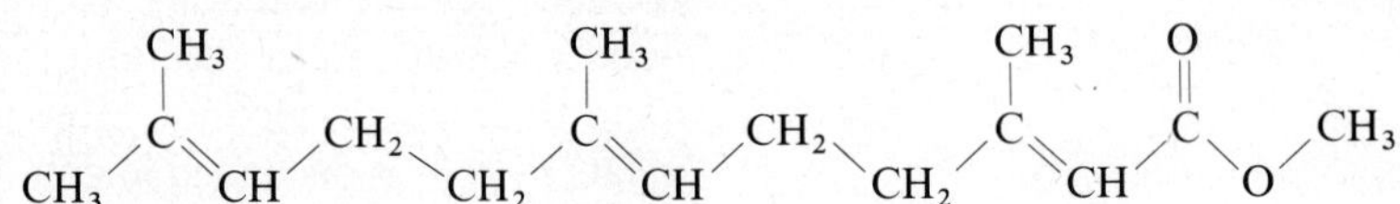

available terpene

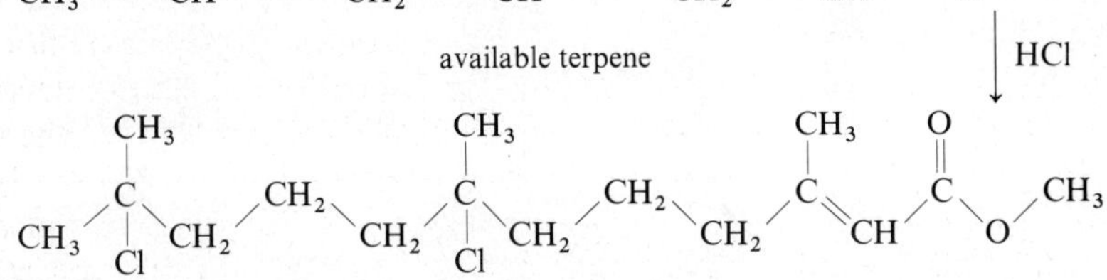

synthetic juvenile hormone

Exercise 3.9

Is the HCl added to the double bonds as Markovikov's rule would predict?

3.12 Polymers

Of the many addition reactions in industry, the most important is the formation of polymers. Polymers are giant molecules assembled by adding together a large number of small units called *monomers.* This process is called *polymerization.*

One class of polymers, the addition polymers, is formed by adding alkenes to *each other* to produce an extremely long chain:

$$\text{a large number of } \underset{\substack{\text{ethene}\\\text{(ethylene)}}}{CH_2{=}CH_2} \longrightarrow \underset{\text{polyethylene}}{\left(CH_2{-}CH_2\right)_{\text{repeated many times}}}$$

Using n to represent an uncertain but large number,

$$n\mathrm{CH_2{=}CH_2} \longrightarrow \mathrm{-\!\!(CH_2{-}CH_2)_{\mathit{n}}\!\!-}$$

The polymerization is initiated by a catalyst that stimulates the formation of either an ionic or a free radical intermediate.

$$\underset{\text{Free radical}}{\mathrm{A}^{\cdot}} + \mathrm{CH_2{=}CH_2} \longrightarrow \mathrm{A\,\dot{\cdot}\,\overset{\odot}{C}H_2{-}\overset{\odot}{C}H_2} \longrightarrow \mathrm{A{-}CH_2{-}CH_2}\cdot$$

The catalytic group is now on the monomer.

$$\mathrm{A{-}CH_2{-}CH_2}\cdot + \mathrm{CH_2{=}CH_2} \longrightarrow \mathrm{A{-}CH_2{-}CH_2\dot{\cdot}\overset{\odot}{C}H_2{-}\dot{C}H_2}$$

$$\downarrow$$

$$\mathrm{A{-}CH_2{-}CH_2{-}CH_2{-}CH_2}\cdot$$

And the process repeats many times.

Because of their high molecular weight, polymers are solids. Most solids you have dealt with in chemistry consist of the individual molecules or ions arranged in a very regular pattern. We call these solids *crystalline*. Polymers cannot assume crystalline patterns because the individual molecules are irregular in shape and size. Within the individual molecule, however, the polymers do have the same structure repeated many times, so an orderly arrangement is possible. Typically in polymeric solids there are regions that are very orderly (crystalline) and regions that are disorganized (amorphous) (Figure 3.4). Such solids are not rigid and brittle the way a crystal of rock salt is. Because of the amorphous region, they can be bent and stretched. The higher the proportion of amorphous area, the more flexible the solid will be. One of the useful properties of polymers is that we can prepare solids with exactly the degree of flexibility or hardness we wish by influencing the ratio of amorphous to crystalline character in the product.

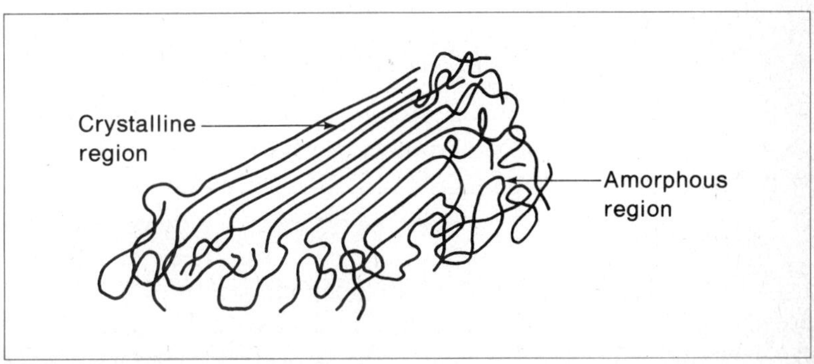

Figure 3.4
Polymeric solids
Long polymer chains form solids in which the chains are neatly organized in rows (crystalline) or twist around one another randomly (amorphous). The proportions of these types of organization determine the properties of the solid.

When heated, the molecules in many polymers eventually twist and vibrate enough to slip by one another. Crystalline regions lose their organized construction, and the polymer can flow. Under these circumstances we can re-form (mold) the shape of the solid into a desired pattern. When it cools, the polymer solidifies and keeps the new shape. Polymers that can be heated and formed in this fashion are called *thermoplastic* ("flexible on heating"). In fact, the general term *plastic* has been applied to all synthetic polymers, even those that do not soften on heating.

Many of our common plastics are addition polymers (prepared by the addition of alkenes to themselves). Examples are provided in

Table 3.6
Some Industrial Addition Polymers

Monomer	*Polymer*	*Common Name*	*Sample Uses*
$CH_2{=}CH_2$	$(CH_2-CH_2)_n$	Polyethylene	Kitchenware, food wrap, bags bottles
$CH_2{=}CH(CH_3)$	$(CH_2-CH(CH_3))_n$	Polypropylene	Kitchenware, milk bottles
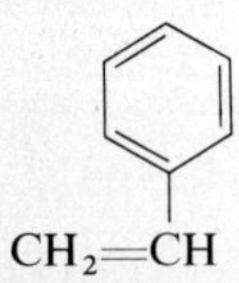	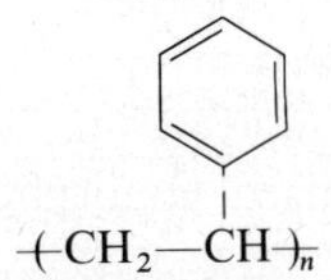	Polystyrene	Toys, insulation (as foam)
$CH_2{=}CH(Cl)$	$(CH_2-CH(Cl))_n$	Polyvinylchloride, vinyl plastic	Imitation leather, rainwear, records
$CH_2{=}CH(C{\equiv}N)$	$(CH_2-CH(C{\equiv}N))_n$	Polyacrylonitrile, Orlon	As a textile fiber in clothing, carpeting
$CH_2{=}C(CH_3)(C(=O)-OCH_3)$	$(CH_2-C(CH_3)(C(=O)-OCH_3))_n$	Polymethylmethacrylate, Lucite, acrylic	Unbreakable windows, paints

Figure 3.5
Some common plastic objects
Many thermoplastic addition polymers are important in our daily lives. Here we see a polypropylene bottle, an acrylic juice glass, a Teflon-coated cookie pan, and a polystyrene egg carton. (Photo by Catherine Feryok, Eastern Michigan University.)

Table 3.6 and Figure 3.5. Once again, *n* indicates a large but unknown number of units in the final molecule. The individual molecules grow rather randomly during polymerization and so vary in size from one another in a single sample. Therefore *n*, if given a value, would represent an average value. How the chain starts and ends contributes very little to the properties of the solid because it is such a small portion of the total molecule, so this is usually not written into the structure.

Additional information is contained in the Appendix 2 article, "Plastics in Cars."

Summary

Alkenes and alkynes are hydrocarbons with double and triple bonds, respectively. In nature the triple bond is rare but the double bond is very common. Terpenes, a class of compounds very common in plants, are made of five carbon units joined together and usually have many double bonds. Fats and oils have double bonds, and the greater the number they have, the lower their melting point. Alkenes and alkynes combine with simple reagents such as H_2, H_2O, HCl, HBr, Cl_2, and Br_2 by a process called the addition reaction. Adding H_2 to liquid oils eliminates double bonds, raising the melting point. Other addition reactions, catalyzed by enzymes, occur in the use of fats by living organisms. In industry the most important alkene reaction is the addition of compounds containing double bonds to themselves, generating very high-molecular-weight molecules called polymers or, more simply, plastics.

Problems

1. Name the following:

a) $CH_3—CH=CH—CH_2—CH_3$

b) 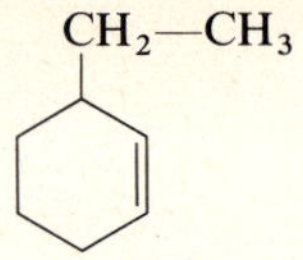

c) 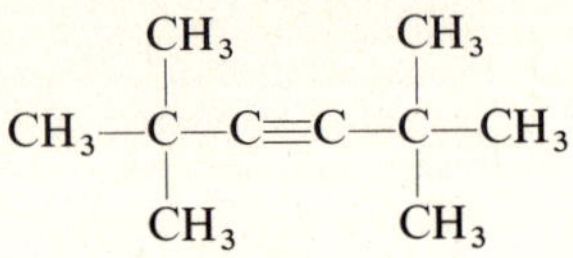

d) 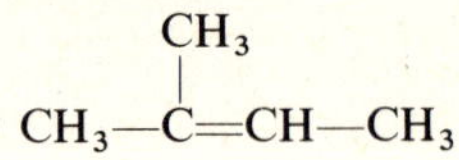

e) 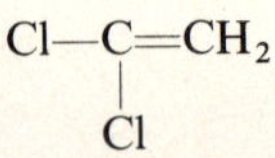

f) 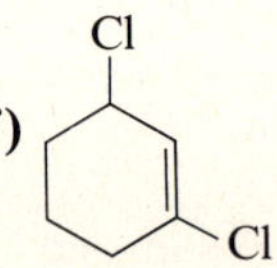

g) $H—C\equiv C—CH_2—CH_2—Br$

h) 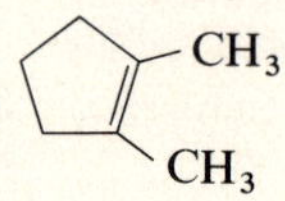

i) 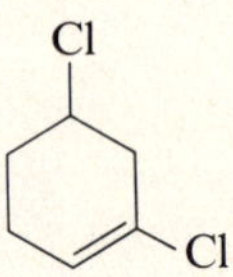

2. Draw structures of the following:

a) 2-methyl-3-hexene

b) 1,2-dichloro-1-propene

c) 1-iodo-2-butyne

d) 3-isopropylcyclopentene

e) 3-ethyl-3-methyl-1-hexene

f) 1,4-dimethyl-1-cyclohexene

g) 4,4-dichloro-2-pentyne

h) 1-isopropyl-3-methyl-1-cyclopentene

3. Look at the covalent bonds in each of the following compounds, and determine for each the total number of sigma electrons and the total number of pi electrons:

a) 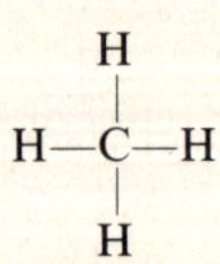

b) 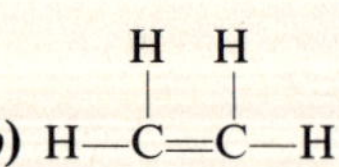

c) 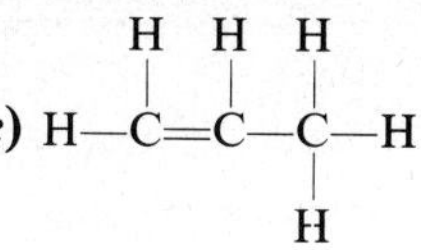

d) H—C≡C—H

e) 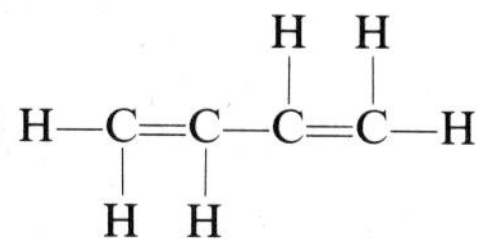

f) 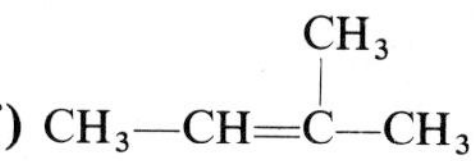

g) H—C≡C—CH=CH_2

h) 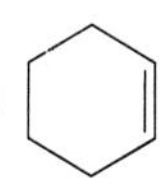

i) $HCCl_3$

j) 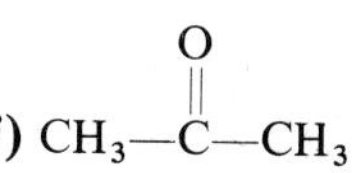

4. Which of the following could display cis–trans isomerism?

a) 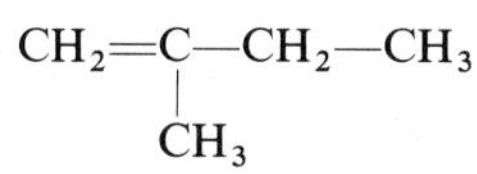

b) Cl—CH=CH—Cl

c) CH_3—C≡C—CH_2—CH_3

d) 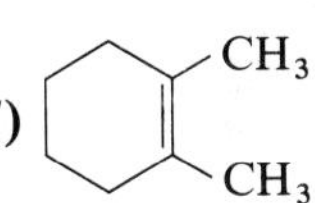

5. Below are some compounds that are isolated from living organisms. Which do you suspect to be isoprenoid?

a) 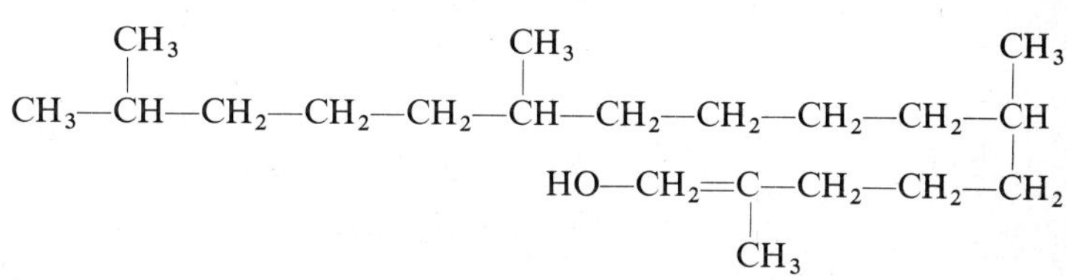

phytol, part of the chlorophyll molecule

b) 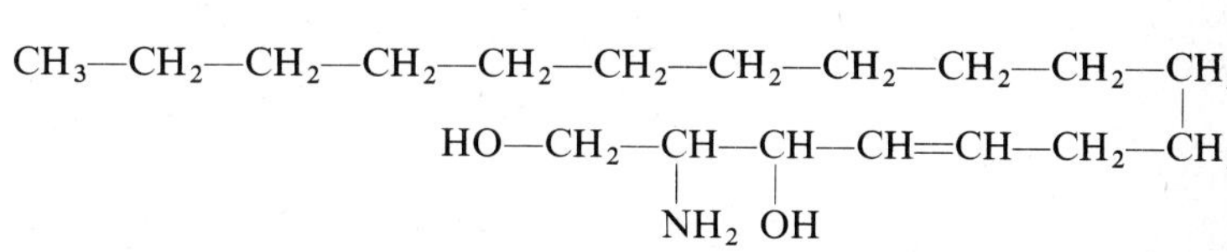

sphingosine, found in brain lipids

c) 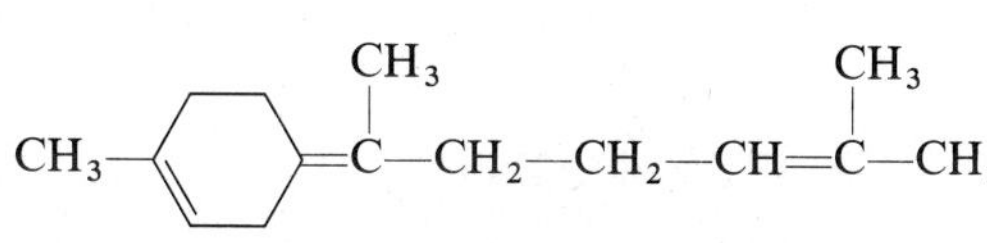

γ-bisabolene, from oil of bergamot

d) 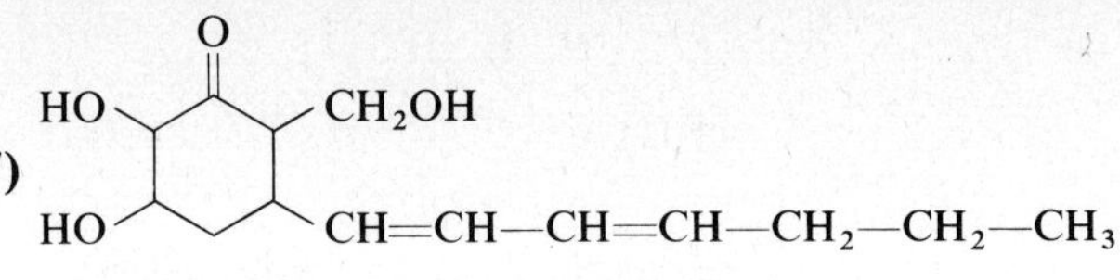

palitantin

6. Refer to Table 3.4, Insect Pheromones.

a) For each compound listed, identify from Table 1.5 the functional group or groups present.

b) For each compound with a double bond, decide whether geometric isomerism could exist.

c) For each double bond that can have geometric isomerism, decide whether the bond is drawn cis or trans. In which cases does this become difficult to decide? Are there any where both isomers are listed?

d) Identify all the compounds that are isoprenoid.

7. If a compound has a double bond for which there are geometric isomers, there are two forms of the compound, cis and trans.

a) If the molecule contains two such bonds, how many isomers are there?

b) How many isomers exist for three double bonds, each displaying geometric isomerism?

c) How many geometric isomers are possible for isoprene (page 70); each of the bee sex attractants (Table 3.4); citronellol, geraniol, farnesol, and vitamin A (Table 3.3); and β-carotene (Table 3.2)?

8. Complete the following reactions:

a) $CH_3-CH=CH-CH_3 + HBr \longrightarrow$

b)  $+ H_2O \xrightarrow{H^+}$

c) $CH_2=CH-CH_3 + Cl_2 \longrightarrow$

d) $CH_3-\underset{\displaystyle CH_3}{\underset{|}{C}}=\overset{\displaystyle CH_3}{\overset{|}{C}}-CH_3 + H_2 \xrightarrow{Pt}$

e) $CH_3-C\equiv CH + \text{limited } Br_2 \longrightarrow$

f) $CH_3-CH_2-C\equiv CH + \text{excess } Cl_2 \longrightarrow$

g) $CH_2=CH-CH_2-CH=CH_2 + \text{limited } H_2 \xrightarrow{\text{catalyst}}$

h) $CH_3-CH_2-CH=CH_2 + HCl \longrightarrow$

i) $CH_3-CH=\underset{}{\overset{CH_3}{\overset{|}{C}}}-CH_3 + H_2O \xrightarrow{H^+}$

j) 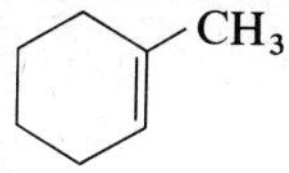 $+ HBr \longrightarrow$

k) $H-C\equiv C-H$ + limited HCl $\longrightarrow$

l) $H-C\equiv C-H$ + excess HCl $\longrightarrow$

m) $CH_2=\underset{CH_3}{\underset{|}{C}}-CH_2-\underset{CH_3}{\underset{|}{C}}=CH_2$ + limited HBr $\longrightarrow$

9. Write equations for the conversion of acetylene ($H-C\equiv C-H$) into each of the following:

a) $CH_2=CH_2$

b) CH_3-CH_3

c) CH_3-CHCl_2

d) $CHCl=CHCl$

e) $CHBr_2-CHBr_2$

f) CH_2Cl-CH_2Cl

g) $CH_2Cl-CHCl_2$

10. This polymer is used in making flexible film for food wraps. What is the monomer?

$$\left(CH_2-\underset{Cl}{\underset{|}{\overset{Cl}{\overset{|}{C}}}}\right)_n$$

11. How would you make the polymer polyvinyl chloride (Table 3.6) starting with acetylene?

12. Suggest how a manufacturer could produce the soft spreadable margarine sold in tubs rather than the hard stick margarine.

13. What do you imagine happens to the thermoplastic properties of a polymer when the individual polymer chains are bridged together as illustrated below:

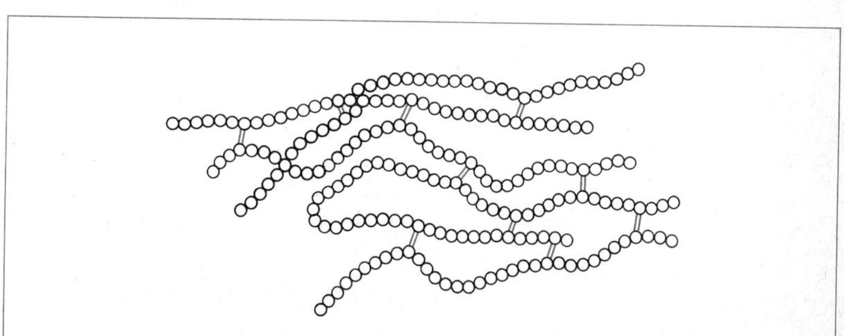

4

Conjugated Double Bonds

Objectives

In this chapter we:

1. *Introduce the concept of conjugated double bonds.*
2. *Learn to name compounds with more than one double bond.*
3. *Study 1,4-addition reactions.*
4. *Survey the chemistry and properties of rubber-type polymers.*
5. *Introduce the concept of aromatic structures.*
6. *Learn to name aromatic structures.*
7. *Study some aromatic substitution reactions.*
8. *Learn about the relationship between chemical structure and color.*

The pi (second pair) electrons in double bonds are able to move about relatively freely. This free movement is important to addition reactions. In fact, pi electrons can have effects that extend beyond the boundaries of the atoms they join. If two double bonds are next to each other, a very reactive system results, the chemistry of which we do not pursue in this text. However, even when the double bonds are separated by a single bond they are capable of interacting, and this is of interest. We call such double bonds *conjugated.*

A conjugated system

$$CH_2{=}CH{-}CH{=}CH_2$$

Double bond	Single bond	Double bond

Exercise 4.1

Which of these double bonds are in conjugation?

a) 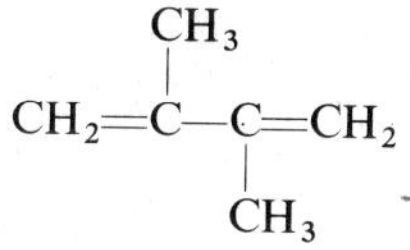

b) 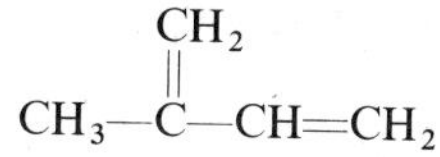

c) $CH_2{=}CH{-}CH_2{-}CH{=}CH_2$

d) 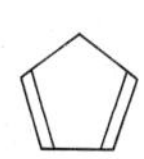

e) 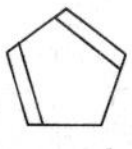

f) 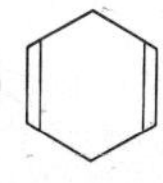

g) 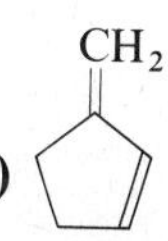

4.1 Naming Compounds with Multiple Double Bonds

The -ene ending of alkene names indicates one double bond. When there is more than one double bond in a molecule, we change the -ene ending to one that tells how many double bonds there are: -diene for two double bonds or -triene for three double bonds. Numbers to indicate the double bond location come immediately before the name of the longest carbon chain.

$CH_2{=}CH{-}CH{=}CH_2$

1,3-butadiene

$CH_2{=}CH{-}CH{=}CH{-}CH{=}CH_2$

1,3,5-hexatriene

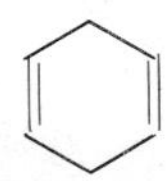

1,4-cyclohexadiene

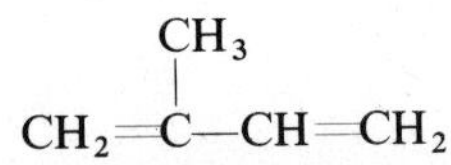

2-methyl-1,3-butadiene
(isoprene)

4.2 Reactions of Conjugated Double Bonds

When a reagent that adds to double bonds is added in limited quantity to a conjugated pair of double bonds, it may add to one of the double bonds in the manner described in Chapter 3. However, it is more likely to add across the entire conjugated system, eliminating one double bond and shifting the remaining double bond to the center of the once conjugated system.

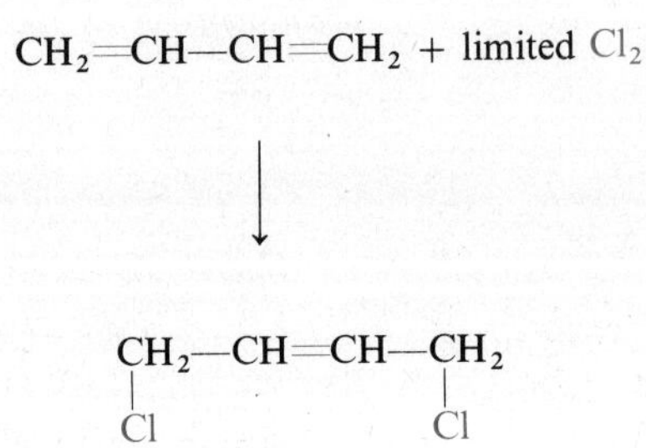

We call this *1,4-addition.*

> ### *Exercise 4.2*
> What is the product if limited HCl is added to 1,3-butadiene?

When a reagent adds to the conjugated system, the shift of the pi electrons occurs as follows:

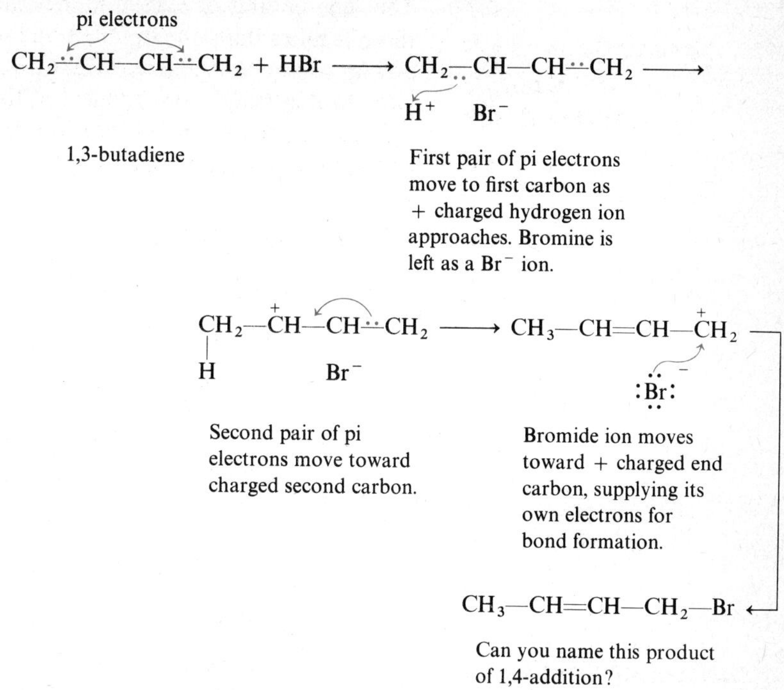

Notice how freely the pi electrons can move about. As a generalization, addition to a 1,3-diene results in addition at carbons 1 and 4, the elimination of the 1 and 3 double bonds, and the generation of a 2 double bond.

4.3 *Rubber, a Product of 1,4-Addition*

Rubber-type compounds, both natural and synthetic, are produced by 1,4-addition reactions. Natural rubber is an isoprenoid compound (Chapter 3). In fact, it is the product that would result from the 1,4-addition of isoprene to itself.

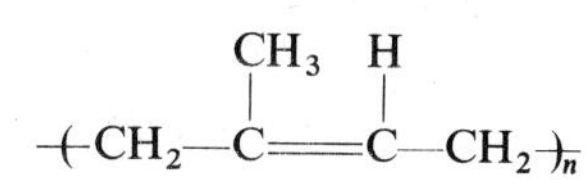

natural rubber

The resulting double bond could have cis and trans isomers, but in natural rubber only cis is found. Biochemical reactions lead to the formation of only one isomer if more than one is possible, and that is true for natural rubber. An all-trans polymer of isoprene called *gutta-percha* also occurs. The difference in properties of the two is an interesting reflection of the difference molecular shape causes. The

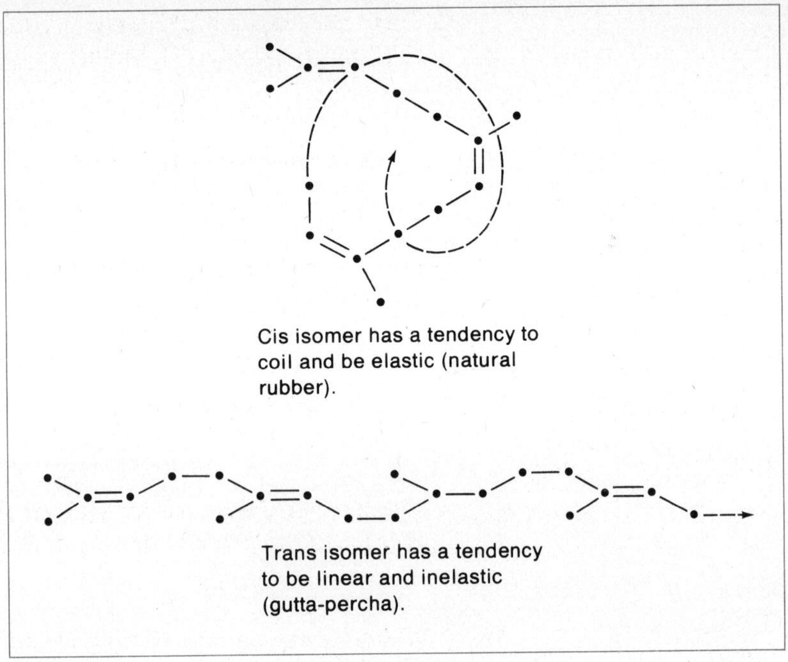

Figure 4.1
Isomers of polyisoprene

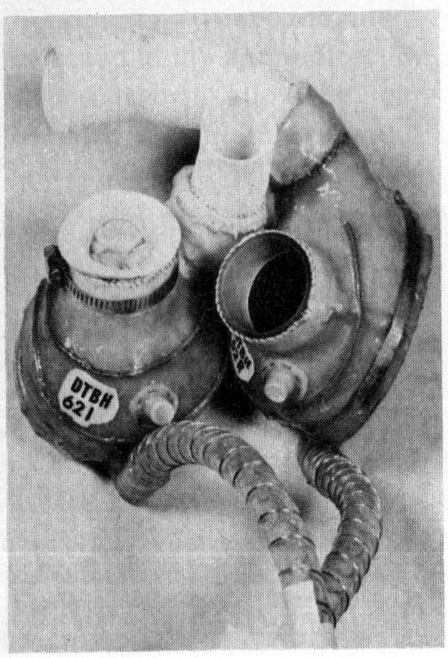

Figure 4.2
Rubber artificial heart
At Cleveland Clinic, Goodyear scientists are developing an artificial heart of rubber. A calf with such a heart survived and grew normally. (From Raymond J. Kiraly, Cleveland Clinic, Cleveland, Ohio.)

cis isomer spontaneously forms long springlike coils that are responsible for the elastic properties of rubber. The trans polymer is linear rather than coiled, with the result that gutta-percha is inelastic and rigid (Figure 4.1). The elastic properties of rubber-type compounds qualify them for uses ranging from tires and rubber bands to artificial hearts (Figure 4.2).

Isoprene can be isolated from petroleum that has been subjected to *cracking*, a process that converts straight-chain hydrocarbons to branched structures. Original attempts to generate rubber by a 1,4-addition of isoprene to itself failed to produce an all-cis product. However, in 1954 a special catalyst, the Zeigler catalyst, caused the formation of an all-cis polyisoprene with properties identical to those of natural rubber.

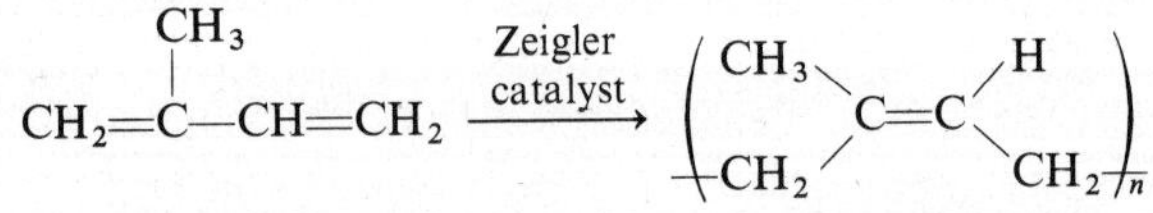

isoprene *cis*-polyisoprene

Neoprene, an important synthetic rubber, is formed by an interesting series of addition reactions from the starting material acetylene. First one acetylene is added to another.

$$2\,HC{\equiv}CH \xrightarrow{\text{catalyst}} H{-}C{\equiv}C{-}CH{=}CH_2$$

Then HCl in limited supply is added to the triple bond to make chloroprene (2-chloro-1,3-butadiene).

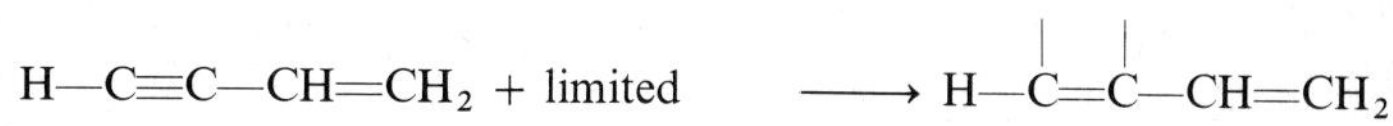

chloroprene

Finally the monomer chloroprene is polymerized by 1,4-addition to produce neoprene.

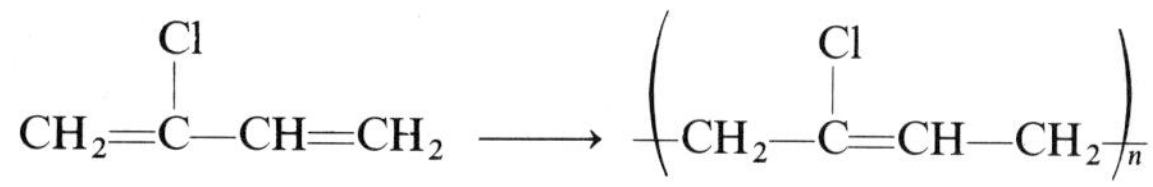

neoprene

Neoprene is often used for seals or tubing where exposure to oil or organic solvents would destroy natural rubber.

4.4 *Aromatic Compounds*

A six-membered ring with three conjugated double bonds is called *benzene*. Benzene, its derivatives, and a number of analogous ring compounds are classed as *aromatic* compounds.

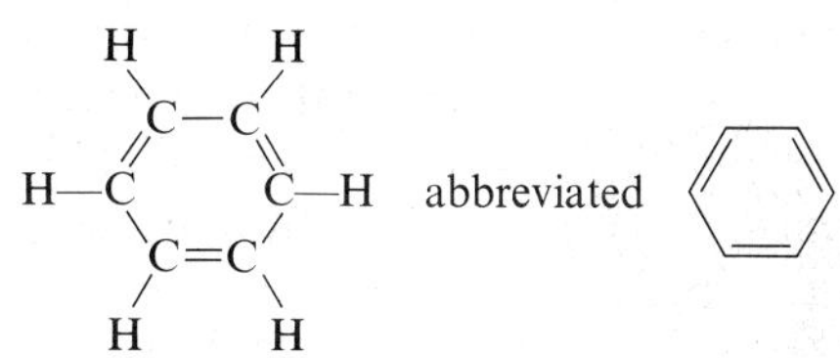

benzene

Aromatic compounds have chemical properties rather different from those of alkenes. Benzene does not undergo addition reactions of the double bond type other than the addition of hydrogen to produce cyclohexane.

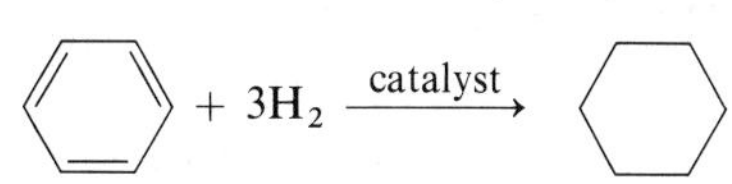

cyclohexane

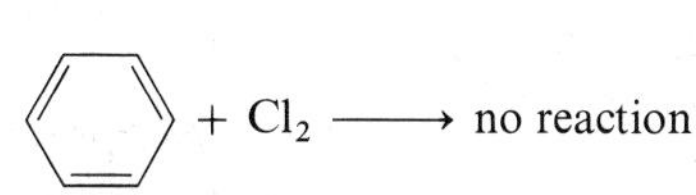

The lengths of bonds in covalent molecules can be measured. Carbon-to-carbon double bonds are shorter than single bonds. The carbon-to-carbon bonds in benzene are all found to be the same length, midway between the length of a single bond and a double bond. We visualize the pi electrons to be spread evenly around the ring, making each carbon-to-carbon bond a sort of "1½ bond."[1]

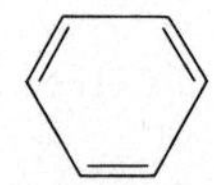

True single and double bonds would generate long and short sides.

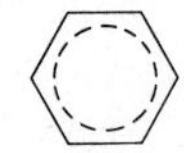

Sides are the same length, and each bond has a share (50 percent) of the pi electrons.

4.5 *Naming Aromatic Compounds*

We have the same kinds of derivatives of benzene that we have of aliphatic hydrocarbons; that is, hydrogen can be replaced with such groups as chloro, bromo, iodo, nitro, and the various functional groups (Chapter 1). Notice that each carbon on a benzene ring has *only one* hydrogen to be replaced, however. The following are some simple benzene derivatives.

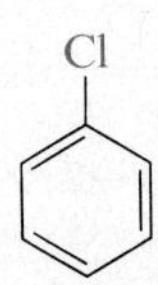

chlorobenzene

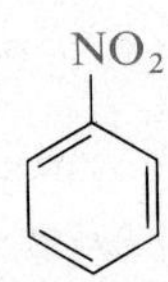

nitrobenzene

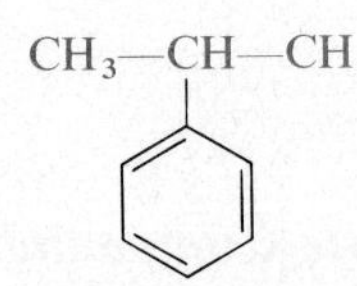

isopropylbenzene (also called cumene)

Exercise 4.3

Name these:

a)

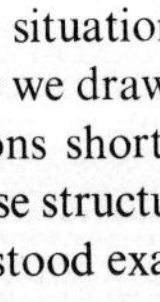

b)

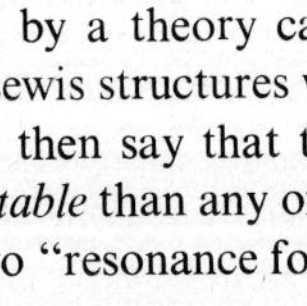

c)

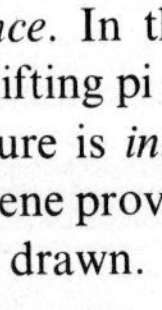

---

[1] This situation is dealt with by a theory called *resonance*. In the resonance theory we draw all the valid Lewis structures we can by shifting pi or unbonded electrons short distances. We then say that the true picture is *intermediate* to all these structures and *more stable* than any of them. Benzene provides an easily understood example where two "resonance forms" can be drawn.

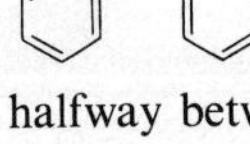

The true picture would then be halfway between the two structures and more stable than either.

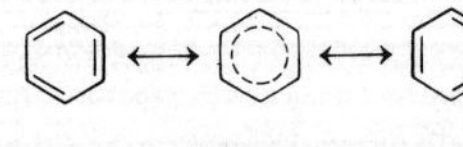

Some of the simple derivatives have special names.

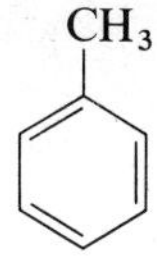

(rather than methylbenzene)

The benzene ring itself can be named as a side group called the phenyl group (just like a methyl or ethyl group). Isopropylbenzene could also be called 2-phenylpropane.

Where there is more than one substitution on benzene, we use numbers just as we would with cyclohexane.

1,2-difluorobenzene

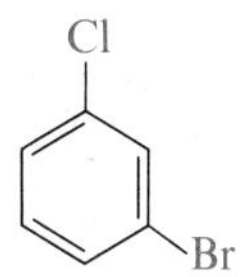

1-bromo-3-chlorobenzene

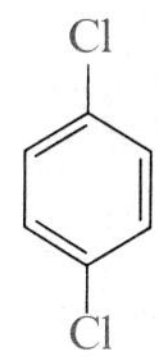

1,4-dichlorobenzene (used to kill moths and caterpillars)

The relationship between two substitutions can also be expressed in the terms ortho, meta, and para. Ortho, abbreviated *o*-, means adjacent. Therefore 1,2-difluorobenzene could also be called ortho difluorobenzene, or simply *o*-difluorobenzene. Meta, or *m*-, infers a 1,3- relationship. Thus 1-bromo-3-chlorobenzene could also be called *m*-bromochlorobenzene. Para, or *p*-, means across the ring from, so 1,4-dichlorobenzene could also be called *p*-dichlorobenzene.

Exercise 4.4

Name the following two different ways, by numbers and by *o*-, *m*-, or *p*- notation:

a) 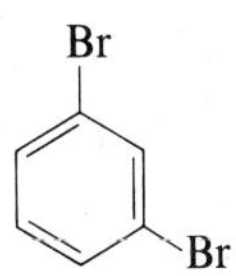

b) 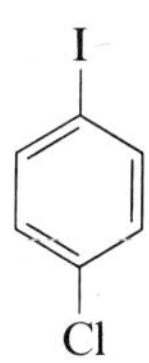

c) 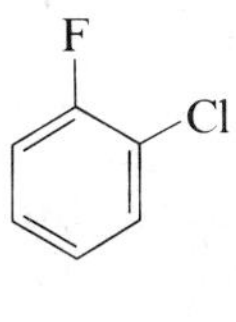

d) 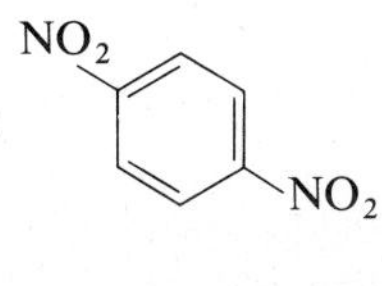

When one of the substitutions gives the compound a special name, that substitution is automatically considered to be at carbon 1.

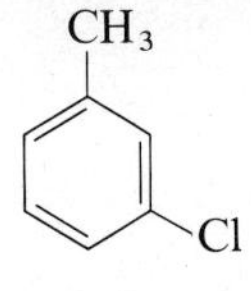

m-chlorotoluene
or
3-chlorotoluene

When there are more than two substitutions, numbers are always used.

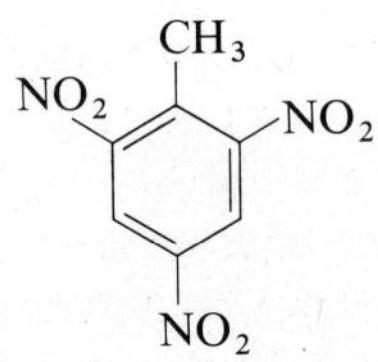

2,4,6-trinitrotoluene
(TNT, the explosive)

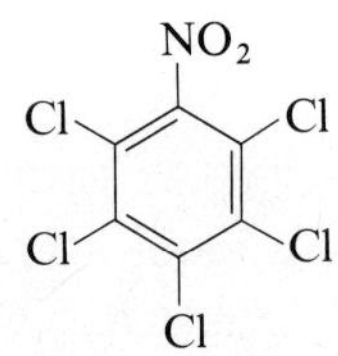

2,3,4,5,6-pentachloronitrobenzene
(a fungicide[2] used to treat seeds)

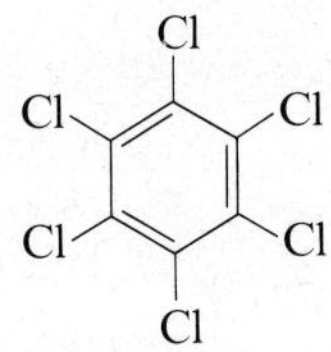

hexachlorobenzene, a fungicide (Why are no numbers necessary?)

Naming can be complicated. For example,

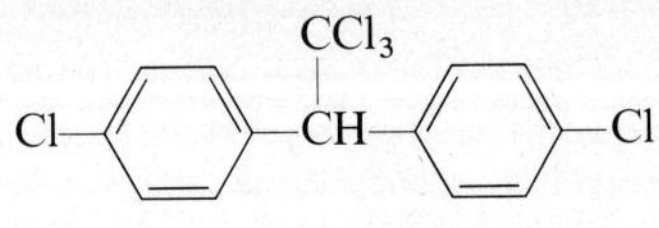

This is 1,1,1-trichloro-2,2-bis- (*p*-chlorophenyl) ethane. Not too surprisingly, the complete correct name is seldom used. Instead it is called DDT (the insecticide)[3]

Other aromatic compounds have more than one benzene ring linked together.

---

[2] A fungicide is a compound that restricts the growth of fungus. Some such agents are used to treat fungal diseases, others to improve the storage characteristics of a wide variety of commodities. Of particular significance is the prevention of the growth of fungi on seeds destined for planting.

[3] DDT is actually an abbreviation for *d*ichloro*d*iphenyl*t*richloroethane, a nonsystematic name for the compound. The term "bis" in the systematic name means *two* (*p*-chlorophenyl) much as "di" in dichloro means *two* chloro.

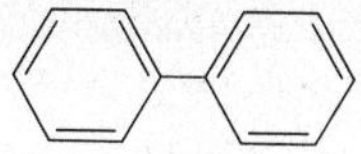

biphenyl

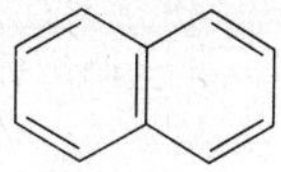

naphthalene

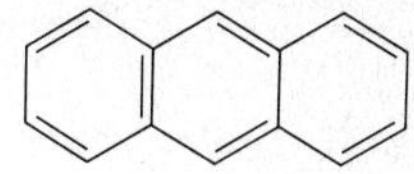

anthracene

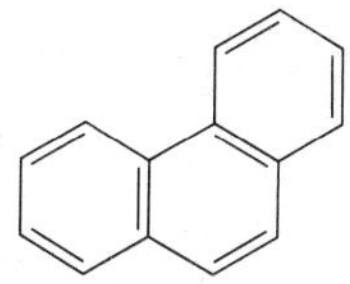

phenanthrene

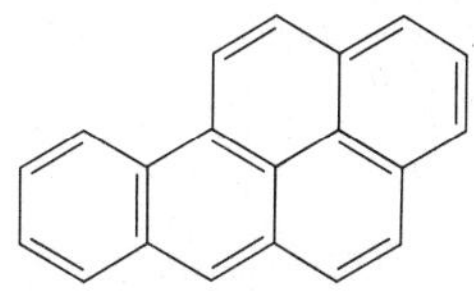

benzopyrene

The more complex polycyclic (multiple-ringed) aromatic structures can be obtained from the breakdown of coal on heating or from the heating of a variety of organic molecules to very high temperatures. Several of these polycyclic aromatics, including phenanthrene and benzopyrene, have been found to be carcinogens. Carcinogens are chemical agents that produce or incite cancer. Such aromatic structures generated by the burning of tobacco may be the cause of lung cancer in smokers.

4.6 Reactions of Aromatic Compounds

As we have already pointed out, the addition reactions presented in Chapter 3 are not characteristic of the behavior of aromatic compounds. Conjugated structures are more stable because of the spreading around of the pi electrons. To perform such an addition reaction requires the elimination of the totally conjugated benzene structure. This is not likely to happen. Instead, aromatic compounds are altered by substitution reactions. The hydrogens attached to the benzene ring are replaced by other groups. Here are some sample reactions ($AlCl_3$ and H_2SO_4 are catalysts):

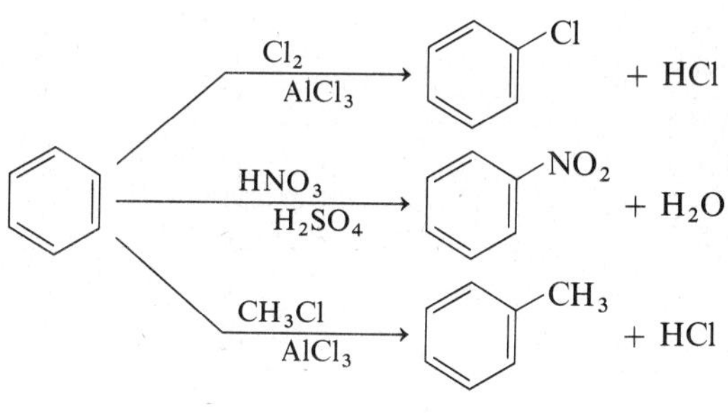

Exercise 4.5

Decide which of the following reactions are likely to be correct and which are definitely wrong or very unlikely:

a) 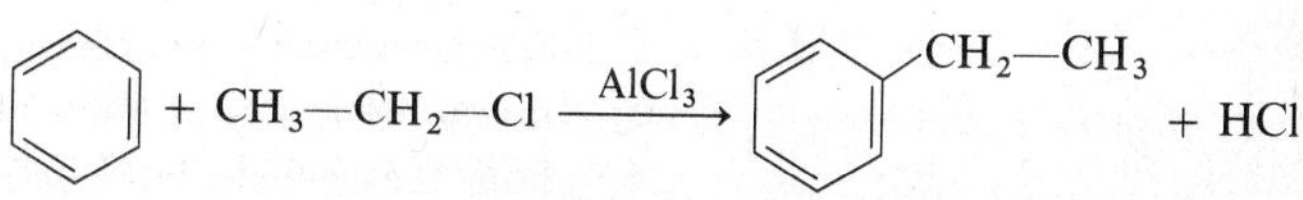

Exercise 4.5 (Continued)

b)
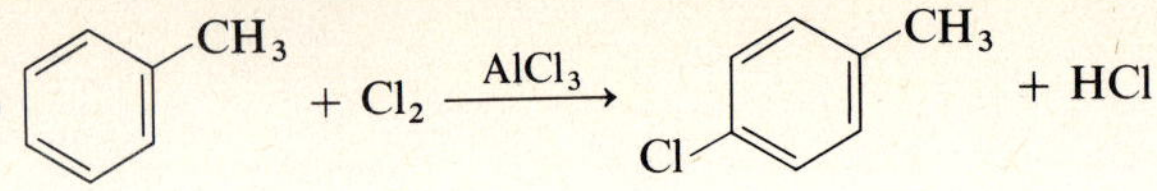

c)
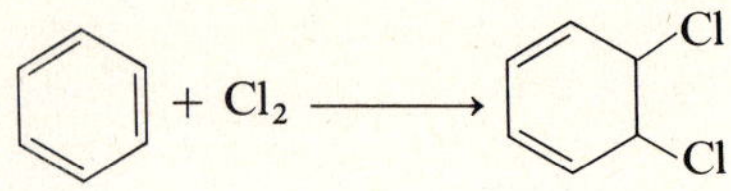

d)
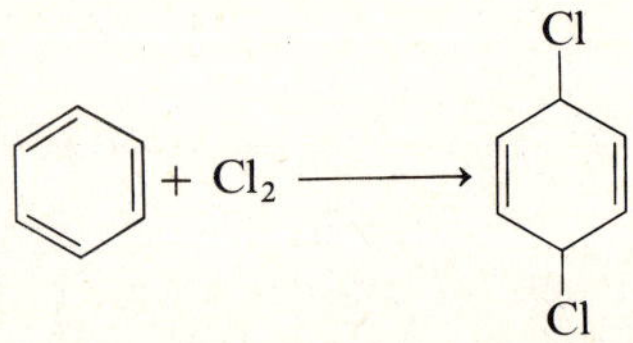

e)
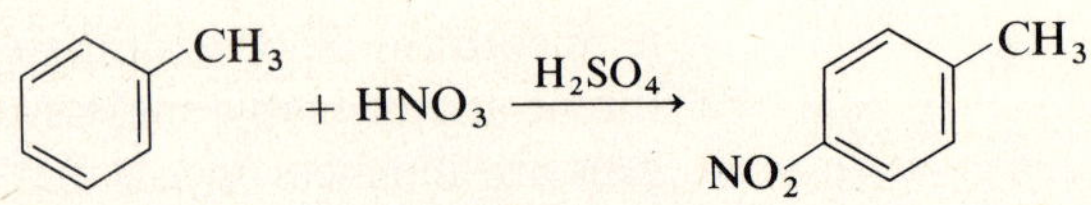

4.7 *Aromatic Compounds of Biochemical Significance*

Most biochemical compounds are not aromatic. However, here are a few very important examples of ones that are.

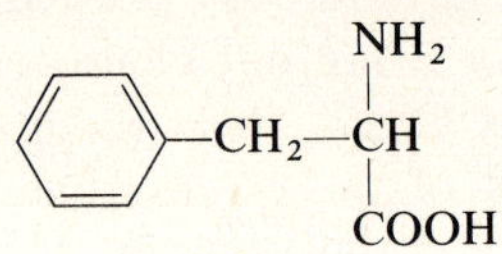

phenylalanine
(component of protein)

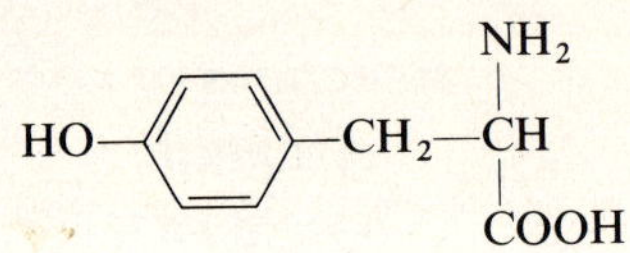

tyrosine
(component of protein)

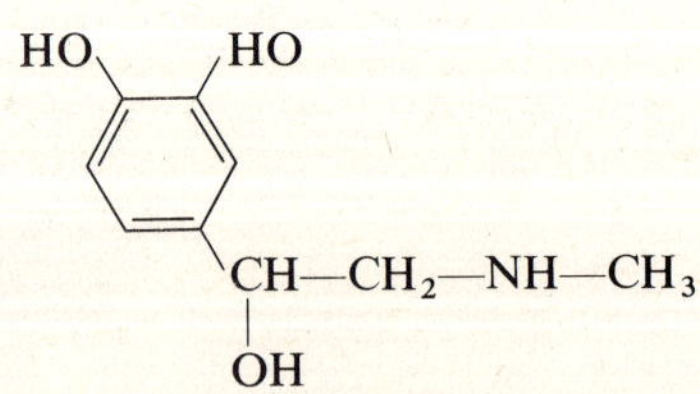

epinephrine
(a hormone)

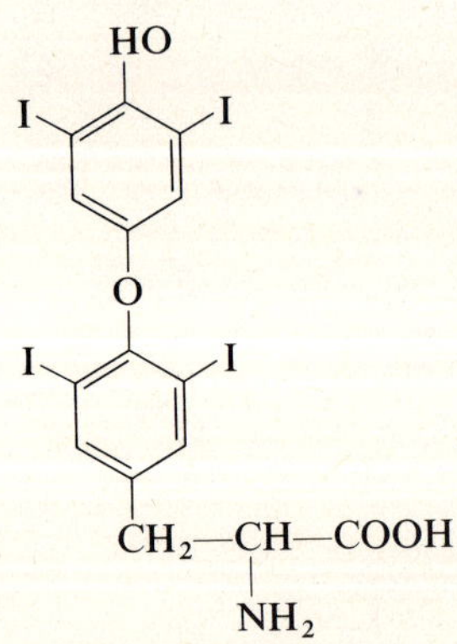

thyroxin
(a hormone)

4.8 *Color in Chemicals: Dyes and Pigments*

If we were to view a large variety of organic compounds, we would soon decide that most were either transparent liquids or white solids. Some do have colors, even very brilliant and intense colors.

Colored compounds that appear in nature were sought out by even very primitive civilization for the purpose of decorations as pigments for pottery or wood carvings or as dyes for fabrics. Perhaps the earliest fabric dye was the plant pigment indigo. In the nineteenth century, vast areas of farmland were devoted to the culture of plants producing this pigment. The German firm BASF spent millions of dollars at a time when research expenditures were relatively small to develop a synthetic indigo. By 1897 they had succeeded in entering the market with such a product. By the start of the twentieth century, the production of natural indigo dropped rapidly in one of the great historic economic readjustments.

What is the relationship between chemical structure and color? A study of colored compounds reveals that conjugated double bonds and atoms with unshared electrons are very common in such structures.

To our eyes, a compound has color when it absorbs some portion of visible light. White light (light including all wavelengths) reflected by or passing through the colored compound loses some of the wavelengths. What we see is the light not absorbed that now does not have all wavelengths and as a result appears colored. The absorbed light energy causes internal changes in the molecule. A very important point must be made at this time. These internal changes involve very exact amounts of energy. Different wavelengths of light (hence different colors) have different amounts of energy. Therefore each specific internal change is caused by the absorption of a specific wavelength of light.

Lower-energy (longer-wavelength) light generally causes parts of the molecule capable of vibrating either to begin vibrating or to vibrate faster. Organic chemists use the absorption of certain wavelengths of infrared light to detect the presence of particular structures in the molecules. Because we cannot see infrared light, a device called an *infrared spectrophotometer* is used to determine exactly what wavelengths of light were absorbed.

Higher-energy (shorter-wavelength) light excites electrons. Certain positions around the individual atomic nucleus or around the several nuclei of covalently bonded atoms are called *energy levels*. Electrons may not exist outside these levels. At room temperature the electrons of molecules are normally in the lowest-energy positions. However, visible or ultraviolet light can cause the electrons to jump to a higher-energy position. The energy of the light must equal exactly the difference in energy between the lower- and the higher-energy state. Ultraviolet light is more energetic than visible light and thus causes electrons to cross between levels that are relatively farther apart. Compounds that are colorless to the eye contain no shifts that require only the limited amount of energy available in visible light.

Table 4.1
Dyes and Conjugated Bond Systems

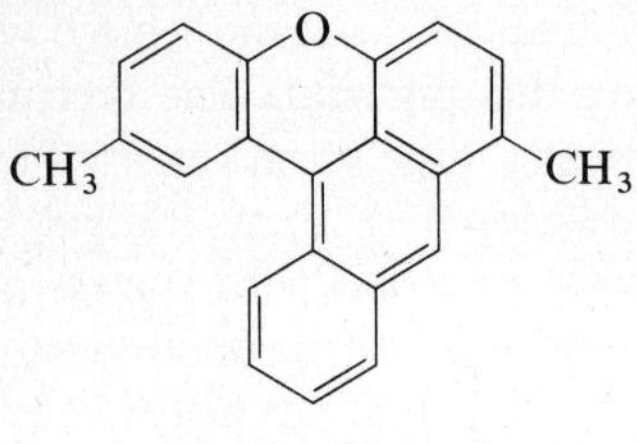

fluorol 5G

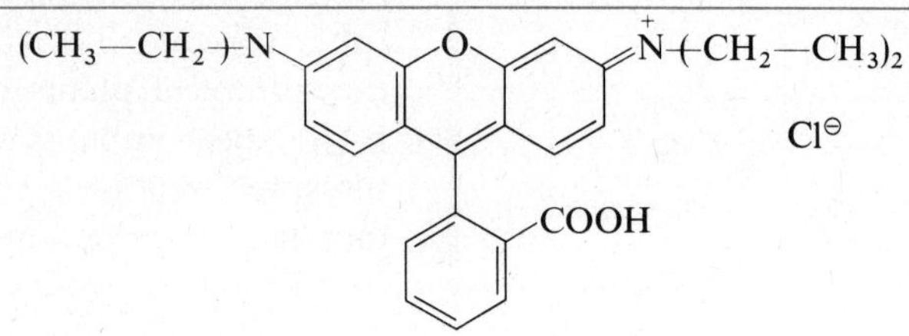

rhodamine B

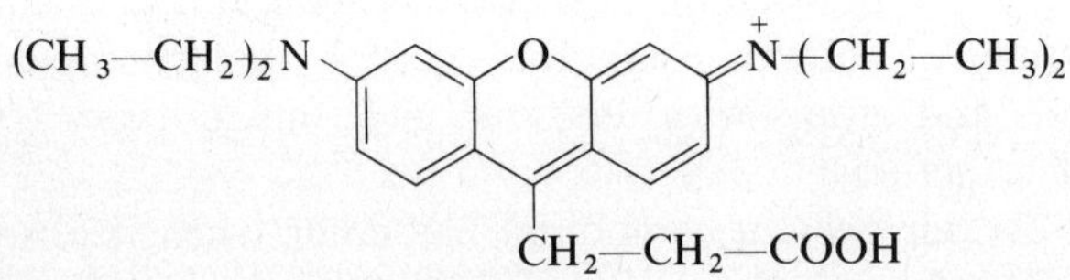

C.I. basic red 11

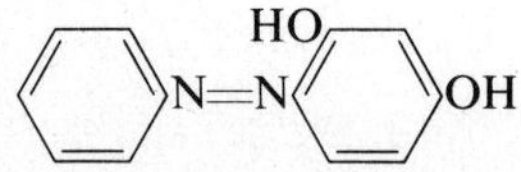

Sudan orange G

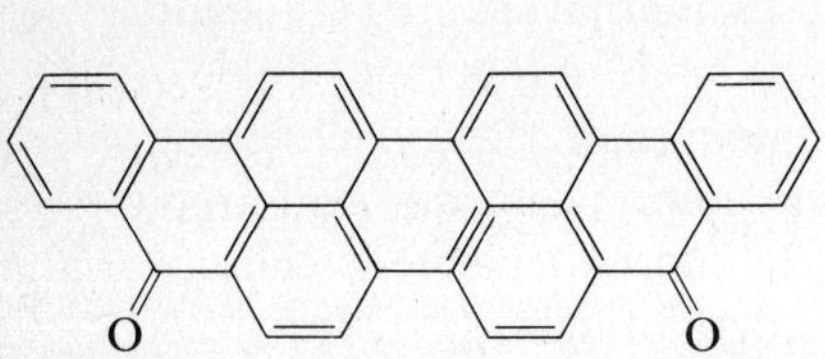

violanthrone

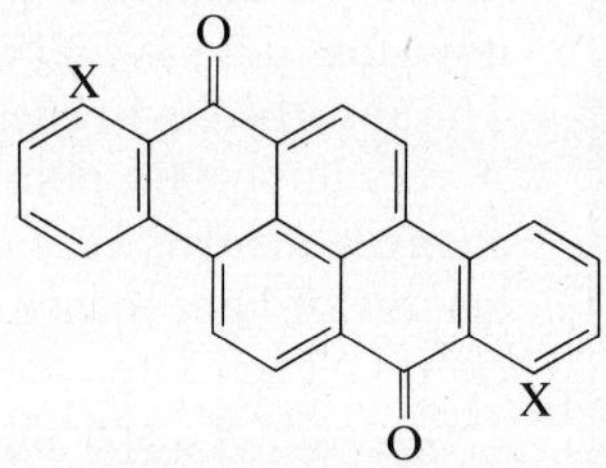

indanthrene golden yellow GK

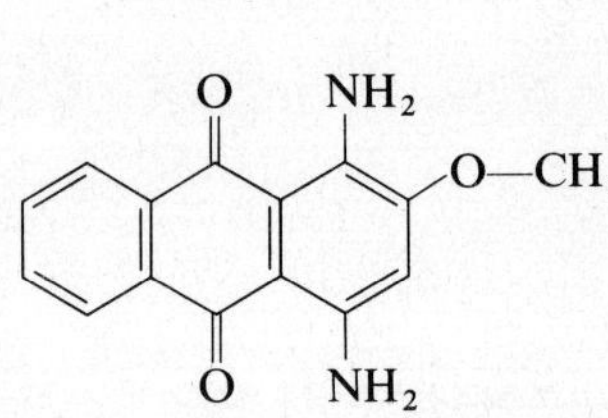

celliton fast red FF 3B

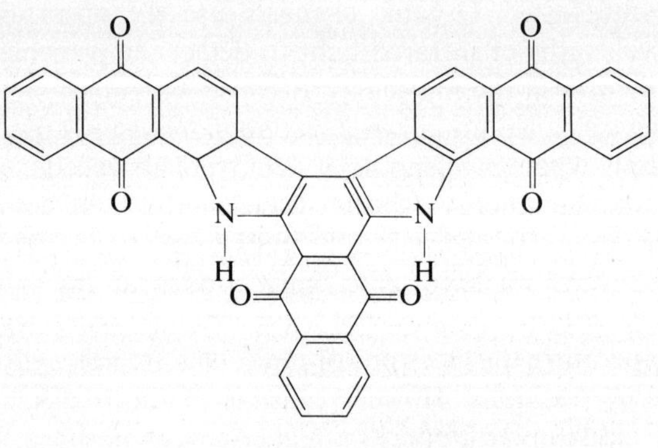

indanthrene brown GR

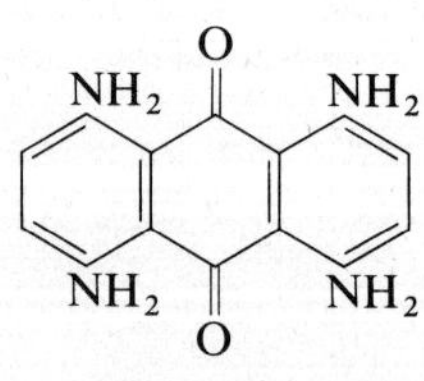

oracet
(sapphire blue G)

Table 4.1 (Continued)

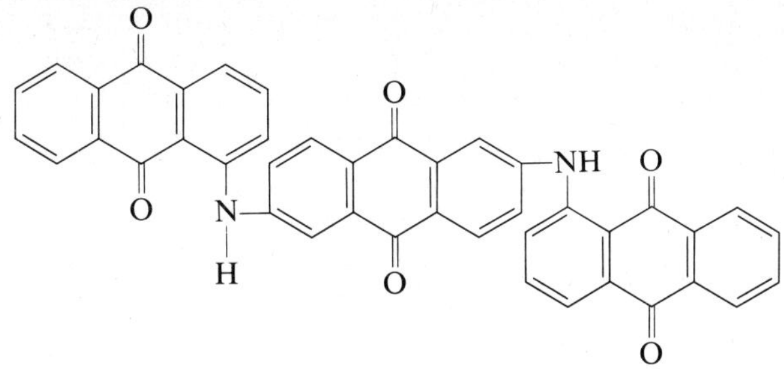

indanthrene orange 7RK

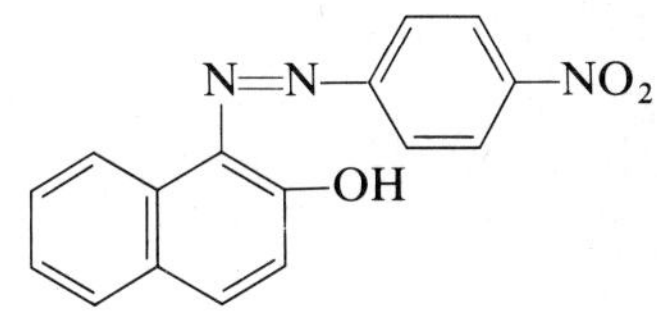

para red

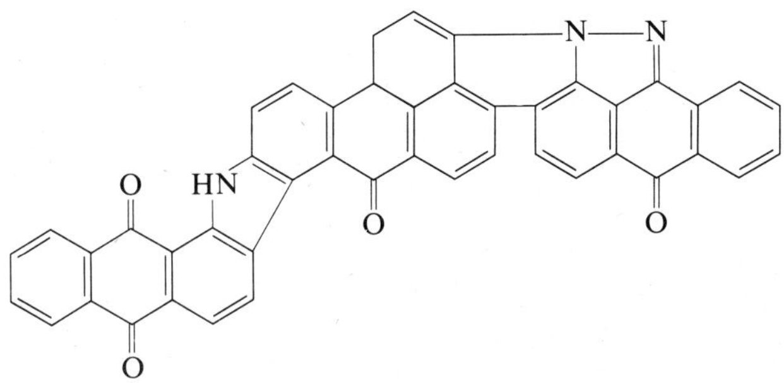

indanthrene gray M

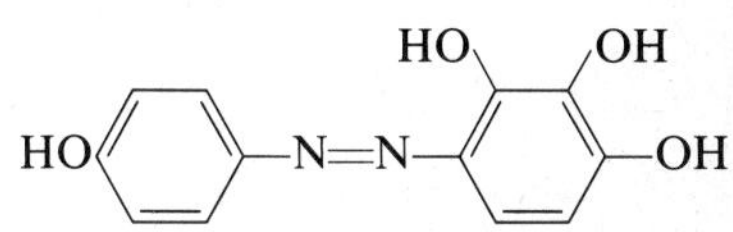

mordant brown 54

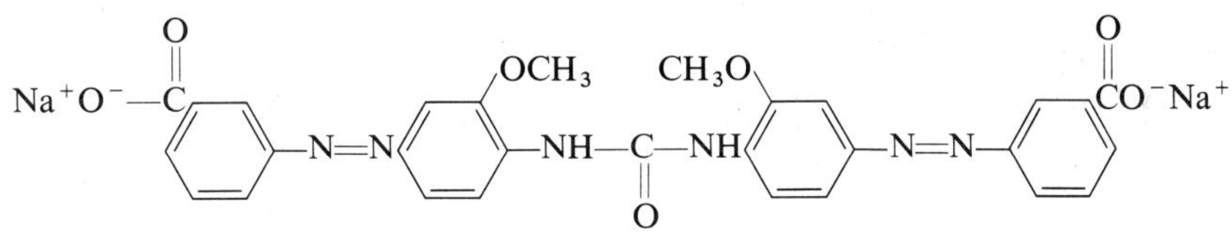

sirius yellow G

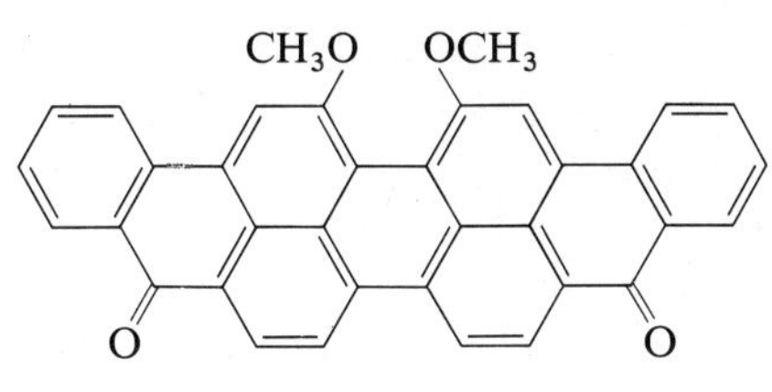

celadon jade green XBN

It is generally true that the exact amount of energy involved in one of these shifts is affected by neighboring parts of the molecule. Conjugating double bonds has an important effect. If we can tie together a chain of conjugated double bonds and atoms with unshared electrons, the pi electrons will have available energy levels that are reached by activation with visible light. A look at a number of dyes confirms the importance of conjugation and such sources of unshared electrons as amine and hydroxy groups (Table 4.1).

Summary

When in compound a pair of double bonds are separated by one single bond, the double bonds are said to be conjugated. Such a pair of double bonds can react by 1,4-addition. In the presence of limited amounts of addition reagent, the two halves of the reagent add to the most distant of the carbons associated with the double bonds. The double bonds both become single bonds, and the central single bond becomes a double bond. Natural and synthetic rubbers are formed by 1,4-addition polymerization.

The benzene ring is a special case of conjugation. In this fully conjugated system the pi electrons are equally distributed about the ring. The molecule does not behave like an alkene, and it undergoes substitution rather than addition reactions.

Organic compounds that absorb light in the visible region contain a long system of double bonds and unbonded electrons that are in conjugation.

Problems

1. Name the following:

a)

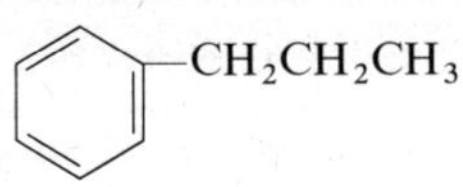

b) $CH_3{-}CH{=}\underset{\displaystyle CH_3}{\overset{|}{C}}{-}CH{=}\underset{\displaystyle CH_3}{\overset{|}{C}}{-}CH_3$

c)

d)

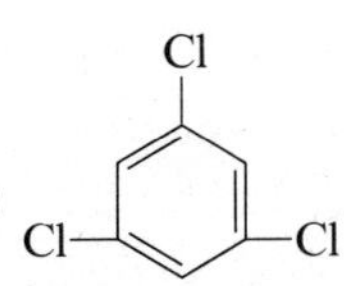

e)

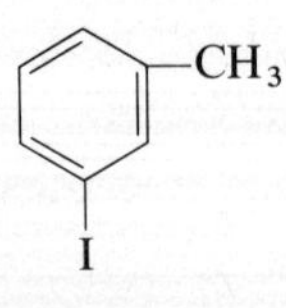

f)

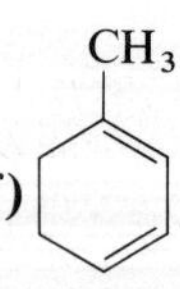

g) 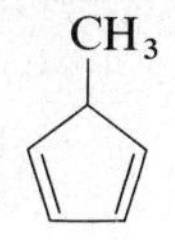

h) 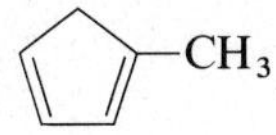

i) 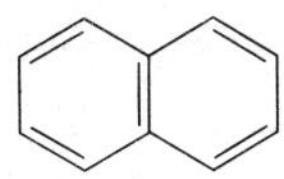

2. Draw the structure of each of the following:

a) iodobenzene

b) 1,5-hexadiene

c) *p*-bromonitrobenzene

d) 2,3-dimethyl-1,4-pentadiene

e) 1-ethyl-1,4-cyclohexadiene

f) 2,4-dinitrotoluene

3. Refer to Tables 3.2, 3.3, and 3.4. For each compound with more than one double bond, tell whether the bonds are conjugated.

4. Which of the structures in 1 and 2 can undergo 1,4-addition?

5. Benzene that has had two methyl groups substituted on it is given the special name xylene. How many different xylene structures are possible? Draw *m*-xylene.

6. How many possible chloronaphthalenes are there? Draw each.

7. Complete the following reactions:

a) $$\begin{array}{ccccc} & H_3C & & CH_3 & \\ & | & & | & \\ CH_2{=} & C & - & C & {=}CH_2 + \text{limited } Br_2 \longrightarrow \end{array}$$

b) Product of part a + Br_2 $\longrightarrow$

c) Product of part a + HBr $\longrightarrow$

d) $CH_2{=}CH{-}CH{=}CH_2$ + limited HCl $\longrightarrow$

e) Product of part d + Cl_2 $\longrightarrow$

f) Product of part d + HCl $\longrightarrow$

g)  + limited H_2O $\xrightarrow{H^+}$

h) 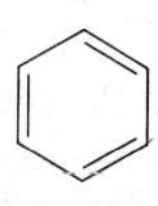 + Cl_2 $\xrightarrow{AlCl_3}$

i) 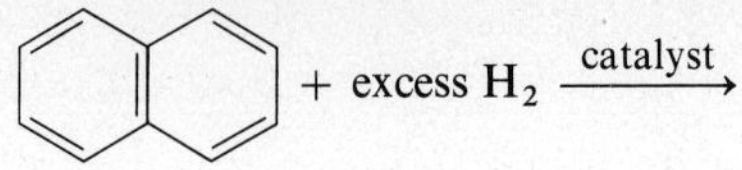

j) 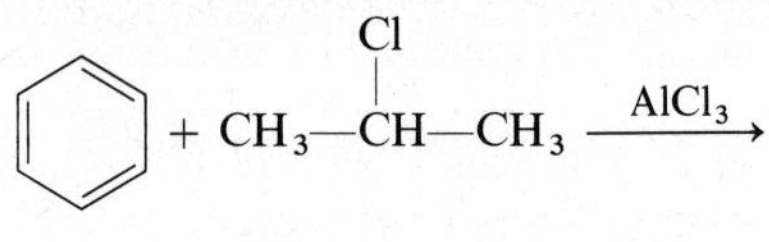

k) 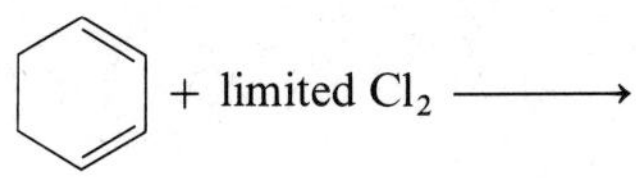

l) Product of part k + HCl ⟶

8. Consider synthetic rubbers. To compensate for a lack of natural rubber, Germany produced a polymer, a segment of which looks like this:

—CH_2—CH_2—CH=CH—CH_2—CH_2—CH=CH—CH_2—CH_2—CH=CH—

This was called Buna rubber. What do you suppose the monomer is? Does this exhibit cis-trans forms?

Challenge Problems

9. We find that when a nitro group is already substituted on a benzene ring, it causes the next group to substitute meta to itself. Complete this reaction:

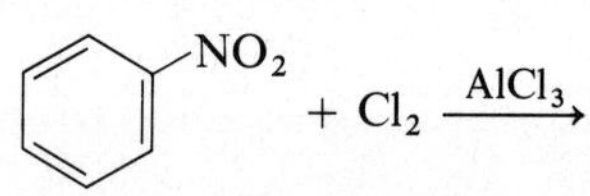

We find that when a methyl or chloro group is already substituted on a benzene ring, the next group to be added will substitute ortho or para. Complete the following:

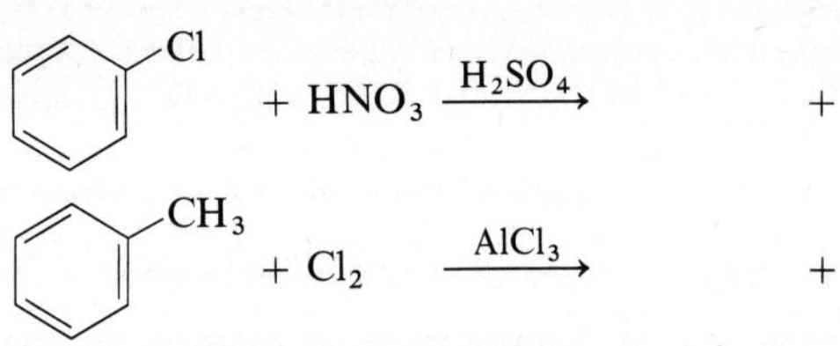

Based on this information, how would you synthesize

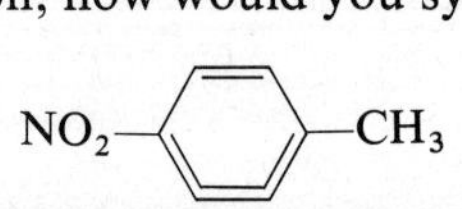

from benzene?

10. On page 77 a few substitution reactions were presented that allow the replacement of a hydrogen on the benzene ring with another group. These catalysts function by helping to generate a *positive ion*, which then displaces H^+ from benzene. Aluminium chloride ($AlCl_3$) served as the catalyst for the substitution of both —Cl and an alkyl group.

a) If the $AlCl_3$ causes Cl_2 to split up to produce a Cl^+ ion, what other ion is formed?

b) If the $AlCl_3$ causes CH_3Cl to split up to produce a CH_3^+ ion, what other ion is formed?

c) The Lewis structure of $AlCl_3$ does not conform to the octet rule for stability:

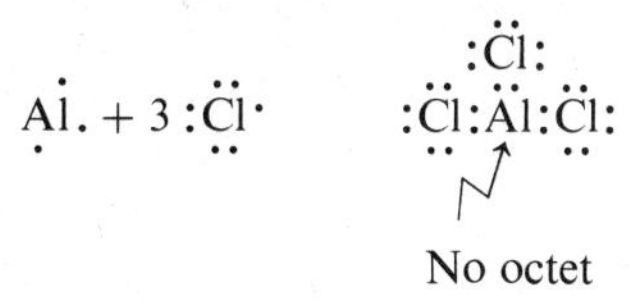

Can you predict the formation of a structure between the catalyst $AlCl_3$ and the half of the reagent left when the positive ion is formed that would be a more satisfactory Lewis structure?

d) What remains when the reactive ion NO_2^+ splits off of HNO_3? This reaction requires the presence of H^+. Of what value is this?

e) H_2SO_4 could assist the substitution of $NO_2{}^+$ on the benzene ring by reacting with the remaining portion of the HNO_3. Write an equation for such a reaction.

5

Alcohols, Ethers, and Thiols

Objectives

In this chapter we:

1. Learn to name alcohols and phenols.
2. Consider the effect of an alcohol group on such properties as water solubility and boiling point.
3. Survey substitution, elimination, and oxidation reactions of alcohols.
4. Learn reactions that synthesize alcohols.
5. Consider a few uses of alcohols and phenols.
6. Learn to name ethers and thiols.
7. Survey the reactions of thiols.
8. Define and illustrate stereoisomerism.
9. Consider the properties and importance of stereoisomers.

The compounds discussed so far have been nonpolar. As a result, they are not water soluble. Because biological systems operate in a water solution, the saturated and unsaturated hydrocarbons are not very significant biologically. Now we shall introduce a very polar functional group, the alcohol group. Just as an interior decorator will assess the effect of a brightly colored accessory to a room or a chef will judge the value of adding a bit of spice to a stew, we shall observe the change in the physical and chemical properties of the hydrocarbons that this addition produces. One change, the increase in water solubility, increases the likelihood of biochemical involvement. In fact, we find alcohol groups to be significant in such biochemically important compounds as carbohydrates, lipids, and proteins.

5.1 Alcohol Nomenclature

Alcohols are organic molecules containing the —OH functional group. They are named by dropping the final -e from the alkane or alkene name and replacing it with an -ol. As with other substitutions, numbers are used to indicate the position of the alcohol group. These numbers are placed immediately before the alkane name.

$CH_3—CH_2—OH$	Start with *ethane;* take away the -e and add -ol, *ethanol.* This is the alcohol of alcoholic beverages.
$CH_3—OH$	*methanol*, sometimes called methyl alcohol or wood alcohol.
$CH_3—CH(OH)—CH_3$	*2-propanol*, sometimes called isopropyl alcohol or rubbing alcohol. Used for massages.

The presence of more than one alcohol functional group is indicated by writing -diol (two —OH) or -triol (three —OH) after the name of the parent compound:

$HO—CH_2—CH_2—OH$

1,2-ethanediol (Also called ethylene glycol. Used as a permanent antifreeze in automobile radiators.)

$HO—CH_2—CH(OH)—CH_2—OH$

1,2,3-propanetriol (Also called glycerine or glycerol. This is a component of fats and is used as a sweetener or solvent in medicines; in cosmetics; and in suppositories.)

As compounds with more than one type of functional group are encountered, attention frequently is centered on a functional group other than the alcohol group. It is useful, then, to name the —OH group in the same manner that we named such substitutions as —Cl (chloro), —Br (bromo), and $—NO_2$ (nitro). The substitution type name for —OH is *hydroxy*. Chemically, the most interesting group in this molecule is the carboxy group.

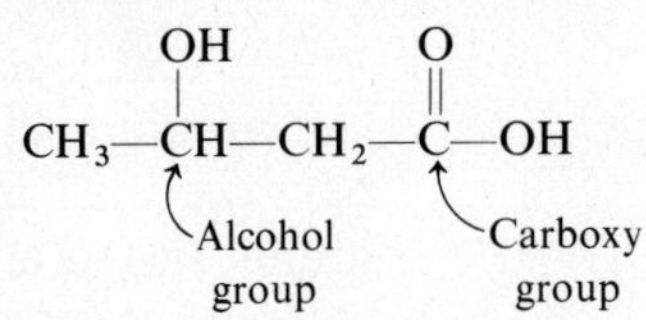

It is most often named 3-hydroxybutanoic acid, focusing attention on the carboxy group. In the example above, glycerol was named 1,2,3-propanetriol. Most chemists would understand you if you said 1,2,3-trihydroxypropane instead.

Aromatic compounds with the hydroxy group are called *phenols*.

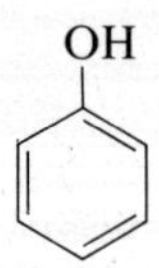

phenol

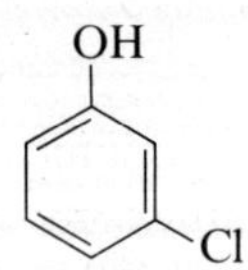

m-chlorophenol

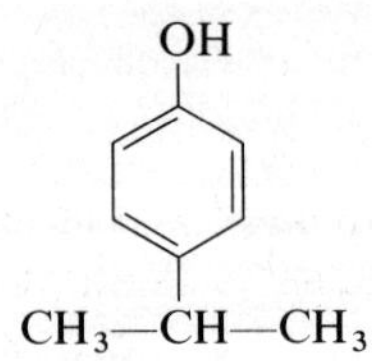

p-isopropylphenol

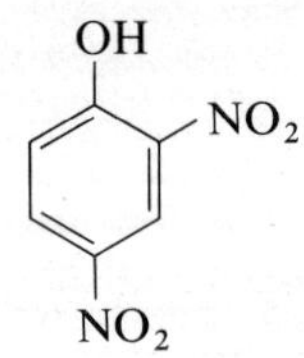

2,4-dinitrophenol

Several phenols have special names.

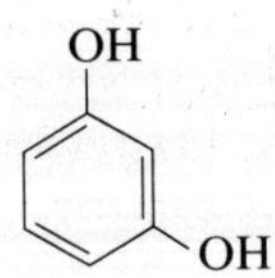

resorcinol

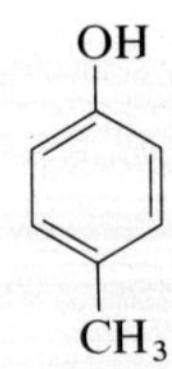

p-cresol

Exercise 5.1

Complete the following names:

a) $CH_3—CH_2—CH_2—CH_2—OH$

1-butan ___

b) 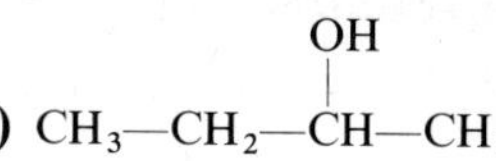

___ -butan ___

c) 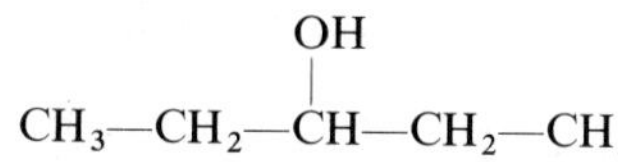

____

d) 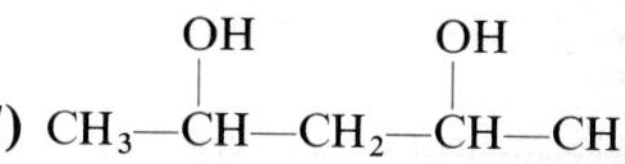

2,4-pentane __________

e) 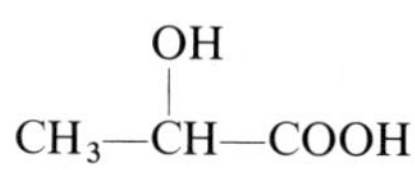

2- ________ -propanoic acid

f) 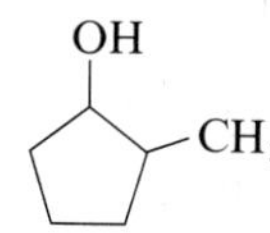

g) 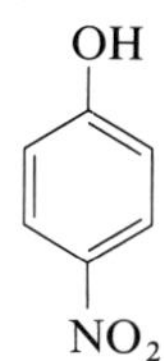

p-nitro ______________

or

___ nitro ______________

h) 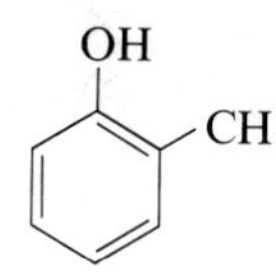

____

or

____

(Special name)

i) 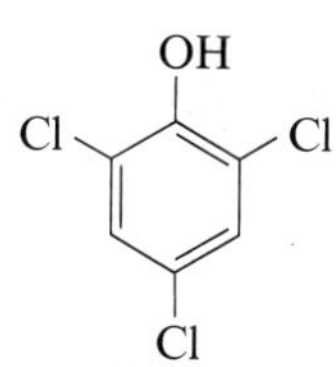

____

5.2 *Properties of Alcohols*

The presence of the very polar alcohol group (Chapter 1) increases the attractive forces between molecules. As a result, alcohols have higher boiling points than compounds of similar molecular weight that lack the polar group. This is because the polar groups attract one another. Molecules that are "sticky" toward one another resist the sort of separation that boiling involves. It requires more heat energy to pull them apart. In alcohols this intermolecular attractiveness goes beyond polar-bond attractions. Because oxygen is electronegative and has unbonded electrons, alcohol groups can hydrogen bond with one another.

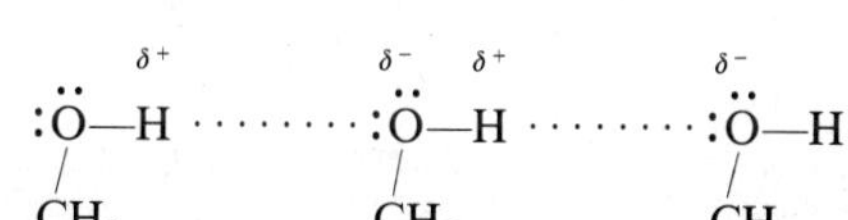

Table 5.1 Water Solubility of Alcohols

Compound	*Solubility in water (g/100g)*
Methanol	Miscible[a]
Ethanol	Miscible
1-Propanol	Miscible
2-Propanol	Miscible
1-Butanol	9
2-Butanol	12.5
1-Pentanol	2.7
2-Pentanol	4
1-Heptanol	0.18
1-Nonanol	Insoluble

[a] Miscible means that the two liquids will mix with one another in all proportions to form a solution.

The alcohol group interacts with water both as a polar group and by hydrogen bonding. As a result, some alcohols are water soluble. However, the water solubility of alcohols illustrates that we must consider the complete structure of the molecule before we predict its properties. As the length of a hydrocarbon chain with a single alcohol group is increased, the importance of the nonpolar hydrocarbon portion in determining the properties of the molecule also increases. Correspondingly, the influence of the alcohol group in allowing the molecule to mix with water decreases (Table 5.1). The effect of one alcohol group among a great amount of hydrocarbon is very small. In methanol, however, its effect is very great because there are few other atoms. Putting it another way, a single alcohol group can carry some but not a great deal of nonpolar molecule into a water solution.

Boiling points reveal the same picture (Table 5.2). Adding an alcohol group to methane increases the attractive forces between molecules enormously, as indicated by the increase in boiling point of 227°C. Once the chain has reached a length of nine carbons, however, the increase in boiling point that results from adding an alcohol group is only 43°C. An alcohol group with its considerable polarity and ability to hydrogen bond has a much greater effect than a less polar group. Chloroalkanes, which are less polar than alcohols and are incapable of hydrogen bonding, show smaller effects but the same trends.

In molecules with more than one alcohol group, these effects are greater, as you would expect. Imagine the effect alcohol groups

Table 5.2 Effect of Alcohol Group on Boiling Points

Compounds and Their Boiling Points (°C)						*Difference Between Boiling Point of Alcohol*	
Hydrocarbon		*Chloro-compound*		*Alcohol*		*and Alkane*	*and Chloroalkane*
Methane	−162	Chloromethane	−27	Methanol	65	227	92
Ethane	−89	Chloroethane	13	Ethanol	78	167	65
Propane	−42	1-Chloropropane	46	1-Propanol	98	140	52
		2-Chloropropane	37	2-Propanol	83	125	46
Butane	0	1-Chlorobutane	78	1-Butanol	117	117	39
		2-Chlorobutane	68	2-Butanol	100	100	32
Pentane	36	1-Chloropentane	108	1-Pentanol	138	102	30
		2-Chloropentane	97	2-Pentanol	119	83	22
Heptane	98	1-Chloroheptane	159	1-Heptanol	175	77	16
Nonane	151	2-Chlorononane	190	2-Nonanol	194	43	4

have on the properties of carbohydrates, where typically all but one of the carbons have alcohol groups. With all these alcohol groups, carbohydrates can hydrogen bond both to water and to each other very strongly. Not surprisingly, carbohydrates are very water soluble and have very high melting points. In fact, many carbohydrates have such high melting points that they decompose (lose their chemical structure because covalent bonds break) rather than melt on heating. This process, called caramelizing, is used in candy making. A small amount of the decomposition product imparts a desirable flavor.

5.3 Reactions of Alcohols

Alcohols can undergo substitution reactions. The most important example is the conversion of an alcohol to a halogen compound using a reagent such as PCl_3, PBr_3, or $SOCl_2$.

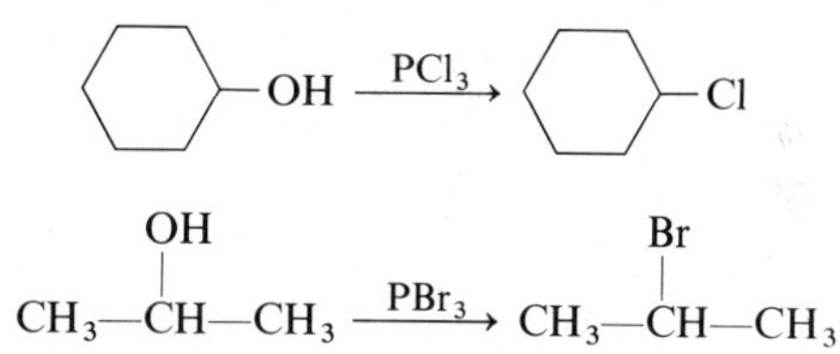

$$CH_3OH \xrightarrow{SOCl_2} CH_3Cl$$

Because there is no double bond in the alcohol functional group, addition reactions are not possible. However, the reaction by which an alcohol is formed from an alkene, the addition of water across the double bond, can be reversed using a concentrated acid such as H_2SO_4 as the reagent. Such a process to form a double bond is called an *elimination reaction.*

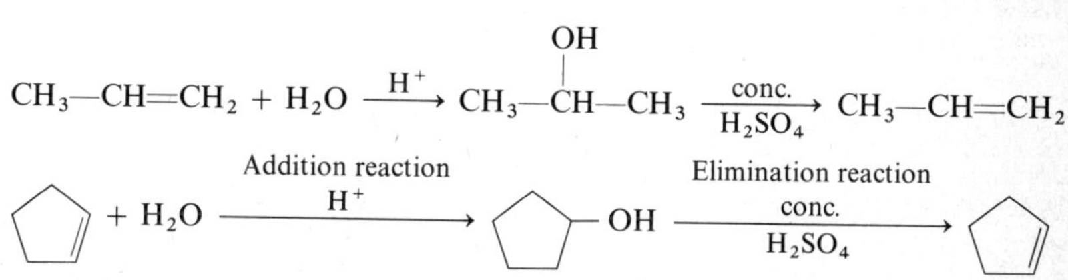

Enzymes also catalyze elimination reactions.

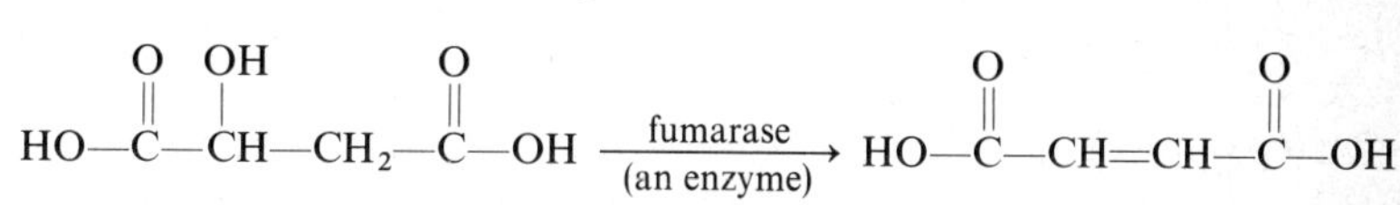

Recall in Chapter 2 that the oxidation of hydrocarbons to CO_2 and H_2O was described and that oxidation or combustion was performed at high temperatures using oxygen. Oxidation of a less dramatic sort can be carried on by use of mild conditions and such reagents as $KMnO_4$ and $Na_2Cr_2O_7$. The carbon to which the alcohol group is attached must have a hydrogen attached for this reaction to be successful, a requirement some alcohols will not meet.

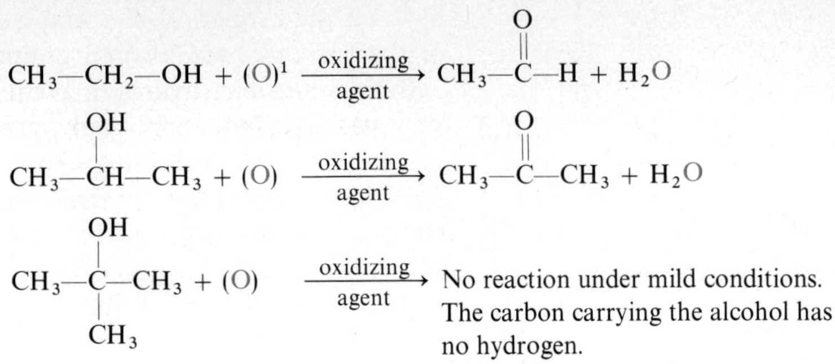

Notice that this could also be described as an elimination reaction in which two hydrogens are removed to create a double bond.

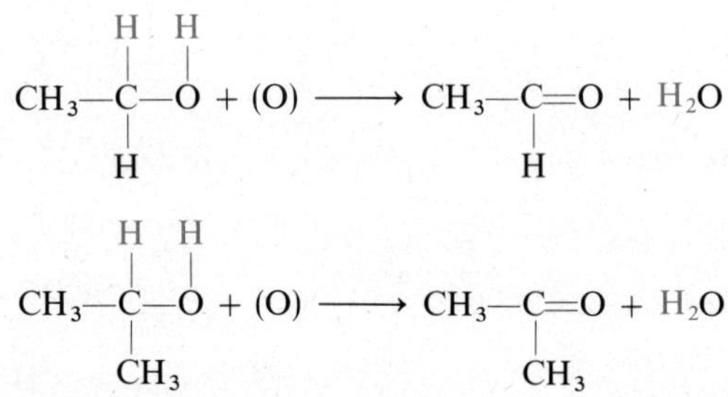

Enzymes also can oxidize alcohols. The following examples are steps in the utilization of fat:

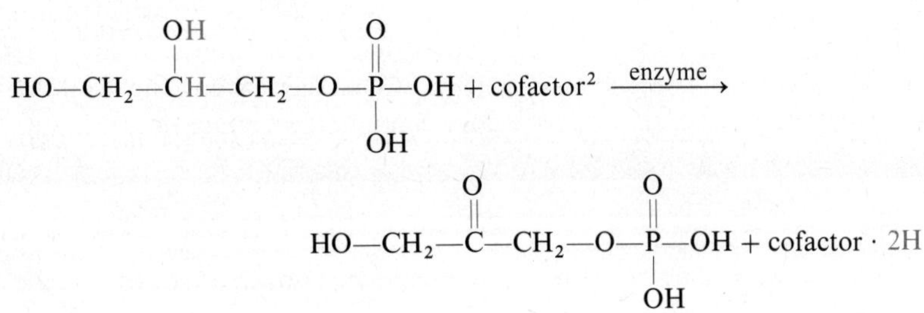

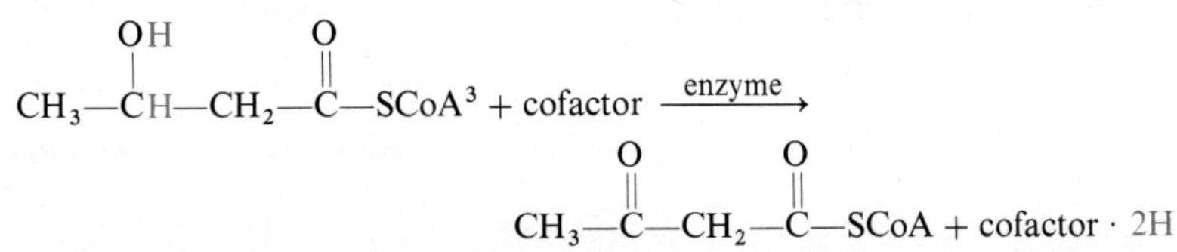

---

[1] (O) here represents oxygen supplied by the oxidizing agent ($KMnO_4$ or $Na_2Cr_2O_7$).

[2] A cofactor is an organic structure that assists an enzyme in catalyzing a chemical change. Here the cofactor accepts the hydrogens eliminated from the alcohol.

[3] CoA is an abbreviation for a complex molecule often attached to carboxylic acids when they are being chemically altered inside a living cell. This is discussed more completely on page 101 and in Chapter 12.

Notice that in the first example one of the alcohols was oxidized but the other was not. Enzymes are capable of this sort of selectivity in the reactions they catalyze. If laboratory reagents had been used, both alcohols would have been oxidized.

Exercise 5.2

Complete these reactions:

a) 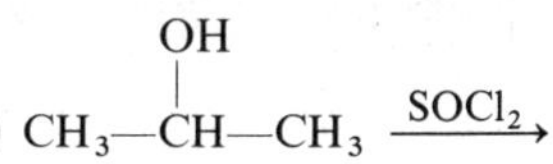

b) $CH_3{-}CH_2{-}OH \xrightarrow{PCl_3}$

c) $CH_3{-}OH \xrightarrow{PBr_3}$

d) $CH_3{-}CH_2{-}OH \xrightarrow[H_2SO_4]{conc.}$

e) 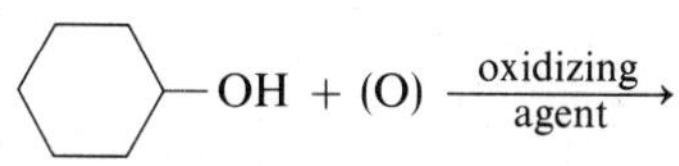

f) 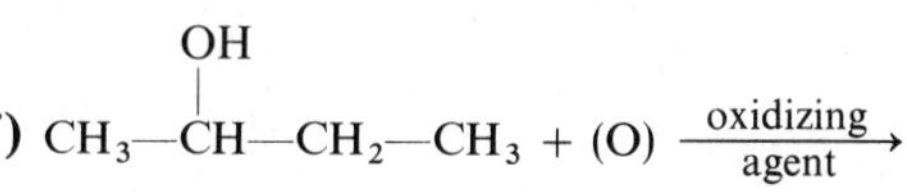

g) $CH_3{-}CH_2{-}CH_2{-}CH_2{-}OH + (O) \xrightarrow[agent]{oxidizing}$

h) 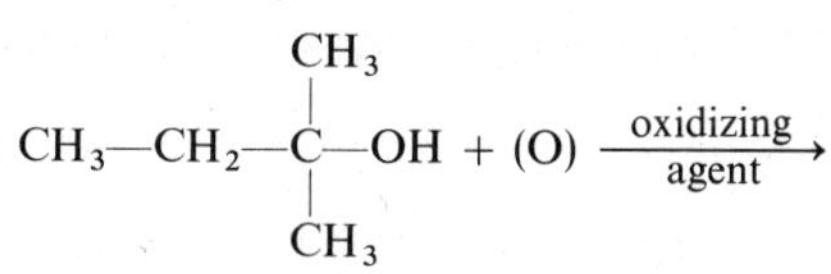

Suggest reagents for performing the indicated chemical changes:

i) 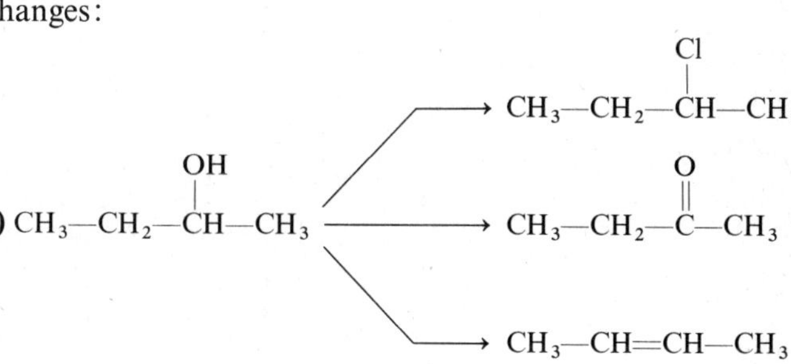

5.4 *Formation of Alcohols*

Alcohols can be formed by reactions that are roughly the reverse of the reactions that alcohols undergo. Obviously the reagents are different.

1. Alcohols can become chloroalkanes

$$CH_3{-}OH \xrightarrow{SOCl_2} CH_3{-}Cl$$

and chloroalkanes can become alcohols

$$CH_3{-}Cl \xrightarrow[\text{NaOH}]{\text{aqueous}} CH_3{-}OH$$

2. Alcohols can become alkenes

$$CH_3{-}\underset{}{\overset{OH}{\overset{|}{C}H}}{-}CH_3 \xrightarrow[H_2SO_4]{\text{conc.}} CH_3{-}CH{=}CH_2$$

and alkenes can become alcohols (see Chapter 3)

$$CH_3{-}CH{=}CH_2 + H_2O \xrightarrow{H^+} CH_3{-}\overset{OH}{\overset{|}{C}H}{-}CH_3$$

3. Alcohols can become aldehydes and ketones

$$CH_3{-}CH_2{-}CH_2{-}OH + (O) \xrightarrow[\text{agent}]{\text{oxidizing}} CH_3{-}CH_2{-}C\begin{matrix}{=}O\\ \diagdown H\end{matrix} + H_2O$$

$$CH_3{-}\overset{OH}{\overset{|}{C}H}{-}CH_3 + (O) \xrightarrow[\text{agent}]{\text{oxidizing}} CH_3{-}\overset{O}{\overset{\|}{C}}{-}CH_3 + H_2O$$

and aldehydes and ketones can become alcohols (this is addition to a double bond)

$$CH_3{-}CH_2{-}C\begin{matrix}{=}O\\ \diagdown H\end{matrix} + H_2 \xrightarrow{\text{catalyst}} CH_3{-}CH_2{-}CH_2{-}OH$$

$$CH_3{-}\overset{O}{\overset{\|}{C}}{-}CH_3 + H_2 \xrightarrow{\text{catalyst}} CH_3{-}\overset{OH}{\overset{|}{C}H}{-}CH_3$$

Exercise 5.3

Complete the following reactions:

a) $CH_3{-}CH_2{-}Cl \xrightarrow{\text{aqueous NaOH}}$

b) 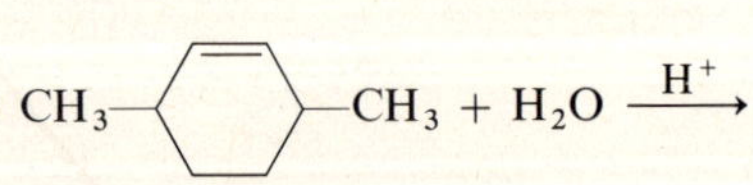

c) $CH_3{-}\underset{CH_3}{\underset{|}{C}H}{-}\overset{O}{\overset{\|}{C}}{-}CH_3 + H_2 \xrightarrow{\text{catalyst}}$

d) 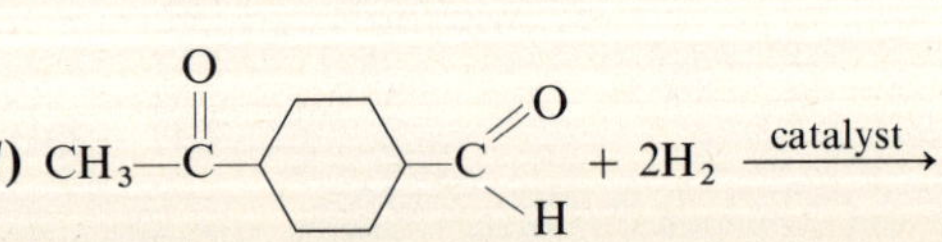

5.5 Uses of Alcohols

Methanol, ethanol, and 2-propanol are used as solvents, sometimes on a very large scale. Ethanol is the alcohol of alcoholic beverages. When ethanol is to be used as a solvent or for other nonconsumption purposes, it is made poisonous (denatured), usually by the addition of methanol. The methanol is both toxic and difficult to separate from the ethanol. Isopropyl alcohol (2-propanol) is also poisonous, but it is nonetheless safe for use as a rubbing alcohol because it is not absorbed through the skin.

Phenol used to be called carbolic acid and was used very early as an antiseptic. In fact, the first sterile surgery was performed in 1867 by Joseph Lister using phenol to kill bacteria. Many phenol derivatives are toxic to bacteria (Table 5.3) and are used today for this purpose. Phenol is used as the standard to measure the effectiveness of a compound for killing bacteria. The phenol index or coefficient (P.I.) is the ratio of those concentrations of phenol and the specific agent that are equally effective in killing bacteria. For example, 2 percent *o*-cresol is as effective as 5 percent phenol, so the phenol coefficient is 5/2, or 2.5. Alcohols and phenols fulfill a variety of other roles in medicine, as shown in Table 5.4.

Notice that in Tables 5.3 and 5.4 the commercial names, the names used by pharmacists, do not follow the IUPAC rules. This practice is especially confusing because the pharmacists make up words that look as though they were following the rules. A name like hexachlorophene may send you scrambling to the nomenclature rules, but it is simply an invention for a molecule that is phenolic and has six chloro groups.

The storability (shelf life) of foods is extended by use of compounds that combat oxidation. Butylated hydroxytoluene (BHT) is such a compound. Notice how often you see this compound listed on the labels of food products.

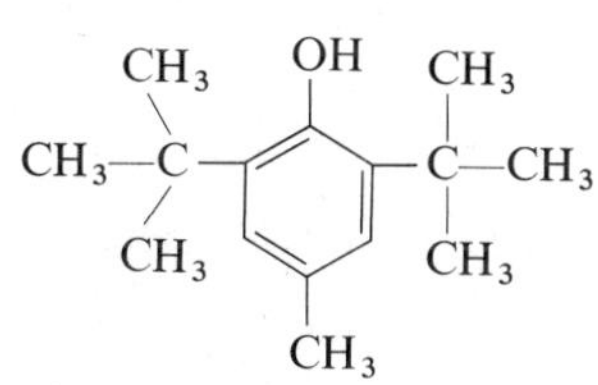

2,6-ditertiary butyl-4-methylphenol
(butylated hydroxytoluene, or BHT)

5.6 Ethers

Ethers have two alkyl groups attached to the same oxygen. They are usually named simply by listing the two groups, followed by the word "ether."

$CH_3—O—CH_2—CH_3$

methyl ethyl ether

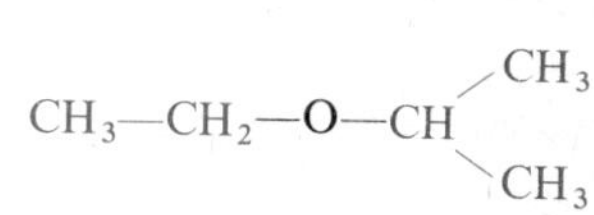

ethyl isopropyl ether

$CH_3—CH_2—O—CH_2—CH_3$

diethyl ether
(the anesthetic ether)

Table 5.3
The Structures of Bacteriocidal and Fungicidal Phenols

1.
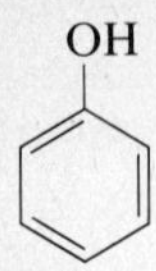
carbolic acid

2.
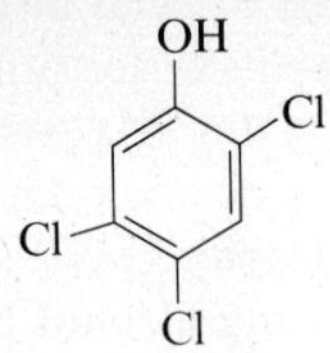
Dowicide 2

3.
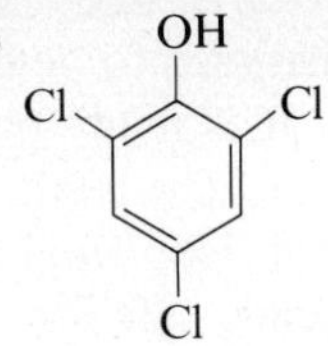
Dowicide 2S

4.
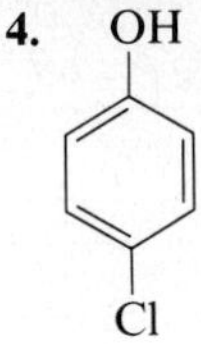

5.
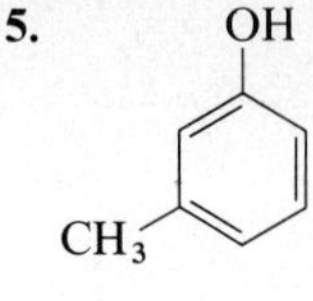

6.
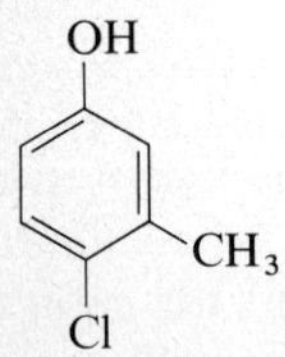

7.
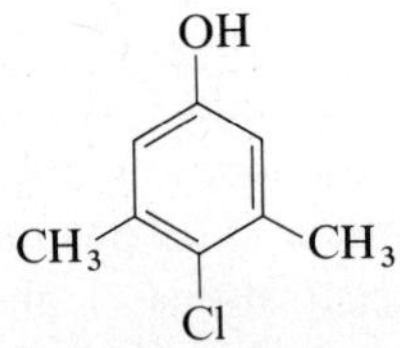

8.
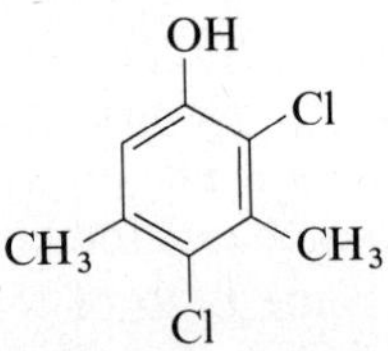
dichloroxylenol

9.
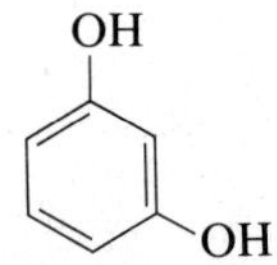

10.
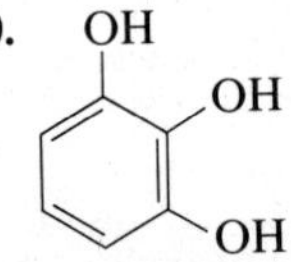
pyrogallol

11.
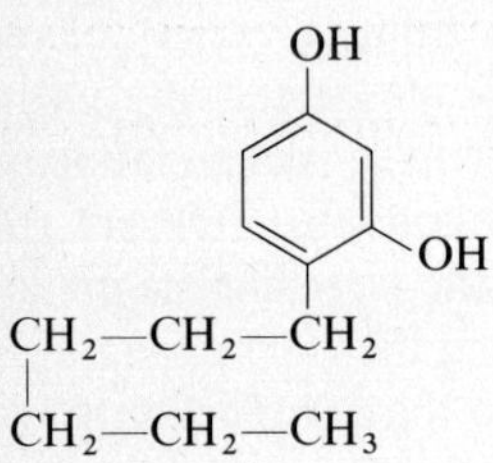
hexylresorcinol

12.
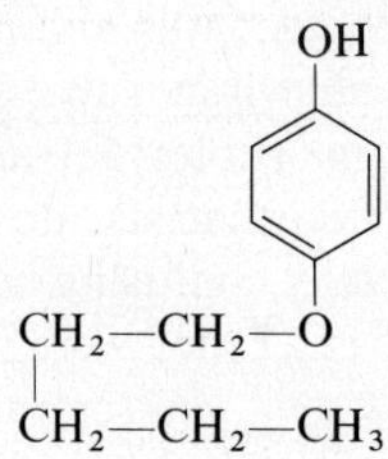

13.
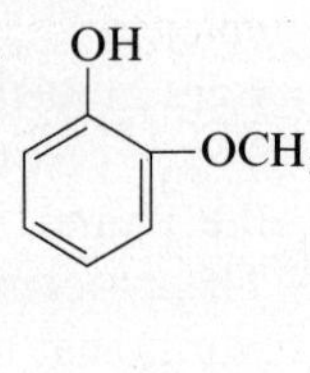
guiacol

14.
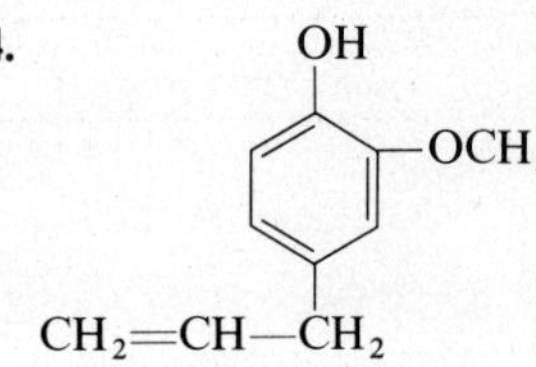
eugenol

15.
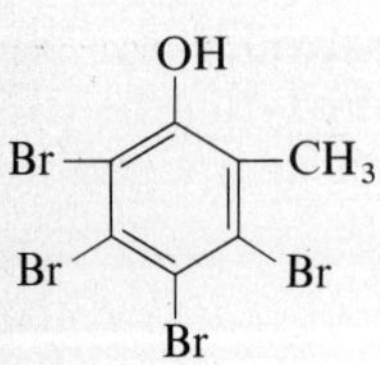

16.
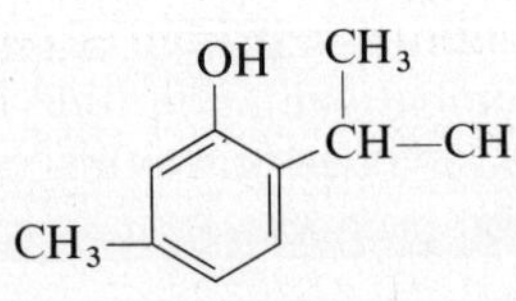

17.
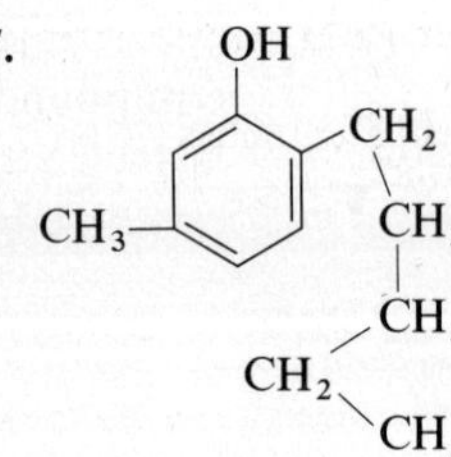

18.
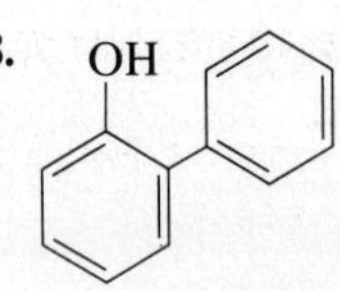
Dowicide 1

19.
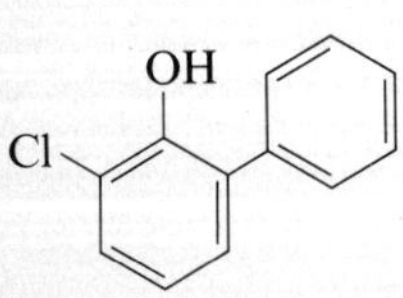

20.
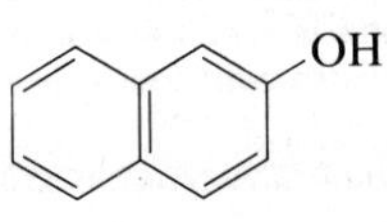

21.
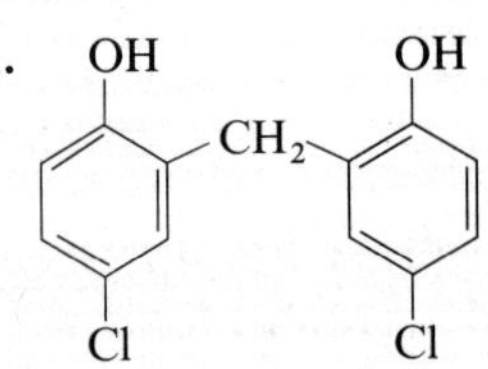
dichlorophene

22.
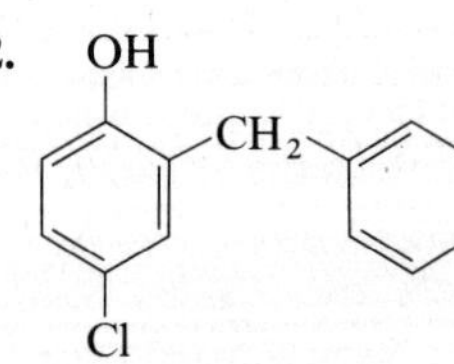
chlorophene

23.
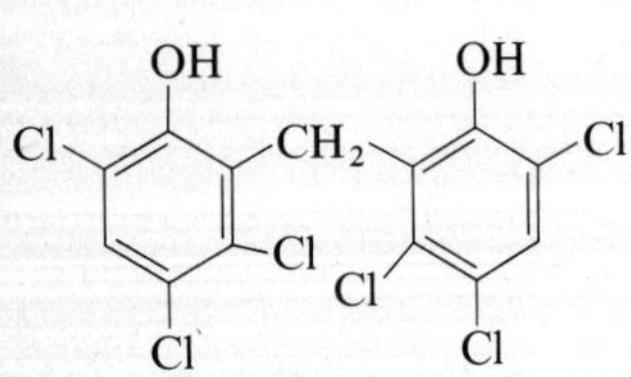
hexachlorophene

Note: Only commercial names are provided. See Problem 16.

Table 5.4
Other Alcohols and Phenols Used in Medicine

Estrogen (Female Sex Hormone) Replacements[a]

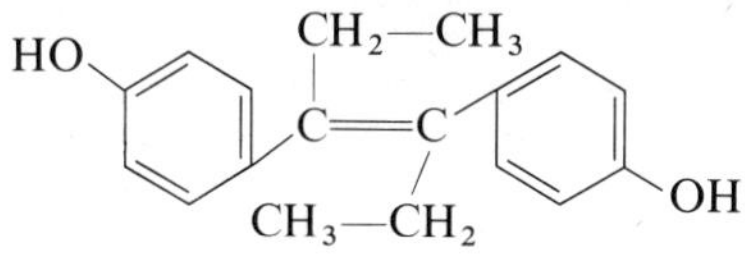

diethylstilbesterol (DES)

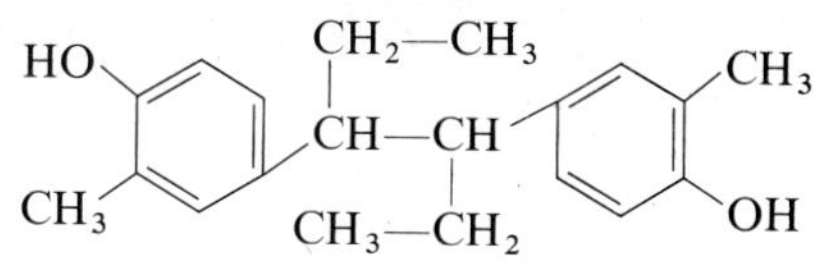

Promethstrol

Sedatives

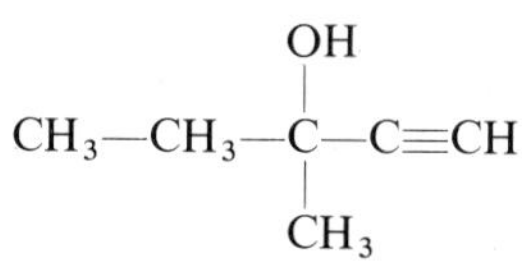

Meparfynol

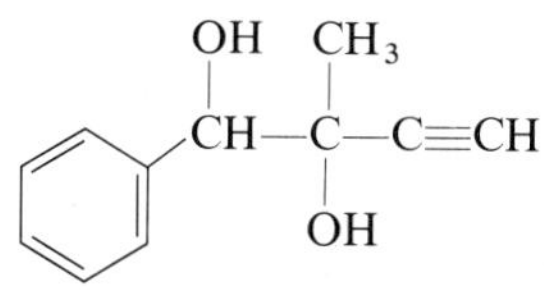

Centalun

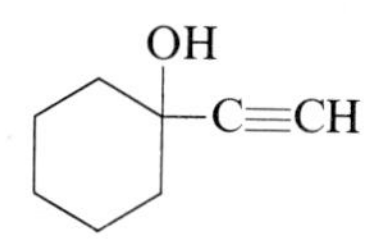

1-ethynylcyclohexanol

Anesthetics

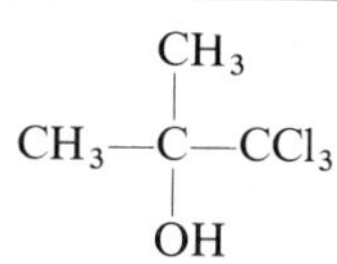

Chlorobutanol

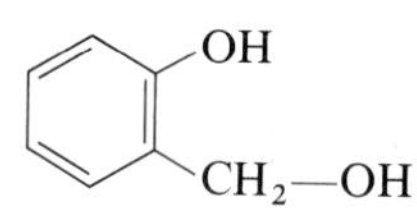

salycyl alcohol

Other Medicinals

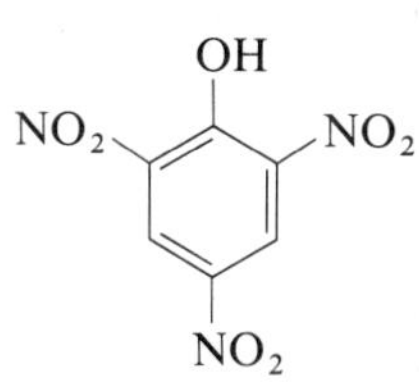

picric acid,
used to seal burns,
prevent fluid loss

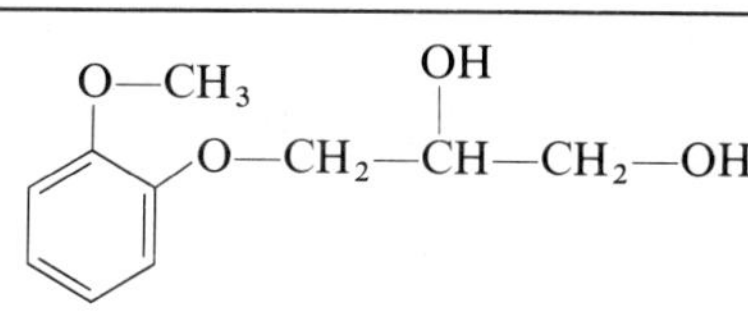

guiacol glyceryl ether,
used in cough syrup

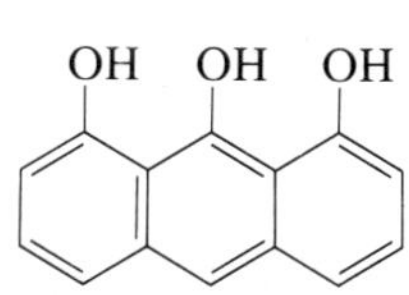

Dithranol,
used against skin problems
such as psoriasis

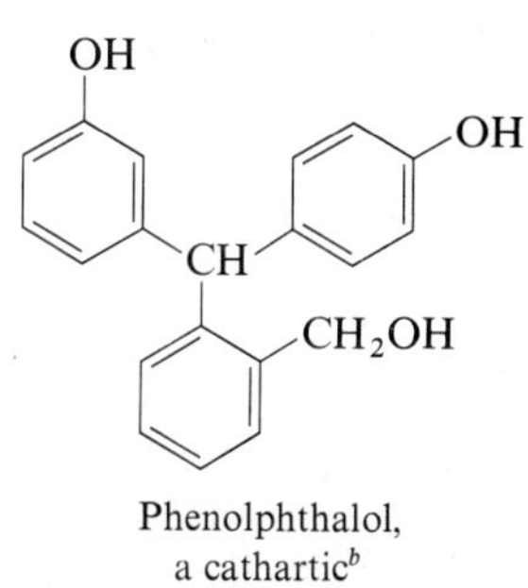

Phenolphthalol,
a cathartic[b]

[a] DES became controversial because of its widespread use and possible carcinogenic properties.
[b] A cathartic is an agent used to stimulate bowel activity, a laxative.

Aromatic ethers may have special names:

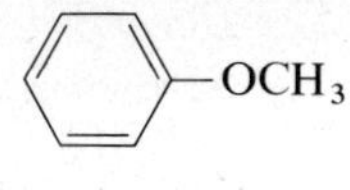

anisole

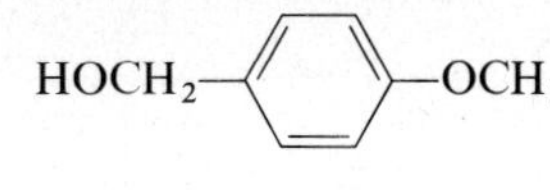

anise alcohol

The $—OCH_3$ of methyl ethers is often called a *methoxy group.*

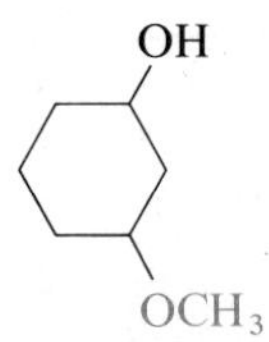

3-methoxy-1-cyclo-hexanol

$CH_3—O—CH_2—CH_2—O—CH_3$

1,2-dimethoxyethane

Exercise 5.4

Name these structures:

a) 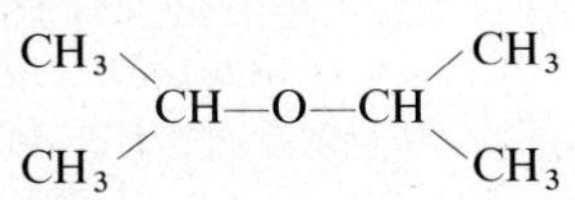

b) 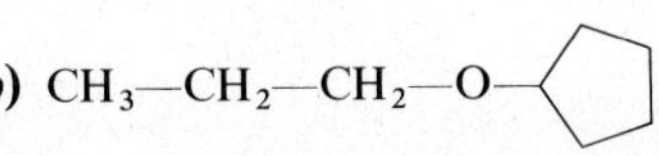

c) 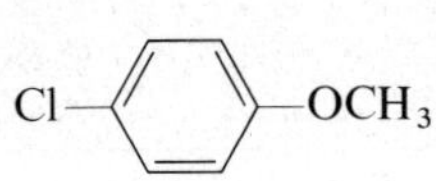

d) 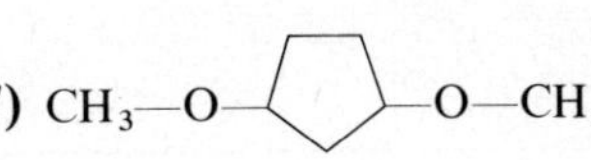

5.7 *Reactions of Ethers*

Ethers are usually chemically inert. This is not surprising, because the functional group does not include the type of polar bonds or the unsaturation associated with chemical reactivity. Diethyl ether is the ether of the anesthetist. Here chemical nonreactivity is an advantage, because a reactive compound might well be more toxic. However, because of its low polarity and low molecular weight, diethyl ether has a low boiling point. Low-boiling-point compounds evaporate rapidly, a property referred to as *volatility.* The hydrocarbon structures (ethyl groups) confer flammability on the compound. Volatility and flammability are a dangerous combination. Diethyl ether must be handled very carefully or fires may result.

5.8 *Thiols*

On the periodic chart, sulfur is directly under oxygen. This means that the outer electron shells of the two elements are the same; therefore their chemistry is rather similar. There are sulfur compounds bonded like alcohols or ethers but with sulfur in place of the oxygen. The alcohol-type structure is called a *thiol* or a *mercaptan,* and the etherlike compound is a *thioether* or *sulfide.*

$CH_3{-}OH$ methanol, or methyl alcohol

$CH_3{-}SH$ methane thiol, or methyl mercaptan

$CH_3{-}O{-}CH_2{-}CH_3$ methyl ethyl ether

$CH_3{-}S{-}CH_2{-}CH_3$ methyl ethyl sulfide

A few thiols and thioethers are found in biological systems. The most important examples are the amino acids cysteine and methioine.

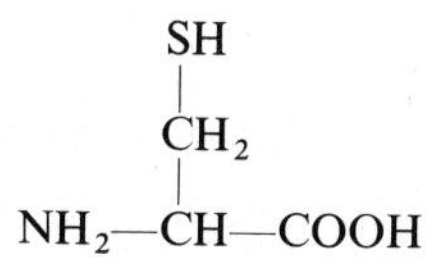
cysteine

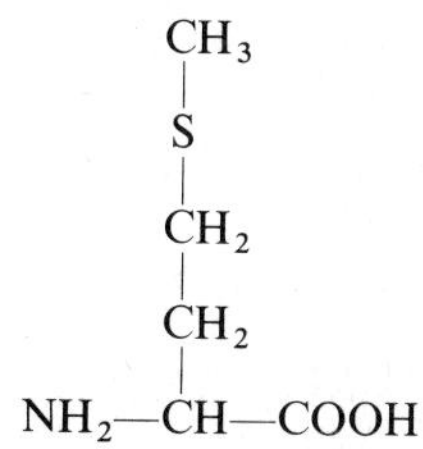
methionine

Thiol groups are important to the activity of a number of enzymes. Most often the source of the thiol group is a cysteine in the enzyme molecule. Heavy metals such as mercury react very readily with the thiol group. If the thiol group takes part in the process of catalysis, the resulting mercury derivative does not function. This explains some of the toxicity of these heavy metals.

There are a number of thiols in nature. One family, including butanethiol, *trans*-2-butenethiol, and 3-methylbutanethiol, would seem relatively unimportant unless you found yourself in a particularly awkward situation. It would be useless information at the time the skunk became unhappy with you, but it might be a source of conversation later to explain that you were reeking of simple mercaptans (Figure 5.1).

$CH_3{-}CH_2{-}CH_2{-}CH_2{-}SH$

butanethiol

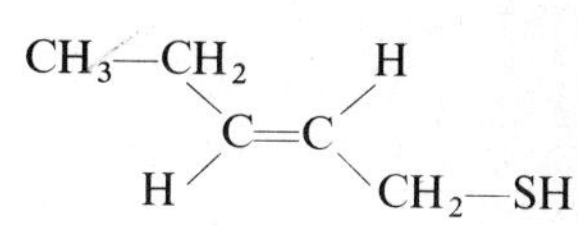
trans-2-butenethiol

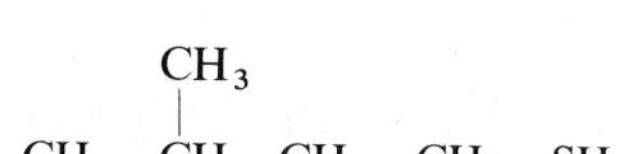
3-methylbutanethiol

Methanethiol is a part of the flavor of cheese, and 1-propanethiol contributes to the flavor of onions.

Figure 5.1
Thiols
Skunks are an unpleasant natural source of thiols. (Photo by John H. Gerard, DPI.)

$CH_3{-}SH$ — methanethiol

$CH_3{-}CH_2{-}CH_2{-}SH$ — 1-propanethiol

One of the thiol compounds that is used medicinally presents an interesting example of administering a compound with which the toxic agent reacts in the same fashion as and instead of the natural compound. British Anti Lewisite (BAL) is a thiol that reacts with poisonous metals like mercury instead of the cysteine of the enzymes.

$$\begin{array}{ccccc} OH & & SH & & SH \\ | & & | & & | \\ CH_2 & - & CH & - & CH_2 \end{array}$$

British Anti Lewisite

Thiols readily undergo an oxidation reaction. Two thiols join together to form a disulfide, transferring two hydrogens to an oxidizing agent. This agent could be oxygen.

$$2CH_3SH + \tfrac{1}{2}O_2 \rightarrow CH_3{-}S{-}S{-}CH_3 + H_2O$$

disulfide bond

$$\underset{\text{cysteine}}{HOOC{-}\underset{\underset{NH_2}{|}}{CH}{-}CH_2{-}SH} + HS{-}CH_2{-}\underset{\underset{NH_2}{|}}{CH}{-}COOH + \tfrac{1}{2}O_2 \rightarrow HOOC{-}\underset{\underset{NH_2}{|}}{CH}{-}CH_2{-}S{-}S{-}CH_2{-}\underset{\underset{NH_2}{|}}{CH}{-}COOH + H_2O$$

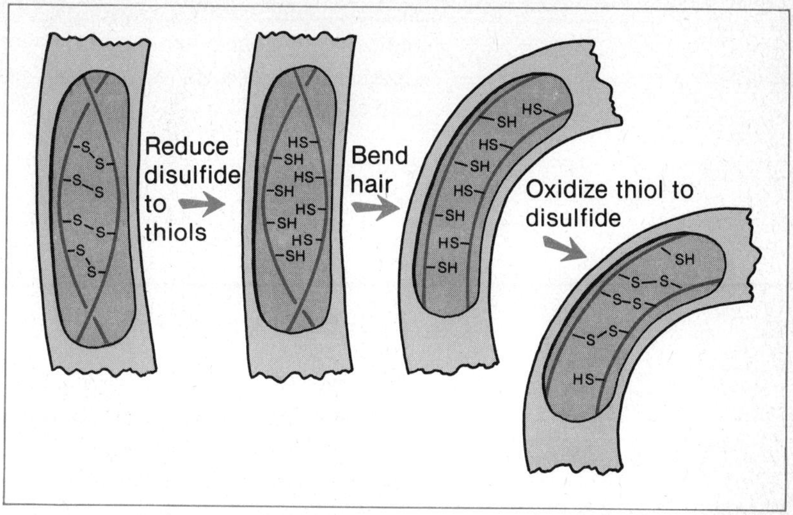

Figure 5.2
Permanent waving of hair
Hair is composed of long, ropelike molecules of fibrous protein that are fixed into a pattern by disulfide cross-links. In permanent waving the disulfide cross-links are broken, the hair is adjusted to a new pattern, and the cross-links are re-formed. The pattern eventually changes back to the old one for two reasons. First, the bonds occasionally break and re-form spontaneously and may well reestablish the natural more stable pattern. Second, the natural pattern is hereditary, and it is the form of the new hair resulting from growth.

This disulfide linkage between two cysteines is important in the structure of many proteins. Hair, which is composed of protein, provides an example. The hair protein is held in a particular shape by a large number of disulfide bonds between cysteines. The chemical agents used in permanent waving of the hair function by reducing the disulfide bonds in hair to thiols. The hair is adjusted to the new configuration; then the thiols are reoxidized to disulfides, holding the hair in the new pattern (Figure 5.2).

This reaction is the manner of function of lipoic acid, an enzyme cofactor (Chapter 12).

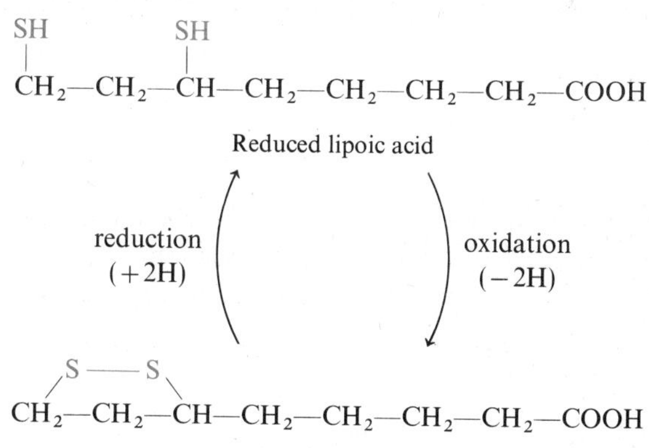

Oxidized lipoic acid

Coenzyme A is another biochemically significant thiol. It has a rather complex structure (Chapter 12). However, it is just the thiol group that is chemically altered as coenzyme A performs its function. For this reason the molecule is often represented simply as CoA-SH. Notice on page 92 that the compound chemically altered in the enzyme-catalyzed example of alcohol oxidation is bound to a molecule of coenzyme A. The combination of that compound and CoA-SH has created a thioester, which is a sulfur analogue of a conventional ester.

ester thioester

Coenzyme A is generally found bound as a thioester when it is in use.

5.9 *Stereoisomerism*

For many years great scientists like Louis Pasteur and Emil Fischer dealt with some observations that were difficult to explain. They found that two isomers seemed to exist for certain molecules, such as glyceraldehyde.

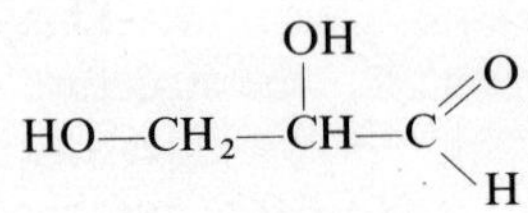

glyceraldehyde

Today, because we understand that carbon with four attached groups is shaped as if it were at the center of a pyramid with the groups at each point, we can explain how these isomers differ. Consider a carbon with groups *A*, *B*, *C*, and *D* attached to it. Notice that it is possible to place these groups around the carbon in two different ways.

Notice in each of these structures that *A* is up and *D* is back, but *B* is to the right in one and to the left in the other.[4] Sometimes we use the term *mirror images* to describe how one isomer relates to the

[4] It is very useful at this point to make two ball-and-stick models using four different colored balls. You will see that no matter how one is turned, it always differs from the other. Try it with two identical groups, and notice that no matter how the models are assembled, they can always be turned to look the same. If ball-and-stick models are not available, colored marshmallows or gumdrops and toothpicks can be used.

other, because one is arranged like a reflection of the other. Another popular analogy is the comparison of your right and left hands. The parts are all the same but the arrangement is exactly reversed. You cannot put a right-hand glove on your left hand. Such mirror image isomers are termed *enantiomers*. Enantiomers are possible only when the carbon has *four different* groups. Such carbons are called *asymmetric*, or *chiral*.

Exercise 5.5

Which of the following structures include an asymmetric carbon?

a)
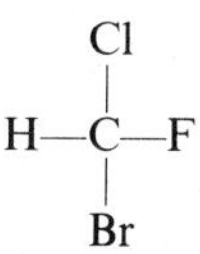

b)
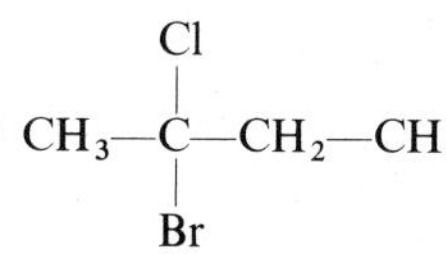

c)
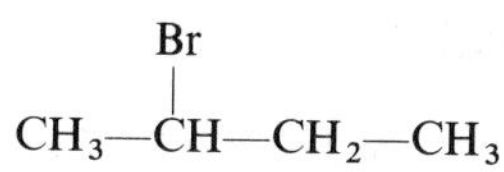

d)
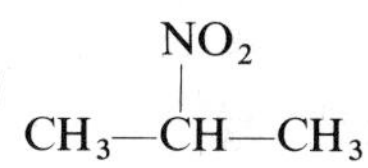

e)
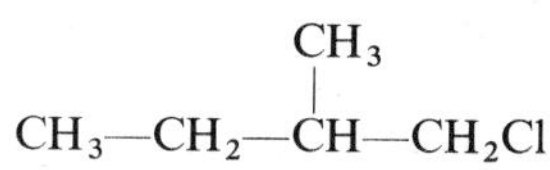

How do enantiomers differ in properties? It has been shown that cis-trans isomers differ in shape, and as a result they differ in such properties as boiling point, melting point, and density. Consider enantiomers. Although the models are visually distinguishable from one another, they actually have the identical shapes in the same sense that the right and left hand do. As a result, with respect to most chemical and physical properties, enantiomers are identical. However, there is a technique able to sort out enantiomers. *Polarizing filters* are able to selectively block components of light oriented in one direction and pass light oriented at 90° to that direction so that the light takes on a directional character. We call this light *polarized.* If we pass light through two polarizing filters, light passing through the first will pass also through the second if the filters are oriented with their polarizing abilities parallel, but if the second filter is turned 90°, then light that is passed by the first filter will be oriented so as to be blocked by the second (Figure 5.3).

Polarized sunglasses provide a commonplace example of this process. Take two pairs of such glasses, and, holding lenses back to back, turn one lens with respect to the other. As you rotate a lens, the view through the lens will dim and brighten. Polarized sunglasses are particularly effective because many surfaces that produce glare, such as road surfaces or automobile finishes, partially polarize the light reflected from them. Try tipping your head parallel to the road while wearing such sunglasses in summer and notice how much more glare you experience.

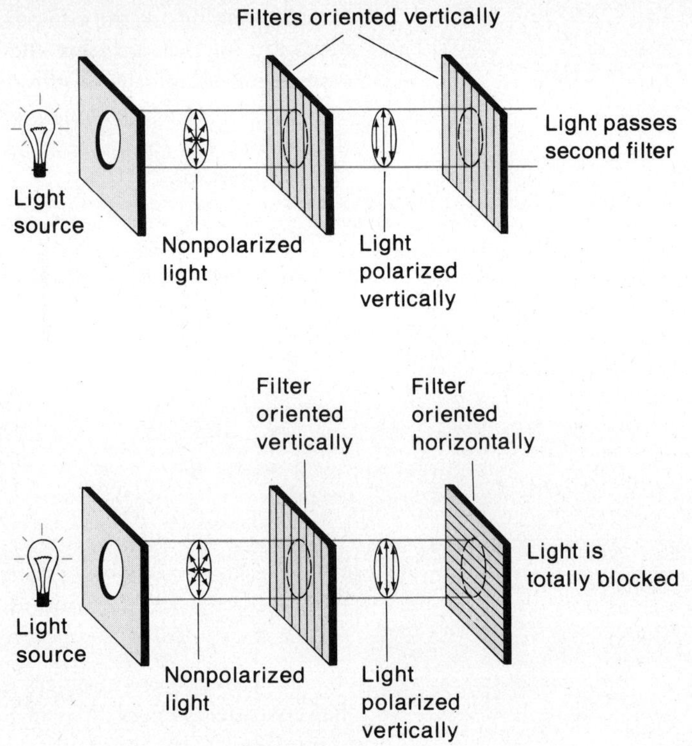

Figure 5.3
Polarizing filters and polarized light
Polarized light behaves as though the polarizing filter had removed all orientations of light but one. A second filter oriented in the same direction as the first will allow the light to pass. Turning that filter 90° will block the light.

Asymmetric carbons have an interesting ability. If you place a container of water between two polarizing filters and locate the exact angle at which the least light passes through the second filter, you assume the polarizing abilities are at 90° to each other. However, as one of a pair of enantiomers is added to the water, the angle necessary for maximum light bloackage is changed (Figure 5.4). The enantiomer twists the light as it passes through. The amount of twisting depends on concentration. The more of the enantiomer you add, the more the light is twisted. Furthermore, if one of the enantiomers twists the light a given amount to the right, the same concentration of the other enantiomer twists it to the left by exactly the same amount. This twisting is called *optical rotation*. Because of this property, compounds with asymmetric carbons are often termed *optically active*.

Just as you could take a barrel of various kinds of gloves and sort them into piles of right-handed and left-handed gloves, all asymmetric carbons can be sorted out into two groups, one D (dextrorotatory) and the other L (levorotatory). For any pair of enantiomers, one is D and the other is L. This does not amount to sorting into the same pile enantiomers that turn light in the same direction. D or L is a basic property relating to how the parts are ar-

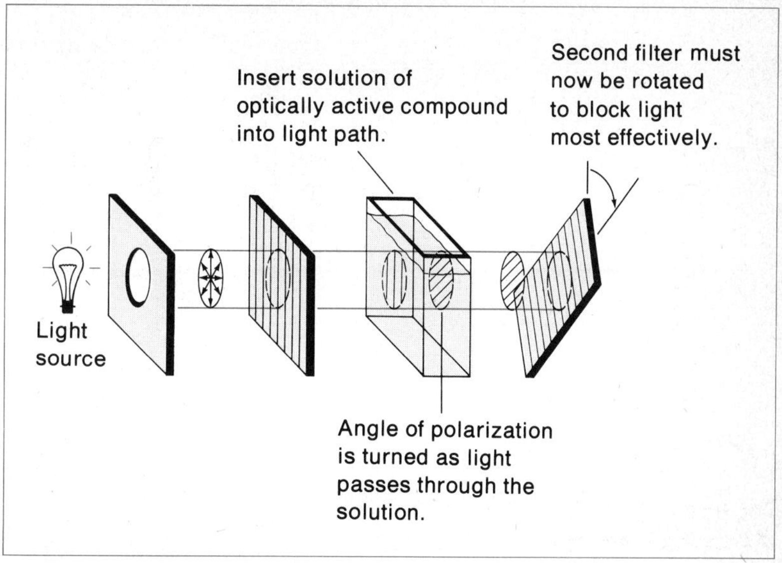

Figure 5.4
Optical rotation by asymmetric carbons
Compounds with asymmetric carbons twist the orientation of polarized light, requiring that the second filter be turned to some angle other than 90° to block the polarized light most effectively.

ranged. But some D compounds twist light one direction while others twist it the other direction. For example, D-glyceraldehyde twists light in the opposite direction from D-lactic acid.

Going back to the two filters with the container of water between them, if we add to the water exactly equal amounts of the D and the L form of a particular compound, however much the D form twists the light in one direction, the L form twists it right back again. The result is *no apparent* optical activity. Such a mixture is called *racemic*.

5.10 Two or More Asymmetric Carbons

Here is a molecule with two asymmetric carbons. Each can be either D or L.

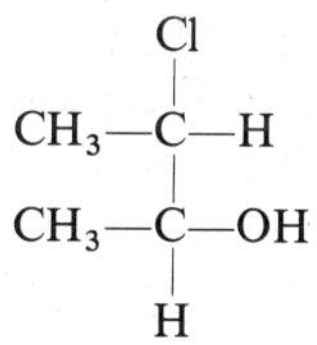

(D or L)

(D or L)

a	*b*	*c*	*d*
D	L	D	L
D	D	L	L

Possible combinations

When the upper (chloro-substituted) carbon is D, the lower carbon can be either D or L, and thus there are two different isomers: $\genfrac{}{}{0pt}{}{D}{D}$ and $\genfrac{}{}{0pt}{}{D}{L}$ Similarly, when the upper carbon is L, the lower carbon again can be either D or L: two more isomers. When all possible combinations of D and L for each asymmetric carbon are considered, there are four different combinations. Furthermore, when there are two asymmetric centers, the mirror image of one isomer has to be opposite in configuration at both centers so the enantiomer of $\genfrac{}{}{0pt}{}{D}{D}$ will be $\genfrac{}{}{0pt}{}{L}{L}$.

Exercise 5.6

a) How many isomers are possible for 2-chloro-3-bromopentane?

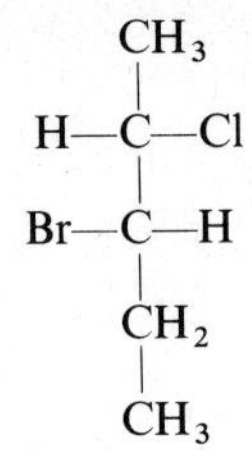

b) What is the enantiomer of the compound in the preceding discussion that is labeled $\substack{\mathrm{L}\\\mathrm{D}}$?

5.11 *Diastereoisomers*

This leaves the question, What is the relationship between $\substack{\mathrm{D}\\\mathrm{D}}$ and $\substack{\mathrm{L}\\\mathrm{D}}$? They are not enantiomers. However much polarized light is twisted one way by $\substack{\mathrm{D}\\\mathrm{D}}$, it will be twisted that amount in the opposite direction by $\substack{\mathrm{L}\\\mathrm{L}}$, but the twisting by $\substack{\mathrm{L}\\\mathrm{D}}$ will be different from either. We say that $\substack{\mathrm{D}\\\mathrm{D}}$ and $\substack{\mathrm{L}\\\mathrm{L}}$ are identical in most properties because they have exactly reversed versions of the same shape. However, $\substack{\mathrm{D}\\\mathrm{D}}$ and $\substack{\mathrm{L}\\\mathrm{D}}$ have different shapes. Isomers of a structure with more than one asymmetric center that are not mirror images are called *diastereoisomers.* Because diastereoisomers have different shapes, they have different properties, just as cis and trans isomers have different properties. As a result, it is much easier to separate diastereoisomers in the laboratory than to separate enantiomers.

If a compound has two asymmetric carbons, we have seen that there are four isomers:

D	D	L	L
D	L	D	L

If we pick one isomer out, say $\substack{\mathrm{D}\\\mathrm{D}}$, there is one enantiomer, $\substack{\mathrm{L}\\\mathrm{L}}$. The other two structures are diastereoisomers. Extending the idea to a molecule with three asymmetric carbons, the following patterns are possible:

D	D	D	D	L	L	L	L
D	D	L	L	D	D	L	L
D	L	D	L	D	L	D	L

If we pick one isomer, say $\substack{\mathrm{D}\\\mathrm{D}\\\mathrm{L}}$, there is one enantiomer, $\substack{\mathrm{L}\\\mathrm{L}\\\mathrm{D}}$. The other six isomers are all diastereoisomers of $\substack{\mathrm{D}\\\mathrm{D}\\\mathrm{L}}$.

Exercise 5.7

What is the enantiomer of $\substack{\mathrm{L}\\\mathrm{D}\\\mathrm{L}}$?

Exercise 5.8

Draw the pattern of possible isomers for a molecule with four asymmetric carbons.

Exercise 5.9

How many asymmetric centers are there in each of the following, and how many isomers are there of each?

a) 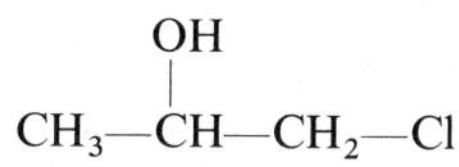

b) 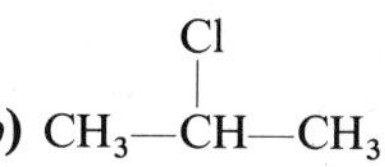

c) 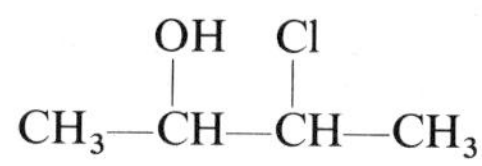

d) 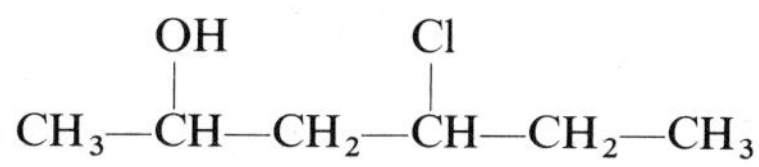

e) 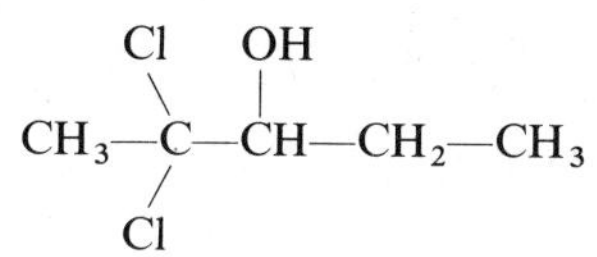

f) 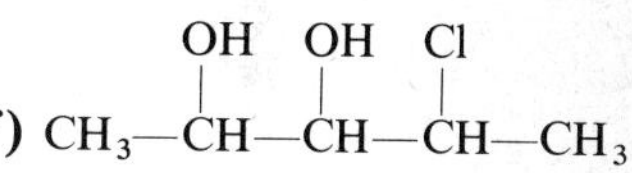

g) 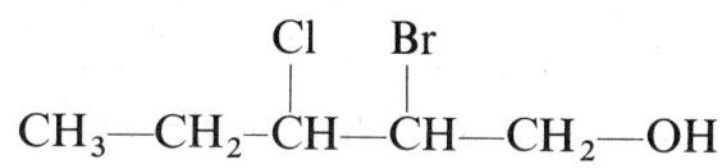

h) 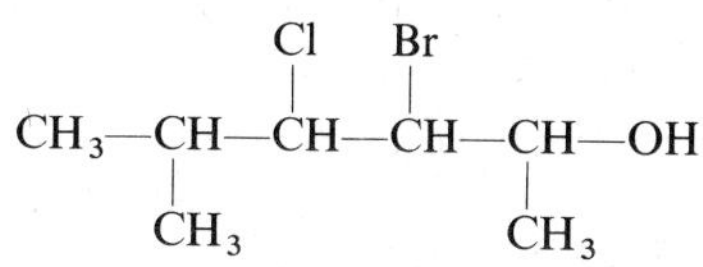

5.12 Optical Isomers in Biochemistry

In biochemistry, optical isomerism is of great importance. Many biologically significant compounds do have asymmetric carbons, sometimes several on one molecule. For example, almost all amino acids, the monomer units of which proteins are built, have an asymmetric carbon. It follows, then, that there is a D and L form of these amino acids. Only the L form is incorporated into protein molecules. Carbohydrates usually have more than one asymmetric carbon, so there are not only enantiomers, there are also diastereoisomers. Two of the most common carbohydrates, D-glucose and D-galactose, are, in fact, diastereoisomers of the same structure. Their properties, metabolism, and biological function are distinctly different in spite of this close structural relationship. In fact, there is a genetic defect encountered in which the individual can utilize D-glucose but not D-galactose, even though they are only slightly different structurally. These individuals at birth cannot dispose of the useless galactose (which is present in their milk diet) rapidly enough, and the accumulated sugar can cause brain damage.

This selectivity toward one of a group of optical isomers is part of a more important general principle in biochemistry. Very frequently biomolecules operate or interact correctly because they fit together exactly. This relationship, termed *complementarity*, includes a fit by shape and a binding by chemical attractive forces. The difference between optical isomers is not one of functional group, it is most clearly one of shape. For complementary fit to occur, this shape must be exactly right. Therefore the difference between enantiomers is a very important one viewed from this standpoint.

Summary

Alcohols, ethers, and thiols are compounds that have oxygen or sulfur functional groups but that lack double bonds.

The alcohol group is very polar and is capable of hydrogen bonding with molecules containing oxygen or nitrogen, including other alcohols and water. The presence of this group in a molecule causes increased boiling point and greater solubility in water when compared with the corresponding molecule lacking the alcohol. Some alcohols can be oxidized to such products as aldehydes, ketones, or carboxylic acids. They can be converted to alkenes by elimination reactions, a reversal of the addition of water to a carbon double bond.

Ethers, because they lack an O—H bond, are not polar like alcohols. They are chemically very unreactive.

Thiols are like alcohols, except that sulfur replaces oxygen. Biologically, the most important reaction of the thiol group is the joining of two thiols with elimination of hydrogen to form a disulfide.

When four different groups are substituted on a carbon, it is called an asymmetric, or chiral, carbon, and the groups can have two different arrangements. A compound including an asymmetric carbon exists as two isomers. The direction of the plane of polarized light passed through a solution of one of these isomers is rotated in one direction while the other isomer rotates the light in the opposite direction. For this reason the isomers are termed optical. When only one carbon in the molecule is asymmetric, there are two isomers that are mirror images or enantiomers of one another. If there are two or more asymmetric carbons, isomers that are exactly opposite are enantiomers. Others are not mirror images and are called diastereoisomers. Biological systems are selective about which of a series of possible optical isomers is produced or can be utilized.

Problems

1. Indicate which of the following are alcohols and which are phenols:

a) 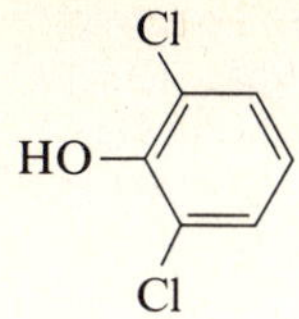

b) 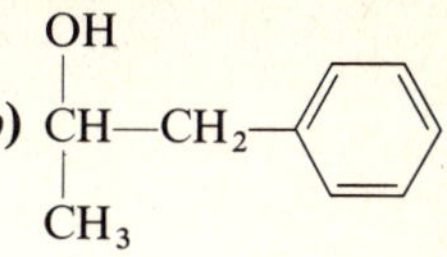

c) 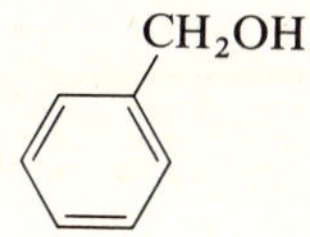

d) 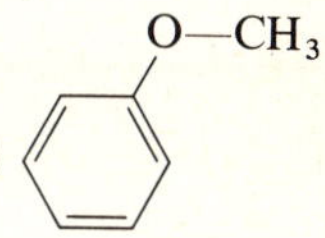

2. Name the following compounds:

a) $CH_3—CH_2—CH_2—CH(OH)—CH_3$

b) 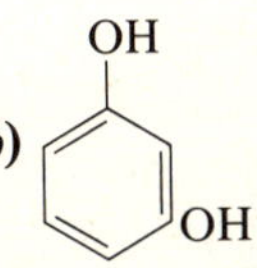

c) $HO—CH_2—CH_2—CH_2—OH$

d) $CH_3—C(CH_3)_2—OH$

e) 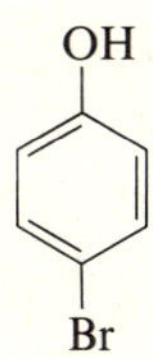

f) $Cl—CH(OH)—CH(Cl)—CH_3$

g) 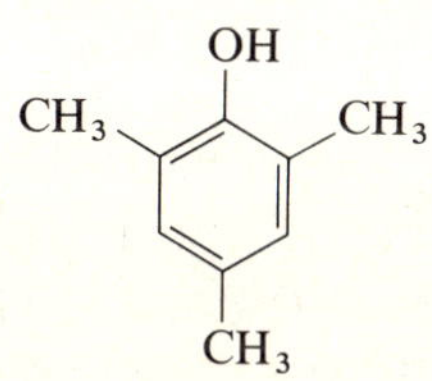

h) 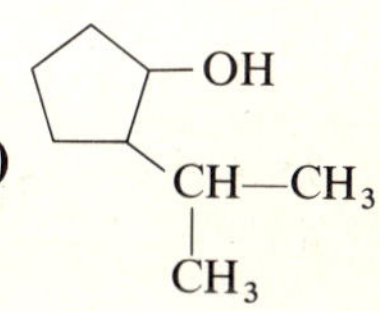

i) 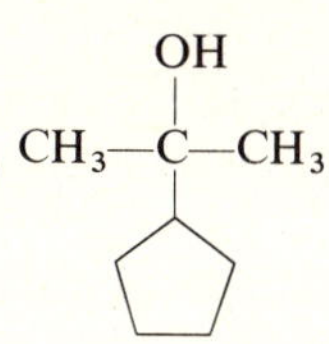

j) 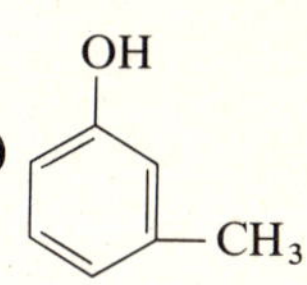

k) $CH_3—CH_2—CH_2—O—CH(CH_3)—CH_3$

l) 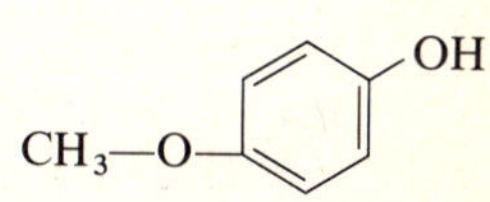

m) $CH_3—CH_2—SH$

n) $CH_3—CH(CH_3)—S—CH(CH_3)—CH_3$

3. Arrange the following in order of anticipated water solubility:

a) 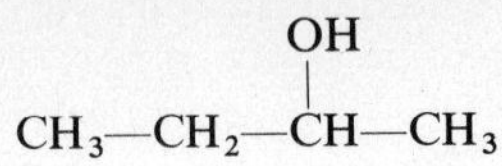

b) $HO—CH_2—CH_2—OH$

c) 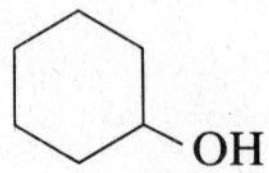

4. Draw structures for a—m:

a) 1-bromo-2-propanol

b) 3-isopropyl-3-hexanol

c) *m*-nitrophenol

d) 1-propanethiol

e) resorcinol

f) methyl isopropyl ether

g) 2-ethylcyclohexanol

h) 1,3-butanediol

i) *m*-cresol

j) *p*-fluorophenol

k) 2-methyl-1,3-cyclopentanediol

l) 2-phenyl-1-butanol

m) *o*-methoxyphenol

n) Name b differently

5. Complete the following reactions:

a) 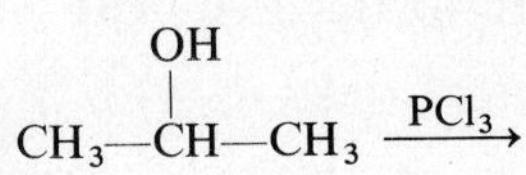

b) 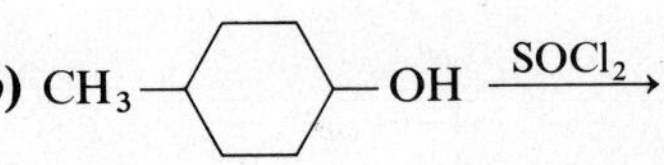

c) $CH_3—O—CH_2—CH_2—OH \xrightarrow{PBr_3}$

d) 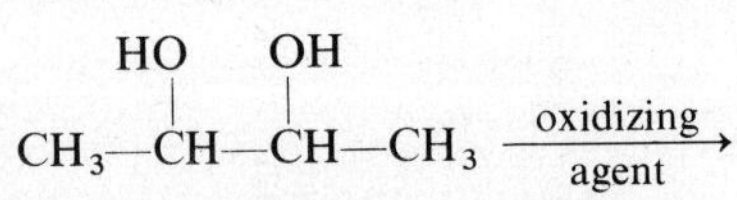

e) 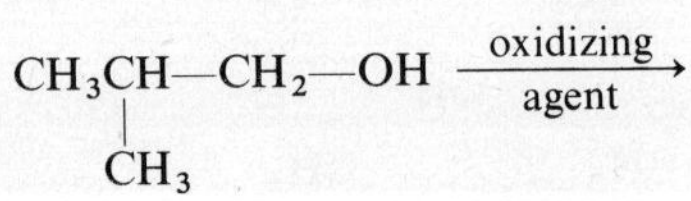

f) 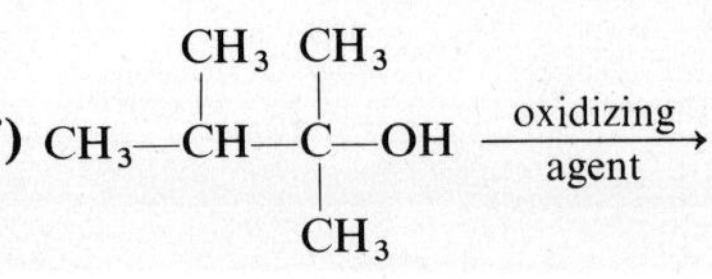

g) 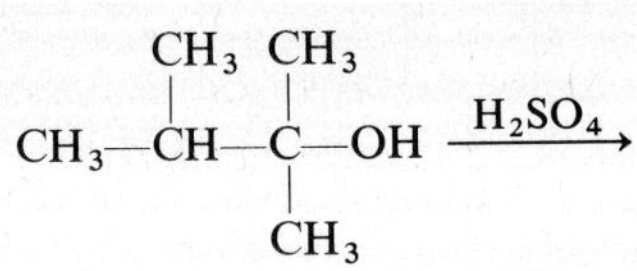

h) $CH_3—CH_2—SH + (O) \longrightarrow$

i) 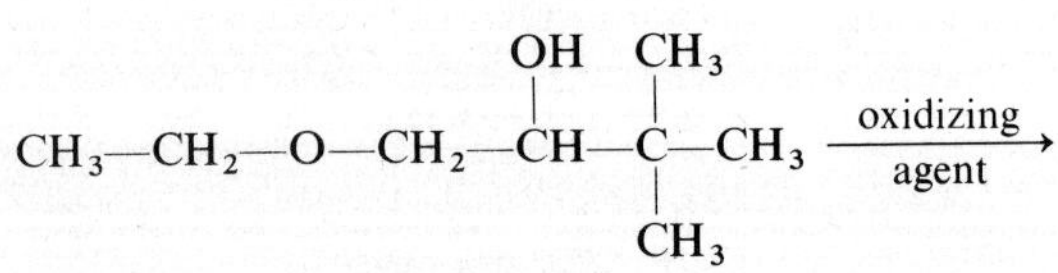

j) 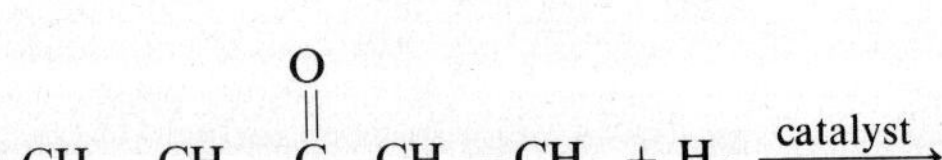

k) 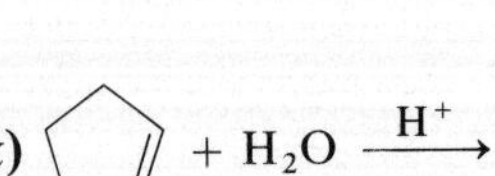

l) 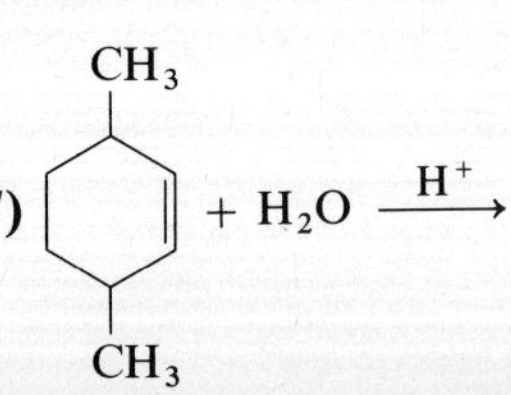

m) 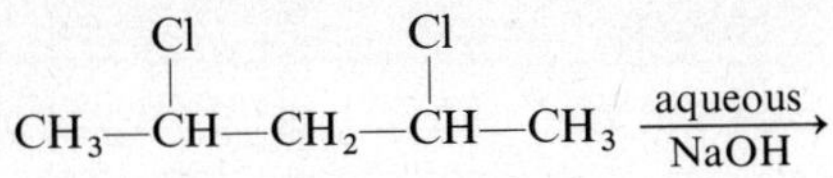

n) 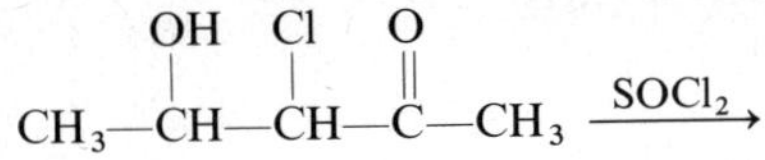

o) 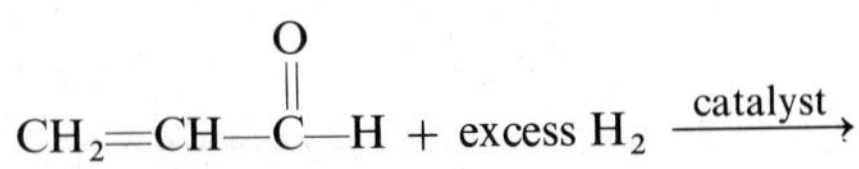

6. Insert the appropriate reagent:

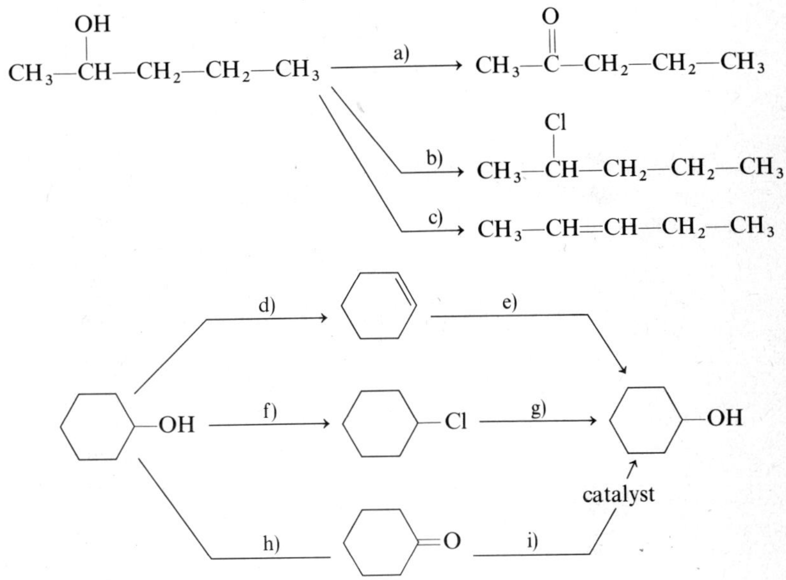

7. Classify each of the six reactions in Problem 6 as substitution, addition, or elimination.

8. Add to your chemical bag of tricks a new elimination reaction:

$$CH_3—CH_2—Cl \xrightarrow[\text{in alcohol}]{\text{KOH}} CH_2{=}CH_2$$

Show the reaction of the following with KOH in alcohol:

a) 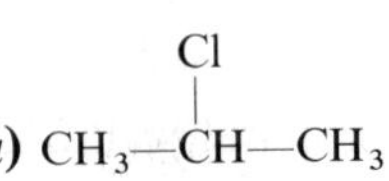

b) 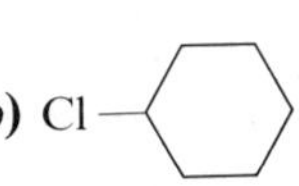

c) 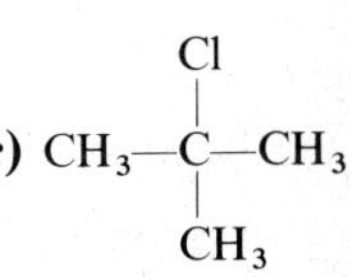

d) 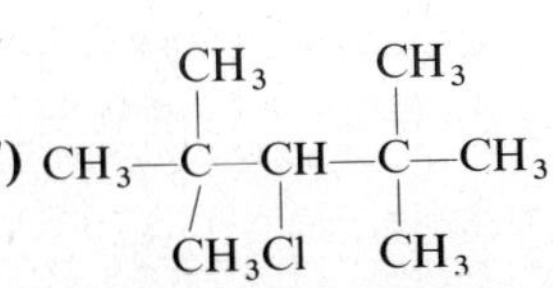 (careful)

Now try these:

e) $Cl{-}CH_2{-}CH_2{-}CH_2{-}CH_2{-}Cl$

f) $Cl{-}CH_2{-}CH_2{-}Cl$ *g)* $CH_3{-}CHCl_2$

9. It is possible to start with a compound (usually some readily available substance) and end up with a different material by putting together a sequence of intermediate reaction steps. In fact, many organic chemists are hired by the chemical industry to design these multistep reaction sequences.

Pretend you are such an organic chemist. Your assignment is to devise several sequences of reactions to make a desired product, then to recommend to the company which one is best to use. Consider each item of the following list, and decide whether it would or would not be important in completing your assignment.

a) For each sequence, what is the cost of the starting material?

b) For each sequence, what are the costs of the reagents?

c) For each starting material, is there a sufficient supply available to handle the proposed level of production should that sequence be selected?

d) For each starting material, where is it obtained? Must it be transported far? Is the source in a politically unstable or unfriendly part of the world?

e) How large is the market for the proposed product?

f) Is the market for the proposed product shrinking, stable, or growing?

g) Are other companies planning new facilities to produce the proposed product?

h) What is the percentage of starting material that actually becomes the desired product? In other words, what is the yield of each sequence?

i) What are the explosion, fire, and health hazards of reagents and intermediates in each pathway?

j) For each pathway that might be selected, what are the waste products or by-products generated that might damage the environment?

k) What is the potential cost of the equipment needed to run each sequence on a commercial scale?

Have any considerations been omitted?

10. Now let us work out some simple pathways from a readily available starting material, ethene ($CH_2{=}CH_2$), obtained from a refinery.

a) Show two pathways to prepare ethanol.

b) Show a pathway to prepare ethanal ($CH_3\overset{O}{\overset{\|}{C}}{-}H$).

c) Show two pathways to prepare chloroethane.

11. Referring to Problem 8, create pathways from ethene to each of the following:

a) $HC{\equiv}CH$

b) $ClCH_2{-}CHCl_2$

c) $Cl_3C{-}CCl_3$

d) $Cl_2CH{-}CCl_3$

12. You are an industrial chemist dealing with two isoprenoid compounds of possible value in perfumes:

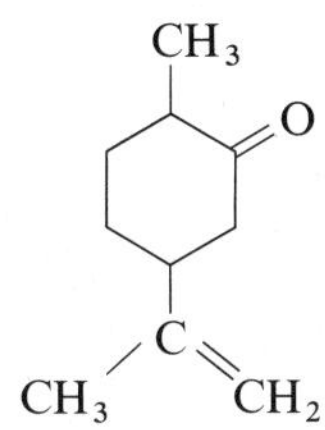

and

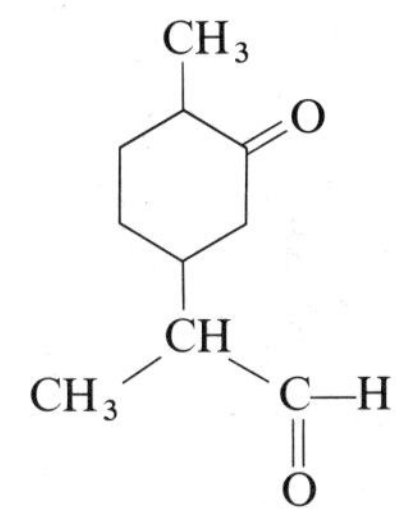

In the laboratory you manage to prepare the following:

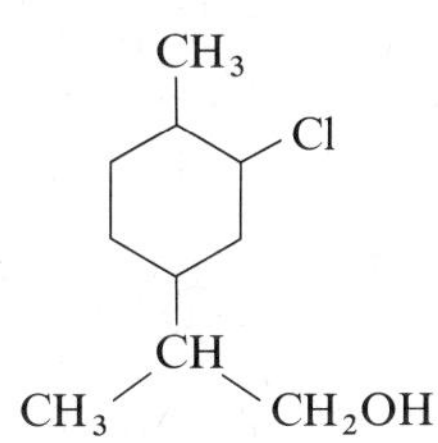

Suggest reactions for converting this intermediate product to each of the products of interest.

13. Locate the asymmetric carbons in the following structures, and place an asterisk at each. How many optical isomers are there of each?

a) 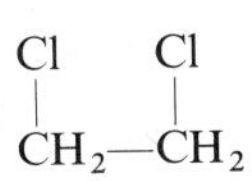

b) 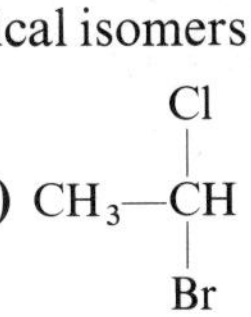

c) 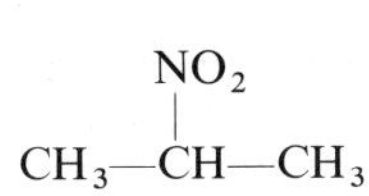

d) 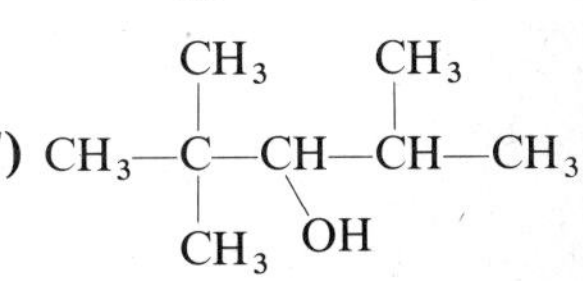

e) 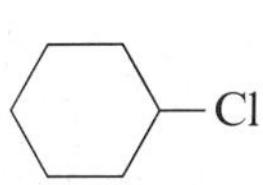

f) 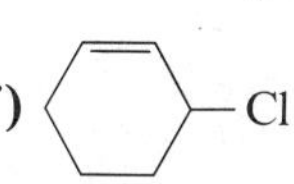

g) 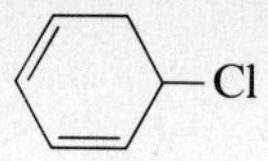

h) 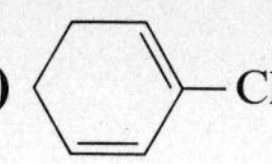

i) 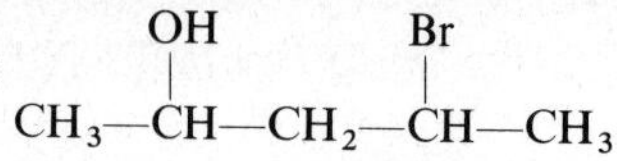

j) 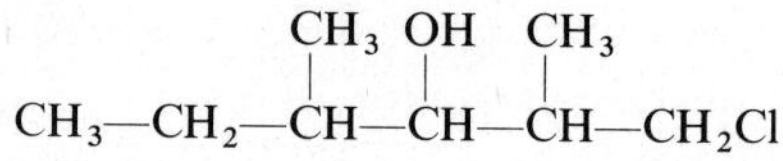

14. Below are drawn the unusual sugars L-fucose and L-rhamnose, which are found in bacterial cell walls. The structures are drawn vertically, representing the D form of the asymmetric carbon by directing the hydroxy group to the right and the L form by directing it to the left.

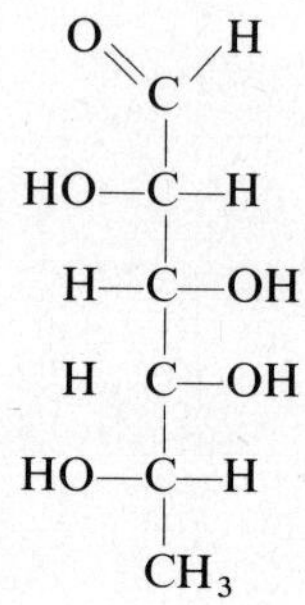

L-fucose

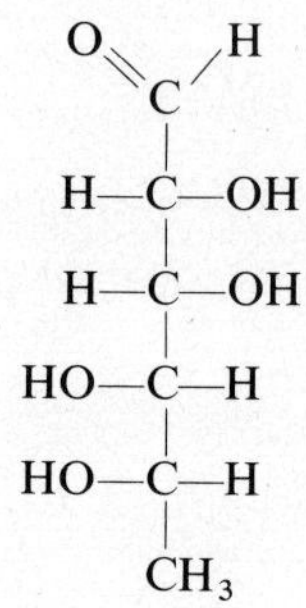

L-rhamnose

a) Number each structure with the aldehyde group as 1 and the methyl group as 6.

b) What number carbons are asymmetric?

c) What number carbons are D in fucose? in rhamnose?

d) What number carbons are L in fucose? in rhamnose?

e) Are these two molecules enantiomers or diastereoisomers?

f) Draw the enantiomer of each.

g) How many optical isomers of this structure are there?

h) Draw a structure that is a diastereoisomer of both of these.

15. Refer to Table 5.3. IUPAC names have been omitted. Provide such names for all but 12, 14, 21, 22, and 23. We must assume from the fact that they appear frequently that some structural features improve germicidal activity. Study the table and make a list of such features. What would you judge the effect of these features to be on water solubility?

Challenge Problems

16. It is possible to have the optical activity of two carbons in the same compound exactly cancel one another. This can happen when a line can be drawn through the center of a compound in such a fashion that one side is the mirror image of the other. Here is an example:

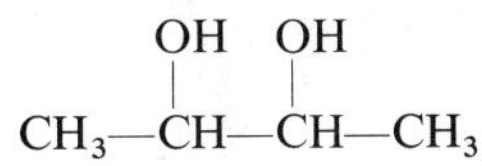

a) Name the structure.

b) Draw a line through the structure as described above to show that one side is in fact the mirror image of the other.

Now let us assign configurations to the two asymmetric carbons to establish what optical isomers are possible:

	Carbon 2	***Carbon 3***
Isomer A	D	D
Isomer B	L	L
Isomer C	D	L
Isomer D	L	D

c) Using the conventions of Problem 14, draw each isomer.

d) Isomers A and B are clearly different from one another. In fact, they are (enantiomers, diastereoisomers).

e) Isomers A and C, A and D, B and C, and B and D are clearly different. In each case, the two isomers compared are (enantiomers, diastereoisomers).

f) Now compare isomers C and D. Turn D end for end and it is identical to C. Are these really two different isomers?

g) A solution of isomer A rotates the plane of polarized light 28° to the left. What will be the optical rotation of an identical solution of isomer B?

h) If isomer A rotates polarized light 28° to the left, then because the two assymmetric carbons are identical we can say that each is responsible for 14° of rotation to the left. On the basis of that reasoning, what rotation would you predict for isomer C (D)? Such an isomer as C (D) is called a *meso* form.

i) Indicate all the compounds below that display meso isomers:

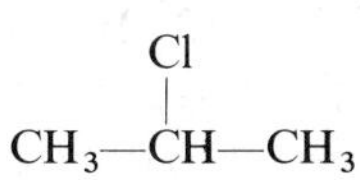

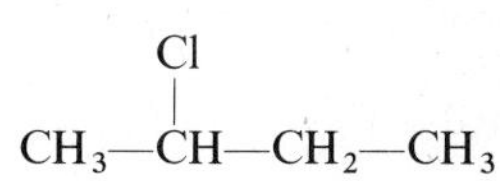

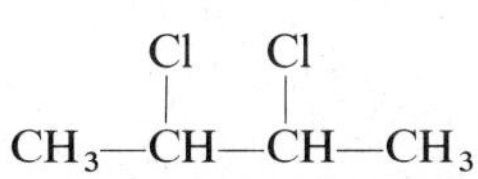

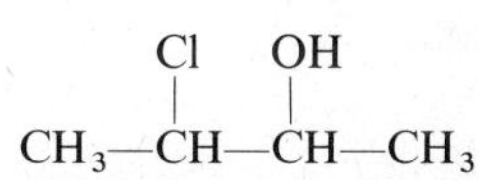

6

Aldehydes, Ketones, and Carbohydrates

Objectives

In this chapter we:

1. Consider the character of the carbonyl group.
2. Learn to name aldehydes and ketones.
3. Survey the terminology, stereochemistry, and nomenclature of carbohydrates.
4. Learn some carbonyl group reactions.
5. Discuss the important disaccharides and polysaccharides.

As chemists, we frequently encounter aldehydes and ketones. They are a component of one of the major groups of biological compounds, the carbohydrates. In addition, carbohydrates have alcohol groups, but the reactions of interest are generally those of the aldehyde and ketone groups. Aldehydes and ketones occur also in steroids and in a variety of natural products, including terpenes and alkaloids.

6.1 *The Carbonyl Group*

The aldehydes and ketones contain a carbonyl group, a carbon-to-oxygen double bond.

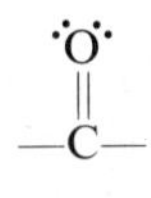

In alkenes, we saw that addition reactions occurred because of the availability of pi electrons. In the carbonyl group, we have a double bond, and hence pi electrons, to encourage addition reactions. However, because oxygen is more electronegative than carbon, the double bond is also polar.

To complete an electron octet, the oxygen requires just the double bond. The carbon, however, requires two more covalent bonds. When hydrocarbon (alkyl) structures provide both of these bonds, the resulting structure is classed as a *ketone*, while bonding to one hydrocarbon and one hydrogen produces the *aldehyde* class of compounds.

$$CH_3-\overset{\overset{\displaystyle O}{\|}}{C}-CH_2-CH_3$$

butanone, a ketone (The carbonyl carbon has completed its octet by sharing with a methyl and an ethyl group.)

$$CH_3-CH_2-\overset{\overset{\displaystyle O}{\|}}{C}-H$$

propanal, an aldehyde (The carbonyl carbon has completed its octet by sharing with an ethyl group and a hydrogen.)

6.2 Naming Ketones and Aldehydes

Ketones are named by determining the length of the longest straight hydrocarbon chain.

$$CH_3-CH_2-\overset{\overset{\displaystyle O}{\|}}{C}-CH_2-CH_3$$

pentane

The -e ending of the hydrocarbon name is replaced by -one.

$$CH_3-CH_2-\overset{\overset{\displaystyle O}{\|}}{C}-CH_2-CH_3$$

pentane becomes pentanone

The position of the carbonyl group is indicated by numbering the hydrocarbon chain and placing the appropriate number immediately before the name.

$$\underset{1}{CH_3}-\underset{2}{CH_2}-\underset{3}{\overset{\overset{\displaystyle O}{\|}}{C}}-\underset{4}{CH_2}-\underset{5}{CH_3}$$

3-pentanone

Exercise 6.1

Complete the following names:

a) $CH_3-\overset{\overset{\displaystyle O}{\|}}{C}-CH_2-CH_2-CH_3$

2 -pentanone

b) $CH_3-CH_2-\overset{\overset{\displaystyle O}{\|}}{C}-CH_2-CH_2-CH_3$

3- hexanone

c) $CH_3-\underset{\displaystyle CH_3}{\underset{|}{CH}}-\overset{\overset{\displaystyle O}{\|}}{C}-\underset{\displaystyle CH_3}{\underset{|}{CH}}-CH_3$

2,4-dimethyl-___-__________

d) $CH_3\text{—}\overset{\overset{\displaystyle O}{\|}}{C}\text{—}CH_3$

____________________ (Is a number necessary?)

e) 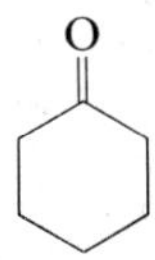

____________________ (Is a number necessary?)

f) $Cl\text{—}CH_2\text{—}CH_2\text{—}\overset{\overset{\displaystyle O}{\|}}{C}\text{—}CH_2\text{—}\overset{\overset{\displaystyle Cl}{|}}{CH}\text{—}CH_2\text{—}Cl$

___ - _______________ - ___ - _______________

Exercise 6.2

What is the shortest possible aliphatic ketone?

As usual, aromatic structures have special names.

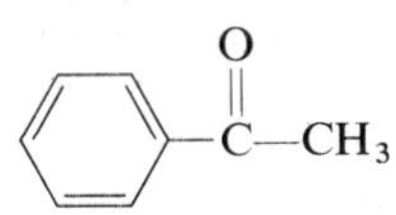

acetophenone
(used in perfume)

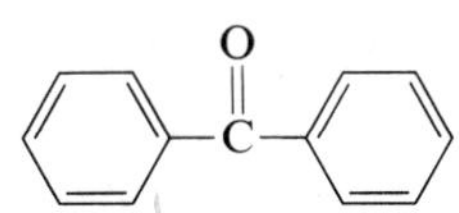

benzophenone
(used in perfume, especially in soaps)

Aldehyde naming also begins with the identification of the longest hydrocarbon chain.

$$CH_3\text{—}CH_2\text{—}CH_2\text{—}\overset{\overset{\displaystyle O}{\|}}{C}\text{—}H$$

butane

The final -e of the hydrocarbon name is replaced with -al.

$$CH_3\text{—}CH_2\text{—}CH_2\text{—}\overset{\overset{\displaystyle O}{\|}}{C}\text{—}H$$

butane becomes butanal

Because the aldehyde group is at the end of the chain, it is designated as carbon 1.

$$\underset{4}{CH_3}\text{—}\underset{3}{CH_2}\text{—}\underset{2}{CH_2}\text{—}\underset{1}{\overset{\overset{\displaystyle O}{\|}}{C}}\text{—}H$$

It is normally not necessary to indicate the position of the aldehyde group by number when naming an aldehyde.

Exercise 6.3

Complete the following names:

a) 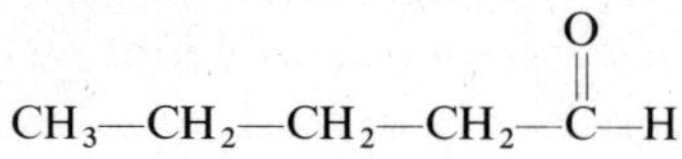

____

b) 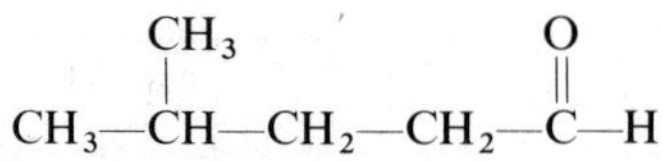

___ -methyl _______________

c) 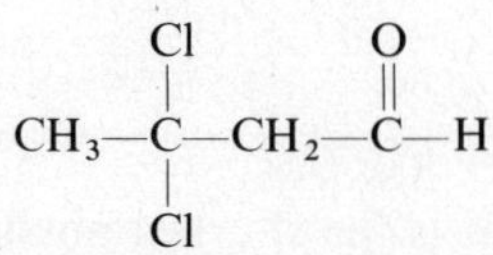

Exercise 6.4

What is the shortest aldehyde chain possible?

It is sometimes useful to name the aldehyde group as a side group when other functional groups are present. In such cases it is called a *formyl* group.

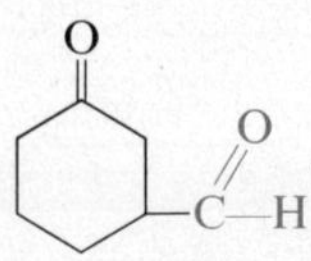

3-formyl cyclohexanone

The simplest aromatic aldehyde is called benzaldehyde.

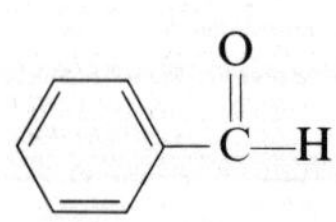

benzaldehyde
(almond flavor, used in perfumes)

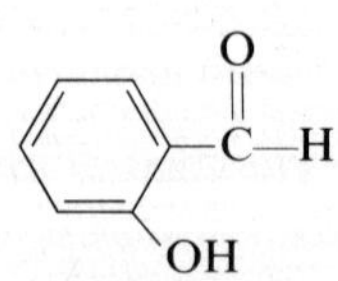

salicylaldehyde, or
2-hydroxybenzaldehyde
(used in perfumes)

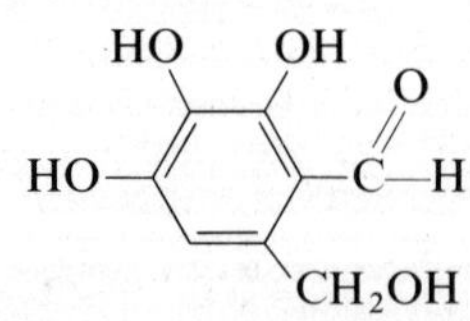

6-hydroxymethyl-
2,3,4-trihydroxybenzaldehyde
(Fomecin A, an antibiotic)

6.3 Carbohydrates

Carbohydrates constitute one of the major classes of compounds occurring in living organisms. Chemically they can be defined as *polyhydroxy* (containing many alcohol groups) *aldehydes* and *ketones*.

Names associated with carbohydrates are common names rather than names conforming to IUPAC rules. These common names often end in -ose. Carbohydrates with an aldehyde group are termed *aldoses* and those with a ketone group *ketoses*. Most of the carbohydrates encountered have either five or six carbons, *pentoses* and *hexoses*, respectively. These terms can be combined. For example, a five-carbon aldehyde sugar would be an aldopentose.

Exercise 6.5

Classify the following as aldo- or keto-pentoses or -hexoses:

a)

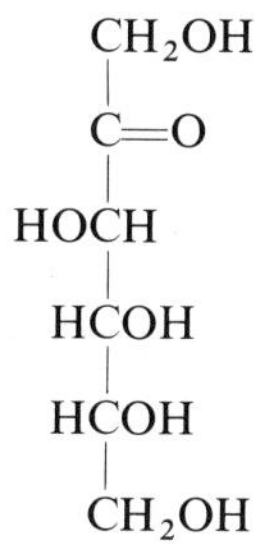

b)

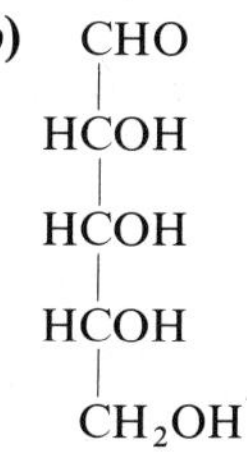

c)

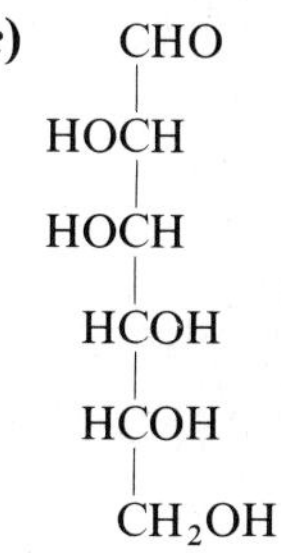

Consider the aldohexose

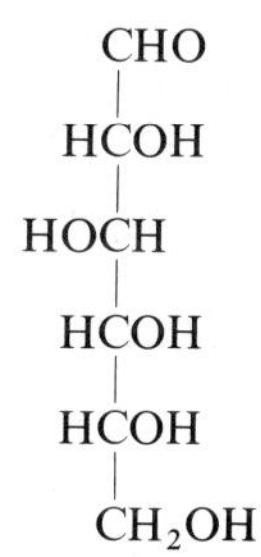

Review if necessary the definition of an asymmetric carbon (Chapter 5). Locate the asymmetric carbons on the aldohexose structure. Each of carbons 2–5 is asymmetric. Four such centers produces 16 different isomers:

1.

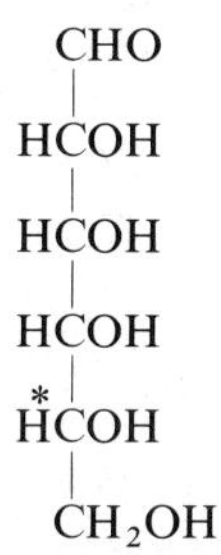

D-Allose

2.

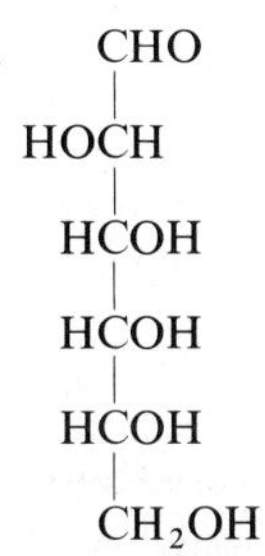

D-Altrose

3.

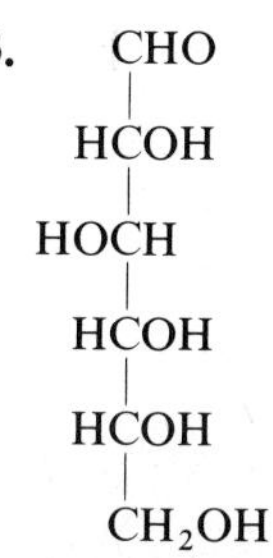

D-Glucose

4.

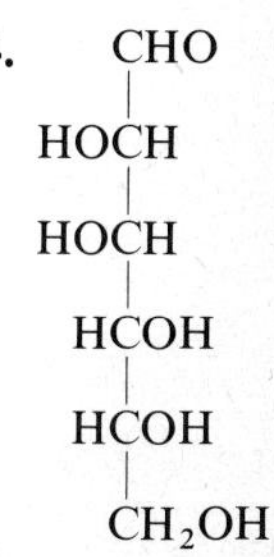

D-Mannose

5. 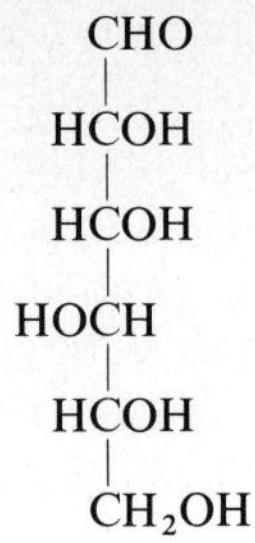

D-Gulose

6. 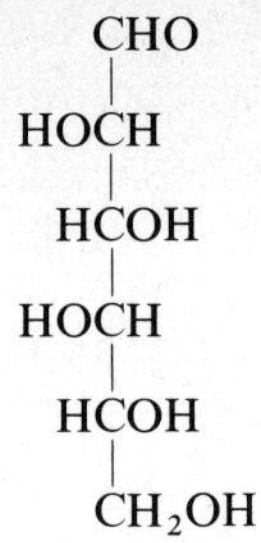

D-Idose

7. 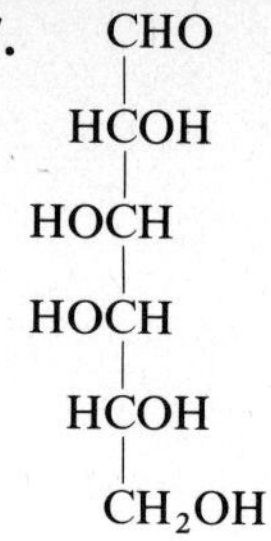

D-Galactose

8. 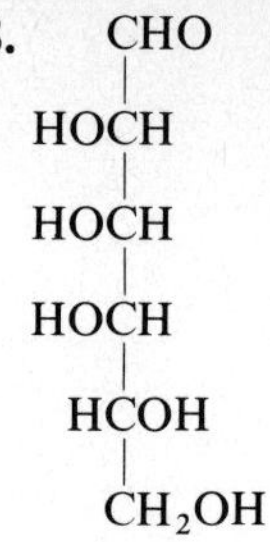

D-Talose

9. 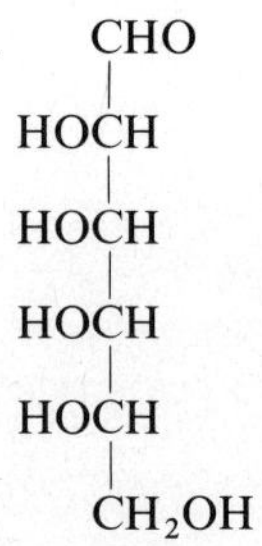

L-Allose

10. 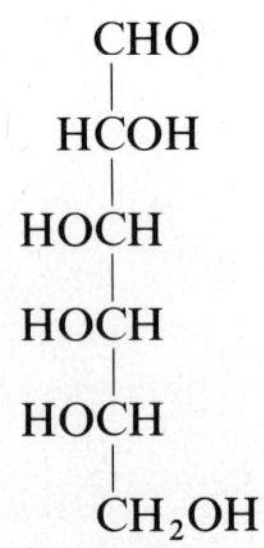

L-Altrose

11. 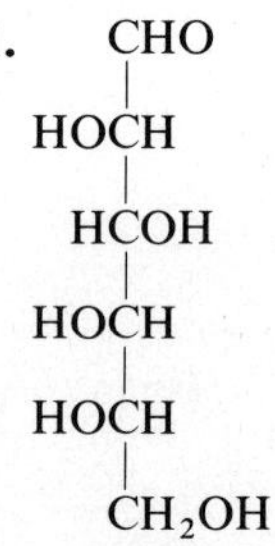

L-Glucose

12. 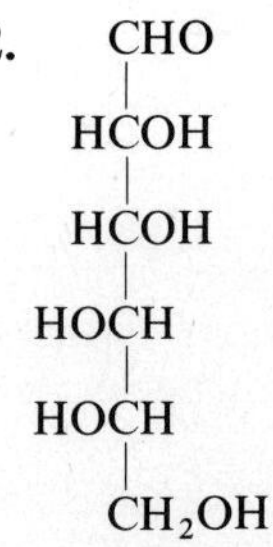

L-Mannose

13. 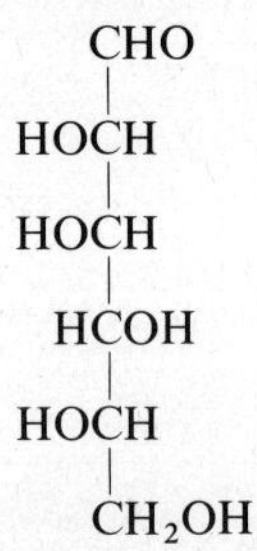

L-Gulose

14. 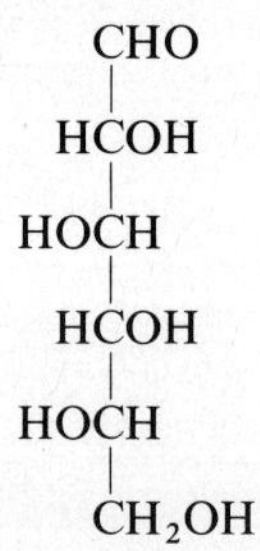

L-Idose

15. 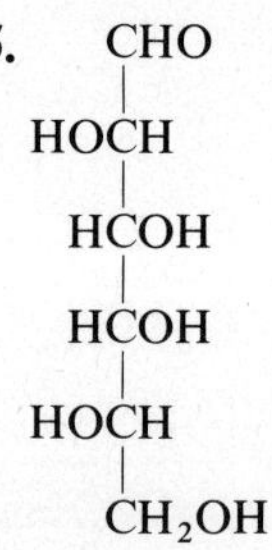

L-Galactose

16. 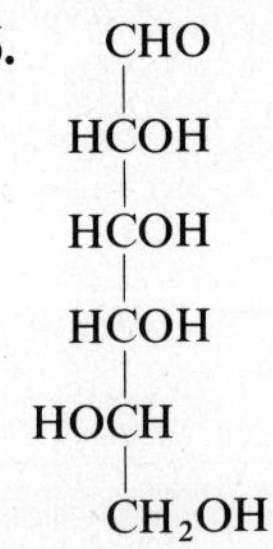

L-Talose

Exercise 6.6

What structure is the enantiomer of 1? Is 2 an enantiomer or a diastereoisomer of 1?

Remember that once we select one of the 16 isomers, only one of the remaining isomers is an enantiomer (mirror image). The other 14 isomers are diastereoisomers. These will have chemical and physical properties different from the reference isomer.

Of the 16 optical isomers, there are 8 pairs of enantiomers. Each pair is given a common name, with one of the pair designated as D and the other as L. We assign the D and L designation according to the configuration of the second highest numbered carbon. In the preceding examples, carbon 5 (starred on D-allose) determines the absolute configuration of the carbohydrate molecule. Thus all the carbohydrates numbered 1–8 are D, and all numbered 9–16 are L. Almost all of the carbohydrates found in nature are D. Three of those in the top (D) set occur commonly: glucose, galactose, and mannose. D-glucose, in fact, is one of the most common organic compounds on earth.

Other common carbohydrates include a ketohexose, D-fructose, and aldopentoses, such as D-ribose and D-xylose.

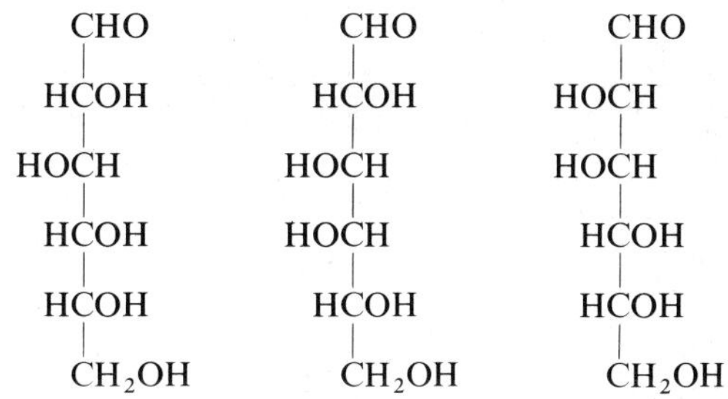

D-Glucose D-Galactose D-Mannose

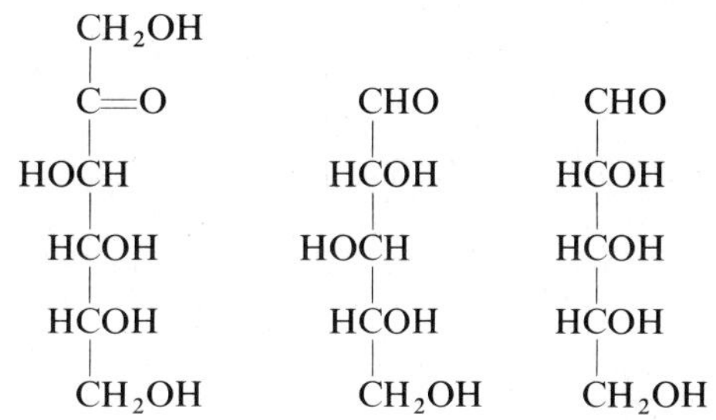

D-Fructose D-Xylose D-Ribose

6.4 *Reactions of Carbonyl Groups: The Addition or Elimination of Hydrogen*

Carbonyl groups contain double bonds and undergo addition reactions just as alkenes do. The addition of hydrogen is exactly the same as with alkenes, except that the product is an alcohol rather than an alkane (see Chapter 5).

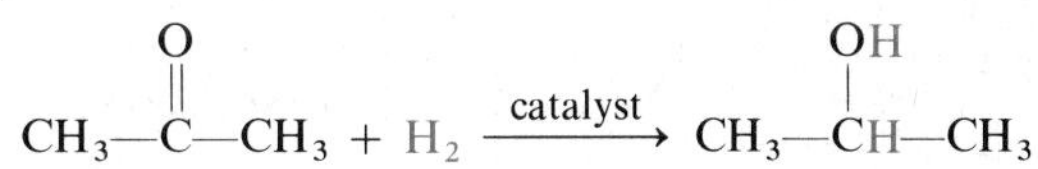

propanone (common name: acetone) 2-propanol

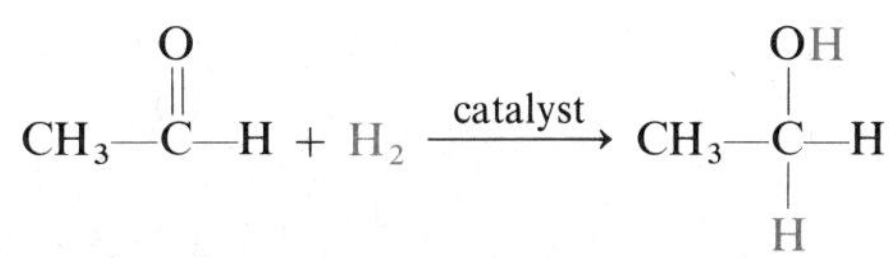

ethanal (common name: acetaldehyde) ethanol

In Chapter 3 the addition of hydrogen was classified as a reduction and its removal as an oxidation. These are therefore *reduction reactions*.

Using appropriate oxidizing agents, these reduction reactions can be reversed. The resulting oxidations are also called *dehydrogenations* (see page 91).

$$CH_3{-}CH_2{-}\overset{\overset{\large OH}{|}}{C}H{-}CH_3 + (O) \xrightarrow[\text{agent}]{\text{oxidizing}} CH_3{-}CH_2{-}\overset{\overset{\large O}{\|}}{C}{-}CH_3 + H_2O$$

2-butanol — butanone

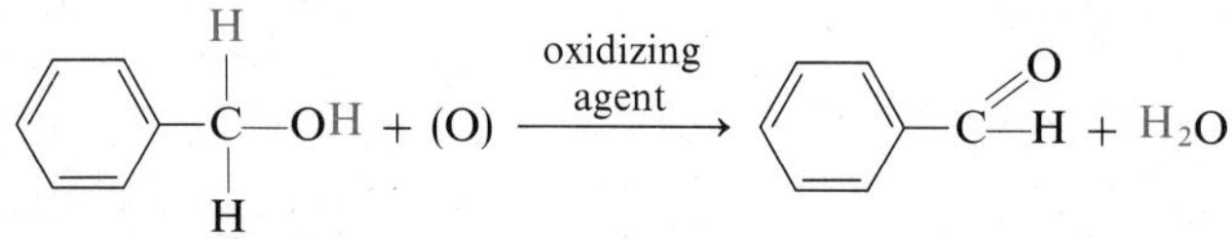

phenylmethanol (benzyl alcohol) — benzaldehyde

Exercise 6.7

Complete the following reactions:

a) $CH_3{-}CH_2{-}\overset{\overset{O}{\|}}{C}{-}CH_2{-}CH_3 + H_2 \xrightarrow{\text{catalyst}}$

b)

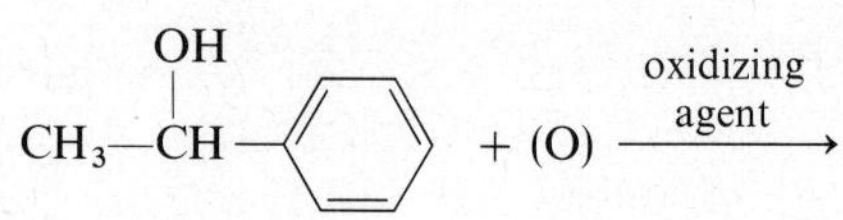

c) $CH_3{-}CH_2{-}CH_2{-}OH + (O) \xrightarrow[\text{agent}]{\text{oxidizing}}$

d)

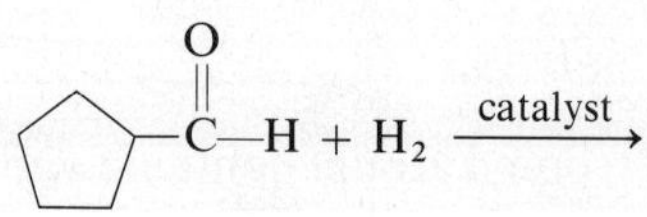

Exercise 6.8

A dehydrogenation is a type of (oxidation, reduction). Indicate in Exercise 6.7 the reactions that are dehydrogenations.

Enzymes catalyze reactions both in forward and reverse directions. There are enzymes whose specific function is to interconvert aldehydes or ketones with alcohols. Thus they catalyze oxidation in one direction and reduction in the other. Such enzymes are classed formally as *oxidoreductases*, or informally as *dehydrogenases*. When they remove hydrogens from an alcohol, they transfer the hydrogen to small organic molecules called *cofactors*. The cofactors are reduced by the addition of the hydrogen at the same time that the alcohol is oxidized by the loss of it. The reverse process occurs when the enzyme converts an aldehyde or ketone to an alcohol. The conversion of ethanal to ethanol is performed by yeast cells as the last step in the fermentation (metabolism in the absence of oxygen) of sugar. Fermentation is the source of alcohol in beer or wine. The total pathway is discussed in Chapter 15.

$$\underset{\text{ethanal}}{CH_3{-}\overset{\overset{\displaystyle O}{\|}}{C}{-}H} + \underset{\substack{\text{reduced}\\ \text{cofactor}}}{\text{cofactor}\cdot 2H} \xrightarrow{\text{enzyme}} \underset{\text{ethanol}}{CH_3{-}CH_2{-}OH} + \underset{\substack{\text{oxidized}\\ \text{cofactor}}}{\text{cofactor}}$$

When ethanol is utilized as a source of energy in aerobic metabolism—that is, metabolism in the presence of oxygen—the first step is to reverse the above reaction using the same enzyme and cofactor combination.

$$CH_3{-}CH_2{-}OH + \text{cofactor} \xrightarrow{\text{enzyme}} CH_3{-}\overset{\overset{\displaystyle O}{\|}}{C}{-}H + \text{cofactor}\cdot 2H$$

Another example of the oxidation of alcohols to ketones in biological systems occurs in the metabolism of fats. Recall from Chapter 3 that an alkane portion of a fat is converted to an alkene. Water is then added to the double bond to create an alcohol. Now we can continue the process one more step. A pair of hydrogens are removed from the alcohol to generate a ketone.

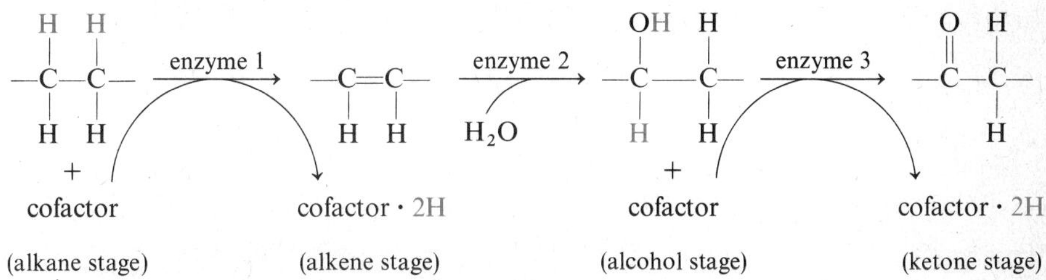

Notice that the conversions both of alkane to alkene and of alcohol to ketone represent the creation of a double bond by dehydrogenation (oxidation) and that both require a cofactor. Further notice that each step is catalyzed by a different enzyme.

In the case of aldehydes, oxidation can proceed one step further. With an appropriate oxidizing agent, an aldehyde is converted to a carboxylic acid.

$$\underset{\text{ethanal}}{CH_3{-}\overset{\overset{\displaystyle O}{\|}}{C}{-}H} + (O) \xrightarrow{\substack{\text{oxidizing}\\ \text{agent}}} \underset{\text{ethanoic acid}}{CH_3{-}\overset{\overset{\displaystyle O}{\|}}{C}{-}OH}$$

It requires careful control of conditions to oxidize an alcohol to an aldehyde but not to go further to a carboxylic acid. Ketones cannot oxidize in this fashion.

Exercise 6.9

Why it is not possible for ketones to be oxidized to carboxylic acids as aldehydes are?

Exercise 6.10

Complete these reactions:

a) 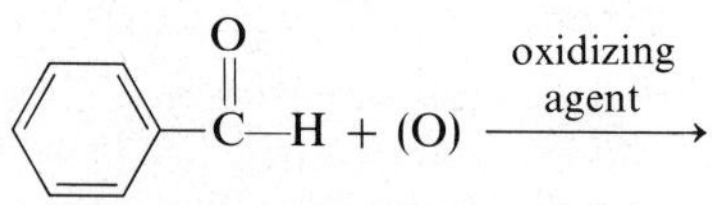

b) 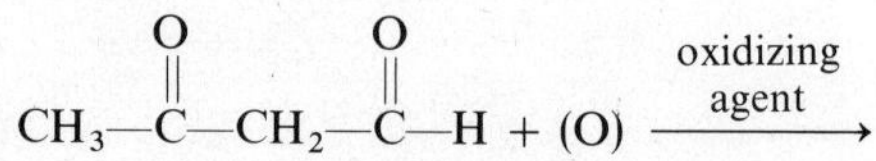

The oxidation of aldehyde is used to test for aldose sugars. The presence of excessive levels of D-glucose in blood or urine is symptomatic of diabetes mellitus. It is therefore routine to check urine specimens for D-glucose by a *reducing sugar test*. The traditional method utilizes a solution of a copper ion complex (Benedict's solution or Fehling's solution). When the Cu^{2+} ion and the aldehyde group react in boiling alkaline solution, the aldehyde is oxidized to a carboxylic acid and the Cu^{2+} becomes Cu^{+}.

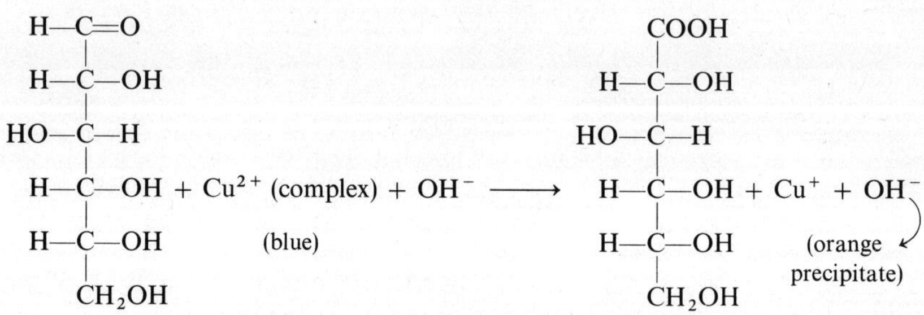

glucose gluconic acid

The solution changes from intense blue through aqua, green, and yellow, to orange. The degree of color change depends on the amount of reducing sugar present. When done carefully, the final color of the solution can be compared with a color chart to estimate D-glucose concentration.

The modern method for glucose analysis also depends on the oxidation of glucose to gluconic acid. However, in place of the copper reagent, the enzyme glucose oxidase is utilized. Hydrogen peroxide is a second product of this enzyme-catalyzed method. The hydrogen peroxide is used to oxidize some compound for which the oxidation involves a color change.

$$\text{glucose} + O_2 \longrightarrow \text{gluconic acid} + H_2O_2$$

$$\text{reduced dye} + H_2O_2 \longrightarrow \text{oxidized dye} + H_2O$$

Paper strips are saturated with these reagents and are dipped into urine to provide a rapid and extremely simple glucose color test.

It is important with newborn infants to test the urine for sugar for more than one reason. It is desirable to detect diabetes early, of course. There is also a congenital problem involving inability to utilize D-galactose, a component of milk sugar. In such cases, D-galactose appears at high levels in the urine and responds to reducing sugar tests in the same way as D-glucose. If not detected early enough, brain damage can occur. Once detected, however, it is necessary only to provide a substitute milk free of galactose to avoid serious consequences.

Numerous enzymatic examples of aldehyde oxidation exist. In the metabolism of carbohydrates the following reactions occur:

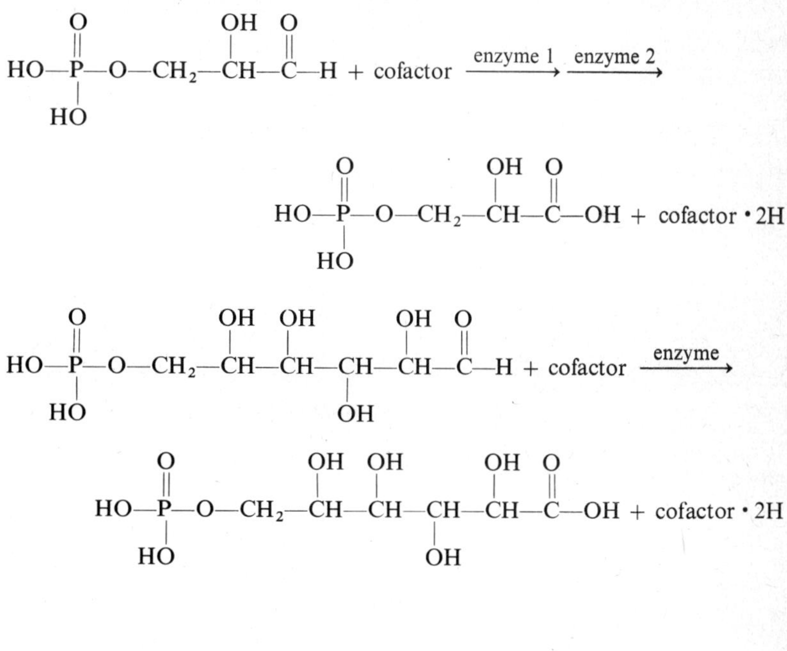

6.5 *Formation of Hemi- and Full Acetals*

Reactions involving addition of other than hydrogen to carbonyls are found. From the standpoint of carbohydrates, the most important example is the addition of alcohols to the double bond. Let us illustrate this first with a simple example.

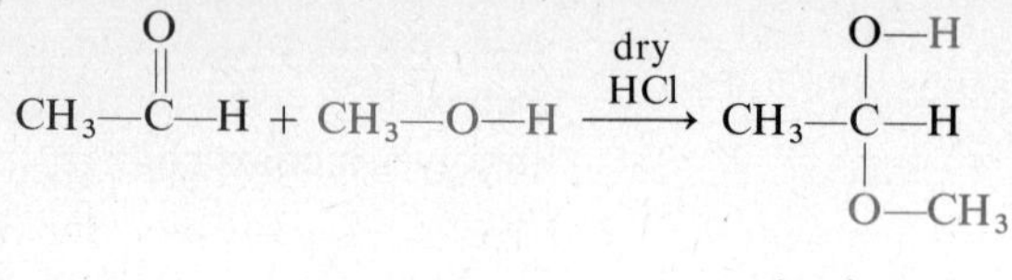

hemiacetal

The product is called a *hemiacetal*. The following are additional examples:

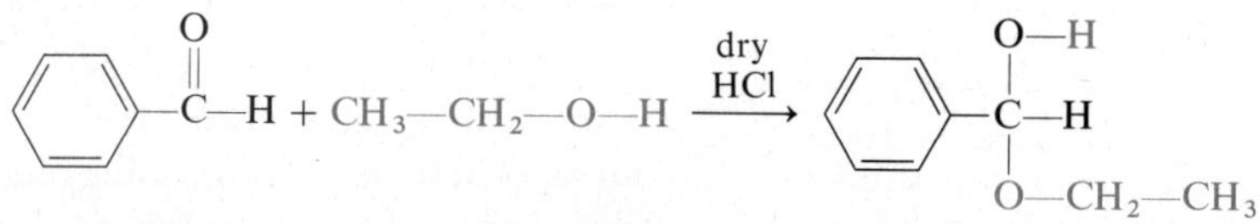

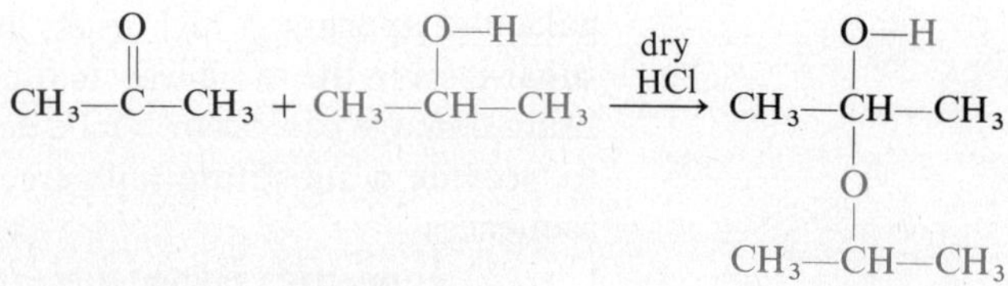

Exercise 6.11

Complete the following reactions:

a) 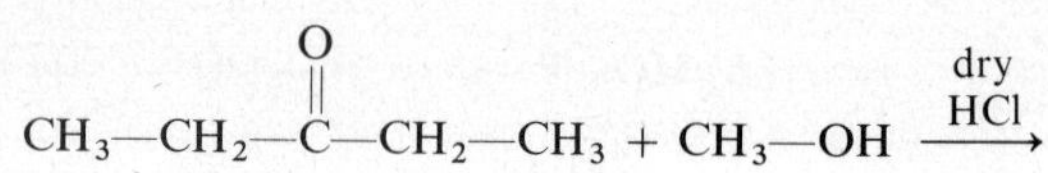

b) 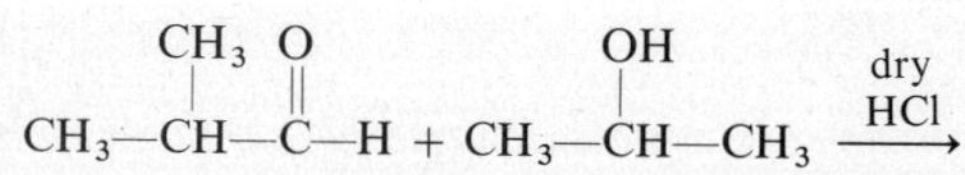

Carbohydrates are a special case. Both the carbonyl and the alcohol group are present on the same molecule. If the carbohydrate molecule can bend so that the two groups are adjacent, an internal hemiacetal can be formed.

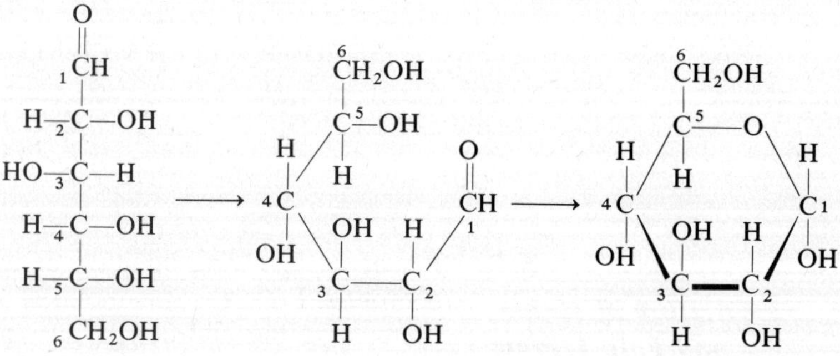

Most often these ring-type hemiacetal carbohydrates are drawn as if they were flat hexagons seen from in front of and slightly above the ring structure. This called a *Haworth projection*. The following are Haworth structures for a variety of carbohydrates. Notice the convention followed in orienting the ring with carbon 1 at the right and carbon 6 at the upper left. Carbon atoms with the D configuration are drawn with the hydroxyl group to the right in the linear vertical presentation of the carbohydrate. That same carbon has the hydroxyl group down in the Haworth projection.

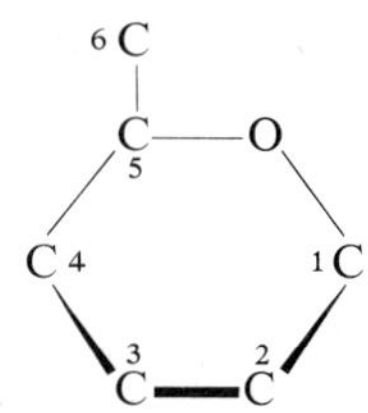

These numbers correspond to the numbers on the free aldehyde form of the carbohydrate.

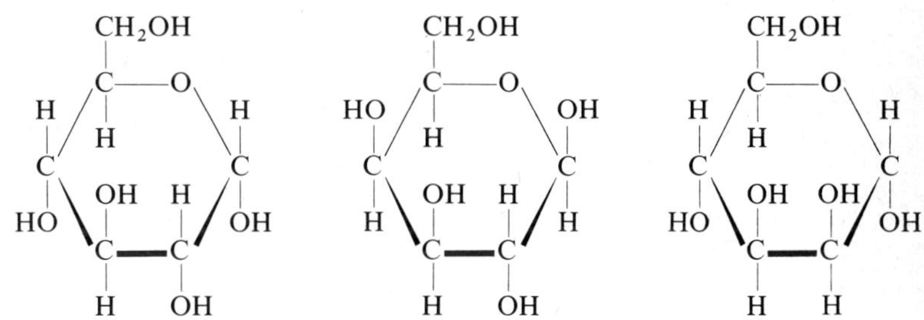

D-glucose D-galactose D-mannose

These are all six-membered rings. We also find five-membered rings.

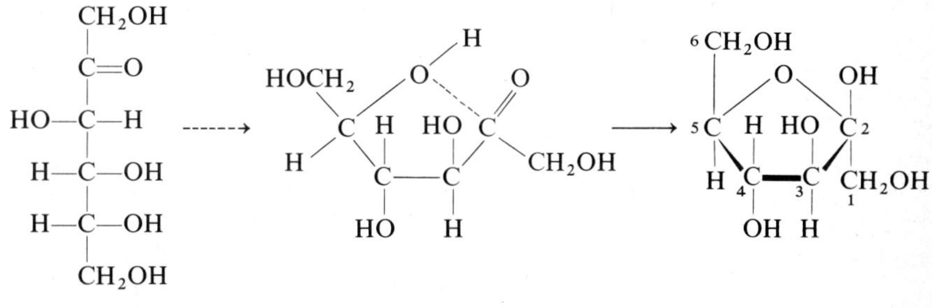

D-fructose

Other five-membered rings are these:

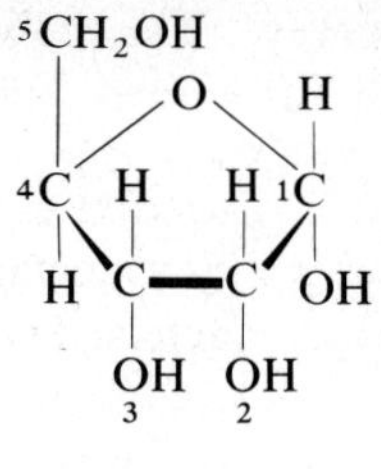

D-ribose

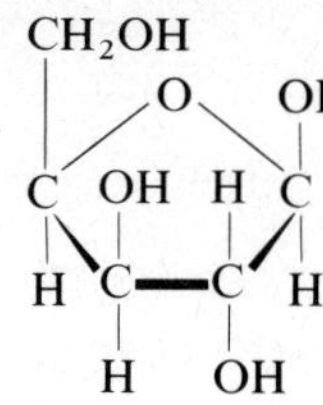

D-xylose

In solution, almost all the molecules of the common carbohydrates are in the form of these internal hemiacetal rings. When we discuss carbohydrate chemistry, we consider primarily the chemistry of these hemiacetals.

Hemiacetals, in turn, can react with alcohols to form *acetals* or *full acetals*.

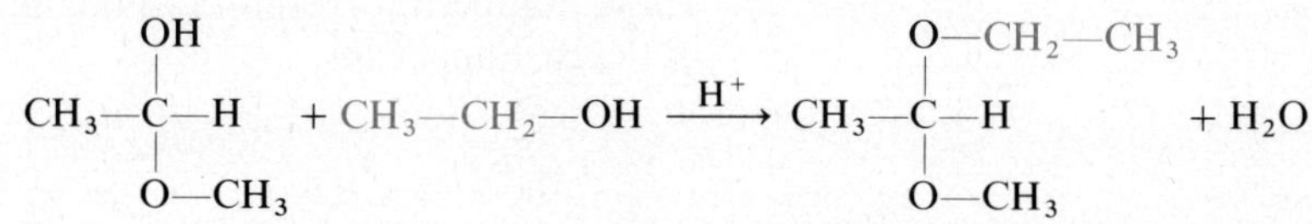

This is a condensation rather than an addition reaction. In such a reaction, the two structures are joined with the elimination of water, part of the water coming from each structure.

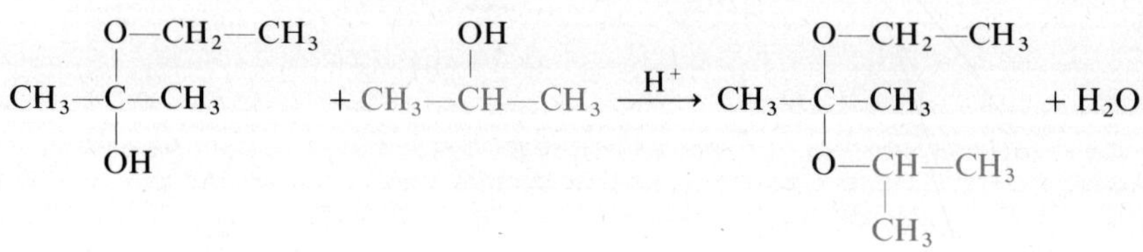

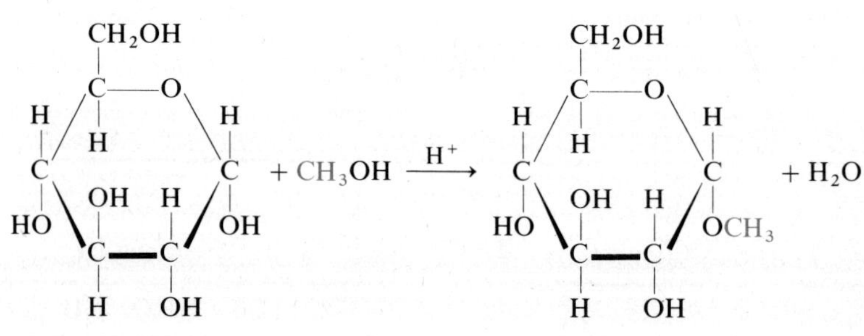

D-glucose methyl glucoside

Notice that in the case of carbohydrates the —OH at carbon 1 looks like an alcohol, but, in fact, it reacts differently than do the —OH groups at other carbons. Only the hemiacetal —OH group reacts with an alcohol to form an acetal.[1]

6.6 Hemiacetals, Acetals, and Optical Isomerism

The carbonyl carbon of a carbohydrate is not asymmetric, but when converted to a hemiacetal or acetal, it has four different groups attached to it and is therefore asymmetric. The introduction of this new asymmetric center means that there are two isomers of each carbohydrate once the ring is formed. These are designated α and β, α being the form with the —OH of the hemiacetal or the —OR of the acetal drawn down and β being the form with the group drawn up.

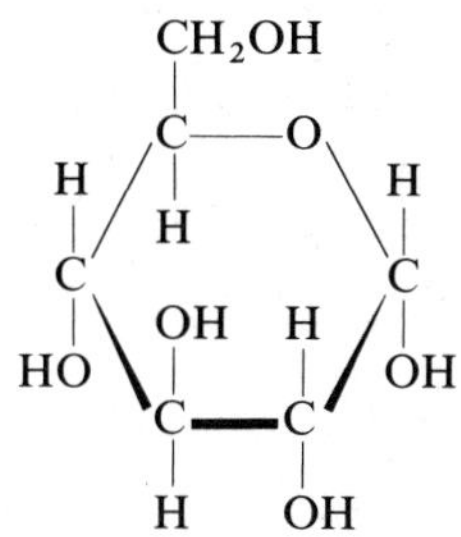

α-D-glucose

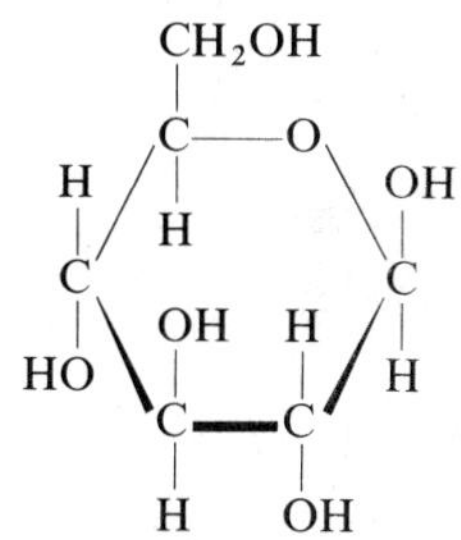

β-D-glucose

In solution, the hemiacetal is in equilibrium with the free aldehyde or ketone.

[1] The structure

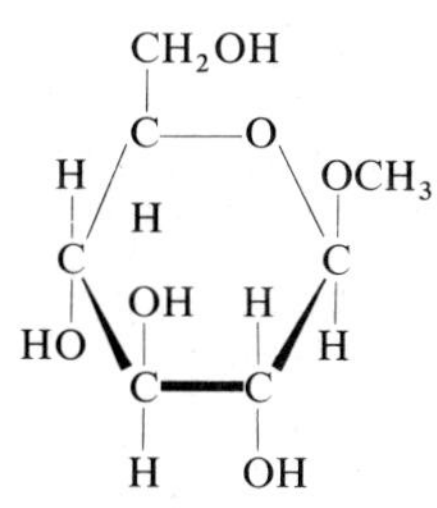

is sometimes represented as

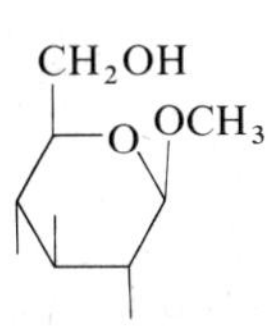

where the lines indicate the orientation of the —OH groups. —H groups are omitted.

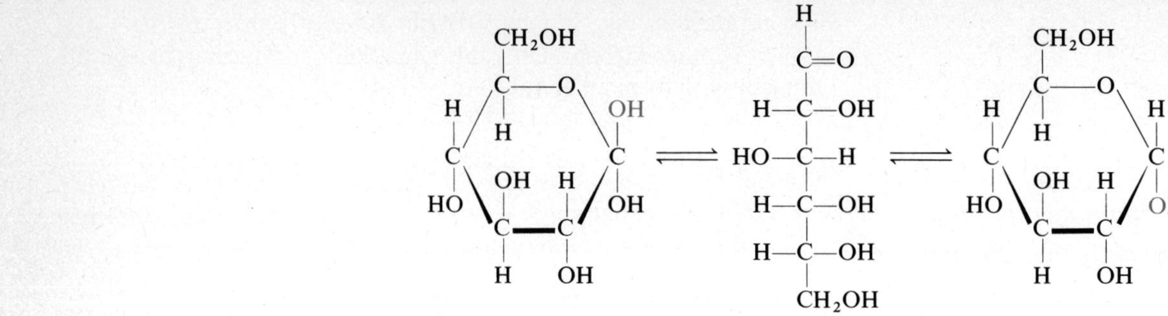

β form D-glucose α-form

As a result, we encounter a mixture of the α and β forms along with a tiny amount of the free aldehyde or ketone. The acetal is not in such an equilibrium. Once we have the α form of a full acetal, it remains α. It is important with full acetals to name the compound as the α or β form.

Exercise 6.12

Study the α and β forms of the methyl full acetal of D-glucose shown above. Are they enantiomers, diastereoisomers, or neither?

Exercise 6.13

Complete reactions a–d, and draw e–g:

a) 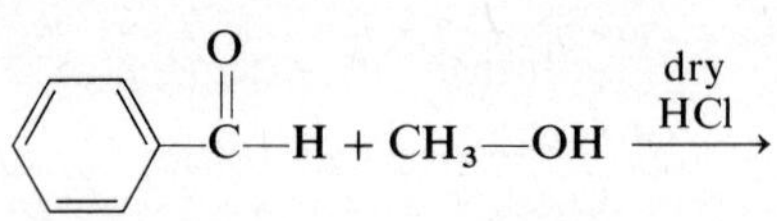

b) 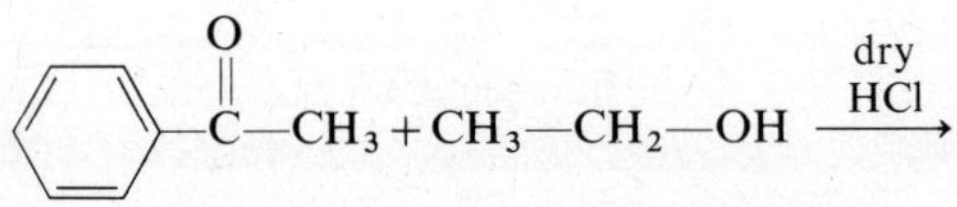

c) 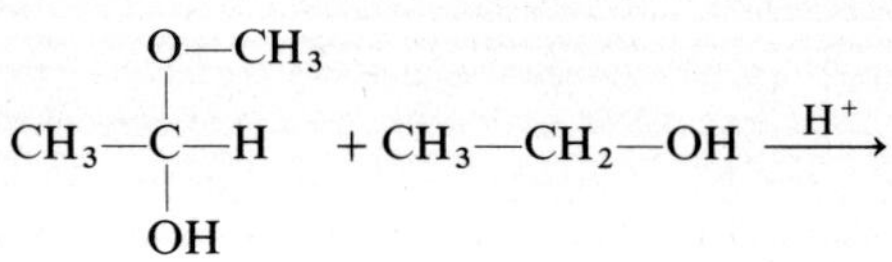

d) 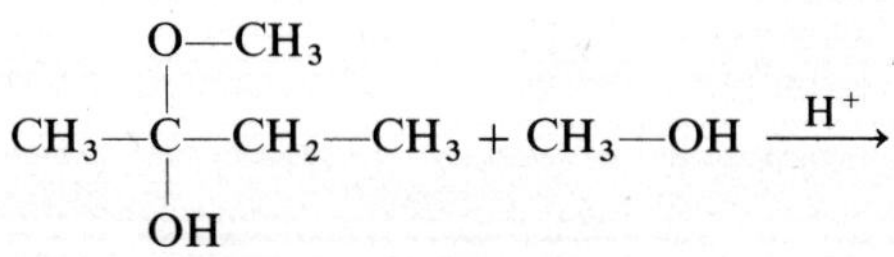

e) The ethyl-β-acetal of D-glucose
f) The isopropyl-α-acetal of D-galactose
g) The methyl-α-acetal of D-fructose

6.7 *Disaccharides*

The most significant example of acetal formation is the case where the hemiacetal group of one carbohydrate molecule reacts with an alcohol group of another carbohydrate molecule to join the two

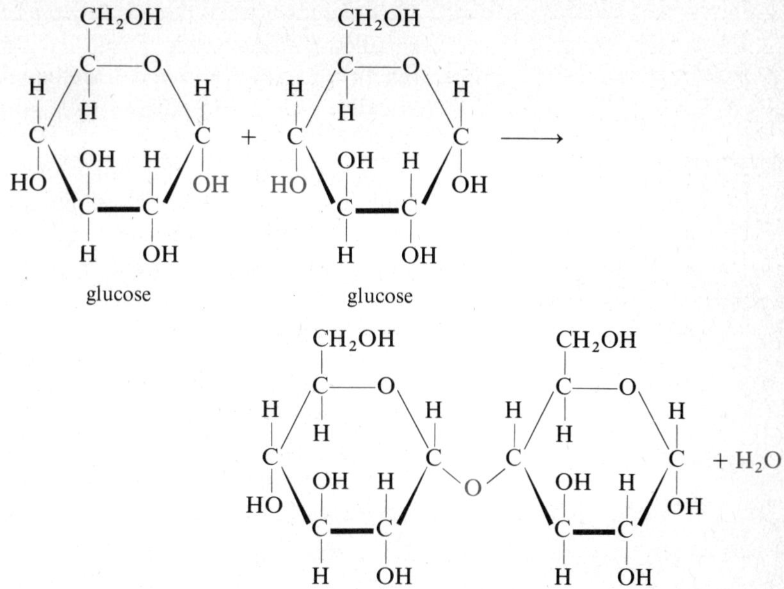

maltose

carbohydrates together. Individual carbohydrate units such as the reactants shown are called *monosaccharides*. When two monosaccharides are joined by the formation of a full acetal as illustrated,

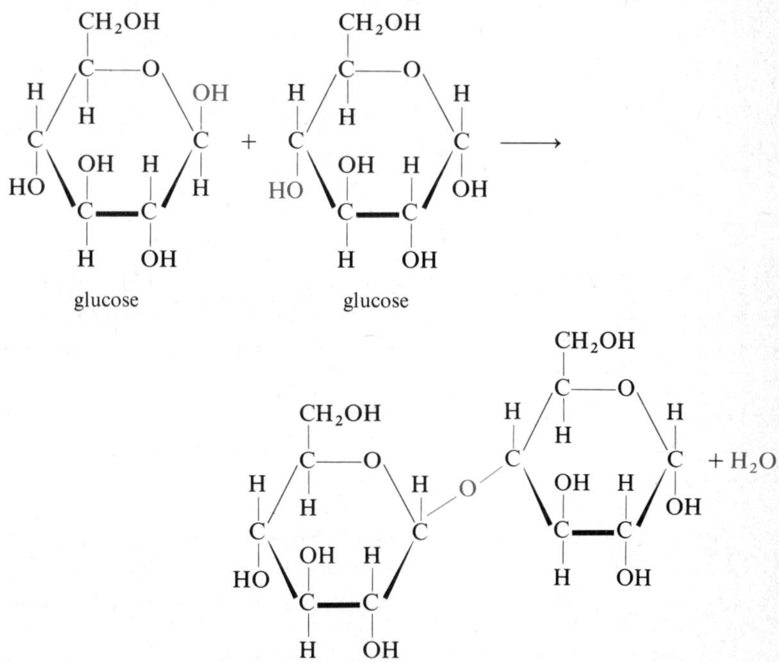

cellobiose

glycosidic linkage with β configuration

the linkage between them is called a *glycosidic linkage*, and the product is called a *disaccharide*. Disaccharides are given common names. The reaction product above is usually called *maltose*. To name it systematically, even using glucose to signify the individual units so joined, is awkward.

Recall that the full acetal carbon is asymmetric. Maltose contains a glycosidic linkage that extends below the level of the ring (the α linkage). There is also an isomer (cellobiose) with the glycosidic linkage extending above the level of the full acetal ring (the β linkage).

Exercise 6.14

Are maltose and cellobiose enantiomers or diastereoisomers?

In each case the glycosidic linkage is between carbon 1 of the left ring and carbon 4 of the right ring. The linkages are differentiated by calling the glycosidic bond in maltose an α-1,4 linkage, because the acetal carbon has an α configuration, and in cellobiose a β-1,4 linkage, because the acetal carbon has a β configuration. These are distinctly different compounds in terms of properties and biological function.

There are monosaccharides and disaccharides in nature that are of nutritional significance. D-glucose and D-fructose are found in foods, especially fruits and honey. Sucrose, however, a disaccharide of frutose and glucose, is more common in our diets (Figure 6.1).

Exercise 6.15

In sucrose what is the orientation, α or β, of the acetal linkage of glucose? of fructose?

Sucrose is table sugar, the sweetener we add to our cereal or coffee in the morning. Notice that sucrose has the hemiacetal of glucose linked to the hemiketal of fructose. Whereas maltose and cellobiose still have a hemiacetal group after reaction, sucrose does not. Although sucrose is found in a large variety of plants, it is particularly abundant in sugar cane, sugar beets, and the saps of sugar maples. Each American consumes about 100 lb of sucrose a year. World production is about 150 trillions lb/yr, of which the United States produces between 7 and 8 percent.

Sweetness is not an inherent carbohydrate property. Many carbohydrates are not sweet at all. Sucrose is sweet, but it is not as sweet as fructose and is sweeter than glucose. Dextrose is another name for glucose, and corn syrup or dextrose syrup is primarily a glucose solution. Treatment of dextrose solutions with the enzyme glucose isomerase results in the conversion of some of the glucose to fructose.

C = 12
H = 22
O = 11

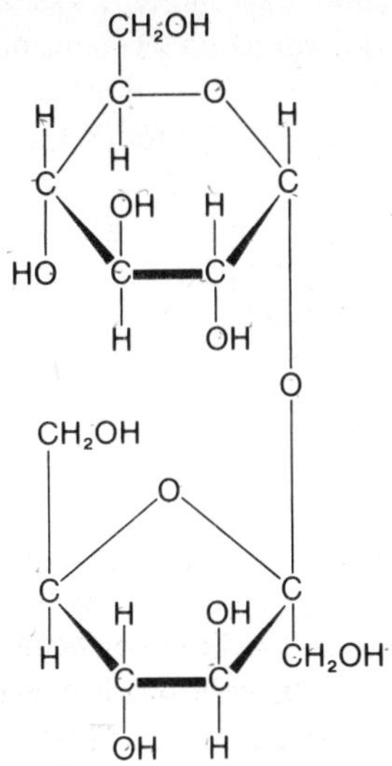

Figure 6.1
Sucrose
The glycosidic linkage is between the hemiacetal of glucose and the hemiketal of fructose. (Photo by Catherine Feryok, Eastern Michigan University.)

RMM sucrose 342g.

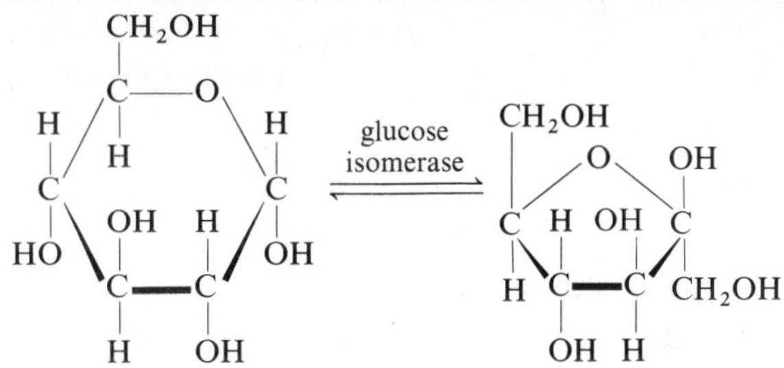

D-glucose D-fructose

A syrup containing both glucose and fructose is comparable in sweetness to a sucrose syrup and is used in canned fruits and soft drinks.

Another disaccharide of nutritional significance is lactose, which is composed of galactose and glucose. Notice that it has a β-1,4 linkage.

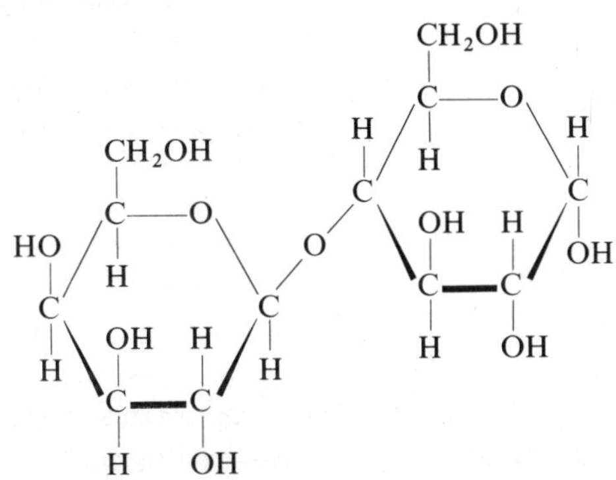

lactose (milk sugar)

Lactose is the carbohydrate found in milk. In fact, it is more than one-third of the solid material found in milk. It is about half as sweet as sucrose. As mentioned earlier, problems can arise when a person is unable to utilize the galactose portion of this molecule. When milk is converted to cheese, the liquid waste, called *whey*, is very rich in lactose (70 percent of the solids). This is a potentially rich food source, perhaps 1.5 billion lb/yr resulting from cheese making in the United States. However, many adults stop producing enzymes necessary for lactose utilization.

Exercise 6.16

Enzymatic digestion of starch produces a small amount of isomaltose, a disaccharide composed of two glucose molecules joined by an α-1,6 linkage. Draw a structure for isomaltose.

6.8 Polysaccharides

Carbohydrates can be joined together by glycosidic bonds to form long chains called polysaccharides. These are actually condensation polymers, a class of compounds discussed in detail in Chapter 7. Polysaccharides are the most frequently encountered types of carbohydrates in nature. Although they perform a variety of functions, most commonly they either serve as food storage or fulfill a structural role.

6.9 Starch

The most important and best-known food storage polysaccharide is starch. Seeds or tubers of plants often store starch as the food reserve to give the plant its first spring growth. Such starch-producing plants as corn, wheat, rice, and potatoes are major components of the human food supply in most of the world. Starch is a polymer of glucose and occurs in two structural forms, amylose and amylopectin. The relative amounts of the two forms vary considerably. For example, "waxy" or glutinous starch from corn used as animal feed is almost completely amylopectin, while sweet corn has a high proportion of amylose.

Amylose is a linear polymer of glucose joined together by α-1,4 glycosidic linkages. Molecular weights are generally around 50,000 or more, or about 300 glucose units per molecule. Amylopectin is a branched polymer. The chains have the same α-1,4 linkage as amylose. However, occasional glucose units, the branch-point units, have a second chain attached to the 6 position. Amylopectin is larger, sometimes having more than 600,000 glucose units per molecule (Figure 6.2).

6.10 Starch Digestion

Starches are digested by two groups of enzymes, the α- and β-amylases. When seeds germinate, these enzymes are produced to release the food reserve. The α-1,4 linkages are attacked at random locations by

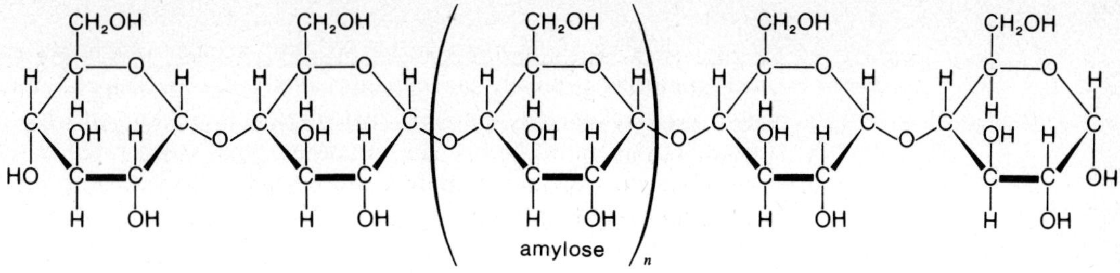

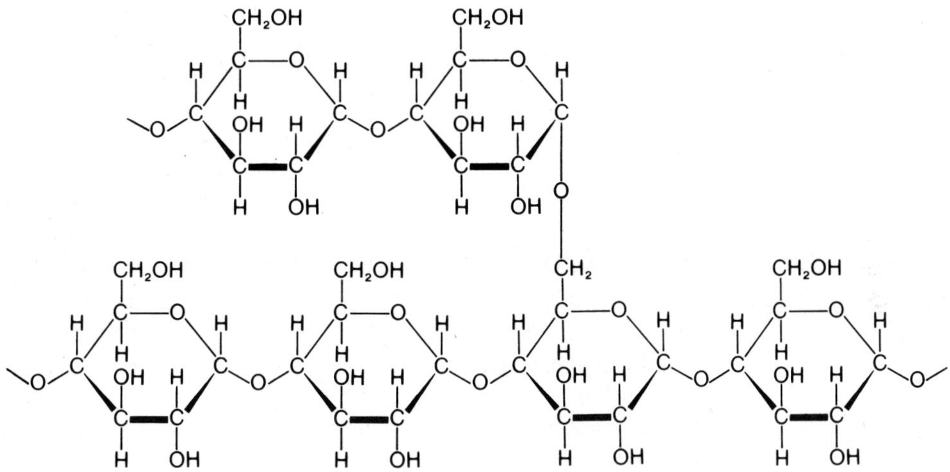

Branched chain of glucose molecules in amylopectin

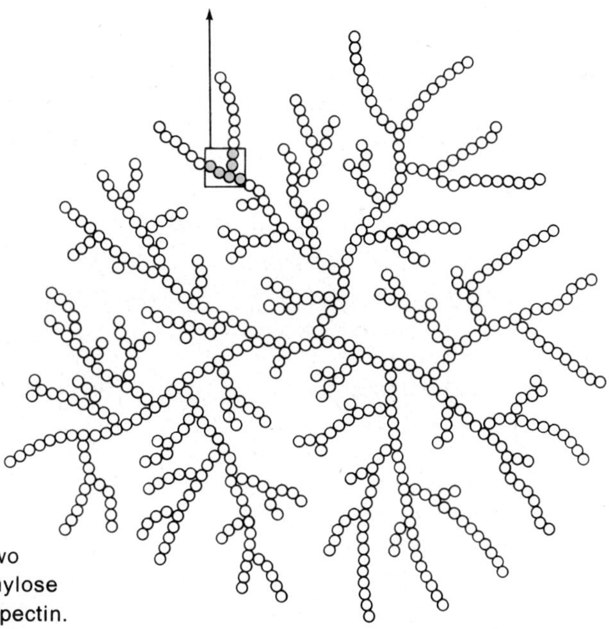

Figure 6.2
Structure of starch
Starch is composed of two structures, the linear amylose and the branched amylopectin.

α-amylase. This serves to convert the high-molecular-weight polymers into numerous relatively short chains called *dextrins*. The end of each chain is attacked by β-amylase hydrolyzing the chain into maltose (dissacharide) units. Notice how these enzymes cooperate. Without α-amylase there would be only a few ends for β-amylase to attack; with it there are many (Figure 6.3).

6.11 Glycogen

Glycogen is a polysaccharide used for food storage by animals. In humans it is found in the liver and the muscle. Primarily it serves as a reserve that can be converted rapidly to glucose in an emergency. It does not represent a massive food reserve to combat a period of fasting. This is the function of body fat.

Glycogen closely resembles amylopectin, having α-1,4 links and α-1,6 branch points, but it is more highly branched. When the body requires additional glucose, this reserve is activated by the enzyme phosphorylase. Phosphorylase is stored with the glycogen in an inactive form. When converted to the active form, it catalyzes the following reaction:

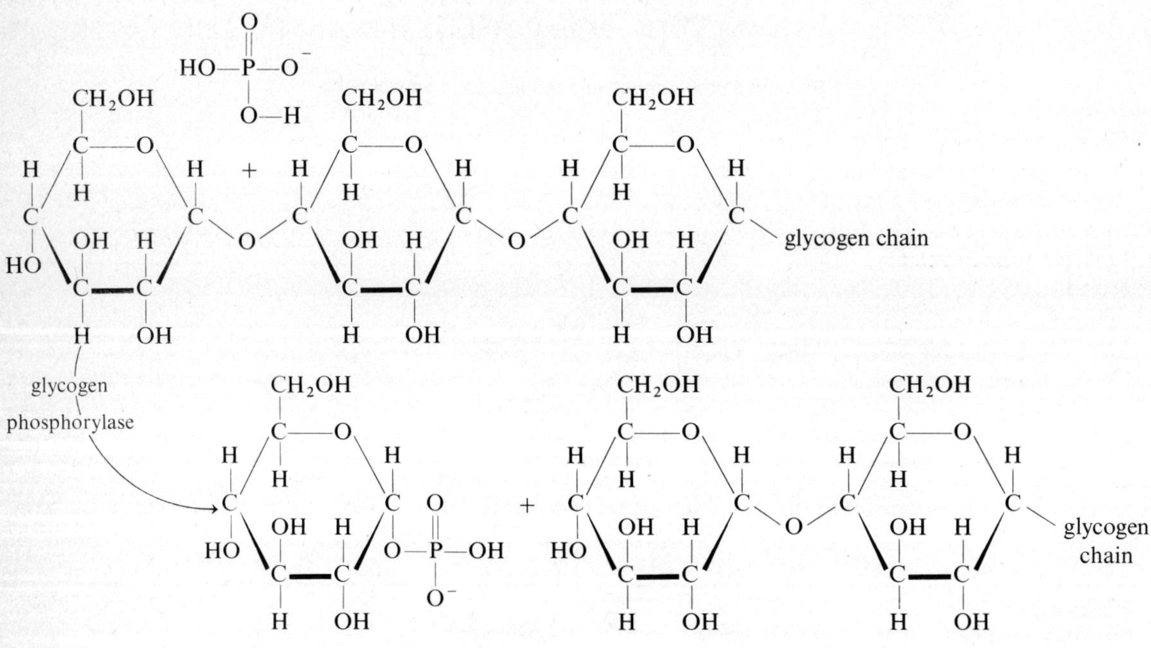

Notice that the product is not free glucose but rather a phosphate derivative of glucose. The glucose-1-phosphate so produced can be broken down for energy directly or can be converted to glucose and transported in the blood for use at a site distant from the place of storage. Phosphorylase attacks the ends of the branch chains and

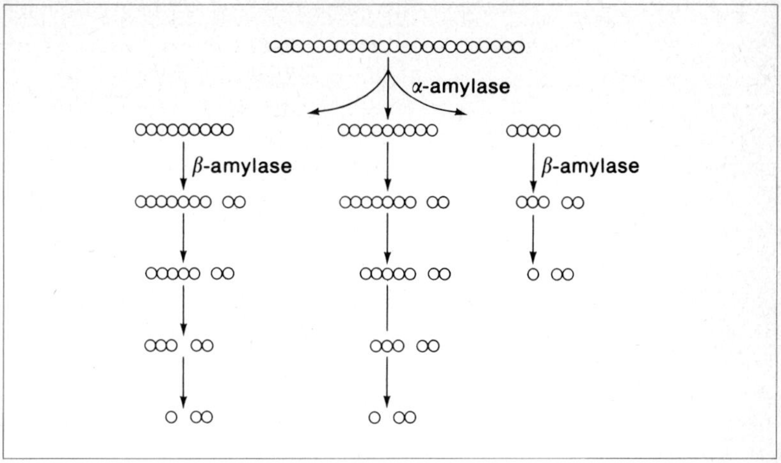

Figure 6.3
Action of the amylase enzymes
The combination of α- and β-amylase rapidly converts amylase into mono- and disaccharides.

can successively remove glucose until the branch point is reached. The enzyme is stopped by the branch point. As a result, only the outer chains of the glycogen molecule are used, not the entire molecule. The core of the molecule, called the *limit dextrin*, remains untouched. The limit dextrin will later be rebuilt by adding glucose back onto the chain ends to restore the original glycogen (Figure 6.4).

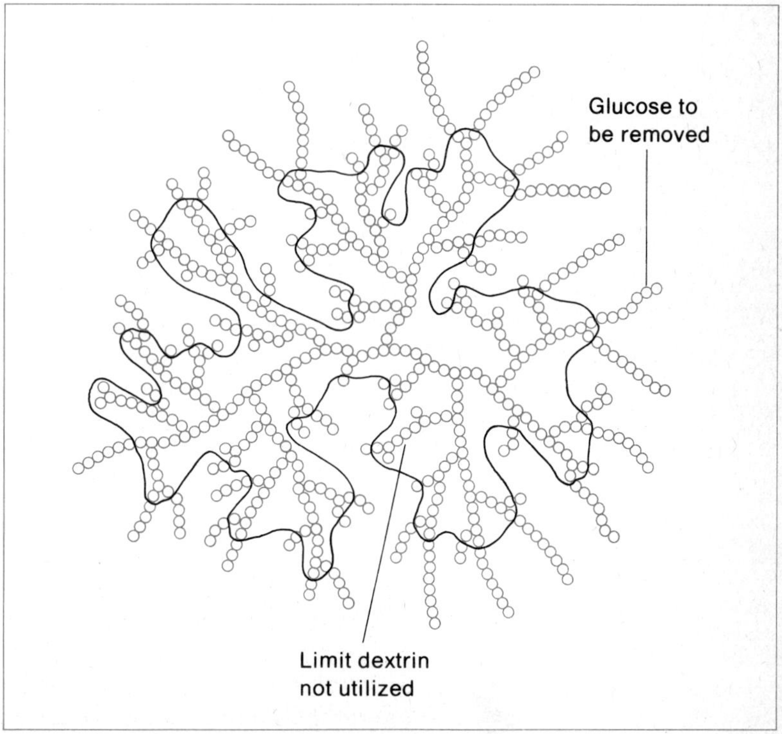

Figure 6.4
Glycogen and limit dextrin
When glucose is removed from glycogen to be metabolized, the enzyme can remove units only until a branch point is reached.

6.12 Cellulose

Cellulose is an example of a structural polysaccharide. It is the major structural material in higher plants, forming the bulk of cell walls. The amount of cellulose produced by plant growth is incredible, one estimate placing annual production at 10^{11} tons. Wood is a combination of cellulose and a less orderly material called *lignin*. Lignin is removed to isolate the cellulose in the process of making paper from wood. Cotton is an example of almost pure cellulose (Figure 6.5).

Cellulose is, like amylose, a linear polymer of glucose units linked by 1,4 glycosidic linkages. However, rather than being an α glycosidic bond, it is a β linkage.

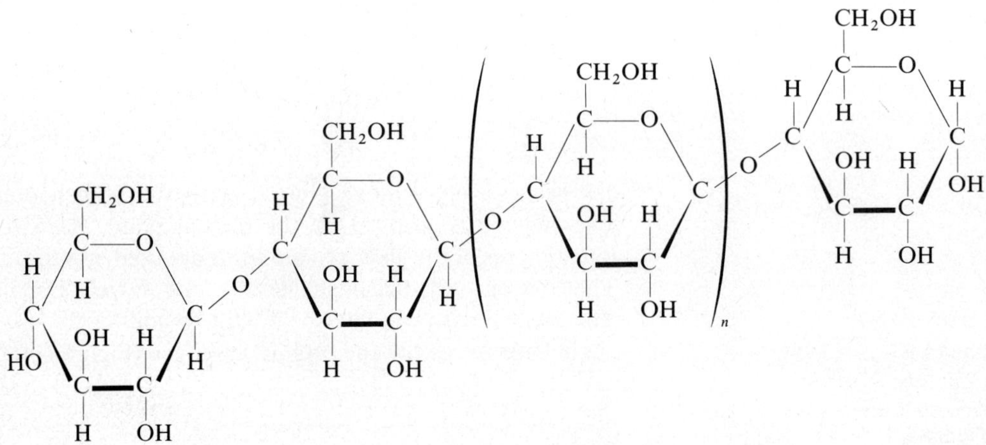

cellulose

With the β linkage, the molecules tend to form long, straight rods that can be stacked very close together. There can be many hydrogen bonds between the glucoses of the adjacent polymer strands forming the cellulose molecules into tight bundles (Figure 6.6).

Water tends to be excluded from these tight bundles. As a result of their strong interstrand binding and repelling of water, cellulose does not dissolve in water the way amylose does. This is just as well when you are caught in the rain in a cotton outfit.

Cellulose is modified chemically for a variety of purposes (Figure 6.7). Cellulose nitrate was used as an artificial ivory for piano keys and to make such items as clip-on collars and cuffs. When the development of motion pictures required the use of a flexible transparent material upon which to coat photographic emulsions, cellulose nitrate was used. This film was used next to a hot, powerful light source in the crowded and often not particularly fireproof early movie theaters. The hazards of this are dramatized by indicating yet another use for cellulose nitrate. It serves in the form of "gun cotton" as the explosive to propel artillery shells. Cellophane was substituted for cellulose nitrate in the development of "safety film" to reduce this fire hazard.

Figure 6.5
Cotton
Cotton is composed almost entirely of cellulose. (Photo by Max and Kit Hunn, Frederíc Lewis Photographs.)

The process used to prepare the rayon textiles involves modifying cellulose so it will dissolve. The strong hydrogen bonds between cellulose molecules must be eliminated. By converting the alcohol groups to acetate esters, hydrogen bonding is prevented. The relatively nonpolar product can be dissolved in acetone. In the xanthate process, the alcohol groups become salt structures and the product is water soluble.

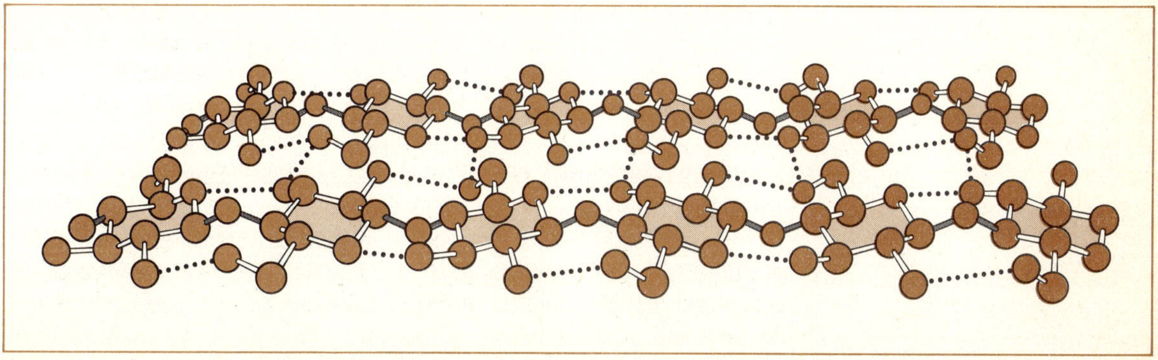

Figure 6.6
Cellulose structure
Adjacent molecules of cellulose are held together by hydrogen bonds, shown here as dashed lines. (Computer-generated structure of cellulose © 1977 by W. G. Davies and J. W. Moore used by permission; reprinted from J. W. Moore, W. G. Davies, and R. W. Collins, *Chemistry*, New York, McGraw-Hill, 1978. All rights reserved.)

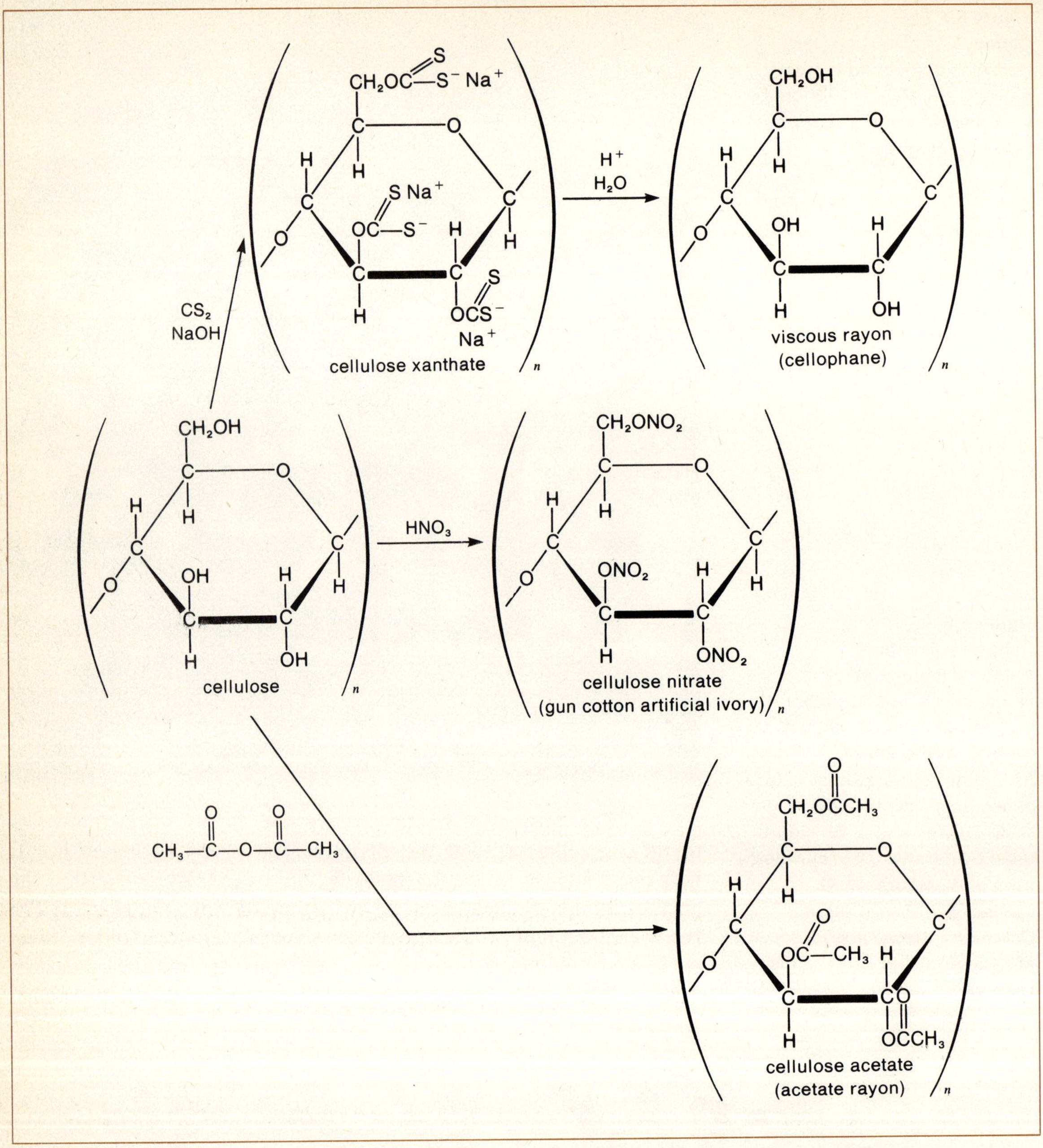

Figure 6.7
Derivatives of cellulose

To prepare a textile fiber, the cellulose solutions are forced through a shower-head-like device called a *spinneret*. Cellulose acetate solutions lose the acetone solvent rapidly in a hot chamber. Cellulose xanthate solutions go into an acid bath, where the cellulose salts are converted back to cellulose. The resulting textile fibers are rayon acetate and viscous rayon, respectively. Rayon can be made from sources of cellulose, such as wood, which are less costly than cotton.

6.13 Derivatives of Monosaccharides

In nature the polysaccharides are sometimes found to contain monomer units that are modified versions of usual monosaccharides. The most common of these carbohydrate derivatives are *amino sugars* and *uronic acids*. Most frequently, D-glucose and D-galactose are the structures modified in this fashion.

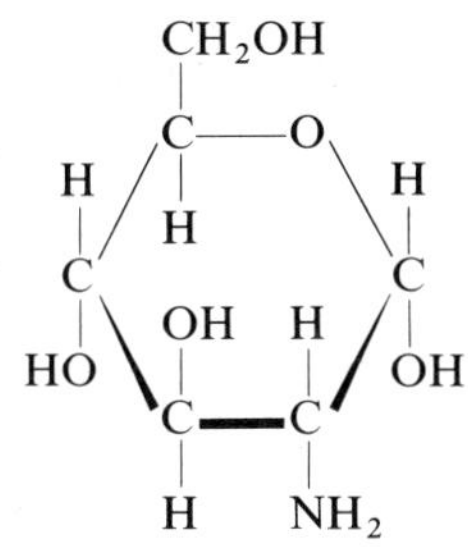

glucosamine

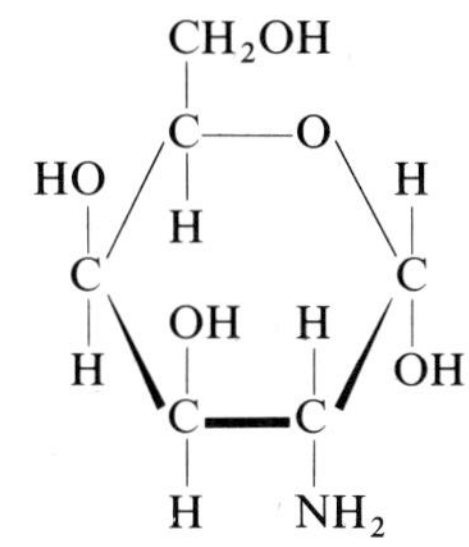

galactosamine

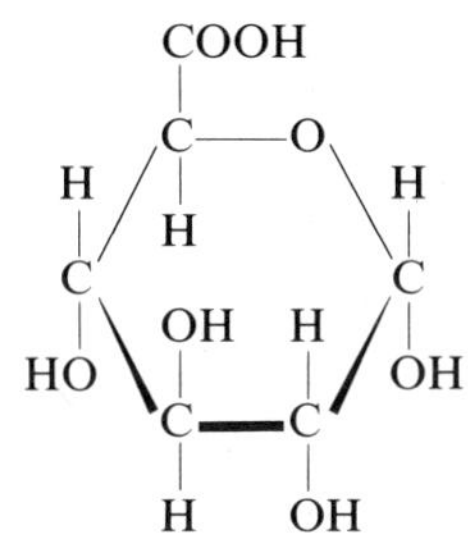

glucuronic acid

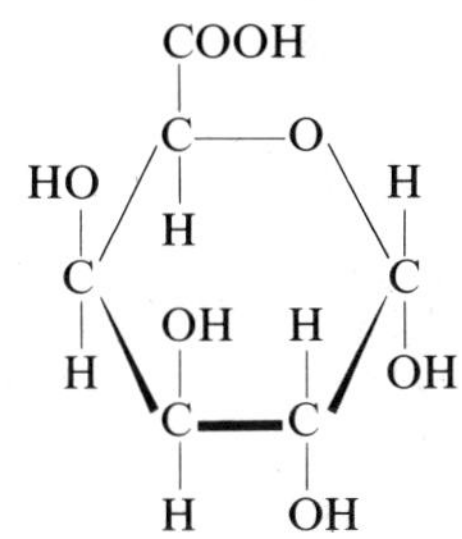

galacturonic acid

Notice that in the uronic acids, carbon 6 has been oxidized to a carboxylic acid. In the amino sugars, the alcohol of carbon 2 has been replaced by an amine group. In nature, the amine group usually has an acetyl group attached to it, so the total structure is as follows:

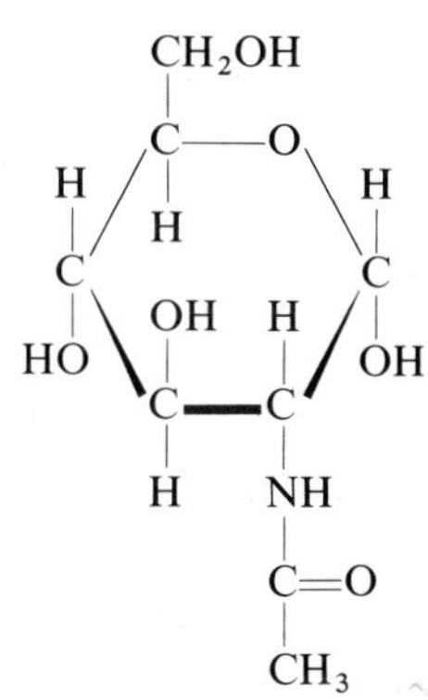

N-acetyl glucosamine

Creatures such as insects, lobsters, crabs and crayfish form a hard outer skeleton from a polymer of glucosamine called *chitin*. All of the amine groups have acetyl groups attached. This is very likely the second most abundant polysaccharide in nature.

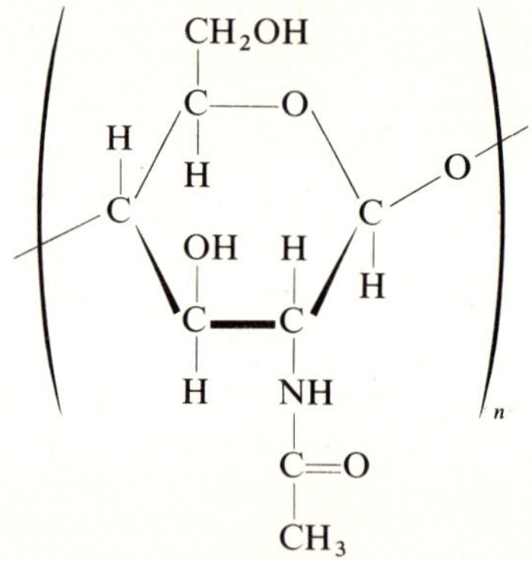

chitin (notice the similarity to

6.14 Hemicelluloses

Structural parts of plants often contain polysaccharides other than cellulose. These are called *hemicelluloses*, and they have a variety of structures, sometimes resembling cellulose itself. Mannans, polymers of mannose, and xylans, polymers of xylose, are important examples.

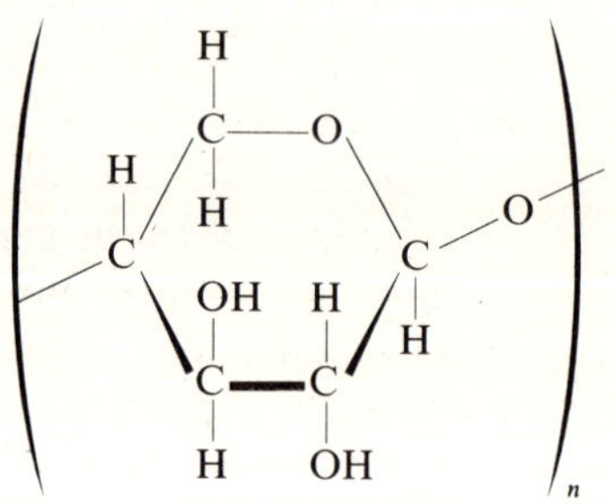

xylan (again note the similarity to cellulose)

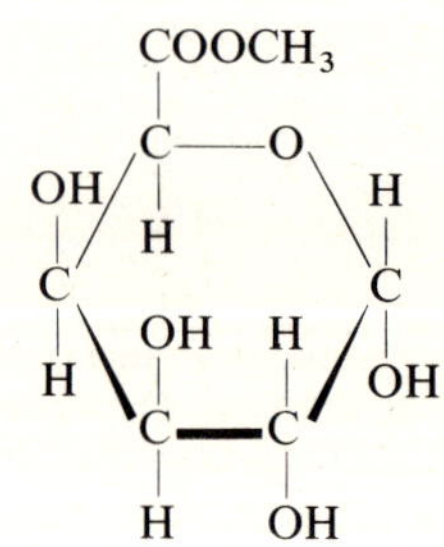

pectin (a polymer of the methyl ester of galacturonic acid)

Polymers of uronic acids are termed *polyuronides*. Pectin is a common example. Pectin is found in fruit skins and is extracted by boiling. When jelly is prepared, the boiling of the fruit dissolves the pectin into the fruit juice. It is this pectin that causes the jelly to set into a semisolid. Citrus rinds are a rich source, containing 30 percent pectin. Commercial pectin preparations sold under such brand names as Sure-Jell or Certo can be added to fruit preparations to increase the tendency to gel.

Table 6.1
Polysaccharide Food Additives

Compound	Composition	Source
Agar	D-Galactose, L-anhydrogalactose	Seaweed
Carrageenan	D-Galactose, sulfate	Seaweed
Alginic acid, alginate salts	D-Mannuronic acid	Seaweed
Furcellaran	D-Galactose, D-mannose, sulfate, other components	Seaweed
Locust bean gum	D-Mannose, D-galactose (20–25% as side groups)	Plant seed
Guar gum	D-Mannose, D-galactose (33–35% as side groups)	Plant seed
Gum arabic (acadia)	Arabinose, D-galactose, rhamnose, uronic acids	Tree gum
Gum tragacanth	D-Galacturonic acid, D-galactose, L-fucose, D-xylose, L-arabinose	Shrub gum
Karaya gum	D-Galacturonic acid, D-galactose, L-rhamnose	Tree gum

6.15 Gums

A large number of polysaccharides in addition to starches are used in foods. Generally termed *gums*, they serve to cause gelling, to thicken, to improve texture, and to stabilize suspensions. The food industry uses billions of pounds of natural gums for these purposes (Table 6.1).

Gums come from a variety of sources. Seaweeds provide agar, carrageenan, furcellaran, and alginic acid. Guar gum is a plant seed product, while gum tragacanth and gum arabic are compounds excreted by trees and shrubs to seal wounds in the bark. The list of specific applications of gums in foods is extensive. In ice cream, for example, gums assure the uniform distribution of fats and inhibit the growth of ice crystals. They affect the texture, smoothness, and solidity of the product. Rapidly prepared puddings, custards, and pie fillings owe their properties to gums. By eliminating eggs from these custards, problems due to bacterial contamination are reduced. Processed cheeses have increased spreadability because of added gums. Meringues, creamed and whipped icings, sundae toppings, salad dressings, and artificial whipped cream are stabilized with gums. Diet foods make extensive use of gums because they add bulk and texture, yet because they are not digested, they add no calories. Drug companies often use gums as excipients (inert agents used to suspend medicines).

6.16 Mucopolysaccharides

In animals we find a group of polysaccharides composed of alternating units, often a uronic acid alternating with an amino sugar. Called *mucopolysaccharides*, they fulfill a variety of functions. Hyaluronic acid, a mucopolysaccharide, is widely distributed, occurring in the fluid of the eye interior and serving as a sort of cement to bind tissues together. When some insects sting, they inject an enzyme called *hyaluronidase*, which breaks the hyaluronic acid linkages. This facilitates the spread of the venom. Concentrated, viscous solutions of hyaluronic acid are found at movable joints such as knees and elbows to serve as lubricant. Heparin is a mucopolysaccharide in blood that inhibits blood clot formation. When a pump for use during heart surgery was being developed, the formation of blood clots on the surfaces of the device was a major problem. Binding heparin to the plastic lining of the pump provided a solution. Chondroitin sulfate, a third mucopolysaccharide, serves a structural role in cartilage, bone, and the cornea of the eye. Mucopolysaccharides are found combined with protein. The combination is termed *mucoprotein*, or *mucin*. Mucins often serve as lubricants, for example as a coating on the esophagus.

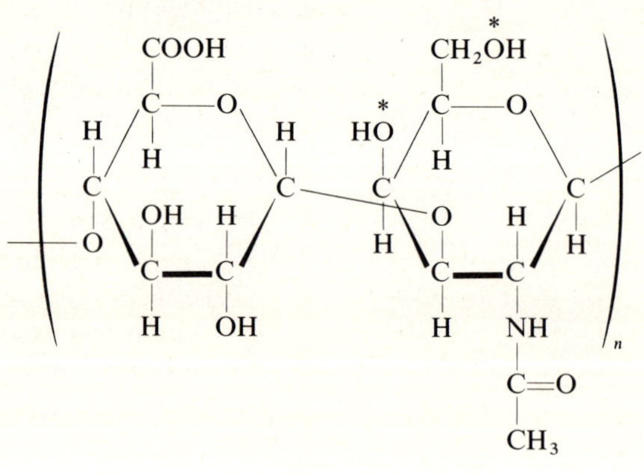

chondroitin sulfate
(the asterisks refer to Problem 13)

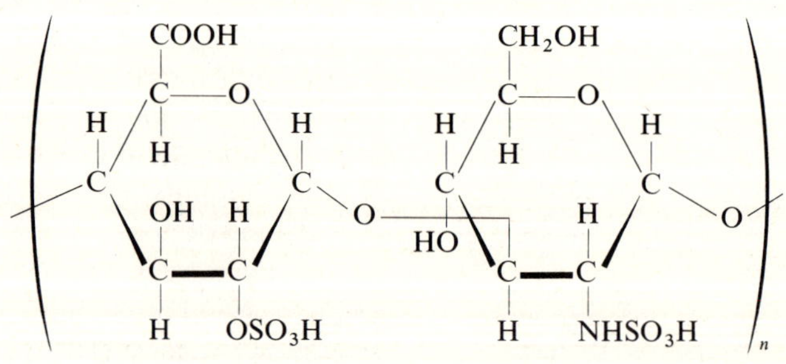

heparin

Summary

Aldehydes and ketones have as their functional group the carbonyl group. In this polar carbon-to-oxygen double bond there are the pi electrons required for addition reactions. Two important reactions are the addition of hydrogen to form alcohol and the addition of an alcohol to form a hemiacetal. Aldehydes may also be oxidized to carboxylic acids. Carbohydrates are polyhydroxy aldehydes and ketones. They usually have several asymmetric carbons, and therefore a number of optical isomers of each basic structure exist. Carbohydrates form rings by internal hemiacetal formation. Hemiacetals can substitute an alcohol alkyl group for an —OH hydrogen, forming a full acetal. A hemiacetal carbohydrate can form a full acetal by reacting with the alcohol group of another carbohydrate. The new bond, a glycosidic linkage, joins the two monosaccharides into a disaccharide. The hemi- and full acetal carbons of carbohydrates are asymmetric, producing two isomers, α and β. A polysaccharide is many monosaccharides joined together by glycosidic links. Starch is a glucose polysaccharide of major importance in the human diet. Cellulose, another glucose polymer, is the chief structural polysaccharide of plants. Other plant polysaccharides include hemicelluloses and gums. Derived monosaccharides, such as amino sugars and uronic acids, are found in many gums. Several gums are used as food additives. Animals have mucopolysaccharides, polymers of these derived monosaccharides, performing structural roles. Examples are hyaluronic acid and chondroitin sulfate. The blood clotting inhibitor, heparin, is a similar structure.

Problems

1. Name the following compounds:

a)
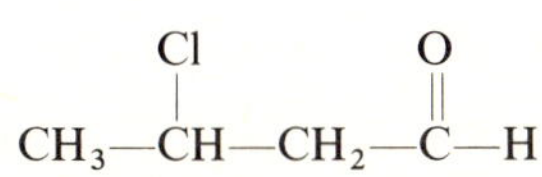

b)
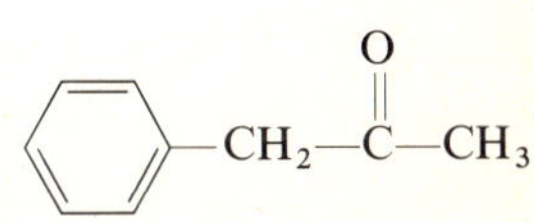

c)
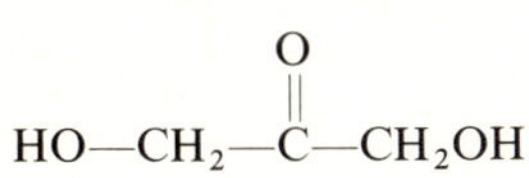

d)
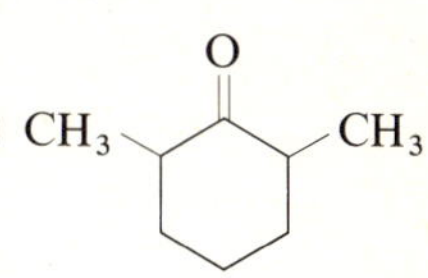

e)
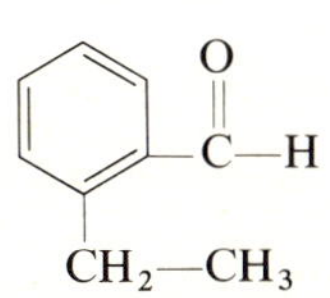

f)
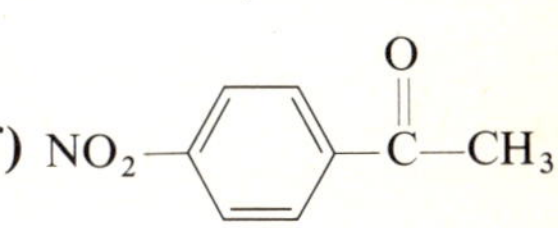

g)
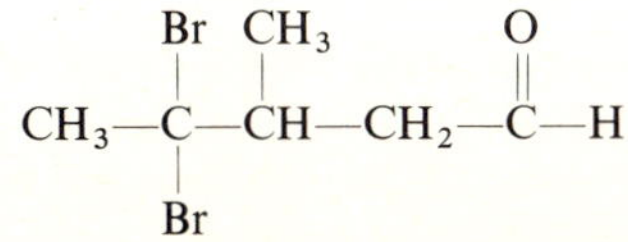

h)
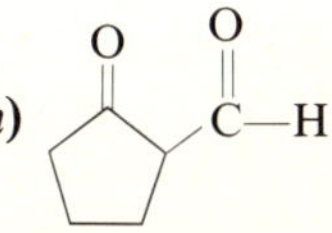

i) 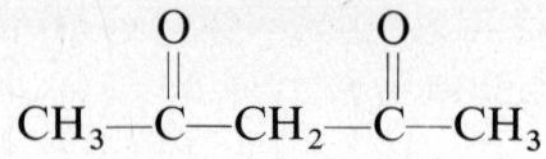

If it gives you trouble, first name

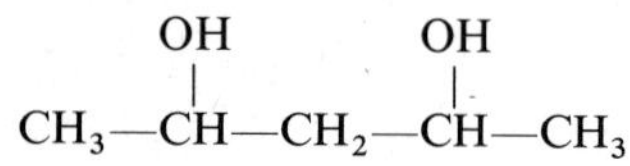

2. Draw structures of the following:

a) 3-methyl-2-butanone
b) acetophenone
c) *m*-chlorobenzaldehyde
d) salicylaldehyde
e) cyclopentanone
f) 3-phenylpropanal
g) 3-ethyhexanal
h) 4-formyloctane
i) 1,1,1,3,3,3-hexachloropropanone

3. Draw the following carbohydrate structures as vertical free aldehydes or ketones:

a) D-glucose
b) L-glucose
c) D-mannose
d) L-fructose

4. Name the following carbohydrates:

a) 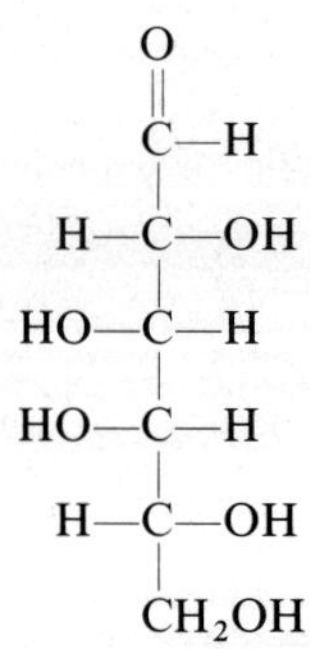

b) 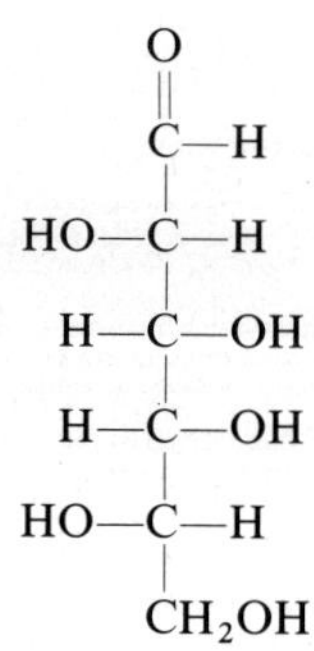

c) 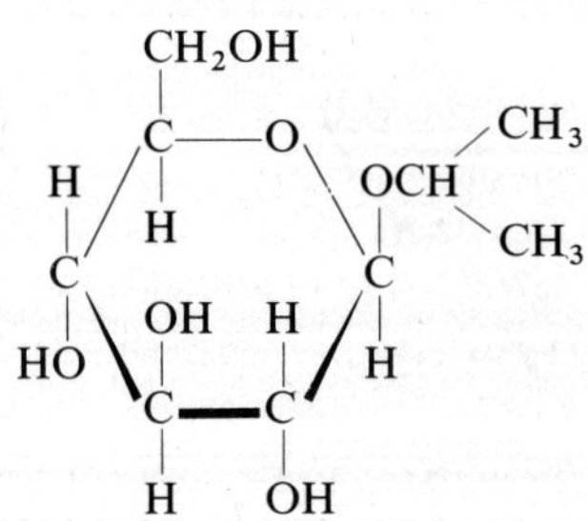

d) 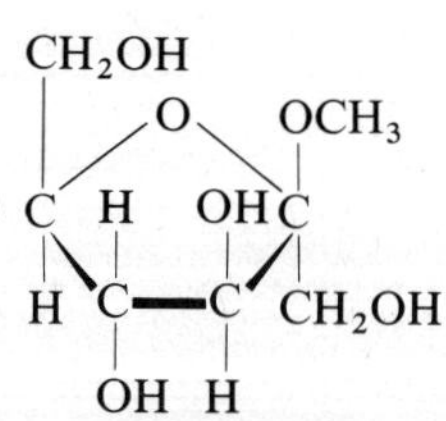

In c and d, number the carbons of the carbohydrate.

5. D-ribose is an aldopentose all of whose asymmetric carbons are D. Draw the structure.

6. If a hexose has six carbons and a pentose has five carbons, what is a triose? Draw all possible aldotrioses and all possible ketotrioses.

7. Identify the following as hemiacetals, full acetals, neither, or both:

a) 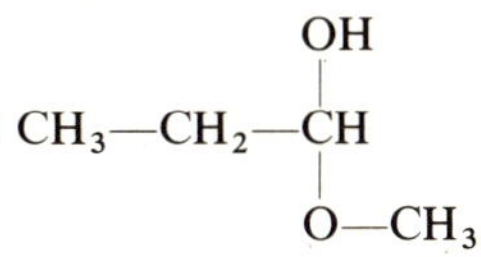

b) 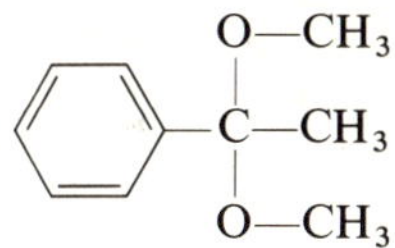

c) 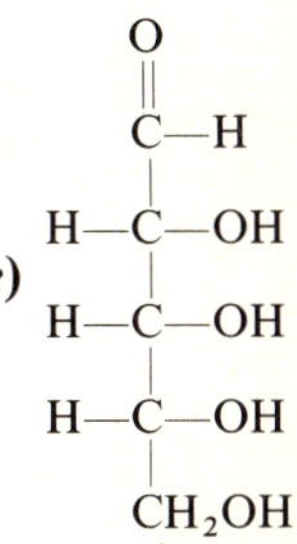

d) 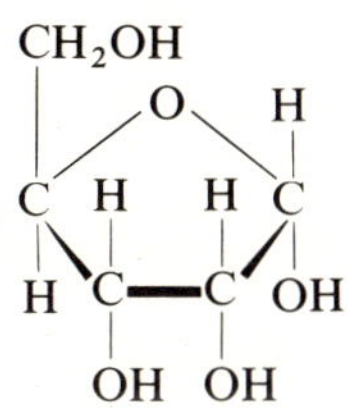

e) 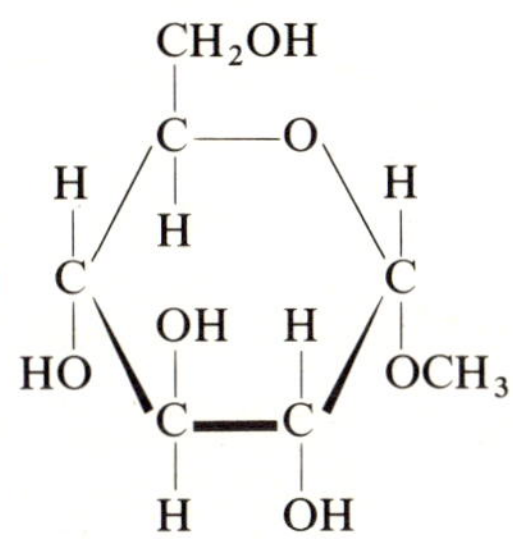

f) 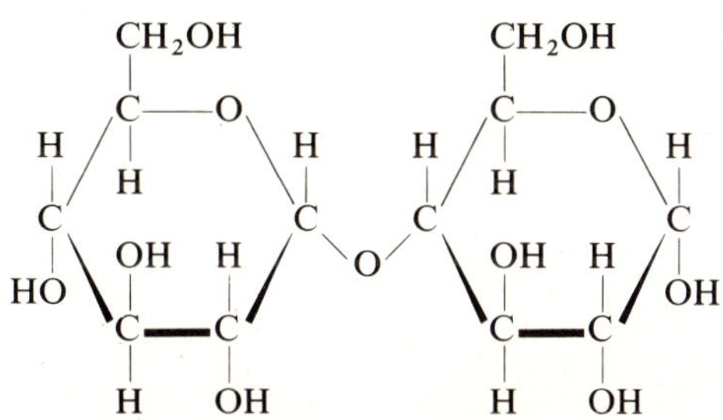

8. Complete the following reactions, indicating which are the addition reactions and which are the elimination reactions:

a) $CH_3-CH_2-\overset{\overset{O}{\|}}{C}-CH_3 + H_2 \xrightarrow{\text{catalyst}}$

b) $CH_3-OH + (O) \xrightarrow[\text{mild conditions}]{\text{oxidizing agent}}$

c) 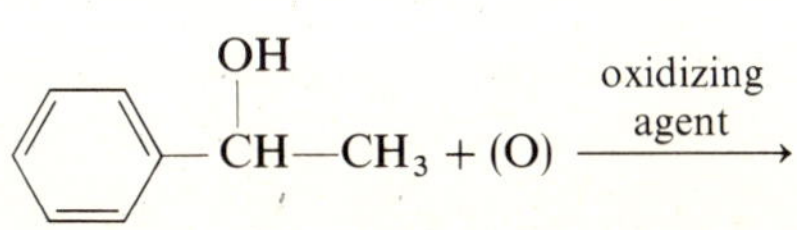

d) $CH_3{-}OH + (O) \xrightarrow{\text{strong conditions}}$

e) 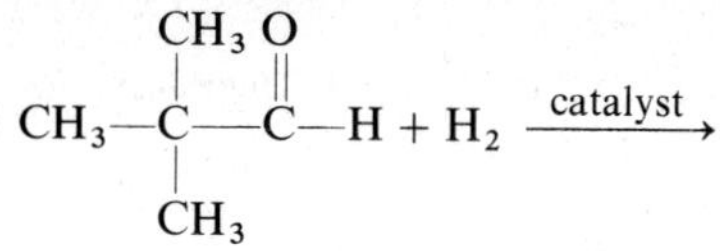

f) 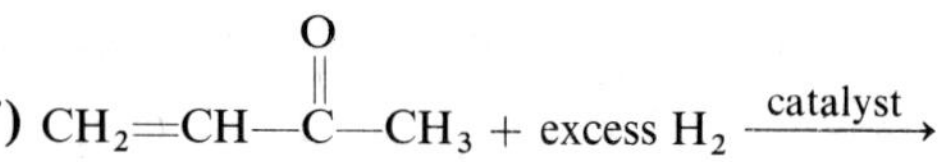

g) 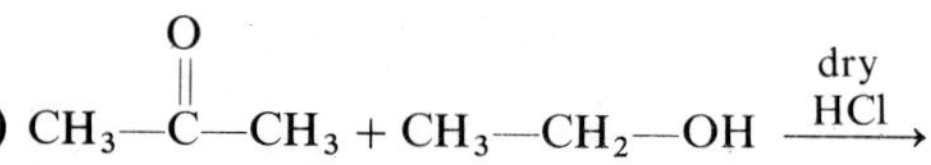

h) 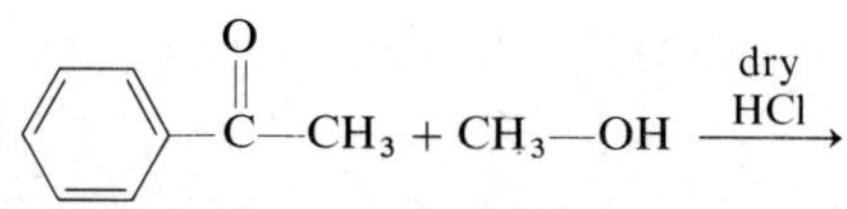

i) 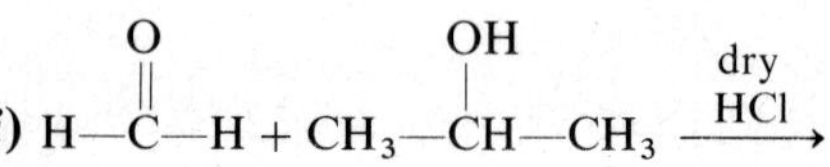

j) 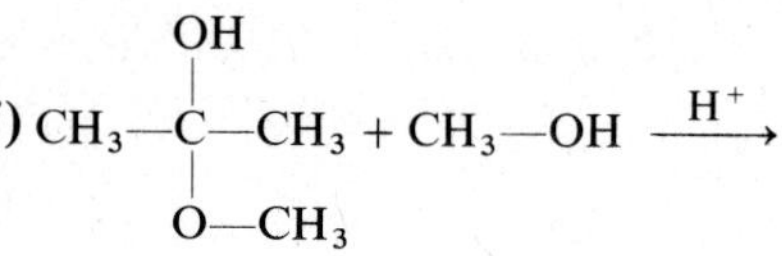

k) 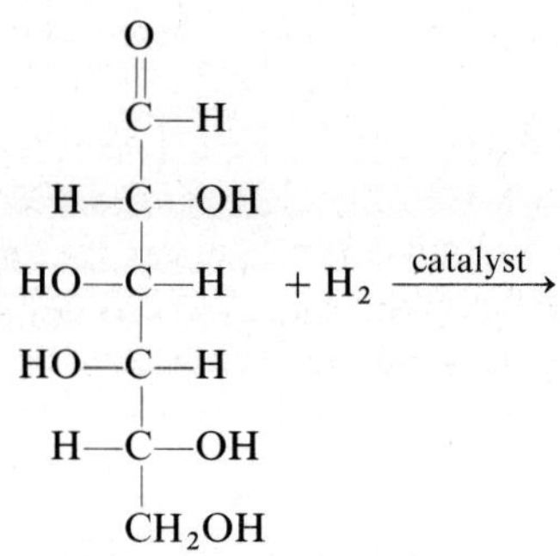

l) 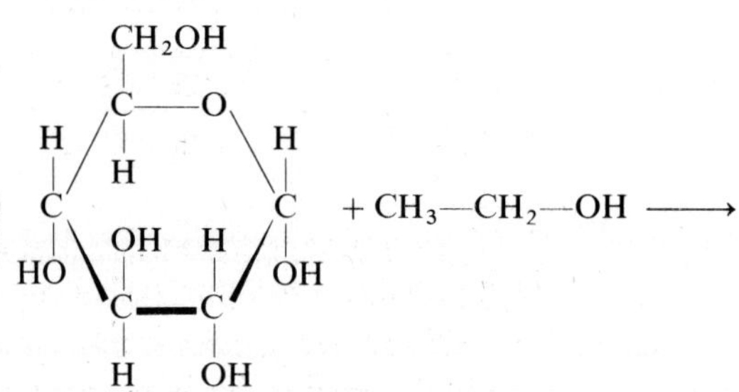

m) 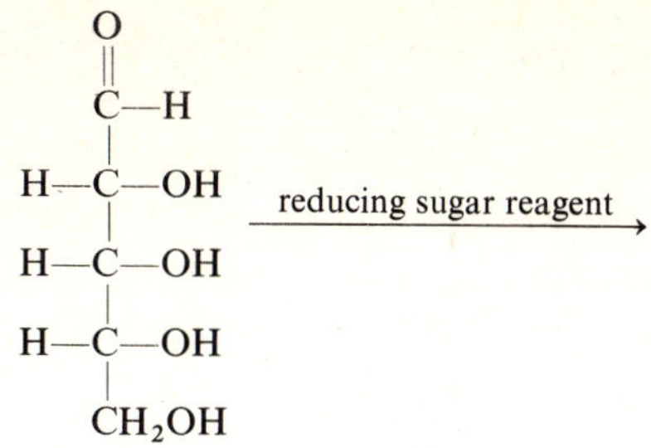

9. Show a series of reactions to perform the following overall conversions. Refer to Chapter 4 where necessary.

a) 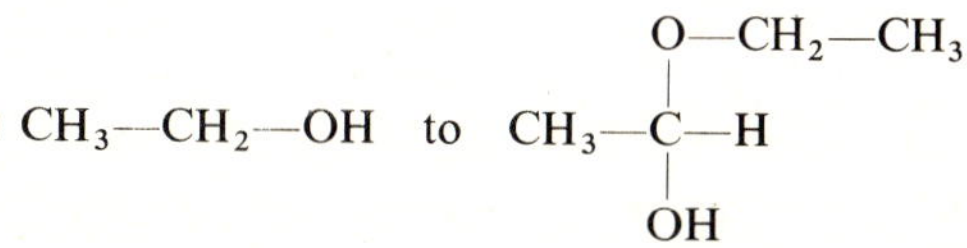

b) 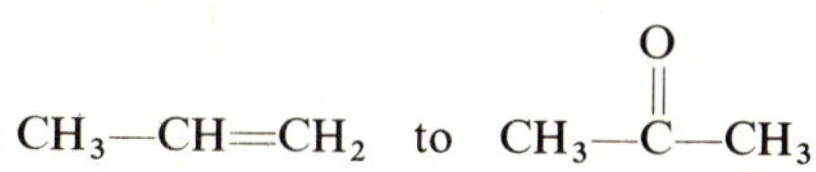

c) 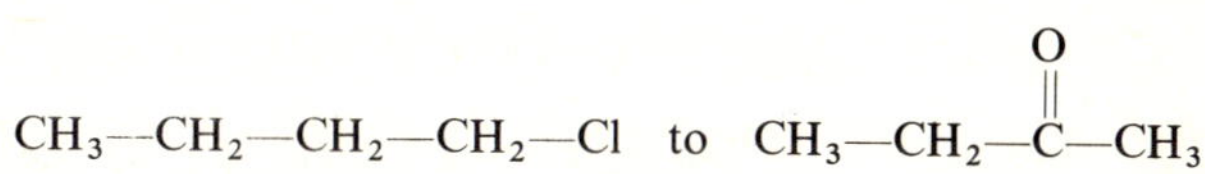

d) 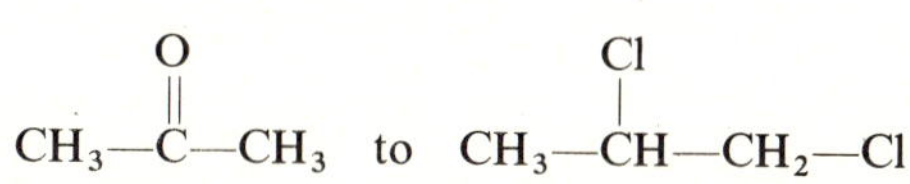

10. Refer to the conversion of the alkane portion of a fat to a ketone structure on page 125. Given that living systems generally perform similar chemical processes by similar chemical pathways, predict the intermediate steps for this conversion that occurs in living cells as part of the Krebs cycle:

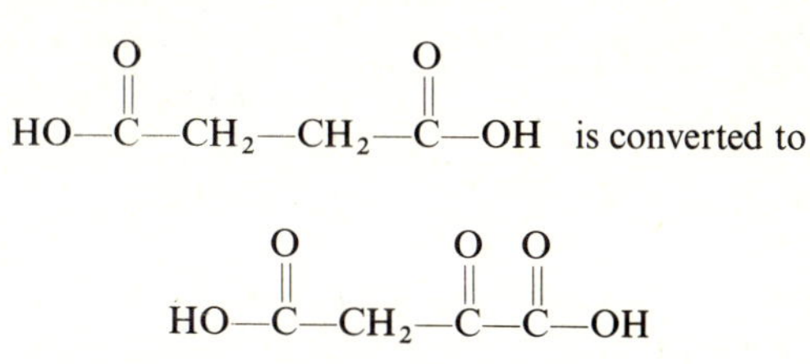

11. A structure composed of three monosaccharide structures joined by glycosidic bonds is called a *trisaccharide*. The most abundant trisaccharide is raffinose, which is found, among other places, in the syrups obtained from sugar beets.

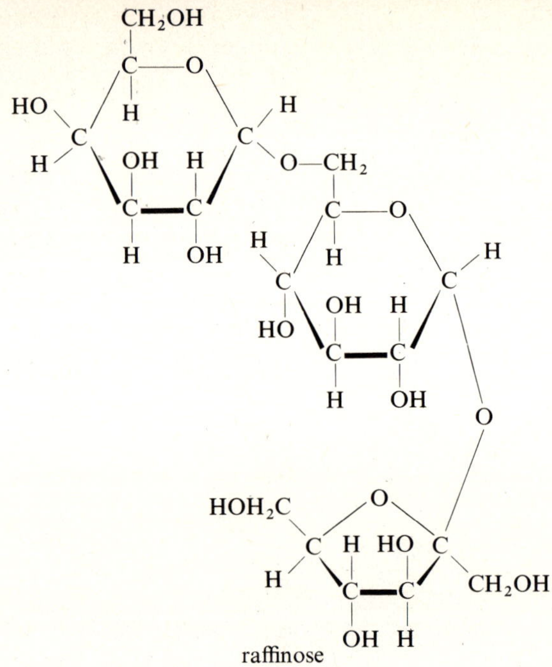

raffinose

a) Identify the three monosaccharides in raffinose.

b) Name the glycosidic bond between the two six-membered-ring monosaccharides.

c) By breaking one of the two glycosidic bonds, you would obtain a very common disaccharide. Which bond would you break, and what is the common disaccharide that would result?

12. Draw the following carbohydrate structures:

a) The β form of D-glucose as a hemiacetal.

b) The ethyl-α acetal of D-galactose.

c) Gentiobiose—a disaccharide composed of two D-glucoses joined β-1,6.

d) The hemicellulose mannan, which is a polymer of mannose units linked β-1,4. Compare the structure with that of cellulose. What is the difference?

e) In some plants (dandelions, dahlias, and artichokes), starch is replaced by inulin as a food storage polysaccharide. Inulin is a β-1,2 linked D-fructose polymer. Draw inulin.

13. What are the differences in structure between heparin and chondroitin sulfate (page 146)? Which of the starred positions in the

chondroitin sulfate would have sulfate esterified in chondroitin-6-sulfate, and which in chondroitin-4-sulfate?

Challenge Problem

14. It has been observed that glucose crystallizes as the hemiacetal, forming crystals that are either pure α or pure β. In the laboratory the optical rotation of a sample of α-D-glucose freshly dissolved at a concentration of 1 g per 100 ml and placed in a 10-cm cell was read. It read $+1.22^\circ$. Fresh β-D-glucose under the same conditions read $+0.187^\circ$. After the tubes stood a while, they read $+0.872^\circ$ for α and $+0.362^\circ$ for β. After a longer time, both tubes read $+0.527^\circ$. Can you explain what happened? This observed change in optical rotation is called *mutarotation*.

7

Carboxylic Acids

Objectives

In this chapter we:

1. *Consider the acidic properties of the carboxyl structure.*
2. *Learn to name carboxylic acids.*
3. *Discuss the relationship between pH and degree of ionization.*
4. *Consider the effects of charged ions on biological molecules.*
5. *Learn about condensation reactions including ester and amide formation.*
6. *Learn about hydrolysis reactions as they relate to the breaking apart of esters and amides.*
7. *Survey medicinal esters and amides.*
8. *Relate flavor to chemical substances.*
9. *Learn the structure and sources of glycerides and discuss rancidity.*
10. *Survey the structure, sources, and uses of waxes.*
11. *Discuss the chemical structure and function of soaps and detergents.*
12. *Learn the structures and survey some functions of complex lipids.*
13. *Discuss the structure and function of prostaglandins.*

Up to this point the compounds discussed have either been nonpolar or have had polar functional groups. With the carboxylic acids we encounter a hydrogen that is very weakly bonded to the rest of the molecule. The hydrogen of the —OH group is easily removed by interaction with a strong base such as hydroxide ion, and is even partly removed by a weak base such as water.

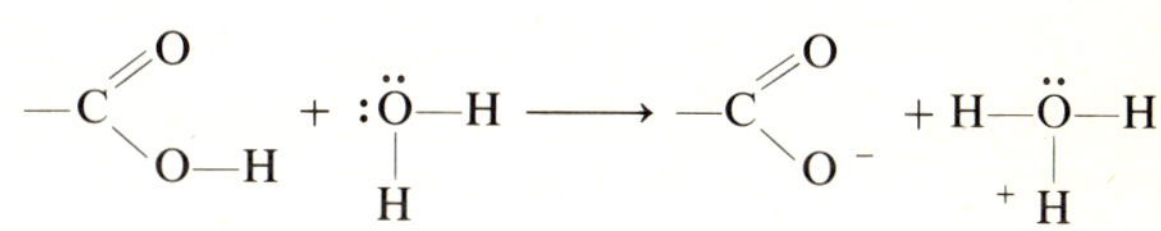

carboxyl group carboxylate ion

Carboxyl groups in water solution thus ionize to some degree. If the H^+ is neutralized with OH^- as it is formed, more ionization occurs. As a result, most carboxyl groups in neutralized water solution are completely ionized. Because the hydrogen leaves as H^+, it leaves behind a negatively charged ion. The charge on this ion influences the properties of the compound considerably. Water solubility is increased, the effect being greater than that of the polar alcohol group (Table 7.1). Even without ionization, the very polar carboxyl group increases boiling points to a greater degree than the alcohol group does. Carboxyl groups are very common in biologically significant molecules, especially in fats and proteins.

7.1 Naming Carboxylic Acids

Carboxylic acids are named by dropping the final -e from the alkane name and adding -oic acid.

Table 7.1
Carboxylic Acids

IUPAC Name	*Common Name*	*Formula*	*Boiling Point (°C)*	*Solubility in Water (g/100 g)*	*Corresponding Alcohol*	*Boiling Point (°C)*	*Solubility in Water (g/100 g)*
Methanoic	Formic	$H—COOH$	100.5	Miscible	Methanol	65	Miscible
Ethanoic	Acetic	$CH_3—COOH$	118	Miscible	Ethanol	78	Miscible
Propanoic	Propionic	$CH_3—CH_2—COOH$	141	Miscible	1-Propanol	98	Miscible
Butanoic	Butyric	$CH_3—CH_2—CH_2—COOH$	164	Miscible	1-Butanol	117	9
Pentanoic	Valeric	$CH_3—CH_2—CH_2—CH_2—COOH$	187	3.7	1-Pentanol	138	2.7
Hexanoic	Caproic	$CH_3(CH_2)_4COOH$	205	1.0	1-Hexanol	157	0.6
Octanoic	Caprylic	$CH_3(CH_2)_6COOH$	239	0.7	1-Octanol	195	0.05
Decanoic	Capric	$CH_3(CH_2)_8COOH$	269	0.2	1-Decanol	228	Insoluble
Dodecanoic	Lauric	$CH_3(CH_2)_{10}COOH$	—	Insoluble	1-Dodecanol	—	Insoluble

$CH_3-CH_2-CH_3$

propane

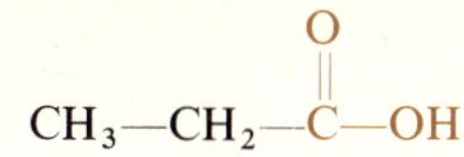

propanoic acid

$CH_3-CH_2-CH_2-CH_2-CH_3$

pentane

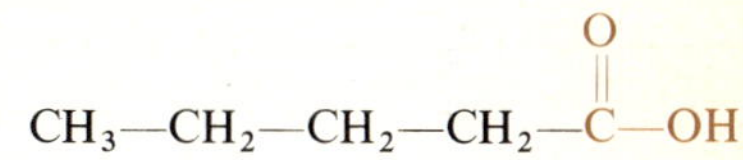

pentanoic acid

Frequently the carboxyl group is abbreviated as —COOH and the carboxylate ion as $-COO^-$. Unfortunately, a number of commonplace (nonsystematic) names have been used with the carboxylic acids, particularly by biochemists and industrial chemists (Table 7.1).

Where the carboxyl group is part of the main chain, it is carbon 1, just as in the case with aldehydes. Other substitutions are indicated as prefixes with numbers in the usual fashion.

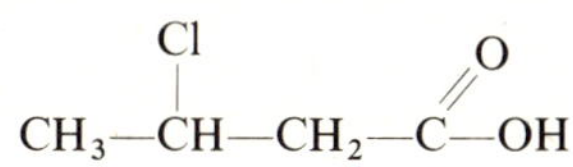

3-chlorobutanoic acid

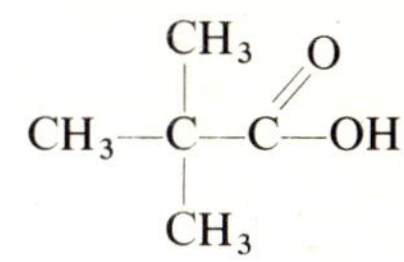

2,2-dimethylpropanoic acid

The aromatic carboxylic acids have special names.

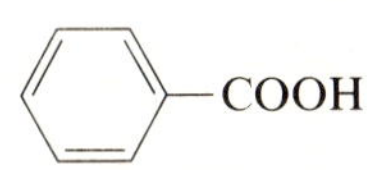

benzoic acid

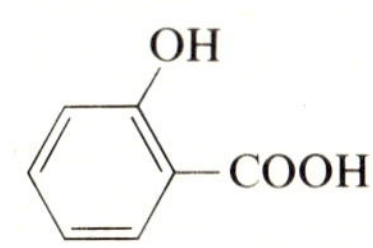

salicylic acid

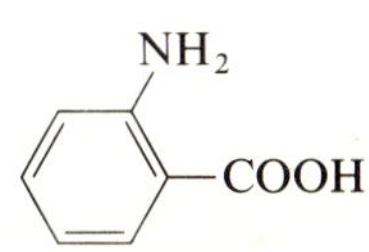

anthranilic acid

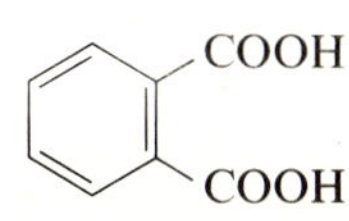

phthalic acid

Exercise 7.1

Name the following:

a) CH_3-COOH

b) 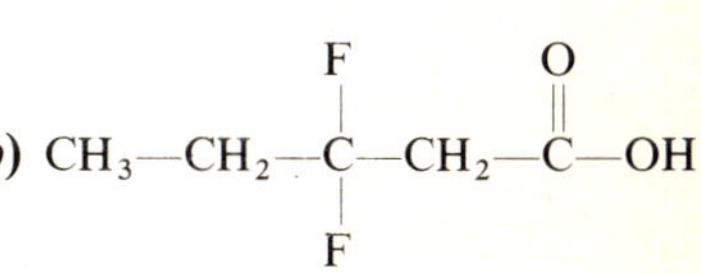

c) 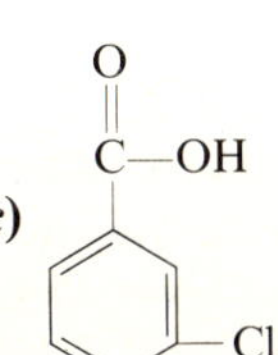

d) 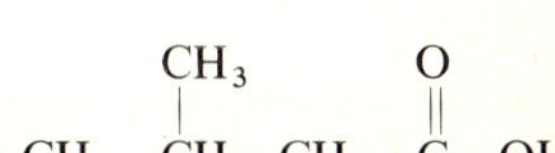

Example 7.1 continued

e) 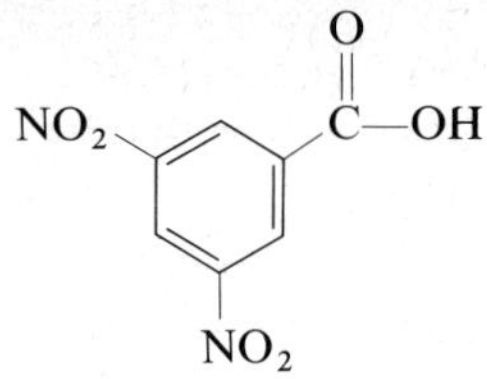

f) $Cl_3C{-}C({=}O){-}OH$

7.2 Acid Properties of Carboxylic Acids

Because these compounds are acids, one of their most significant reactions is with bases to form salts and water in the same fashion as mineral acids do.

$$HCl + NaOH \longrightarrow NaCl + H_2O$$

$$CH_3COOH + NaOH \longrightarrow CH_3COO^-Na^+ + H_2O$$

The salt forms of the carboxylic acids are named differently from the free acid forms. The -ic ending is replaced by -ate.

$CH_3CH_2CH_2C({=}O){-}O^-\ K^+$

potassium butanoate

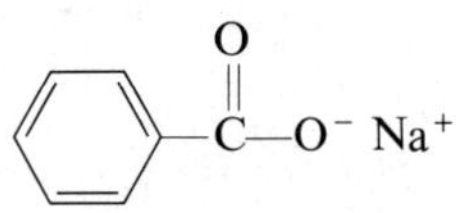

sodium benzoate
(a food preservative;
read your catsup label)

Exercise 7.2

Name these:

a) $CH_3CH_2C({=}O){-}O^-\ Na^+$

b) 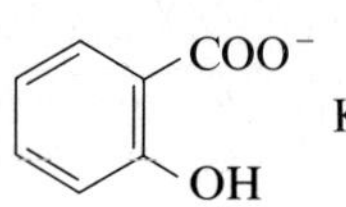

Recall that a common way to indicate the acidity or basicity of a solution is by its pH. The pH is calculated from the concentration of hydrated H^+ (H_3O^+) in the solution.

$$pH = -\log(H_3O^+)$$

Remember these important facts about pH:

1. In acid solutions the pH is less than 7.0.
2. A change in pH of one unit means a tenfold change in the acid concentration. To change the pH from 4.0 to 3.0 requires that the acid concentration be increased tenfold. To go from 4.0 to 2.0 requires that it be increased 100-fold (tenfold × tenfold).

3. In the neutral region, which is around pH 7.0, the pH value is very sensitive to exceedingly small changes in acid concentration. Going from pH 7.0 to 6.0 is accomplished by making a neutral solution one-millionth (10^{-6}) molar in acid. One drop of concentrated HCl is enough to adjust 60 liters of pure water. The pH scale is very useful for biochemists, because biological systems are generally in the neutral region and are also very sensitive to small changes in pH.

Structures like the carboxyl group that can lose and regain protons in water do so under specified conditions. If we start at pH 7.0, almost all carboxyl groups have lost their hydrogen. As we lower the pH by adding an acid such as HCl to the solution, we reach a pH where there is a high enough concentration of H_3O^+ in solution that the equilibrium shifts in favor of the carboxyl form rather than the carboxylate ion form. In a sense, the hydrogen is forced back onto the carboxyl group by the pressure of H_3O^+ concentration. With changing pH, the hydrogens add at first gradually to the carboxylate ion, then at a constantly quickening pace, until half of the carboxylates have added a hydrogen. The rate of addition with decreasing pH then lessens, and the last few hydrogens are slow to add back. If we look at ethanoic acid, or acetic acid as it is more commonly termed, the pattern is apparent.

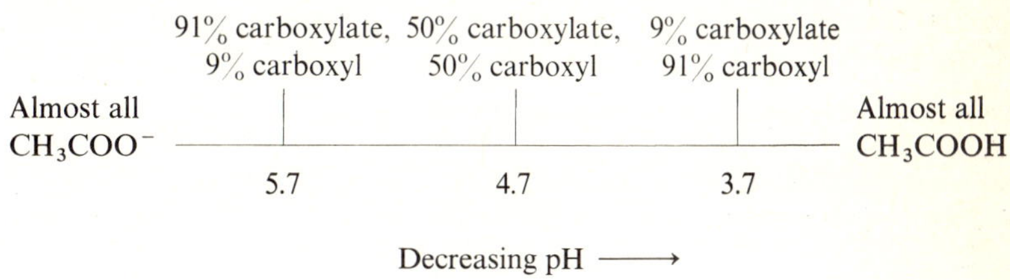

We focus special attention on the pH at which half the hydrogens have been added back, in this case pH 4.7. This always occurs at pH 4.7 for acetic acid and is called the p*K* of acetic acid. Each acid has its own characteristic p*K*. The following points are particularly important when considering acidic groups.

1. Acidic groups do not lose or regain hydrogen all at once but rather across a range of solution pH values.
2. The losing or regaining of hydrogen is most active in the midpoint of the pH range.
3. The pH at the exact midpoint (where half the groups have a hydrogen and half don't) is a characteristic physical constant called the p*K*.

7.3 Biochemical Effects of Acid Properties

The acidic properties of carboxylic acids have important implications for living systems. The most fundamental and obvious is that carboxylic acids change the pH of a solution to which they are added. When the

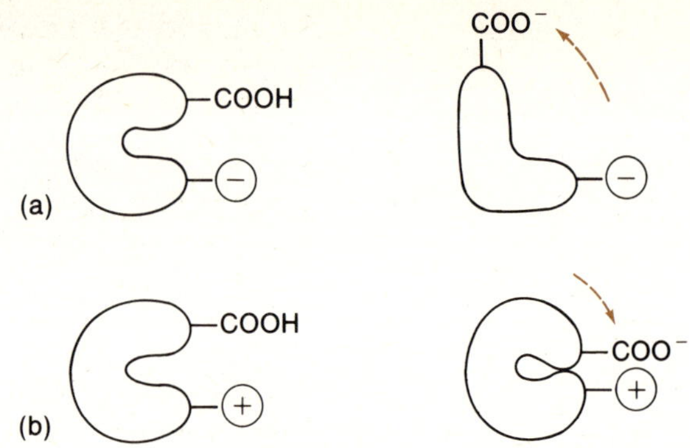

Figure 7.1
Charge and protein shape
(a) A newly ionized carboxyl group moves away from a negative charge. (b) Ionization produces a negative charge, which is attracted to the existing positive charge.

carboxyl group reacts with water, the H_3O^+ produced lowers the pH. In most regions of a living organism, control of pH to a narrow specific range is vital. Taking carboxylic acids into the body or forming them in the body from nonacidic substances lowers the pH. An example of the latter is the enzymatic formation of carboxylic acids from neutral aldehydes (Chapter 6). Although the appearance of acid is quite unpredictable, the organism must have systems to immediately counterbalance its formation. This problem is discussed fully in Chapter 22.

Another change occurs when the carboxyl group ionizes in water. Before reaction the group is uncharged; afterward it has a negative charge. In fact, the reaction of the acid water is referred to as *ionization*, emphasizing the charge aspect. Chapter 2 introduced the concept of complementarity, that biological molecules often must fit together closely and accurately. Shape is important to complementarity, but charge can also be significant. Clearly, even if two surfaces fit exactly, if they both have negative charges, they will resist approaching one another. Conversely, if the charges are opposite (+ and −), they will be eager to join. If one of the molecules to be joined is a protein, charge can exert an effect on shape. Proteins are large, folded molecules. Often the pattern of folding can be altered, changing the molecular shape (Chapter 11 and Figure 7.1).

Exercise 7.3

Is the concentration of (H_3O^+) higher at pH 2.1 or 3.6? at pH 7.0 or 8.0?

Exercise 7.4

In a solution at pH 3.2, the concentration of H_3O^+ is (twice as great, half as great, 3.2/4.2 times as great, ten times as great, one-tenth as great) as a solution at pH 4.2.

Exercise 7.5

You are told that the pK of a carboxylic acid is 3.6. You dissolve 0.1 mole of this acid in a liter of water and adjust the pH to 3.6. The solution will then contain which of these:

a) 0.1 mole of carboxylic acid
b) 0.05 mole of carboxylic acid and 0.05 mole of carboxylate ion
c) 0.01 mole of carboxylate ion
d) none of these

7.4 Other Reactions of Carboxylic Acids: Ester Formation

Acids and alcohols react to form *esters*. In this process the hydrogen of the alcohol and the —OH of the acid combine to form water. This process is sometimes called a *condensation reaction*. The reaction is speeded by an acid catalyst.

$$CH_3-\overset{\displaystyle O}{\overset{\|}{C}}-O-H + CH_3OH \xrightarrow{H^+} CH_3-\overset{\displaystyle O}{\overset{\|}{C}}-O-CH_3 + H_2O$$

The reaction can be reversed simply by heating the ester with dilute acid. Because breaking the ester apart requires that water be added back to the ester bond, the reaction is termed *hydrolysis* (*hydro*, water, and *lysis*, breaking).

$$CH_3-\overset{\displaystyle O}{\overset{\|}{C}}-O-CH_3 + H_2O \xrightarrow{H^+} CH_3-\overset{\displaystyle O}{\overset{\|}{C}}-O-H + CH_3OH$$

Hydrolysis is also accomplished using a hydroxide ion as catalyst, but the product is then the salt of the acid rather than the un-ionized acid.

$$CH_3-\overset{\displaystyle O}{\overset{\|}{C}}-O-CH_3 + NaOH \longrightarrow CH_3-\overset{\displaystyle O}{\overset{\|}{C}}-O^-\ Na^+ + CH_3OH$$

7.5 Naming Esters

Esters are named by listing the alkyl group of the alcohol, then the name of the ionized form of the acid. Generally the common rather than the correct names of the acids are used. The compound in the examples should correctly be

$$CH_3-\overset{\displaystyle O}{\overset{\|}{C}}-O-CH_3$$

methyl ethanoate

However, we almost always hear it called methyl acetate.

Exercise 7.6

Name the following esters:

a) 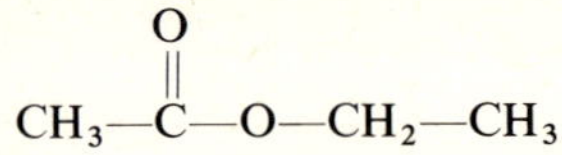

b) 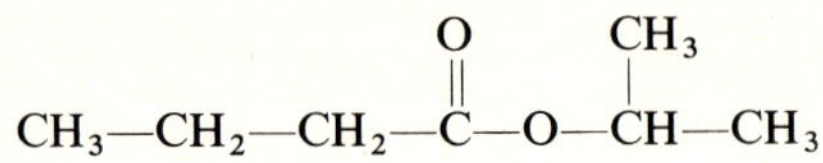

c) 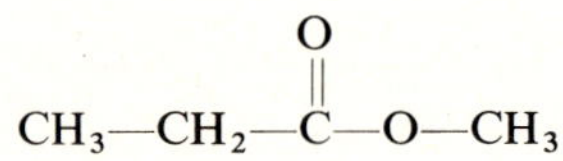

d) 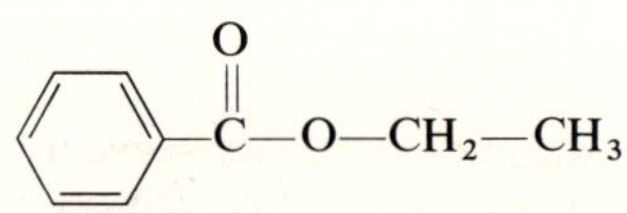

e) 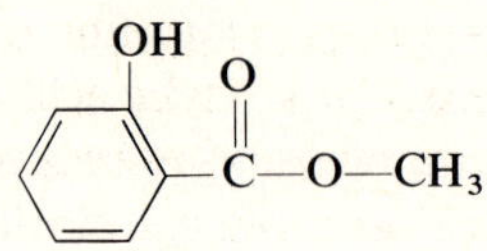

f) 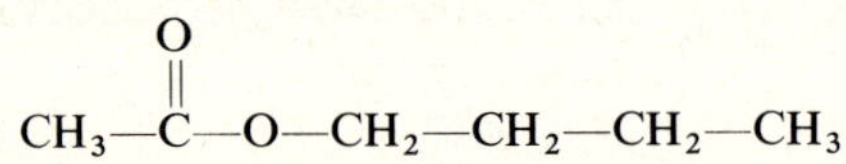

Because both the acid and the alcohol have lost polar —O—H groups in forming the ester, esters are rather nonpolar. Most simple esters are liquids and have pleasant, fruity odors.

7.6 *Esters of Mineral Acids*

In biological systems we find alcohols bound in the ester fashion to mineral acids. The most common mineral acid encountered is phosphoric acid, with sulfuric acid occurring occasionally. In metabolism, for example, carbohydrates are converted first to phosphate esters. Below are common examples of such carbohydrate phosphates.

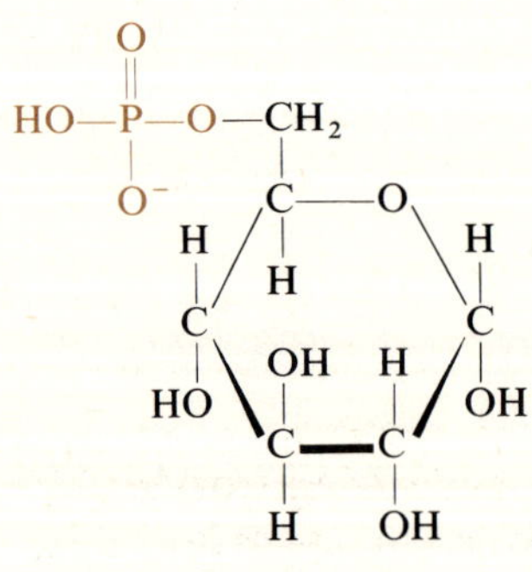

glucose-6-phosphate

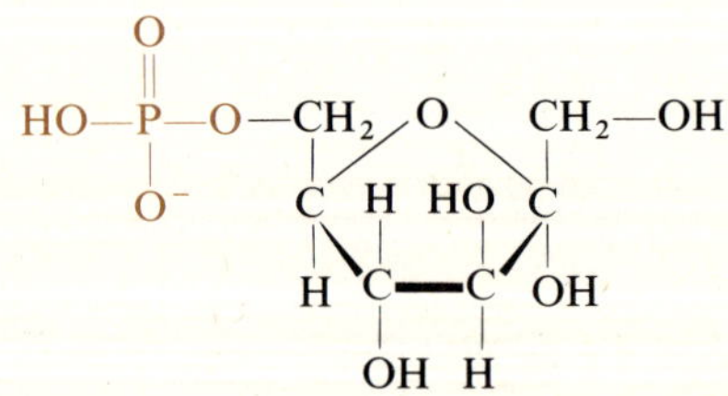

fructose-6-phosphate

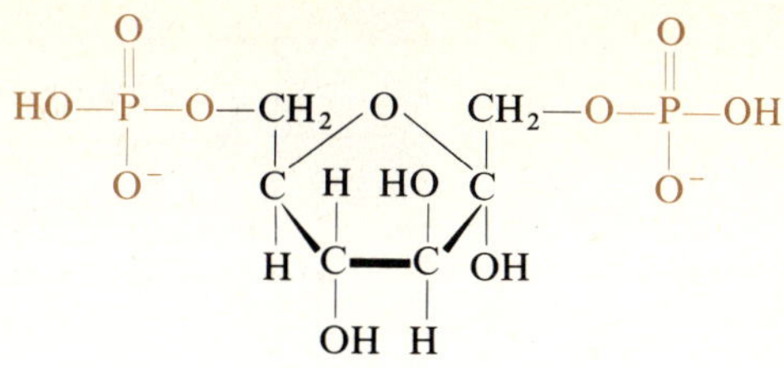

fructose-1,6-diphosphate

More examples of mineral acid esters are found in the lipids section of this chapter.

7.7 Medicinal Esters

A number of esters serve as medicinal compounds. Esters of salicylic acid have been particularly significant. Perhaps the best-known, most often used nonprescription drug in the world is aspirin. Aspirin is the acetate ester of salicylic acid and is used as an analgesic (pain reliever) and antipyretic (fever reducer). Large doses of aspirin are used to relieve the problems of arthritis. Other salicylates, such as acetylsalicylsalicylic acid and phenylsalicylate, are used similarly (Figure 7.2).

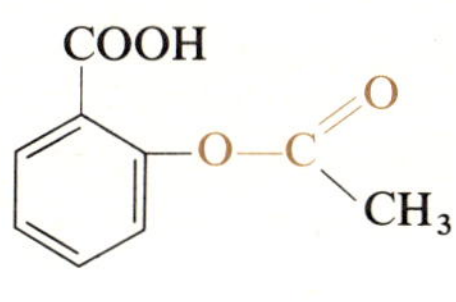

acetylsalicylic acid
(aspirin)

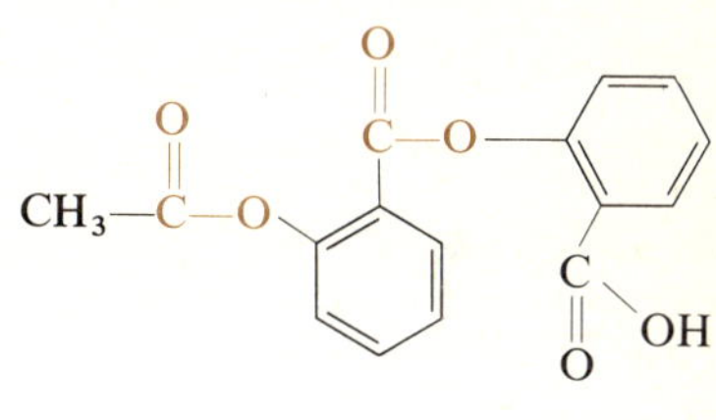

acetylsalicylsalicylic acid

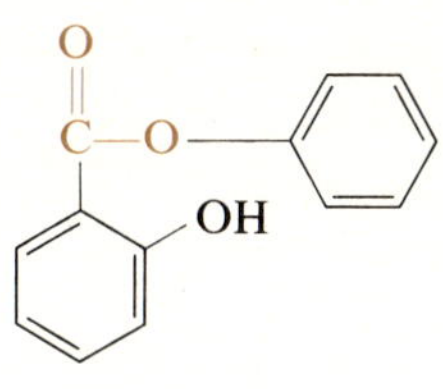

phenylsalicylate

Esters with anesthetic properties include propyl aminobenzoate and methyl-3-amino-4-hydroxybenzoate (Orthocaine). One of the reasons tuberculosis is so much less threatening today than many years ago is the ester antibiotic phenylaminosalicylate (PAS).

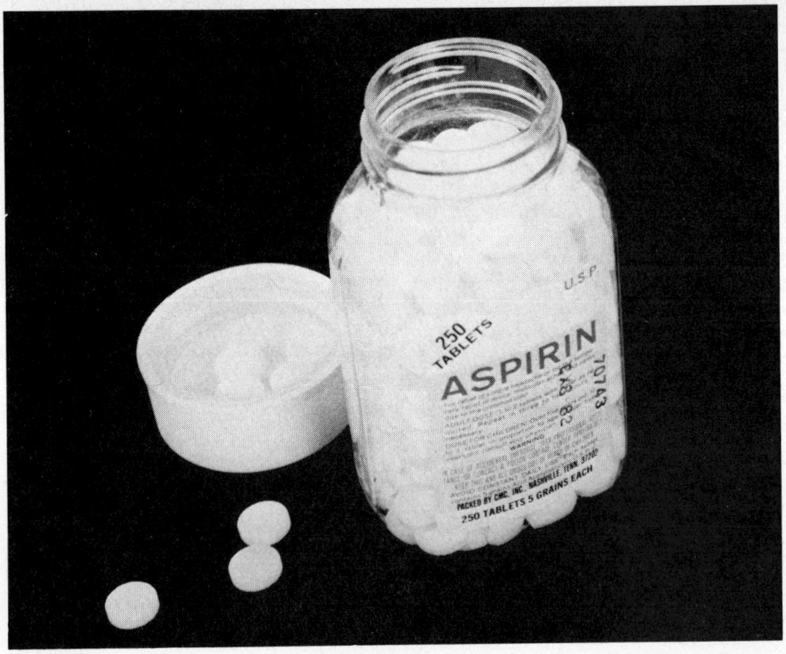

Figure 7.2
Aspirin
(Photo by Catherine Feryok, Eastern Michigan University.)

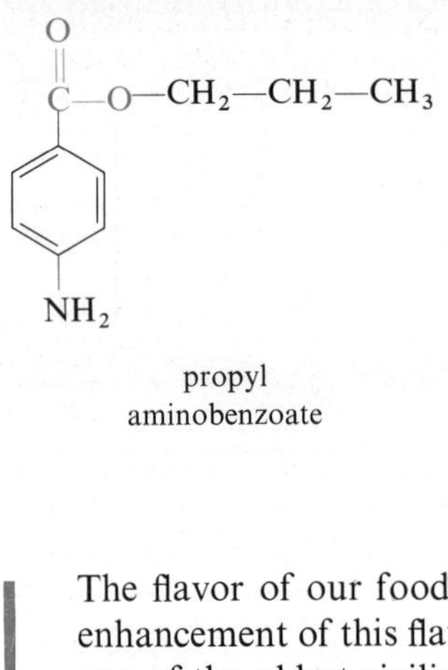

propyl aminobenzoate

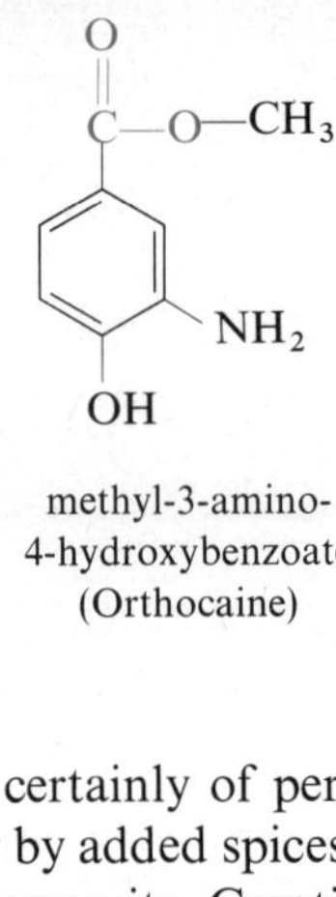

methyl-3-amino-4-hydroxybenzoate (Orthocaine)

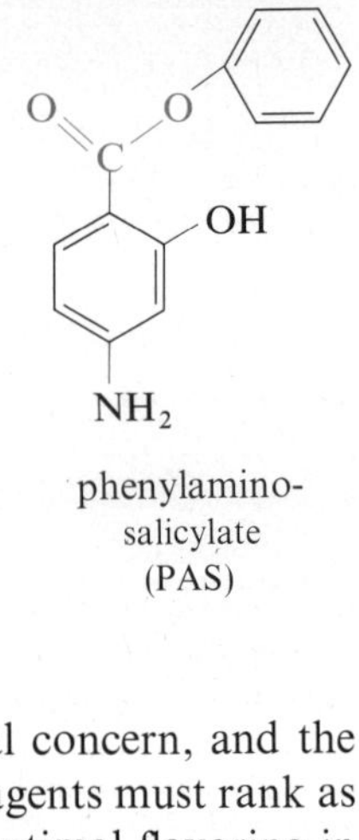

phenylaminosalicylate (PAS)

7.8 Flavor

The flavor of our food is certainly of personal concern, and the enhancement of this flavor by added spices or agents must rank as one of the oldest civilized pursuits. Creating optimal flavoring in commercial foods is a large industry, and the testing and approval of agents added to foods for this purpose represents a major effort by the Food and Drug Administration. Are the agents added to alter food flavor totally different from natural flavor agents? In an environment of suspicion about additives, should we use only the natural compounds? To deal with these questions, we need to know more facts.

Some of what we describe as flavor represents the response of sensors on the tongue. These sensors give us sensations of sweet, sour, bitter, and salty. The more subtle aspects of flavor are the

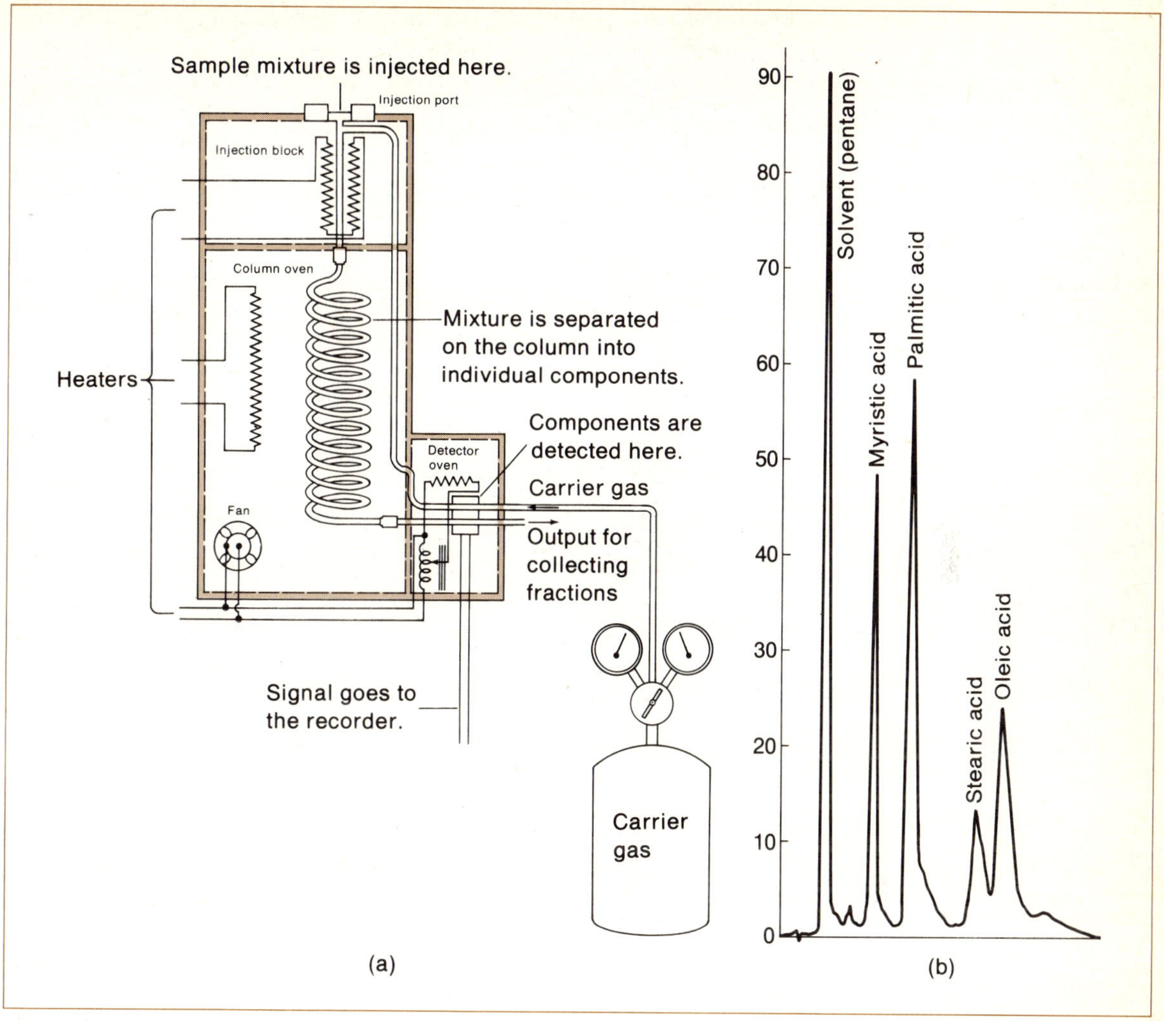

Figure 7.3
Gas chromatograph
(a) The gas chromatograph heats liquid mixtures to convert them to the vapor state, separates the mixture into its components, detects them, and indicates their presence as a peak on recorder chart paper. (b) A student prepared the methyl esters of the fatty acids in a sample triglyceride, separated them on the gas chromatograph, and obtained the above results.

result of our sense of smell detecting volatile food components. Recall how flat food tastes during a head cold. Notice that the flavor is detected because the compound is volatile. Because flavor elements are volatile, on standing or particularly on processing at elevated temperatures these compounds are likely to be lost. It's true that food fresh from the garden tastes better and that excessive cooking can spoil the flavor. As you smell the coffee percolating or the strawberries being boiled for jam, reflect that you are enjoying the smell because those pleasant flavor elements are no longer in the food but rather in the air of the kitchen. Some food processors trap these volatile compounds by condensing them from the air over the cooking vat and returning them to the food. Freezing foods to preserve them is an excellent way of avoiding the loss of volatile flavor elements; these elements are trapped at the low temperatures. Freeze-dried coffee also retains more of the fresh flavor than conventional instant coffees. Unfortunately, it is often necessary to partially cook (blanch) most vegetables before freezing in order to inactivate enzymes that would slowly degrade the food in the frozen state, so a total avoidance of heating is not possible.

As the sophistication of chemists in separation and analysis improves, the ability to determine exactly what gives each food its peculiar flavor is heightened. Gas chromatography is a very significant technique for this purpose. In the gas chromatograph, mixtures of volatile material are injected into a hot chamber. They are then carried as gases through a tube packed with a substance that attracts and so retards the flow of the compounds. The retarding is greater for the more polar compounds. The result is that the components of the mixture are separated. The compounds then flow through a sensitive detector that indicates how much of each compound is present (Figure 7.3).

Figure 7.4
Artificial flavor
Compounds are added to a variety of foods to provide or enhance the flavor. (Photo by Catherine Feryok, Eastern Michigan University.)

By this technique and others, the complex mixtures that comprise the flavor of foods are studied. Table 7.2 shows the flavor constituents of strawberry and pineapple. Notice that the pineapple components are esters, alcohols, aldehydes, and ketones. Is this also true for strawberry? Notice, too, that by contrast to some of the other biochemically significant molecules you have seen, these are often very simple compounds.

Because the natural flavor components are difficult to retain, or because the product being prepared is totally synthetic, perhaps hard candy or a soft drink, the food industry adds Food and Drug Administration approved compounds whose smell resembles that of the natural food (Figure 7.4).

Table 7.3 lists esters that are used in this fashion. Compare this list with Table 7.2 and you will see that some of the artificial flavor elements are also components of the natural food flavor.

7.9 *Amides*

Carboxylic acids also react with ammonia to produce *amides*. At first the ammonia as a base simply forms an ammonium salt with the acid. On heating, however, water is lost to produce an amide. This, too, is a condensation reaction.

$$CH_3{-}\overset{\overset{\displaystyle O}{\|}}{C}{-}OH + NH_3 \longrightarrow \underset{\text{a salt}}{CH_3{-}\overset{\overset{\displaystyle O}{\|}}{C}{-}O^-\ NH_4^+} \xrightarrow{\Delta} \underset{\text{an amide}}{CH_3\overset{\overset{\displaystyle O}{\|}}{C}{-}NH_2} + H_2O$$

Amides usually are named by dropping the -ic ending from the common name of the acid and substituting -amide in its place (see page 172).

Table 7.2
Flavor Constituents

Volatile Flavor Constituents of the Crushed Strawberry (ppm)			
More Than 1	***0.1–1***	***0.01–0.1***	***Less Than 0.01 (Traces Only)***
Hexene-2-al (6.35)	Hexanal	Diacetyl	Acrolein
	Hexene-2-ol *trans*	Pentene-2-al	Crotonaldehyde
Acetone (3.0)	Hexanol	*cis*-Hexene-3-al	Propanal
Acetaldehyde (1.8)	Methyl butyrate	Heptanone-2	Pentanone-2
	Ethyl butyrate	Methanol	Propyl acetate
Ethanol (1.51)	Methyl caproate	Butanol	
	Ethyl caproate	Pentene-1-ol-3	
	Hexyl acetate	Heptanol-2	
		Ethyl acetate	
		Ethyl propionate	
		Butyl acetate	
		Isopropyl butyrate	
		Ethyl crotonate	
		Heptanone	
		Hexene-2-yl acetate	
		Hexene-2-yl butyrate	
		Hexyl butyrate	
		Isoamyl acetate	

Volatile Constituents of Pineapple Flavor	
Esters	Acids
Methyl acetate	Acetic
Ethyl acetate	
Ethyl acrylate	Alcohols
Methyl *n*-butyrate	Methanol
Ethyl *n*-butyrate	Ethanol
Methyl *n*-valerate	*n*-Propanol
Methyl isovalerate	Isobutanol
Ethyl lactate	*n*-Pentanol
Ethyl isovalerate	
Methyl *n*-caproate	Carbonyl compounds
Methyl isocaproate	Acetone
Ethyl *n*-caproate	Formaldehyde
Methyl *n*-caprylate	Acetaldehyde
Ethyl *n*-caprylate	Diacetyl
n-Amyl *n*-caproate	Furfural
Methyl β-methylthiopropionate	Methyl-*n*-propylketone
Ethyl β-methylthiopropionate	5-Hydroxy-2-methylfurfural

Source: From Schultz, Day, and Libbey, *The Chemistry and Physiology of Flavors*, Avi Publishing Company, Westport, Connecticut. Reprinted by permission.

Table 7.3
Esters Approved as Flavor Agents in Food

Ester	*Name*
1. 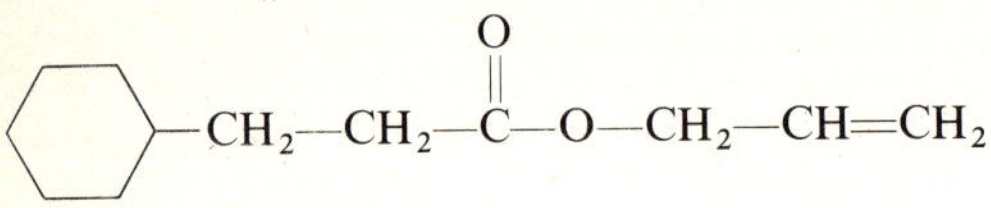	Allyl-3-cyclohexylpropionate (pineapple odor)
2. 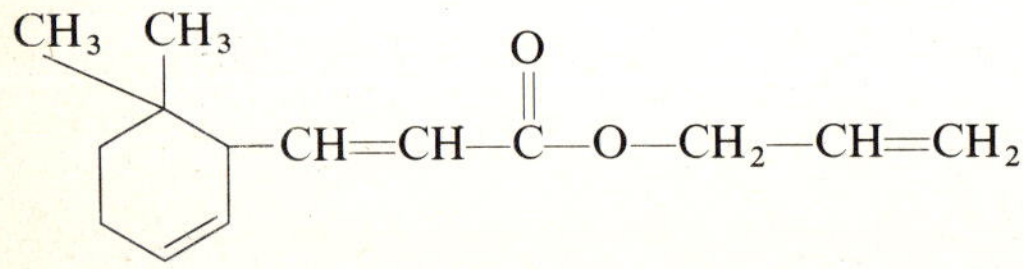	Allyl ionone (pineapple odor)
3. $CH_3—CH_2—CH_2—\overset{O}{\overset{\|}{C}}—O—CH_2—CH_2—CH_2—CH_3$	Butyl butyrate (pineapple)
4. 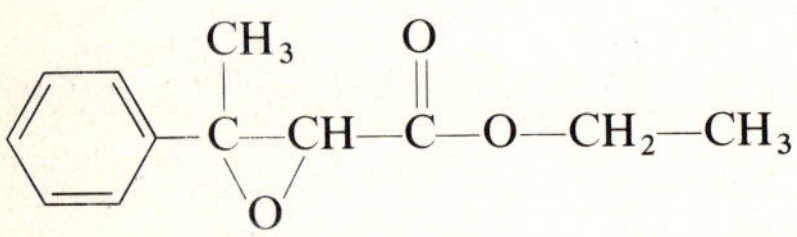	Ethyl methylphenylglycidate (strawberry)
5. 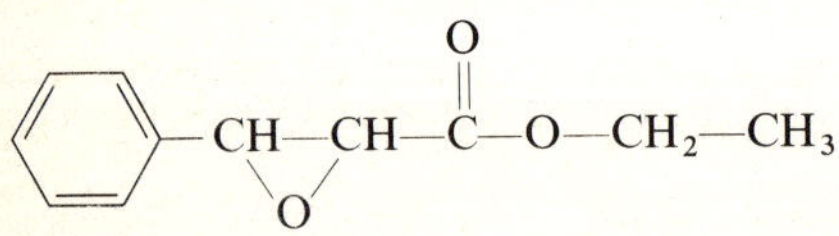	Ethyl phenylglycidate (strawberry)
6. $H—\overset{O}{\overset{\|}{C}}—O—CH_2—CH_2—CH_2—CH_2—CH_3$	Amyl formate (plum odor)
7. 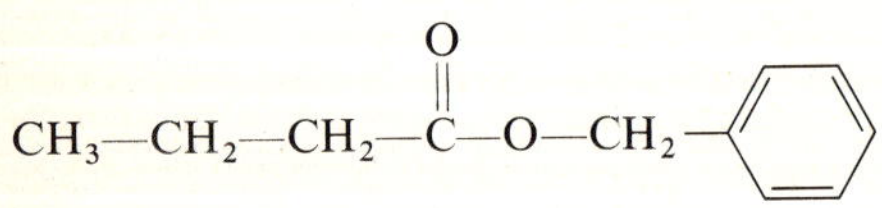	Benzyl butyrate (plum)
8. 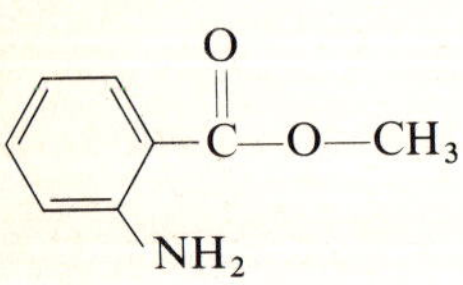	Methyl anthranilate (grape)
9. 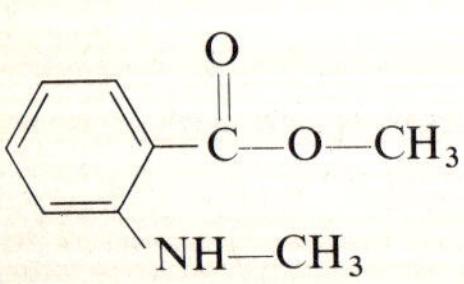	Methyl N-methylanthranilate (grape)

Table 7.3 ***(Continued)***

Ester	*Name*
10. $CH_3-CH_2-\overset{\overset{\displaystyle O}{\|}}{C}-O-CH_2-CH_3$	Ethyl propionate (rum)
11. 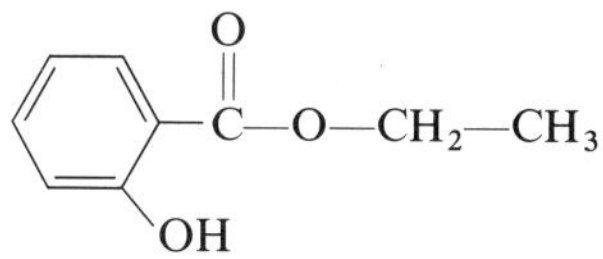	Ethyl salicylate (wintergreen)
12. 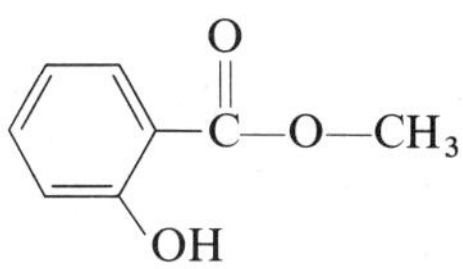	Methyl salicylate (wintergreen)
13. 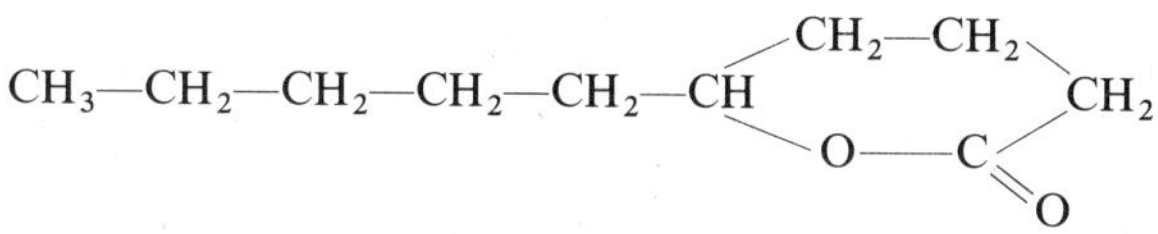	Δ-Decalactone (coconut)
14. 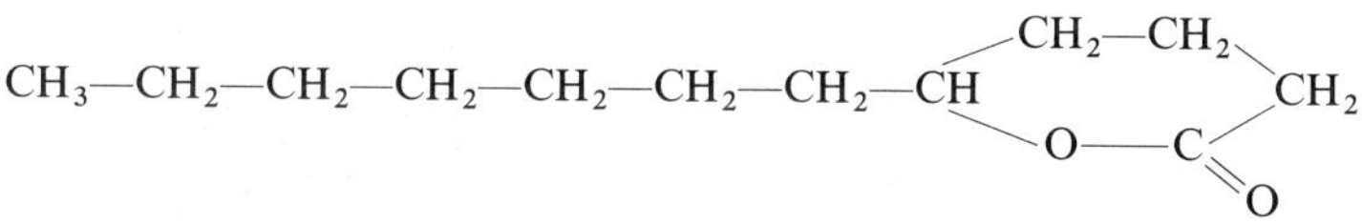	Δ-Dodecalactone (coconut)
15. 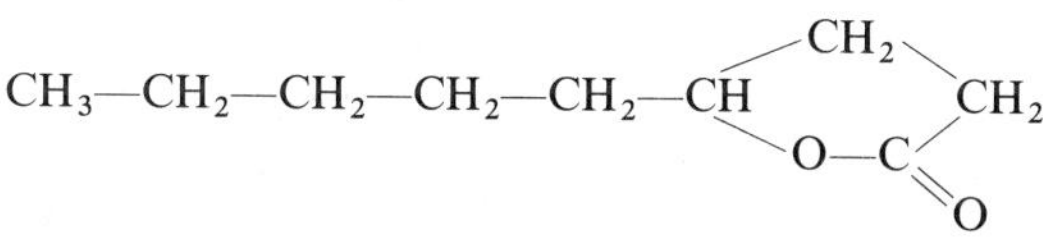	γ-Nonalactone (coconut)
16. $CH_3-\overset{\overset{\displaystyle CH_3}{\vert}}{CH}-CH_2-\overset{\overset{\displaystyle O}{\|}}{C}-O-CH_2-CH_3$	Ethyl isovalerate (apple)
17. $CH_3-\overset{\overset{\displaystyle CH_3}{\vert}}{CH}-CH_2-CH_2-\overset{\overset{\displaystyle O}{\|}}{C}-O-CH_2-CH_2-\overset{\overset{\displaystyle CH_3}{\vert}}{CH}-CH_3$	Isoamyl isovalerate (apple)
18. 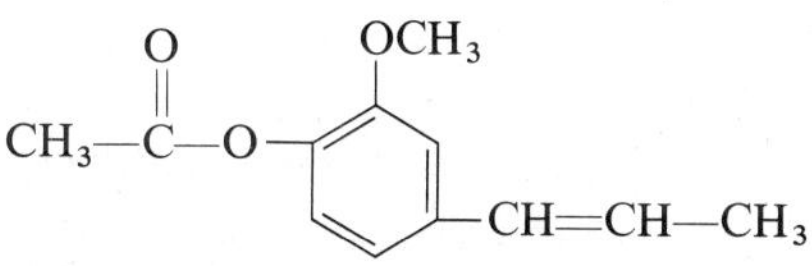	Isoeugenyl acetate (clove, spicy)

Table 7.3 **(*Continued*)**

	Ester	*Name*
19.	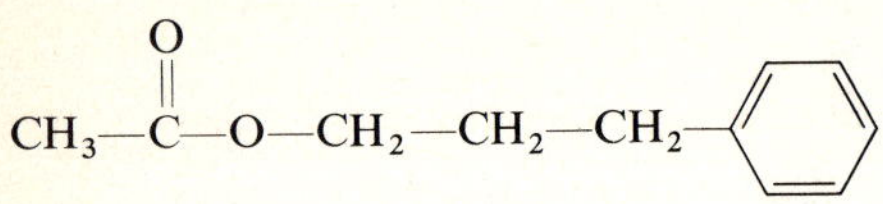	3-Phenylpropyl acetate (spicy)
20.	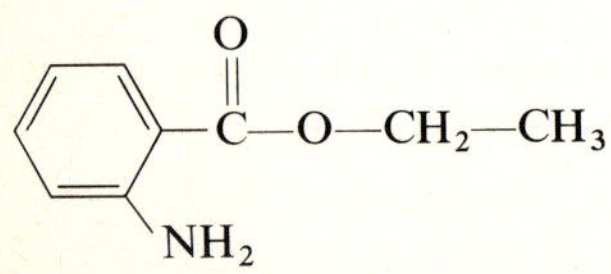	Ethyl anthranilate (orange blossom)
21.	$CH_3-C(CH_3)=CH-CH_2-CH_2-C(=O)-O-CH_3$	Methyl heptenoate (citrus)
22.	$CH_3-C(=O)-O-CH_2-CH_2-CH_2-CH_2-CH_2-CH_2-CH_2-CH_3$	Octyl acetate (orange, jasmine)
23.	$CH_3-CH_2-CH_2-C(=O)-O-CH_2-CH_2-CH(CH_3)-CH_3$	Isoamyl butyrate (fruity)
24.	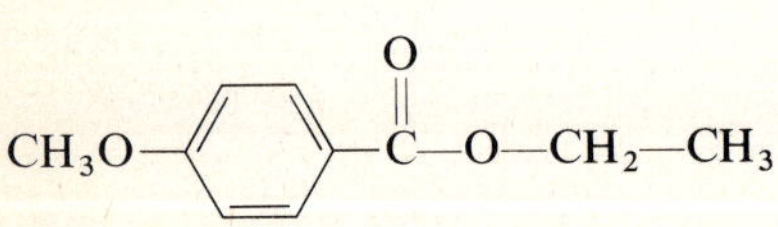	Ethyl anisate (anise, fruity)
25.	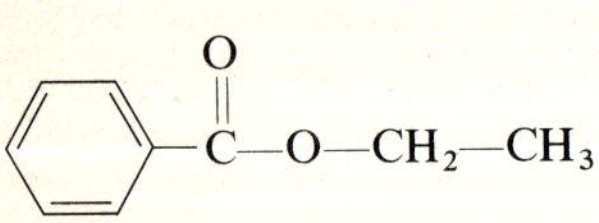	Ethyl benzoate (fruity)
26.	$CH_3-CH_2-CH_2-C(=O)-O-CH_2-CH_3$	Ethyl butyrate (fruity)
27.	$CH_3-CH_2-CH_2-CH_2-CH_2-C(=O)-O-CH_2-CH_3$	Ethyl caproate (fruity)
28.	$CH_3-CH_2-CH_2-CH_2-CH_2-CH_2-C(=O)-O-CH_2-CH_3$	Ethyl heptanoate (fruity-wine)

Table 7.3 (*Continued*)

Ester	*Name*
29. $CH_3-CH_2-CH_2-CH_2-CH_2-CH_2-CH_2-CH_2-C(=O)-O-CH_2-CH_3$	Ethyl pelargonate (fruity-cognac)
30. $CH_3-C(=O)-O-CH_2-CH(CH_3)-CH_3$	Isobutyl acetate (fruity)
31. $H-C(=O)-O-CH_2-CH_2-CH_2-CH_2-CH_2-CH_2-CH_2-CH_3$	Octyl formate (fruity)
32. 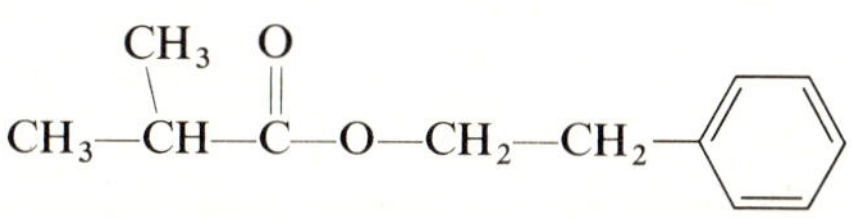	2-Phenylethyl isobutyrate (fruity)
33. 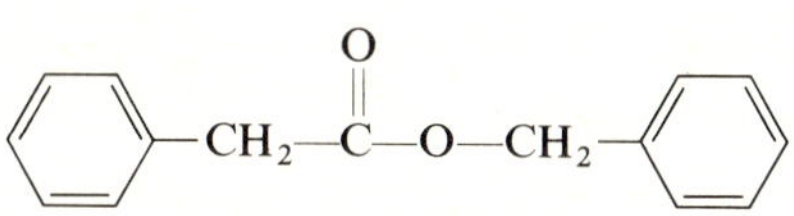	Benzyl phenylacetate (floral, honey)
34. 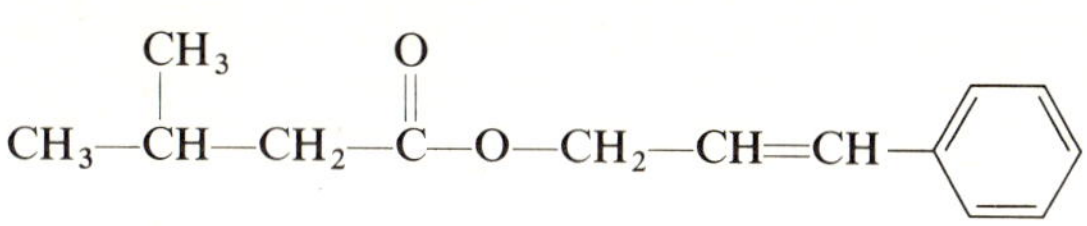	Cinnamyl isovalerate (spicy, fruity)
35. 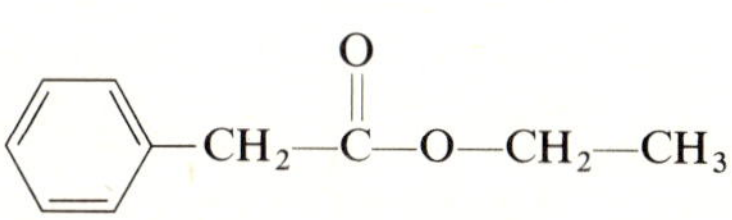	Ethyl phenylacetate (honey, sweet)
36. 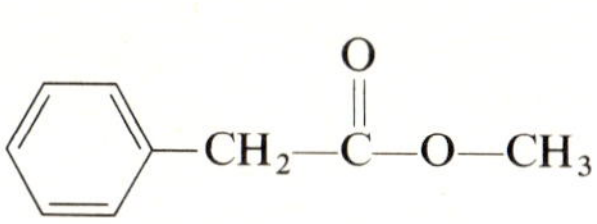	Methyl phenylacetate (honey)
37. 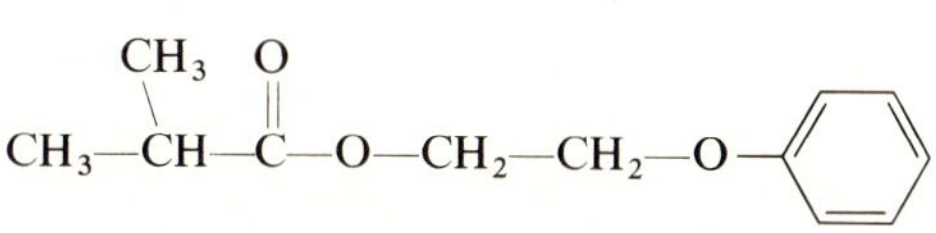	Phenoxyethyl isobutyrate (honey, rose)

acetamide propionamide

7.10 Reactive Acid Derivatives

There are derivatives of acids that are more reactive than acids in the formation of esters or amides. Just as we could use $SOCl_2$, PCl_5, or PCl_3 to replace the —OH of an alcohol with —Cl, we can replace the —OH of a carboxylic acid in the same way. The product is called an *acid chloride.*

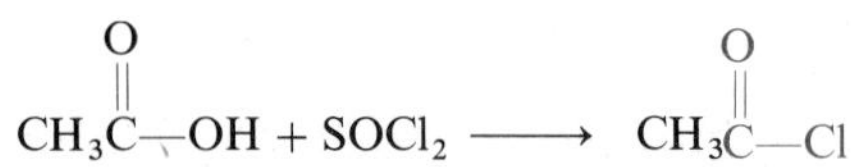

acetyl chloride

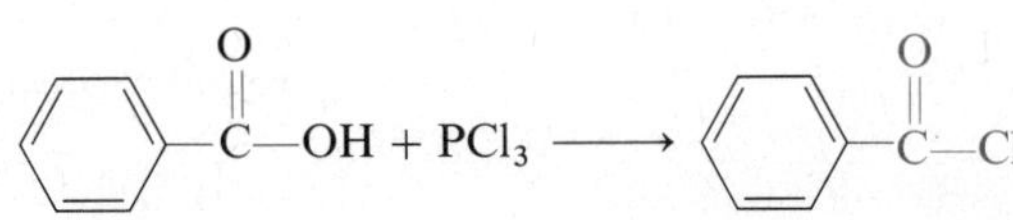

benzoyl chloride

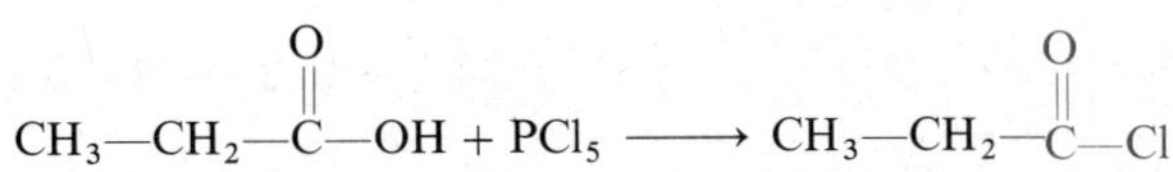

propionyl chloride

These acid chlorides are usually named by replacing the -ic of the common acid name with -yl chloride.

Another active derivative of carboxylic acids is the *acid anhydride.* It is formed by the condensation of two carboxyl groups. The most common example is acetic anhydride.

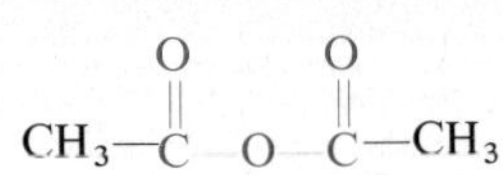

acetic anhydride

Acid chlorides and acid anhydrides react with alcohols and ammonia to form esters and amides, just as the carboxylic acids do. However, they do so under very mild conditions. Often preparing an ester or an amide is done more successfully by first converting the acid to one of these reactive forms.

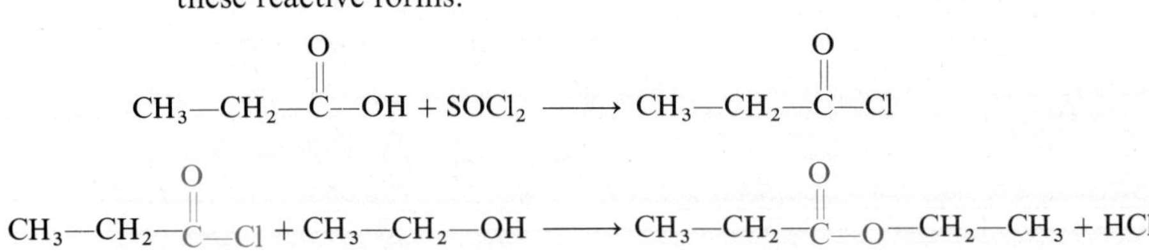

ethyl propionate

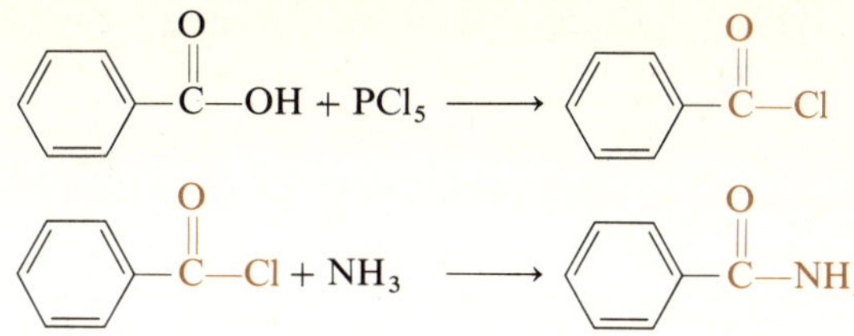

benzamide

$$CH_3-\overset{O}{\overset{\parallel}{C}}-O-\overset{O}{\overset{\parallel}{C}}-CH_3 + CH_3OH \longrightarrow CH_3-\overset{O}{\overset{\parallel}{C}}-O-CH_3 + CH_3-\overset{O}{\overset{\parallel}{C}}-OH$$

methyl acetate

By reacting an acid chloride with an amine, an amide is formed with substitutions on the nitrogen.

$$CH_3-\overset{O}{\overset{\parallel}{C}}-Cl + CH_3-NH_2 \longrightarrow CH_3-\overset{O}{\overset{\parallel}{C}}-NH-CH_3$$

N-methylacetamide

Exercise 7.7

Complete the following reactions:

a) $CH_3-CH_2-\overset{O}{\overset{\parallel}{C}}-OH + PCl_3 \longrightarrow$

b) $CH_3-CH_2-\overset{O}{\overset{\parallel}{C}}-OH + CH_3OH \xrightarrow{H^+}$

c) $CH_3-CH_2-\overset{O}{\overset{\parallel}{C}}-OH + NH_3 \xrightarrow{\Delta}$

d) $CH_3-CH_2-\overset{O}{\overset{\parallel}{C}}-Cl + CH_3-\overset{OH}{\overset{|}{CH}}-CH_3 \longrightarrow$

e) 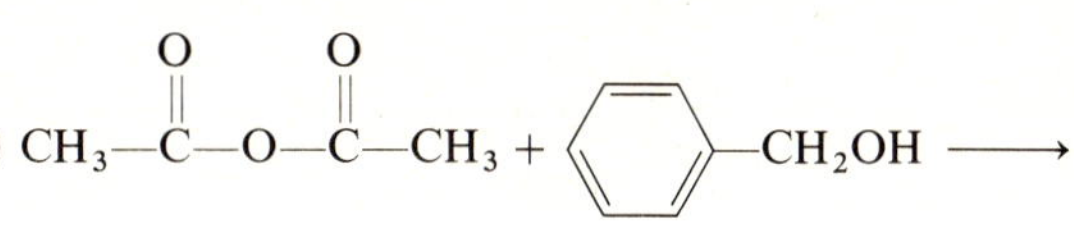

f) $CH_3-\overset{O}{\overset{\parallel}{C}}-O-CH_2-CH_3 + H_2O \xrightarrow{H^+}$

g) $CH_3-CH_2-\overset{O}{\overset{\parallel}{C}}-Cl + CH_3-CH_2-NH_2 \longrightarrow$

h) $CH_3-CH_2-\overset{O}{\overset{\parallel}{C}}-O-CH_2-CH_3 + H_2O \xrightarrow{NaOH}$

7.11 Medicinal Amides

Acetoaminophen (the amide of acetic acid and *p*-amino-phenol) is often used as an aspirin substitute. Just as the esters of salicylic acid served as analgesics, so do salicylamide and N-isopropyl salicylamide. Salicylanilide is an antifungal agent, and oxaphenamide (the amide of salicylic acid and *p*-amino phenol) is used to affect bile secretion. Oxyclozanide is used in veterinary medicine to eliminate intestinal worms.

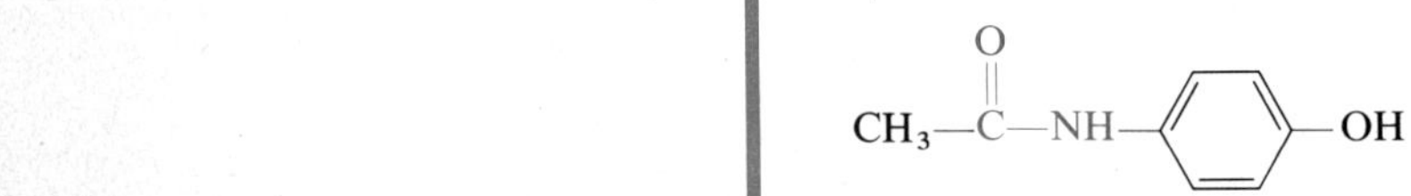

acetoaminophen

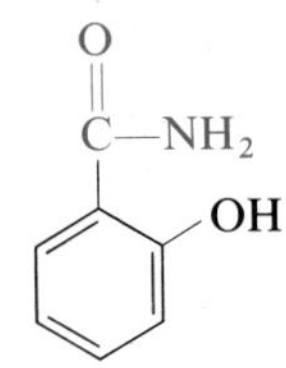

salicylamide

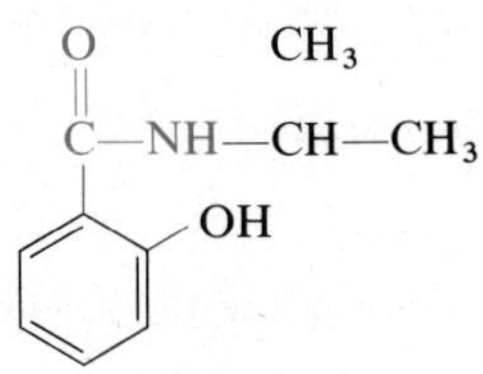

N-isopropyl
salicylamide

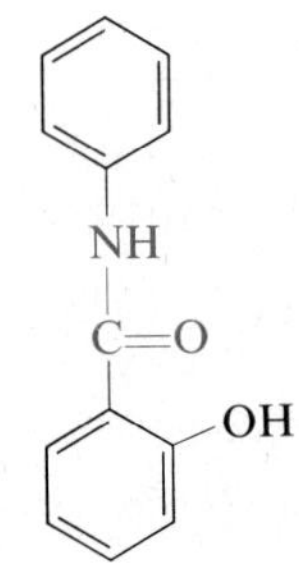

salicylanilide

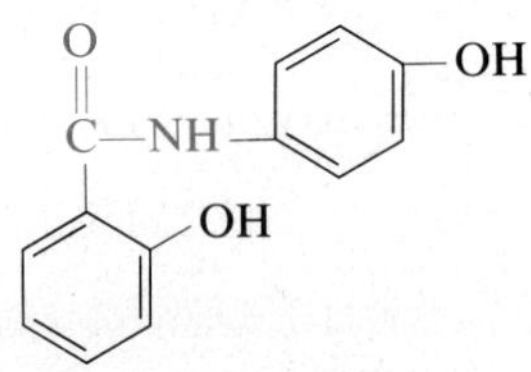

oxaphenamide

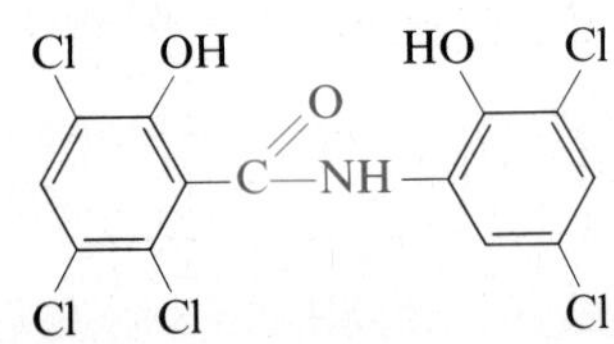

oxyclozanide

Other amides used in veterinary medicine include 3,5-dinitrobenzamide (nitromide) and 2-chloro-4-nitrobenzamide (aklomide), both antibiotics. Derivatives of phthalic acid are used in agriculture. N-1-naphthylphthalamic acid is a pre-emergence herbicide and 3-methylphthalanylic acid (Duraset) is used to prevent early dropping of fruit from trees.

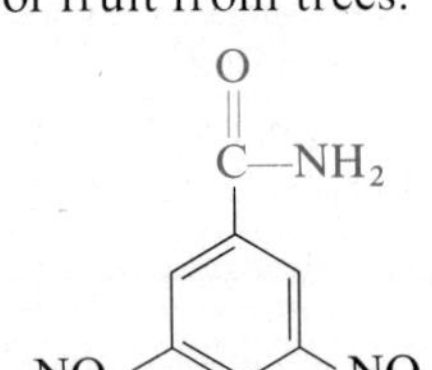

3,5-dinitrobenzamide
(Nitroamide)

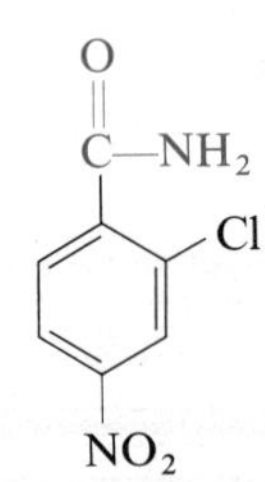

2-chloro-4-nitrobenzamide
(Aklomide)

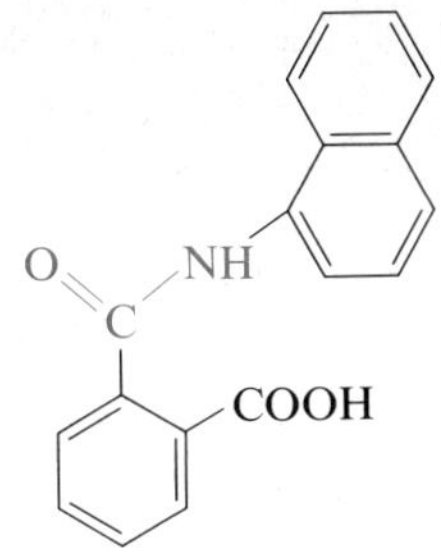

N-1-naphthylphthalamic acid

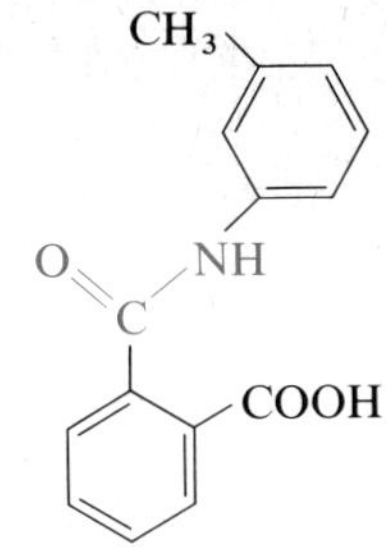

3-methylphthalanylic acid (Duraset)

Although few amides are found as flavor agents, the "hot" of hot pepper is the amide capsaicin.

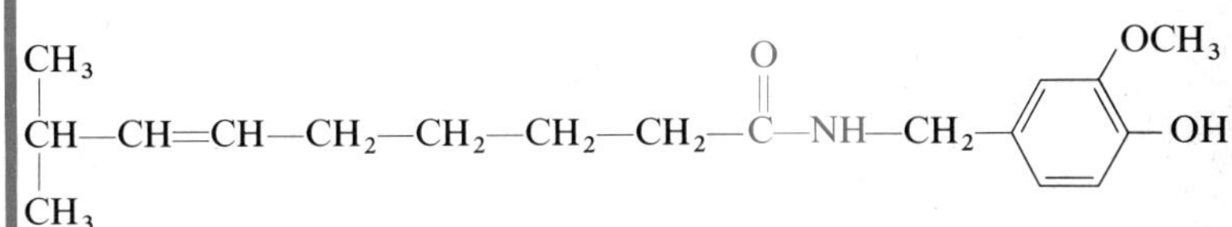

capsaicin

7.12 Condensation Polymers

Polymers are high-molecular-weight molecules prepared by adding a large number of small units, monomers, together. In the addition polymer (Chapter 3) the units combined by addition reactions. It is possible to prepare polymers by other types of reactions. Condensation reactions provide a variety of polymers second only to those produced by the addition reactions.

Obviously, a simple carboxylic acid and a simple alcohol can combine only to form an ester, which is larger than either reactant but is clearly not a polymer. However, suppose we start with an acid that has two carboxyl groups and a "diol" type alcohol:

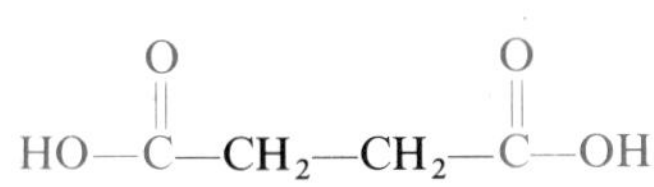

succinic acid

$HO—CH_2—CH_2—OH$

ethylene glycol

When one carboxyl group has reacted with one alcohol group, an ester linkage has been generated.

But the ester formed still has a carboxyl group that can react wit another alcohol.

$$HO{-}CH_2{-}CH_2{-}OH + HO{-}\overset{\overset{\large O}{\|}}{C}{-}CH_2{-}CH_2{-}\overset{\overset{\large O}{\|}}{C}{-}O{-}CH_2{-}CH_2{-}OH$$

$$\downarrow$$

$$HO{-}CH_2{-}CH_2{-}O{-}\overset{\overset{\large O}{\|}}{C}{-}CH_2{-}CH_2{-}\overset{\overset{\large O}{\|}}{C}{-}O{-}CH_2{-}CH_2{-}OH$$

Now a carboxylic acid can react with either end to extend the chain still further. This can continue until the supply of either alcohol or acid is exhausted. The final product is represented by bracketing a portion of polymer chain that includes one acid unit and one alcohol unit.

$$-\!\!\left(O{-}CH_2{-}CH_2{-}O{-}\overset{\overset{\large O}{\|}}{C}{-}CH_2{-}CH_2{-}\overset{\overset{\large O}{\|}}{C}{-}O\right)_{\!n}\!\!-$$

Such a polymer is classed as a *polyester*. Dacron is the best-known polyester.

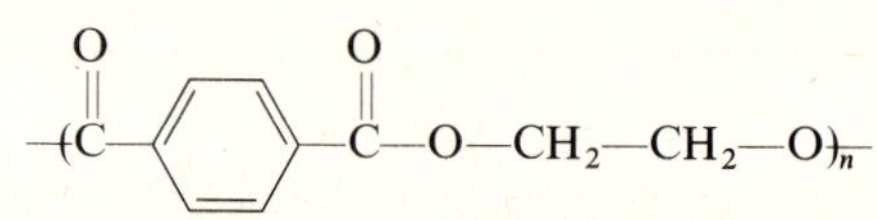

Dacron (What were the monomers?)

Polyamides, condensation polymers where the monomers are joined by amide linkage, are also possible. These are also called Nylons and are produced in a variety of structural variations. See Table 7.4.

7.13 Simple Lipids

Lipids are one of the largest classes of biologically significant molecules we encounter. Originally the name "lipid" meant simply everything that would extract from a biological sample with nonpolar solvents. This is still not a bad place to start, but we have become much more sophisticated about the categories and structural types of lipids encountered. Lipids are not usually water soluble.

Lipids we commonly refer to as fats or oils are ester structures. They all have the alcohol glycerol.

$$\begin{array}{l} CH_2{-}OH \\ | \\ CH{-}OH \\ | \\ CH_2{-}OH \end{array}$$

glycerol (1,2,3-propanetriol)

One two, or three carboxylic acids can be esterified (formed into an ester) with the alcohol groups.

Table 7.4
Some Common Condensation Polymers

Polymer	Monomers	Name
Polyamides		
$\left(NH-(CH_2)_5-\overset{O}{\overset{\|}{C}} \right)_n$	$NH_2-(CH_2)_5-\overset{O}{\overset{\|}{C}}-OH$	Polycaprolactam, Nylon-6, Perlon
$\left(NH-(CH_2)_6-NH-\overset{O}{\overset{\|}{C}}-(CH_2)_4-\overset{O}{\overset{\|}{C}} \right)_n$	$NH_2-(CH_2)_6-NH_2$ + $HOOC-(CH_2)_4-COOH$	Nylon-6,6
Polyesters		
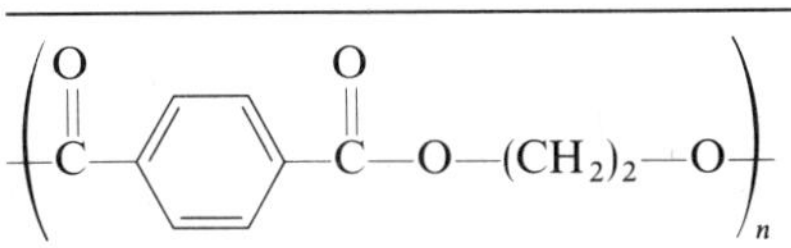	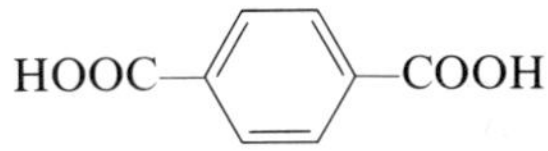 + $HO-CH_2-CH_2-OH$	Dacron, Terylene, Mylar
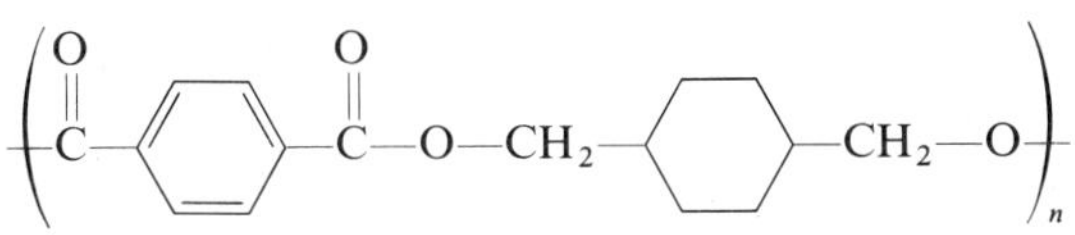	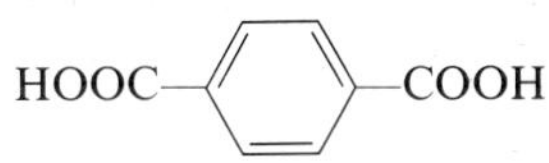 + 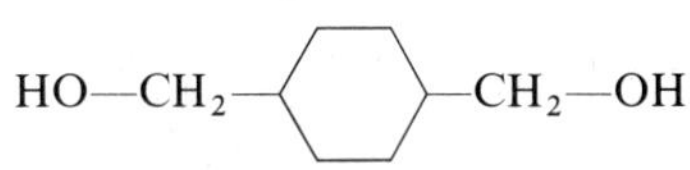	Kodel
Polycarbonate (similar to polyester)		
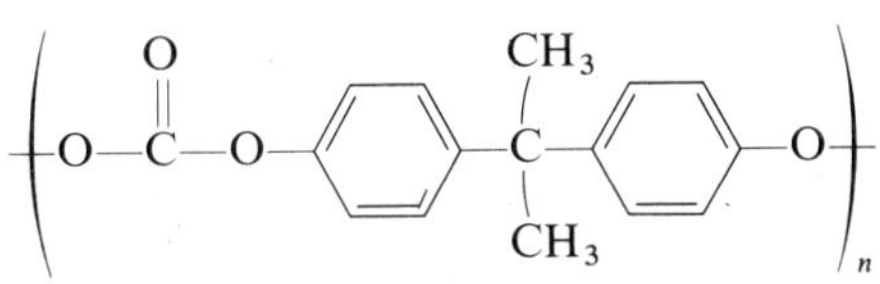	$HO-\overset{O}{\overset{\|}{C}}-OH$ + 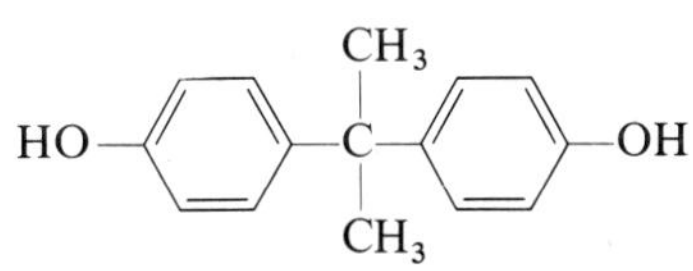	Lexan

Most fats and oils have all three alcohols in the form of ester structures, although we commonly find small quantities with one or two. Glycerol lipid structures in general are called *glycerides.* A *monoglyceride* has one acid esterified, a *diglyceride* two, and a *triglyceride* three. Representing the variable part of the molecule by —R, we have these structures:

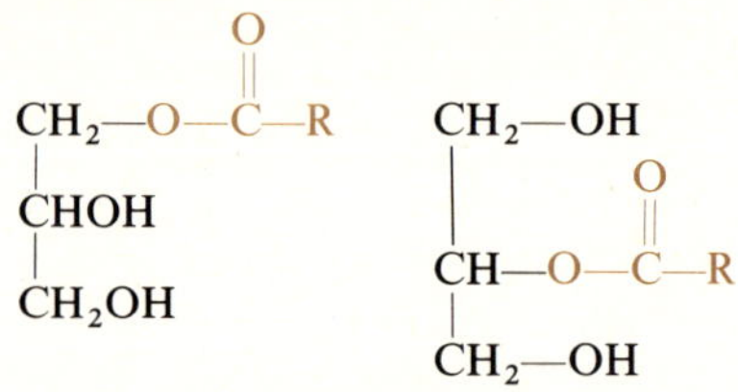

monoglycerides

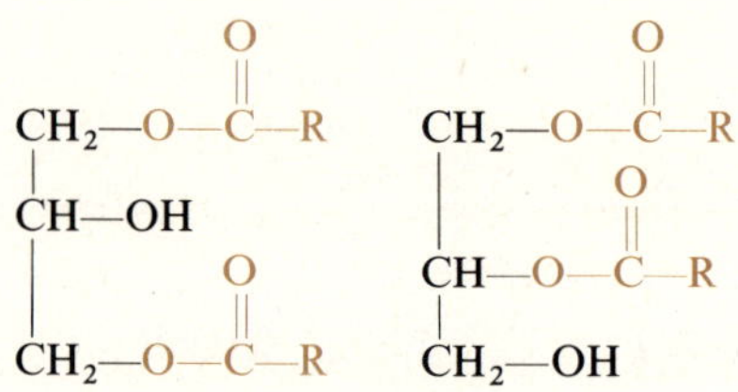

diglycerides

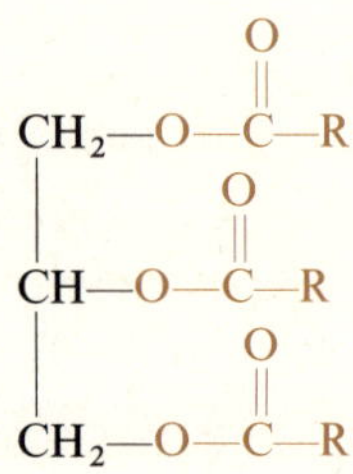

triglyceride

Exercise 7.8

Are there other possible arrangements for mono- or diglycerides?

The carboxylic acids of the glycerides are referred to as *fatty acids*. This does not mean they are something altogether new and different. They are a group of rather ordinary looking long-chain carboxylic acids. However, they do conform to a set of rules. Although every one of these rules is broken in some lipid, most fatty acids share these characteristics.

1. Fatty acids have rather long unbranched hydrocarbon chains.
2. Fatty acids have even numbers of carbons. A chain of 18 carbons is most common, with 16, 14, and 12 carbons also being encountered frequently.
3. Fatty acids may be saturated (no double bonds), unsaturated (one double bond), or polyunsaturated (more than one double bond).
4. The double bonds are cis.
5. Where more than one double bond is present, these bonds are not conjugated.

Table 7.5 shows the structures and names of the most common fatty acids. Do they fit the rules?

Table 7.5
Common Fatty Acids

	Name	*Formula*	*Melting Point (°C)*
Saturated	Stearic acid	$CH_3(CH_2)_{16}COOH$	70
	Palmitic acid	$CH_3(CH_2)_{14}COOH$	63
	Myristic acid	$CH_3(CH_2)_{12}COOH$	54
Unsaturated	Oleic acid	$CH_3(CH_2)_7—CH{=}CH—(CH_2)_7COOH$	16
Poly	Linoleic acid	$CH_3(CH_2)_4—CH{=}CH—CH_2—CH{=}CH—(CH_2)_7COOH$	−12
Unsaturated	Linolenic acid	$CH_3CH_2—CH{=}CH—CH_2—CH{=}CH—CH_2—CH{=}CH—(CH_2)_7—COOH$	

The terms *fat* and *oil* both refer to triglycerides from natural sources. Most samples of fat or oil from nature are not samples of a single molecular species but rather are mixtures of triglycerides with a variety of fatty acids present. Fats are generally solids, their melting point being above room temperature. Run through a list of fats in your mind—lard, tallow, butter—and you realize that they generally come from an animal source. Most oils are liquids at room temperature, and a list of such oils—peanut, corn, soya, linseed, cottonseed—reveals that they are usually derived from plants.

Fats and oils function primarily for food storage. In animals, body fat also serves as an insulation against the cold and as a cushion for some internal organs. Seed oils are a more compact food source than starch (Chapter 6). We extract oil for food from seeds in large quantities by pressing it out (Figure 7.5).

What is the structural difference between fats and oils that leads to the difference in melting point? For the most part the difference lies in the number of double bonds in the fatty acids. Oils have more double bonds than fats do. Notice in Table 7.5 how sharply the melting point drops as the number of double bonds increases. Our taste in foods sometimes dictates the use of solid triglycerides rather than liquids in our cooking. Traditionally, lard or butter served such functions, but oils pressed from plant seeds are less expensive. So that we may enjoy these cheaper products, unsaturated oils are reacted with hydrogen using a catalyst (Chapter 3). The oils are "hardened" into solids for use as margarine or vegetable shortenings by lowering the number of double bonds.

7.14 Rancidity of Fats and Oils

When fats or oils are to be used as food, care must be taken with regard to both the manner and the length of time of storage. Two classes of chemical problems plague fats and oils. The first of these, called *hydrolytic rancidity*, is the breakdown of the triglycerides by the hydrolysis of ester linkages to produce free fatty acids. This is particularly obnoxious when the triglyceride contains short-chain fatty acids, as in the case with butter or coconut oil. These short-chain (therefore low-molecular-weight) fatty acids are more volatile and have the undesirable rancid smell. Comparing samples of butanoic and stearic acids emphasizes the comparative unpleasantness of the short-chain compound. The hydrolysis of the ester is usually catalyzed by an enzyme called *lipase*. Lipase can be present as a result of the processing of the oil-bearing seed, because of mixing with other food components such as milk or milk powder, or as a consequence of the presence of bacteria. Water must also be present to perform the hydrolysis. Keeping triglycerides dry or adding a drying agent (0.1 percent lecithin is sometimes used) retards hydrolytic spoilage.

The second class of chemical problems of fats and oils is

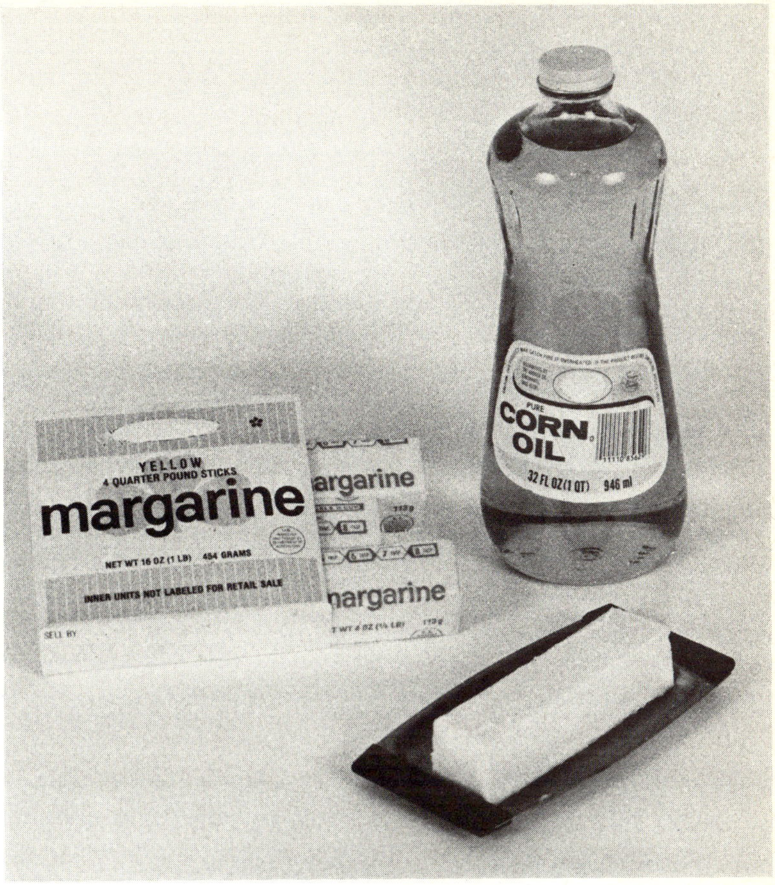

Figure 7.5
Commercial fats and oils
Fats and oils are mixtures of triglycerides extracted from a variety of plant and animal sources. (Photo by Catherine Feryok, Eastern Michigan University.)

termed *oxidative rancidity*. Oxygen reacts in the vicinity of double bonds in the fatty acid to produce aldehydes, peroxides, ketones, short-chain carboxylic acids, and oxyacids (acids with ketone groups nearby). The reaction is catalyzed by light or by metals. The products can be poisonous. Vitamin E (Table 3.3), which is often a natural component of a seed oil preparation, reacts with oxygen to remove it from the oil. This provides some protection against oxidative rancidity. The problem is further reduced by storing oils in tightly closed containers to exclude oxygen. The containers should be nonmetallic and tinted to reduce the likelihood of catalysis of the reaction.

7.15 Waxes

Waxes are products of a broad variety of living creatures and are composed primarily, if not entirely, of esters. They are usually inert, water-insoluble, soft solids. As the term *wax* is used in an everyday sense, it refers to any compound with waxlike properties. This would include, for example, very long-chain hydrocarbons from petroleum (paraffin wax) and such synthetic polymers as silicones used to perform the traditional commercial functions of natural waxes. Natural waxes have been collected and used commercially for a long time for candles, as agents to provide a smooth waterproof finish or coating, and for carving and modeling objects in an artistic sense. In early times the practice of using wax to hide flaws in pottery or ceramics generated the word "sincere" (*sine*, without, and *cera*, wax) to describe a sound, genuine item. Today waxes are also used in cosmetics and as a base for medicinal ointments.

Figure 7.6
Waxes
Large quantities of wax are used in candles and in polishes. (Photo by Catherine Feryok, Eastern Michigan University.)

In plants and animals the waxes often serve as inert waterproof coatings; they retard the evaporation of water from land creatures. Plant leaves and fruit can often be polished to give a lustrous shine because of the waxes present. The comb of honey bees is an example of structural modeling in wax. Some aquatic mammals use waxes as a food reserve in place of triglycerides. The blubber (body fat) of the sperm whale is composed primarily of wax-type esters.

Typical wax esters are composed of fatty acids of the same character as those in triglycerides. However, the chain length is likely to be longer, extending to over 30 carbons rather than halting at the 18 carbons typically encountered in fats and oils. The short-chain fatty acids found in some fats are generally missing from waxes. The alcohol of the wax ester is most often a very long-chain compound with a single alcohol located at the end of the chain. Structurally, it closely resembles the fatty acid.

Commercial waxes come from a variety of sources (Figure 7.6). Beeswax is the wax most often used in candles, cosmetics, and medicinals. It is actually only about 70 percent wax esters. Free acids and simple hydrocarbons are also important components (10–15 percent each). Lanolin or wool wax is extracted from wool. Besides regular wax esters it has esters of sterols, more complex ring-structure alcohols (Chapter 8). Lanolin, too, is used in ointments and cosmetic preparations. A high-quality wax for floor and furniture polish is obtained from leaves of the carnauba tree in Brazil. Much of the commercial drive for whaling in earlier times was the market for candle wax. The spermacete from the sperm whale is about 90 percent cetyl palmitate.

$$C_{15}H_{31}\overset{\overset{\displaystyle O}{\|}}{C}{-}O{-}C_{16}H_{33}$$

cetyl palmitate

7.16 *Soaps and Detergents*

Water-insoluble (nonpolar) materials present special problems in the cleansing of clothing, walls, floors, and our skin. Besides esthetic considerations, removal of such impurities fulfills a hygienic function, because harmful bacteria are removed at the same time. Nonpolar compounds can be removed by nonpolar solvents, which is exactly what we do when we dry-clean fabrics. However, the most common

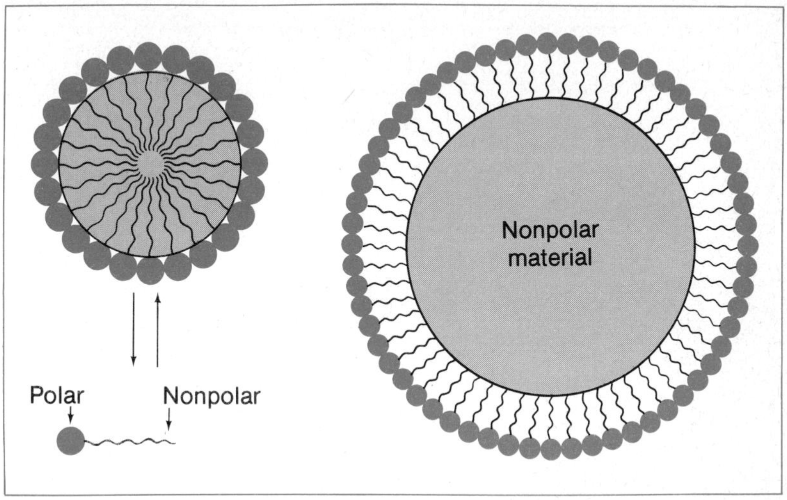

Figure 7.7
Micelles
Droplets of nonpolar material can be suspended in water by coating the surface of the droplet with a surface active agent.

solvent available to us is water, and it is possible to suspend nonpolar compounds in water by use of *surface-active agents.*

In order to suspend nonpolar materials in water, it is necessary to form *micelles.* A micelle is a body of nonpolar material coated with a surface-active agent. Soaps and detergents are the most common examples of surface-active agents. Chemically, surface-active agents act as a bridge between the nonpolar compounds in the body of the micelle and the very polar solvent water. A large portion of the molecule is nonpolar, allowing it to dissolve in the nonpolar substance. At least one structure on the molecule is very polar, usually ionic. Because the polar portion does not dissolve in the body of the micelle, it remains at the surface. If enough surface-active agent is present, the micelle becomes completely coated with the polar groups. This allows water to interact with the surface of the particle so that it may be suspended in a stable fashion in the water. The smaller the particle, the more likely stable suspension becomes. In soap the polar group is ionic, carrying a negative charge. When two micelles are coated with a negative charge in this fashion, they repel each other electrostatically. As a result, they are very unlikely to collide and merge (coalesce) (Figure 7.7).

If we mix oil and water, the oil forms a separate distinct layer on top of the water. Even after we shake the mixture to break the oil into droplets, the droplets coalesce into a separate layer again. If we add increasing amounts of a surface-active agent, the coalescing of the layers goes more slowly, until at some point the oil remains dispersed in the water. Such a stabilized mixture is called an *emulsion.*[1] Adding

[1] An example of an emulsion is homogenized milk. The fat is distributed throughout the milk in fine droplets. The surface-active agent is probably largely milk proteins.

more surface-active agent increases the stability and uniformity of the emulsion. This is because the micelles become smaller and less likely to coalesce. With enough surface-active agent the micelles reach an optimum and roughly uniform small size.

Animal fats were an early source of surface-active agents. By extracting wood ash, potassium hydroxide is obtained. When fats are boiled with potassium hydroxide, the esters are hydrolyzed or saponified to release glycerol and salts of the fatty acids.

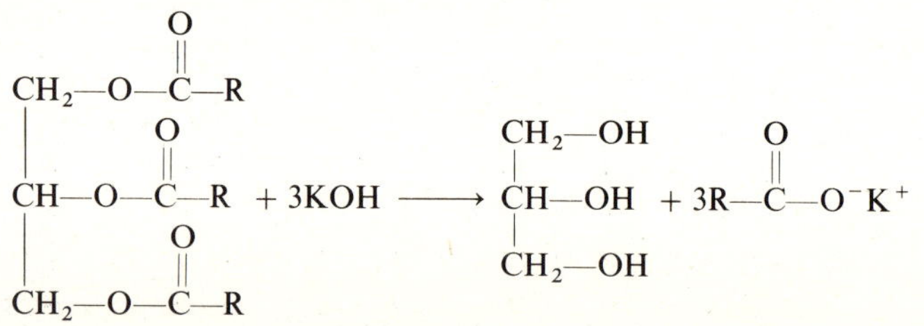

The fatty acid salts are *soap*. Today sodium hydroxide is more commonly used than potassium hydroxide but the process is the same. Notice how nicely these salts fit our requirements for a surface-active agent. The long hydrocarbon chain can associate with the nonpolar particle, leaving the ionized carboxyl group at the surface to interact with the water. Soaps have advantages due to their simple natural origin. They are unlikely to produce toxic or allergic response. However, they also have disadvantages. They form insoluble salts with a number of divalent metal ions. These include Ca^{2+} and Mg^{2+}, which are common constituents of tap water, especially where wells are the water source. These insoluble salts result in the ring around the bathtub and can deposit in clothing to dull the fabrics. Once in the form of insoluble salts, the soap is no longer able to emulsify nonpolar material. Additional soap must then be added until there is an excess of soap over divalent ions.

A series of compounds were synthesized that provide a combination of polar and nonpolar structure like that of soaps, but which do not precipitate in the presence of the divalent ions found in water. These synthetic detergents (syndets) have largely replaced soap for laundry purposes. They are not without their own problems, however. The compounds first employed were effective surface-active agents but created difficulties once they were emptied down the drain. Where waste soaps are degraded by bacteria in the sewage treatment process, these early synthetic detergents are not. There followed a nightmare of rivers with long-lasting islands of suds. The key was to change the detergent formula to one the bacteria would attack, a so-called biodegradable formula.

7.17 Complex Lipids

The lipids found in living organisms include compounds more complex than the triglycerides. We shall consider two groups of these, the phosphoglycerides and the sphingolipids.

Phosphoglycerides are derivatives of phosphatidic acid. Phosphatidic acid is a triglyceride derivative in which one of the fatty acid ester structures has been replaced by a phosphate ester structure. Phosphoric acid has three removable hydrogens, only one of which is used to form the ester linkage. Of the others, one or both are ionized at neutral pH. Phosphoglycerides are therefore ionic molecules.

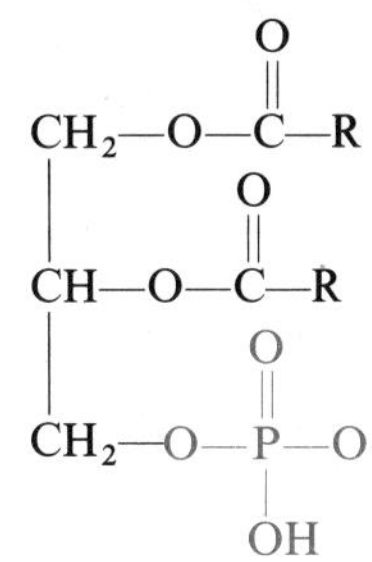

phosphatidic acid

Phosphatidic acid itself is found in nature but is not a phosphoglyceride of major importance. The variety of derivatives of it where the second hydrogen on the phosphate is also esterified are commonplace, however. Most common are phosphatidyl ethanolamine, -serine, -choline, and -inositol.

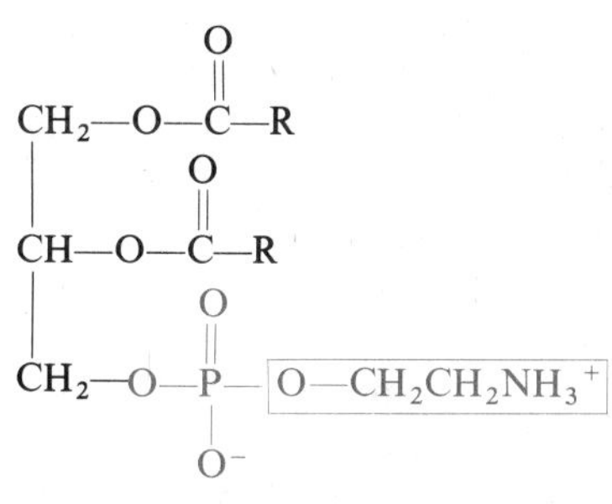

phosphatidyl ethanolamine

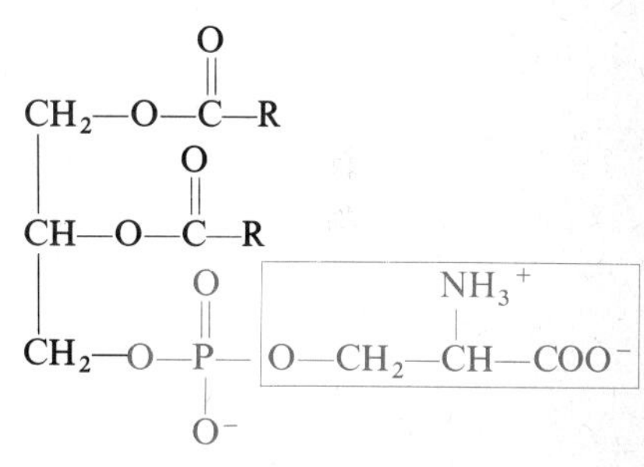

phosphatidyl serine

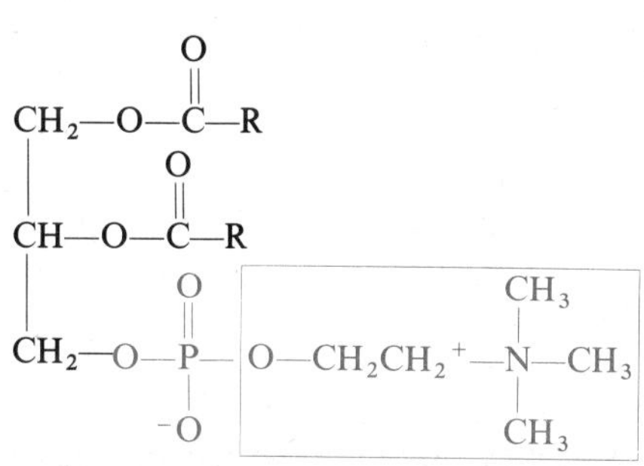

phosphatidyl choline (lecithin)

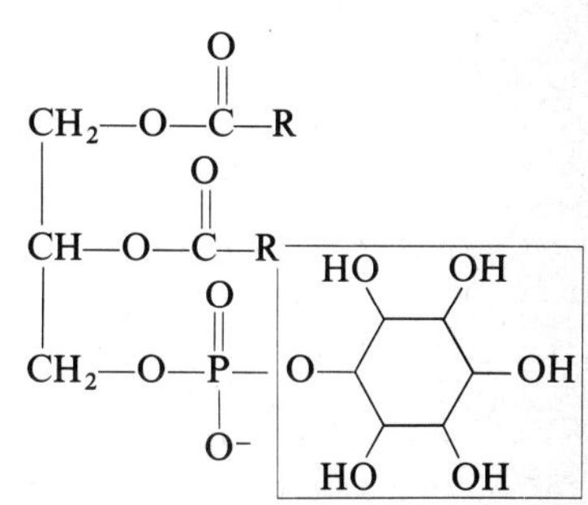

phosphatidyl inositol

7.18 Phosphoglycerides as Surface-Active Agents

Notice that these molecules meet the structural requirements to be detergents. The fatty acids have long hydrocarbon groups to dissolve into nonpolar material. Even the ester structures do not interfere with the low polarity of the alkyl groups. Without a polar —O—H group, ester functional groups are actually rather nonpolar themselves. At the other end is the charged phosphate group and polar or charged substitutions on the phosphate. The biological functions of these molecules often relate to their ability to serve as surface-active agents.

These compounds are utilized outside of living systems as detergents. For example, egg yolk is a rich source of phospholipids. Adding egg to a cake batter allows the nonpolar shortening to be suspended together with flour and milk proteins and carbohydrates in a water-solvent system. Phosphoglycerides, especially lecithin, are extracted commercially from sources such as soybeans and are used to stabilize emulsions in cosmetics and food preparations. Notice how often you see lecithin listed as an additive in a processed food.

In the living system, micelles are the form in which triglycerides are transported by the circulatory system. Phosphoglycerides are important as surface-active agents to stabilize these micelles.

7.19 Membranes

Membranes are the most interesting and important phosphoglyceride structures. Cells, particularly those of higher plants and animals, have a variety of membranes that serve to separate the cell interior from the outside environment and that subdivide the cell interior into functional compartments. The fundamental structure of these membranes is a bilayer of phosphoglyceride molecules where the nonpolar regions mingle with one another in the center of the membrane, and the polar groups align themselves at the surfaces of the membrane facing the water solutions of the cell interior or exterior. Protein molecules are also part of this structure and are concerned with the selective transporting of materials across the membrane (Figure 7.8).

7.20 Sphingolipids

Sphingolipids all include the compound sphingosine.

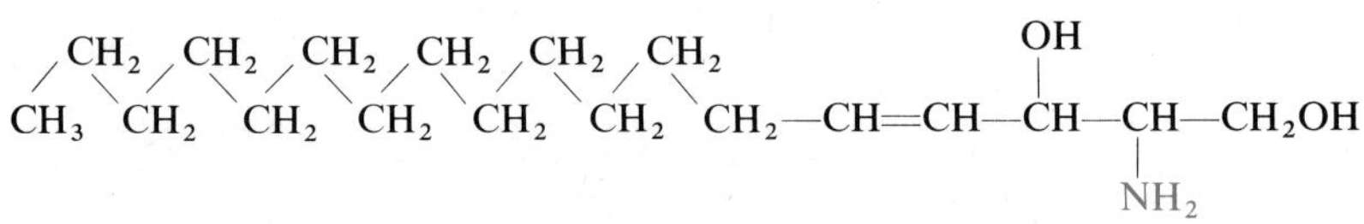

sphingosine

The sphingolipids have a fatty acid attached by an amide linkage at the amine group of sphingosine. Compounds with this as the only substitution are called *ceramides*. They are the simplest sphingolipids. The more complex sphingolipids have a variety of groups attached to the terminal alcohol group of sphingosine.

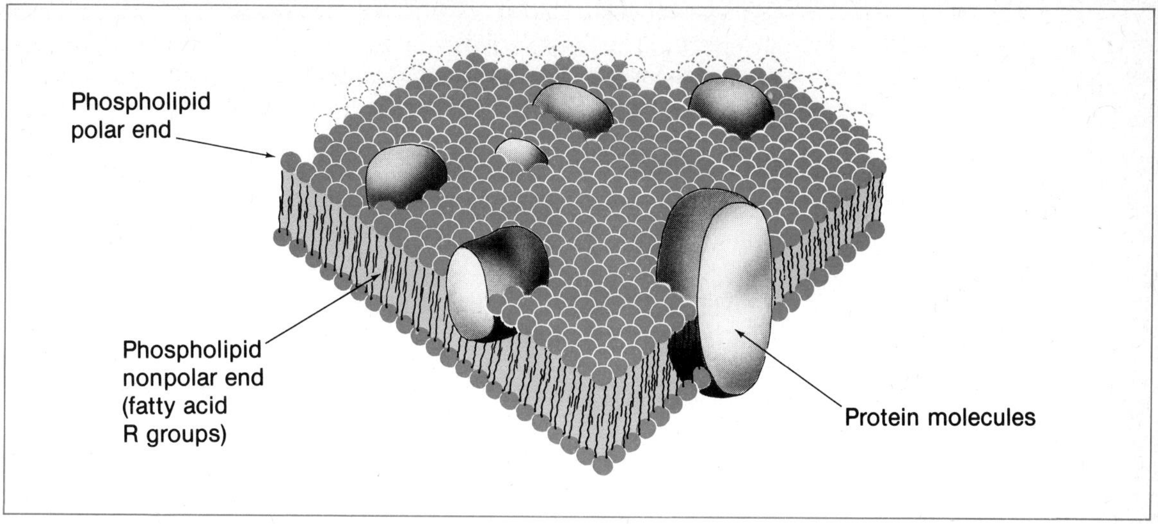

Figure 7.8
Fluid mosaic model of the cell membrane
The most likely model of membrane structure pictures phospholipids oriented with their fatty acid R groups at the center and the phosphate esters at the surface. Proteins are embedded throughout the membrane. (From S. J. Singer and G. L. Nicolson, *Science* 175 (1972): 723. Copyright 1972 by the American Association for the Advancement of Science.)

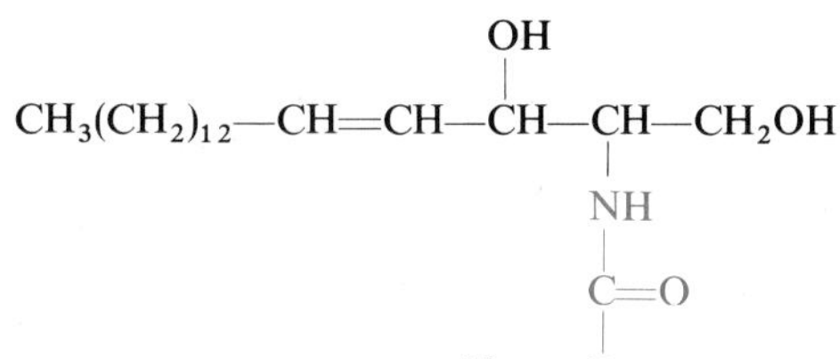

a ceramide

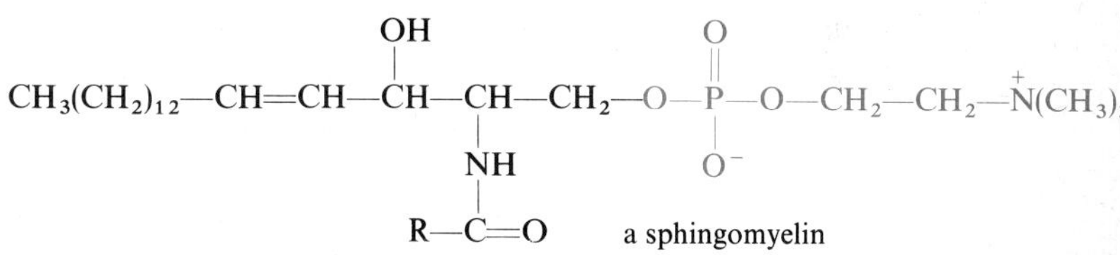

a sphingomyelin

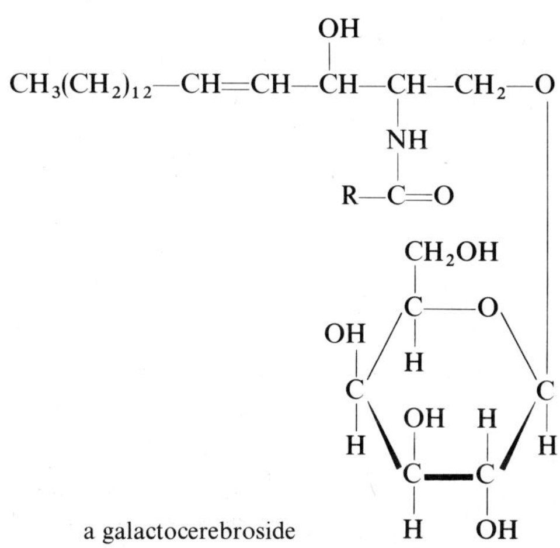

a galactocerebroside

Like the phosphoglycerides, sphingolipids are very widely distributed in cells. Both groups are found in highest concentration in brain and nervous tissue. Exact functions for these compounds have not yet been established.

7.21 *Prostaglandins*

Some fatty acids, arachadonic and linoleic acids specifically, must be present in the diet to maintain good health. They are sometimes referred to as the *vitamin F* compounds. These essential fatty acids are used to synthesize the prostaglandins. As the name infers, prostaglandins were first observed as a product of the prostate gland but have since shown to be the product of many tissues and to be very widely distributed. Chemically, there are several prostaglandin variations, but all are based on the 20-carbon prostanoic acid. They are all designated by PG, followed by A, B, E, or F and a subscript.

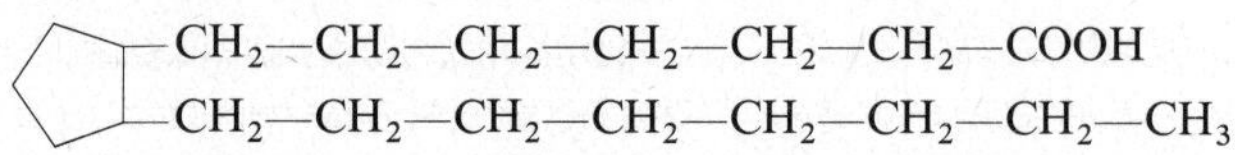

prostanoic acid

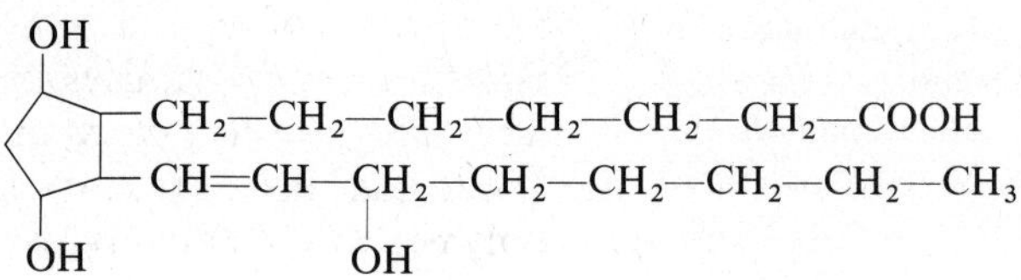

prostaglandin $F_{1\alpha}$($PGF_{1\alpha}$)

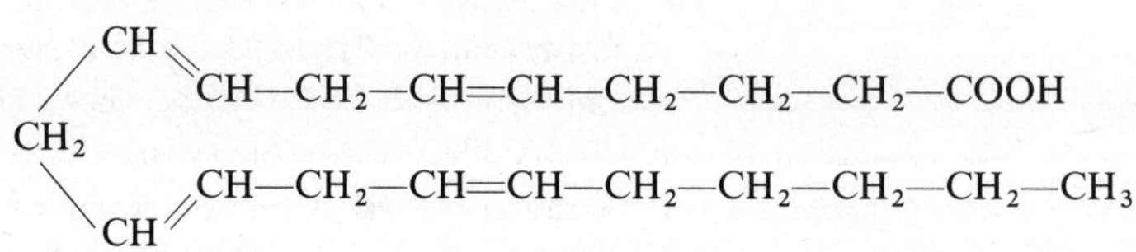

arachadonic acid (prostaglandin precursor)

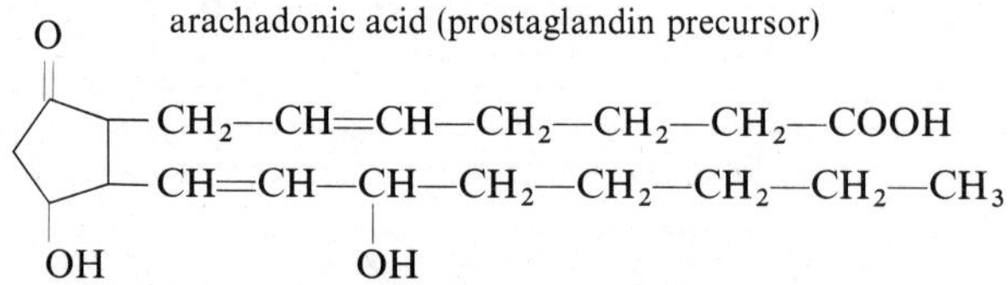

prostaglandin E_2 (PGE_2)

The prostaglandins are not truly hormones, because they do not arise in a single specialized gland. They are sometimes classed with hormones, however, because of their hormonelike effects. They affect the levels of a regulatory substance, cyclic AMP (Chapter 10), and stimulate the release of some hormones, the catecholamines (Chapter 8). More broadly, prostaglandins induce the blood vessels to dilate, increasing blood flow and decreasing blood pressure. They cause membranes to pass water and salts more readily, increasing urinary volume. Prostaglandins sedate and tranquilize, and they reduce gastric secretion and mobility.

Aspirin interferes with the synthesis of prostaglandins. As prostaglandin levels are lowered, the ability of membranes to pass fluids is reduced. Inflammation is caused by the movement of fluids into tissues. Hence, the reduction of inflammation caused by aspirin is the result of lowered prostaglandin levels.

Summary

The carboxyl functional group has a weakly bonded hydrogen that is readily removed by strong bases and is at least partly ionized in water solution. The reaction of carboxylic acids with bases produces salts. Because of the polar character of the un-ionized carboxyl group and the charge of the salt, both acids and salts tend toward higher melting points or boiling points and are relatively water soluble. The pH at the midpoint of the titration of a carboxylic acid is the pK, an important constant. Carboxylic acids condense with alcohols to form esters, with amines to form amides, and with one another to form acid anhydrides. Substitution of the carboxyl —OH with —Cl produces an acid chloride. Specific esters and amides are important as medicines, and esters make a contribution to natural and artificial flavors. Either ester or amide bonds can link monomers into the long chains called condensation polymers. Acid anhydrides and acid chlorides are used primarily as reactive intermediates in the synthesis of esters and amides. In simple lipids the triol glycerol is esterified by carboxylic acids of similar structural characteristics called fatty acids. Salts of fatty acids are soaps and are models for detergents that can stabilize suspensions of nonpolar material in water. Waxes are esters of long-chain alcohols and fatty acids. Phosphoglycerides are simple lipids in which a fatty acid has been replaced by substituted phosphoric acid. These also serve as detergents and are the chief structural components of cell membranes. Sphingolipids include the compound sphingosine to which a fatty acid attaches by an amide linkage, and which is possibly substituted further by carbohydrates or esterified phosphoric acids. Prostaglandins are hormonelike control substances synthesized from certain fatty acids.

Problems

1. Name the following compounds:

a) $\mathrm{H{-}C({=}O){-}OH}$

b) $\mathrm{CH_3{-}CH(CH_3){-}COOH}$

c) $\mathrm{CH_3{-}CCl_2{-}CH_2{-}COOH}$

d) $\mathrm{F{-}CF_2{-}COOH}$

e) 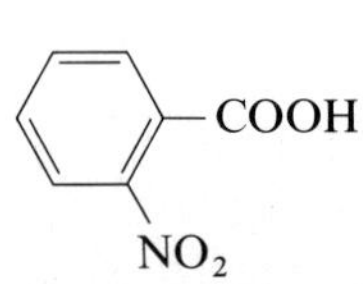

f) 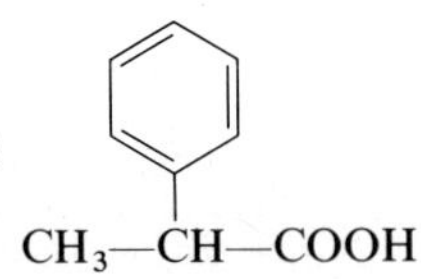

g) 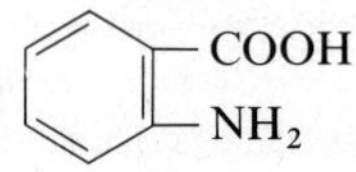

h) 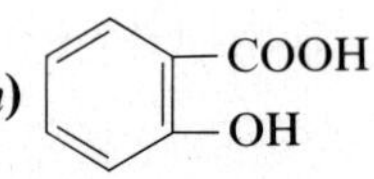

i) 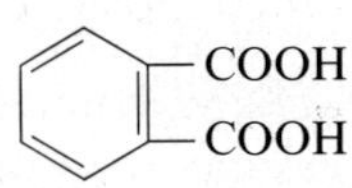

j) $\mathrm{CH_3CH_2CH_2CH_2{-}C({=}O){-}O^-Na^+}$

k) 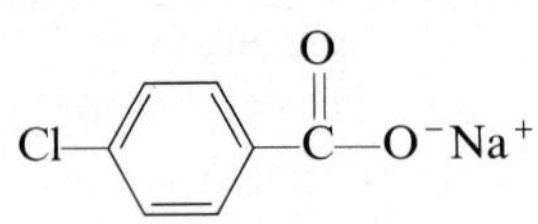

l) 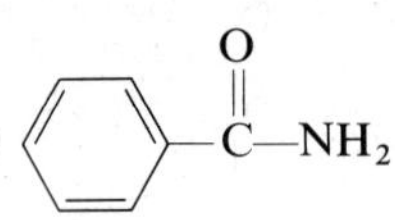

m) $\mathrm{CH_3{-}CH_2{-}C({=}O){-}O{-}C({=}O){-}CH_2{-}CH_3}$

n) $\mathrm{CH_3{-}C({=}O){-}O{-}CH(CH_3){-}CH_3}$

2. Draw structures for the following:

a) chloroethanoic acid

b) chloroacetic acid

c) acetyl chloride

d) potassium benzoate

e) 3-ethylhexanoic acid

f) methyl anthranilate

g) *p*-nitrobenzamide

h) sodium stearate

i) methyl oleate

j) a triglyceride of palmitic acid

k) phosphatidic acid with two oleic acids

l) a wax where both hydrocarbon portions have 20 carbons

***m*) phosphatidyl choline with one stearic acid and one linoleic acid

***n*) a ceramide with myristic acid ***o*) galactose-6-phosphate

***p*) fructose-1-phosphate

3. Complete the following reactions:

a) 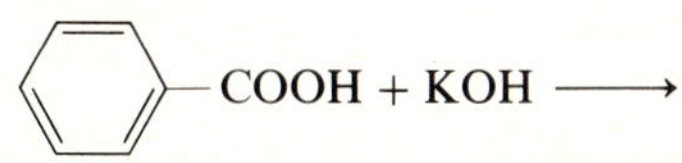

b) $CH_3—CH_2—COOH + CH_3—CH_2OH \xrightarrow{H^+}$

c) 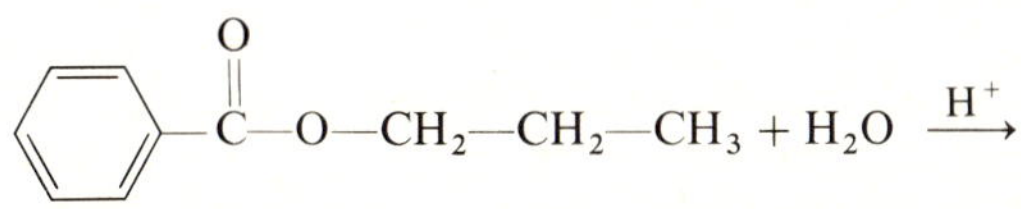

d) $CH_3—CH_2—\overset{O}{\overset{\|}{C}}—OH + NH_3 \xrightarrow{\Delta}$

e) 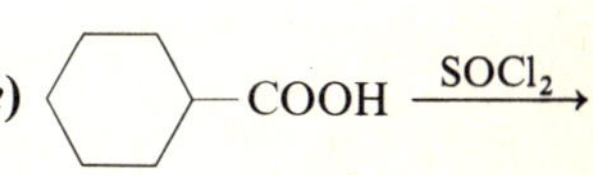

f) $CH_3—\overset{O}{\overset{\|}{C}}—Cl + CH_3—\overset{OH}{\overset{|}{C}H}—CH_3 \longrightarrow$

g) $CH_3—CH_2—\overset{O}{\overset{\|}{C}}—O—\overset{O}{\overset{\|}{C}}—CH_2—CH_3 + CH_3OH \longrightarrow$

h) 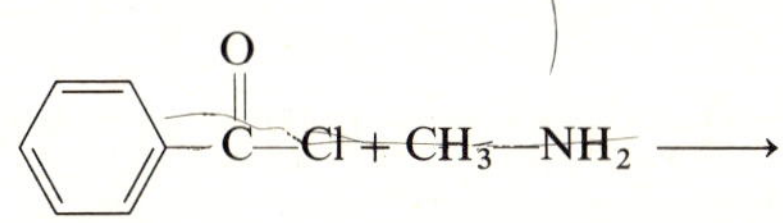

i) $Cl—\overset{O}{\overset{\|}{C}}—CH_2—CH_2—\overset{O}{\overset{\|}{C}}—Cl + \overset{OH}{\overset{|}{C}H_2}—\overset{OH}{\overset{|}{C}H_2} \longrightarrow$

j) $H—\overset{O}{\overset{\|}{C}}—O—CH_2—CH_3 + H_2O \xrightarrow{NaOH}$

k) 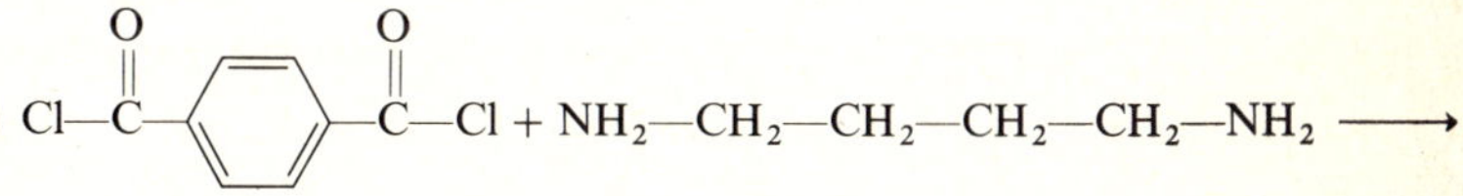

4. Look at the structure below. In what class of those discussed in this chapter does it belong? What would you suppose is the starting material to make this structure? Would you expect the structure to be more or less reactive than the starting material? Try to name the structure.

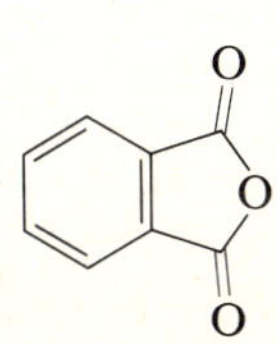

5. To gain some practice in recognizing lipid structures, match the following names with the structures below:

_____ ***a*) diglyceride**

_____ ***b*) phosphatidyl inositol**

_____ ***c*) sphingosine**

_____ ***d*) triglyceride**

_____ ***e*) ceramide**

_____ ***f*) phosphatidyl choline**

_____ ***g*) monoglyceride**

***h*) phosphatidyl serine**

_____ ***i*) prostaglandin**

_____ ***j*) cerebroside**

_____ ***k*) sphingomyelin**

***A*)**

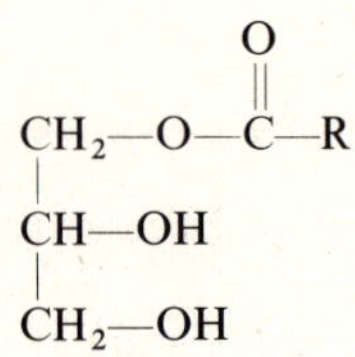

***B*)**

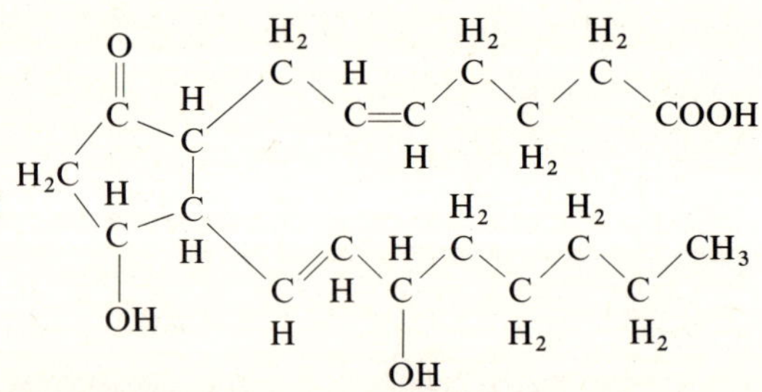

***C*)**

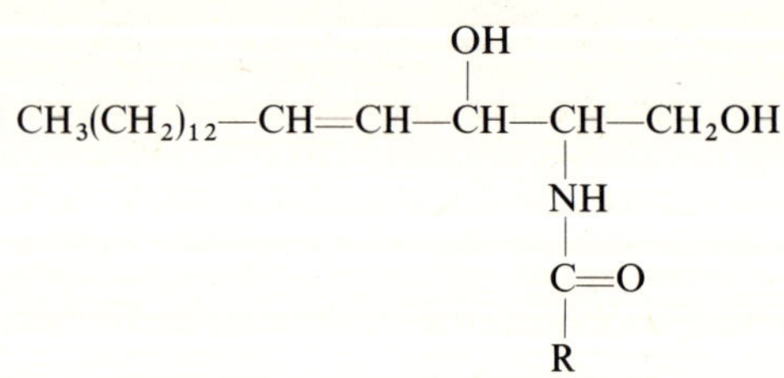

***D*)**

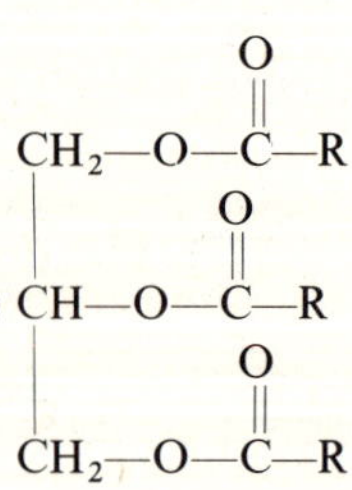

***E*)**

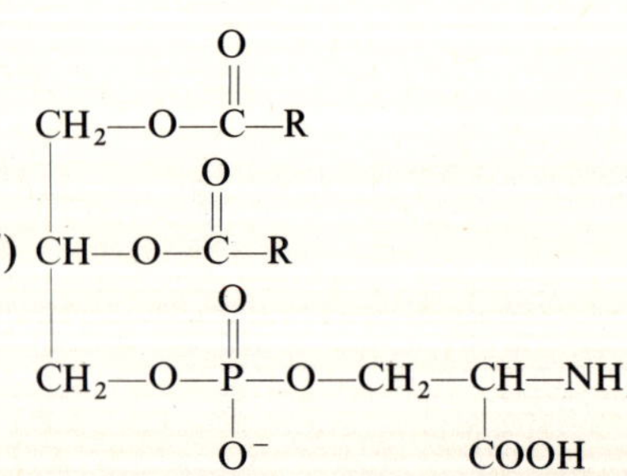

F)
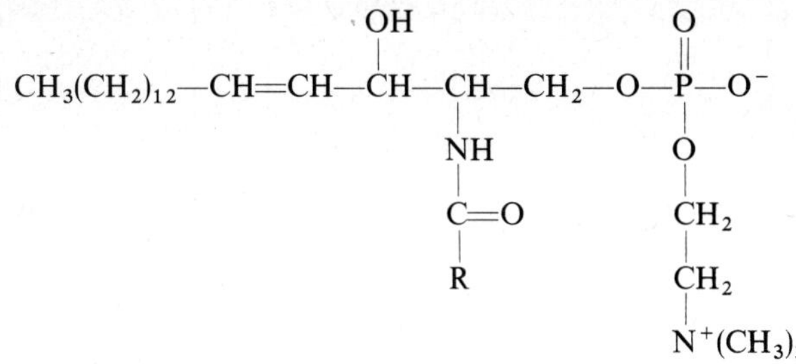

G)
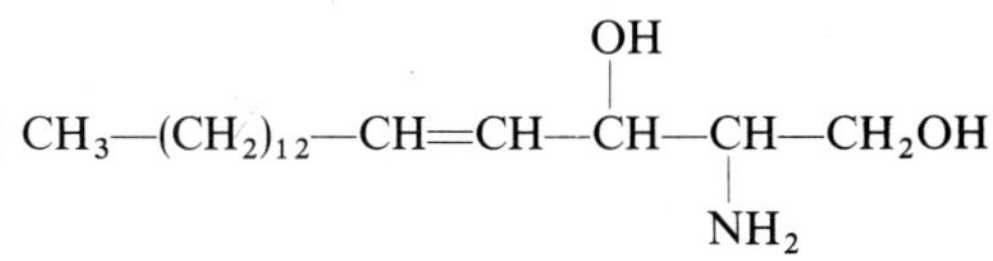

H)
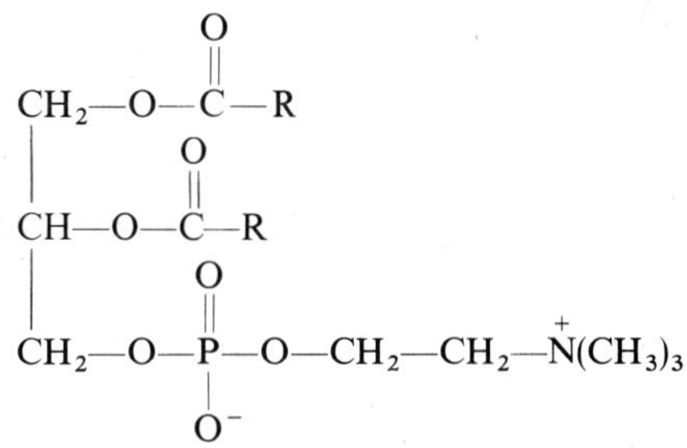

I)
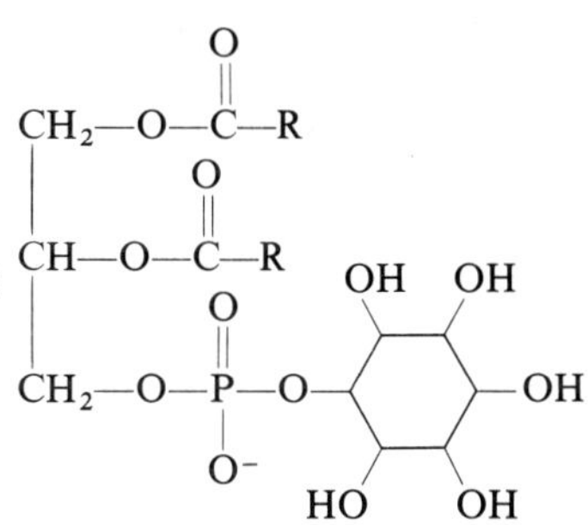

J)
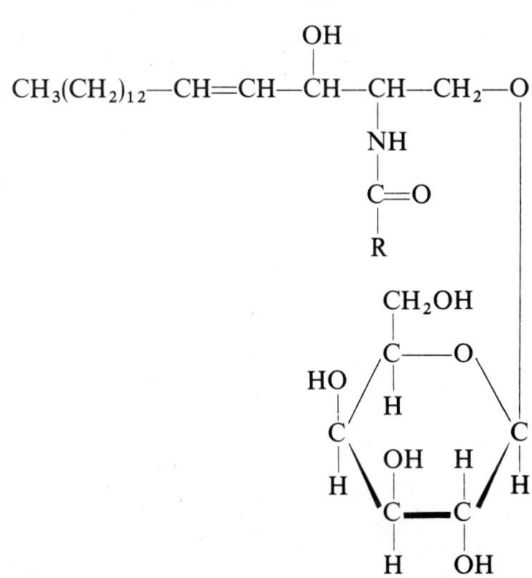

K) 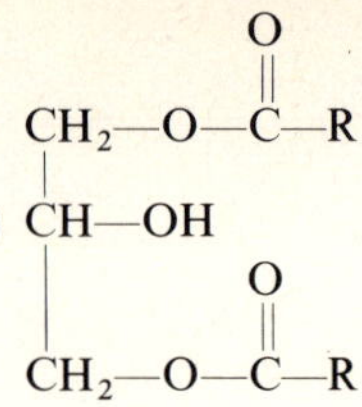

6. Which of the following can form condensation polymers, which may have difficulty with the formation of rings, and which cannot form a polymer?

a) 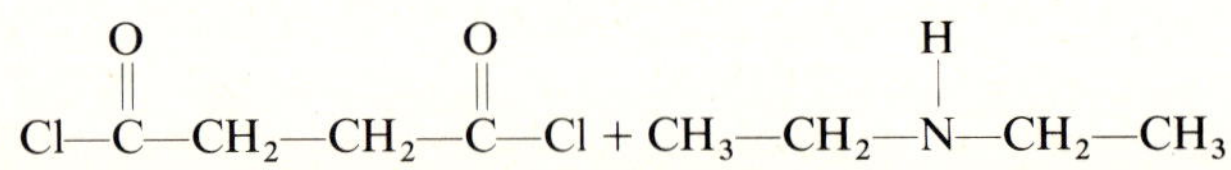

b) 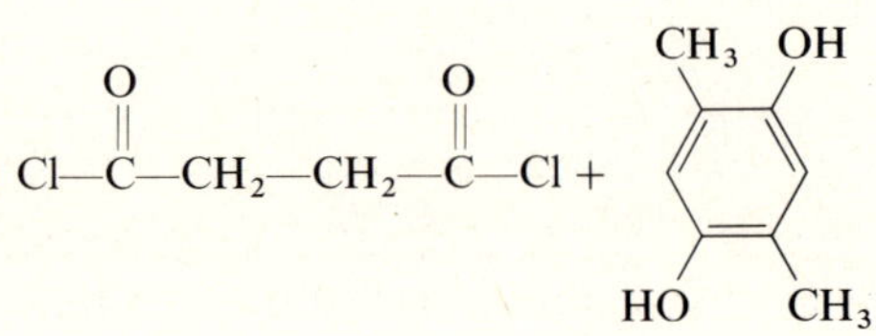

c) 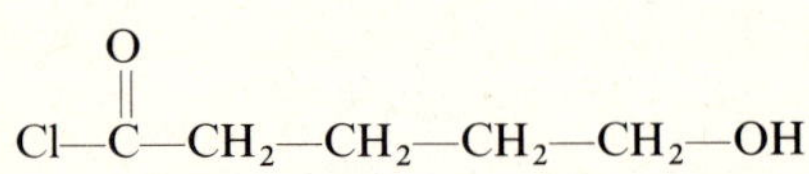

d) 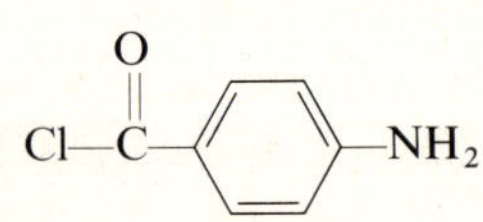

e) 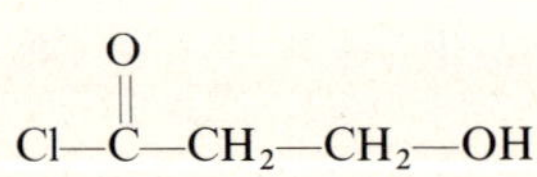

7. Using these starting materials to form a condensation polymer,

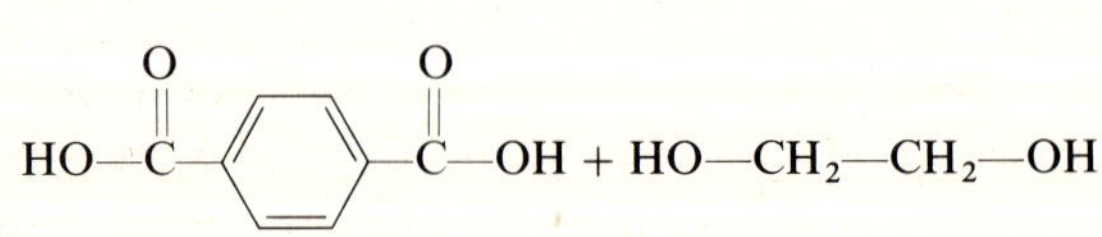

you would encounter difficulty because as the reaction proceeds, water is a product. The water is trapped in the polymer and can adversely affect its properties. Water requires extreme conditions, including high temperatures, for its removal. How might this problem be avoided? What is the comparable situation with addition polymers (Chapter 3)?

8. Comment on the desirability of each of the following for use as a surface-active agent:

a) 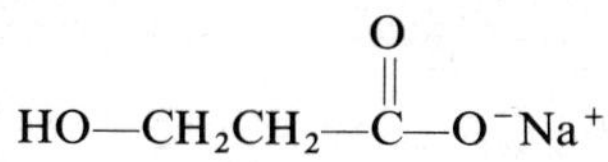

b) 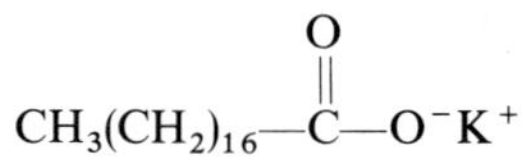

c) 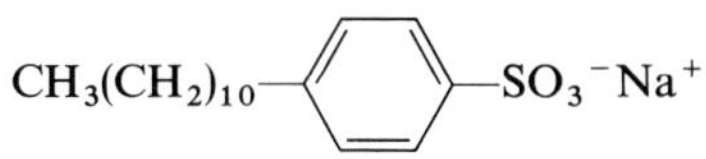

d) 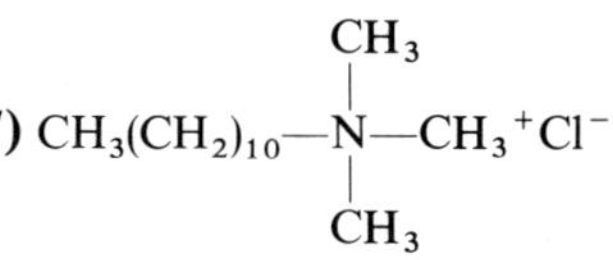

e) 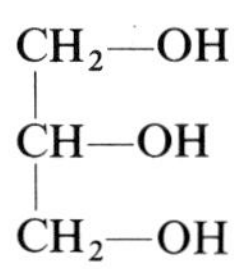

9. If you had a set of triglycerides with the following fatty acid composition, arrange them in what you would expect to be the order of their melting points:

a) Three stearic acids

b) Three oleic acids

c) One linolenic acid, two stearic acids

d) Two oleic acids, one linolenic acid

10. You are in the business of manufacturing margarine from corn oil. How might you deal with the problem that in summer the margarine melts at room temperature whereas in winter it is so hard that it tears the bread in spreading?

11. You are asked by your boss, the director of the labs of Fine Flavor Enhancers, Inc., to prepare a new ester. In order to cause the reaction to occur rapidly enough, you raise the temperature to a level that causes decomposition of the reactants. After that they do not taste very good. How can you avoid this problem?

12. In the following lipid structures, locate all asymmetric carbons:

a) 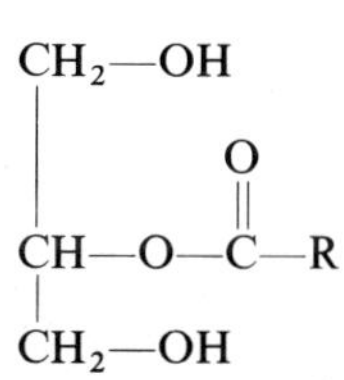

b) 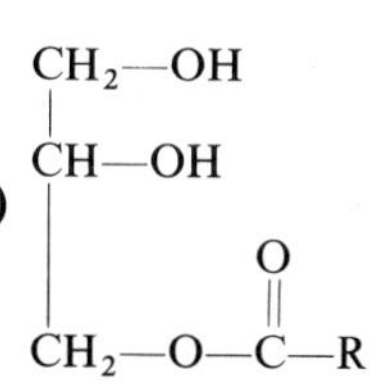

c) 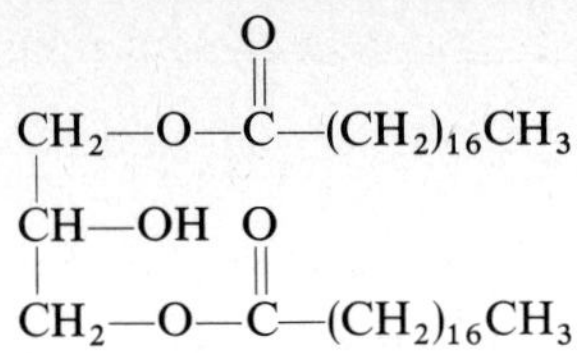

d) 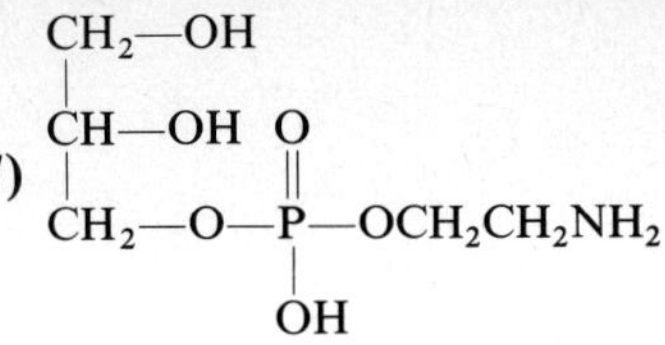

e) 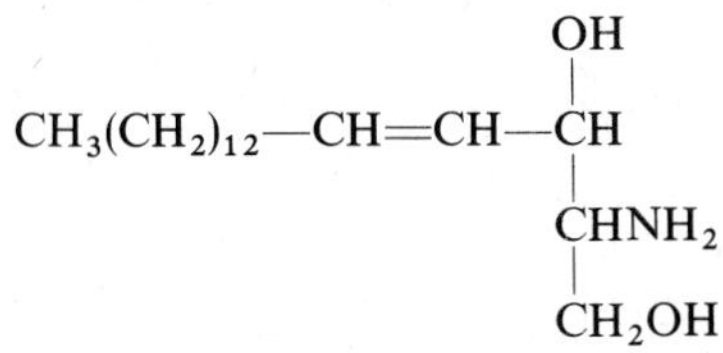

13. Arrange in order of water solubility:

a) 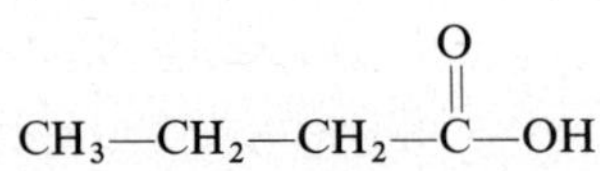

b) 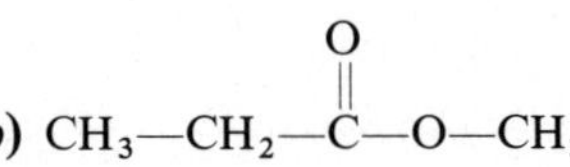

c) $CH_3{-}CH_2{-}CH_2{-}CH_2{-}OH$

Challenge Problems

14. a) The strength of a carboxylic acid depends on its willingness to release the hydrogen. The more willing the acid is, the (stronger, weaker) it is as an acid.

b) Suppose you added side groups to the acid and tested the effect on acid strength. You would observe a change in p*K* as the acid changed strength. The stronger the acid became, the (higher, lower) the p*K* would be.

c) Starting with propanoic acid, which would you expect to increase the acid strength—a positive, or electron-attracting, group or a negative, or electron-donating, group?

$CH_3{-}CH_2{-}COOH$ (propanoic acid) $\quad CH_3{-}\overset{+}{C}H{-}COOH \quad CH_3{-}\overset{-}{C}H{-}COOH$

In answering this question, consider the shift of electrons all the way to the hydrogen.

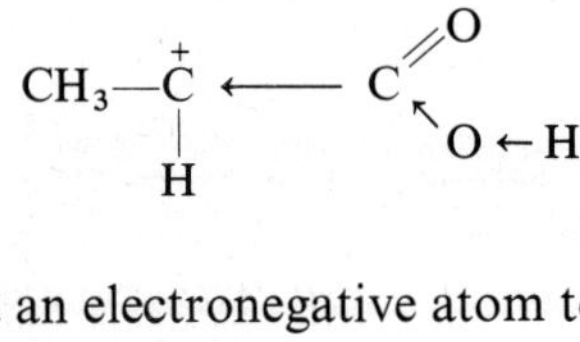

d) Would you expect an electronegative atom to be electron-donating or electron-attracting?

e) Indicate the order of acid strength in the following series:

CH_3COOH $\quad$ $Cl{-}CH_2COOH$

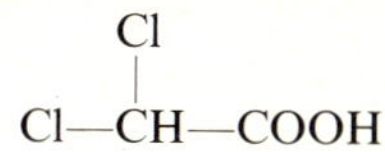

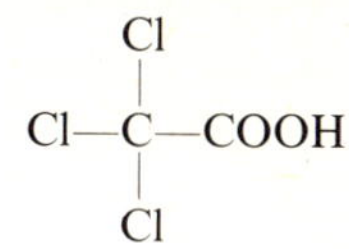

15. Refer to Table 7.3.

a) Write the structures of the acids and alcohols that result from the acid hydrolysis of the flavor agent esters in 1, 2, 3. and 4.

b) What is unusual about 13, 14, and 15? Such structures are termed *lactones.* Notice that 13 and 14 are called δ (delta) lactones, while 15 is called γ (gamma) lactone. Study the structures and decide what this means. What would you expect an ε (epsilon) lactone to look like?

c) What is unusual about 4 and 5? Such a ring is called an *epoxide.* To what functional group class do epoxides belong? These are named as alkene oxides, for example,

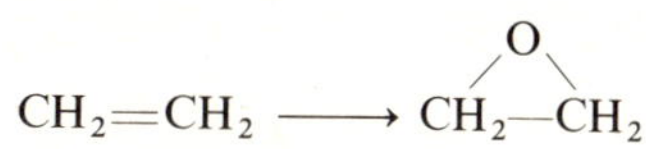

ethylene oxide

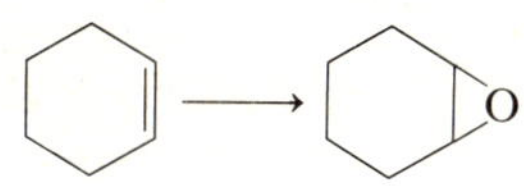

cyclohexene oxide

d) Provide IUPAC names for the acid structures 1, 3, 10, 11, 16, 17, 21, 24, 27, 29, 32, and 33.

e) Consider the following boiling points: acetic acid, 118°C; ethanol, 79°C; ethyl acetate, 77°C. Can you explain in terms of structure why the boiling point of ethyl acetate is lower than those of acetic acid and ethanol in spite of the fact that the molecular weight of ethyl acetate is much higher? Why does having a lower boiling point contribute to the properties of esters as flavor agents?

16. Here are the molecular weights, melting points, and boiling points of a series of compounds. In terms of attractive forces between molecules (polar bonds, ionic electrostatic attractions, van der Waals forces, hydrogen bonds), why do the melting and boiling points vary as they do?

	Molecular Weight	***Melting Point (°C)***	***Boiling Point (°C)***
$CH_3-\overset{\overset{\displaystyle O}{\|}}{C}-CH_3$	58	−95	57
$CH_3-\overset{\overset{\displaystyle O}{\|}}{C}-OH$	60	17	118
$CH_3-\overset{\overset{\displaystyle O}{\|}}{C}-Cl$	78	−112	51
$CH_3-\overset{\overset{\displaystyle O}{\|}}{C}-NH_2$	59	81	222
$CH_3-\overset{\overset{\displaystyle O}{\|}}{C}-O-CH_3$	74	−99	57

8 Steroids

Objectives

In this chapter we:

1. *Learn the basic structure of steroids.*
2. *Discuss the structure and function of cholesterol, vitamin D, and bile salts.*
3. *Introduce concepts of regulation by hormones.*
4. *Learn the structures and functions of corticosteroids and sex hormones.*
5. *Study two specific hormonal systems: control of corticosteroid concentrations, and regulation of the menstrual cycle.*

In Chapter 3 the terpenes were introduced. The assembly of the isoprene units can generate large and complex ring systems. One of the ring systems so generated, the steroid system, is so important that we are devoting a separate chapter to it. Steroids serve as hormones, are involved in calcium metabolism, are surface-active agents in the intestine, and are used as medicinals. Plants produce a variety of steroids, some of which have medicinal value and some of which are very poisonous. A great deal of concern and research is directed at the steroid cholesterol as it occurs in the diet, in the blood, and in arterial walls.

Steroids share this four-ring structure:

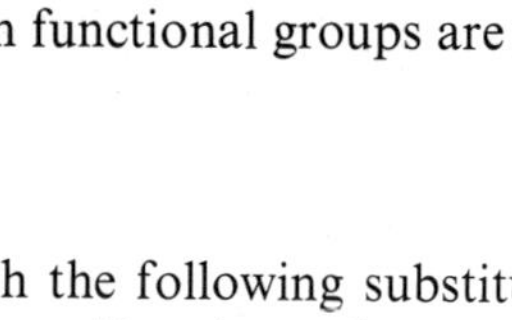

The standard pattern for numbering is shown above. Most steroids have methyl groups at carbons 10 and 13. If there is a side chain, it will be at carbon 17. Oxygen functional groups are most likely to occur at 3, 7, 11, 12 and 17.

Exercise 8.1

Draw a steroid with the following substitutions: 3-keto, 10, 13-dimethyl, 17-hydroxy. You have drawn the male sex hormone dihydrotestosterone (page 206).

The four rings could join together in a number of ways to produce a steroid molecule with a great variety of shapes. However, most steroids are assembled so that the molecule is roughly flat. When chemists have tried to synthesize steroids in the laboratory, it has been difficult to orient all the bonds properly to obtain this flat shape. Without the correct shape the molecules do not perform the desired biological function. Steroids are synthesized in cells by forming the ring structure from the terpene squalene, producing lanosterol.

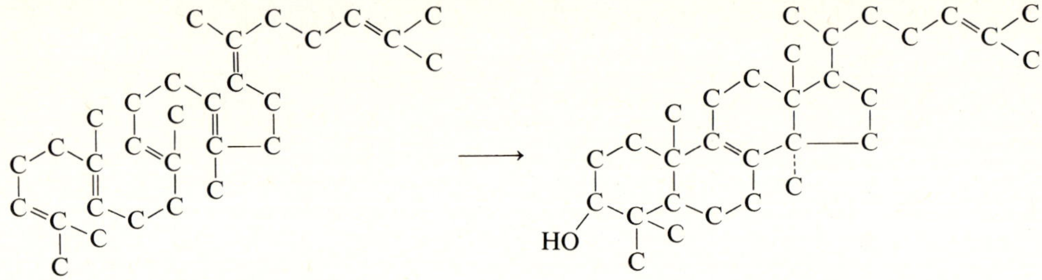

squalene lanosterol

8.1 Cholesterol

Lanosterol is an intermediate in the formation of the most common and most discussed steroid, cholesterol. In turn, most other steroids are synthesized from cholesterol. Cholesterol is found in blood, in bile, and in every cell to some degree. It is transported in the blood along with triglycerides as a component of the protein-coated micelles. Both free cholesterol and cholesterol esters are found in these micelles. The esters are formed between the cholesterol and fatty acids, usually saturated fatty acids.

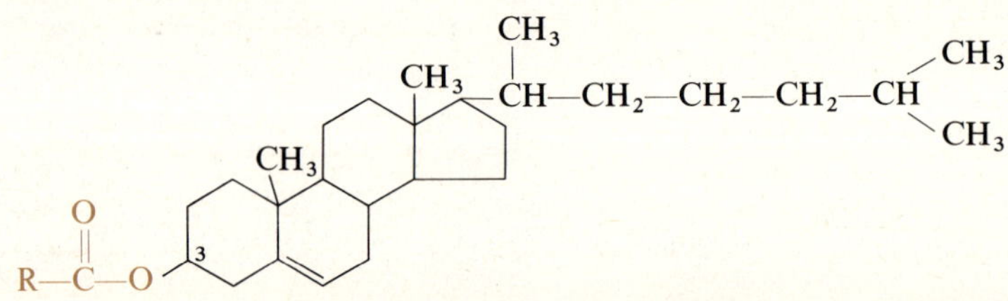

cholesteryl ester

In many individuals, cholesterol deposits along with other serum lipids in arterial walls, generating a problem termed *arterio-sclerosis*. Flow through the vessels is restricted and the walls become less flexible (Figure 8.1). This deposition of cholesterol and other lipids is more likely to occur when serum lipids are at high levels. Levels of cholesterol in the diet seem not to have a significant effect on deposition, but a high ratio of saturated to unsaturated fatty acids in the diet favors deposition. Exercise reduces both serum cholesterol and serum triglycerides.

8.2 Vitamin D

An insufficiency of vitamin D produces poorly formed and easily broken bones, a disease called *rickets*. The precursor of vitamin D is the steroid ergosterol. In the skin, light catalyzes the conversion of ergostercol to calciferol.

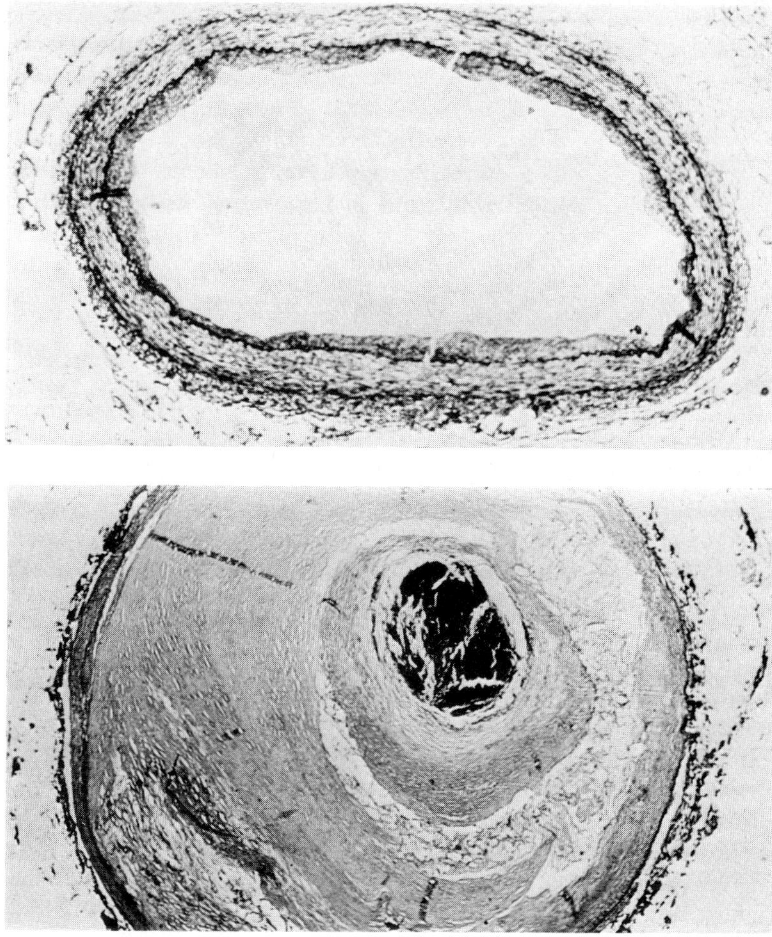

Figure 8.1
Normal and arteriosclerotic blood vessels
Deposits of lipids narrow the passage through a blood vessel in arteriosclerosis. (Photos courtesy of American Heart Association.)

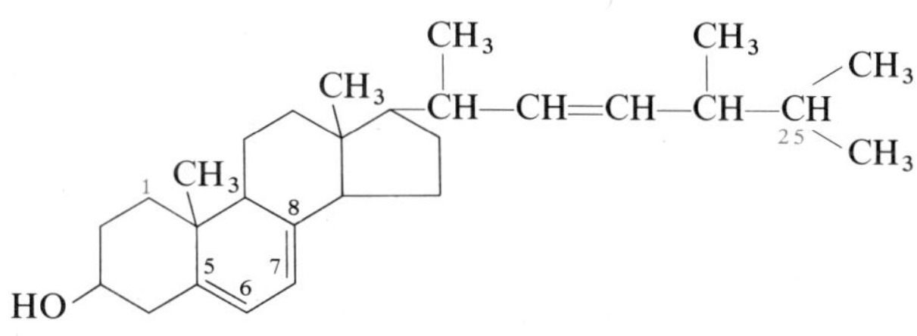

ergosterol

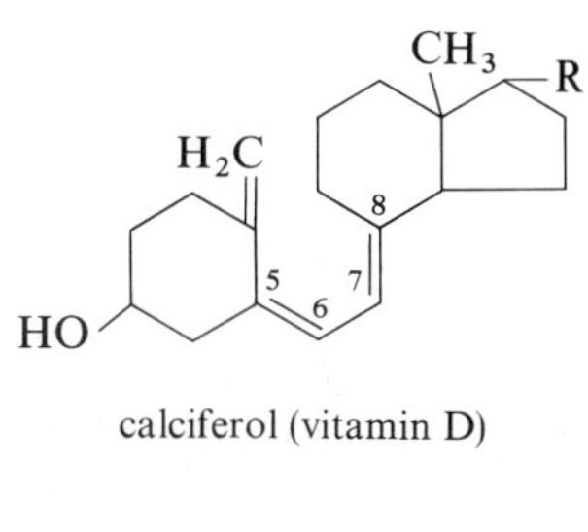

calciferol (vitamin D)

In the liver and kidneys, the calciferol is then hydroxylated at positions 1 and 25. The final products are involved in the uptake of Ca^{2+} from the intestine and in the mobilization of Ca^{2+} from bones.

8.3 Bile Salts

Next to the liver there is a sac called the *gall bladder* that stores a fluid called *bile*. When food transfers from the stomach to the intestine, the gall bladder is stimulated to contract. This sends the bile through a duct into the intestine.

Four steroids, called bile acids, are found in human bile. The most abundant is cholic acid. The carboxylic acid group of the bile acids forms an amide linkage to either glycine or taurine. These products are called *bile salts* and are the useful form of the steroids.

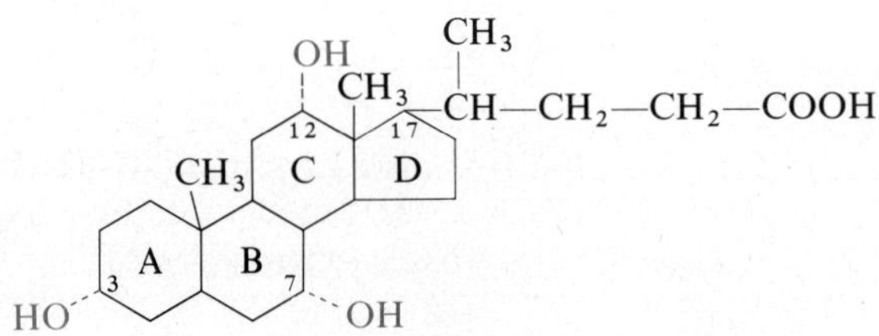

cholic acid

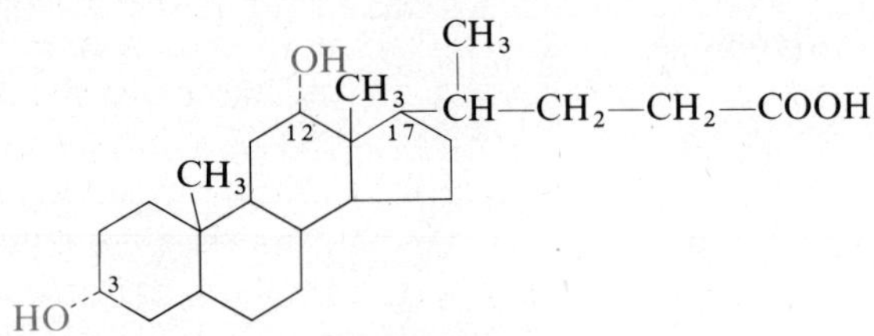

deoxycholic acid

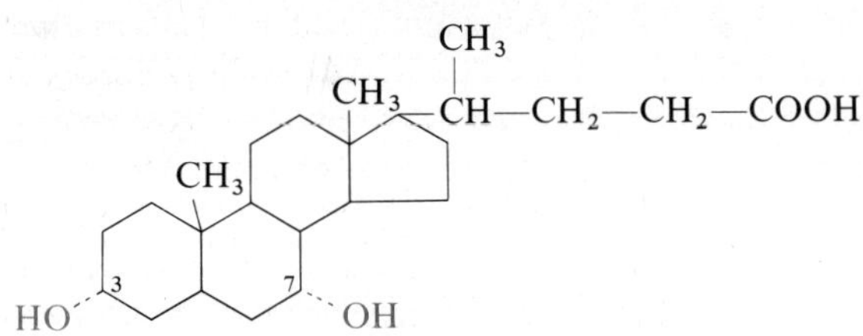

chenodeoxycholic acid

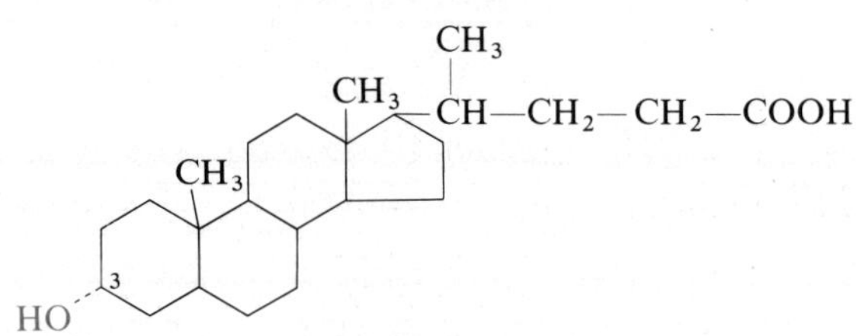

lithocholic acid

bile acids

$$\left.\begin{array}{c} C_{23}H_{26}(OH)_3\overset{\overset{\displaystyle O}{\|}}{C}-\overset{H}{N}-CH_2-COOH \\ \text{glycocholic acid} \\ \\ C_{23}H_{26}(OH)_3\overset{\overset{\displaystyle O}{\|}}{C}-\overset{H}{N}-CH_2-CH_2-SO_3H \\ \text{taurocholic acid} \end{array}\right\}\ \text{bile salts}$$

The bile salts serve as detergents to emulsify triglycerides in the intestine. Later the steroids are reabsorbed from the intestine, removed from the blood by the liver, and collected in the gall bladder for reuse.

In the liver, cholesterol is also transferred from the blood to the bile. Cholesterol when present at moderate levels dissolves in the bile. At high concentrations cholesterol can form crystals that grow into gallstones, which can cause serious problems if they block the bile duct. Gallstones are the most frequent reason the gall bladder needs to be surgically removed.

8.4 Hormones

Let us deal now with a sophisticated regulatory mechanism. A compound is produced in a specialized organ, is transported by the blood circulation, and has a stimulatory or inhibitory effect on one or more "target" tissues. The organs are called *glands*, and the compounds produced by glands are termed *hormones*.

In mammals these glands, the endocrine system, produce a variety of hormones having complex, overlapping, and counterbalancing functions. Early study of the endocrine system was stimulated by the interest and needs of the medical profession. Malfunction of particular glands produces characteristic diseases or syndromes. Surgical removal of gland tissue or injection of crude gland extract into animals produced the same symptoms, thus identifying the malfunctioning gland. However, the isolation of the chemical agent responsible, the hormone, was a more difficult task. Hormones are very potent and so are produced only in small quantities. Furthermore, the hormone could only be detected and measured by means of its physiological effect on a living organism, which is an awkward and often inexact method of assay.

8.5 Steroid Hormones

Humans have two sources of steroid hormones. In males, steroid hormones originate in the testes and the adrenal cortex, and in females in the ovaries and the adrenal cortex. Interestingly, all the steroid-synthesizing glands derive from the same tissue during embryonic development. The effect of steroid hormones is to stimulate the rate of synthesis of specific proteins in a target tissue.

8.6 Sex Hormones

Male sex hormones, or androgens, are produced primarily in the testes. Testosterone and dihydrotestosterone are the principal examples.

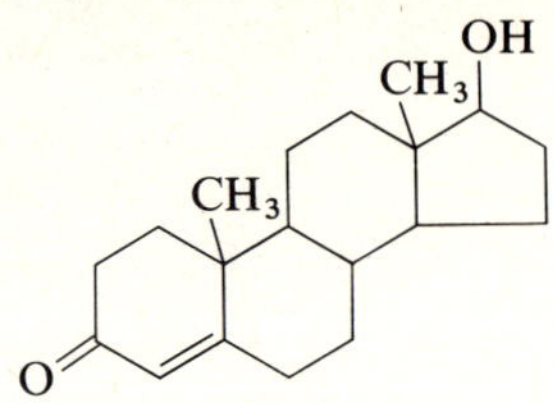

testosterone

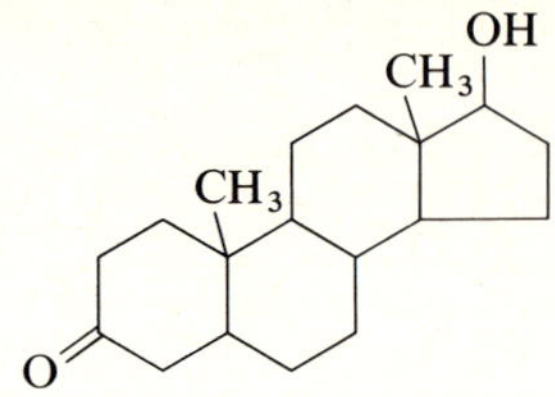

dihydrotestosterone

During puberty, androgen levels increase. They stimulate the development of the male sex characteristics, such as facial and body hair, deepening of the voice, and an increase in muscular strength. In the absence of androgens these changes fail to occur, and body fat accumulates in a more feminine distribution. Loss of androgens (as by castration) after puberty results in some loss of male sex characteristics, a situation that can be reversed by administering additional androgens.

Estrogens (female sex hormones) are produced primarily by the ovaries. Three estrogens are observed in humans, all having a benzene-type ring with a hydroxy group at carbon 3.

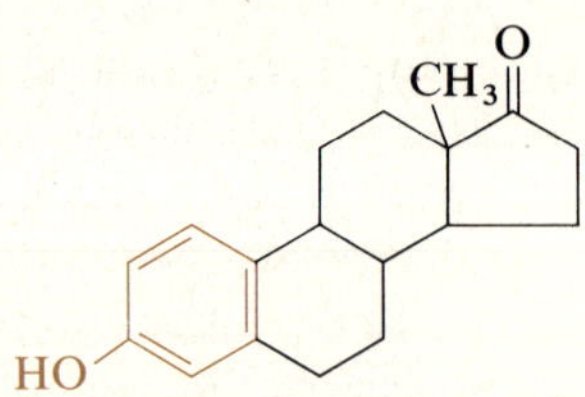

estrone

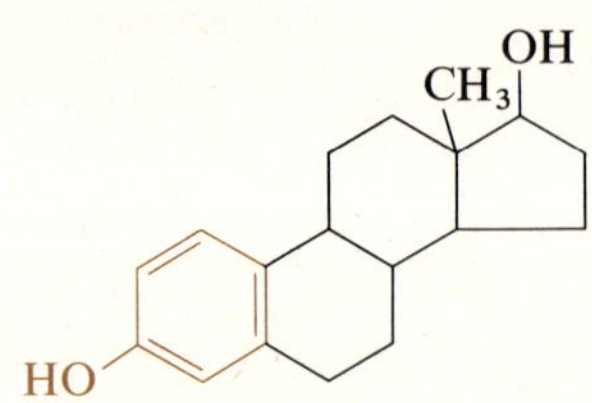

estradiol

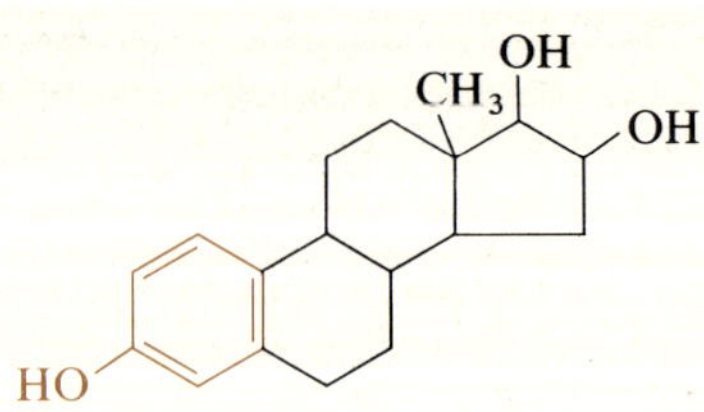

estriol

Like androgens, the estrogens are responsible for the development of female sex characteristics during puberty. These include a broadening of the pelvis, development of the breasts, and growth of pubic hair. The reproductive system matures, and the menstrual cycle begins.

The synthesis of androgens and estrogens can be traced from cholesterol. In the testes, cholesterol is converted to testosterone. The ovaries also produce testosterone, which is then converted to estradiol, the most potent of estrogens.

When androgens or estrogens are used medicinally, they are most effective when injected. Taken orally, much of the potency is lost in the digestive process. However, synthetic analogues of the sex hormones survive digestion and therefore can be used orally. Methyl testosterone is an oral androgen, and 17α-ethynylestradiol is an oral estrogen.

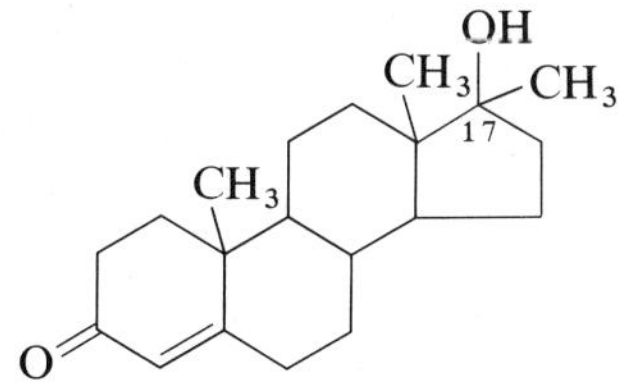

methyltestosterone

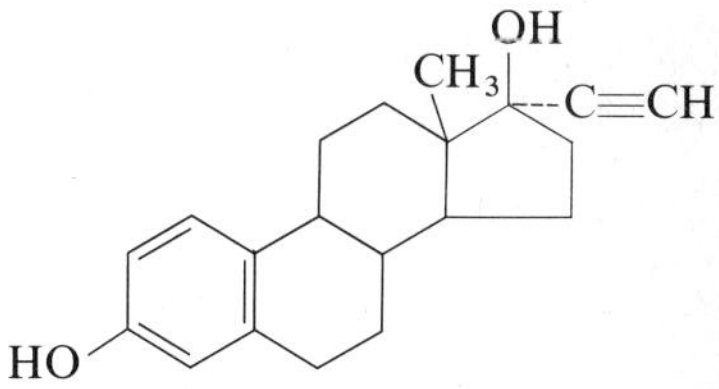

17α-ethynylestradiol

Diethylstilbesterol has also been used as an oral estrogen, but it has been found to have harmful side effects.

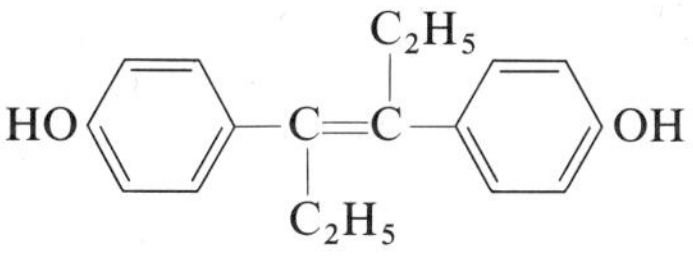

diethylstilbesterol, or DES

8.7 Progesterone

Progesterone is produced in the ovaries by a structure called the *corpus luteum*, as well as in adrenals and in the placenta. Medicinally, progesterone can be used to treat menstrual disorders and to prevent abortion of a developing fetus. As with androgens and estrogens, progesterone is rapidly metabolized if taken orally. Artificial analogues of progesterone, called *progestins*, are available and can be taken orally.

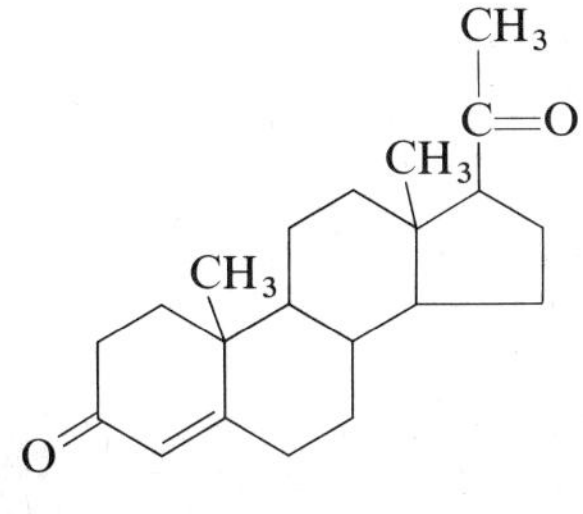

progesterone

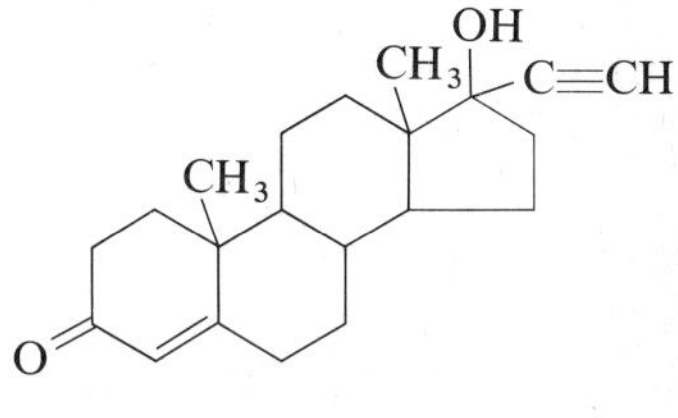

17α-ethynyltestosterone

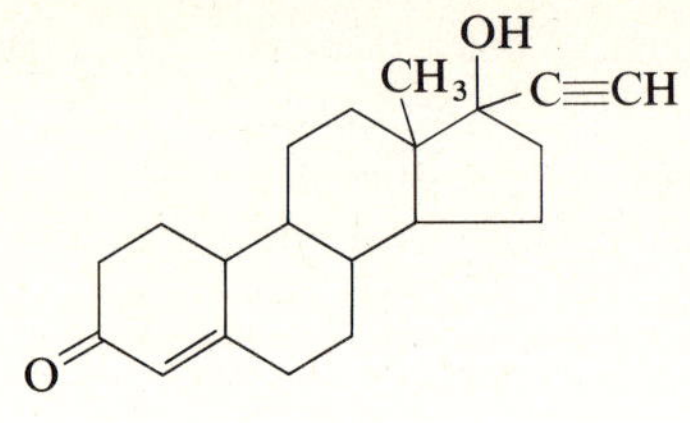

17α-ethynylnortestosterone

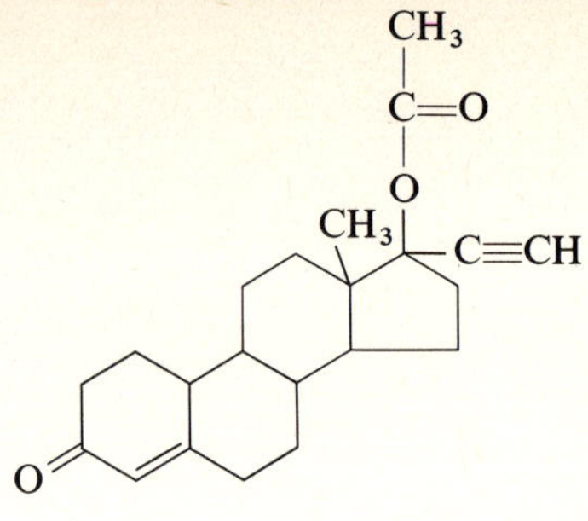

nor-ethisterone acetate

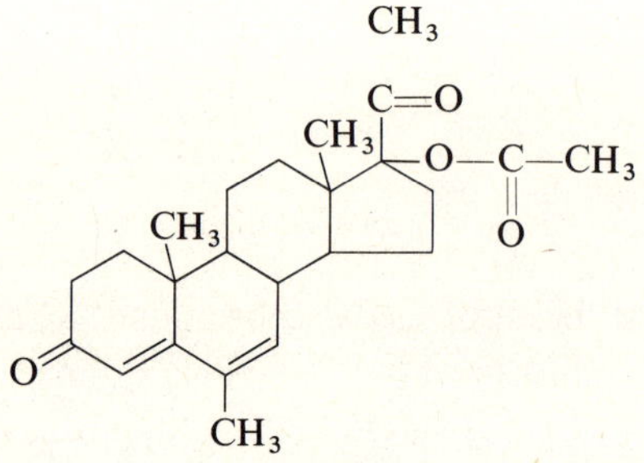

megestrol acetate

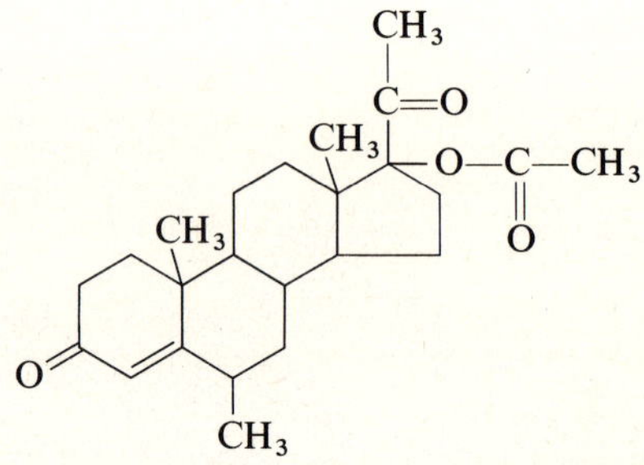

medroxy progesterone acetate

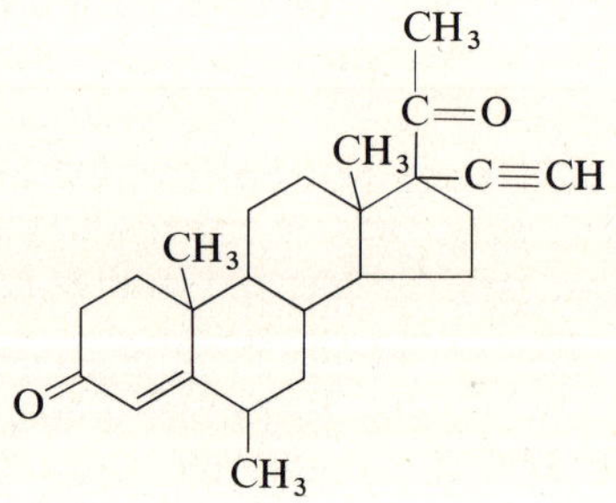

dimethistrone

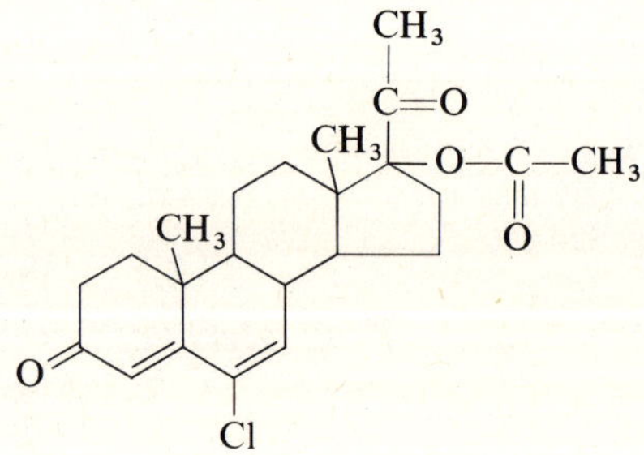

chlormadinone acetate

Progesterone and progestins prevent ovulation and so are used in conjunction with estrogens as oral contraceptives.

8.8 *Corticosteroids*

A large variety of steroids are found in the adrenal cortex. The three important steroids secreted are cortisol, corticosterone, and aldosterone.

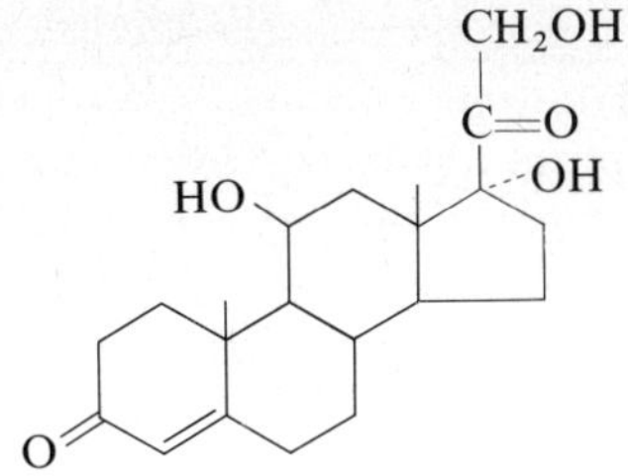

cortisol

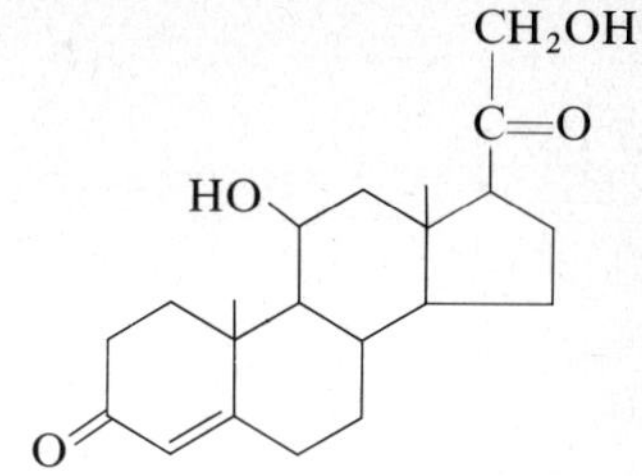

corticosterone

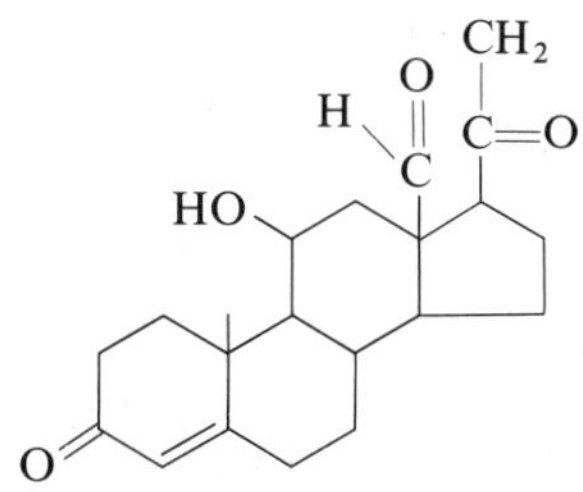

aldosterone

These hormones affect mineral and water balance. Specifically, they stimulate water and Na^+ retention. Excess corticosteroid leads to an increase in blood pressure as extra fluid is retained. Insufficient corticosteroids lead to a loss of fluids and a disruption of the correct balance between K^+ and Na^+. The latter is a factor in nerve and brain function. In the water–mineral effects, aldosterone is particularly effective (Chapter 22).

Carbohydrate levels in the body are affected by the corticosteroids, particularly by cortisol. Under the influence of cortisol, carbohydrate synthesis and storage is stimulated, while carbohydrate breakdown for energy decreases.

Lipids are released to the blood from body fat storage by these hormones.

Corticosteroids combat inflammation such as the swelling and irritation accompanying injury, infection, and allergic response (Chapter 11). This effect may be utilized medicinally by injecting or applying the hormones or their analogues either directly at the inflamed site or so as to raise the total body level. In this regard, cortisone, the 11-keto derivative of cortisol, was observed to have a dramatic effect on rheumatoid arthritis, and hope rose that an effective treatment for this widespread disease had been found. However, the reduction of inflammation reverses when treatment with the hormone is stopped. Continued use of cortisone leads to problems with water and salt retention (edema and high blood pressure) and to emotional disturbance.

Excessive levels of corticosteroids, a condition termed Cushing's syndrome, results from an adrenal tumor. The same side effects that appear during cortisone administration are observed. In addition, blood sugar is high, and secondary sexual characteristics may be overstimulated (resulting in early puberty in the case of children). Two diseases are described that result from low levels of corticosteroids, Addison's disease and Simmond's disease. In each syndrome there is a loss of salt and water, which causes muscular weakness, and low blood sugar. Addison's disease results from damage to the adrenals and is treated by administering corticosteroids or their analogues. Simmond's disease results from a failure in the system controlling output of adrenal hormones, but the adrenals are normal. Administration of ACTH, the adrenal control substance, relieves Simmond's disease. How ACTH operates is the subject of Section 8.10.

8.9 Control Mechanisms for Corticosteroids

Aldosterone production is directly governed by levels of sodium ion in the serum. Low serum Na^+ stimulates aldosterone production, which in turn improves the retention of sodium ion by the kidney. This is an example of a direct feedback control. The synthesis of a compound is regulated by the level of the compound whose retention it is designed to control.

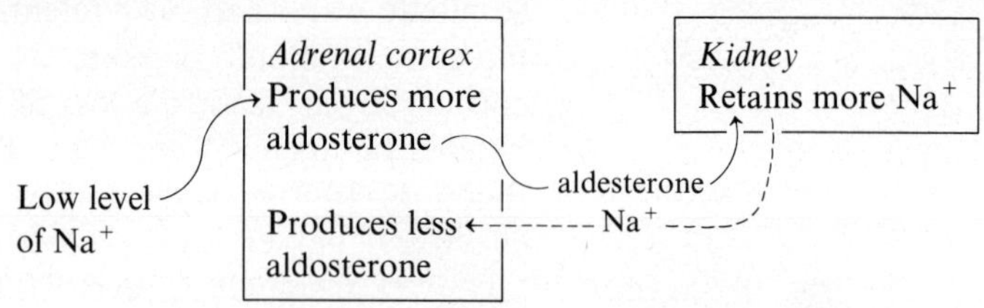

A feedback control system

8.10 The Adenohypophysis

There is a more sophisticated feedback control system than the one shown above. Just under the brain is found the hypophysis (the pituitary gland). The gland has two lobes, the anterior and the posterior, and it is with the anterior lobe, or adenohypophysis, that we are concerned. The peptide hormones secreted by this "master gland" provide a mechanism for the regulation of some of the other gland systems. Corticosteroids are controlled by the adrenocorticotropic hormone (ACTH). This is a peptide of 39 amino acids whose sequence is known. Corticosteroids exert a feedback control on ACTH production, assuring a constant level of hormones. If the level of corticosteroids drops, the hypophysis produces additional ACTH. This stimulates the synthesis of corticosteroids, thereby restoring the correct level (Figure 8.2).

There is a control that can override the self-regulation. There are neurohumeral substances, *corticotropin-releasing factors*, released

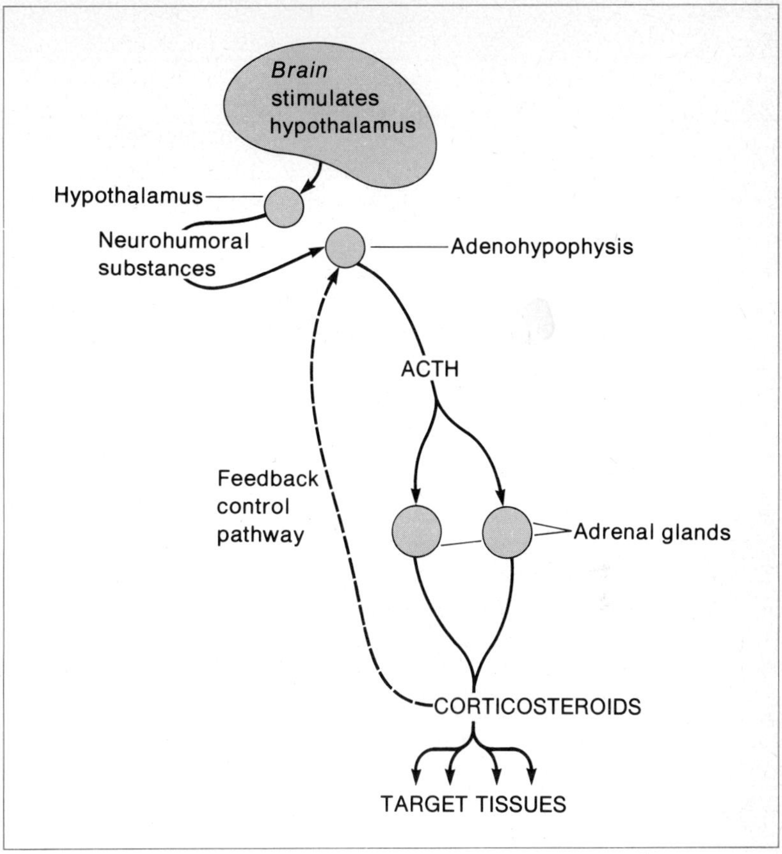

Figure 8.2
Regulation of corticosteroid levels
Corticosteroid levels are regulated by the adenohypophysis by a feedback control mechanism. This self-regulation can be overridden by stimulation of the hypothalamus by the brain.

by another gland, the hypothalamus, in response to nervous stimulation from the brain. These factors are small peptides that need travel only a short distance from the hypothalamus to the hypophysis. The corticotropin-releasing factors stimulate the hypophysis to produce ACTH, even in the presence of adequate levels of corticosteroids. A broad variety of factors stimulate the hypothalamus to release corticotropin-releasing factors, including stress, trauma, and drugs.

8.11 Self-Regulation of a Cyclic Process: The Menstrual Cycle

The menstrual cycle presents an excellent example of coordination of a physiological process by hormones. The cycle does the following:

1. It generates and makes available an egg.
2. It prepares the uterus for implantation of the egg if fertilization should occur.
3. If a fertilized egg is implanted, it maintains the uterus in the desired condition for embryonic development until the embryo is capable of interrupting the cyclic pattern.
4. It terminates the preparation of the uterus and begins a new cycle of egg production if fertilization does not occur.

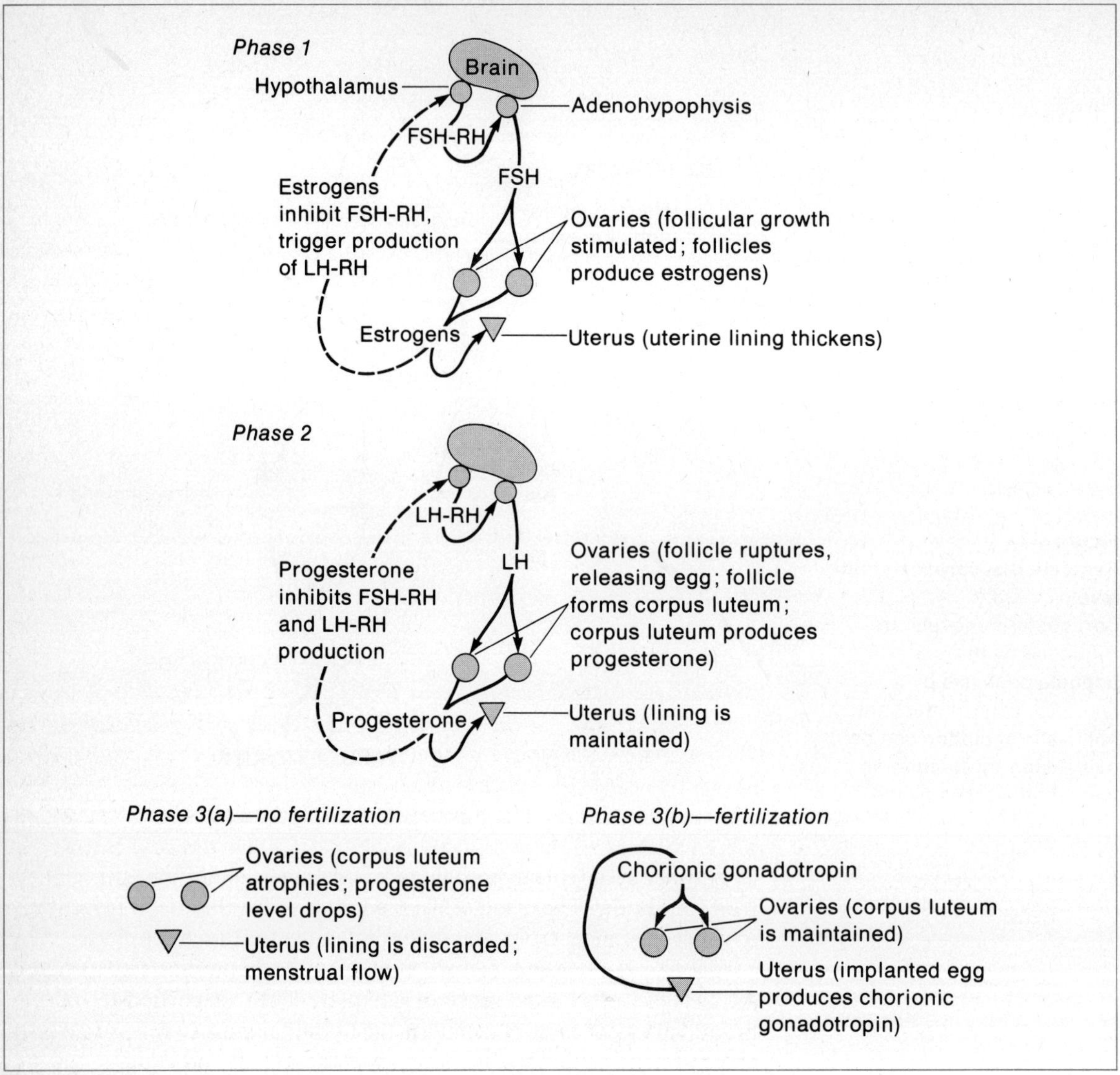

Figure 8.3
The menstrual cycle
The menstrual cycle represents a self-regulating interruptible cyclic process.

In dating events of the menstrual cycle, the beginning of menstrual flow, the discarding of the uterine endometrium (lining) from the last cycle, is considered to be day 1. Presumably in response to a signal from the brain, the hypothalamus produces *follicle-stimulating hormone release hormone* (FSH-RH). This is carried to the adenohypophysis, where it brings about the release into the bloodstream of *follicle-stimulating hormone* (FSH). FSH in turn stimulates follicles, vessels which include maturing eggs, to grow in the ovaries. The follicles are the source of *estrogens*, so as the follicles mature, the estrogen levels increase. Estrogens stimulate the thickening of the uterine endometrium, the lining of the uterus into which the egg, if fertilized, will implant. Estrogens also inhibit the formation of FSH-RH and as a consequence reduce the levels of FSH in the bloodstream. This first

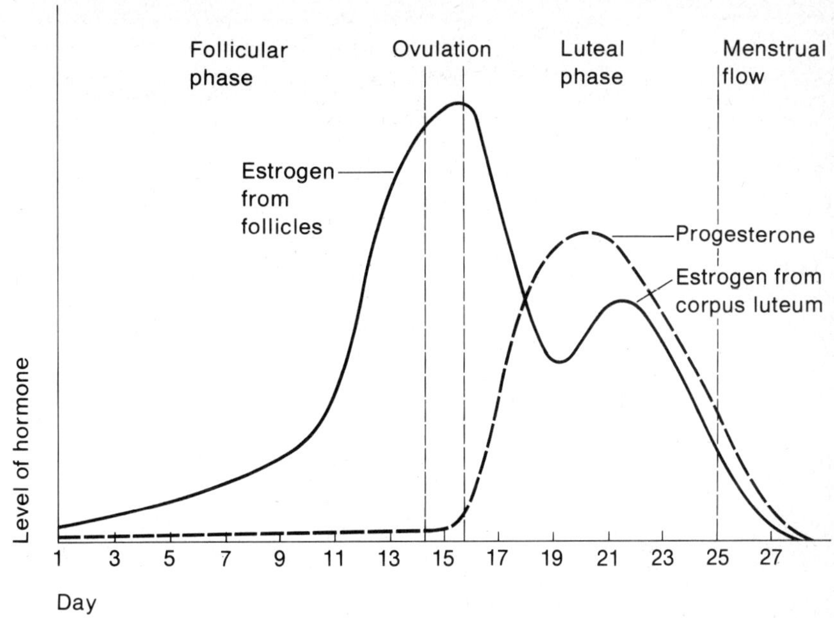

Figure 8.4
Hormone levels during menstrual cycle
Hormone levels vary markedly as the menstrual cycle proceeds. This explains the periodic changes in water retention and correlates with periodic emotional highs and lows.

period in the cycle is called the *follicular*, or *proliferative*, phase. In response to a signal from the ovulation center of the hypothalamus, this thought to be triggered by the high estrogen levels, the hypothalamus releases a quantity of *luteinizing hormone releasing hormone* (LH-RH). This travels to the adenohypophysis and causes a sharp rise in the release of *luteinizing hormone* (LH) into the bloodstream. LH, in turn, triggers ovulation, the rupturing of a mature follicle to release the egg. This marks the end of the follicular phase and ushers in the luteal, or secretory, phase.

Under the influence of luteinizing hormone, the follicle transforms into the corpus luteum, a glandular body secreting *progesterone* and a lesser amount of estrogen. Although progesterone is generally antagonistic to estrogens, it does continue the job begun in the follicular phase by the estrogens of building the uterine endometrium. High progesterone levels inhibit the production of LH-RH and FSH-RH, so levels of both FSH and LH drop. Eventually the corpus luteum atrophies, progesterone and estrogen levels drop as a result, and the cycle ends. The uterine endometrium is no longer maintained by the steroid hormones and is eroded and lost. In the absence of progesterone and estrogen, the levels of FSH and LH again increase and a new cycle begins.

Suppose, however, that a fertilized egg is implanted. The implanted embryo produces *chorionic gonadotropin*, which prevents the regression of the corpus luteum, thus assuring high progesterone levels and a continuation of maintenance of the uterine endometrium. Eventually the placenta becomes a major source of progesterone itself (Figures 8.3 and 8.4).

Summary

Steroids are isoprenoid compounds sharing a four-ring system. The most prevalent steroid is cholesterol, which is found throughout the body. It is a starting point for the synthesis of other steroids. Cholesterol can form gallstones as it crystallizes out of bile, or it can be an important constituent in fatty deposits in artery walls. The extensiveness of the latter problem, termed arteriosclerosis, depends on diet. Vitamin D, which is involved in calcium utilization, is a steroid. Bile salts are steroids that act as detergents in the intestine, stabilizing fatty micelles. Hormones are compounds produced in glands and transported by the blood circulation to target tissues where they regulate some aspect of the target tissue function. Steroid hormones are produced by the adrenal cortex and by the testes or ovaries. Androgens are male sex hormones and estrogens are female sex hormones. These are associated with the occurrence and maintenance of sexual maturation. Corticosteroids are produced in the adrenal cortex. They are important in mineral and water balance and in carbohydrate and lipid utilization. A variety of feedback mechanisms control hormone levels. Some respond directly to the levels of the substance they regulate. The adenohypophysis regulates levels of several hormones by releasing hormones that stimulate production of the target hormone. The adenohypophysis is in turn regulated by the hypothalamus, which responds to signals from the brain. The menstrual cycle represents an example of the self-regulating cyclic process involving the hypothalamus, the adenohypophysis, the ovaries, and the uterus.

Problems

1. For each statement on the left, list as many items from the list on the right as are appropriate.

______ ***a***) Is a precursor of cholesterol but is not a steroid	***A***) Bile salt ***B***) Estradiol ***C***) Cortisone
______ ***b***) Assists in the utilization of calcium	***D***) Aldosterone ***E***) FSH
______ ***c***) Helps emulsify fats in the intestine	***F***) Squalene ***G***) Lanosterol
d) Male sex hormone	***H***) Androgen ***I***) Vitamin D
______ ***e***) Female sex hormone	***J***) Estrone
______ ***f***) Prevents ovulation	***K***) Testosterone ***L***) Progesterone
______ ***g***) Is produced by the corpus luteum	***M***) Estrogen
______ ***h***) Stimulates water retention	***N***) Corticosteroids ***O***) Glycocholic acid
______ ***i***) Combats inflammation	***P***) LH ***Q***) Taurocholic acid

_______ *j*) Production stimulated by ACTH

_______ *k*) Stimulates follicles to grow

_______ *l*) Hormone produced by follicles

2. How many carbons are there in squalene? How many isoprene units are there in squalene? This means that steroids are (mono-, di-, tri-, or hexaterpenes).

3. Take the lid off the water tank of the toilet at home. Flush the toilet and watch what happens in the tank. Compare this water-level control system to the system by which aldosterone maintains the Na^+ concentration.

a) The water level in the tank is analogous to the Na^+ concentration in a person. How does the tank know when it is full? Where is that detection system in a person?

b) How does the control system cause the water level to be increased? Where is the machinery for raising the Na^+ concentration in a person?

c) How is the level detector connected to the level "increaser" in the tank? How is this principle accomplished in a person?

Challenge Problems

4. Steroids are named as derivatives of a series of "parent hydrocarbons," much as the names of simple aliphatic compounds are based on methane, ethane, propane, and so forth. The following is a steroid parent hydrocarbon:

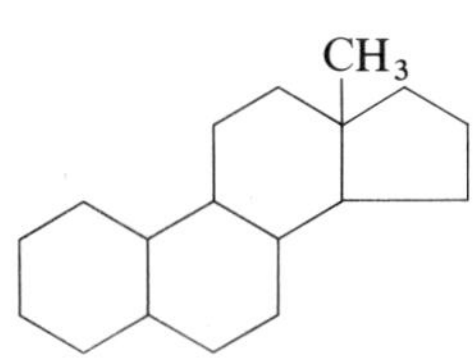

estrane

a) Number the ring carbons.

b) If a double bond connects two carbons that are not sequentially numbered, the higher number is placed in parentheses and is written immediately after the lower number. According to that rule, draw the structures of estra-1,3,5-triene and of estra-1,3,5(10)-triene. In which case are the double bonds conjugated? In which case is an aromatic structure produced?

c) If we attach a group to an aromatic ring, how many different orientations in space can it have?

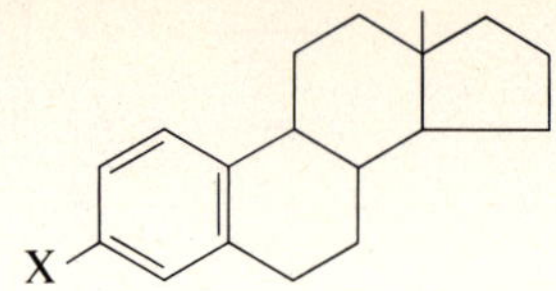

If we attach a group to a cycloalkane ring system, it could be above the plane of the ring or below it. This might, but wouldn't always, generate two isomers. It is conventional in steroid chemistry to indicate that a group is below the plane of the ring by using a dashed line bond and calling the group α. Groups above the plane of the ring are designated by solid line bonds and are called β. Where else have you used α and β in exactly the same sense?

***d*)** Which of the following are truly different isomers?

***A*)**

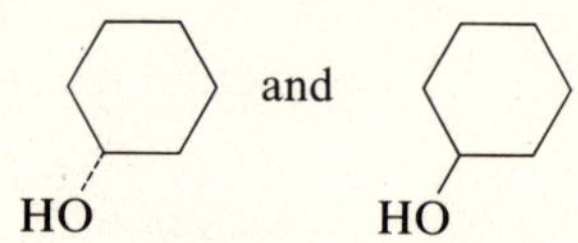

***B*)**

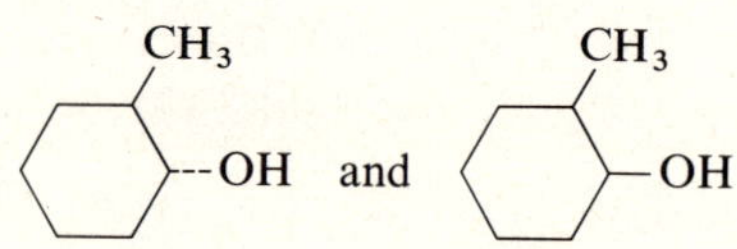

***C*)**

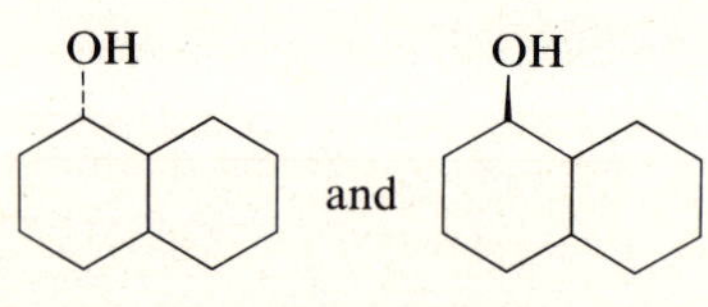

***e*)** This structure would be a 3α-hydroxyestrane:

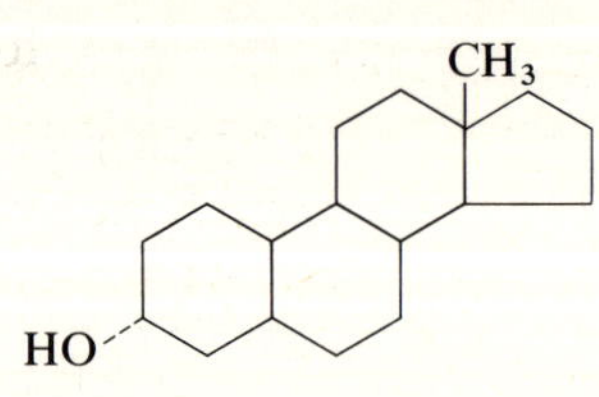

Draw 3β-hydroxyestrane.

***f*)** Name the following:

***A*)**

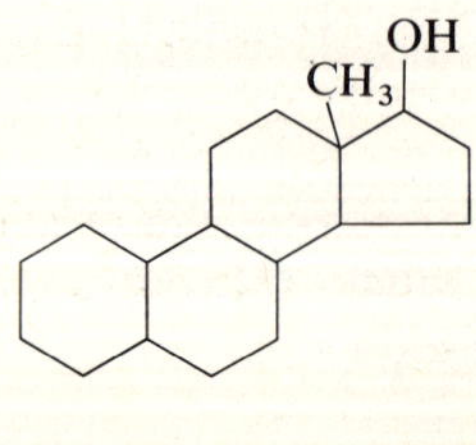

***B*)**

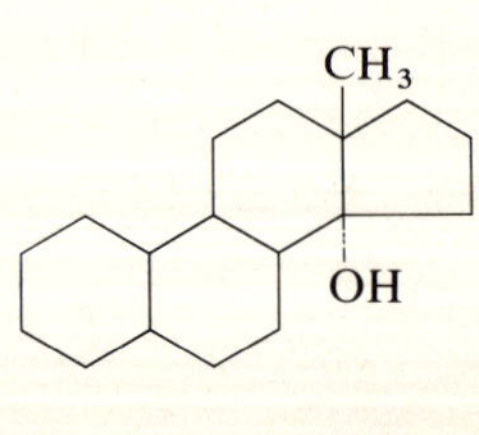

C)
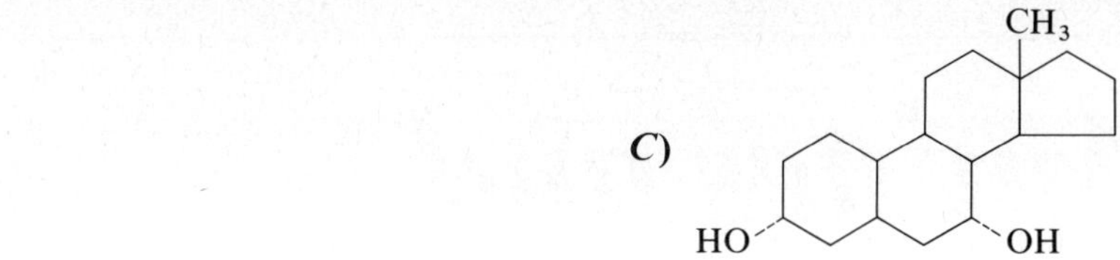

D)
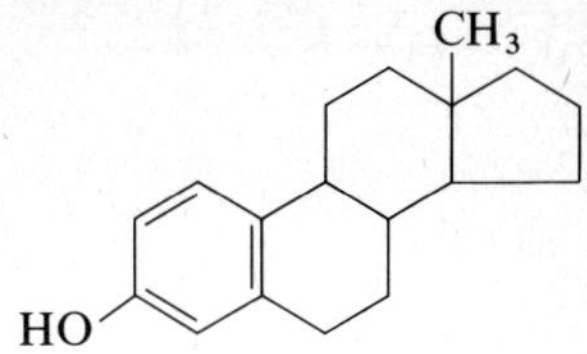

g) Draw the following, then identify each by its trivial name as presented in the chapter:
A) Estra-1,3,5(10)-triene-3,17β-diol
B) Estra-1,3,5(10)-triene-3,16α,17β-triol
C) 3-Hydroxyestra-1,3,5(10)-triene-17-one

5. Given the following parent hydrocarbons, draw the structures from the formula names, and identify as many as possible by trivial names from the chapter:

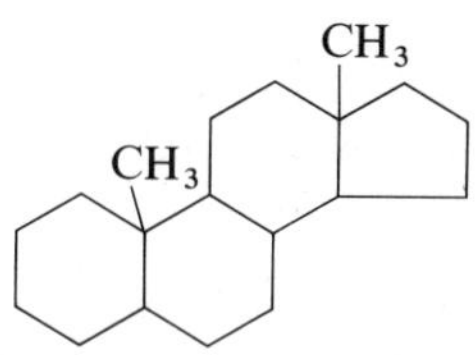

androstane

(How is this different from estrane?)

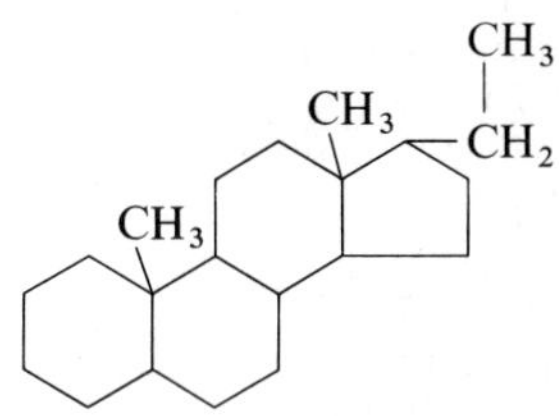

pregnane

a) 17β-Hydroxyandrost-4-en-3-one
b) Preg-4-ene-3,20-dione
c) 11,17,21-Trihydroxypregn-4-ene-3,20-dione
d) 11,21-Dihydroxypregn-4-ene-3,20-dione

6. Make the following substitutions to the basic steroid ring system:

a) 10,13-Dimethyl (What parent hydrocarbon do you now have?)
b) 2β,3β,14α-Trihydroxy
c) 7-Ene
d) 6-One
e) A side chain at position 17 that is 3,6-dihydroxy-6-methyl heptane and is bound to steroid position 17 by carbon 2 of the side chain.

You have just drawn the structure of ecdysone, the moulting hormone of insects. This hormone is involved, as is the juvenile hormone in Chapter 3, in the process of maturation of insects from larva to pupa to adult insect.

7. At first glance, diethylstilbesterol and an estrogen look very different. Study the two structures carefully, and describe the structural similarities.

9

Amines and Amino Acids

Objectives

In this chapter we:

1. Learn to name simple amines.
2. Learn the structures of the common amino acids.
3. Discuss the importance of amino acids in the diet.
4. Study the acid–base properties of amines and amino acids.
5. Learn how amino acids are bonded together to form peptides and proteins.
6. Discover the importance of protein sequence to function, the variability of sequence between and within species, and the relationship between mutation and the protein sequence.
7. Study the separation technique called electrophoresis and how this technique depends on acid–base chemistry.
8. Learn to name quaternary amines, and survey some uses of these compounds.
9. Discuss catecholamines and their relationship to amphetamines.

We shall now focus attention on the organic chemistry of nitrogen. Nitrogen is rather special. It has five electrons in its outer shell, which means that it requires the formation of three covalent bonds to complete an octet.

ammonia

Like oxygen, not all the electrons are involved in a covalent bond. Two nitrogen electrons are still unbonded. We shall return to those two electrons when we discuss the acid–base properties of amines.

9.1 Amine Structures and Names

Amine structures are generated by replacing the hydrogens of ammonia with alkyl groups. Replacing one hydrogen creates a *primary amine*. These amines are named simply by prefixing the word "amine" with the name of the alkyl group.

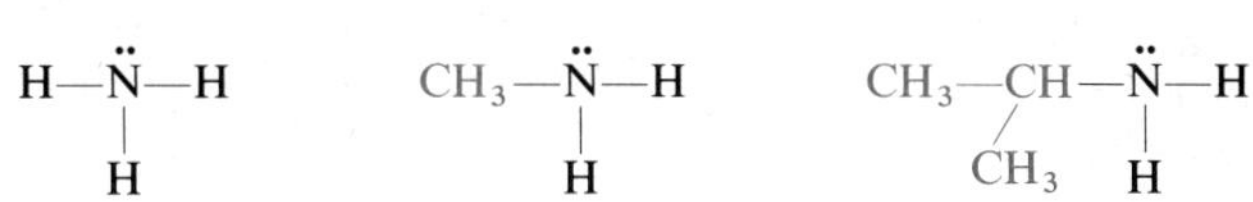

ammonia　　methyl amine　　isopropyl amine

Aromatic amines can have special names.

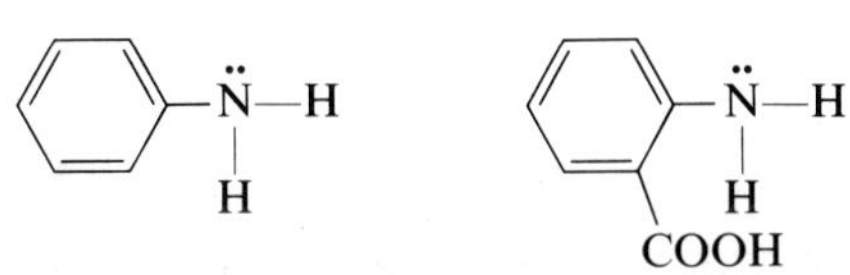

aniline　　anthranilic acid

Replacing two hydrogens with alkyl groups produces *secondary amines.*

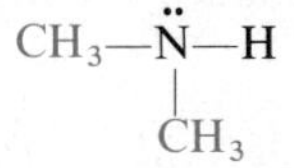

dimethyl amine

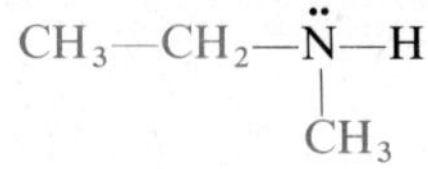

ethyl methyl amine

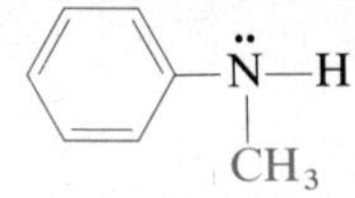

methyl phenyl amine
(N-methyl aniline)[1]

And replacing all three hydrogens generates *tertiary amines.*

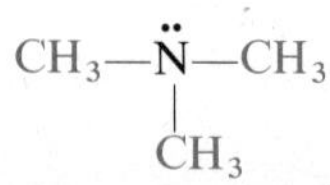

trimethyl amine

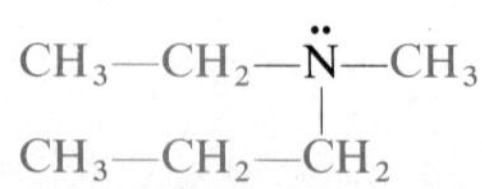

ethyl methyl *n*-propyl amine

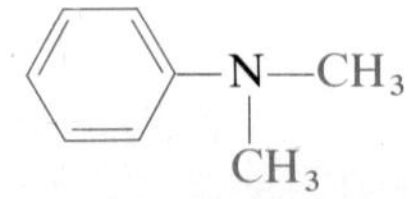

dimethyl phenyl amine
(N,N-dimethyl aniline)

The —NH_2 group can be named as a substituted group—the *amino* group—when present in a structure with other functional groups.

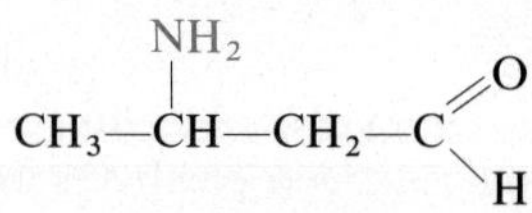

3-aminobutanal

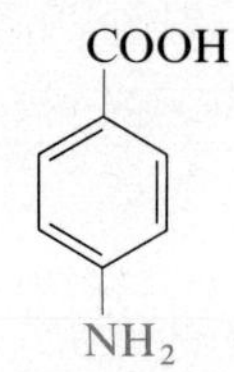

p-amino benzoic acid
(a growth factor for some organisms)

Exercise 9.1

Name the following:

a) $CH_3—CH_2—NH_2$

b) 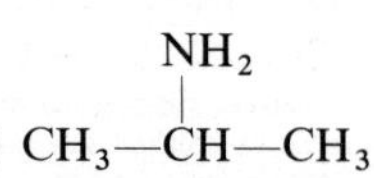

c) 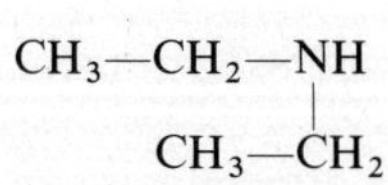

d) 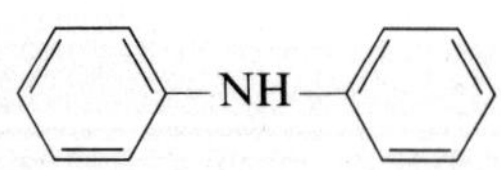

---

[1] The compound name is prefixed N-methyl. This locates the methyl group on the nitrogen rather than on the benzene ring. This convention in naming amines is sometimes used when one of the groups attached to the nitrogen is much more important than the others, especially if it provides a unique name. In this case, phenyl amine is usually named aniline, and the presence of the aromatic structure has a considerable effect on the properties of the amine.

e) 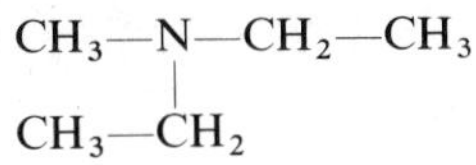

f) 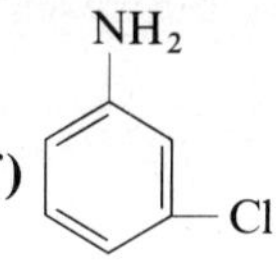

g) 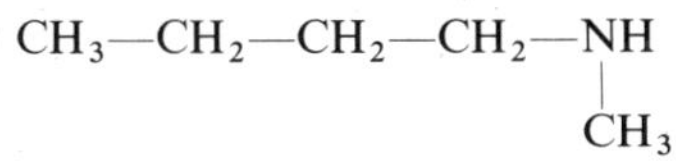

h) 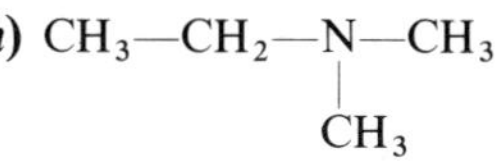

i) 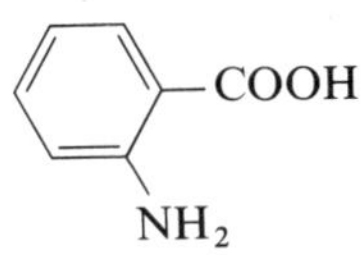

j) 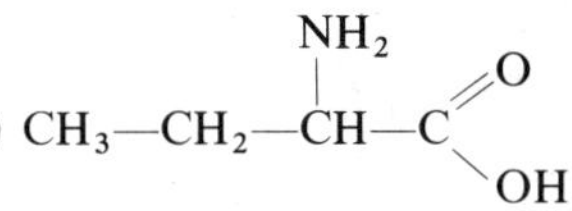

k) 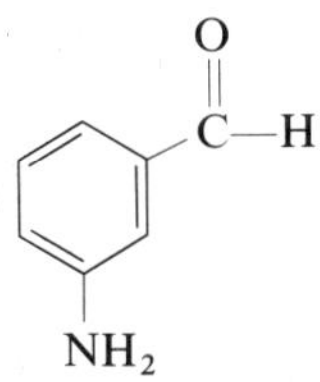

9.2 Amino Acids

Among the most interesting structures containing amine groups are the *amino acids*. Primarily they serve as the building blocks of *proteins*. They share this structure:

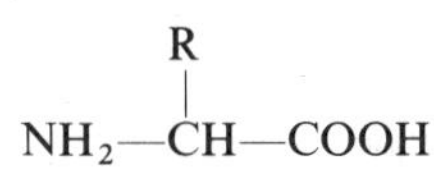

an amino acid

There are a wide variety of amino acids in nature, differing from one another in the composition of the R group. In proteins, however, only twenty different structures are found (Table 9.1). Notice that the R groups are classified as nonpolar, polar uncharged, and polar charged. The properties of a protein depend heavily on the particular assortment of R groups present.

All of the amino acids found in protein except glycine are asymmetric at carbon 2. As is typical of living systems, only one isomer is synthesized. In the case of every asymmetric amino acid, the natural isomer is the L form. We do find D amino acids in nature, for example in certain antibiotics. However, the amino acids in proteins are invariably L.

9.3 Protein Nutrition

Protein is an essential component of the diet. However, proteins once eaten are not used directly to perform a biological function, even if the dietary protein functions in the same way as a protein

Table 9.1
The Amino Acids of Protein

Nonpolar R Group

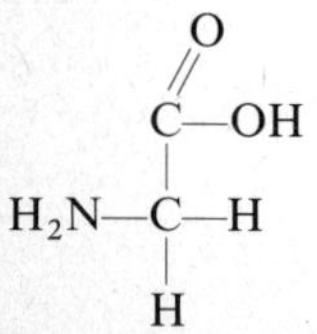
Glycine (Gly)

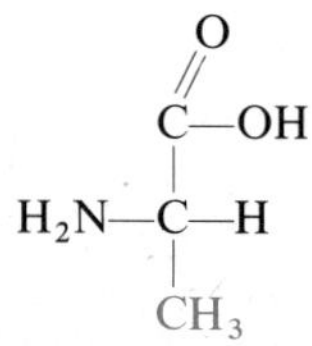
Alanine (Ala)

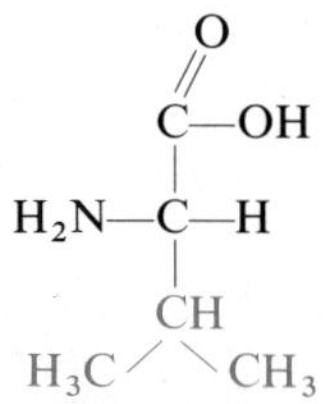
Valine (Val)[a]

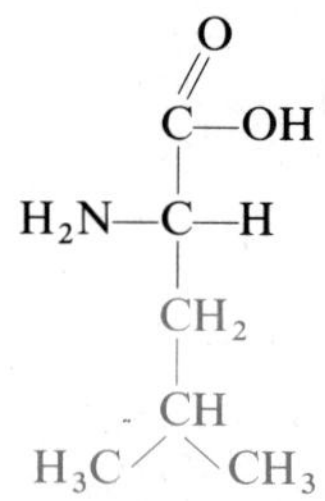
Leucine (Leu)[a]

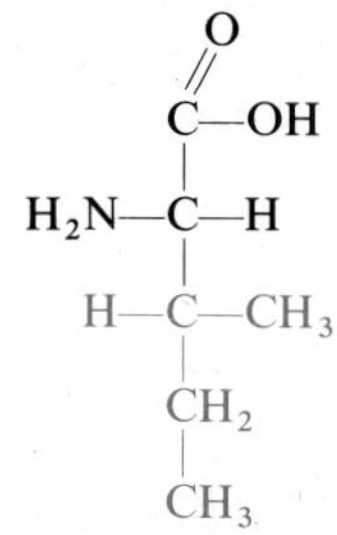
Isoleucine (Ile)[a]

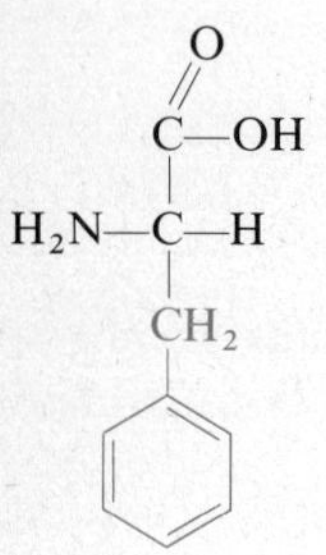
Phenylalanine (Phe)[a]

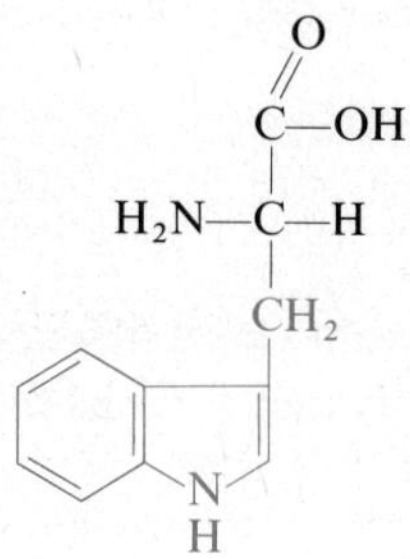
Tryptophan (Trp)[a]

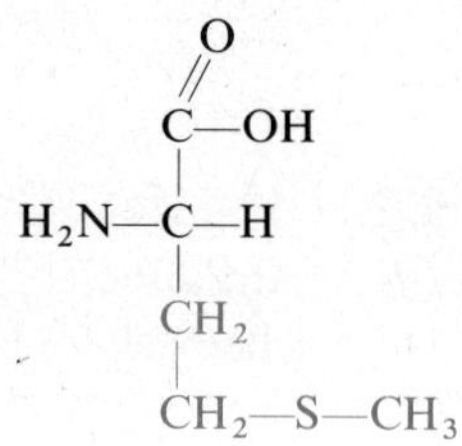
Methionine (Met)[a]

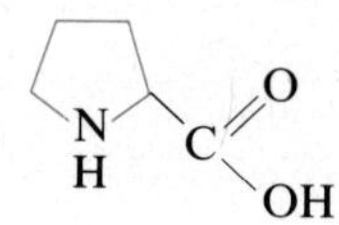
Proline (Pro)

Polar Uncharged R Group

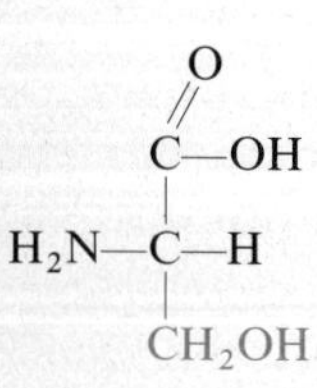
Serine (Ser)

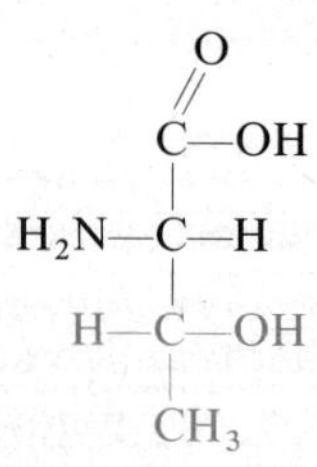
Threonine (Thr)[a]

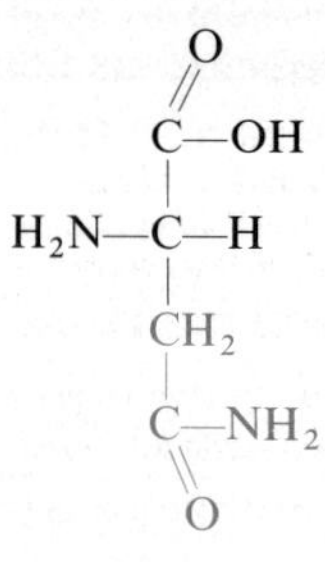
Asparagine (Asn)

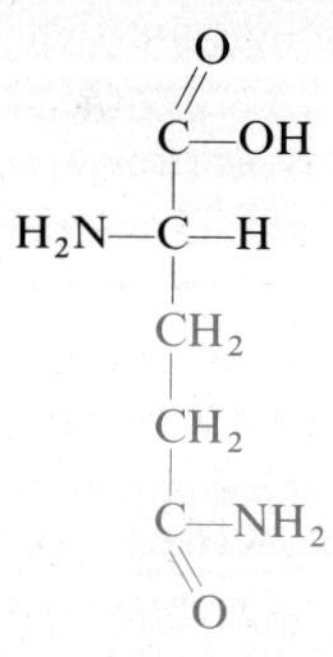
Glutamine (Gln)

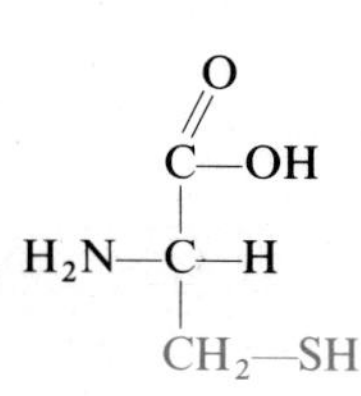
Cysteine (Cys)

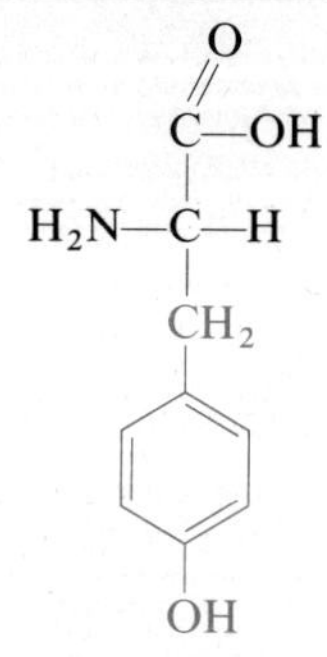
Tyrosine (Tyr)

Table 9.1 (Continued)

Polar Charged R Group				
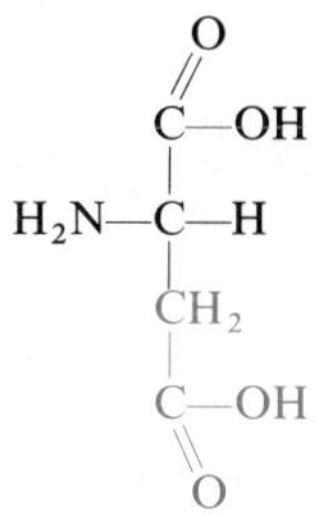	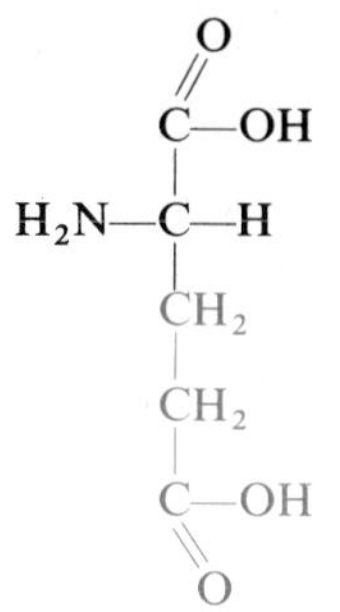	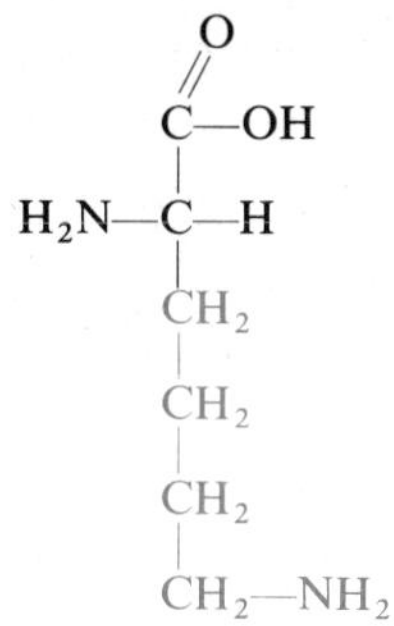	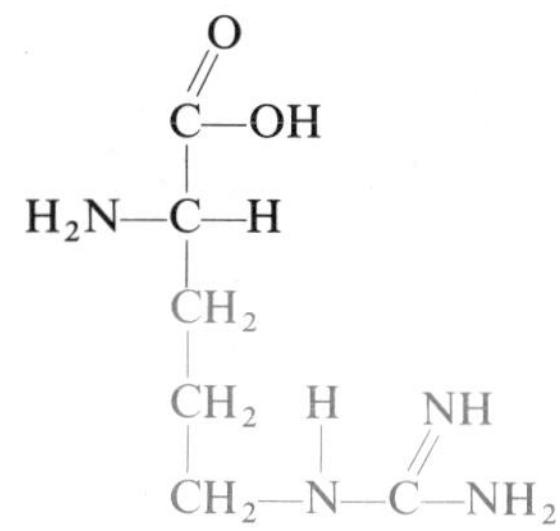	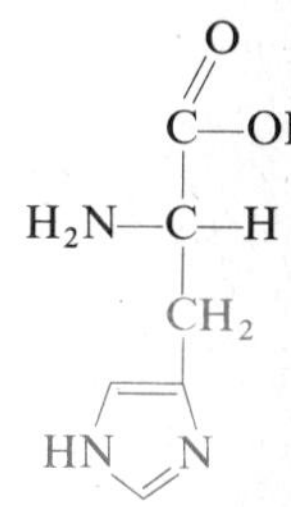
Aspartic acid (Asp)	Glutamic acid (Glu)	Lysine (Lys)[a]	Arginine (Arg)	Histidine (His)

[a] "Essential" amino acids.

we require. Instead, the proteins in the diet are digested into individual amino acids before they are absorbed from the intestine into the blood. The individual amino acids are then transported to the cells, where they are synthesized into whatever proteins the cell needs at that moment.

If our diet does not contain a sufficient supply of certain amino acids, others can be altered into the needed structures. Other amino acids can be synthesized from nonprotein sources in the diet. However, human beings have lost the ability to synthesize some amino acids in any fashion. These *essential* amino acids must be present in our diet in adequate supply. What appears to be critical in the diet is not a specific total amount of protein. Rather it is a sufficient supply of each of the essential amino acids. The lack of a single essential amino acid, even in the presence of a more than adequate total supply of protein, leads to a state of malnutrition. Furthermore, we do not store amino acids in the sense that we store body fat. Our ability to utilize a particular day's supply of amino acids for synthesis of new protein is restricted to the quantity of the least adequately supplied essential amino acid. Beyond this, the amino acids are metabolized for energy, even the essential amino acid that might be limiting at the next meal. If there is a continued shortage of an essential amino acid, body proteins beginning with liver and muscle protein are broken down to supply the scarce amino acid. Obviously this situation leads to serious problems if allowed to continue.

Recently, concern has been expressed about the increase in the earth's population and the ability of human beings to continue to feed themselves. Frequently, the wastefulness of feeding grain

Figure 9.1
Who eats the grain?
Do we harvest our fields to feed livestock, or should we use the grain we grow directly for human consumption? Understanding the complexities of this question is important as we face the challenge of feeding an increasingly hungry world. [Photos by Bob Taylor, DPI (top); Jerry Frank, DPI (middle); Joel Gordon, DPI (bottom).]

to animals, then consuming the animals, is pointed out. Without question we lose a large percentage of the total calories of the grain in this fashion. Fewer acres of land would be required to provide the minimum diet if we consumed the grain ourselves (Figure 9.1). The equation is not as simple as it is often presented to be, however. First, many of the grasslands upon which grazing animals are raised are of borderline value for the planting of grain. More importantly, vegetable protein does not provide as satisfactory a distribution of amino acids as does animal protein, such as meat, eggs, fish, and milk. A person who would meet his or her amino acid requirements with about 20 g of meat or egg would require about 70 g of whole wheat bread to supply the same level of amino acid nutrition.

Maintenance of an adequate protein nutrition is most important in children. Perhaps the greatest nutritional problem in the world is the disease often termed *kwashiorkor*. The problem occurs when infants are prematurely weaned onto a low-protein cereal diet. This diet has less than half the protein per calorie as mother's milk. It is often particularly short of lysine. As the child runs short of essential amino acids, tissue protein is sacrificed. In extreme cases the production of digestive enzymes is halted, destroying the ability to digest what protein is present in the diet. High mortality also results from infection or dehydration because of diarrhea. Addition of lysine to the diet or the introduction of new strains of corn very high in lysine are employed to combat kwashiorkor.

9.4 Reactions of Amines

The product of the addition of an amine to a carbonyl group is a Schiff's base. The reaction can be visualized as happening in two steps. The first is addition of the amine to the double bond of the carbonyl, and the second is the elimination of water to form a new double bond.

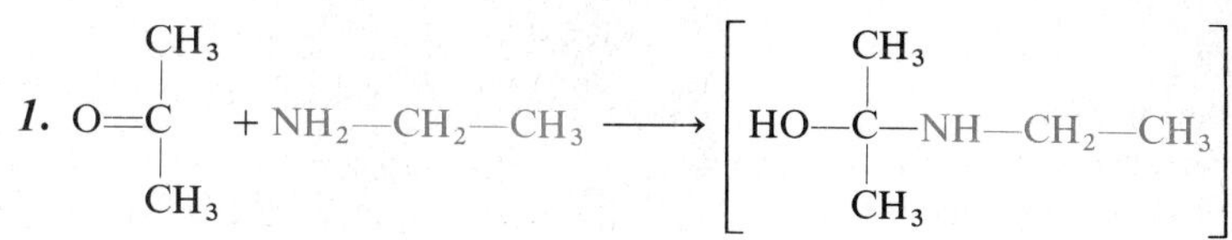

unstable intermediate

This intermediate is not stable enough to isolate.

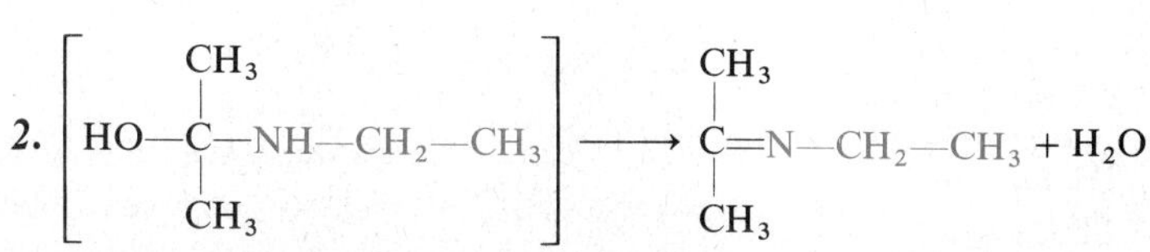

Schiff's base

Other examples are

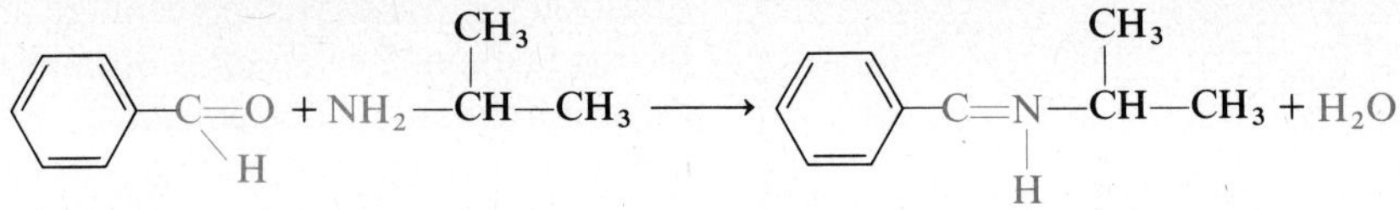

and

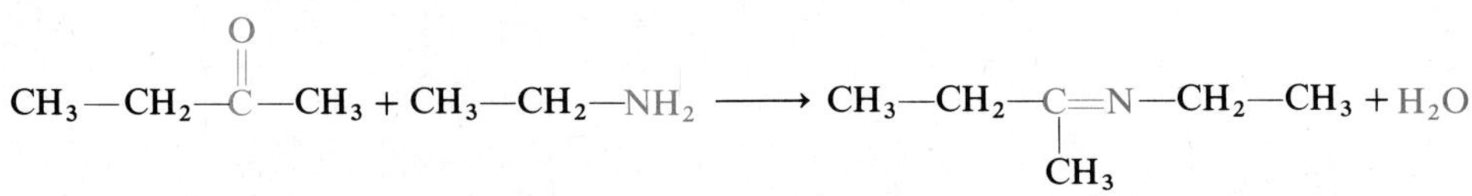

It is fairly easy for a hydrogen ion and the pi electrons of the double bond to shift to effectively move the double bond to the other side of the nitrogen. This sort of shift involving just hydrogen ion and electrons is called a *tautomeric shift*.

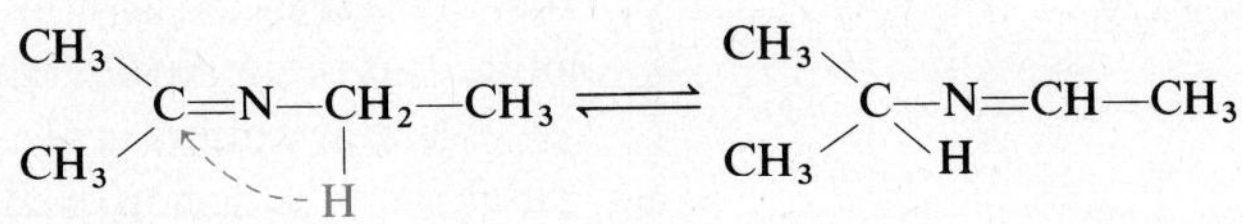

Exercise 9.2

Write the reaction for the formation of a Schiff's base between each of the following:

a) 3-pentanone and 3-aminopentane
b) pentanal and 3-aminopentane

This reaction is important in enzyme chemistry. For example, the R group amine of lysine on an enzyme sometimes forms a Schiff's base with an aldehyde or ketone as part of the enzyme function.

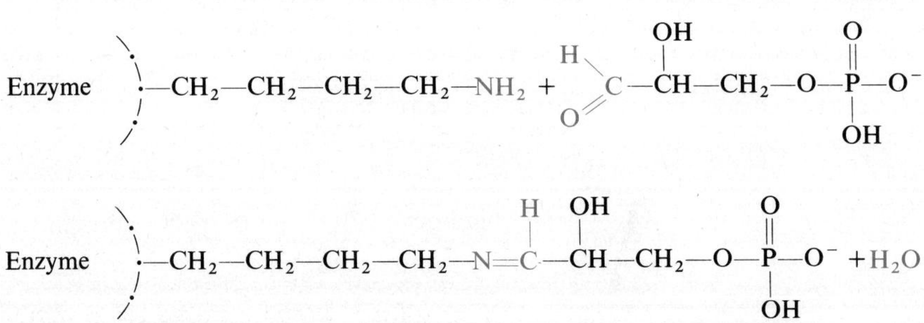

The cofactor pyridoxal (which is attached to the enzyme) has an aldehyde group that forms a Schiff's base with the amino group of an amino acid during the course of its normal functioning.

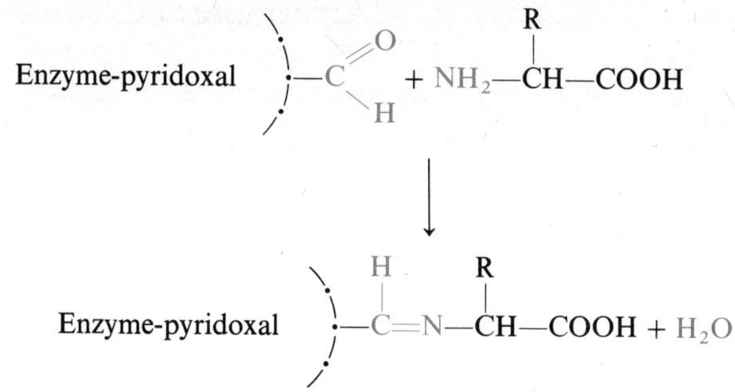

This is discussed in more detail in Chapter 12.

The most important reaction of amines for our purposes is their ability to act as bases.

9.5 *Acid–Base Properties of Amines*

Carboxylic acids function as acids because they include a hydrogen that can be attracted away readily. A hydrogen atom has two particles, a proton and an electron. Because the hydrogen was attracted away without its electron, that is, simply as a proton, the acid was a *proton donor*. Using the definition that an acid is a proton donor, a base is exactly the opposite; a base is a *proton acceptor*. Amines are bases. The unused pair of electrons can accept a proton.

Acid (Donates a hydrogen ion):

$$CH_3-\overset{\overset{\displaystyle O}{\|}}{C}-OH \rightleftharpoons CH_3-\overset{\overset{\displaystyle O}{\|}}{C}-O^- + H^+$$

Acid in water (Donates a hydrogen ion to water):

$$H_2O + CH_3-\overset{\overset{\displaystyle O}{\|}}{C}-OH \rightleftharpoons CH_3-\overset{\overset{\displaystyle O}{\|}}{C}-O^- + H_3O^+$$

Base (Accepts the hydrogen ion):

$$H^+ + CH_3-\underset{\underset{\displaystyle H}{|}}{\ddot{N}}-H \rightleftharpoons CH_3-\overset{\overset{\displaystyle H}{|}}{\underset{\underset{\displaystyle H}{|}}{N}}\!\overset{+}{-}H$$

Base in aqueous acid:

$$H_3O^+ + CH_3-\underset{\underset{\displaystyle H}{|}}{\ddot{N}}-H \rightleftharpoons H_2O + CH_3-\overset{\overset{\displaystyle H}{|}}{\underset{\underset{\displaystyle H}{|}}{N}}\!\overset{+}{-}H$$

In each case the product of the reaction, whether we started with an acid or a base, has a charge.

Acid–base reactions are reversible. In a water solution, the condition that determines whether an acid has lost its proton or a base has gained it is the pH of the solution (review Chapter 7). Whereas the process of gaining or losing a proton occurs in acidic (low pH) solutions for carboxylic acids, it occurs for amines in basic (high pH) solutions.

As with acids, amines do not gain or lose protons all at once but do so across a pH range. The center of this range is the pH value called the p*K* for the amino group.

$$CH_3\text{—}NH_3^+ \underset{+H^+}{\overset{-H^+}{\rightleftharpoons}} CH_3\text{—}NH_2$$

form in acid solution	changes in region of p*K*	form in basic solution

Amino acids have both acidic and basic groups and so assume a number of different charge patterns as the pH is changed. Suppose the general amino acid structure shown above had a carboxyl p*K* of 2.5 and an amino p*K* of 9.5. If we started in a very acidic solution, both of these groups would have hydrogens on them. This would result in a positive ion form of the amino acid.

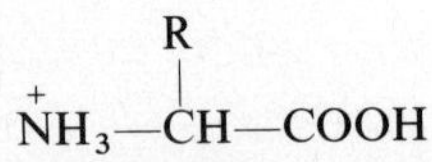

form of amino acid at very low pH

As we add base to this solution, the pH increases, and in the region of the carboxyl p*K*, the hydrogen is ionized from the carboxyl group. This changes the group from being neutral to being negatively charged. The amino acid now has both a positive and a negative charge.

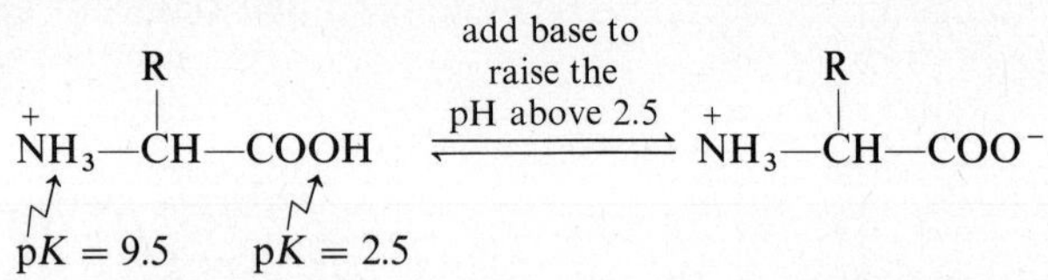

positive ion in very acid solution — *dipolar ion*

Notice that in these acidic solutions the extra proton on the amine group is not disturbed. At pH 2.5 the solution contains 50 percent positive ion and 50 percent dipolar ion.

The *dipolar ion* is neutral in overall charge. In fact, it does have two charges, so its properties are very different from what they would be if both functional groups were neutral. The dipolar ion is water soluble and forms a very high-melting crystalline solid.

The dipolar ion form predominates from pH 2.5 to pH 9.5. As we come near pH 9.5, the amine group begins to lose its extra proton. At pH 9.5 it is half lost, and above pH 9.5 the predominant form is a *negative ion*.

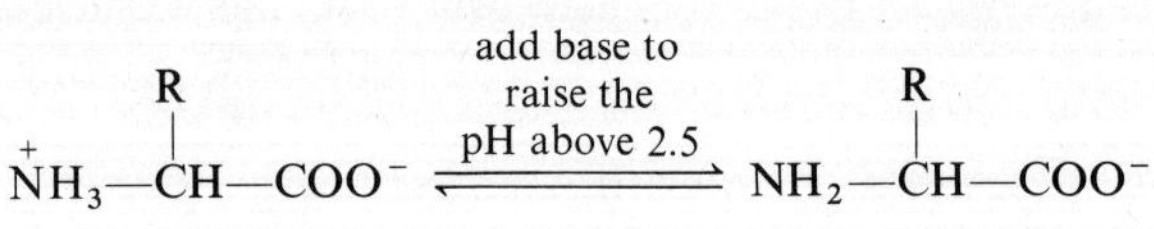

dipolar ion — *negative ion* in very basic solution

Notice that as base is added, protons are removed, and the structure moves from more positive to more negative charge.

Recognize that the carboxyl group p*K* varies somewhat from one amino acid to another, but not by as much as one pH unit. The same is true for the p*K* of the amino group. All amino acids, then, at neutral pH (7.0) should have a negatively charged carboxylate ion and a positively charged ammonium ion.

Exercise 9.3

The structures in Table 9.1 are not correct for pH 7, because they show neutral amine and carboxyl groups. Would structures ever exist as shown in Table 9.1? Redraw the carboxyl and amine groups of glycine, alanine, and serine as they would really be at pH 7.0.

Some of the amino acids have acidic or basic R groups, complicating the picture further. Aspartic acid, glutamic acid, lysine, histidine, and arginine all have R groups that can gain or lose a hydrogen ion. Below are the acidic and basic forms of each R group and the approximate p*K* for the hydrogen ion gain or loss.

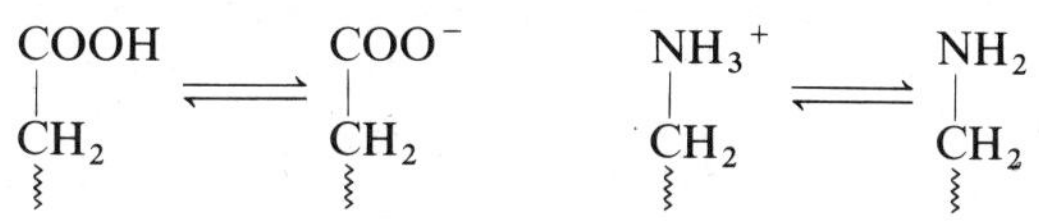

aspartic acid, 3.9
glutamic acid, 4.3

lysine, 10.5

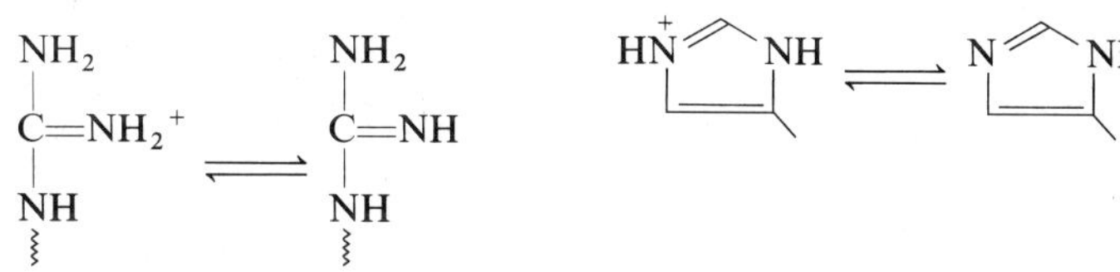

arginine, 12.5

histidine, 6.0

Exercise 9.4

Draw the complete structure with correct charges at each functional group for each of the preceding amino acids at pH 7.0.

9.6 Proteins: The Peptide Bond

Amino acids are joined to one another by *peptide bonds* to form linear polymers. The peptide bond is an amide linkage between the carbon 1 carboxyl of one amino acid and the carbon 2 amino of the next.

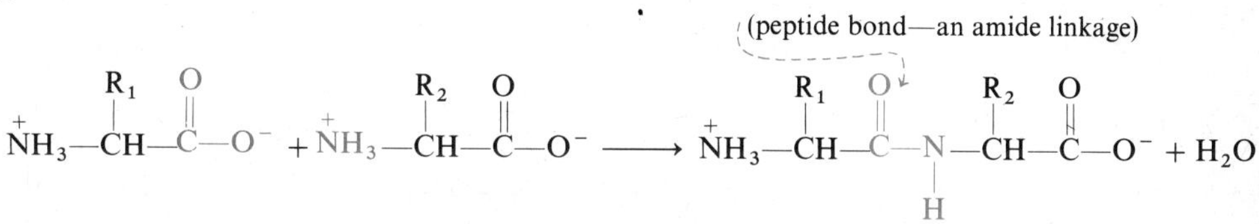

a peptide, specifically
a dipeptide

This structure, which contains two amino acid residues, amino acids that have lost water due to the condensation reaction, is called a *dipeptide*. It has an unreacted functional group at each end by which the chain can be extended to any desired length. Short polymer chains are called *peptides* or *polypeptides*, and longer chains are *proteins*. The size boundary between peptides and proteins is around 45–50 amino acids per molecule. Typical proteins have from 100 to several hundred amino acids per molecule.

Exercise 9.5

a) Write the structure of the dipeptide (two amino acid peptide) containing two alanines.

b) How many different dipeptides can be made from one alanine and one glycine?

9.7 The Primary Structure of Proteins

Unlike industrial polymers, or even the polysaccharides found in living systems, peptides or proteins do not occur in random chain lengths. Each molecule of a particular kind of protein in a given living organism has exactly the same amino acid composition. Furthermore, the amino acids of the protein are assembled in exactly the same order every time. This sequence of amino acids is called the *primary structure of protein*.[2] The sequence is usually written left to right starting with the amino acid whose amine group is not included in a peptide bond (N-terminal amino acid) and ending with the amino acid whose carboxyl group is not involved in a peptide bond (C-terminal amino acid).

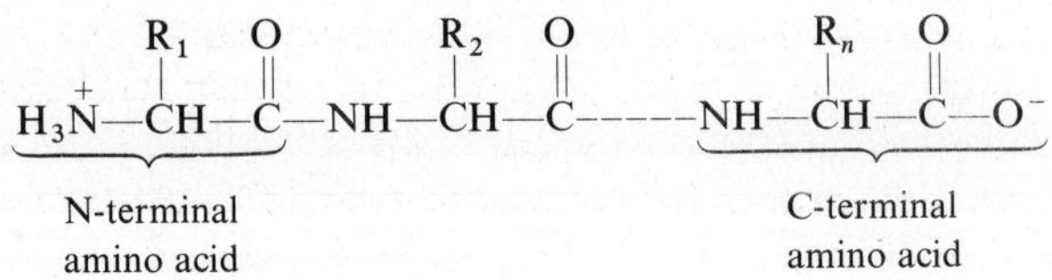

As we have indicated, all the molecules of a specific protein from the same individual have the same primary structure. Furthermore, the primary structure of samples of the same protein from different individuals of the same species is also very likely to be the same. However, if we compare that primary structure with that of the functionally equivalent protein from a different species, it will be

[2] In fact, the primary structure of all the proteins of the organism is the information that is contained in the genes. Because every cell has the same set of genes, and because the sets of genes are handed on to the next generation, the universality of the protein sequence in an individual or a species is not surprising.

different. Each species has unique primary structures for its set of proteins (Figure 9.2). In other words, the amino acid sequence of any sample of hemoglobin from the blood of a given cow is always the same; it is probably the same as the hemoglobin of another cow; and it is almost certainly not the same as the hemoglobin of any given dog.

9.8 Mutant Protein Sequences

Occasionally, when comparing samples of a specific protein from a number of individuals of the same species, we do find an individual whose protein has an unusual primary structure. There almost certainly is only one amino acid different from the normal sequence for that species. The change of an amino acid in the sequence is attributed to a genetic change called a *mutation*. The change in sequence does not occur by exchanging amino acids at some position after the protein is synthesized. Instead the genetic information directing the sequence of that protein has been altered. This has two important implications. First, because the genetic information is the same in every cell of an organism, the altered sequence appears in every cell. Second, because the composition of the genetic information of progeny is a compilation of the information provided by the parents, the message for an altered sequence is inherited by the progeny.

A mutant protein is seldom better than the normal protein. Usually the change caused by the amino acid replacement is either harmless or a minor disadvantage. In cases where the change severely damages the ability of the protein to function, the individual either can survive without the protein, possibly at a lowered level of efficiency, or in fact does not survive. These problems are often termed *congenital* or *inborn defects*, or *genetic diseases*. In the enzymes of amino acid metabolism alone, more than a dozen genetic diseases are observed, including phenylketonuria and alkaptonuria. These are discussed more completely in Chapter 19.

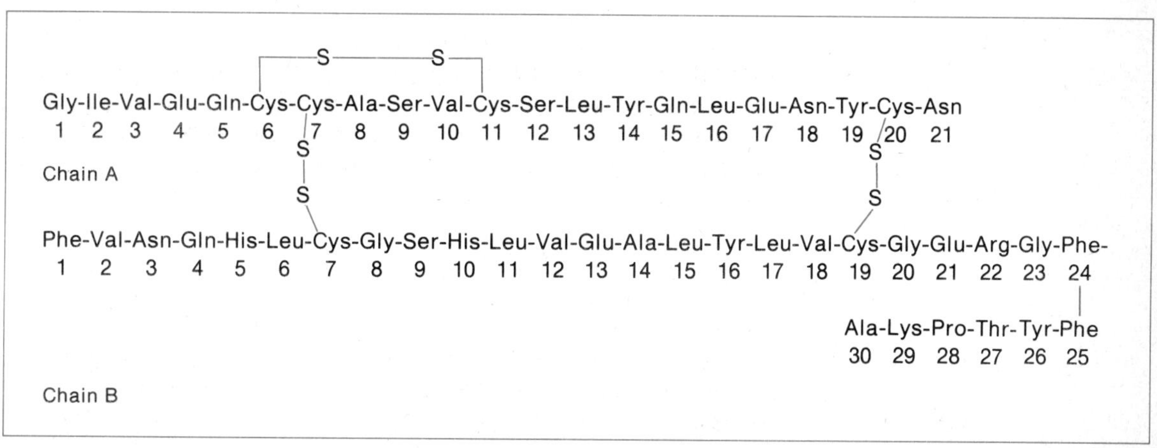

Figure 9.2
The sequence of the protein insulin
Insulin was the first protein for which the amino acid sequence was determined. This was accomplished by Dr. F. Sanger in 1953. He was later awarded the Nobel prize for this achievement. Notice that the protein is composed of two chains joined by disulfide linkages.

9.9 Hemoglobin

One of the most thoroughly studied proteins is human hemoglobin. Studies of hemoglobin sequences indicate that more than 99 percent of all humans have the same hemoglobin. However, about 150 different kinds of abnormal or mutant hemoglobin have been identified. On determining the sequence of these abnormal hemoglobins it is found that one amino acid differs from the normal sequence. These are often named after the city of their discovery. For example, in Hemoglobin Norfolk an aspartic acid has replaced a glycine, and in Hemoglobin Ypsilanti a tyrosine has replaced an aspartic acid.

On comparing variations in the primary structure of hemoglobin from humans and from other animals it is found that certain positions in the sequence always have the same amino acids, while at other positions a variation in the sequence is tolerated. The interpretation is that the properties of the protein depend on the sequence. The various amino acids in the sequence are not uniformly important in determining the properties, however. In the less important portions of the sequence, a change generated by mutation can occur without serious difficulty. Presumably, if change occurred in the more important parts of the sequence such that the functioning of the molecule was impaired, the fetus would not survive.

The most studied of the mutant hemoglobins is Hemoglobin S, the cause of sickle cell anemia. An individual who has received this trait from one parent may suffer no serious disadvantage and is said merely to have sickle cell trait. Half of the individual's hemoglobin is normal and half is Hemoglobin S. One who receives the trait from both parents has only Hemoglobin S and therefore has sickle cell anemia. Red blood cells containing Hemoglobin S are more fragile and therefore have a shortened useful life. This condition results in low red cell counts, a problem termed *anemia*. In a circumstance where oxygen supplies are low, the hemoglobin molecules are largely in the oxygen-free state. In this state the molecules form long fibers. These fibers distort or "sickle" the cell. The distorted cells clump together, blocking capillaries. Once the capillary is blocked, new supplies of oxygen-rich red blood cells cannot enter, so the oxygen concentration drops still further. This causes more sickling of the cells. The tissue destruction that results generally leads to the early death of the victim. In the United States about 10 percent of the black population have sickle cell trait, and 0.25 percent have sickle cell anemia.

Why did a mutation that seems to have such negative characteristics survive? One would expect it to have disappeared from the population because its carriers would be less likely to survive. In Africa, where the mutant originated, Hemoglobin S conferred an advantage on the individual. People with Hemoglobin S are more resistant to malaria, which has always been a major killer.

Why do the fibers form? When sickle cell hemoglobin is sequenced, it is found to have a valine in place of a glutamic acid. This amino acid R group is on the surface of the hemoglobin. The nonpolar valine is a hydrophobic bonding (greasy) spot on the hemoglobin surface. When the hemoglobin is in the nonoxygenated folding pattern, such hydrophobic sites line up and stick together. The hemoglobins then polymerize into long, straight chains or fibers that distort the cell.

9.10 Protein Charge and Electrophoresis

Because proteins are all composed of the same set of twenty amino acids, they are very similar in properties. As a result, they are very difficult to separate. Yet we are discussing separating proteins differing by only one amino acid in a long sequence.

Earlier in the chapter we saw the change of charge that occurs as the pH of an amino acid solution changes. Amino acids go from positively charged molecules in acid solution, through a region of counterbalanced positive and negative charge, finally to negatively charged molecules at basic pH. The protein molecule has a free carboxyl group at one end, a free amino group at the other end, and the R groups of aspartic acid, glutamic acid, lysine, arginine, and histidine between. Because these groups all gain and lose hydrogen ions, proteins change charge with changing pH just as do individual amino acids. At any pH the total charge of two proteins is different if their content of acidic and basic R group amino acids differs. We take advantage of this variation in charge for separation of proteins in a technique called *electrophoresis* (Figure 9.3).

An electrophoresis apparatus has a buffer-saturated support medium, at each end of which are electrodes connected to a high-voltage power supply. The medium can be paper, cellulose acetate, starch gel, agar gel, or acrylamide gel. On such a medium the charged protein moves toward the oppositely charged electrode. There is for

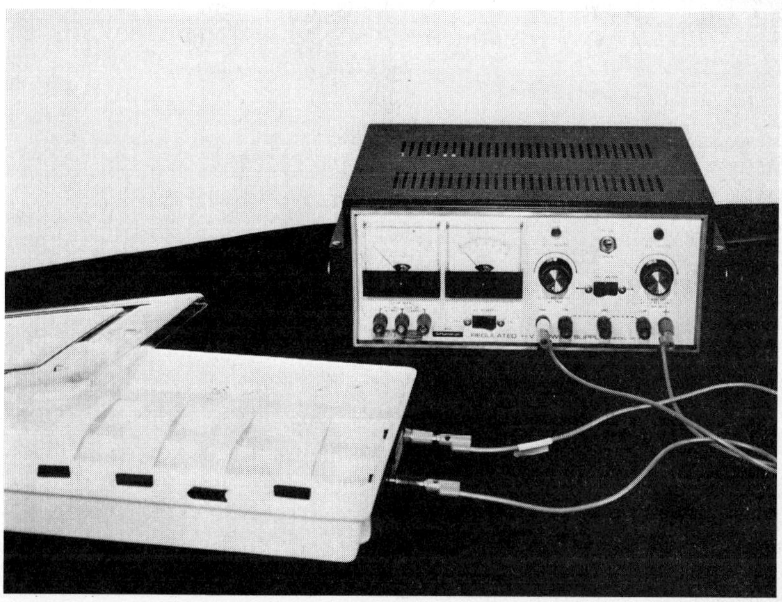

Figure 9.3
An electrophoresis apparatus
Electrophoresis separates molecules by charge. Applying the technique to proteins provides valuable information for researchers and for clinical analysts. (Photo by Catherine Feryok, Eastern Michigan University.)

each protein a pH at which the total + and − charges are exactly equal. At this pH the protein does not move when the power supply is on. This pH is termed the *isoelectric point* of the protein.

To perform a separation, the medium is soaked with a buffer of some desired pH. The mixture of proteins is placed in a narrow band or small spot on the medium. The proteins gain or lose hydrogen ions to establish the charged pattern appropriate to the pH of the buffer. The power supply is turned on, and the protein moves toward the oppositely charged electrode. After a period of time, the power supply is disconnected, and the medium is stained for detection of the protein. Patterns obtained in this fashion, for example from blood plasma, can be very powerful diagnostic tools.

Exercise 9.6

Below are some of the substitutions that occur in hemoglobin mutant types. Indicate which you would expect would easily separate by electrophoresis from normal hemoglobin and which would not. Remember, electrophoresis detects differences in charge.

Hemoglobin type	***Normal residue***	***Replacement***
M Boston	Histidine	Tyrosine
G Philadelphia	Asparagine	Lysine
S (Sickle cell)	Glutamic acid	Valine
M Milwaukee	Valine	Glutamic acid
E	Glutamic acid	Lysine
Koln	Valine	Methionine

9.11 *Quaternary Amines*

Just as the unused pair of electrons in a tertiary amine can bind a hydrogen ion to form a positive ion, they can also bind an alkyl group to form a positive ion.

$$(CH_3)_3N: + H^+ \rightleftharpoons (CH_3)_3\overset{+}{N}H$$

$$(CH_3)_3N: + CH_3{}^+ \longrightarrow (CH_3)_4\overset{+}{N}$$

Such a structure is called a *quaternary amine*. As a positive ion, it is found in conjunction with a negative ion such as Cl^- or Br^-. Quaternary amines are named by listing the groups substituted on the nitrogen just as was done with other amines. Instead of ending with the word "amine," however, the word "ammonium" is used. Then the negative ion associated with the positively charged quaternary structure is named.

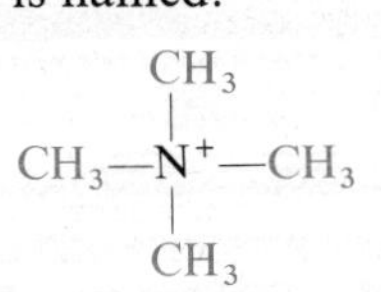

tetramethyl ammonium chloride

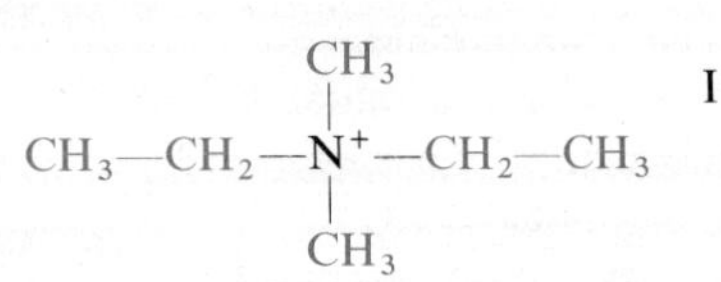

dimethyl diethyl ammonium iodide

Exercise 9.7

Name these structures:

a) 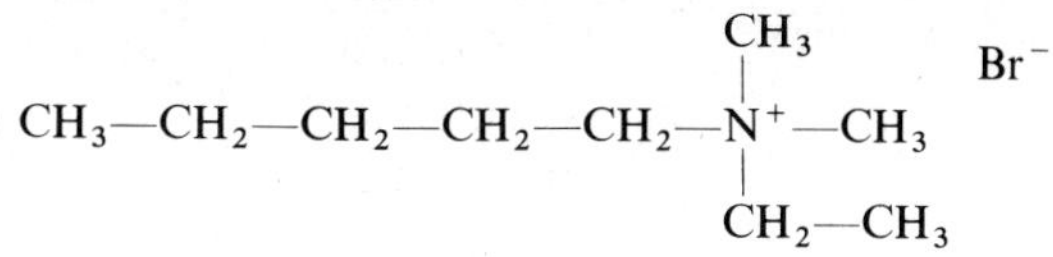

b) 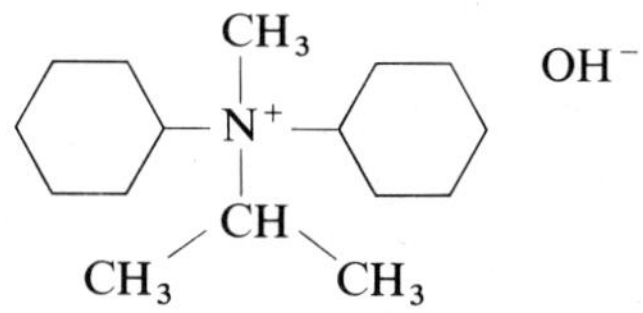

9.12 Uses of Quaternary Amines

Quaternary amines have been used for some time as germicidal detergents in food processing and medicine. A quaternary amine with one large alkyl group meets the requirements for detergent properties (Chapter 7). It has a nonpolar position, the alkyl group, to dissolve into the nonpolar substance to be suspended, and a charged group, the quaternary amine, to interact with the water. Because they are positively rather than negatively charged, quaternary amine detergents are often called *invert soaps*.

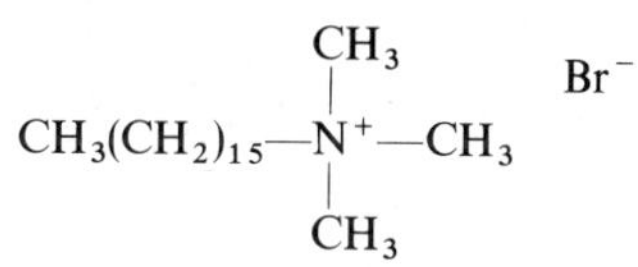

cetyl trimethyl ammonium bromide (Cetavlon)

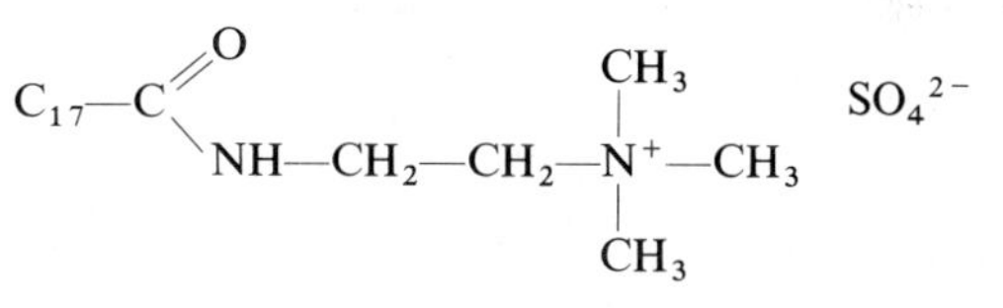

Sapamine

9.13 Choline

A biologically important quaternary amine has already been encountered in the discussion of phosphoglycerides (Chapter 9). The choline of phosphatidyl choline is not just a lipid component but rather part of other significant structures, for instance the acetyl choline involved in the transmission of nerve impulses (Chapter 21).

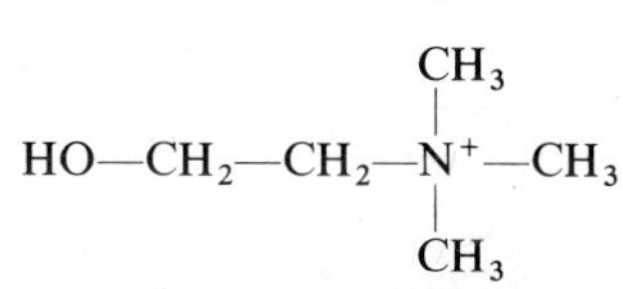

choline

Table 9.2
Amphetamines

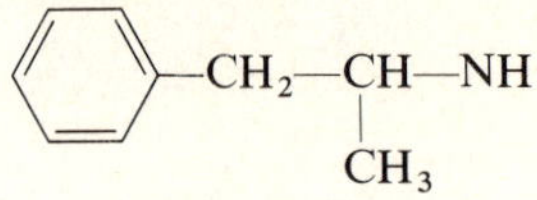

amphetamine
(Benzedrine)

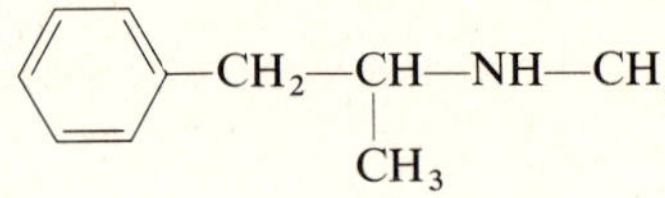

N-methyl amphetamine
(speed)

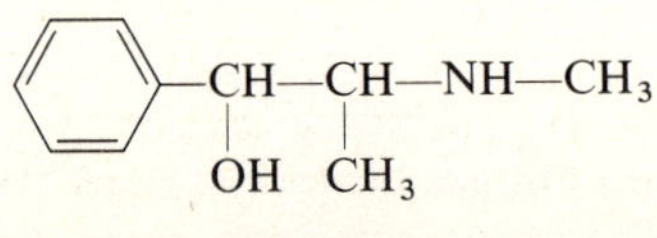

Ephedrine

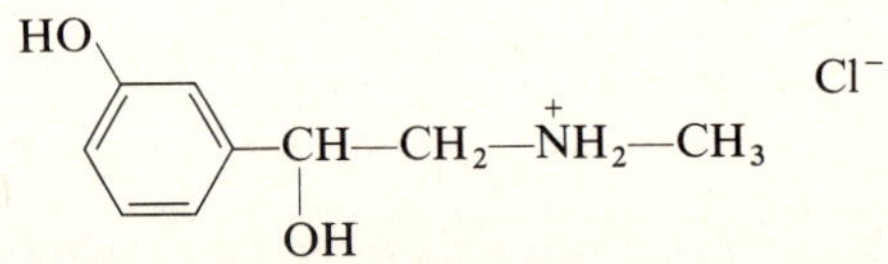

Phenylephrine
(Neosynephrine)

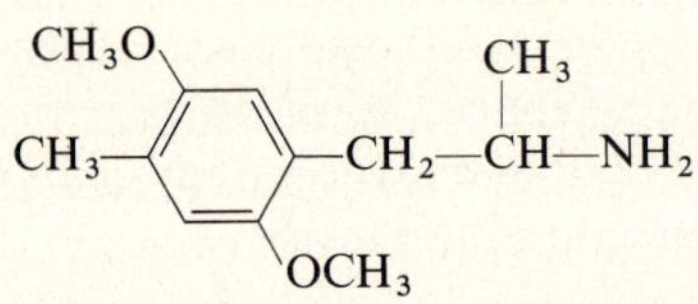

2,5-dimethoxy-4-methylamphetamine
(STP)

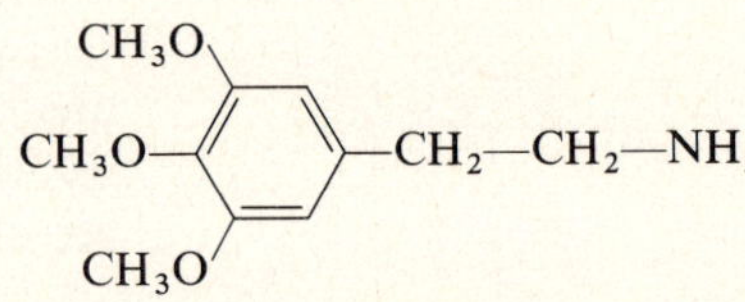

mescaline
(from peyote cactus)

9.14 Catecholamines

Epinephrine (adrenaline) and norepinephrine are amine structures produced by the adrenal medulla. They are sometimes collectively called the *catecholamines*. Elevating levels of these hormones in the blood produces an increase in blood pressure and heart rate and an increase in blood flow to the skeletal muscles. The musculature of the digestive tract relaxes. Blood sugar and fatty acid levels are elevated. In simple terms, the person is prepared for battle or flight.

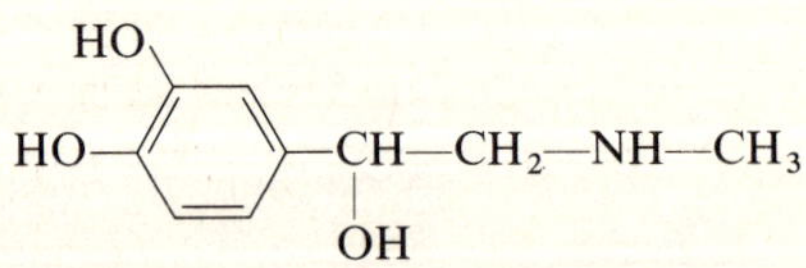

epinephrine

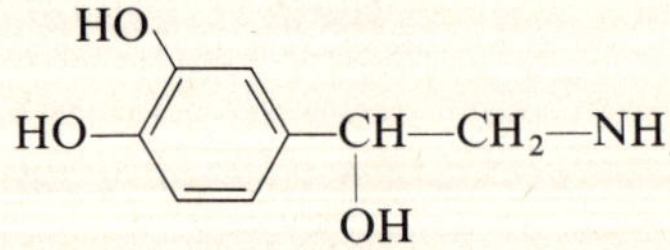

norepinephrine

These hormones are strikingly similar in structure to amphetamines, a family of synthetic drugs. Amphetamines are stimulants and have been used as appetite suppressants. Misuse and overuse of these drugs has led to serious problems. Furthermore, they have become an important part of the illegal drug scene as "uppers" (Table 9.2). How such drugs exert their physiological effect is discussed in Chapter 21.

Summary

Nitrogen can obtain an electron octet by forming three covalent bonds. This leaves two unbonded electrons. Amines are structures wherein the covalent bonds formed by nitrogen include at least one alkyl group. Amines with one alkyl are termed primary, with two secondary, and with three tertiary.

Amino acids are carboxylic acids with an amine group at carbon 2. They are used chiefly to build protein polymers. Of the twenty amino acids found in protein, all but glycine are asymmetric at carbon 2, and all asymmetric carbons are in the L form. Dietary protein is digested into individual amino acids before absorption. Several amino acids, termed essential, cannot be synthesized by humans and so must be in the diet in adequate amounts.

Amines are bases, being able to bind a hydrogen at the unused pair of electrons. Amino acids, therefore, have both an acid and a basic group, which causes them to have charged structures regardless of pH. If instead of a hydrogen a tertiary amine has an alkyl group bound to the fourth pair of electrons, a positively charged quaternary amine is formed. When one of the alkyl groups is large, quaternary amines can serve as detergents.

Proteins are formed by linking the carboxyl of one amino acid to the amine of the next in an amide linkage called the peptide bond. The sequence of amino acids in a specific protein from a particular organism is always the same and is called the primary structure of the protein.

The sequence of proteins varies from one species to another. Some amino acids in a particular sequence are important to function and never vary; others seem less critical. Sequence also varies occasionally within a given species, such as the human population. Such variations are due to genetic alterations and are termed mutations. The most clinically important abnormal hemoglobin is the one associated with the hereditary disease sickle cell anemia. As with all abnormal hemoglobins, only one amino acid is different from the normal sequence. Victims of sickle cell anemia have weakened red blood cells and thus are anemic. The variant amino acid also triggers a polymerization of the deoxygenated form of the hemoglobin, which distorts the cells into a sickle shape, blocking capillaries and causing tissue damage.

Problems

1. Name the following:

a) 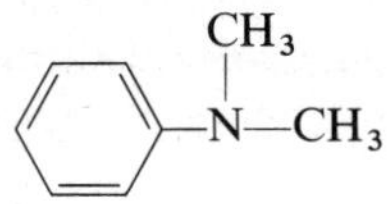

b) 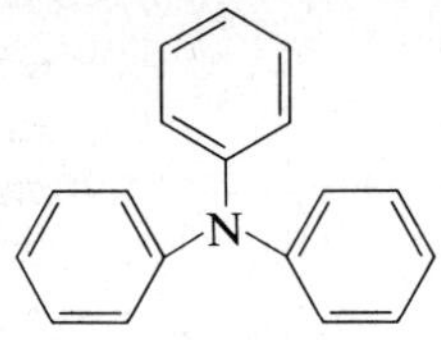

c) 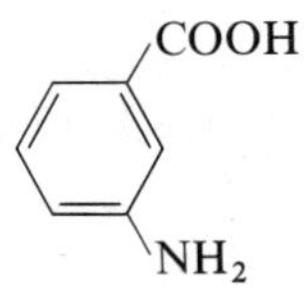

d) $CH_3-CH(NH_2)-CH(NH_2)-CH_3$

e) $CH_3-CH(NH_2)-COOH$

f) $CH_3-CH_2-CH_2-CH_2-N^+(CH_3)_2-CH_3 \quad Cl^-$

2. Draw structures for the following:

a) diethylamine
b) N,N-diethylaniline
c) 2,6-diethylaniline
d) N-methylanthranylic acid
e) proline
f) tetraethylammonium bromide

3. In Problems 1 and 2, indicate whether each amine is primary, secondary, tertiary, or quaternary.

4. Draw the following:

a) Given that the p*K* values for valine are 2.3 (—COOH) and 9.6 ($-NH_3^+$), draw the most prevalent structure of valine at pH 1.0, at pH 6.0, and at pH 11.0.

b) Given that the p*K* values of aspartic acid are 2.0 (—COOH), 3.9 (R group—COOH), and 10.0 ($-NH_3^+$), draw the structure of aspartic acid that would be most prevalent at pH 1.0, 3.0, and 6.0. At neutral pH, toward which pole of an electrophoresis apparatus would the compound migrate?

c) Given that the p*K* values for lysine are 2.2 (—COOH), 9.0 ($-NH_3^+$), and 10.5 (R group $-NH_3^+$), what are the two forms most prevalent at pH 9.0, and what percentage of each would be found?

d) Draw the structure of a dipeptide of glycine and serine, with the glycine being N-terminal.

e) Draw the structure of the tripeptide alanine (N-terminal)-aspartic acid—lysine. Place all the charges on the molecule that you might expect at pH 7.0. What would you predict about its behavior in an electrophoresis apparatus?

5. You separate three proteins at neutral pH: hemoglobin, insulin, and papain (a plant enzyme). The separation appears as follows:

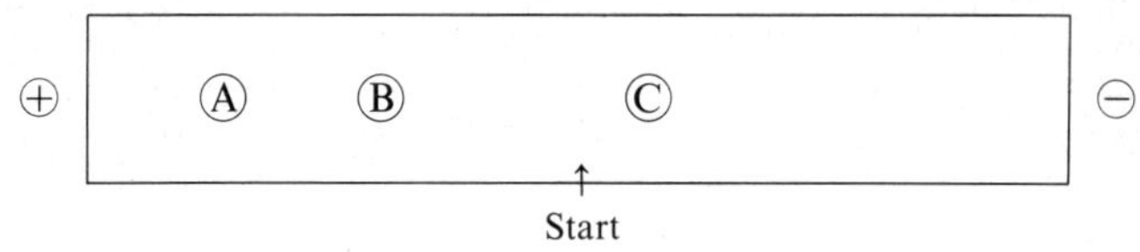

Given the following information about the amino acid composition of each protein, identify the three spots.

	Percentage of Each Amino Acid				
Protein	*Aspartic acid*	*Glutamic acid*	*Lysine*	*Arginine*	*Histidine*
Hemoglobin	10.0	7.4	10.6	3.5	8.5
Insulin	7.0	18.0	2.6	3.0	5.4
Papain	9.8	10.9	5.0	7.0	0.8

6. You separate three proteins at neutral pH by electrophoresis. The proteins and their isoelectric points are given:

Protein	***Isoelectric Point***
Pepsin	1.1
Hemoglobin	6.7
Lysozyme (an enzyme that destroys bacteria)	11.0

Assign the spots below to the appropriate protein.

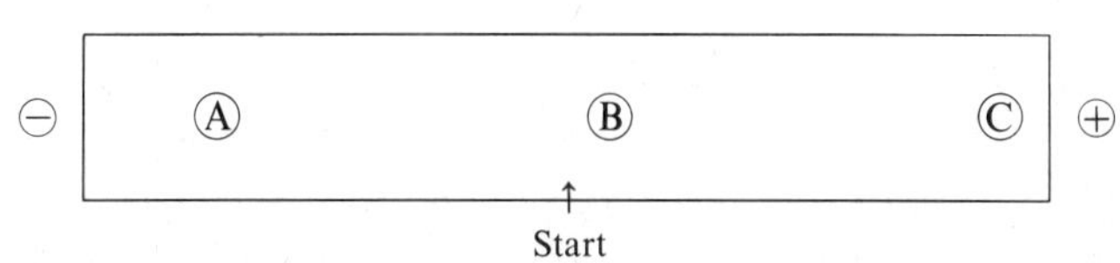

7. Comparing the information in Problems 5 and 6, what would you predict about the amino acid composition of pepsin and lysozyme?

8. What amino acid(s):

a) Doesn't conform to the general structure on page 221?

b) Have more than one asymmetric carbon?

c) Is a substituted phenol? (Is the substitution *o*, *m*, or *p*?)

d) Contain an amide structure?

e) Contains a thioether?

f) Have either a functional group or a ring structure substituted on alanine?

g) Has no asymmetric carbon?

9. Glutathione is a tripeptide very widely distributed in living organisms. It acts in conjunction with a number of enzymes. In one interesting example, some flies produce an enzyme, DDT dehydrochlorinase, that attacks the DDT structure with the help of glutathione. Study the structure. What amino acids are present? Indicate how it is a very unusual peptide.

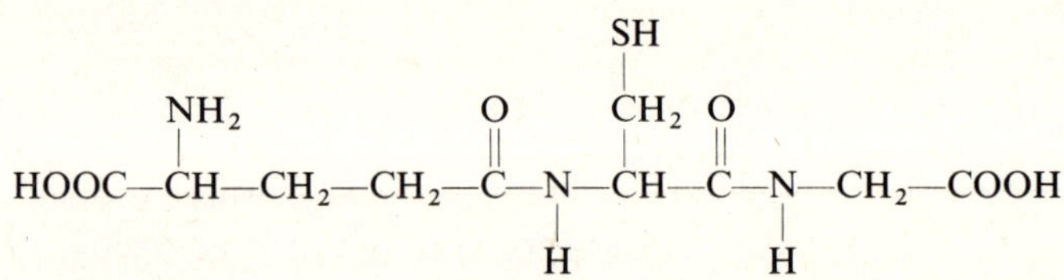

10. Considering what has been said in this section about mutation, what do you think is the likelihood of genetic damage from radioactivity producing giant grasshoppers, lizards, spiders, or any of the other unfriendly monsters created by the moviemakers?

10
Heterocycles

Objectives

In this chapter we:

1. Survey the structures, names, and simple chemistry of some heterocyclic structures.
2. Learn of the source and structure of a variety of alkaloids.
3. Discuss drugs of abuse.
4. Survey some heterocyclic medicinals, both natural and synthetic.
5. Learn the structures and functions of nucleosides and nucleotides.

To this point, most of the compounds discussed have been acyclic (no rings) or homocyclic (rings formed by carbon atoms only). A large and important area of chemistry is concerned with cyclic compounds where the ring includes atoms other than carbon. In nature some of the most important structures, including the nucleic acids, carriers and translators of genetic information, and many of the vitamins, include nitrogen heterocycles. In fact, you already have seen some heterocycles. The hemiacetal form of carbohydrates is an oxygen heterocycle, and the R group of histidine contains the imidazole ring, a nitrogen heterocycle.

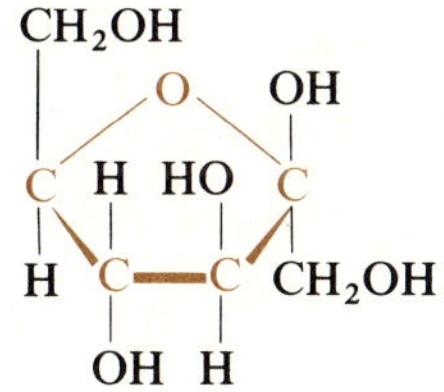

β-D-fructose

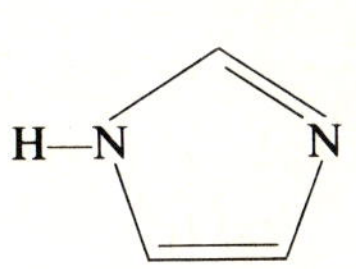

imidazole ring
of histidine

To begin consideration of heterocycles, look at the structure and names of the more common ring systems (Table 10.1).

Exercise 10.1

Using the information in Table 10.1, identify the heterocyclic ring system in each of the following compounds:

Table 10.1
Common Heterocyclic Ring Systems

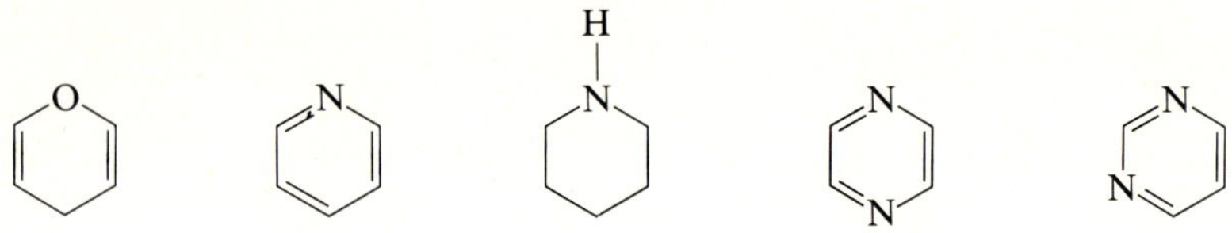

pyran | pyridine | piperadine | pyrazine | pyrimidine

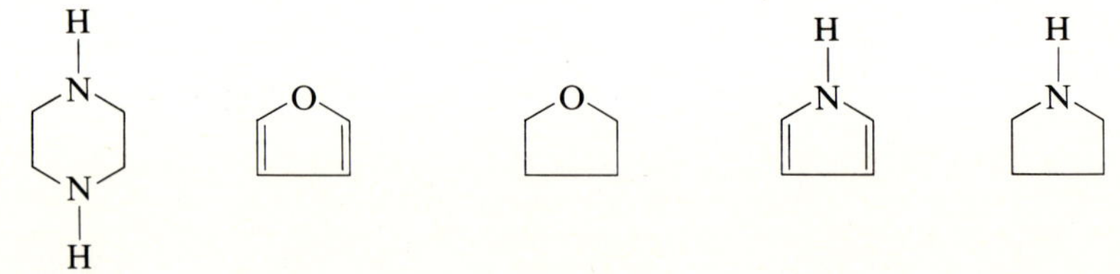

piperazine | furan | tetrahydrofuran | pyrrole | pyrrolidine

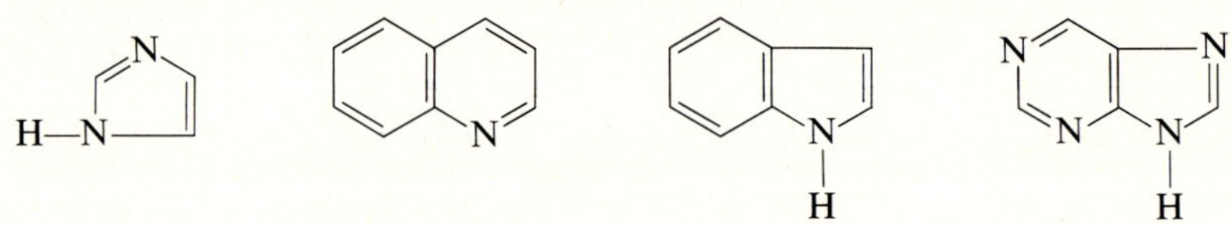

imidazole | quinoline | indole | purine

Exercise 10.1 continued

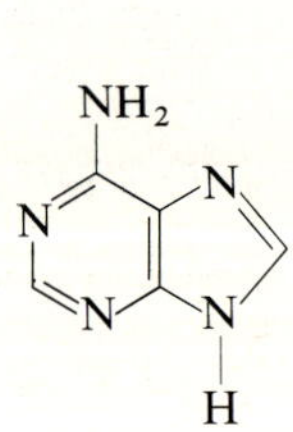

a) adenine

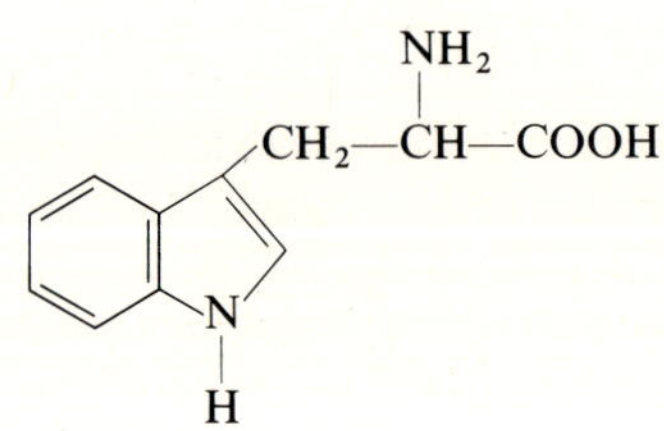

b) tryptophan

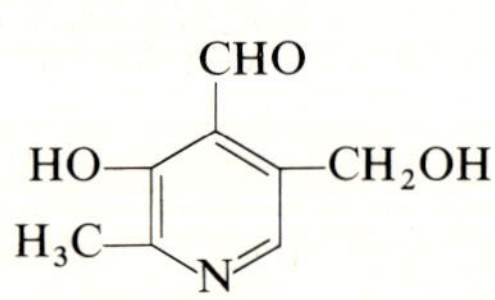

c) pyridoxal (vitamin B_6)

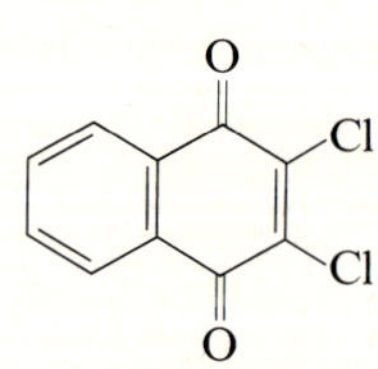

d) 2,3-dichloro-1,4-naphthoquinone (a fungicide)

Chemically, do these ring compounds belong to any of the functional group classes? Oxygen heterocycles might be ethers, esters, or acid anhydrides. Consider the following examples:

cyclic ether, tetrahydrofuran (furan with four added hydrogens)

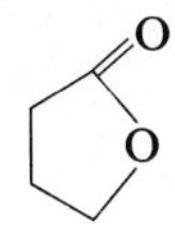

cyclic ester (the lactone of 4-hydroxy butanoic acid)

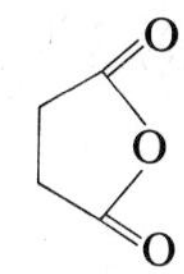

cyclic acid anhydride (succinic anhydride)

Exercise 10.2

In each of the following structures, locate and identify examples of cyclic ethers, lactones, and cyclic anhydrides:

a)

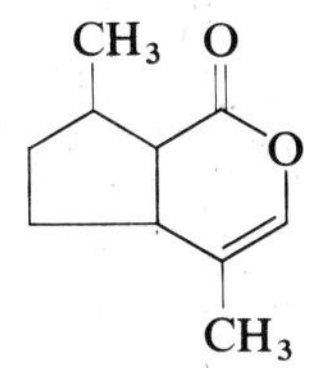

nepetalactone (chief component of oil of catnip)

b)

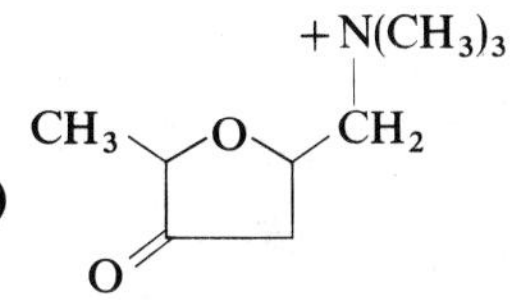

muscarine (poisonous mushroom alkaloid—used as a cholinergic)

c)

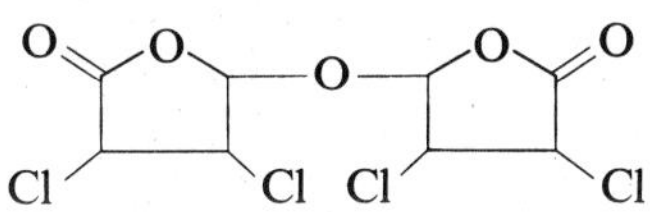

mucochloric anhydride (used as a fungicide)

d)

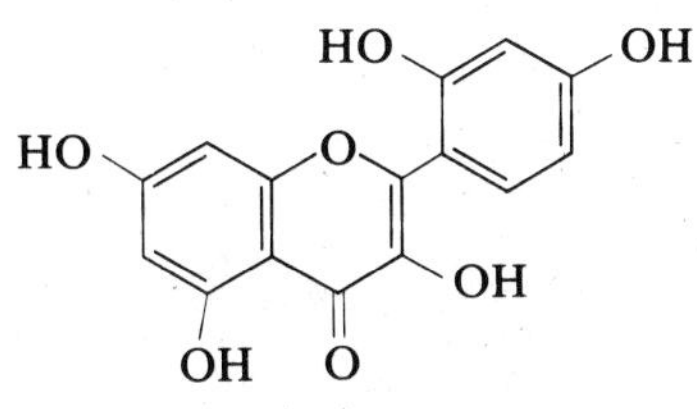

morin (yellow dye derivative in flower pigments)

e)

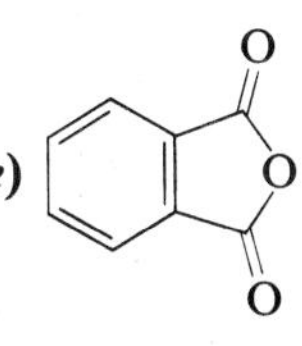

phthalic anhydride (plastics monomer)

The chemistry of such cyclic structures is the same as that of the noncyclic equivalents. Similarly, the nitrogen heterocycles may be chemically classed as amines or as amides.

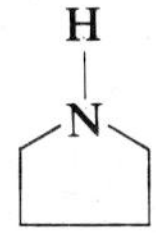

a cyclic amine, pyrrolidine

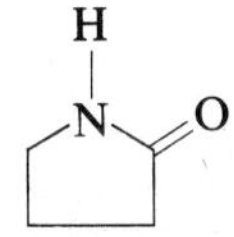

a cyclic amide, 2-pyrrolidone

Exercise 10.3

In each of the following heterocyclic structures, identify examples of cyclic amines and cyclic amides:

a)
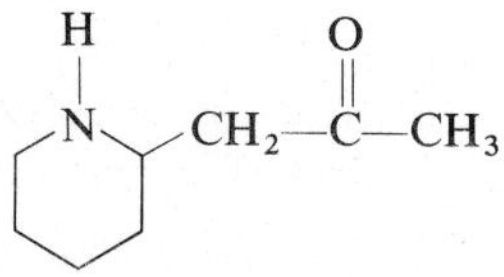

pelletierine (an agent for use against internal worms)

b)
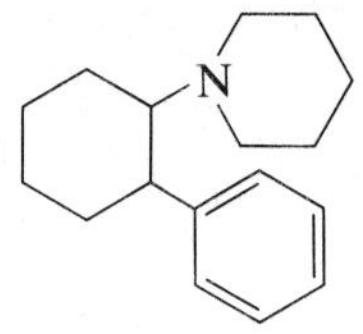

phencyclidine (an anesthetic)

c)
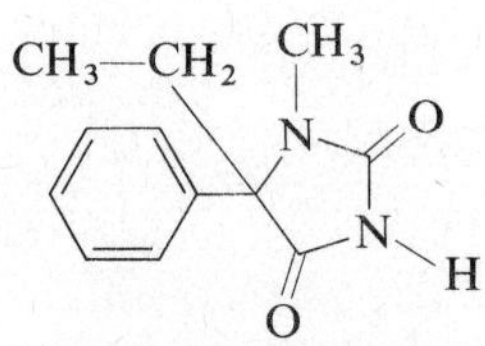

methetoin (an agent for use against convulsions)

d)
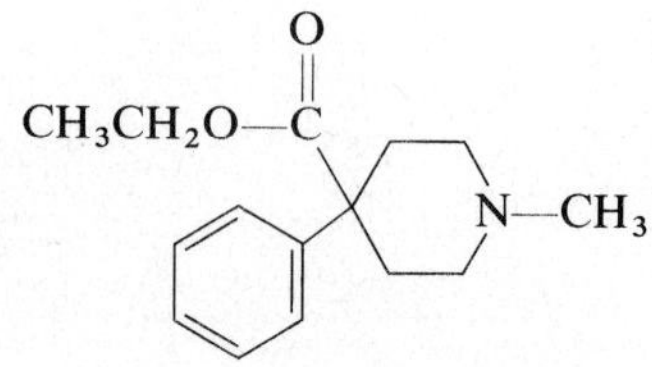

meperidine (an analgesic)

Naming Heterocycles. Derivatives of heterocyclic ring systems often are complex structures and are referred to by short, nonsystematic names for convenience. It is possible to provide systematic names using the name of the ring system. Substituents are named by listing the substituted group preceded by the number of the position of substitution. It is necessary to find the ring-numbering pattern in a reference to do this correctly, because each heterocycle has a standard numbering pattern. Usually the ring is numbered by assigning 1 to a noncarbon ring atom, then numbering in such a direction as to give other noncarbon ring atoms as small numbers as possible.

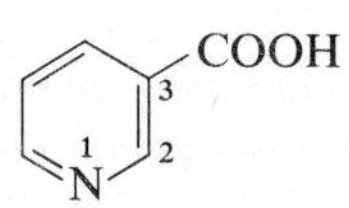

pyridine-3-carboxylic acid (used by veterinarians to treat pellagra)

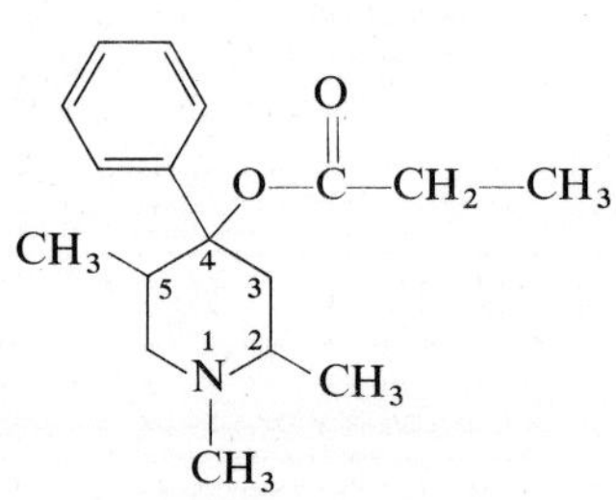

1,2,5-trimethyl-4-phenyl-4-proprionoxy-piperadine (used for certain heart problems)

The earlier convention that rings are numbered so that the substituents have the smallest set of numbers is followed wherever possible, as has been done here.

> ### *Exercise 10.4*
> Refer to the structure of pyridoxal in Exercise 10.1. Fill in numbers in the blanks provided to generate a correct name for pyridoxal: __-formyl-__-hydroxy-__-hydroxymethyl-__-methylpyridine.

10.1 Alkaloids

Plants produce a variety of nitrogen heterocycles that are referred to as *alkaloids*. These tend to be complex structures. Often our understanding of alkaloids has progressed only to the stage of isolation and structural analysis. Alkaloids occasionally have a trivial function or importance. For example, a few flavors relate to alkaloid structures. Piperine is an alkaloid in black pepper.

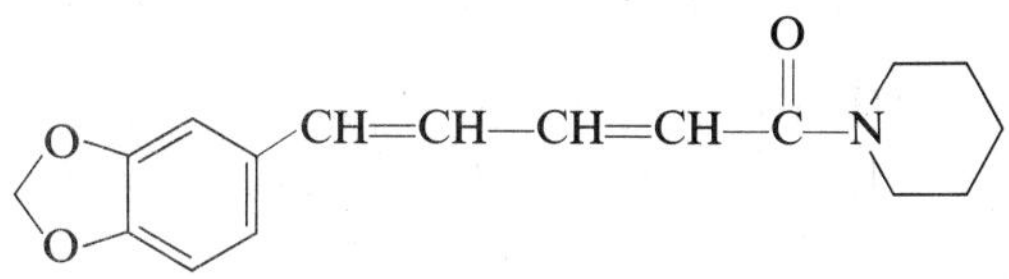

piperine
(from black pepper, used to flavor brandy and to kill insects)

Alkaloids have been successfully studied for uses in medicine. It is easy to generate the attitude that all alkaloids have powerful physiological or psychological effects on humans. In truth, most alkaloids are innocuous, but it is those that are not that are at the center of scientific and often public attention. Not surprisingly, we seem to know much more about the effects of alkaloids on humans than we do about their usefulness to the plant that produces them.

Quinine, extracted from cinchona bark, is used to combat malaria, which is still one of the most serious health problems in the world. The quinine structure has two linked heterocyclic systems.

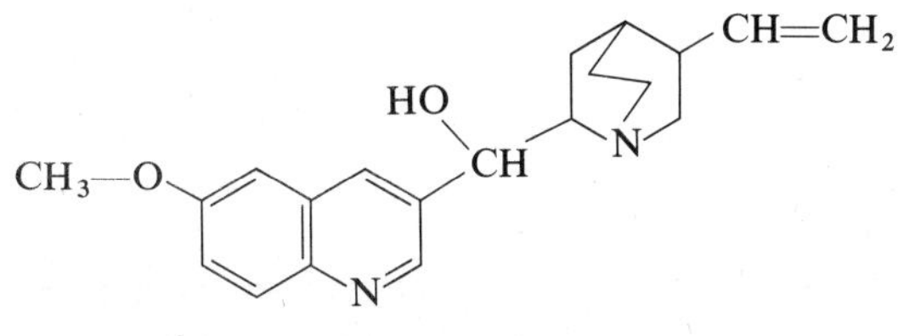

quinine
(antimalaria drug)

Rauwolfia roots yield reserpine, a compound that sedates or tranquilizes.

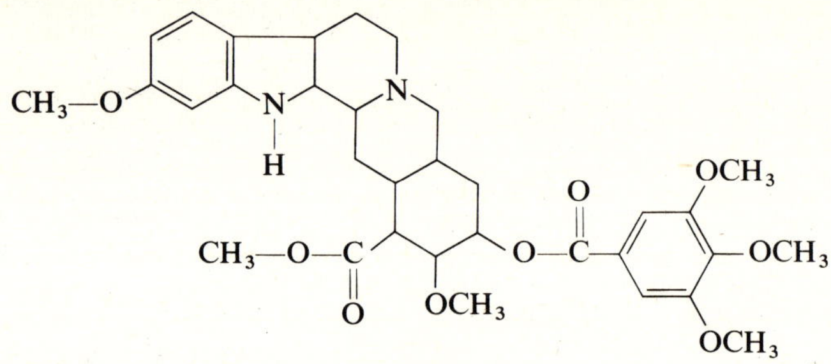

reserpine
(a tranquilizer)

Several antibiotics have heterocyclic structures. The penicillins, for example, have a complex heterocyclic ring system.

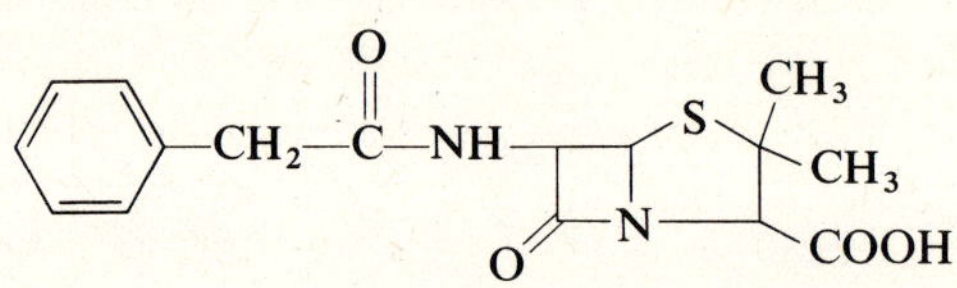

penicillin G
(an antibiotic; notice the four-membered ring)

Many of the alkaloids that we think of negatively because of their poisonous nature are, in the proper dosage and with controlled use, valuable medicinals. Curare is used by some South American Indians to poison arrows. The active component is *d*-tubocurarine, a powerful muscle relaxant that has been used in small doses for this purpose in medicine (Figure 10.1). Succinyl choline iodide (Chapter 8) is similar in effect and is more commonly used.

Other poisonous alkaloids include coniine. This is extracted from hemlock bark and is believed to be the substance that killed

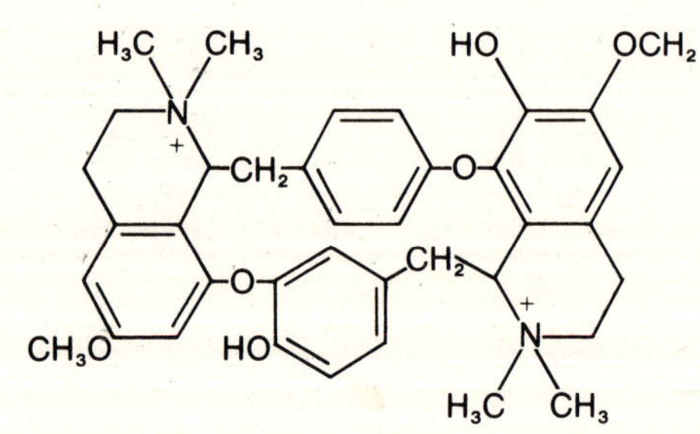

Figure 10.1
Tubocurarine chloride
This is a curare alkaloid that serves as a muscle relaxant and can cause death by respiratory paralysis.

Socrates. Nicotine, less toxic, represents about 75 percent of the total alkaloid in tobacco.

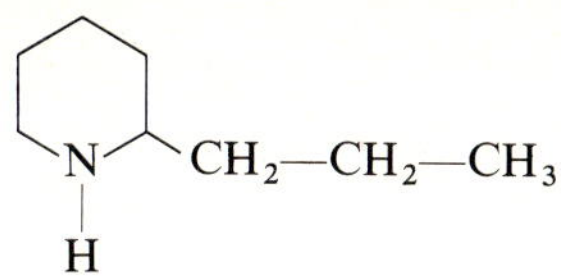

coniine
(hemlock bark alkaloid)

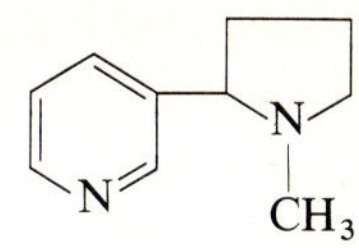

nicotine
(tobacco alkaloid)

10.2 Alkaloid Drugs of Abuse

Several alkaloid compounds produce psychological effects and have been used illegally to induce these effects. Ergot (a cereal fungus) alkaloids produce derivatives of lysergic acid. These compounds, particularly the diethyl amide (LSD), are hallucinogens. They may function by disturbing the natural serotonin (depressant)–epinephrine (stimulant) balance in the brain. Lysergic acid has an indole structure present, as does serotonin. Bufotenine, obtained from toadstools (and toads, interestingly) is very similar to serotonin and is also a hallucinogen. The ability of LSD to produce hallucinations long after use hints at incorporation into or permanent alteration of some brain structure.

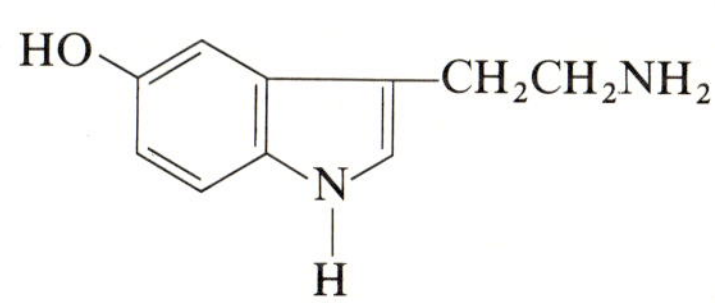

serotonin

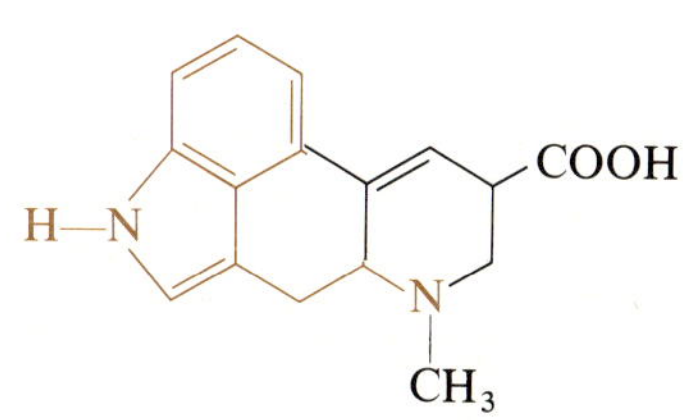

lysergic acid
(ergot alkaloid)

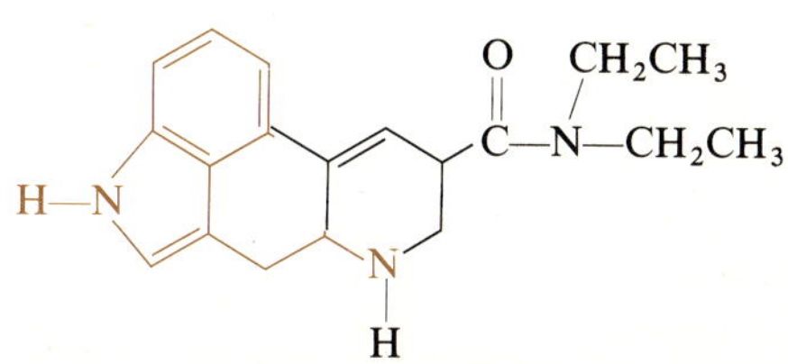

lysergic acid diethylamide
(LSD)

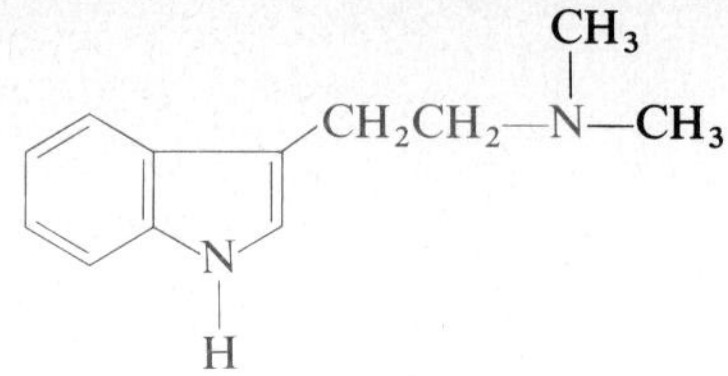

bufotenine
(toadstool alkaloid)

Opium, obtained from poppies, is a mixture of over two dozen alkaloids. These include narcotic substances of varying effectiveness. Codeine, for example, is used in cough remedies and as a pain reliever of greater strength than aspirin. Care must be exercised in its use medicinally, because it can be addictive. When addicted, the individual experiences a need for a compound for the continuation of normal function. A serious hazard lies in the need for continuously expanded dosage to alleviate withdrawal symptoms. As dose increases, so do harmful side effects, with an often fatal consequence. Morphine, about 10 percent of the total opium alkaloid, has been used to relieve pain. Its overuse for this purpose has led to unintended addiction. Heroin, one of the most serious drugs of abuse, is the diacetate of morphine. Narcotine and papaverine are structurally similar. Papaverine has been used as an antispasmodic.

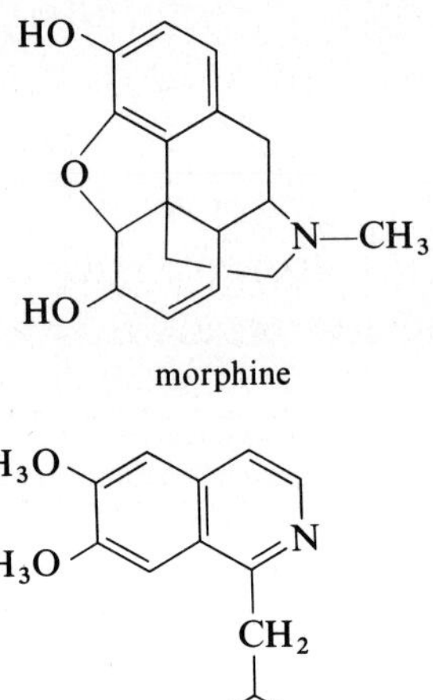

morphine

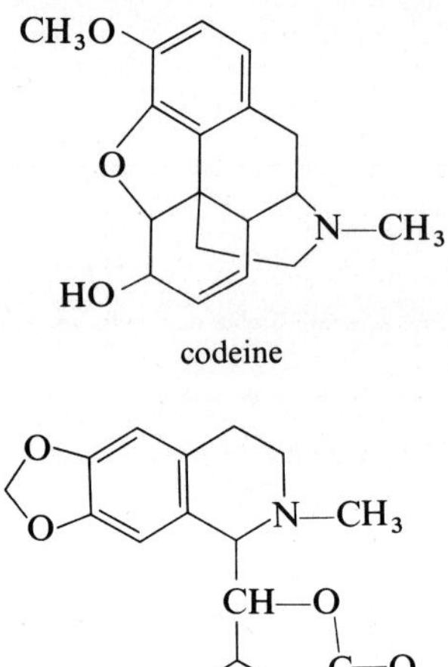

codeine

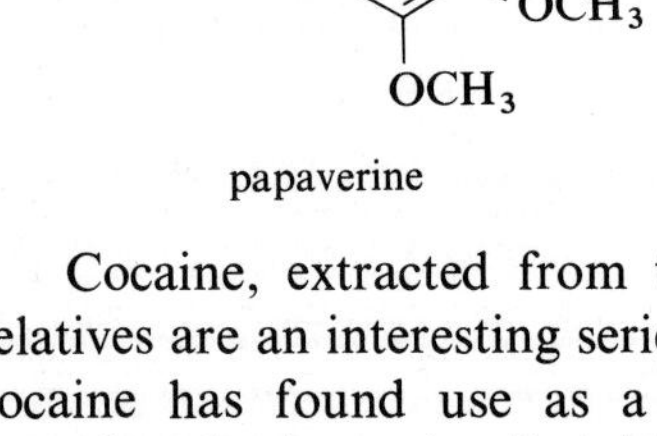

papaverine

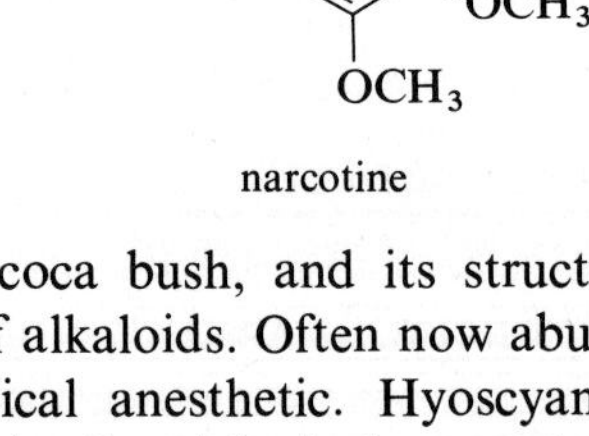

narcotine

Cocaine, extracted from the coca bush, and its structural relatives are an interesting series of alkaloids. Often now abused, cocaine has found use as a topical anesthetic. Hyoscyamine (atropine) is the toxic alkaloid of deadly nightshade, mandrake, and henbane. It is used medicinally as an "anticholinergic," that is, a drug antagonistic to nerve impulse transmission (Chapter 21). These sorts of compounds are used in sleeping pills. Scopolamine has a reputation as a truth serum. It is, in fact, a central nervous system depressant and hallucinogen. Medicinally it has been used as an anti-Parkinsonism agent and to prevent motion sickness.

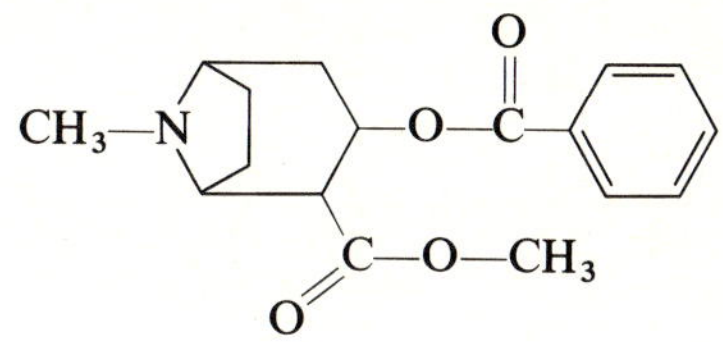

cocaine

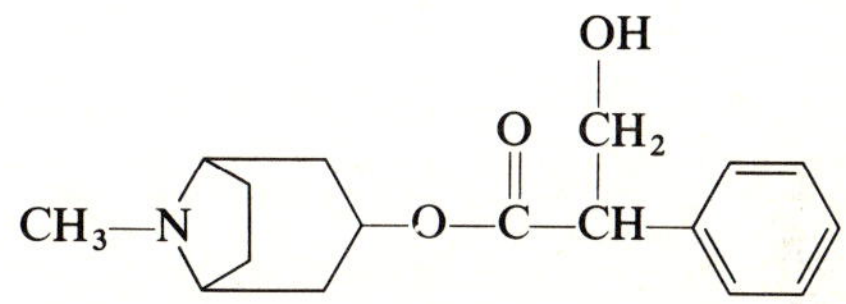

hyoscyamine
(atropine)

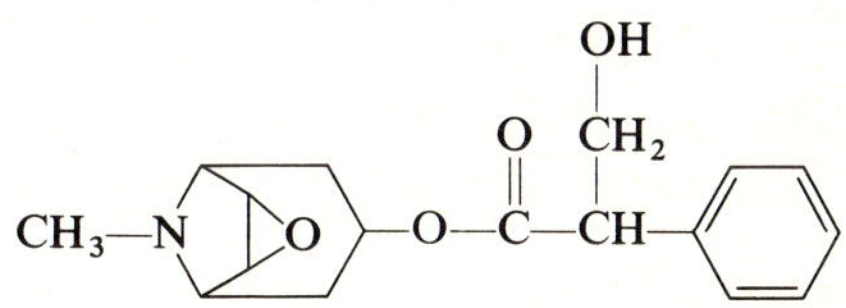

scopolamine

Mescaline, from peyote cactus, is a hallucinogen, but it is not a heterocycle.

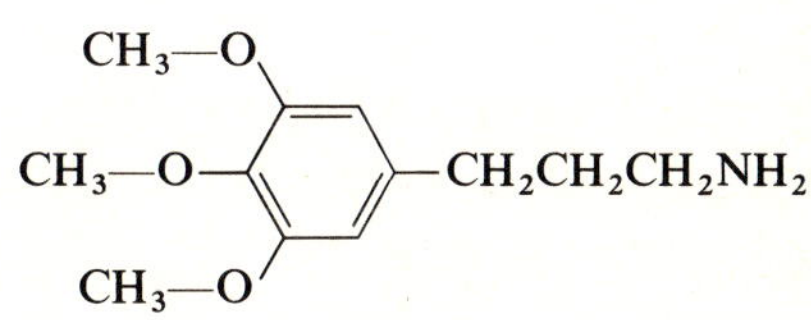

mescaline

The active principles of marijuana, tetrahydrocannabinol and related structures, are also heterocycles.

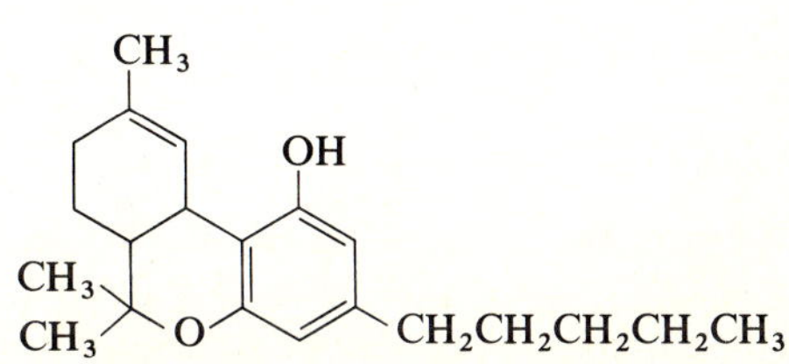

tetrahydrocannabinol (from marijuana)

10.3 Synthetic Heterocyclic Medicinals

Studies of molecules with physiological effects continue to be a major field of research. Organic chemists try to synthesize new compounds that have structural features suspected of relating to effectiveness in known medicinals. It is hoped that scientists can create compounds capable of treating ailments but displaying fewer undesirable side effects and lowered toxicity. These compounds must then undergo exhaustive testing to establish effectiveness and toxicity before approval is obtained for manufacture and sale. The sophistication and expense of these procedures can seldom be borne by other than the relatively few large pharmaceutical companies in the world.

Often the compounds produced synthetically as medicinals are heterocycles. For example, synthetic tranquilizers include the following:

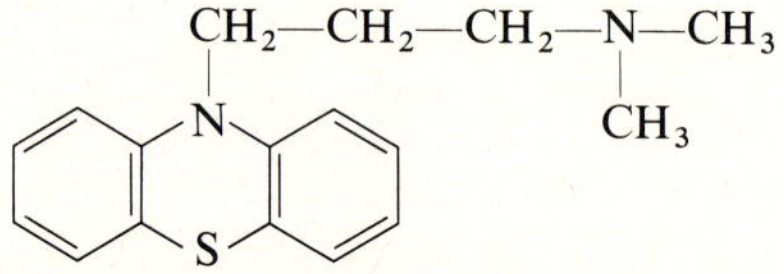

chlorpromazine
(thorazine)

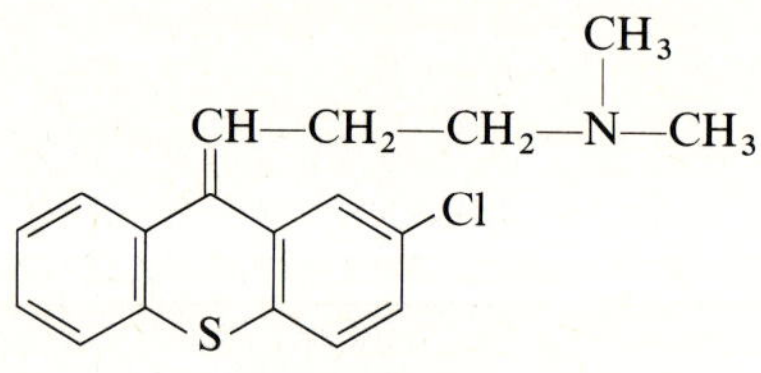

chlorprothixene

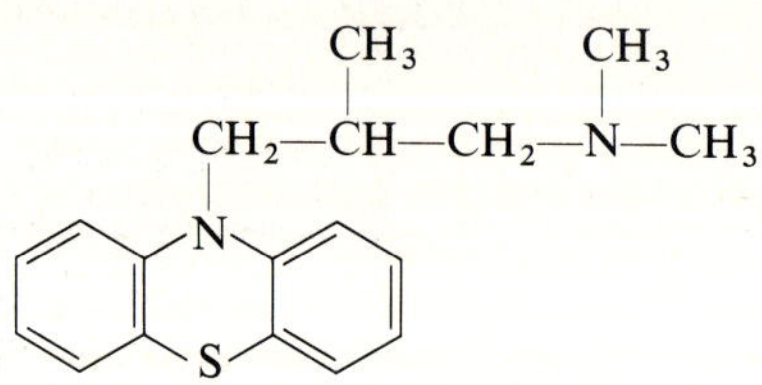

cyamepromazine
(also serves as antihistamine)

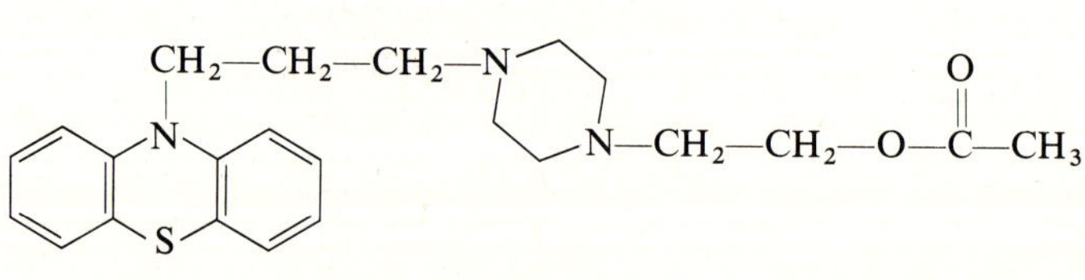

thiopropazate
(Dartal)

Many antihistamines (Chapter 11) are heterocycles. Chlorpheniramine and tripelennamine are examples.

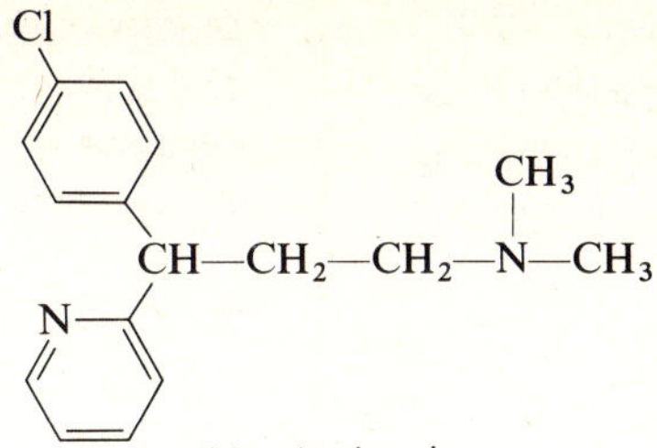

chlorpheniramine
(Chlor-trimeton)

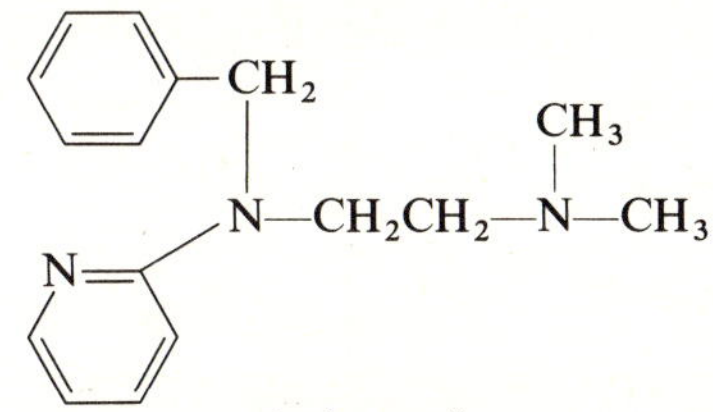

tripelennamine
(pyribenzamine)

Barbiturates are produced as sedatives and are used in sleeping pills.

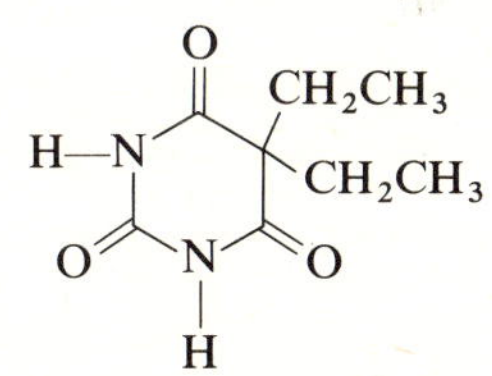

diethylbarbiturate
(Barbital, Vernol)

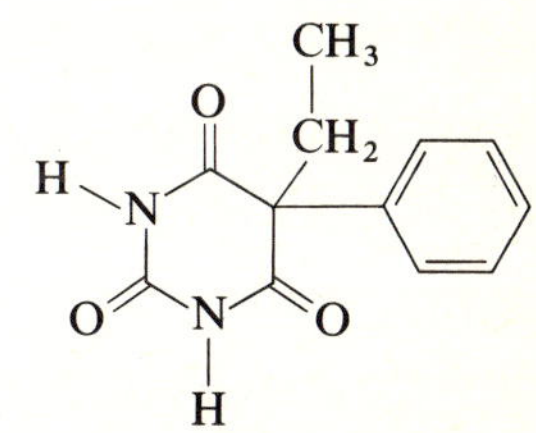

ethyl phenylbarbiturate
(Phenobarbital, Luminal)

Unfortunately, these two barbiturates under the slang term "downers" have also become drugs of abuse. A tragic situation develops when individuals become dependent on amphetamines (Chapter 9) to be stimulated and barbiturates to be relaxed. Barbiturate dependence is very difficult to terminate because of severe, sometimes fatal, physiological trauma.

Another heterocycle, isonicotinic acid hydrazide (Isoniazid), is one of the antibiotics that has emptied our tuberculosis hospitals of patients.

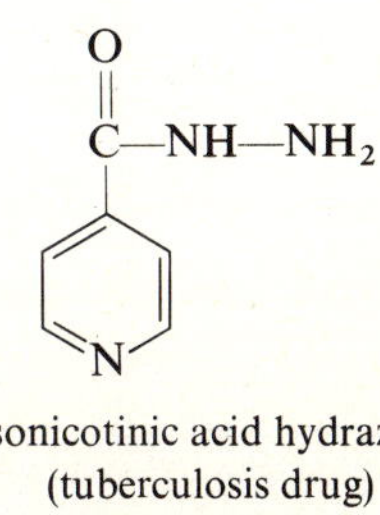

isonicotinic acid hydrazide
(tuberculosis drug)

10.4 Heterocyclic Dyes

Heterocycles have also been synthesized for use as fabric dyes. Colored compounds can be used in foods, in beverages, and as medicinals, as long as they are soluble in the substance and are safe to ingest. However, to color a fabric the dye must attach to that fabric in some lasting fashion. We say that the color must be fast. This may involve some interesting chemical interaction between the dye and functional groups on the fabric. Other dyes depend on a physical entrapment in the fabric, particularly in fabrics that lack reactive functional groups.

Primulin Red is an example of an azo dye. It is used as an ingrain dye, that is, a dye formed in the cloth and thus trapped physically. The cloth is saturated with one portion of the dye molecule, which is soluble. The second portion of the dye is added, reacting with the first portion to form the complete dye molecule. The complete dye is insoluble and is ensnared in the fabric.

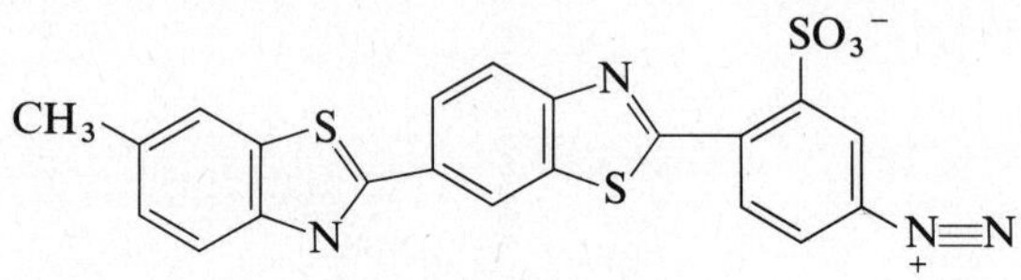

Saturate the fabric with this.

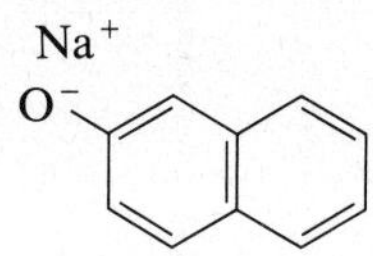

Add this next.

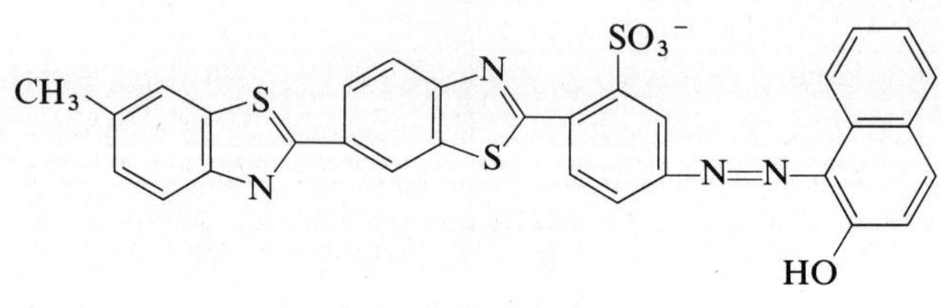

Primulin red: The completed dye is ensnared in the fabric.

Heterocyclic dyes have a wide range of structures, as shown below. Indigo was discussed in Chapter 4.

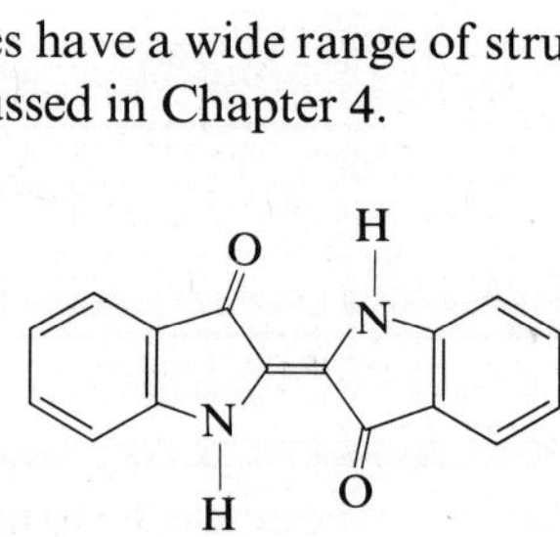

indigo

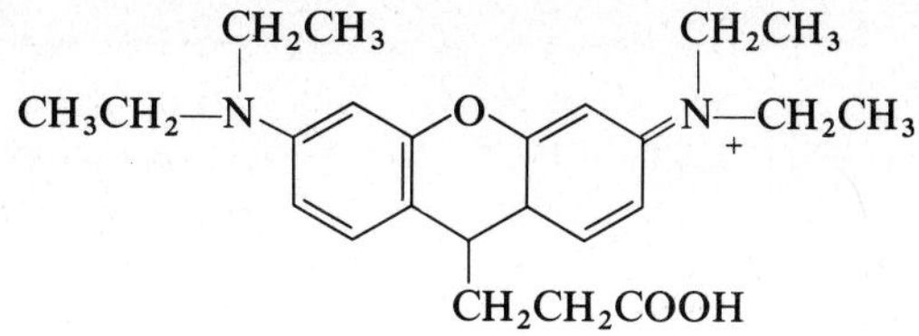

basic red 11

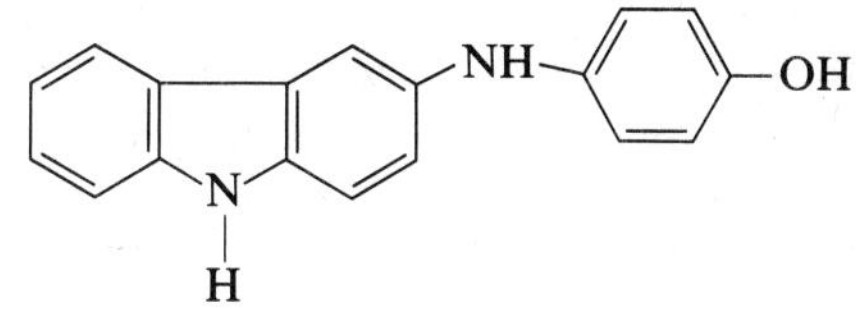

hydron blue R

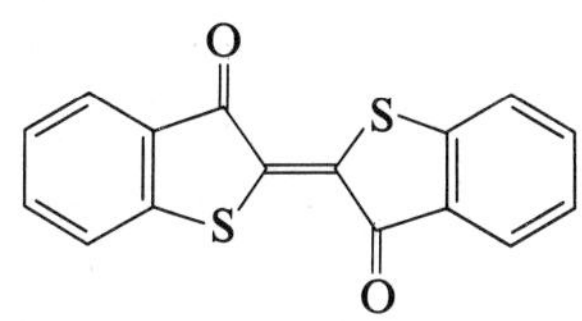

thioindigo red B

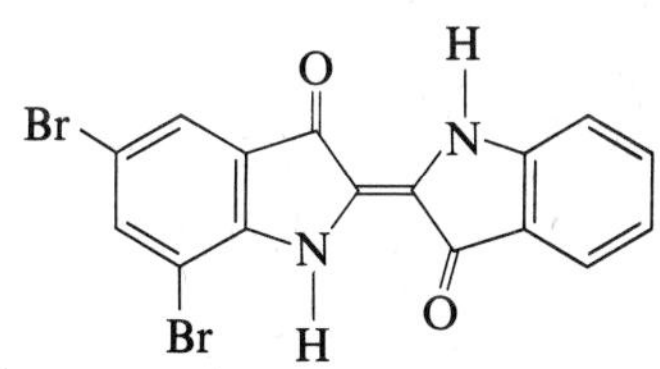

vat blue 5

Review the discussion in Chapter 4 relating color to conjugated double bonds; then study the preceding structures.

10.5 Nucleosides and Nucleotides

Several important compounds in biological systems contain derivatives of purine or pyrimidine ring systems. Two specific purines are widespread in occurrence, adenine and guanine.

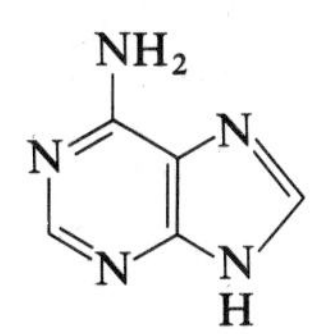

adenine

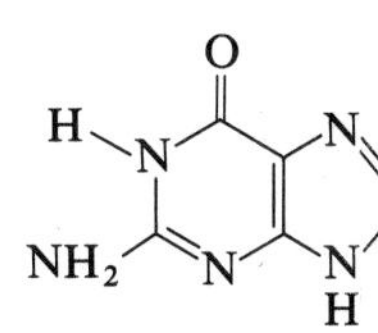

guanine

Similarly, we frequently encounter three pyrimidines, cytosine, uracil, and thymine.

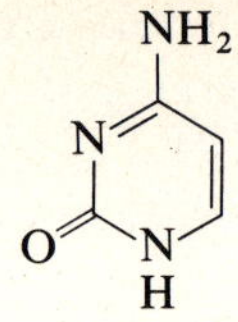

cytosine

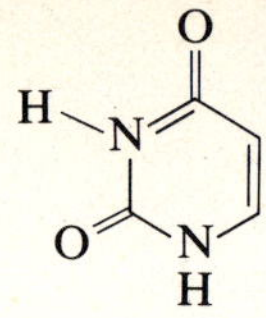

uracil

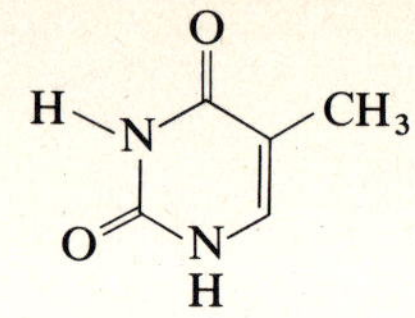

thymine

We seldom see the heterocycles by themselves. They most often occur attached to the carbohydrate D-ribose. Such a combination is termed a *nucleoside*.

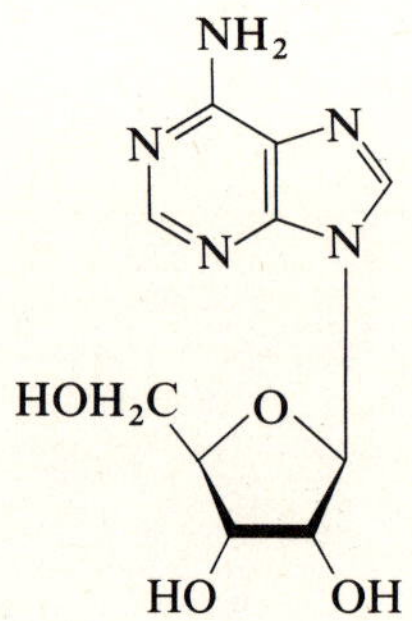

adenosine

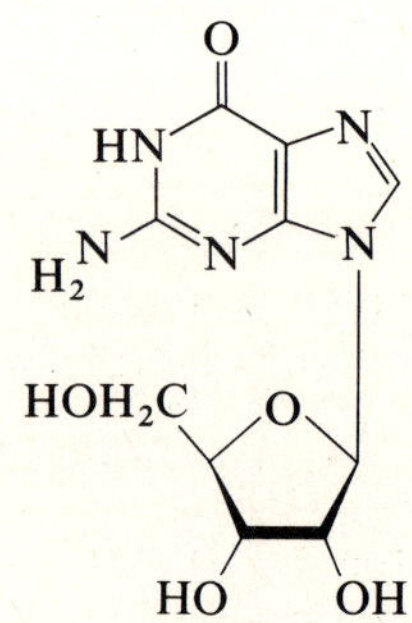

guanosine

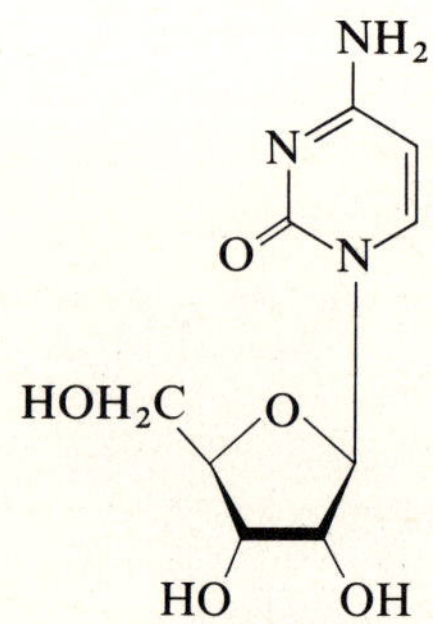

cytidine

Notice that in the nucleoside the carbohydrate has three alcohol groups available for reaction. One of these is likely to be esterified by the phosphoric acid, generating a nucleotide. The position of the phosphate group is indicated by the number of the alcohol group substituted. This number is in front of the name of this nucleotide and has a prime (′) mark to differentiate the carbohydrate ring numbers from the purine or pyrimidine ring number.

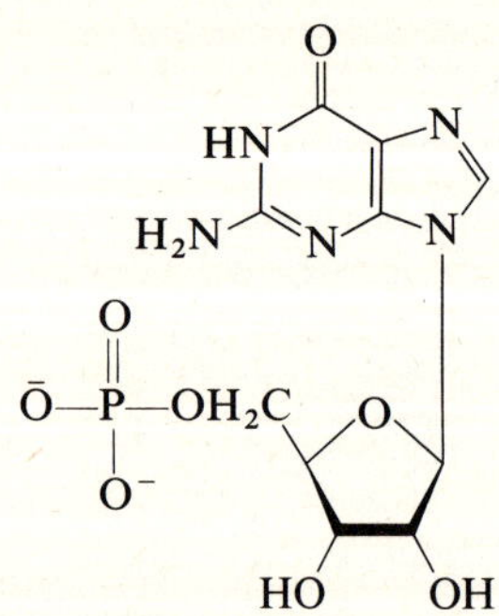

guanosine-5′-monophosphate

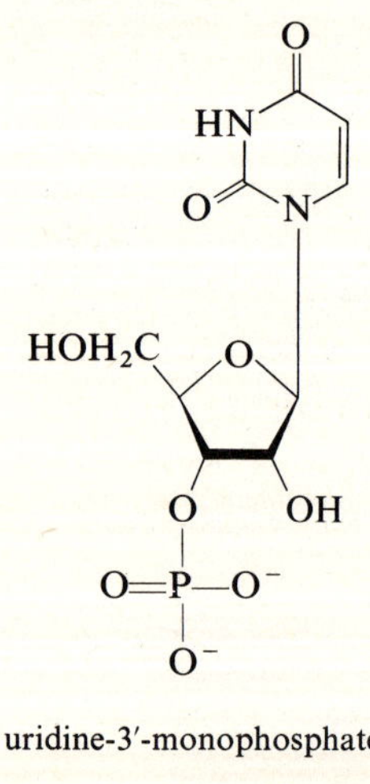

uridine-3′-monophosphate

Many nucleotides have important biological functions. The best known depend on the binding of extra phosphates. The single phosphate nucleotides are often named as nucleoside *monophosphates* to emphasize that they do not have extra phosphate. The nucleoside names are then abbreviated to single letters (A, G, C, U, and T), the monophosphate to MP.

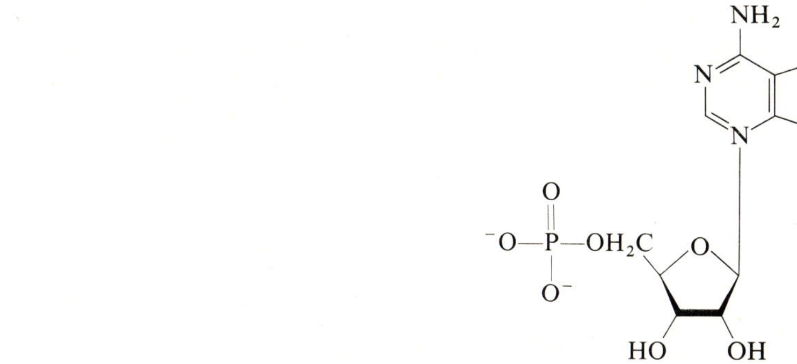

adenosine-5′-phosphate
5′-adenosine monophosphate
5′-AMP

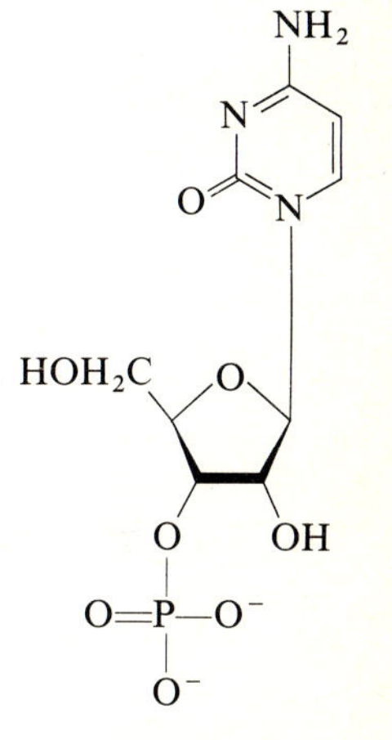

cytidine-3′-phosphate
3′-cytidine monophosphate
3′-CMP

Attaching a second phosphate to the first produces nucleoside diphosphates, with DP in the abbreviation.

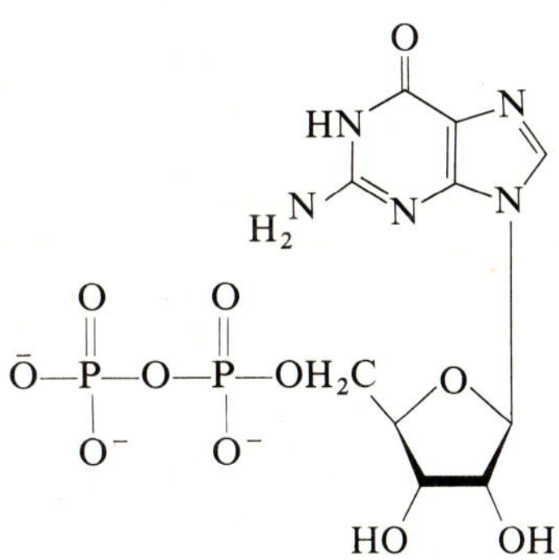

5′-guanosine diphosphate
5′-GDP

Finally, a third phosphate may be added, resulting in nucleoside triphosphates, with TP in the abbreviation.

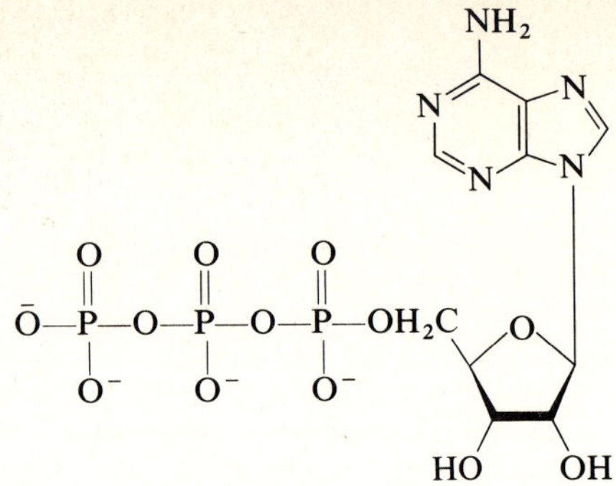

5′-adenosine triphosphate
5′-ATP

The bonds connecting the phosphate groups store an unusually large amount of energy. Therefore, they are often called *high-energy bonds*. When we break down food for energy, it is largely in the phosphate-to-phosphate bonds of ATP that the energy is captured (Chapter 18).

Exercise 10.5

Draw structures for the following:

a) 5′-ADP ***b)*** 3′-UMP ***c)*** 5′-CTP ***d)*** 5′-GMP

10.6 Cyclic Nucleotides

Recently, two important nucleotides, cyclic AMP (cAMP) and cyclic GMP (cGMP), have been discovered and studied. In each of these, a single phosphate is bonded by ester linkages to both the 3′ and the 5′ hydroxy group of the ribose.

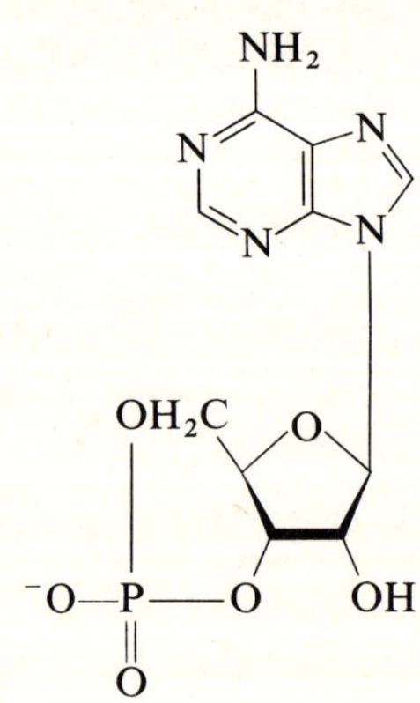

cyclic AMP

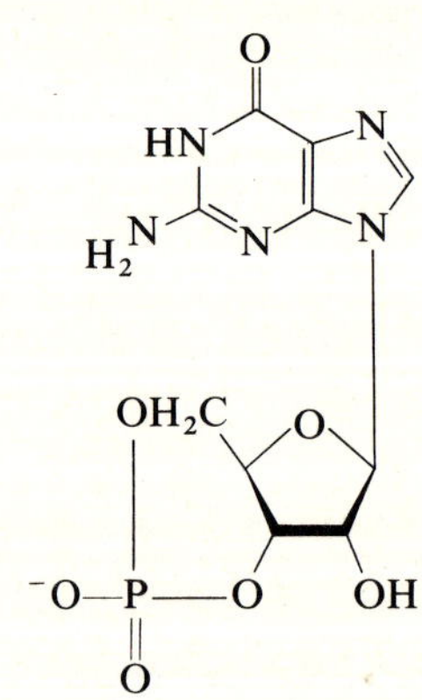

cyclic GMP

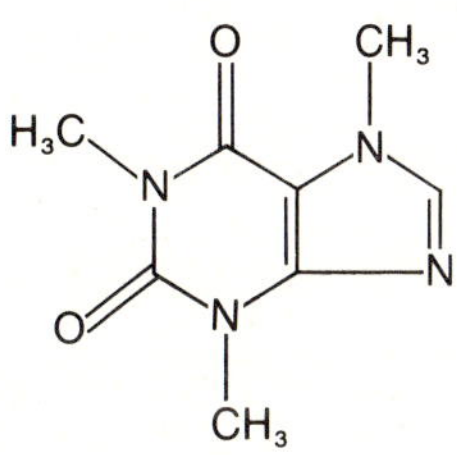

Figure 10.2
Caffeine
Coffee contains the alkaloid caffeine. Of which heterocyclic ring system is caffeine a derivative? (Photo by Catherine Feryok, Eastern Michigan University.)

These nucleotides are sometimes called the "second messengers" of hormones. Hormones in the blood regulate the synthesis of these cyclic nucleotides inside the cell. For example, the hormones epinephrine and glucagon activate the system that synthesizes cAMP in the cell. The cAMP in turn affects the activity of certain enzyme systems in the cell, producing the desired change in cell chemistry. When the hormones drop in concentration, the synthesis of cAMP slows. An enzyme in the cell inactivates the cAMP, which soon causes the concentration of cAMP to drop. The hormone insulin inhibits the synthesis of cAMP but stimulates the synthesis of cGMP. This produces a quite different change in cell chemistry.

The stimulating effect of the alkaloid caffeine found in coffee, tea, and cola-based soda pop may relate to cAMP levels (Figure 10.2). Caffeine inhibits the enzyme that inactivates the cAMP. As a result, the levels remain high even after the epinephrine concentration has dropped.

10.7 Final Comment

The most important role of the nucleotide structures is as the polymers RNA and DNA. These are discussed in detail in Chapter 13.

Summary

Heterocyclic compounds are ring structures that include both carbon and other kinds of atoms, particularly oxygen and nitrogen in the ring. The ring systems have special names. Naming the total structures can be very complex, particularly because the structures themselves are often very complicated. Chemically, the oxygen heterocycles are usually cyclic ethers, cyclic esters (lactones), or cyclic acid anhydrides. The nitrogen heterocycles are usually either cyclic amines or cyclic amides. Plants produce a large variety of heterocycles called alkaloids. These have been of greatest interest because of their physiological effects, either for use as medicinals or as drugs of abuse. Chemists have similarly produced a variety of synthetic heterocycles in the hopes of producing useful compounds.

Nucleosides and nucleotides are structures composed of purines or pyrimidines linked to the carbohydrate ribose. In the case of nucleotides, phosphate is also part of the structure. When more than one phosphate is bonded, the compounds serve as reservoirs of chemical energy. Cyclic phosphate esters are also found that cooperate with hormones in a regulatory role.

Problems

1. Referring to the tables or pages listed, indicate the heterocyclic ring systems in the following compounds:

a) Muscarine—Exercise 10.2, page 245

b) Pelletierine—Exercise 10.3, page 246

c) Phencyclidine—Exercise 10.3, page 246 ***d)*** Nicotine—page 249

e) Bufotenine—page 250 ***f)*** Chlorpheniramine—page 253

g) Uridine-3′-monophosphate—page 256

2. Referring to the structures in Table 10.1, decide the following:

a) How many different derivatives could you make that consist of a single methyl group substituted on the basic ring system?

b) Name any methyl derivative you select of each of the following:

1) Pyran 2) Pyrrol 3) Pyridine
4) Piperadine 5) Imidazole

3. Draw the structure of the heterocyclic compound that could be prepared in one step from the following:

a) 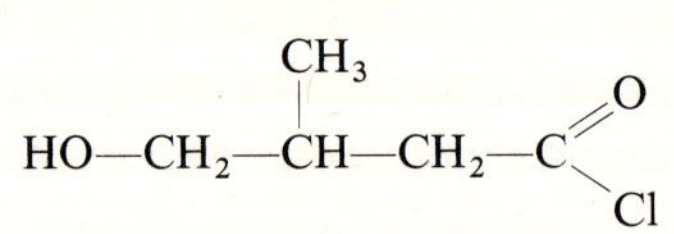

b) $NH_2-CH_2-CH(Cl)-CH(Cl)-CH_2-C(=O)Cl$

c) $Cl-C(=O)-CH_2-C(=O)-Cl + NH_2-CH_2-NH_2$

4. As we view a series of compounds that exert similar effects biologically, a common structural feature becomes apparent.

a) What are the basic features of a barbiturate structure? Look at Barbital, Phenobarbital, and the following:

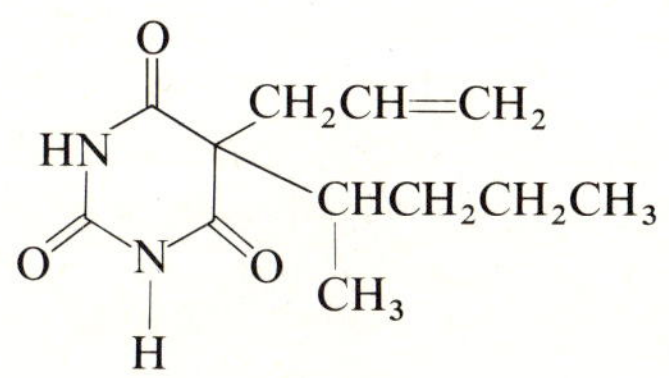

Secobarbital

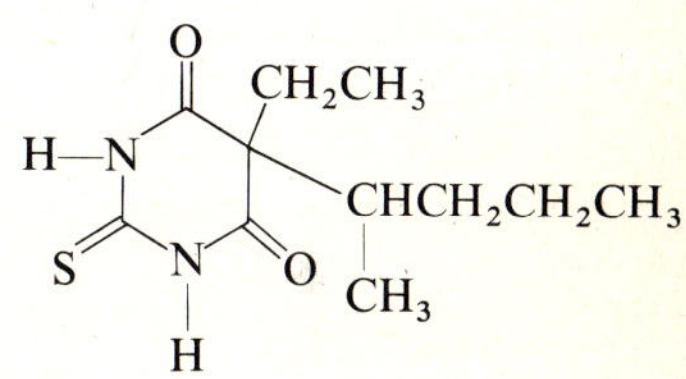

Pentothal

b) What are the common features of the synthetic tranquilizers (page 252)?

c) How do you explain the above correlations of structure and function?

11
Protein Shape and Function

Objectives

In this chapter we:

1. Obtain a concept of the size and shape of protein molecules.
2. Survey the structure and properties of fibrous proteins, with special emphasis on secondary structure.
3. Enumerate the forces involved in the maintenance of protein three-dimensional shape.
4. Consider the importance of binding sites on globular proteins as exemplified by antibody–antigen interactions.
5. Discuss the mechanisms of allergy.
6. Present transport proteins as another example of globular proteins with binding sites.
7. Learn about the transport protein hemoglobin in detail.

What is a protein really like? We can make models of monosaccharides or simple organic molecules and imagine that we had somehow shrunk ourselves to molecular size and were looking at the actual molecule. To be more realistic, we would have to visualize the parts of the molecule spinning and vibrating because of heat energy. That wouldn't be too hard to do. But a protein is so big that it is too much to visualize at once. When a molecule gets that big, certain characteristics and properties exist that may be lacking in smaller, simpler structures. With simpler molecules, the functional groups were the chemical center of attention. Is this still true in proteins, or has something else become more important?

Proteins fulfill a broad range of functions; therefore, they take on a variety of physical appearances. A structural protein whose function is to hold skin or tendon together in a tough but flexible fashion might look like a long strand of rope to us. However, most proteins function by binding other structures to themselves, either to carry them around or to modify them somehow. These molecules tend to be relatively spherical. If we were to shrink to molecular dimensions, these proteins would resemble huge individual dirigibles. The surfaces of these great masses are sparked with a number of charged groups, and as the pH of the water changes, these charges wink off and others appear. These charged groups attract water and oppositely charged ions. Numerous polar groups attract water and hold it to the surface of the protein, keeping the protein suspended in the water around it. Whenever the protein shifts about, masses of water trail along with it, attracted to the surface. If we stayed shrunken long enough to watch our typical protein at work, we would see it attract and bind other molecules to hollows or valleys in the protein surface. Just as we relate the chemical activity of small molecules to a functional group, the protein has a pattern of functional groups in this cavity, or binding site, on its surface. Portions of the binding site might also appear greasy or sticky with hydrocarbon groups—certainly not as attractive to water as the rest of the surface. There might also be small nonprotein units at the binding site providing some functional groups the protein doesn't have.

As we watch the protein bind the other molecule, we become aware of massive shifts and movements in the whole protein—not the spinning methyl groups or vibrating bonds of small molecules, but great shape changes of the protein mass. As we swim around to watch other proteins performing the same function as the one we first observed, we are aware that the intricate folding and twisting of the long chain of amino acids from which the protein is assembled is exactly the same for every protein of the same type.

11.1 Protein Function

What is it, then, that these giant molecules do? We have alluded to a structural role. The ropelike molecules performing that function are termed *fibrous* proteins. The more spherical proteins with the binding sites are termed *globular*. Function is a question of what is bound, and what is done with what is bound. Some proteins transport compounds; they serve a bit like delivery vans. Oxygen, metals, fatty acids, and hormones are among the substances so transported. Antibodies bind foreign proteins or polysaccharides either by themselves or as part of the surface of invading organisms entering the blood. Here the function is to neutralize this threatening invader. Muscle proteins provide mechanical movement, but they do so by a process involving binding. Enzymes bind compounds, then modify them chemically while they are bound.

There is, finally, a group of proteins that do none of the above. They are simply present as a food reserve. The ovalbumin of egg white and the casein of milk are such proteins. Ovalbumin is intended simply to be degraded for use by the growing chick; casein, similarly, nourishes the newborn mammal. Such molecules are generally globular but lack a binding site.

11.2 Protein Folding

As discussed in Chapter 9, the sequence of amino acids in a particular protein is characteristic for the protein and is called the *primary structure*. Why is assembling the amino acids in the correct order so important? We would not expect our long, stringlike amino acid polymer to bend and twist in a random fashion. Because of the sequence, specific attractive and repulsive forces between the amino acids cause the molecule to fold a particular correct way. Because the sequence is always the same, the folding is always the same. This process results in the final molecules of a given protein all having identical shapes. This correct pattern of folding is called the *tertiary structure*.

Having this exact shape is essential for the protein to function properly. The forces holding the protein in this shape are only moderately strong, so the shape can be distorted. Such changes as adjusting the protein solvent to an acidic or basic pH, adding organic solvents to the protein solution, or heating the protein solution can cause this distortion in shape. We refer to the change in shape of a protein as *denaturation*. You see denaturation happen when you cook an egg. Egg white is essentially a water solution of the protein ovalbumin. In its correctly folded form it is water soluble, but as we heat the solution, the ovalbumin unwinds and becomes water insoluble. In this example the denaturation is, as it often is, irreversible. In some cases the forces that normally hold the protein in its correct shape can restore it to that shape after denaturation by a process called *renaturation*.

Unless the function is very mundane, say food storage, the shape of a protein and the biological function of that protein are closely related. Imagine what unfolding does to the accurately assembled binding site. As a result, denaturation destroys the biological function.

11.3 Structural Proteins: Keratins

Some proteins fulfill a structural role. Skin, hoof, hair, horn, beak, nail, claw, feather, scale—all are made of structural proteins. The fibrous structural proteins are insoluble in water. The fibrous proteins of the structures given are called *keratins*. There are two subclasses, the α-keratins and the β-keratins.

The α-keratins have quite a range of properties. They may be hard and brittle, as in horn and nail, or they may be flexible, as in skin and hair. The difference in rigidity is related to the amount of the amino acid cysteine present. Cysteine residues join to one another by disulfide bonds (Chapter 5). The bonds join neighboring protein chains together, resulting in a more rigid structure. The more cysteine the keratin has, the more rigid it is.

11.4 The α Helix

The α-keratins were studied by a technique called *X-ray crystallography*. This method reveals the actual arrangement in space of the atoms of a molecule. It is a very difficult technique, so only a few proteins have been studied. The few results we have obtained, however, are the most revealing and useful information we have about proteins. The studies of α-keratin indicate that the structure has a very highly

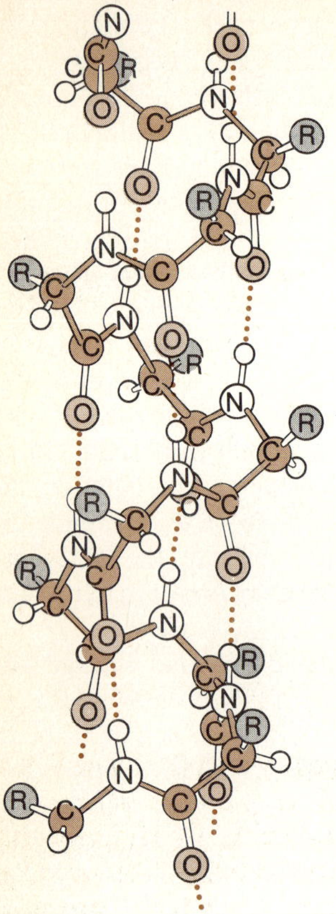

Figure 11.1
The α helix
The peptide chain coils, and the turns of the coil are held together by hydrogen bonds, shown here as dotted lines. (From G. H. Haggis, *Introduction to Molecular Biology*, 2nd ed., London, Longman, 1964.)

organized pattern. Large portions of the protein chains are arranged in a coiled, or helical, pattern. Of the two directions the helix can coil, only one, the "right-handed" helix, is present. Each turn of the helix contains 3.6 amino acid residues. That means that if we started with the $-\overset{\overset{\displaystyle O}{\|}}{C}-$ of a particular amino acid and followed around the helix until we were directly above that $-\overset{\overset{\displaystyle O}{\|}}{C}-$, counting the amino acids we passed as we went around, we would have passed three complete amino acids and six-tenths of another. The pattern has been called the α helix (Figure 11.1).

We can see from the model above that it is possible to coil the protein chain into such a helix. We can coil a cooked spaghetti strand into a helix, too, but it would not stay coiled. There must be a force to stabilize the helix once it forms. Notice in the illustration that the oxygen of the carboxyl residue is directed toward the hydrogen of an amine residue one turn of the helix away. As indicated by the dotted lines, the oxygen and the amine are linked by a hydrogen bond. In fact, the reason the α helix has 3.6 residues per turn is that the peptide bonds are then lined up to permit hydrogen bonding to occur. As you recall, hydrogen bonds are individually rather weak, but in this case we have one formed for every peptide bond in the protein chain. The sum of all these hydrogen bonds becomes a significant force, the force stabilizing the α helix. This sort of regional orderliness or organization within the total folding pattern of the protein is called the *secondary structure*. The α helix is found in most protein molecules somewhere in the overall folding of the molecule.

The α-keratins have a large fraction of their amino acids organized into the α-helix patterns. This amounts to building up a structure from long coil springs. It is not surprising that many α-keratins can be stretched, but they return to the original arrangements when the tension is released. Cross-linking with disulfide bonds reduces this ability to be stretched by holding the coils together with inflexible linkages.

11.5 β-Keratins and the β Structure

X-Ray studies of β-keratins reveals that they too have secondary structure but that it is not the α helix. Instead the protein chains are stretched out straight, actually not arrow-straight, but with a sawtooth pattern common to linear hydrocarbon structures. In such an arrangement the R groups are directed up or down from tips of the sawteeth, while amine residue hydrogen and the carboxyl residue oxygen are directed at right angles to the sawteeth. If we line two such strands up side by side, matching the positions of the sawteeth, the amines and carboxyls are side by side, and hydrogen bonds can form between the strands.

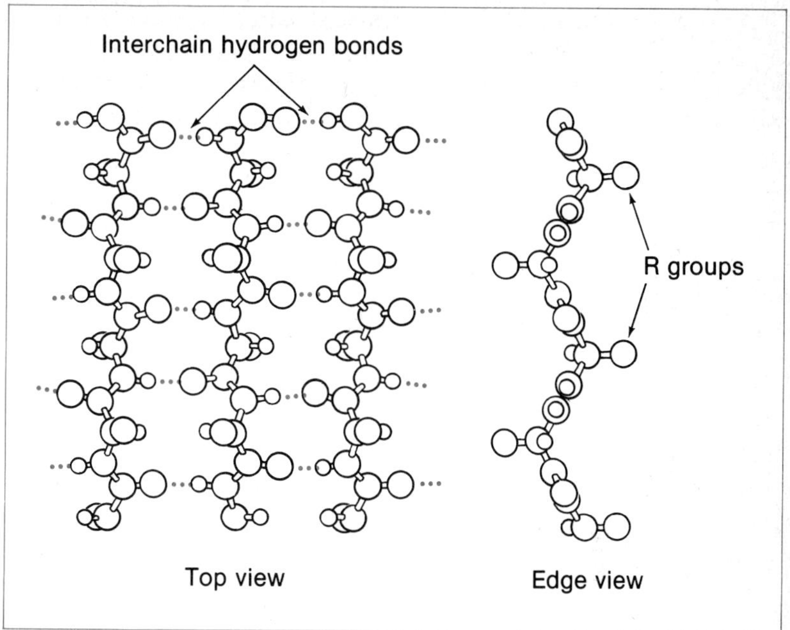

Figure 11.2
The β structure
The peptide chains runs side by side, and these neighboring strands are held together by hydrogen bonds, shown here as dotted lines. (From H. D. Springhall, *The Structural Chemistry of Proteins*, London, Butterworths, 1954.)

If we repeat this step time after time, we build a sheet that is held together side to side by hydrogen bonds. Because the sawteeth are lined up, the surface of the sheet is corrugated or pleated. For this reason, this type of secondary structure is often called the *pleated sheet*. It is also termed the *β structure*. There can be two types of β structure. If all the protein chains are aligned with their amine terminal ends in the same direction, the structure is "parallel." If instead they alternate so that at one end the chain ends are amine, then carboxy, then amine, then carboxy, the structure is called "antiparallel" (Figure 11.2).

Silk is a β-keratin protein. Portions of the silk protein are in β structure and portions are rather randomly folded. The higher the proportion of β structure, the lower the elasticity and flexibility of the silk. This reflects the rather rigid character of the β-structure type of secondary structure. There are no "coil springs."

11.6 Structural Proteins: Collagen

The connective tissue in skin, tendons, and bones is a very strong but inelastic fibrous protein structure. Upon study by X-ray crystallography, it is shown to have a helical secondary structure different from the α helix. This structural protein type is called *collagen* (Figure 11.3). Collagen is composed of bundles of collagen fibrils (fiberlike strands). These fibrils are built from small units, called *tropocollagen*, lined up side by side. Tropocollagen units are connected to one another by cross-linking bonds between the lysine R groups of neighboring units. Tropocollagen consists of three protein chains that wind around one another so that each strand is formed into a helix. This helix is much

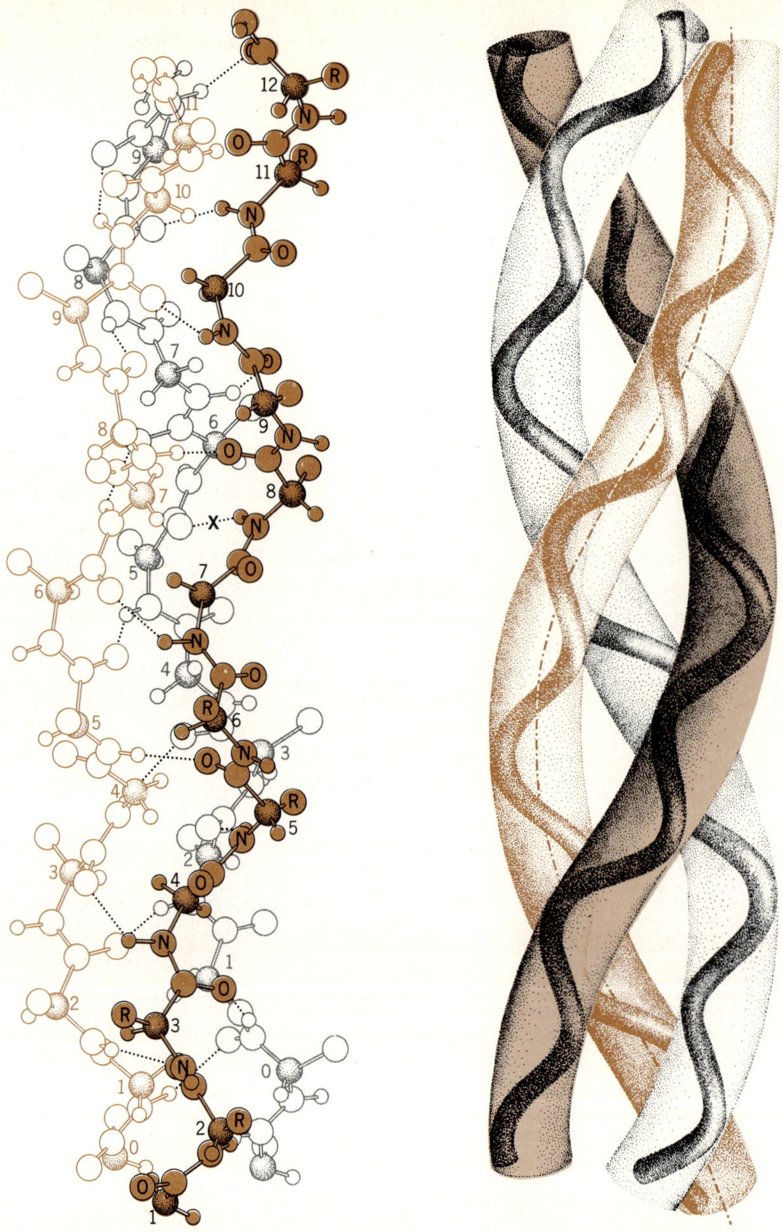

Figure 11.3
Collagen
Three peptide chains, each in a helical form, wrap around one another. The structure is stabilized by hydrogen bonds between chains. (From R. E. Dickerson and I. Geis, *The Structure and Action of Proteins*, New York, Benjamin-Cummings, 1969.)

more open or stretched out than the α helix is. The collagen secondary structure, just like the other secondary structures, is stabilized by hydrogen bonds between carboxyl residue oxygens and amine residue N—H groups. Instead of connecting one turn of the helix to the next, as happens in the α helix, these hydrogen bonds are between the three protein strands. The amount of hydrogen bond stabilization of the secondary structure is smaller than with either the α helix or the β structure. In those cases, virtually every peptide bond is involved in hydrogen bonding. In collagen this is not the case. In boiling water, the collagen structure unfolds, resulting in a mixture of individual soluble protein strands. Denatured collagen is generally called *gelatin* and is used in a number of food products.

11.7 Globular Proteins

Most proteins are not long strands. Instead they are folded into a roughly spherical shape. These spherical, or *globular*, proteins are more compact molecules than the fibrous proteins (Figure 11.4). In fibrous proteins the hydrogen bonds of the secondary structure as well as covalent cross-linkages hold the molecule in its overall correct shape. These same forces help to stabilize the globular protein shape. Myoglobin has about three-quarters of the amino acids arranged in the α-helix secondary structure. Lysozyme has less α helix, but it does have some β structure. Ribonuclease has a very high proportion of β structure, but the α helix is completely absent. There are hydrogen bonds that are not part of a secondary structure. For example, in ribonuclease three tyrosines hydrogen bond to three carboxylate groups.

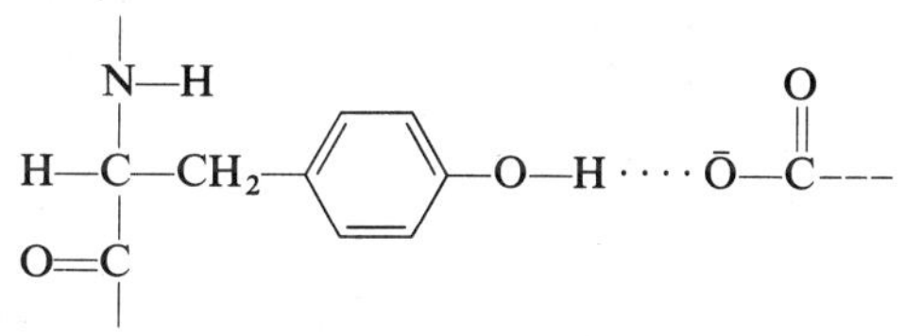

Both lysozyme and ribonuclease have cysteine-to-cysteine disulfide cross-links, which provide very strong shape stabilization (Chapter 5).

Other forces contribute to the maintenance of protein shape. At neutral pH, aspartate and glutamate carboxylate groups lose their protons and so have a negative charge. The basic lysine, arginine, or histidine R groups gain a proton and so have a positive charge. Where a positively and a negatively charged group can approach one another, the *electrostatic attractive force* helps to hold the two charged groups together.

Another type of force, *hydrophobic bonding*, is more difficult to understand. Polar or charged groups attract one another and will mix. The nonpolar structures are excluded from this mixing. However, because of a small force all atoms can exert on one another if they can approach closely enough (van der Waals forces), nonpolar structures attract one another. As a result, amino acids with polar R groups tend to locate at the surface of the protein where they can interact with water. The nonpolar amino acids tend to locate in the interior and associate with one another (Figure 11.5). Globular proteins studied so far have a

Figure 11.4
Tertiary structures of proteins
The twisting lines represent folding of the peptide backbones of a few globular proteins. The folding pattern is unique for each protein and displays a variety of kinds and degrees of secondary structure. (From R. E. Dickerson and I. Geis, *The Structure and Action of Proteins*, 2d ed., in preparation, New York, Benjamin-Cummings.)

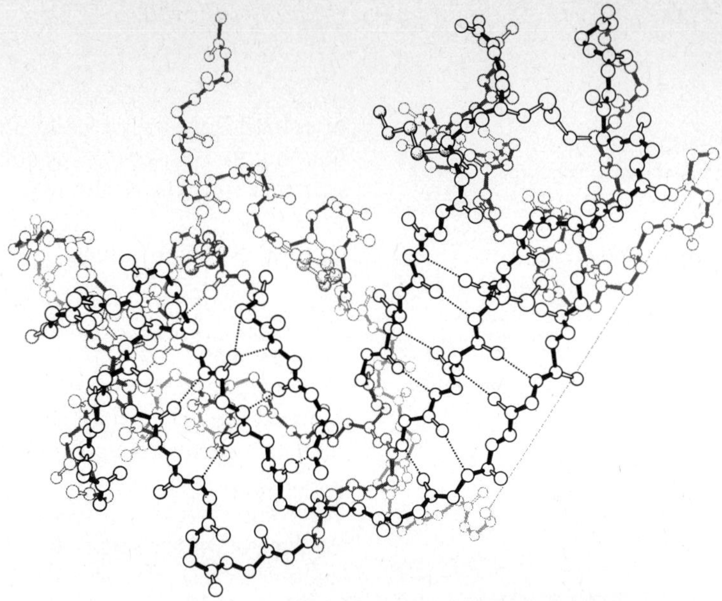

Ribonuclease

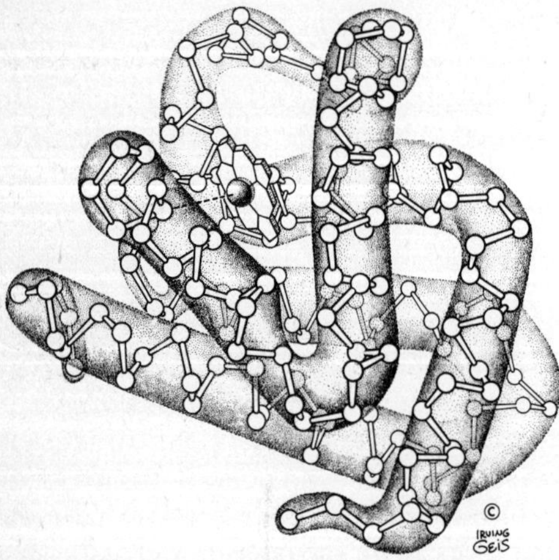

Myoglobin

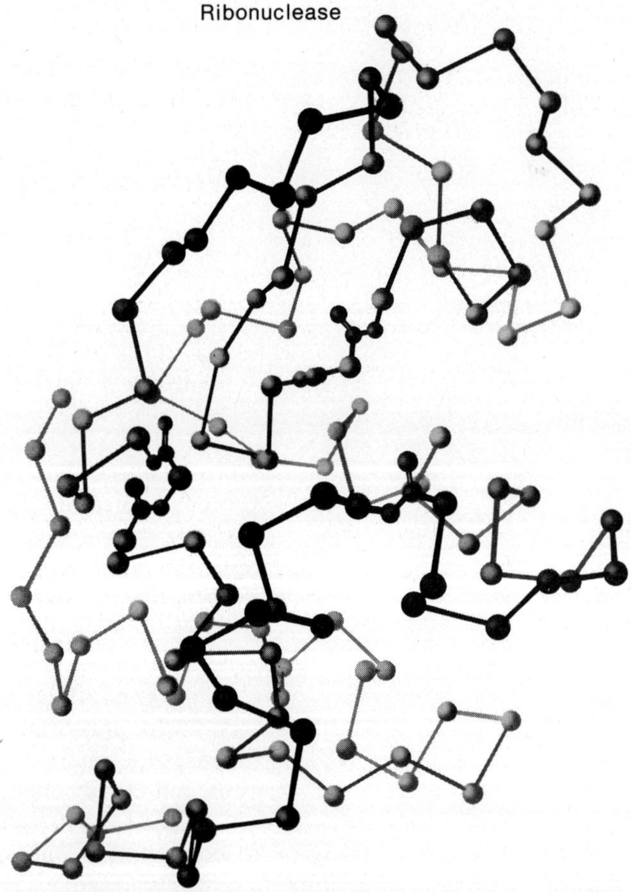

Lysosyme

Figure 11.5
Hydrophobic bonding
Polar portions of flexible large molecules such as proteins locate at the surface of the molecule and interact with water. Nonpolar portions inhabit the interior and interact with one another.

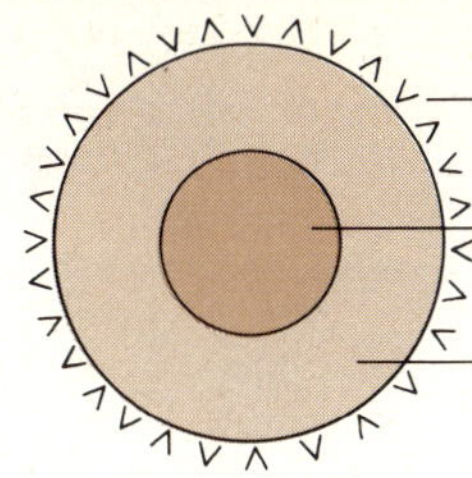

nonpolar interior. Such an arrangement contributes significantly to stability.

11.8 Bonding Forces at the Binding Site

This same list of attractive forces is involved in an exceedingly important aspect of protein function. As we study the globular proteins, we find that they are generally involved with binding something to themselves. This binding occurs at a specific place on the surface of the protein, the binding site. The forces discussed above that keep the protein in a correct shape—hydrogen bonding, electrostatic attraction, covalent bonding, and hydrophobic bonding—also serve to bind other molecules to the protein.

11.9 Antibodies

The process by which we do battle with invaders of our circulatory system provides an excellent example of protein binding sites. A bacterium, virus, or foreign protein entering the blood is a potential threat to health. Such invaders are called *antigens*, and their presence triggers the production of specialized protein molecules called *antibodies*. Antibodies are moderately large proteins composed of four smaller proteins linked together by disulfide bonds (Figure 11.6). As is generally true of the proteins of the blood, antibodies have carbohydrate molecules bonded to their surfaces. The four smaller proteins include two identical smaller, or "light," chains and two identical longer, or "heavy," chains. When the molecule is assembled completely, there are two identical binding sites, each assembled in part from one heavy and one light chain.

When an antigen, such as a bacterial cell, enters the blood, it encounters a type of leukocyte (white cell) called the *lymphocyte*. Each lymphocyte produces only one kind of antibody. These antibodies are bound to the surface of the lymphocyte. Attachment of the antigen to these antibodies stimulates rapid division of the cell, all daughter cells also being specialized in production of the particular antibody. In this fashion the levels of a particular antibody in the blood rise.

The blood proteins can be separated by electrophoresis (Chapter 9). When this is done, the antibodies (immunoglobulins) are

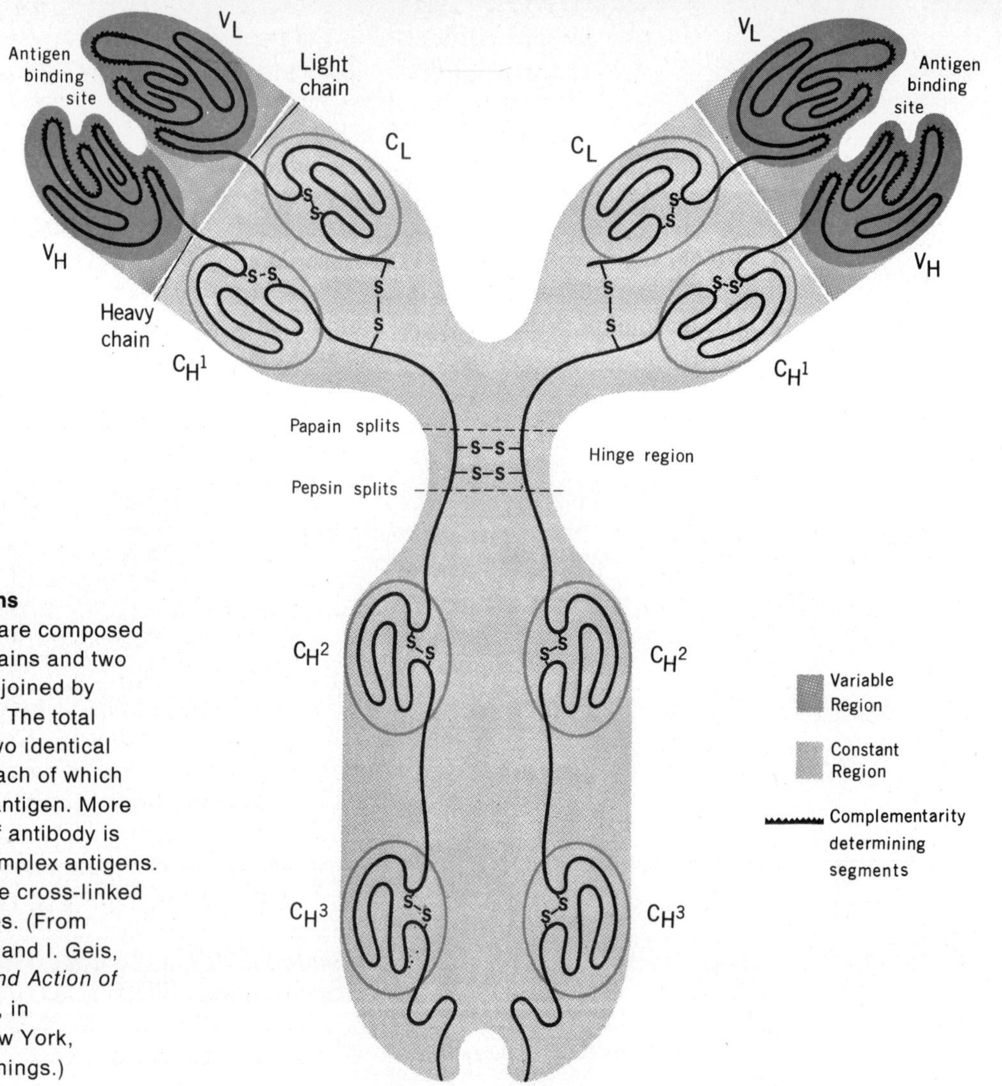

Figure 11.6
Gamma globulins
These proteins are composed of two heavy chains and two light chains, all joined by disulfide bonds. The total molecule has two identical binding sites, each of which can bind to an antigen. More than one type of antibody is produced for complex antigens. The antigens are cross-linked by the antibodies. (From R. E. Dickerson and I. Geis, *The Structure and Action of Proteins*, 2d ed., in preparation, New York, Benjamin-Cummings.)

found largely in a fraction called the γ globulins. There are five classes of immunoglobulins based on structure: IgG, IgM, IgA, IgD, and IgE. An infection produces dramatic changes in the character of the γ-globulin fraction.

Babies do not have a functioning system for antibody synthesis for several months. During this period they are protected by antibodies from the mother's blood. This type of immunity, wherein the antibodies are received pre-formed, is called *passive immunity*. We impart passive immunity when we inject antigen into another animal, such as a horse, then draw blood, collect the antibodies, and inject these antibodies into people. Another approach to giving people immunity is to kill or weaken pathogenic organisms so they are not dangerous but still have characteristic sites (antigenic determinants) for selective antibody attachment. The lymphocytes producing the appropriate antibody proliferate, and the individual is prepared for an invasion of the active pathogen, should it happen.

11.10 Allergy

Another aspect of the antibody system is what is called *allergic response*. A person is sensitive or allergic to some specific substance, whether it be ragweed pollen, penicillin, or cat fur, because he or she produces an antibody to the substance. The antigen (allergen) may be airborne, producing a response in the eyes and respiratory tract, or it may be ingested with the food, producing gastrointestinal or skin reactions. The antibodies involved belong to the IgE (immunoglobulin class E) subclass of antibodies. People with allergies have larger than usual amounts of the IgE antibodies.

Allergic reactions may be immediate (within hours) or delayed (requiring days). The mechanisms of these two reactions are thought to be somewhat different. Of the two, the immediate reaction is the better understood (Figure 11.7). The allergen on first exposure generates IgE antibodies. On later exposures the allergen is bound by the IgE antibodies, which in turn bind to the surface of specific cells. This binding results in the cells releasing a variety of substances including histamine, seratonin, kinins, and SRS-A (a slow reacting substance of anaphylaxis). These compounds cause such responses as smooth muscle contraction and increases in capillary wall permeability. As a result of capillary wall changes, leukocytes enter surrounding tissues. These disintegrate, releasing enzymes, especially enzymes that hydrolyze protein. These events produce the inflammation we observe in tissues displaying an allergic response (Figure 11.7).

The delayed response allergy differs in that the IgE antibodies remain attached to leukocytes (Figure 11.8). Upon reexposure, the allergens bind to the IgE–leukocyte complex. This complex releases a factor (migratory inhibiting factor) that attracts other leukocytes. These invade the surrounding tissue with the effects noted above. Asthma, with its breathing difficulties, the itching and sneezing of hay fever, and the swelling and itching of hives are examples of allergic response.

Drugs are used to combat the effects of allergy (Figures 11.9 and 11.10). Antihistamines operate by blocking the binding sites for histamine on the smooth muscle or capillary walls. They do not affect the rate of histamine production. Antihistamines may fail if histamine production is so great that the antihistamines are overpowered by quantity. Because histamine is not involved, antihistamines are not effective against delayed allergy. Corticosteroids operate by preventing increases in capillary permeability and by stabilizing the leukocytes against the release of the destructive

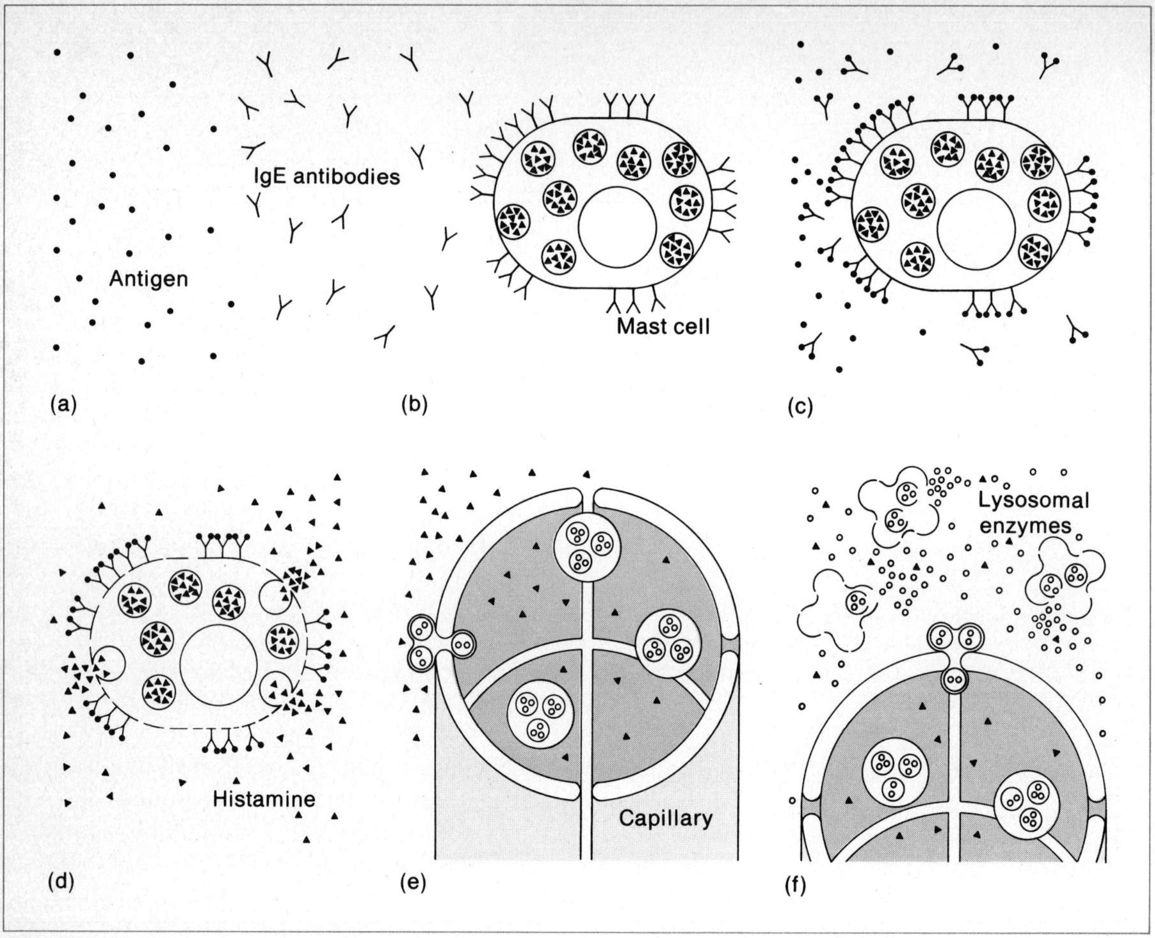

Figure 11.7
History of an immediate type allergy reaction
(a) Antigens generate IgE antibodies, which (b) bind to specific cells. Later exposure to the antigen (c) results in the release of substances such as histamine (d). Histamines increase capillary wall permeability, allowing white blood cells (leukocytes) to invade the surrounding tissues, (e) where they disintegrate, releasing hydrolytic enzymes that damage tissue (f). (From "Cellular Aspects of Allergy and Inflammation," The Upjohn Company, Kalamazoo, Michigan.)

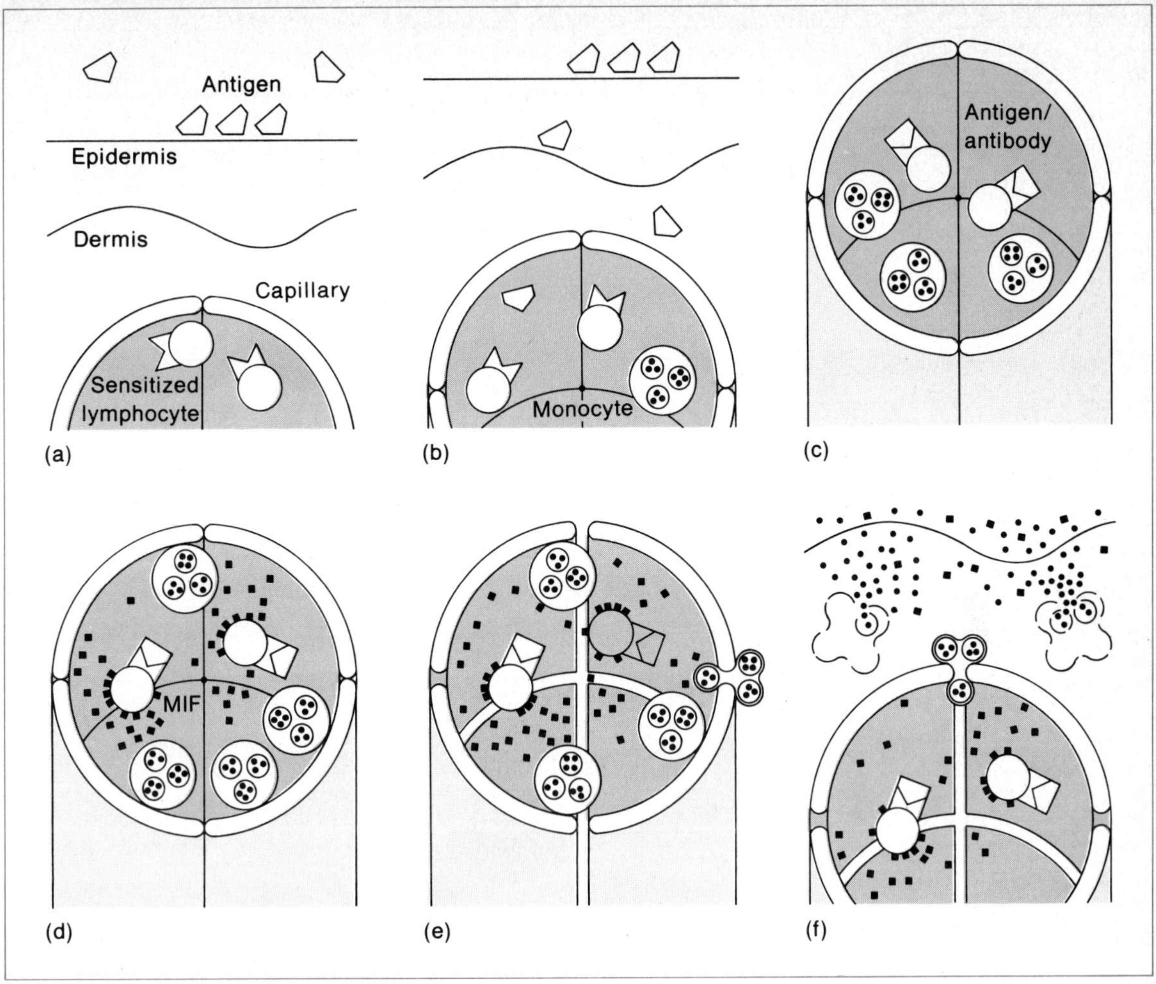

Figure 11.8
History of a delayed type allergy reaction
Exposure to the antigen generates antibodies that remain attached to the lymphocyte (a). Reexposure to the antigen (b, c) results in the release of migration inhibitory factor (MIF, d). MIF leads to damage to the capillary (e) and eventual invasion of surrounding tissues by lymphocytes, which disintegrate (f), releasing hydrolytic enzymes that damage tissue. (From "Cellular Aspects of Allergy and Inflammation," The Upjohn Company, Kalamazoo, Michigan.)

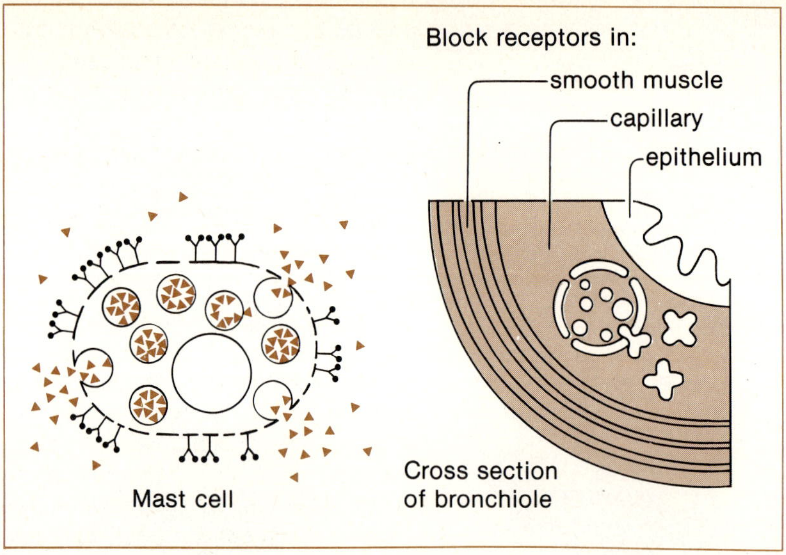

Figure 11.9
Drugs affecting allergic reactions: antihistamines
Antihistamines block the binding sites for histamines and so prevent the dilation of the capillaries. Histamine production is not blocked. (From "Cellular Aspects of Allergy and Inflammation," The Upjohn Company, Kalamazoo, Michigan.)

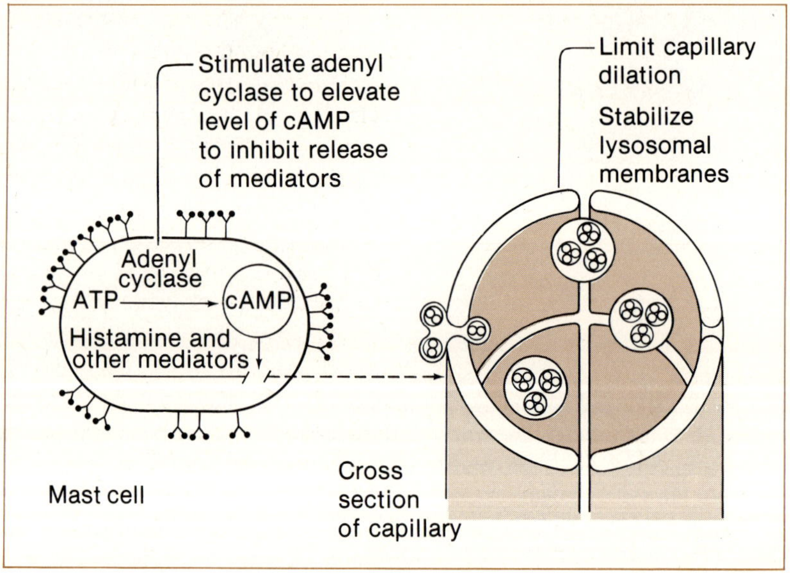

Figure 11.10
Drugs affecting allergic reactions: corticosteroids
Corticosteroids raise the level cAMP and so prevent the release of histamines and the disintegration of leukocytes in the tissue. (From "Cellular Aspects of Allergy and Inflammation," The Upjohn Company, Kalamazoo, Michigan.)

enzymes. Catecholamines inhibit the release of histamines and SRS-A (Figure 11.8).

11.11 Transport Proteins: Hemoglobin

Transport proteins are another example of globular proteins with binding sites. In the blood, a number of substances are transported attached to a specific protein. These include metal ions, hormones, fatty acids, and oxygen. This is discussed further in Chapter 22. The transport protein has a binding site that is specific for the particular substance being transported.

Of all the transport proteins, for many decades the greatest interest has been in hemoglobin, the molecule that transports oxygen. Hemoglobin is made up of more than one protein chain; in fact, four protein chains are present. On studying the chains it was found that there are two each of two kinds. One pair is called α chains, the other pair the β chains. This sort of protein, with more than one complete folded protein structure in the functional molecule, is said to have a *quaternary structure*. Quaternary structures can have as few as two subunits or as many as dozens. These are not just fragments of crystal lattice. Proteins that have a quaternary structure usually function differently or not at all when the subunits are separated.

The α and β subunits of hemoglobin are very similar. Each subunit has a heme group attached to it. The heme unit is a *porphyrin ring* with an Fe^{2+} ion complexed in the center (Chapter 23).

It is the heme structure that actually carries the oxygen. Each subunit is capable of transporting one oxygen molecule. In the capillaries of the lungs, where the concentration of oxygen is very high, oxygen binds to the heme. Out on the capillaries of the tissues the oxygen concentration is low, and the equilibrium shifts the other way—the oxygen is released. Notice that it is not desirable to have the hemoglobin bond oxygen very tightly. If it did so, although the heme would quickly saturate with oxygen in the lungs, it would be unwilling to release the oxygen in the capillaries of the tissues. On the other hand, if oxygen were bound very weakly, it might not be bound in the lungs in the first place. As with any biochemical system involving reversible binding, just the right binding strength is called for.

With this in mind, oxygen binding in fetal blood presents a special problem. Before birth, the source of oxygen is the mother's hemoglobin in the placenta. A hemoglobin adapated to binding oxygen at levels normal for the lungs would be poorly suited to taking oxygen from other hemoglobin molecules. Fetus hemoglobin, therefore, has two γ, or fetal, hemoglobin subunits and two α subunits. The fetal hemoglobin subunits bind oxygen more strongly than the maternal blood. They are therefore successful at taking it from the maternal blood. After birth, β subunits are synthesized in place of γ subunits.

Hemoglobin illustrates another aspect of the importance of the shape of protein molecules. Hemoglobin is one of the few proteins whose complete tertiary and quaternary structure is known. The determination of shape has been done on hemoglobin both with and without oxygen. Hemoglobin changes shape when it binds oxygen. Furthermore, the shape change corresponds to a change in the tightness of oxygen binding. When oxygen is plentiful, as in the lungs, hemoglobin assumes the "tight binding" shape and readily saturates with oxygen. Later the oxygen is released to the tissues. The hemoglobin then changes back to the "loose binding" shape and releases the last of the oxygen more readily. The shape change is triggered by the occupying of the binding sites by oxygen.

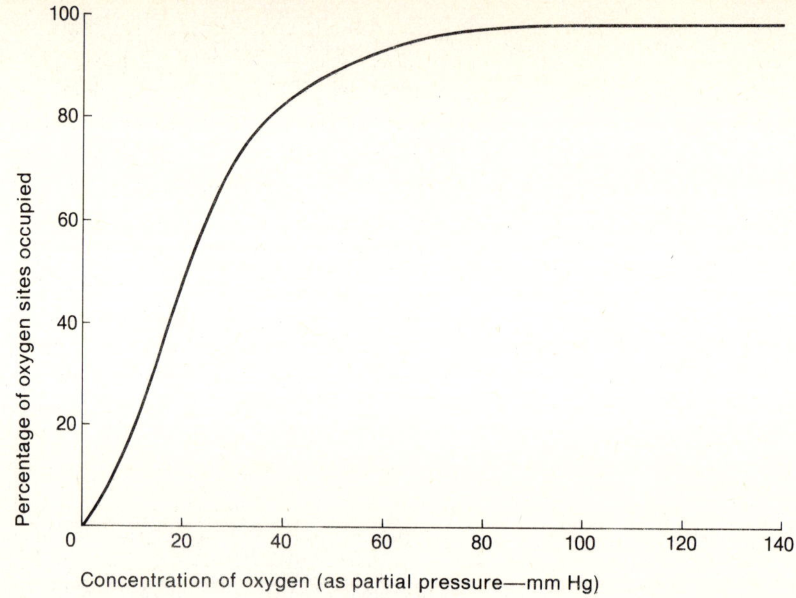

Figure 11.11
The binding of oxygen by hemoglobin
The shape of this curve, called sigmoidal, reflects the weak binding of oxygen by hemoglobin when oxygen is scarce and the strong binding when it is plentiful.

This can be expressed very compactly on a graph (Figure 11.11). If we plot the percentage of the hemoglobin binding sites carrying oxygen versus the concentration of oxygen in solution, the curve has a sigmoidal shape. This is interpreted to mean that at low oxygen concentration the hemoglobin has assumed the weak binding shape. As the concentration of oxygen increases, some binds to hemoglobin in spite of the weak binding conformation. Once some has bound, it triggers the molecule to shift to the strong binding conformation. Thereafter, as reflected by the steep slope, hemoglobin binds oxygen eagerly.

Is this a useful effect? Consider where the oxygen concentration is low—out in the tissues. There we don't want the hemoglobin to hold the oxygen, we want the oxygen released. On the other hand, where is the concentration of oxygen high?—in the lungs. There we want the hemoglobin to bind oxygen, which it is prepared to do.

Summary

In addition to a common sequence or primary structure, protein molecules of a given type all have the same folding pattern or tertiary structure. If this shape is changed, as it can be by a variety of environmental factors such as heat, pH change, and solution in organic liquids, it is said to be denatured. Within the folded pattern are found orderly arrangements of the chains held in place by peptide bond hydrogen bonding and called secondary structures. Types of secondary structure include the α helix, the β structure, and the collagen structure.

Structural proteins are fibrous, have a high degree of secondary structure, and often contain interchain cross-links. Examples include the α- and β-keratins found in hair, hoof, horn, skin, feather, beak, and nail, and the collagen found in skin and tendon.

Other proteins are globular and are held in shape by hydrogen bonding, electrostatic attractions, some internal cross-linking, and hydrophobic bonding. Because of hydrophobic bonding, most proteins have their nonpolar amino acids in the interior of the structure and their polar or charged amino acids on the surface.

Globular proteins usually function by having a portion of the protein surface shaped in such a fashion and composed of such an array of amino acids that they can bind some other molecule. Usually they are very specific about what they bind.

Antibodies are an example of globular proteins with binding sites. These molecules are produced to bind foreign molecules called antigens that enter the blood. Each antibody has two binding sites and so can join two antigens together, eventually clumping large numbers of antigens into a kind of polymer. Antibodies are specific for a particular antigen and are produced by lymphocytes. The presence of antigen stimulates a sharp increase in antibody production. The system represents a major defense against bacterial or viral disease.

Hemoglobin is an example of a transport protein. It carries oxygen from the lungs to the tissues. The hemoglobin molecule is composed of four separate chains and thus is said to have a quaternary structure. Each chain binds a heme structure, which is the actual oxygen binding site. Shape changes in the molecule, triggered by the binding of oxygen, cause the molecule to vary from tight oxygen binding in oxygen-rich regions to weak oxygen binding in oxygen-deficient regions.

Problems

1. Indicate to which class of proteins each of the examples below belongs.

Class of proteins

A) Structural
B) Enzyme
C) Antibody
D) Transport protein
E) Nutrient protein

***a*)** Ovalbumin ______

***b*)** β-Keratin ______

***c*)** Hemoglobin ______

***d*)** Collagen ______

***e*)** IgE ______

***f*)** α-amylase (page 136) ______

***g*)** The only class that would not be globular ______

***h*)** The classes that would have a binding site ______

2. Identify each of the following specific observations about protein structure as being aspects of the 1°, 2°, 3°, or 4° structure.

***a*)** Amino acids 14 through 38 in the sequence are arranged in an α-helix pattern.

***b*)** When the protein was enzymatically broken apart, one of the fragments was found to be the following:

leucine–tyrosine–glycine–alanine–lysine

***c*)** The entire molecule is a distorted sphere 1.3 times as long in one dimension as its diameter.

***d*)** Detergent was added to a water solution of the protein. The molecular weight study before the addition indicated a single protein with a molecular weight of 150,000. After the addition, two proteins were present with molecular weights of 25,000 and 50,000.

3. For each statement below, indicate the bonding force involved.

Bonding force

A) Electrostatic attraction
B) Hydrogen bonds
C) Hydrophobic bonding
D) Covalent bonding

***a*)** The α helix is held together by ______

***b*)** van der Waals forces make a major contribution to ______

______.

c) Two cysteines could create a ______________________ type of stabilizing force.

d) A glutamic acid and a lysine, both as part of the protein chain, would stay close to one another because of ____________________.

e) As a consequence of ______________________, the polar amine acids are at the surface of a protein and the nonpolar ones are in the interior.

f) All secondary structures are stabilized by ________________ ____.

g) The N-terminal amine group might bend around close to an aspartate in the chain in response to ________________________.

4. Denaturation, the changing in shape and consequent change in properties of a protein, occurs when one or more of the forces holding the protein in shape is weakened or overcome. Below are listed some conditions that cause denaturation. Try to decide which force primarily is failing in each case.

Protein is denatured by:	***The force that is failing in the protein is:***
a) Addition of acid to the water solution of protein.	*A*) Hydrogen bonding
b) Addition of a relatively nonpolar solvent to the water solution of protein.	*B*) Electrostatic attraction
c) Heating of the protein solution.	*C*) Hydrophobic bonding
d) Addition of detergent to the water solution of protein.	

5. In each of these statements, what type of secondary structure is being described?

a) The protein chain extends in a line, then turns and runs back past itself, hydrogen bonding between peptide bonds of the two strands holding the structure together.

b) Three protein chains twist around one another in a helical pattern, hydrogen bonds from peptide bonds of one chain to those of another chain holding it together.

c) A single protein chain coils into a helical structure, hydrogen bonds between peptide bonds of adjacent turns of the helix holding it together.

6. Look at the two patterns obtained by electrophoresis of plasma proteins.

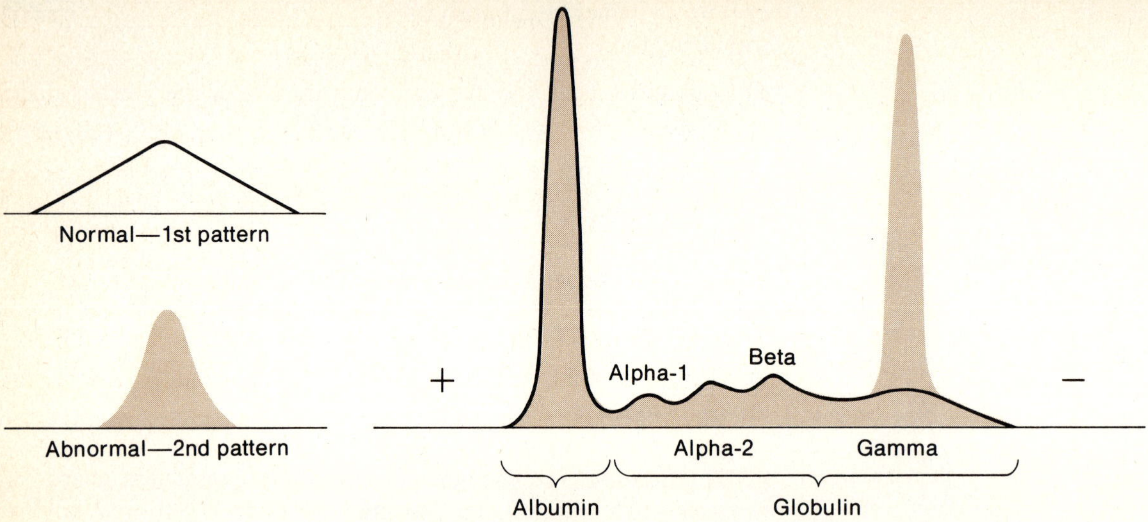

a) Locate the globulin bands in the two and compare them.

b) Can you suggest what might have happened to the pattern in the second case to explain why it is different from the first?

c) Why does neither pattern show a band for hemoglobin even though it is a major blood protein?

Challenge Problems

7. The sigmoidal curve showing oxygen binding (page 278) is shifted as pH changes. Notice that the sigmoidal shape does not change.

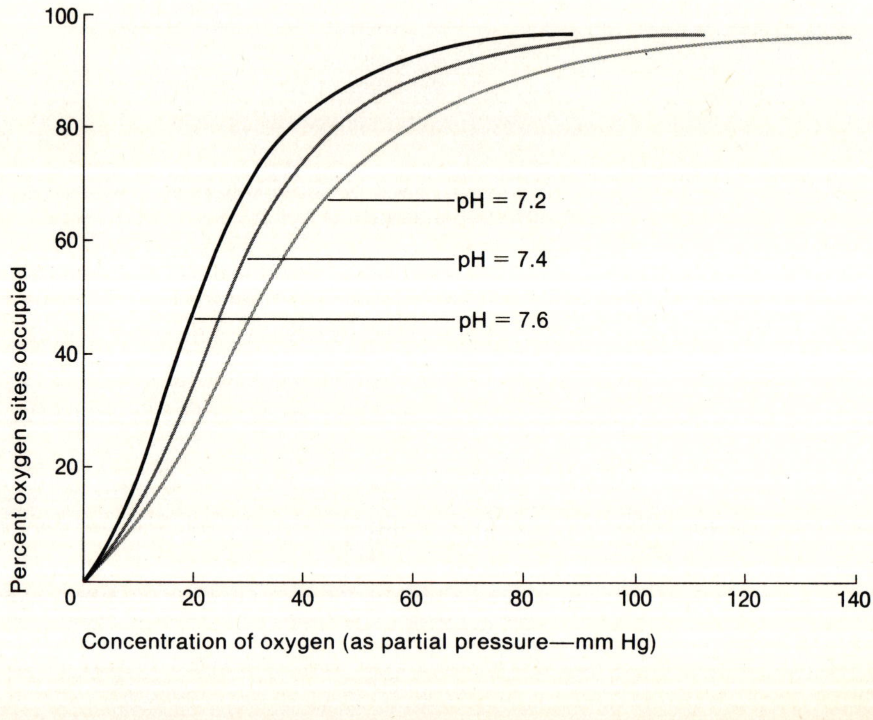

a) Draw a vertical line at $P_{O_2} = 20$ mm Hg. What is the percent saturation with hemoglobin at that concentration of oxygen at each pH?

b) Respiring tissues use oxygen and release carbon dioxide. What is the product of carbon dioxide plus water? Where will the blood be more acidic, in the lungs where CO_2 is leaving and oxygen is entering the blood, or in a distant capillary where CO_2 is entering the blood and O_2 is leaving?

c) Is the hemoglobin going to be more prone to release oxygen, in other words, to have a lower percent saturation of oxygen, at low or high pH? Does all this chemistry assist, hinder, or have no effect on the delivery of oxygen to tissues and the binding of oxygen in the lungs?

12 Enzymes

Objectives

In this chapter we:

1. Study the binding of the substrate to the enzyme.
2. Learn about the relationship of enzymes activity to pH.
3. Study structure and function of the enzyme cofactors and their relationship to the B vitamins.
4. Learn how enzymes are named.
5. Study the digestion of carbohydrates and proteins as simple examples of enzyme-catalyzed processes.
6. Learn about control of proteolytic enzyme activity by synthesizing the enzymes as inactive enzyme precursors.
7. Study the system of blood clotting as an example of enzyme function, of multiplication of rate of an event by use of enzymes activating enzymes, and of control of an event needed only in emergencies.
8. Survey the relationship of enzymes to the flow of chemicals from dietary intake to new cell material or energy production in a cell.
9. Learn about the control of cell chemistry by the use of allosteric enzymes.
10. Study custom-tailored isozymes that differ with the type of tissue, and the usefulness of being able to detect these isozymes to diagnose an illness.
11. Survey the present and future uses of enzymes in food processing.

One way to look at a living system is as a complex array of chemical reactions rushing back and forth to satisfy needs, to adjust to changing circumstances, and to grow. These reactions occur at bewildering speeds almost always to produce just the right amount of only the correct product. In the laboratory we seldom approach the reaction rates achieved in living systems, and in the laboratory quantities of unwanted side products are often produced. Behind the high efficiency of living systems is a series of protein molecules called *enzymes*. Enzymes speed up (catalyze) chemical reactions. For each chemical reaction there is a specific enzyme, and that enzyme usually catalyzes only one reaction. Enzymes are not used up by the reaction. As true catalysts they accelerate reaction after reaction without themselves being changed. How do enzymes perform these chemical wonders?

12.1 The Active Site

Emil Fischer, a chemist of the late 19th century whose thinking was far ahead of his time, puzzled why only a single specific compound was selected by an enzyme. He proposed that the specific molecule altered by the enzyme, the *substrate*, exactly fit into a region on the enzyme. This fit was like two pieces of a picture puzzle joining, or like a key fitting a lock (Figure 12.1). In fact his theory was called the *lock-and-key theory of enzyme specificity*. If a potential substrate has the correct shape, it can attach to the enzyme; if not, it is rejected. This lock-and-key concept is, with slight modification, still the way we explain enzyme specificity toward substrate. This region of the enzyme surface for attachment of the substrate we call the *active site*. The active site is the

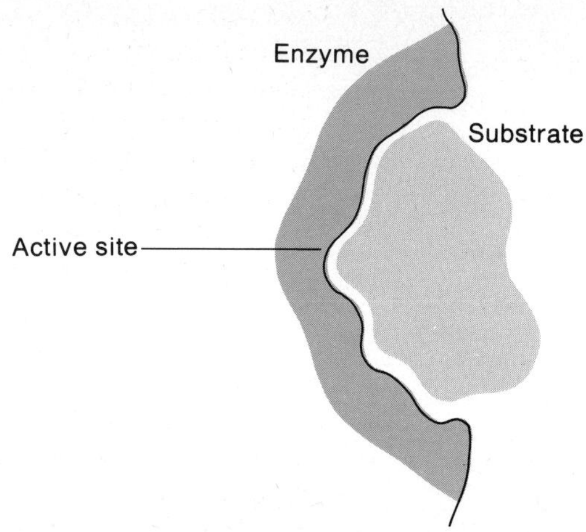

Figure 12.1
The enzyme active site
Emil Fischer proposed that the active site of the enzyme was shaped such that bumps on the substrate would fit into depressions on the active site, much as a key fits into a lock.

enzyme version of the binding site of the transport proteins or antibodies.

This lock-and-key character of the active site explains much of what was said in earlier chapters about the importance of shape in biochemical compounds. For example, even though the chemical properties of enantiomers are identical, the shapes are not. Attaching a D-amino acid to an active site intended for an L-amino acid would be like putting a right hand in a left-handed glove.

The shape of the active site is the result of the folding pattern of the enzyme. Anything that disturbs this folding pattern—that is, anything that denatures the protein—causes the shape of the active site to be incorrect for the substrate so that the enzyme no longer functions.

Saying that a substrate must have the correct shape to fit the active site is really saying only why the wrong substrate is rejected. If the correct substrate has moved into place on the active site, what holds it in place? The same forces that explain how the protein holds its shape (Chapter 11) explain the binding. Sometimes a covalent bond forms between enzyme and substrate. More often we explain the binding by attractions between oppositely charged functional groups or hydrogen bonds between the substrate and the active site. Hydrophobic bonding, the attraction of nonpolar regions for one another, is also important. As stated in the discussion of hemoglobin (Chapter 11), it is not the strongest possible bond between enzyme and substrate that is desired. Then the substrate would never leave. There is a certain strength of binding that is correct, strong enough to hold the substrate but not so strong as to keep it.

Once the substrate is bound to the active site, the parts of the substrate that are to be chemically changed are held next to functional or charged groups. These groups encourage the desired chemical change to occur. We have altered Fischer's original concept of a rigid active site to one that can change shape to assist the binding and catalysis to occur. Sometimes the groups holding the substrate onto the active site also perform the catalysis; at other times different groups are involved. Through study of all the factors that influence the binding and catalysis, we better understand the chemistry of enzymes.

12.2 Effect of pH on Enzymes

Charged structures such as carboxyls and amines are important to bind the substrate, to catalyze chemical change, and to maintain the protein in the correct configuration. These groups gain or lose hydrogen and charge as the pH changes. It is not surprising that enzymes show changes in catalytic effectiveness as the pH of the solution changes. For example, the binding to the enzyme of a substrate containing a carboxylate ion might well involve the pairing of that negative charge with a positively charged lysine R group at the enzyme active site. Therefore making the solution either acid enough to put a hydrogen ion onto the carboxylate ion or basic enough to take a hydrogen ion away from the lysine ammonium ion prevents binding of the substrate to the enzyme active site. For most enzymes there is a range of pH, the *pH optimum*, at which they are most effective. Reduction or complete loss of activity occurs when the pH goes beyond this range. This is an important reason for careful regulation of pH in cells and most body fluids (Chapter 23) (Figure 12.2). Most enzymes have a pH optimum near neutrality. Pepsin is an interesting exception. It operates in the stomach, where the pH is very low, and the enzyme has a correspondingly low pH optimum. At neutral pH, not only is pepsin inactive, it also rather quickly denatures.

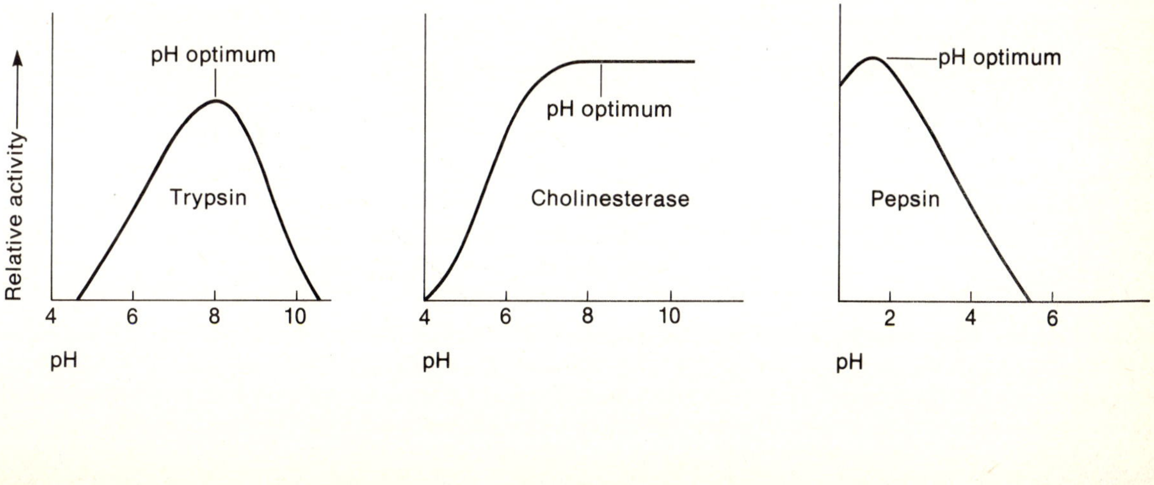

Figure 12.2
pH and enzyme activity
The pH of the enzyme–substrate solution affects the rate at which the reaction is catalyzed. (From A. L. Lehninger, *Biochemistry*, 2nd ed., New York, Worth, 1975, p. 196.)

12.3 The Mechanism of Catalysis

How does catalysis occur? Frankly, we don't know the whole answer to this question. One part of the answer involves bringing very close to one another groups that are intended to react. Where the reaction involves two or more substrate molecules, they are held very close to one another when attached to the active site in the correct position for rapid reaction. The chemical groups of the active site are located next to the substrate. These groups include the R groups of the amino acids of the portion of the protein chain forming the active site. In this fashion carboxyl groups, alcohols, thiols, or amines can be found very close to reactive parts of the substrate.

The reasons for the amazing effectiveness of enzymes as catalysts are much more complex than just the correct placement of reactive groups. For example, enzymes are known to change shape during catalysis; this could bend or stretch bonds in the substrate to make them more reactive. It could also move a functional group into place only when it is needed. However, we are aware only that the enzyme does change shape. What the actual change is, or how it relates to the reaction, is not known, merely predicted. A great deal remains to be learned about enzyme function.

12.4 Cofactors

In addition to amino acid functional groups, we may find nonprotein agents at the active site. These cofactors are of two classes: metal ions and organic molecules. Not all enzymes have cofactors. Some require metal ions, some organic cofactors, some neither, and some both. The functions of the cofactor include helping to bind the substrate to the active site and assisting in catalyzing the reaction. Where such nonprotein structures are associated with an enzyme, the enzyme will not catalyze its reaction without them. If they are removed from the enzyme, activity is lost, but full activity is usually restored when the missing agent is restored.

12.5 Metal Ion Cofactors

The metal ions may bind to both the substrate and the active site, helping to hold the two together (Figure 12.3). Forming such bonds can also contribute to the catalytic process. This is discussed in more detail in Chapter 23.

12.6 Organic Cofactors

The organic cofactors perform simple chemical tasks necessary to catalysis. They may accept and carry a piece that has been removed from the substrate, supply that piece to another substrate, or help bind the substrate to the active site.

12.7 Cofactors and Vitamins

In many cases, our cells have lost the ability to synthesize these organic cofactors. We depend on a ready-made supply of them either from our

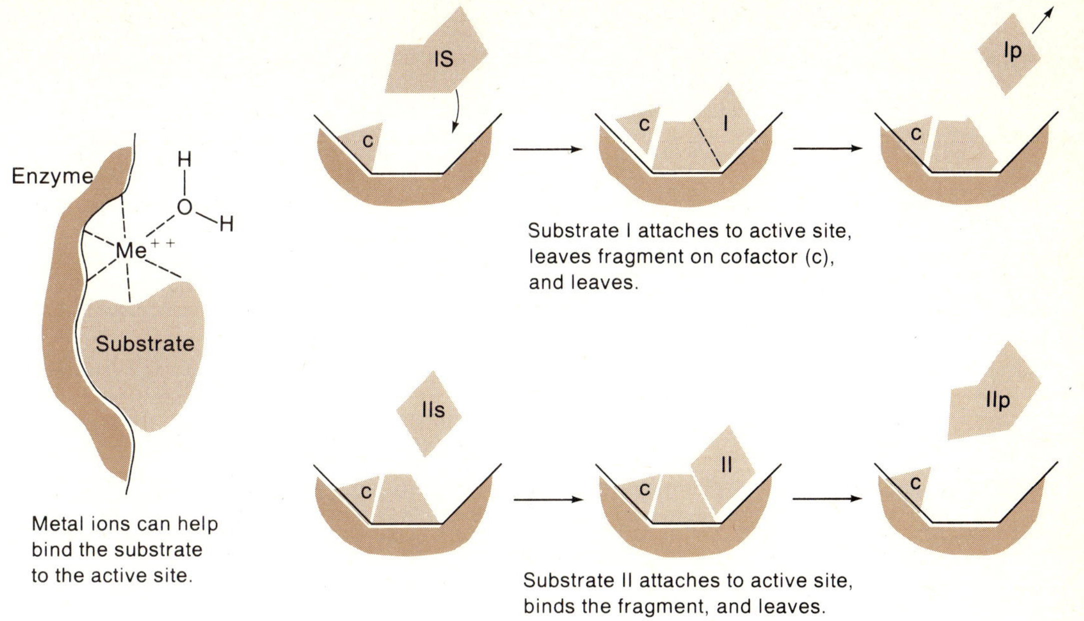

Figure 12.3
Metal ion cofactors

diet or from our intestinal bacteria. Enzymes are not used up during catalysis. Proteins, however, eventually require replacement. The supply of these ready-made cofactors need not be large, only great enough to equip the new enzymes synthesized to replace those that have denatured. What we are describing are vitamins, ready-made molecules in small quantities that must be in the diet to maintain good health. The cofactor-type vitamins are generally water soluble. Many are classed together as B vitamins. In the discussion of the individual cofactors that follows, those of which we must obtain all or part as vitamins are specified.

12.8 Pyridine Nucleotides

First consider cofactors whose function is the carrying of a pair of hydrogens. Enzymes that remove or add substrate hydrogens are oxidizing or reducing that substrate, respectively. They are formally called *oxidoreductases*, and are often informally termed *dehydrogenases* (Chapter 6). The most common example of an oxidoreductase cofactor is nicotinamide adenine dinucleotide, usually abbreviated NAD.[1]

---

[1] An older name, still encountered, is diphosphopyridine nucleotide, or DPN.

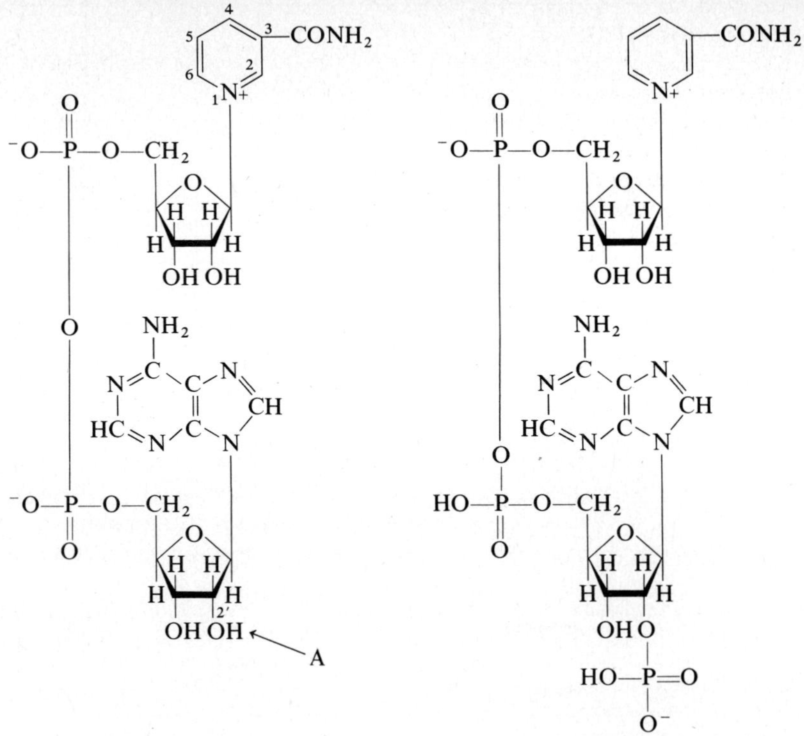

nicotinamide adenine dinucleotide (NAD)

nicotinamide adenine dinucleotide phosphate (NADP)

NAD has a very close cousin, nicotinamide adenine dinucleotide phosphate, or NADP [2] The structures differ only in that NAD has a hydrogen at position A while NADP has a phosphate group ($—OPO_3H^-$). In both forms the reaction performed by the cofactor is exactly the same. The pyridine ring carries a pair of hydrogens. Actually the hydrogen on the pyridine ring nitrogen is very easily removed (acidic). The overall reaction is generally written to reflect this.

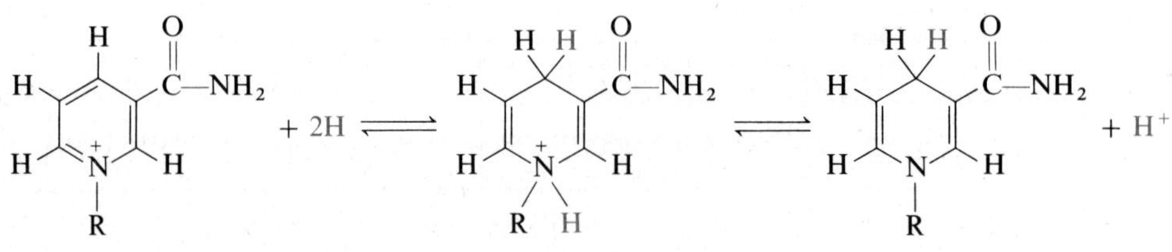

where R represents the rest of the molecule.

[2] An older name is triphosphopyridine nucleotide, or TPN.

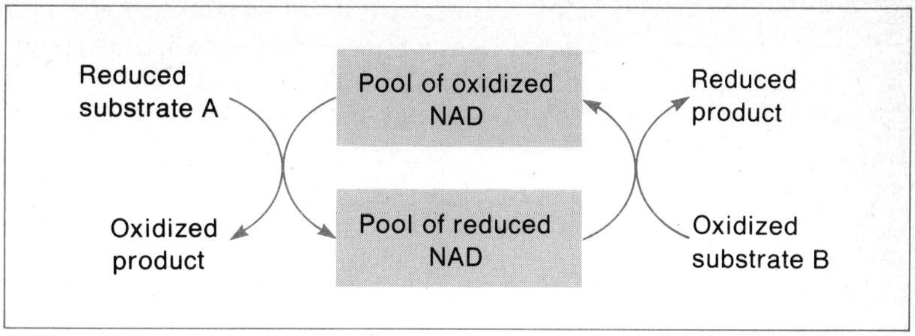

Figure 12.4
The recycling of NAD
During the normal course of cell metabolism, some enzymes usually oxidize NAD and others generally reduce it. The NAD moves about from one enzyme to another being oxidized one time and reduced another.

NAD and NADP are cofactors that do not bind tightly to the enzyme active site. Instead they are attached at about the same time the substrate is attached and leave when the reaction is complete. In reality they are a "second" substrate for the reaction. If the reaction being catalyzed is the oxidation of a substrate, the cofactor is required in its oxidized form. During the reaction it accepts the hydrogens removed from the substrate, the substrate thus becoming oxidized and the cofactor reduced. The oxidized substrate leaves.

$$\underset{\text{lactic acid}}{CH_3\text{—}\overset{\overset{\displaystyle OH}{|}}{C}H\text{—}COOH} + NAD^+ \xrightleftharpoons[]{\text{lactate dehydrogenase}} \underset{\text{pyruvic acid}}{CH_3\text{—}\overset{\overset{\displaystyle O}{\|}}{C}\text{—}COOH} + NADH + H^+$$

Notice that the cofactor, now reduced, is in the wrong form to catalyze the reaction again. So the cofactor also leaves to find an acceptor elsewhere for the hydrogens. It is replaced by a new cofactor, this one oxidized, to allow the enzyme to convert a new substrate to product (Figure 12.4).

Some enzymes specifically use NAD as cofactor, some specifically use NADP. Very few enzymes use them interchangeably. In general, reactions concerned with breaking dietary material down into simpler, more oxidized materials use enzymes specific for NAD. Reactions where more complex new chemicals are being created from simple raw material would be expected to involve enzymes with NADP as cofactor.

We are not able to produce the pyridine ring portion of the cofactor ourselves. This portion of the structure is the B vitamin niacin. A deficiency of niacin in the diet leads to the disease pellagra. In the United States, pellagra was a particularly serious problem in the south through the early 20th century. By 1920 the U.S. Public Health Service had established that pellagra was a dietary problem, not an infectious disease. A diet including meat, eggs, and milk was found to prevent the problem. Although NAD was known as a cofactor at that time, it was not known until the late 1930s that there was a relationship between prevention of pellagra and the cofactor.

12.9 Flavin

Next most important as hydrogen-carrying cofactors are the flavin derivatives: flavin mononucleotide, or FMN, and flavin adenine dinucleotide, or FAD. Flavin is the color portion of the structures below.

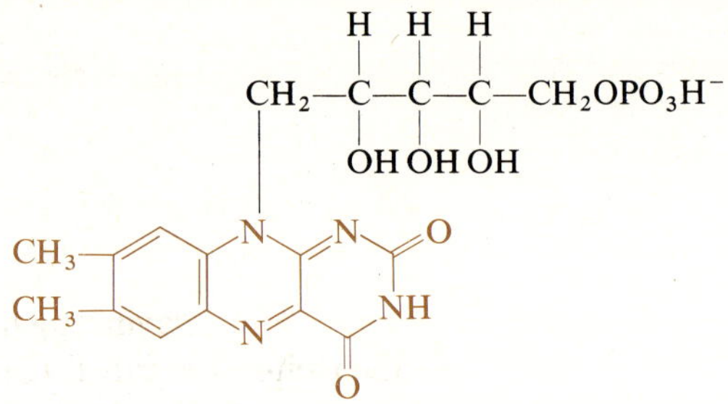

flavin mononucleotide (FMN)

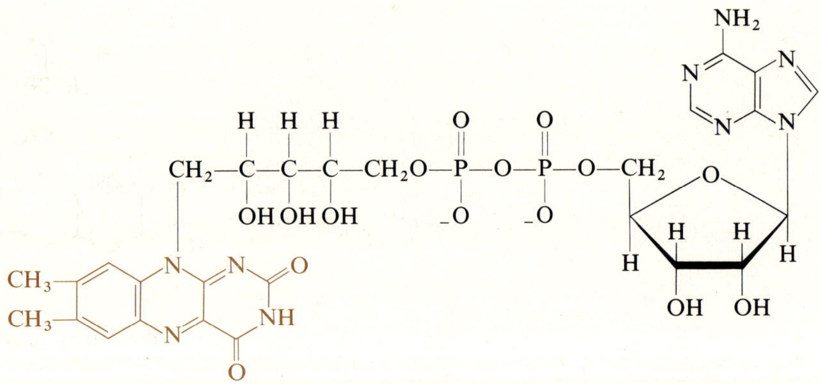

flavin adenine dinucleotide (FAD)

Notice that the carbon structure attached to adenine is ribose, but the one attached to flavin is a "reduced" sugar or polyalcohol called *ribitol.*

As with NAD and NADP, a pair of hydrogens is carried. In contrast to the nicotinamides, where one of the hydrogens ionizes from the reduced cofactor, both hydrogens remain firmly bound to the flavins.

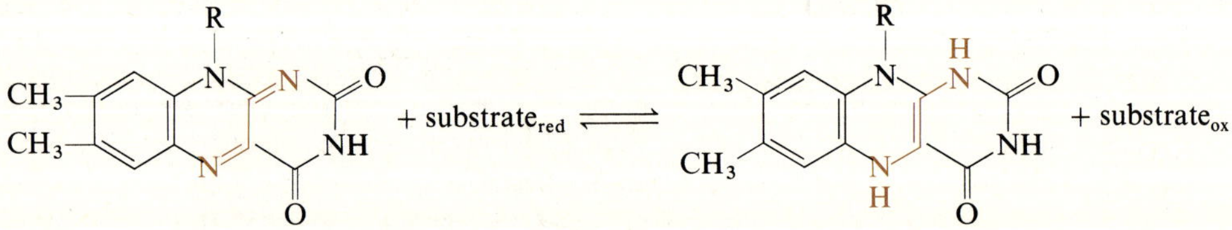

Also in contrast to the nicotinamides, FMN and FAD remain firmly bound to the enzyme molecule. Such enzymes in fact are called *flavoproteins*. Once flavin is reduced in a flavoprotein enzyme, it needs to find an acceptor for the hydrogens. Commonly, then, flavoproteins associate both with sources of hydrogen and receivers of hydrogen. These may well be quite different compounds.

$$NH_2\text{—}\overset{R}{\overset{|}{C}}H\text{—}COOH + H_2O + \text{enzyme-FMN} \rightleftharpoons O\text{=}\overset{R}{\overset{|}{C}}\text{—}COOH + NH_3 + \text{enzyme-FMN} \cdot 2H$$

(NH_2: source of hydrogen)

$$\text{enzyme-FMN} \cdot 2H + O_2 \rightleftharpoons \text{enzyme-FMN} + H_2O_2$$

(O_2: receiver of hydrogen)

The flavin ring must be supplied in our diet, and together with the ribitol it is called *riboflavin* or *vitamin* B_2. Deficiency of riboflavin leads to a variety of skin problems. Riboflavin is very stable and is common in meats, green vegetables, and whole grains.

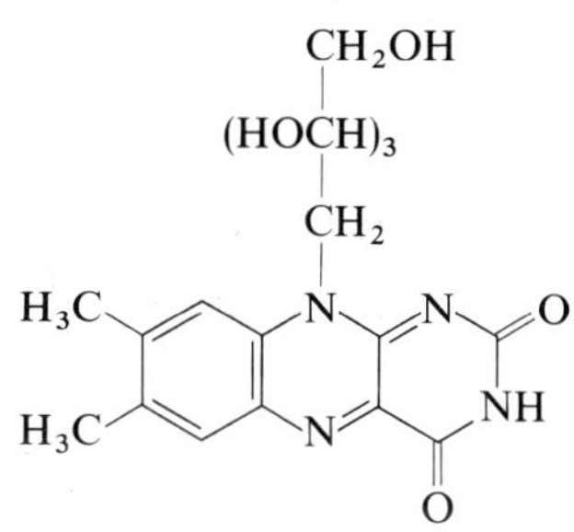

riboflavin or vitamin B_2

12.10 Lipoic Acid

A hydrogen carrier of very limited occurrence is lipoic acid. It is a very simple structure that carries or provides hydrogen by the thiol–disulfide reaction already described for the amino acid cysteine. Lipoic acid is firmly bound to the enzyme by an amide likage between its carboxyl group and the R group amine of lysine at the protein active site.

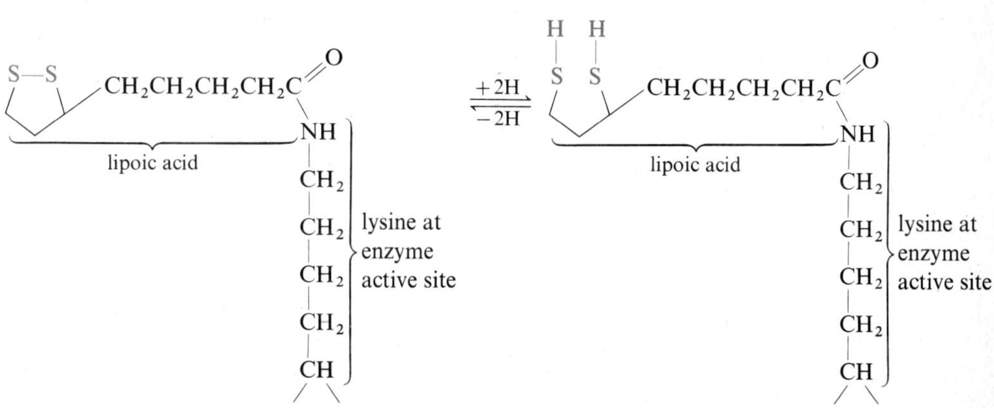

The very long flexible chain carrying the functional part of lipoic acid when it is attached to the protein is thought to allow it to swing easily back and forth between different substrates on the enzyme.

12.11 Tetrahydrofolic Acid

Some cofactors carry carbon structural units. Typically pieces one or two carbons in size are transferred by such structures. Tetrahydrofolic acid is an example. It is found to carry $—CH_3$, $—CH_2OH$, $—\overset{\overset{O}{\|}}{C}—H$, $—CH_2—$, and $—CH{=}NH$ groups. Two nitrogens on the molecule are involved, nitrogens 5 and 10.

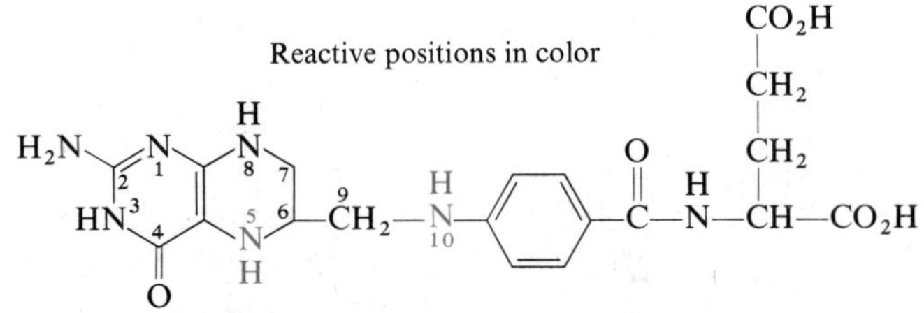

tetrahydrofolic acid (THFA)

Intestinal bacteria provide a supply of this cofactor. A deficiency, usually termed a *folacin deficiency*, produces anemia. A deficiency is more likely to occur during treatment with antibiotics, particularly with sulfa drugs. The sulfa drugs are particularly effective because their mechanism of action is to prevent synthesis of tetrahydrofolic acid. The sulfa drug is chemically similar to the *p*-aminobenzoic acid portion of the cofactor. It is mistakenly incorporated by the microorganism, producing a nonfunctional cofactor.

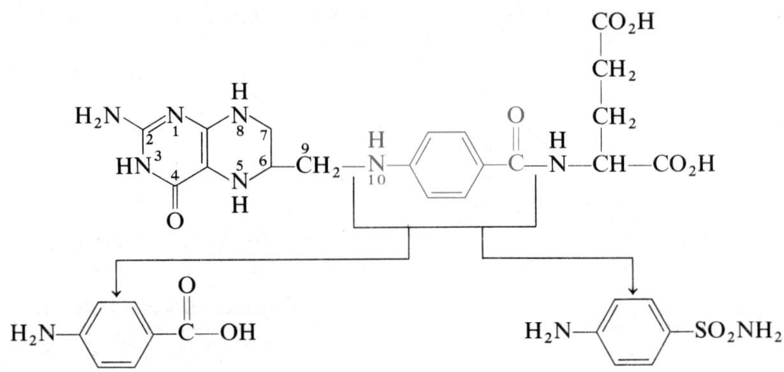

p-aminobenzoic acid

p-aminosulfonamide, a sulfa drug

Some sample reactions involving THFA are as follows:

$$\underset{\text{serine}}{NH_2\text{—}\overset{\overset{\displaystyle CH_2OH}{|}}{C}H\text{—}COOH} + \text{THFA} \rightleftharpoons \underset{\text{glycine}}{NH_2\text{—}CH_2\text{—}COOH} + N^5,N^{10}\text{-methylene-THFA} + H_2O$$

$$N^5\text{-methyl THFA} + \underset{\text{homocysteine}}{HS\text{—}CH_2\text{—}CH_2\text{—}CH(NH_2)\text{—}COOH} \rightleftharpoons \text{THFA} + \underset{\text{methionine}}{CH_3\text{—}S\text{—}CH_2\text{—}CH_2\text{—}CH(NH_2)\text{—}COOH}$$

12.12 Biotin

Biotin also transfers a one-carbon structure, a carboxyl group. The cofactor binds tightly to the enzyme in the same fashion as lipoic acid, by an amide linkage to lysine.

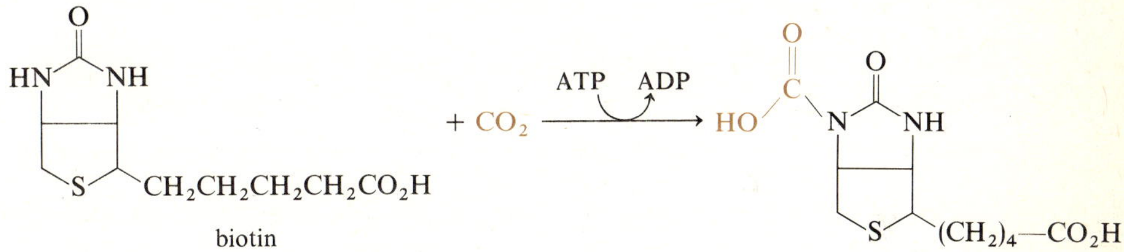

biotin

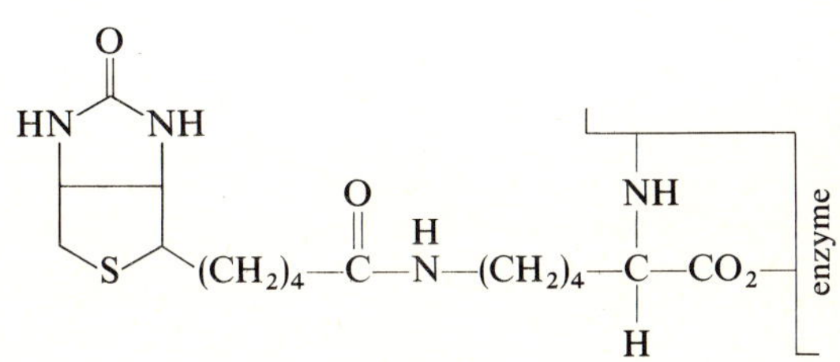

biocytin

Once again the long flexible "stem" on which the cofactor is attached allows it to move about from one site to another without being lost.

Biotin is supplied by the intestinal bacteria, so a deficiency disease is very seldom encountered. However, raw egg has a protein, avidin, that binds biotin very strongly. A prolonged diet of raw egg can cause a biotin deficiency. In cooked egg the avidin is denatured and causes no problem.

Metabolic energy supplied by ATP is necessary to attach the CO_2 to biotin. A sample reaction involving biotin follows:

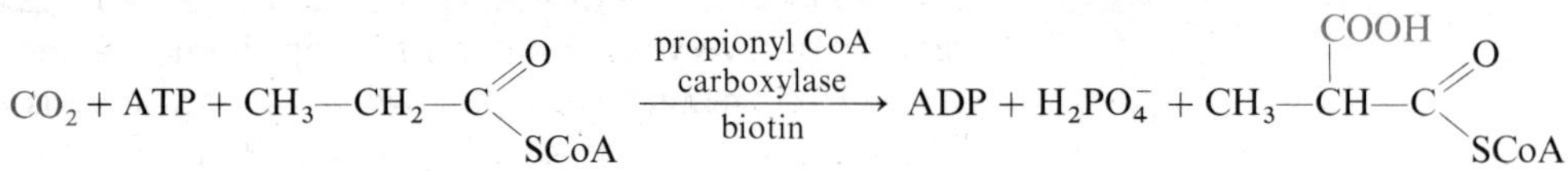

propionyl CoA — methylmalonyl CoA

This reaction is involved in the metabolism of fats (Chapter 16).

12.13 Coenzyme A

Coenzyme A functions differently from the cofactors so far introduced. It attaches to a substrate and goes with that substrate from one enzyme to another, rather than remaining associated with a single enzyme.

The structure is complex, but the functional portion is simply the terminal thiol group. In fact, the importance of the thiol group is emphasized by abbreviating the structure as CoA—SH.

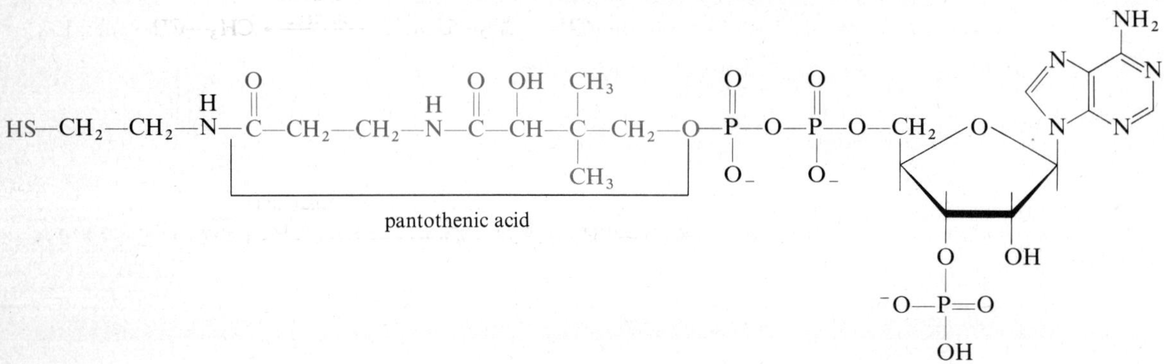

coenzyme A or CoA—SH

The pantothenic acid portion is a vitamin. As with the other B vitamins, meat, milk, and eggs are rich sources.

Coenzyme A attaches to carboxylic acid to form a thioester. The most common such structure, acetyl CoA, is one of the most important intermediates in metabolism. During reactions the substrate may undergo extensive modification, but the CoA is unchanged. Examples have been provided in the sections on biotin and cobamide and in Chapters 4 and 5.

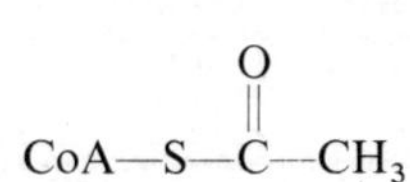

acetyl CoA

12.14 Cobamide

Cobamide, or vitamin B_{12}, is the most complicated of the structures in this section (see Chapter 23), and its chemistry is the least well understood. It is synthesized by certain microorganisms. Humans require only very small amounts of the vitamin. A deficiency produces pernicious anemia due to failure of red blood cells to mature. The deficiency may be the result of an inability to absorb the vitamin from the intestine even when it is present in the diet. Absorption requires that the cobamide be bound to a glycoprotein called *intrinsic factor*.

Vitamin B_{12} helps in reactions where a group is shifted from one place to another on a molecule. Examples follow:

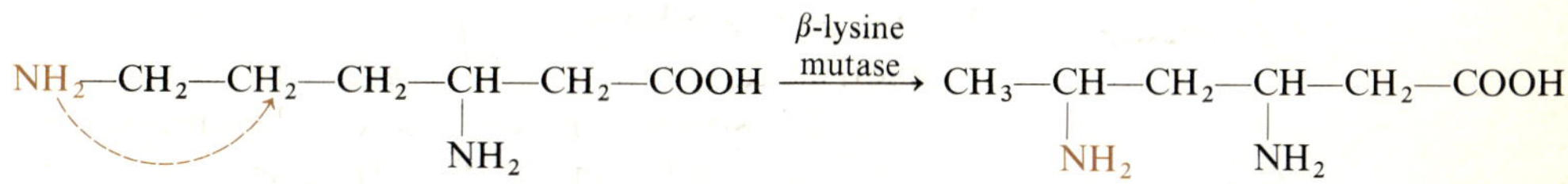

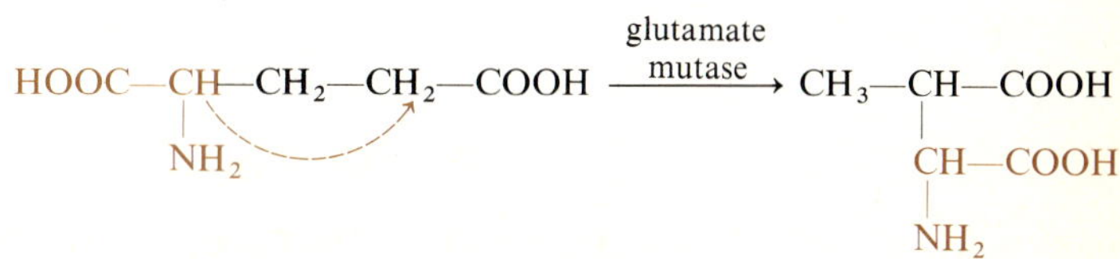

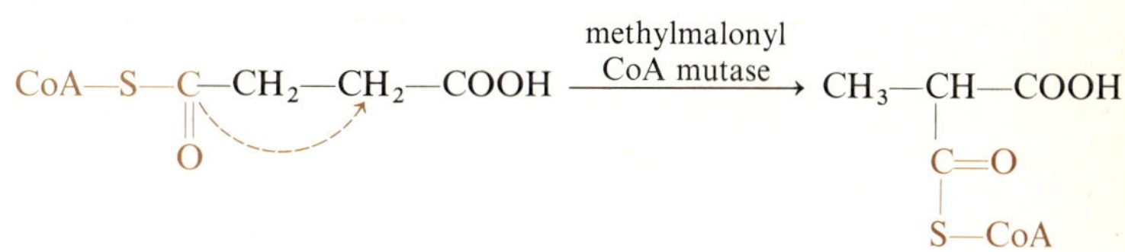

12.15 Thiamine

Thiamine pyrophosphate (TPP) is also called vitamin B_1. A deficiency of vitamin B_1 produces the disease beriberi. Thiamine pyrophosphate is involved in the catalysis of a variety of reactions, a common feature of which is the binding of the entire substrate or a fragment of it to the carbon between the N and the S of the five-membered ring. A fragment so removed may be transferred to another substrate.

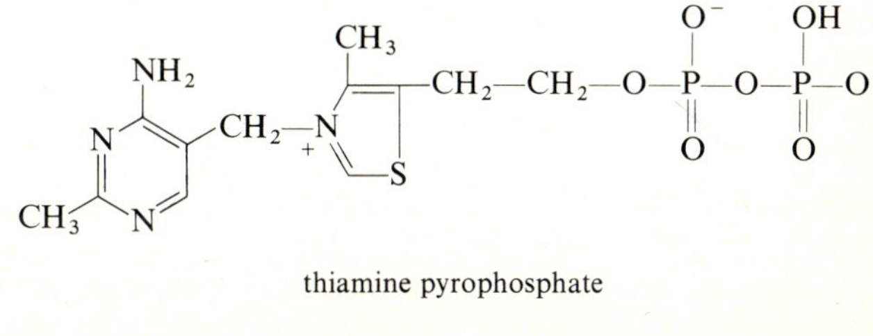

thiamine pyrophosphate

The following are two examples of reactions involving TPP:

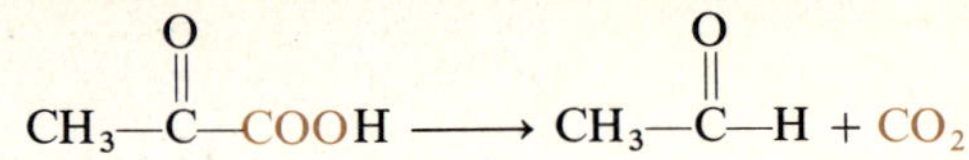

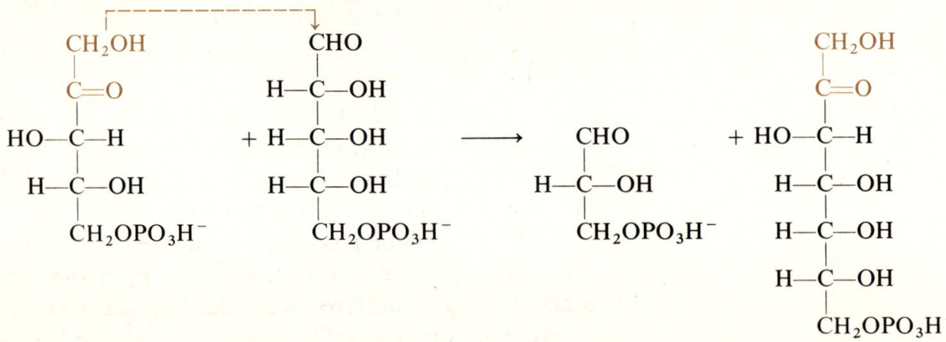

12.16 Pyridoxal

Pyridoxal, or vitamin B_6, assists in the catalysis of several kinds of reactions, including the removal of a carboxyl group, conversion of a single optical isomer of an amino acid to a racemic mixture, and transfer of an amine group. Cases of a deficiency disease have been observed in children but are unusual in adults. A person under treatment for tuberculosis and receiving isoniazide (Chapter 10) can develop a B_6 deficiency. Examples of B_6-assisted reactions follow:

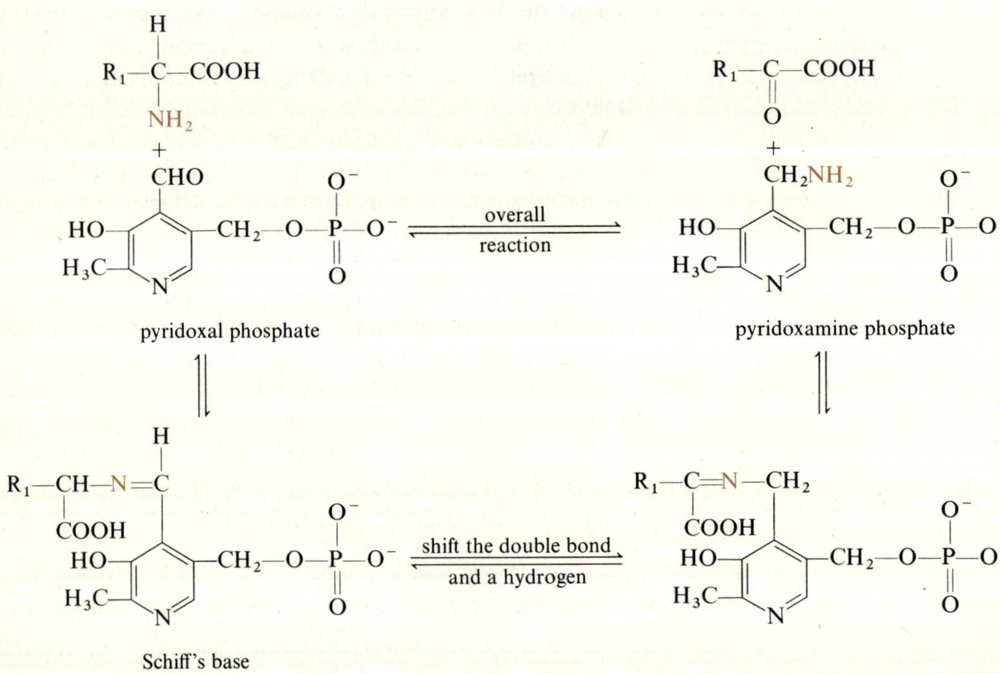

Suppose the above reaction proceeded only as far as the shift of the electrons and hydrogen. At the beginning, carbon 2 of the amino acid was asymmetric. Now it is not. If the reaction is now reversed to restore the amino acid, but is reversed without regard to D or L at carbon 2, a racemic mixture is produced. In this fashion a pure D amino acid would become a racemic mixture. The enzyme is called a *racemase*.

12.17 Naming Enzymes

Enzymes at first were given nonsystematic names. Gradually, as more were discovered, some general practices developed. It became customary to put -ase at the end of the enzyme name. To relate the name to the function, the substrate name was used, giving it an -ase ending. For example, an enzyme modifying fumarate is called *fumarase*, and one acting on aspartate is *aspartase*. To be more descriptive, the type of chemical action catalyzed was added to the name. The enzyme that removes two hydrogens from lactate is therefore *lactate dehydrogenase*. The International Union of Biochemistry (IUB) proposed a set of rules for nomenclature that provides more completely descriptive names and a system of cataloging the enzymes. By these rules lactate dehydrogenase is considered a usable but "trivial" name. The enzyme is correctly named L-lactate: NAD oxidoreductase. The enzymes were all classified by a numerical code. Lactate dehydrogenase is given the catalogue number 1.1.1.27. For simplicity we will use the trivial names rather than those proposed by the IUB.

12.18 Simple Enzyme Examples

Perhaps the best-known examples of enzymes are those involved in digestion. At the same time they are somewhat atypical because they do not display the total specificity toward a single substrate that most enzymes do.

Digestion is primarily a process by which polymers in the diet are converted to monomers. Only as monomers are they capable of crossing the mucosa of the gastrointestinal tract. In a chemical sense, compounds have not really entered the body until they do cross this barrier because they are not available to body cells. The gastrointestinal tract is therefore a tubelike extension of the outdoors providing compartments in which the diet can be mixed with enzymes and from which nutrients can be absorbed.

Consider carbohydrate digestion. Nutritionally our most important dietary carbohydrate is starch. Starch is first digested by saliva in the mouth. An α-amylase (amylose-attacking enzyme) hydrolyses the α-1,4 glycosidic linkages of starch randomly. In this fashion, a large polymer is very rapidly converted to a number of relatively small fragments. Perhaps in an earlier science class you were asked to hold a piece of bread in your mouth and notice that it became sweet because the starch was converted to glucose. (I could never get it to work myself; I always swallowed the bread.) By mixing starch and saliva you can watch the iodine color, and therefore the high-molecular-weight starch, disappear. To catalyze this reaction, the starch must be attached to the active site by reactive amino acid R groups of the enzyme. It must be held so the glycosidic bond is readily hydrolyzed by water.

Carbohydrate digestion stops in the stomach. Enzymes are sensitive to pH, and the salivary α-amylase, which operates well in the neutral environment of the mouth, stops completely in the very acid environment of the stomach. The small intestine is neutral, and another α-amylase is present, produced by the pancreas, to continue the hydrolysis of the polymer.

Dietary disaccharides are also hydrolyzed in the intestine. Specific enzymes are available for each disaccharide.

$$\text{Sucrose} \xrightarrow{\text{sucrase}} \text{fructose} + \text{glucose}$$

$$\text{Lactose} \xrightarrow{\text{lactase}} \text{galactose} + \text{glucose}$$

$$\text{Maltose} \xrightarrow{\text{maltase}} 2\ \text{glucose}$$

The monosaccharides are then absorbed through the mucosa. When infants are breast fed, lactose is the chief dietary carbohydrate. As the infant grows older, lactose can reach high levels in the diet. It sometimes exceeds the ability of intestinal lactase to hydrolyze it. Unhydrolyzed, it is not absorbed from the intestine and generates an undesirable bacterial growth in the large intestine. Some groups of people, principally of Middle Eastern or African origin, stop producing lactase after they are weaned. Milk is undesirable in the diet of these people thereafter because it is low in nutritional value and is intestinally upsetting.

A number of polysaccharides common in the diet are not nourishing because there are no enzymes to digest them. Cellulose is the best example. The α-amylase attacks the α-1,4 linkage but not the β-1,4 linkage. Grazing animals derive nourishment from cellulose even though they also lack a correct digestive enzyme. They depend on intestinal bacteria that do have a cellulase to make the glucose available to them. Other plant structural polysaccharides made of pentoses are not digestible by humans. Bacteria in the large intestine attack them to produce organic acids and CO_2. The acids stimulate peristalsis (intestinal muscle action). The combination of this stimulation and the roughage supplied by cellulose explains the usefulness of indigestible plant material as a laxative.

12.19 Protein Digestion

To be useful in the diet, proteins must first be converted to individual amino acids. The digestion of proteins begins in the stomach. Pepsin, the gastric *proteolytic* (protein-hydrolyzing) enzyme, is an example of an *endopeptidase*. It attacks peptide bonds at specific points in the middle of the peptide chain. This hydrolysis of scattered peptide bonds

Table 12.1
Specificities of Some Endopeptidases

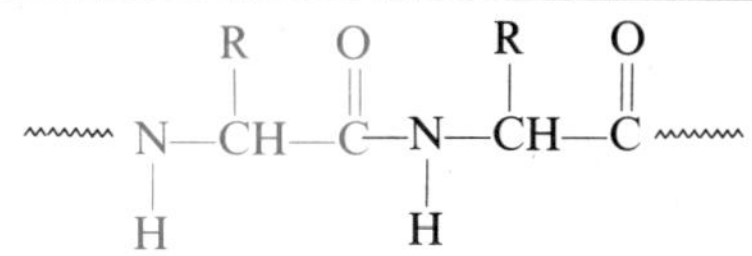

Peptide Bond Hydrolyzed

Enzyme name	*Target amino acids (shaded)*
Pepsin (stomach)	Tryptophan, phenylalanine, tyrosine
Trypsin (intestine)	Arginine, lysine
Chymotrypsin (intestine)	Tryptophan, phenylalanine, tryosine

quickly converts a few long chains into a number of short ones but produces few free amino acids. In the small intestine, more endopeptidases, trypsin and chymotrypsin, continue the conversion of long to short chains. These endopeptidases are all somewhat specific about the peptide bond they hydrolyze. These specificities are summarized in Table 12.1. The conversion of protein to individual amino acids is completed by *exopeptidases*, enzymes that attack the end of the peptide chain. Carboxypeptidase removes whatever amino acid is at the free carboxyl end of the peptide chain. Aminopeptidase does the same thing at the free amine end of the chain. Notice how the actions of endo- and exopeptidases complement one another. Endopeptidases create a very large number of "ends" for the exopeptidases to attack. Once individual amino acids are formed, they can be absorbed through the intestinal mucosa either to be reassembled into new protein or to be further broken down for energy.

12.20 Zymogens

Proteolytic enzymes present a problem to the body. It is desirable to produce these enzymes ahead of time and to store them until they are needed. Otherwise a great deal of very rapid protein synthesis would be called for every time you snack. However, in the protein-lined world of the animal interior, where does one store an enzyme that digests protein? It's a bit like the old puzzle of where to store a universal solvent.

The answer lies in producing the enzymes in an inactive or zymogen form. By hydrolyzing a few peptide bonds on the zymogen, the active site can be uncovered or the configuration changed so that it is converted quickly to an active enzyme. For example, the enzymatic machinery to activate the zymogens is found in the digestive tract, so the enzymes are stored without hazard until they are needed. Then in the place of use they are quickly activated.

In summary, synthesizing enzymes as inactive precursors or zymogens provides an important mechanism for control of where and when the enzyme is active.

12.21 Blood Clotting—A Cascade Mechanism

The mechanism for the forming of a blood clot is an interesting special application of enzymes. The use of enzymes allows this emergency process to be performed very quickly. In addition, the speed of the process is increased even more by the cascade effect. Finally it is a system over which we must keep very close control. Random formation of clots can block circulation in critical areas such as the heart or brain, with lethal consequences.

A blood clot is composed of a protein molecule called *fibrin.* Clot formation occurs in three stages, two of which require enzyme catalysis.

Stage 1: Formation of fibrin. The blood does not normally contain fibrin as such. Instead there is about 3 g of a fibrin precursor, fibrinogen, in every liter of blood. The first step in clot formation is the conversion of fibrinogen to fibrin. Fibrinogen has a quaternary structure consisting of four subunits. A peptide is hydrolyzed away from each subunit by the enzyme thrombin. Once the peptides are removed, the fibrinogen has become fibrin.

Stage 2: Polymerization of fibrin. The removal of peptides from fibrinogen unmasks sites at which the fibrin molecules bind to one another. The resulting "polymer" of fibrin units is called a *soft clot.*

Stage 3: Cross-linking of the fibrin. The enzyme thrombin also hydrolyzes peptides away from another protein in the blood, called *fibrin-stabilizing factor*, converting it into an active enzyme called *fibrinase*. Fibrinase catalyzes the formation of an amide linkage between glutamic acid R group carboxyls of one fibrin and lysine R group amines of another fibrin. The fibrins are now tightly bound together. This joining together of polymer strands is called *cross-linking* (Figure 12.5). This much more rigid and less soluble cross-linked structure is called the *hard clot*. The reaction requires calcium to be present.

If you were watching this process closely, you realized that the key to clot formation was the enzyme thrombin. If thrombin enzymetically stimulates clot formation, and if the raw material fibrinogen is plentiful, why do we not have clots forming continuously? Clot formation can be a problem. Sometimes clots do form in blood vessels and move until they lodge in a narrow vessel and block the flow of blood to tissues. The tissue is starved for oxygen and the cells die. If the tissue so destroyed is in the brain, we term the situation a *stroke*. When heart muscle tissue is killed, the problem is called a *myocardial infarction*, or *coronary throm-*

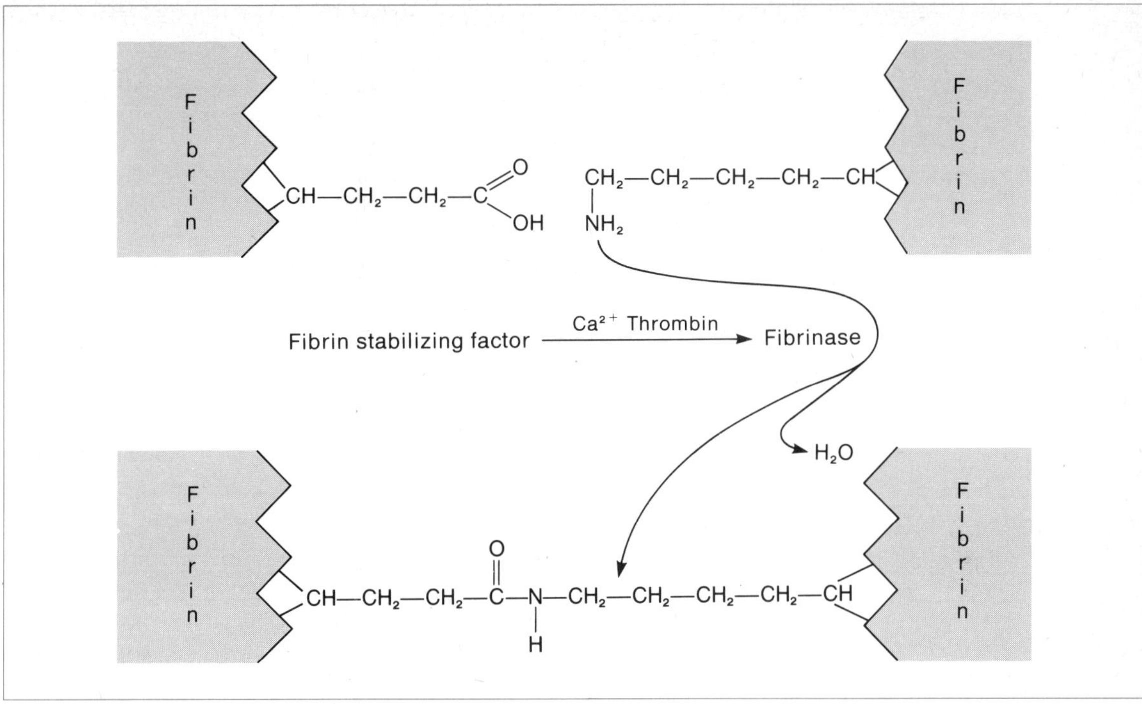

Figure 12.5
The hard clot
Large fibrin molecules are cross-linked by fibrinase to create the huge mass we call a clot. The same enzyme (thrombin) that converts fibrinogen to fibrin activates the fibrinase to cross-link the fibrin polymer.

bosis. Because these are serious health problems, more complete consideration of the clotting mechanism is justified.

Like fibrinase, thrombin is normally present as a zymogen. The thrombin zymogen is called *prothrombin*. Prothrombin is converted to thrombin by action of another proteolytic enzyme, thromboplastin, in the presence of calcium ion, phosphoglyceride, and Stuart factor. Thromboplastin can be supplied by tissues, being released when tissue is damaged, or by blood platelets (cells that circulate in the blood) that accumulate and rupture at the site of blood vessel damage. The massed platelets themselves can serve as the nucleus of the fibrin clot.

Platelets have an important role in clot formation. When drawing blood, the likelihood of clots forming is reduced by using silicone or teflon-coated vessels to which platelets don't stick, or by centrifuging platelets out of the blood.

The mechanism as presented so far consists of proteolytic enzymes that specifically activate other proteolytic enzymes. This results in a system that responds quickly to damage because of a cascade effect. A cascade effect can be understood best from an example. Compare two systems. One has an enzyme that by itself catalyzes a desired change. The other has a series of enzymes, each activated by the one above it in the sequence. For simplicity, pretend that in one second any enzyme converts 100 substrate molecules to product. As you see in Table 12.2, at the end of 5 seconds the direct catalysis of product has produced 500 molecules of Y,

Table 12.2
The Cascade Effect in Enzyme Activation

Direct Activation	Cascade Mechanism
$X \xrightarrow{\text{enzyme 1a}} Y$	$\text{Proenzyme 2} \xrightarrow{\text{enzyme 1}} \text{enzyme 2}$
	$\text{Proenzyme 3} \xrightarrow{\text{enzyme 2}} \text{enzyme 3}$
	$X \xrightarrow{\text{enzyme 3}} Y$

In each case, start with 1 molecule of enzyme 1.

Time (sec)	Direct	Total Yield	Cascade	Total Yield
1	$100X \rightarrow 100Y$	$100Y$	$100 \text{ pro 2} \xrightarrow{\text{enz 1}} 100 \text{ enz 2}$	$0Y$
2	$100X \rightarrow 100Y$	$200Y$	$100 \text{ pro 2} \xrightarrow{\text{enz 1}} 100 \text{ enz 2}$	$0Y$
			and	
			$10^4 \text{ pro 3} \xrightarrow{100 \text{ enz 2}} 10^4 \text{ enz 3}$	
3	$100X \rightarrow 100Y$	$300Y$	$100 \text{ pro 2} \xrightarrow{\text{enz 1}} 100 \text{ enz 2}$	$10^6 Y$
			$2 \times 10^4 \text{ pro 3} \xrightarrow{200 \text{ enz 2}} 2 \times 10^4 \text{ enz 3}$	
			$10^6 X \xrightarrow{10^4 \text{ enz 3}} 10^6 Y$	
4	$100X \rightarrow 100Y$	$400Y$	$100 \text{ pro 2} \xrightarrow{\text{enz 1}} 100 \text{ enz 2}$	$4 \times 10^6 Y$
			$3 \times 10^4 \text{ pro 3} \xrightarrow{300 \text{ enz 2}} 3 \times 10^4 \text{ enz 3}$	
			$3 \times 10^6 X \xrightarrow{3 \times 10^4 \text{ enz 3}} 3 \times 10^6 Y$	
5	$100X \rightarrow 100Y$	$500Y$	$100 \text{ pro 2} \xrightarrow{\text{enz 1}} 100 \text{ enz 2}$	$10^7 Y$
			$4 \times 10^4 \text{ pro 3} \xrightarrow{400 \text{ enz 2}} 4 \times 10^4 \text{ enz 3}$	
			$6 \times 10^6 X \xrightarrow{6 \times 10^4 \text{ enz 3}} 6 \times 10^6 Y$	

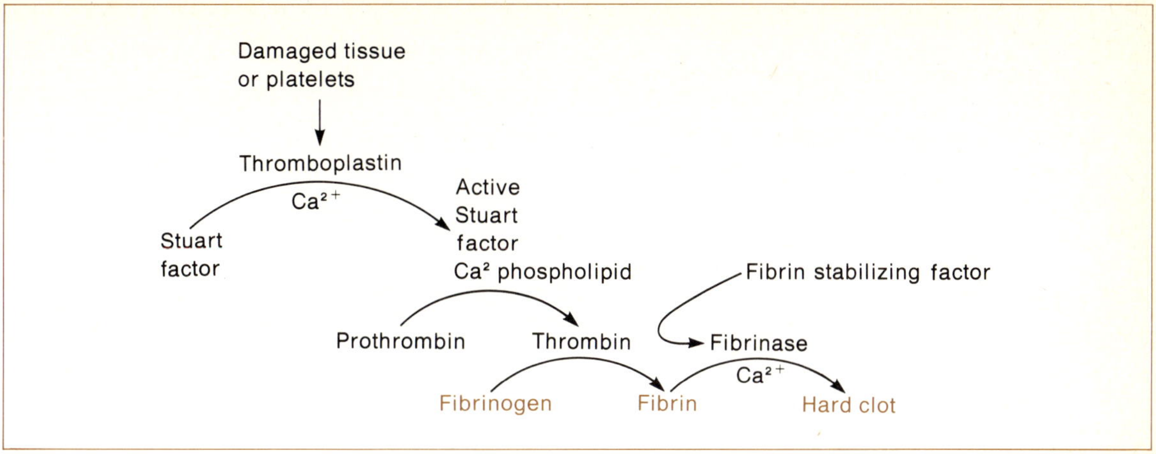

Figure 12.6
The blood-clotting cascade

while the cascade mechanism has produced 10 million and is still accelerating its rate of production.

The cascade in blood clotting is as in Figure 12.6.

Some individuals have blood that does not clot. This may be due to a shortage of platelets that can result from exposure to radiation or to certain chemicals, or from a variety of illnesses. However, some people cannot form clots from birth. The problem is called *hemophilia*. Hemophilia is an inherited trait and is sex linked; that is, it is transmitted only through the mother. In studying blood from various patients with hemophilia, it was found that although the blood samples would not clot by themselves, sometimes two samples mixed together would clot. This inferred that in each case some part of the clotting machinery was missing, but the piece missing was different in each of the two samples that clotted on mixing. When they were mixed, each supplied the other with the missing factor. Analyses of these blood samples revealed factors for clotting in addition to those already mentioned.

The existence of these factors indicates that there is a pathway for clot formation different from that described above. The proposed second pathway is also a cascade of proteolytic enzymes arranged as in Figure 12.7. Notice that some factors are in both pathways and that this second pathway is longer.

Blood contains a polysaccharide called *herapin* (Chapter 6) that inhibits the cascade, preventing clotting.

The synthesis of prothrombin and some other factors by the liver requires an adequate supply of vitamin K (Chapter 3). In its absence, the clotting mechanism will not operate, resulting in hemorrhaging upon injury. In cattle this hemorrhaging results from clover consumption. The problem was traced to the presence in clover of dicumarol, which is similar in structure to vitamin K. Presumably such an analogue is accepted by the liver in place of vitamin K, but it does not function. In cases where the inopportune

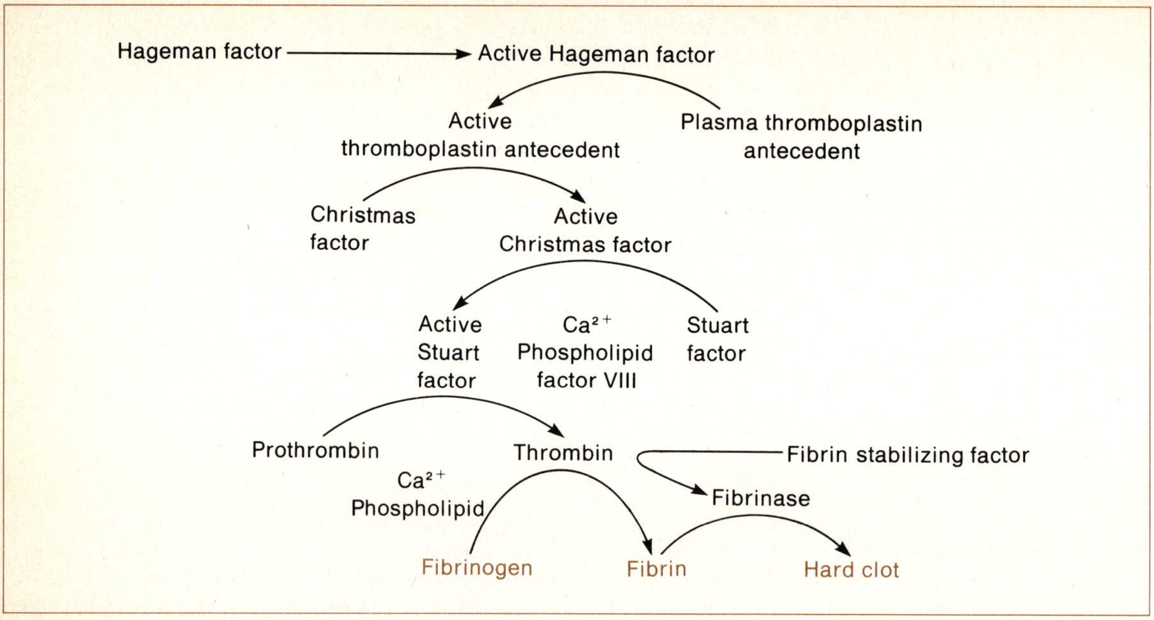

Figure 12.7
The secondary path for blood clotting

formation of a clot is of danger to a person, the danger of a stroke or myocardial infarction can be lessened by dicumarol or warfarin, another vitamin K analogue. For short-term treatment, heparin may be administered.

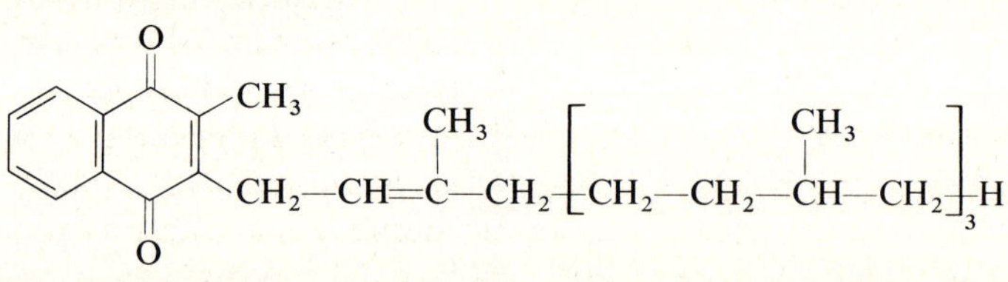

vitamin K_1

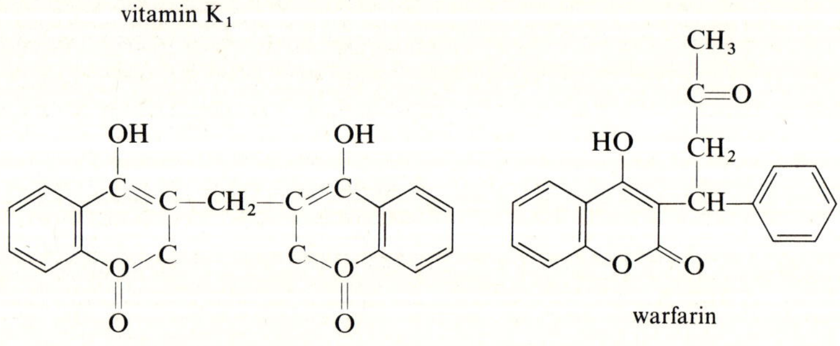

dicumarol

warfarin

When blood is drawn, clot formation may be prevented by adding heparin. Calcium ion is required in several places in this mechanism. Removal of Ca^{2+} can therefore prevent clotting. This is accomplished by adding oxalate or citrate to blood, either one of which traps the Ca^{2+} in a complex.

12.22 Enzymes and Metabolism

Metabolism is the sum of the various chemical reactions performed in a cell. It is the pattern by which nutrients from digestion eventually become new cell material or a source of energy. We are at first bewildered by the complexity of the pattern as we now know it to be. On closer observation, however, pathways leading to specific goals can be seen. Metabolism viewed all at one time on one of the big metabolic charts you can obtain from chemical or equipment companies looks like a series of interconnected river systems, several streams sometimes arising from common headwaters, and occasional whirlpools spinning the streams around. The individual streams on close observation prove to be rows of chemical reactions, each step being enzyme-catalyzed. Part of the complexity arises because enzymes do not individually perform miraculous transformations. For example, glucose can be used as a source of energy, and we could write the overall reaction as follows:

$$C_6H_{12}O_6 + 6O_2 \longrightarrow 6CO_2 + 6H_2O + \text{energy}$$

However, this change is not catalyzed by a single enzyme. Instead the total process involves a very large number of individual steps. Over twenty different enzymes are used just to generate all the CO_2. At that point the water and a useful form of the energy are yet to come. Each enzyme has a single small change to catalyze, and the metabolic pathway is the long sequence of these individual steps.

Because the enzymes are impressively effective catalysts, the process, complex as it is, can run very rapidly. The specificity of the enzymes is even more important. Starting with glucose, an aldehyde with five alcohol groups, a large number of reactions could happen—oxidations, reductions, rearrangements, etc. However, the reaction we desire, the addition of phosphate to carbon 6, is the only one that will occur to any important degree. This is because the enzyme hexokinase is specific in binding glucose, has a single change that it is capable of catalyzing, and performs that change so rapidly that nothing else has time to happen. The product, glucose-6-phosphate, is the substrate specifically for phosphoglucoisomerase, which rapidly converts it to fructose-6-phosphate. In other words, of all the diverse changes that could occur in metabolism, only those for which there are enzymes occur. The direction of metabolism is decided by the set of enzymes available.

12.23 Allosteric Enzymes

Going back to our analogy of metabolism resembling a complex river system, given the effectiveness of enzymes as catalysts these could be very fast-flowing rivers indeed. I am confident that calculations would support the idea that our enzymes are capable of turning us into solid blocks of aspartic acid or triglyceride very quickly. We need to have some dams on our river, some way to control the flow so that we produce exactly the amount we need of some products and no more.

By controlling the catalytic activity of some of the enzymes, we can maintain close control of the amount of product each metabolic pathway creates. How are levels of enzyme catalytic activity controlled? We don't need to control them all. We need to control only the enzymes in key positions in the metabolic pathway. The activity of these key enzymes is regulated by a compound related to the pathway and called an *effector*. Such a controlled enzyme is called *allosteric* and has a separate binding site for the effector. Very often the first enzyme unique to a pathway is allosteric. Allosteric enzymes may be turned either on or off by effectors. In fact, there are examples of enzymes with several effectors, some of which activate and some of which inhibit the enzyme. *Feedback inhibition* is a useful example of the allosteric type of control. In this pattern the allosteric enzyme is the first in a series of enzymes whose function is to produce a specific product. The final product of the pathway is an effector that inhibits the activity of the enzyme.

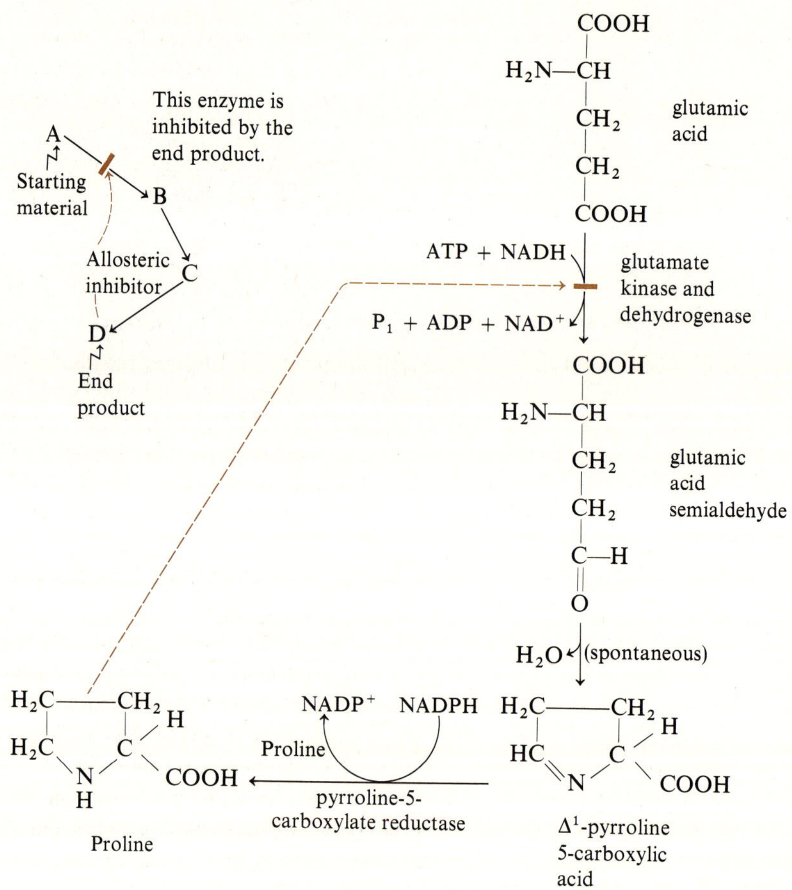

As excess quantities of the effector are produced, they inhibit the enzyme and thus halt the production of a further excess. As the effector product is used up, the enzyme will again be active. Notice that this prevents accumulation of all the intermediate compounds (B and C in the illustration).

12.24 Isozymes

Sometimes more than one type of enzyme molecule is produced to catalyze a particular reaction. Such multiple forms of enzymes are called *isozymes*. Generally isozymes are different in certain characteristics, such as the reactivity toward substrate or the particular allosteric effectors that can exert control over them. In this fashion the enzyme can be custom tailored to the needs of a particular tissue. Liver and muscle cells might well contain different isozyme forms of an enzyme, for example, each having a form best suited to the chemistry going on in the particular type of cell. Being able to separate and measure relative amounts of isozymes can be useful in clinical diagnosis. For example, lactate dehydrogenase exists as five different isozymes. When cells die, the enzymes appear in blood serum. Normal patterns of cell death result in a characteristic mixture of lactate dehydrogenase isozymes in the blood. A pathological condition, for example, a myocardial infarction, leads to an unusually high concentration of lactate dehydrogenase in the blood. Furthermore, the isozymes characteristic of heart muscle are then present in unusually high concentration. The isozymes can be separated from one another by electrophoresis. The occurrence and severity of the heart damage can be detected by a measurement of the relative amounts of lactate dehydrogenase isozymes. In Figure 12.8 we see typical patterns for lactate dehydrogenase isozymes separated by electrophoresis. Both the normal pattern and the pathological patterns are presented.

Creatine phosphokinase (CPK) is another enzyme used in this fashion. There are three isozyme forms of this enzyme. Only one, CKP_3, is normally seen in the blood. CPK_1 is the brain isozyme. Brain and nerve cells do not normally die, so this isozyme does not appear in the blood. The isozyme CPK_2 is in muscle and is released by a myocardial infarction or muscular dystrophy. In the case of myocardial infarction, the enzyme disappears in about 72 hours, while muscular dystrophy patients produce CPK_2 continuously. Patterns are shown in Figure 12.8.

Alkaline phosphatase has five isozymes. These are named according to the tissue source: $liver_1$ (fast liver), $liver_2$ (slow liver), bone, placenta, and intestine. Normal samples show only $liver_2$. Liver and bone disorders can be detected by unusual patterns (Figure 12.8).

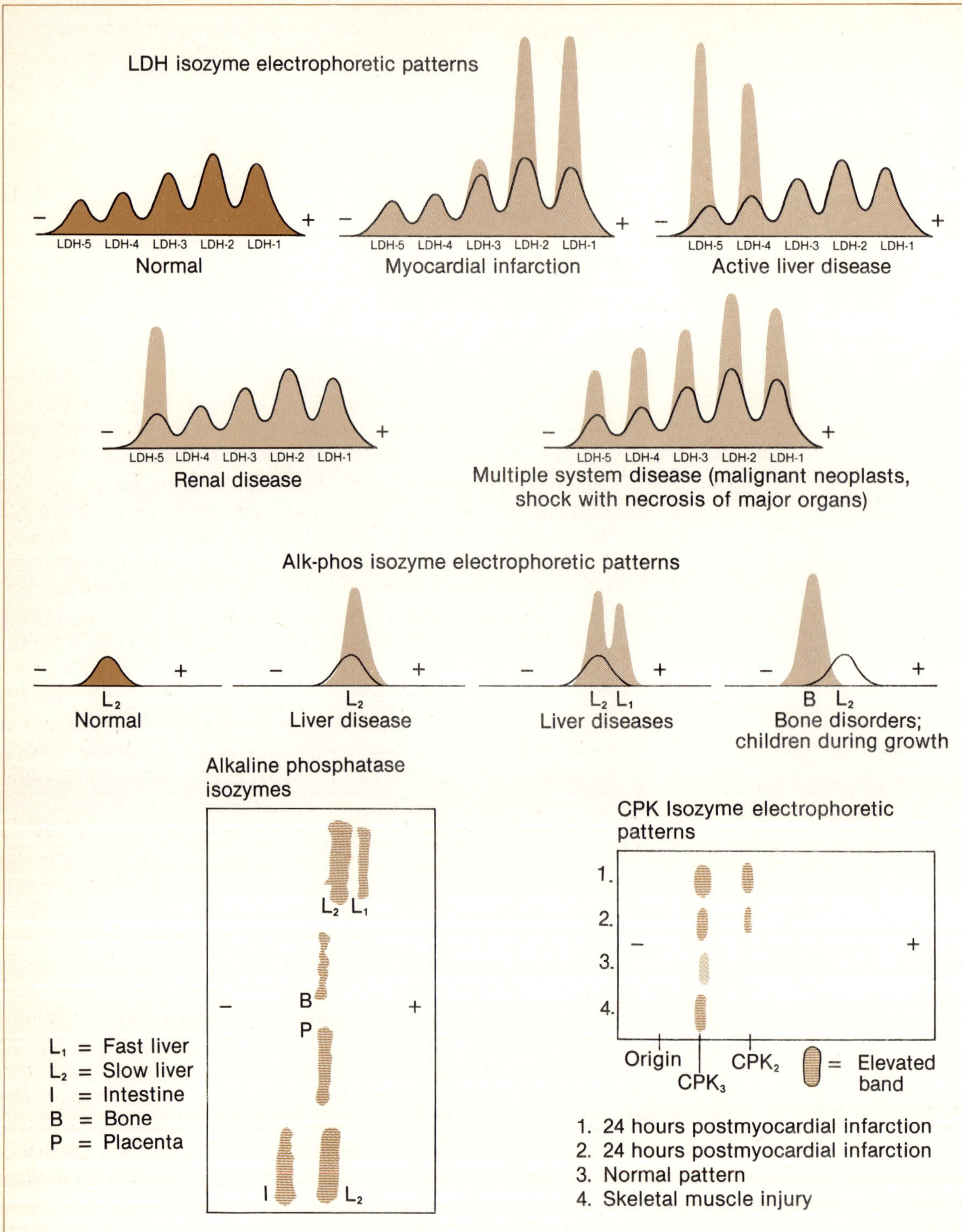

LDH isozyme electrophoretic patterns
− LDH-5 LDH-4 LDH-3 LDH-2 LDH-1 +
Normal
− LDH-5 LDH-4 LDH-3 LDH-2 LDH-1 +
Myocardial infarction
− LDH-5 LDH-4 LDH-3 LDH-2 LDH-1 +
Active liver disease
− LDH-5 LDH-4 LDH-3 LDH-2 LDH-1 +
Renal disease
− LDH-5 LDH-4 LDH-3 LDH-2 LDH-1 +
Multiple system disease (malignant neoplasts, shock with necrosis of major organs)
Alk-phos isozyme electrophoretic patterns
− + L_2
Normal
− + L_2
Liver disease
− + L_2 L_1
Liver diseases
− + B L_2
Bone disorders; children during growth
Alkaline phosphatase isozymes
L_2 L_1
B
P
− +
I L_2
L_1 = Fast liver
L_2 = Slow liver
I = Intestine
B = Bone
P = Placenta
CPK Isozyme electrophoretic patterns
1.
2.
3.
4.
− +
Origin
CPK_3
CPK_2
= Elevated band
1. 24 hours postmyocardial infarction
2. 24 hours postmyocardial infarction
3. Normal pattern
4. Skeletal muscle injury

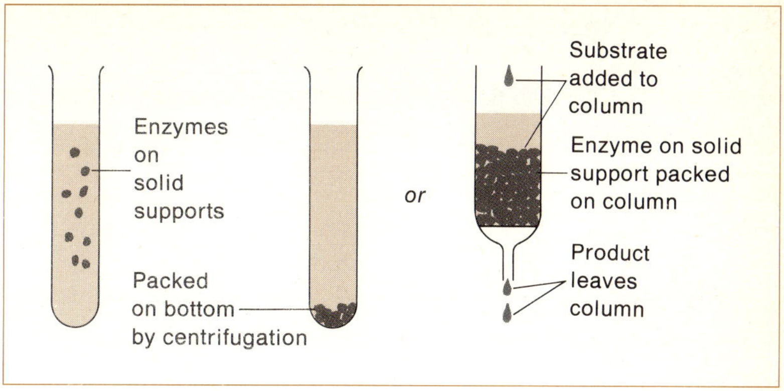

Figure 12.9
Enzymes in food processing
Successful use of enzymes in food processing requires that they be removed from the reaction vessel and reused. Enzymes bound to a solid support can be centrifuged out of solution, or they can be packed on a column through which the reactant is run.

12.25 Enzymes in Food Processing

As indicated, enzymes catalyze chemical changes with a speed and precision that turn a laboratory worker green with envy. One of the really exciting practical research endeavors today is the application of enzymes to industries such as food processing. And why not? Why should this be difficult? In one sense we have been doing this at least since the start of recorded history. Conversion of carbohydrates to alcohol by fermentation certainly takes advantage of yeast enzymes. In this case, however, the organism providing the enzyme remains intact and alive. We might say the yeast cells are providing a life support system for the enzymes of fermentation. However, what is really needed is individual specific enzymes to do just what is wanted and nothing more. Enzymes are generally rather difficult to isolate in pure form. Then after we take the trouble to purify them, they spontaneously denature. Finally to be used, they need to be mixed with the substrate, after which it may be very difficult to get the enzymes back, even though they are perfectly reusable.

A variety of techniques are solving these problems. Scientists are adding cross-links to the enzyme molecule to hold the enzyme more strongly in the correct molecular pattern. This prevents denaturation, extending the useful life of the enzyme. In another method the enzymes are attached to a large, insoluble particle. The enzymes can be added to substrate, then easily centrifuged away from the product later. It would also be possible to pack the solid supports to which enzymes were attached into a column, then pour the substrate through Figure 12.9.

Enzymes are being used already. Probably the most important application is the generation of sweet syrups from cornstarch. D-glucose is not as sweet as sucrose, but D-fructose is sweeter (Chapter 6). A mixture of D-glucose and D-fructose is used in place of sucrose in canned fruits, fruit drinks, and carbonated beverages. This mixture is prepared by use of three enzymes. α-amylase converts starch to shortened units called *dextrins*. These are hydrolyzed

Figure 12.8 (opposite)
Patterns obtained for isozymes in blood and the relationship to disease
(From Gelman Instrument Company, Ann Arbor, Michigan.)

to glucose by glucoamylase. Finally the glucose is converted to fructose by glucose isomerase.

$$\text{cornstarch} \xrightarrow{\alpha\text{-amylase}} \text{dextrin} \xrightarrow{\text{glucoamylase}} \text{glucose} \xrightarrow[\text{isomerase}]{\text{glucose}} \text{fructose}$$

More than a billion pounds of these high fructose syrups are produced annually.

There are several other applications of enzymes in food processing. The first step in cheese production is the precipitation (coagulation) of casein, the chief milk protein. This is done with the enzyme rennin obtained either from calf stomach or more recently from a microorganism. In another example, a problem exists because the unwanted trisaccharide raffinose (Chapter 6, Problem 11) interferes with the crystallization of sucrose from beet sugar syrups. Raffinose increases in concentration when the beets are stored in cold weather. An α-galactosidase is used to convert the raffinose to sucrose and galactose in the syrups. A third example is provided by the processing of coffee. During processing it is necessary to remove a polyuronide (Chapter 6) impurity, here termed a *mucilage*. This mucilage is hydrolyzed by "pectinolytic enzymes," a group of enzymes specific for polyuronides. These enzymes also serve to remove polyuronides from fruit juices and wines in order to remove the cloudy appearance such polysaccharides produce in the juice.

The future potential for such processing is impressive. Consider just one possibility. If a stable cellulase preparation were prepared, the cellulose of plants could be converted to glucose to be used as food. The enormous quantities of discarded cellulose in the form of paper, cotton, wood, garden debris, agricultural wastes, and food processing wastes could be converted to glucose. This glucose could then be fermented to produce ethanol. Every 12.8 lb of glucose produces a gallon of alcohol. Such ethanol could be produced cheaply enough to be considered as a supplement to or replacement for petroleum fuels in automobiles. This process would help solve both an energy problem and a trash disposal problem.

Summary

An enzyme is a protein that catalyzes a specific chemical change in a molecule called the substrate. The catalysis occurs at a region of the protein molecule called the active site. Only the correct substrate has the right shape to attach to the active site. Catalysis occurs because certain functional groups of the enzyme protein are placed very close to reactive parts of the substrate. This catalysis may be assisted by metal ions or organic cofactors that also attach to the active site. The organic cofactors usually serve as carriers of chemical fragments being transferred from one place to another. We have lost the ability to synthesize many of the cofactors and depend on a supply of them in our diet. Several are called B vitamins. Enzymes are named by listing substrate and type of chemical change catalyzed, then ending the name in -ase. Enzymes are very effective catalysts. Because enzymes are synthesized to accelerate only certain reactions, other possible reactions do not occur to a significant degree. Thus the long series of chemical changes compounds undergo in the body is the result of the existence of a set of enzymes selecting the desirable reactions. Some enzymes are controlled so that the correct amount of a product results. When the binding of molecules from the surrounding medium exerts this control, the enzymes are called allosteric.

Carbohydrate digestion begins with the hydrolysis of starch in the mouth. In the intestine, starch hydrolysis continues. A series of specific enzymes hydrolyze disaccharides to produce monosaccharides, which can then be absorbed across the mucosa.

Protein digestion begins in the stomach and continues in the intestine with a series of endopeptidases, enzymes that hydrolyze peptide bonds at specific points in the middle of the chain. The peptide fragments so produced are converted to individual amino acids by exopeptidases, which remove an amino acid at a time from the end of the chain.

Blood clotting involves the conversion of fibrinogen, a protein, to fibrin. Fibrin then spontaneously polymerizes into a soft clot, which is then cross-linked into a hard clot. The fibrinogen conversion is catalyzed by a proteolytic enzyme that, in turn, is converted from inactive to active form by another proteolytic enzyme. In fact, the clotting mechanism as we understand it is a cascade of enzymes activating enzymes, which results in a very rapid massive response. Clotting is prevented by removing Ca^{2+}, by adding heparin, or on a long-term basis by inadequate supply of vitamin K. The latter is necessary for the synthesis of some blood-clotting factors in the cascades.

Isozymes are a family of enzymes different in structure but all catalyzing the same reaction. They often represent adaptations of the enzyme to the needs of different tissues. Extensive use is made of the relative amounts of isozymes of a particular enzyme that appear in the blood due to cell death. The pattern indicates the identity of the tissue destroyed.

Enzymes are being used increasingly in food processing. There is an impressive potential for conversion into useful products of materials presently wasted.

Problems

1. We know the enzyme binds the substrate at the active site. Suppose you were designing your own enzyme. Would you:

***a*)** Bind the substrate to the enzyme very strongly, perhaps by several covalent bonds?

***b*)** Bind the substrate to the enzyme moderately strongly, relying largely on hydrogen bonds, electrostatic attractions, and hydrophobic bonding?

***c*)** Bind the substrate very weakly, perhaps by a few hydrogen bonds?

2. After the word in the left-hand column place the letter of *every* statement in the right-hand column that is appropriate.

***a*)** Enzyme active site

***b*)** Endopeptidase

***c*)** Substrate

***d*)** Zymogen

***e*)** Isozyme

***f*)** Allosteric effector

***g*)** Cofactor

***h*)** Chymotrypsin

***A*)** This word refers to a single specific enzyme.

***B*)** This is a molecule that attaches to some or all kinds of enzymes.

***C*)** This is where molecules bind to the enzyme.

***D*)** This is an inactive enzyme.

***E*)** This is a member of a group, many of which are vitamins.

***F*)** This controls enzyme activity.

***G*)** This is one of a group of different proteins that all exhibit the same enzyme activity.

***H*)** This attacks peptide bonds in the middle of the chain.

***I*)** This attacks peptide bonds at the end of the chain.

***J*)** This term is associated with every enzyme.

3. After the cofactor in the left-hand column, place the letter of every statement in the right-hand column that applies.

a) NAD
b) FMN
c) Pyridoxal
d) Thiamine pyrophosphate
e) Coenzyme A
f) FAD
g) Biotin
h) NADP
i) Tetrahydrofolic acid

A) Has AMP as part of its structure.
B) Has a vitamin as part of its structure.
C) Is an oxidation-reduction cofactor.
D) Is a cofactor that binds a carbon fragment of some sort to itself.
E) Binds to the substrate before the substrate approaches the enzyme.
F) Is involved with a one-carbon fragment.

4. After class you are hungry and snack on a glass of milk and a piece of bread with jam. The chief foods you have taken are as follows:

- Milk: lactose, casein (protein)
- Bread: starch, gluten (protein)
- Jam: sucrose

Trace the course of this snack from your mouth to your stomach and intestine, listing in each location the enzymes that act on each food component and the changes the enzymes produce.

5. The science fiction novel *The Andromeda Strain* by Michael Crichton describes a disease arising from the space program in which the victims' blood is very rapidly and completely clotted. Explain why this feature of the plot has a degree of scientific credibility.

6. Clot formation in freshly drawn blood may be prevented by adding one of the following agents. Why in each case is clot formation prevented?

a) Oxalate *b*) Heparin *c*) Citrate

7. You have heard that administering dicumarol or warfarin to a person prevents clot formation in that person. Yet when you add these to freshly drawn blood, it clots anyway. Explain this discrepancy.

8. Below are a series of reactions that are catalyzed by enzymes that require cofactors. Suggest which cofactor would be best suited to assist each reaction.

a) 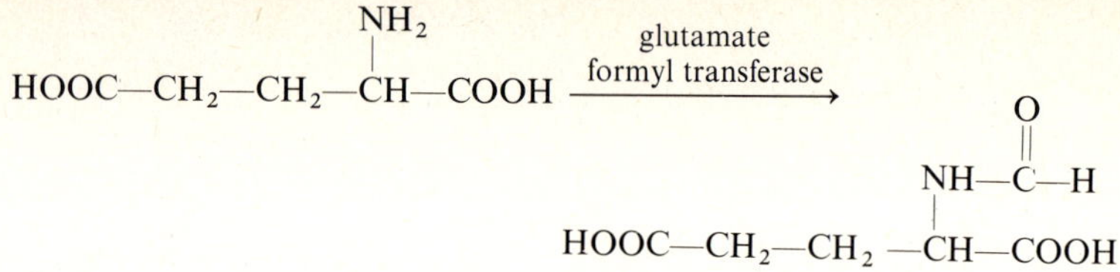

b) 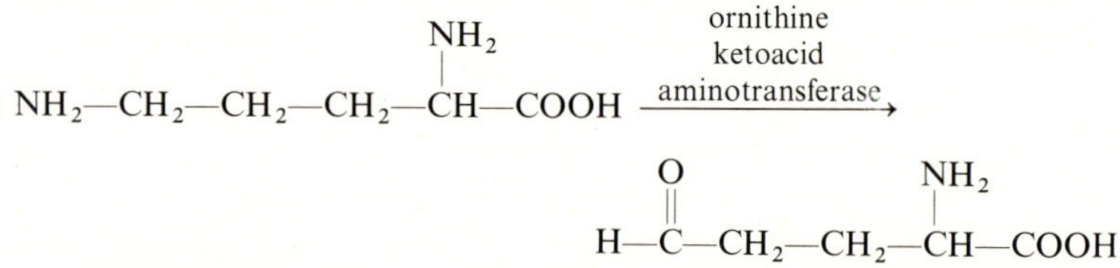

c) 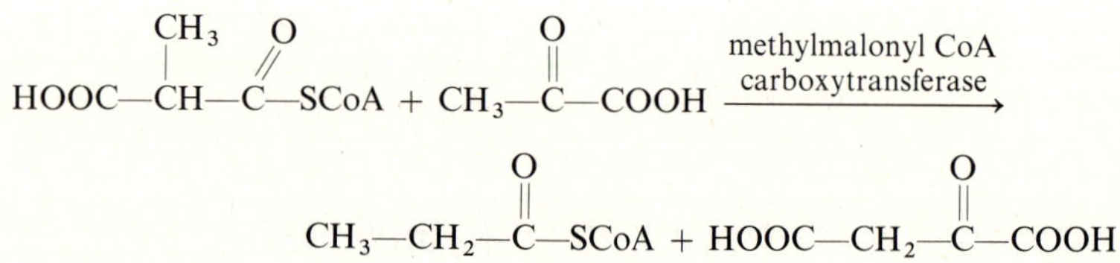

d) 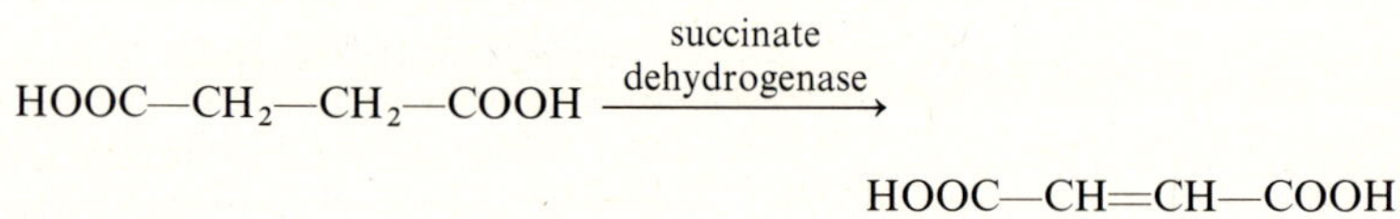

e) 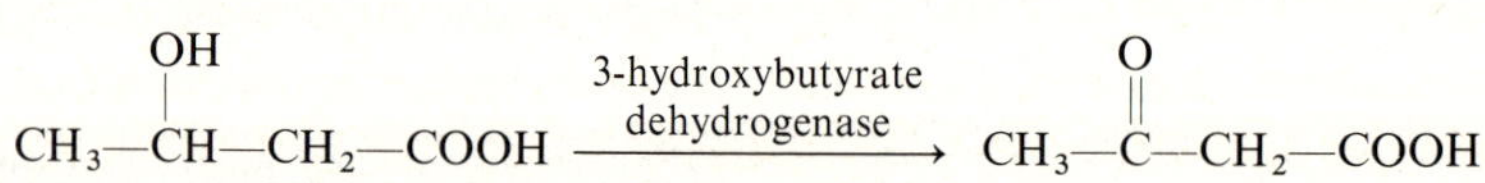

f) 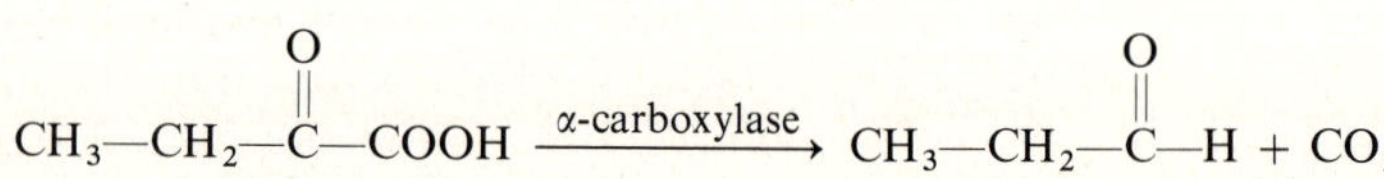

g) 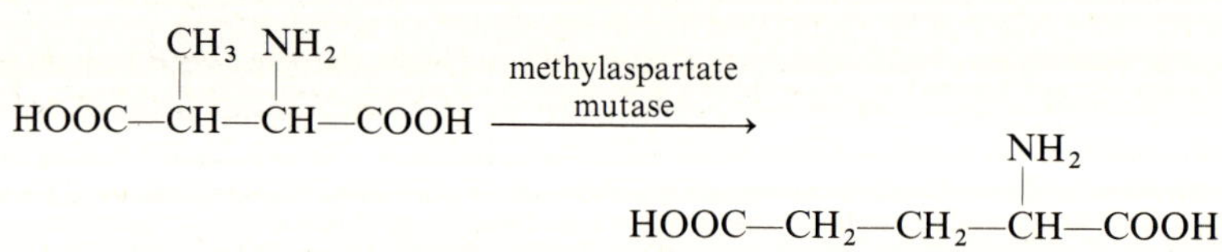

9. You are studying an enzyme (aspartate transcarbamylase) that converts aspartic acid and carbamyl phosphate to carbamyl aspartate and phosphate ion.

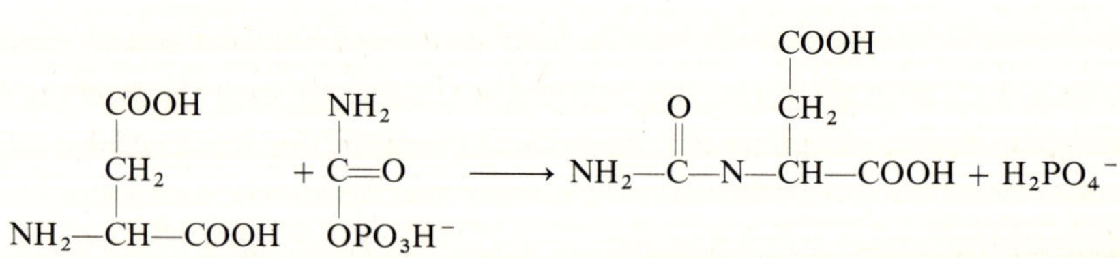

You obtain the following data:

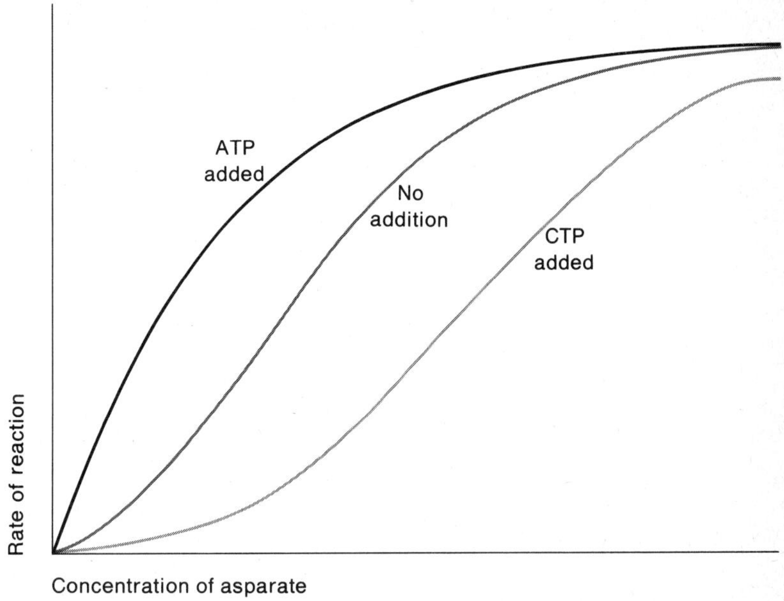

a) If we assume this is an allosteric enzyme, then describe ATP and CTP as either positive or negative effectors.

b) For which of the two allosteric effectors would you suspect this enzyme to be the first synthetic step?

Challenge Problems

10. Consider these facts:

> ***1.*** The substrate binds to the active site, is chemically altered, and is released. All this must take a certain length of time.
> ***2.*** If the concentration of substrate is doubled, it should become twice as likely that the substrate will collide with the active site.

Now predict behavior from these statements:

a) These curves show a relationship between substrate concentration and reaction rate. Which one is the behavior predicted by statement 2 only?

b) As a consequence of statement 1, visualize a very high substrate concentration, so high that there is a new substrate ready to be bound at the active site just as soon as product is released. Will doubling the substrate concentration double the rate under these circumstances?

c) Which curve predicts that behavior?

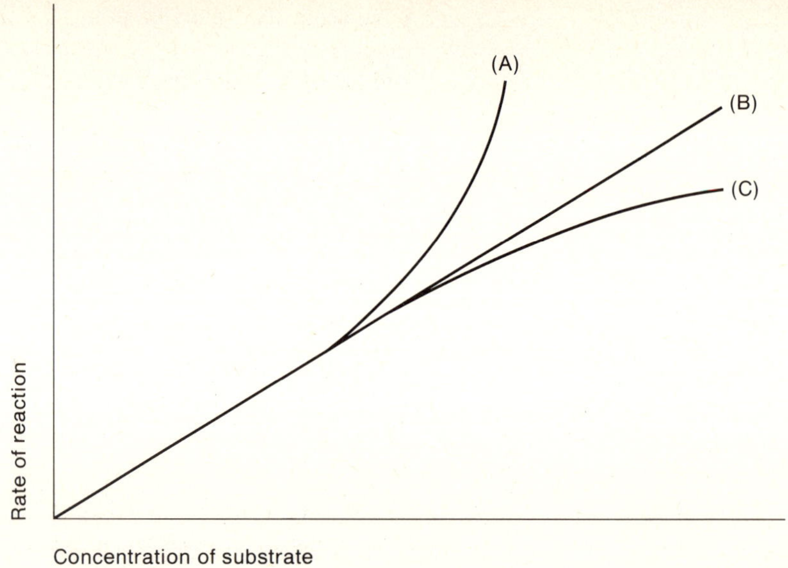

pH	*Relative Rate*
4.5	4
5.5	15
5.75	22
6.0	50
6.5	85
6.75	110
7.0	120
7.5	135
8.0	140

11. An enzyme has been studied with respect to its activity versus pH. Some values obtained are provided here.

a) Plot the data on a piece of graph paper, marking pH from 4.5 to 8.0 on the x axis and relative rate from 0 to 150 on the y axis.

b) Let us assume that the activity drops as the solution becomes acid because a group that can be titrated is in fact having a hydrogen ion added to it. The pK of that group is the pH at which the group is half titrated. Find a point on the curve at which the activity is *half maximum*. At what pH does that occur? Call that the pK.

c) If we assume that the group being titrated is on the enzyme, let us refer to the following table of pK values of amino acid R groups as they would be in the protein and guess what group is probably being titrated.

Amino Acid	*Approximate Range of pK*
C-Terminal carboxyl, aspartate, or glutamate	3–5
Histidine	5.5–7
Lysine or N-terminal amine	7.5–10.5
Tyrosine	10–10.5
Arginine	11.5–12.5

d) Now look at the pH curve for trypsin, cholinesterase, and pepsin (page 287). What is the *pH optimum* for trypsin and pepsin? In each case using the table in part c of this question, decide what amino acids may be operating at the active site. (Note that such decisions are only guesses; other facts are involved.)

12. You are an emzymologist studying an enzyme you have isolated. You find that 1 micromole (10^{-6} mole) of the enzyme converts 60 millimoles (mmoles) (60×10^{-3} moles) of substrate to product every minute.

a) How many μmoles are converted to product per minute? (1 mmole= 1000 μmole)

b) How many molecules of substrate are being converted to product each minute by a typical molecule of enzyme in that reaction study? We sometimes call this the enzyme turnover number, or molecular activity.

c) In order for this to be happening, what is the maximum length of time substrate could be attached to the active site during catalysis (expressed in seconds)?

13. You are a brilliant researcher who has isolated a peptide hormone from beetle brain. Let us begin by pretending that you know the total structure:

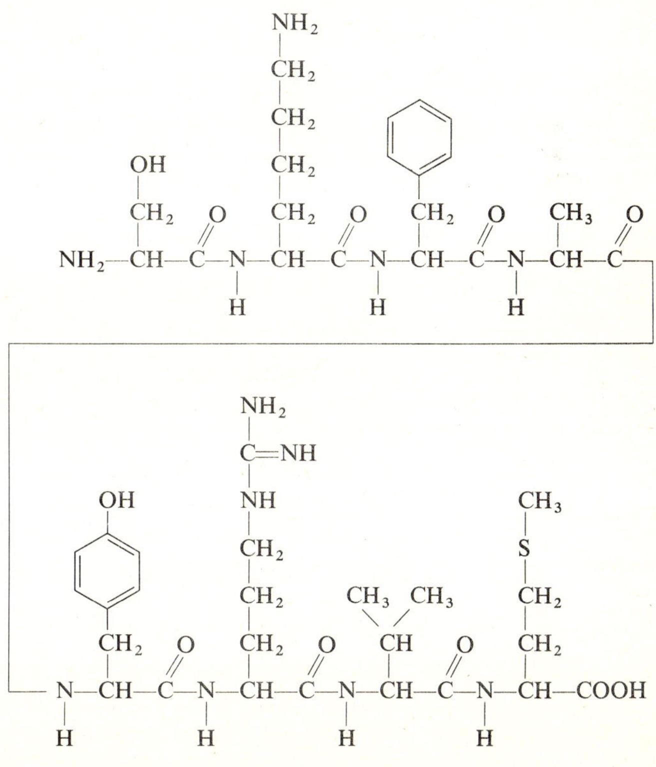

a) Name the amino acids in the sequence.

b) Write the structure of the peptides that would be produced by hydrolysis with trypsin.

c) Write the structure of the peptides that would be produced by hydrolysis with chymotrypsin.

The remainder of this question is more difficult. It illustrates the reasoning used by scientists to determine the sequence of a peptide or protein molecule.

d) Suppose you had obtained a peptide fragment from some other protein by hydrolysis with trypsin. You then hydrolyze that peptide by boiling it in 6*M* hydrochloric acid. This converts it into a mixture of its component amino acids. By paper chromatography of this hydrolysate, you learn that the peptide contains two amino acids, glycine and lysine. Which of the following was the peptide?

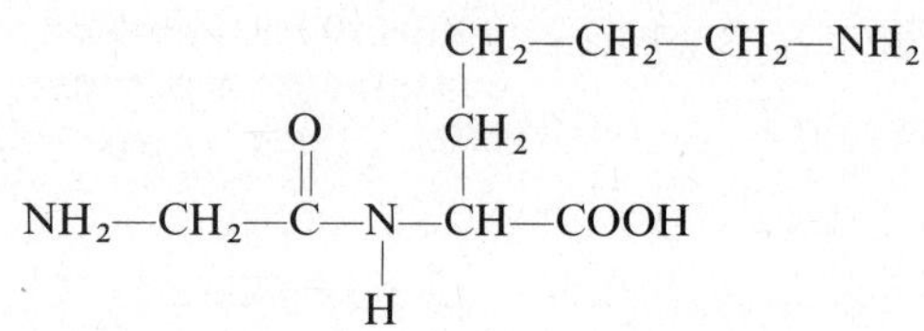

or

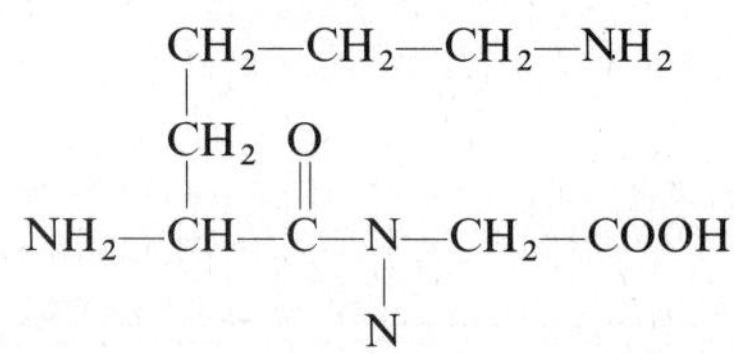

Hint: Remember that it was obtained by hydrolysis with trypsin.

Now let us pretend you do *not* know the total sequence of the original beetle brain peptide.

e) Return to your answer to b. If you had been given the three fragments and told that you were to decide which belonged at the carboxy terminal end of the original structure *without knowing* the total structure, how would you make that decision?

f) Apply the same reasoning to the fragments from c.

g) Now pretend that all you knew about these two carboxy terminal fragments, one from b and one from c, was that one (b) contained methionine and valine, while the other (c) contained methionine, valine, and arginine. In the tripeptide, what would have to be the location of the arginine?

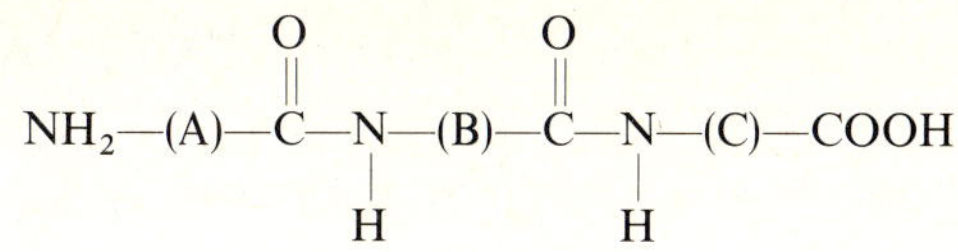

h) In b, what would have to be the order of the dipeptide containing serine and lysine if all you knew was the composition?

i) In c, you obtained a tripeptide that contained lysine, serine, and phenylalanine. This obviously overlaps the peptide mentioned in h, because both contain lysine and serine. Given your answer to h, what has to be the sequence of the tripeptide?

j) In c, you obtained a dipeptide that contained alanine and tyrosine. How would you know the sequence if all you were told was that it had been obtained using chymotrypsin and contained those amino acids?

k) In b, you obtained a tetrapeptide containing phenylalanine, alanine, tyrosine, and arginine. Suppose you knew only the composition; in which position would arginine have to be?

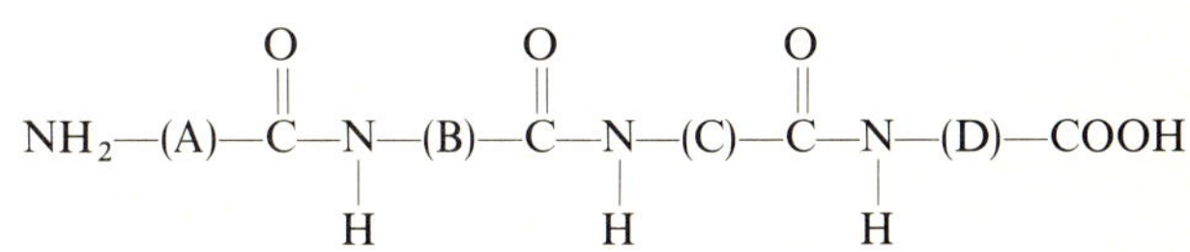

Now given the answer to the tripeptide of part i, where would phenylalanine have to be? That leaves alanine and tyrosine to be located. Given your answer to j, where would they have to be?

l) If you had not known the sequence originally, you would by now have put together almost the entire sequence. Review what you have put together and you can see that only one portion of the sequence has not been solved: Is valine or methionine the carboxy terminal amino acid? Carboxypeptidase is an enzyme that removes the amino acids one at a time from the carboxy terminal end. On brief incubation of the original peptide with carboxypeptidase, which amino acid would appear most quickly in solution?

m) Other tools are used in sequencing. The amine terminal amino acid reacts with 2,4-dinitrofluorobenzene (DNFB) to form a tight bond

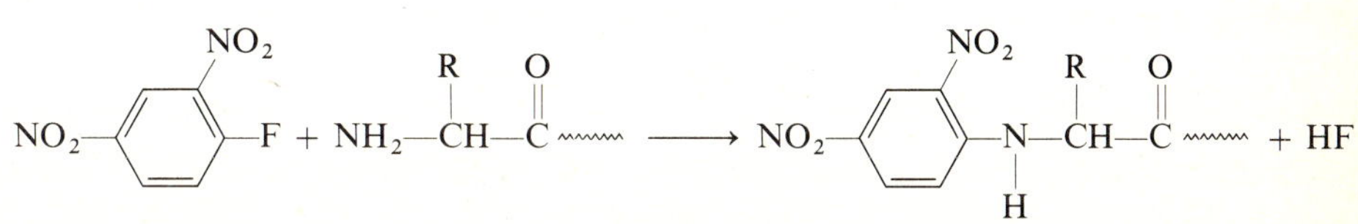

This dinitrophenyl "label" remains attached to that amino acid even when the peptide is broken down into amino acids by heating in 6*M* hydrochloric acid. This identifies the amine terminal amino acid for us. These labeled amino acids show plainly as yellow spots on the paper chromatogram. What amino acid would label if the original peptide had been treated with DNFB? What amino acid would have been labeled in each of the trypsin hydrolysis fragments? What amino acids would have been labeled in each of the chymotrypsin fragments?

The amine group of the lysine R group also forms a derivative, but this is easily distinguished from the amine terminal group.

14. Now try your ability to sequence given the tools of Problem 7.

***a*)** A peptide is labeled with DNFB and is hydrolyzed in 6*M* hydrochloric acid. The amino acids as identified by paper chromatography are (alphabetically): alanine, arginine, aspartic acid, glutamic acid, glycine, leucine, lysine, phenylalanine, and dinitrophenyl serine. Fill in as much as you can of the sequence from that:

_____ _____ _____ _____ _____ _____ _____ _____ _____

***b*)** On hydrolysis with trypsin, three fragments were obtained. Each was labeled with DNFB, hydrolyzed, and chromatogramed. They are *not* in sequential order:

1. Dinitrophenyl aspartic acid, arginine, glycine, phenylalanine
2. Dinitrophenyl alanine, glutamic acid
3. Leucine, lysine, dinitrophenyl serine

These three fragments are amine terminal, carboxy terminal, and central. Which is which? How much can you say about the sequence of each fragment?

1. _____ _____ _____ _____

2. _____ _____

3. _____ _____ _____

The summary of everything we know to this point is:

_____ _____ _____ _____ _____ _____ _____ _____ _____

***c*)** The original peptide is hydrolyzed with chymotrypsin to produce two fragments whose composition is:

4. Aspartic acid, leucine, lysine, phenylalanine, serine
5. Alanine, arginine, glutamic acid, glycine

What is the total peptide sequence?

15. As presented in the chapter, lactate dehydrogenase exists as five different isozyme forms. We recognize that the enzyme has a quaternary structure and that the isozymes are really different combinations of some number of basic subunits. If a mixture of lactate dehydrogenase isozymes is treated so that the subunits separate, and then the subunits are analyzed by electrophoresis, we find that there are only two different subunits in all the isozymes. Now consider that if the quaternary structure consisted of two subunits joined together, there would be three possible isozymes. Labeling the subunits A and B, the isozymes would be:

AA AB BB

***a*)** How many different isozymes can be assembled from two different subunits if the quaternary structure is composed of three subunits joined together?

***b*)** How many subunits must be present in the quaternary structure of lactate dehydrogenase to explain five isozymes?

***c*)** What can we assume about the differences in amino acid composition of the two subunits when we are able to separate them by electrophoresis readily?

***d*)** In the lactate dehydrogenase assay, total blood plasma is analyzed by electrophoresis, not purified lactate dehydrogenase. The strip that is scanned to produce the diagrams in Figure 12.8 ignores all the other proteins and sees only lactate dehydrogenase. Can you imagine how this is accomplished?

13 Nucleic Acids: Genes and Protein Synthesis

Objectives

In this chapter we:

1. Learn the structure of DNA.
2. Relate the structure to the process of making exact copies of DNA during cell division and of repairing damage to DNA.
3. Study the synthesis, structure, and varieties of RNA.
4. Discuss the genetic code.
5. Consider each of the four steps of protein synthesis (amino acid activation, initiation, elongation, and termination) in detail.
6. Survey some antibiotics that function by interfering with the processes discussed in this chapter.
7. Survey viruses briefly.

• • • *genetics deals with phenomena that require . . . the deterministic, exact maintenance of all the millions of genes in every cell. This mechanism requires a degree of order of organized information, that is found nowhere else in nature, not even in the movement of the planets.*

Dr. S. E. Luria, Nobel Laureate
from *Life, the Unfinished Experiment*

In the late 19th and early 20th centuries, scientists who were interested in unraveling the complex chemical secrets of the life process operated largely at the level of isolating and identifying the chemicals found in cells. Without modern techniques such as chromatography, this was a giant challenge. Obviously, before we could do anything more, we had to know what chemicals were present. The next question to be answered—How did those chemicals become part of the cell? Some bacterial cells prosper on glucose and a few mineral salts, obviously generating the multitude of chemicals in a cell from this single source of carbon. In the 1930s and 1940s, the major thrust of biochemistry was to discover the pathways of chemical synthesis—metabolism. This will be the focus of Chapters 14–19.

However, there was a central question in biochemistry. Living cells are differentiated from a complex process of chemical synthesis in the laboratory by what had to be the largest milestone in the origination of life, the ability of cells to reproduce. In reproducing they pass to the daughter cells their accumulated chemical wisdom, the message for the complete cell pattern, fine tuned and adjusted by long evolution. This message is the genetic information. Solving the riddle of how this is accomplished was not the only biochemical frontier of the 1950s and 1960s, but it was certainly the most exciting one. It was fascinating to watch the vocabulary of the geneticist and the terminology of the biochemist become interconvertible as an understanding was reached of what a gene, a mutation, or a chromosome is in terms of organic chemicals.

In Chapter 10 we introduced nucleotides. Now see how nucleotides are polymerized to produce ribonucleic acid (RNA) and deoxyribonucleic acid (DNA). These compounds function to store the genetic information and to convert this stored genetic information into the synthesis of new protein. *The essential functions of cells and of whole organisms are so completely directed and accomplished by protein molecules that control of protein synthesis is equivalent to control over the cell destiny*. This protein synthesis machinery operates so efficiently that mistakes in the genetic message and its conversion to protein sequence are very rare.

13.1 Deoxyribonucleic Acid: DNA

Pneumococci, the organisms that cause pneumonia, normally have a polysaccharide coating. With this coating they are called type S because they have "smooth" colonies on the agar plate. A nonvirulent form does not have the polysaccharide coating and is called type R because the colonies are "rough." In the early 1940s, Oswald Avery showed that DNA extracted from type S and added to type R resulted in conversion of the type R to type S. Furthermore, in the generations that followed, the organisms continued to be type S. This experiment established that genetic information, in this case the message to create the polysaccharide coating, was associated with DNA.

Bacterial cells have only one molecule of DNA, that molecule being incredibly large. *E. coli*, the most studied microorganism, has a single DNA molecule with a molecular weight of over 2 billion. The nucleotide monomers of this molecule are paired, and this molecular weight represents over 3 million pairs of nucleotide bases. This molecule is in the form of a closed loop—so-called *circular DNA*. This circular pattern for the DNA is common. The molecule is so large that with the electron microscope it can be seen and its circular character readily observed.

Large as the DNA of *E. coli* is, more complex organisms have in fact a great deal more DNA. In these organisms the DNA is in a membrane-enclosed compartment, the nucleus. The DNA is present generally as several separate DNA molecules, or *chromosomes*. In bacterial cells there is one set of genetic information, that is, one message for each protein. In higher plants and animals there is a double set of DNA, one set from each parent. Thus more complex organisms not only store more information in the genes, they store two sets of it. Finally, in higher organisms there is a considerable amount of DNA that does not code for protein. The function of that DNA is under investigation.

13.2 DNA: Primary Structure

Chemically, DNA is a polynucleotide. The polymer has a backbone of phosphate and deoxyribose (ribose with no —OH group at the 2 position).

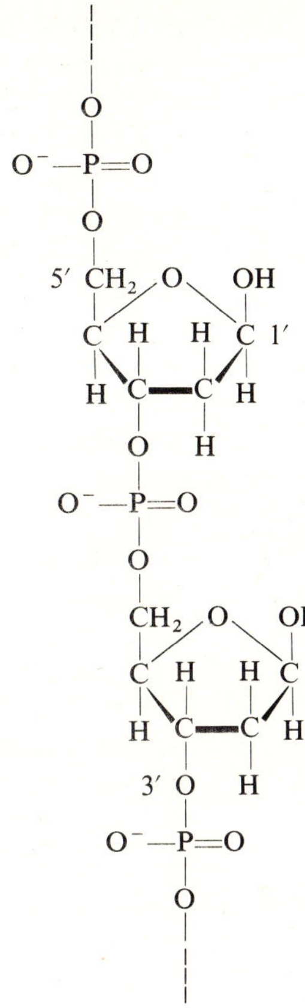

At the 1 position of the deoxyribose are found either the purines adenine or guanine, or the pyrimidines thymine or cytosine (Figure 13.1).

The following are some important points about the DNA primary structure.

1. The purines and pyrimidines (often referred to as the "bases") are not part of the polymer chain. They are side groups sticking out from the chain.

2. The polymer chain has phosphate groups that have been esterified at two of the three possible positions. The third position is quite acidic (p*K* 2) and is therefore ionized at neutral pH. In other words, the DNA polymer strand is ionic and negatively charged.

3. The two ends of the DNA polymer are different, just as in the protein chain where there are the N-terminal and C-terminal amino acid ends. At one end the last deoxyribose connects to the main

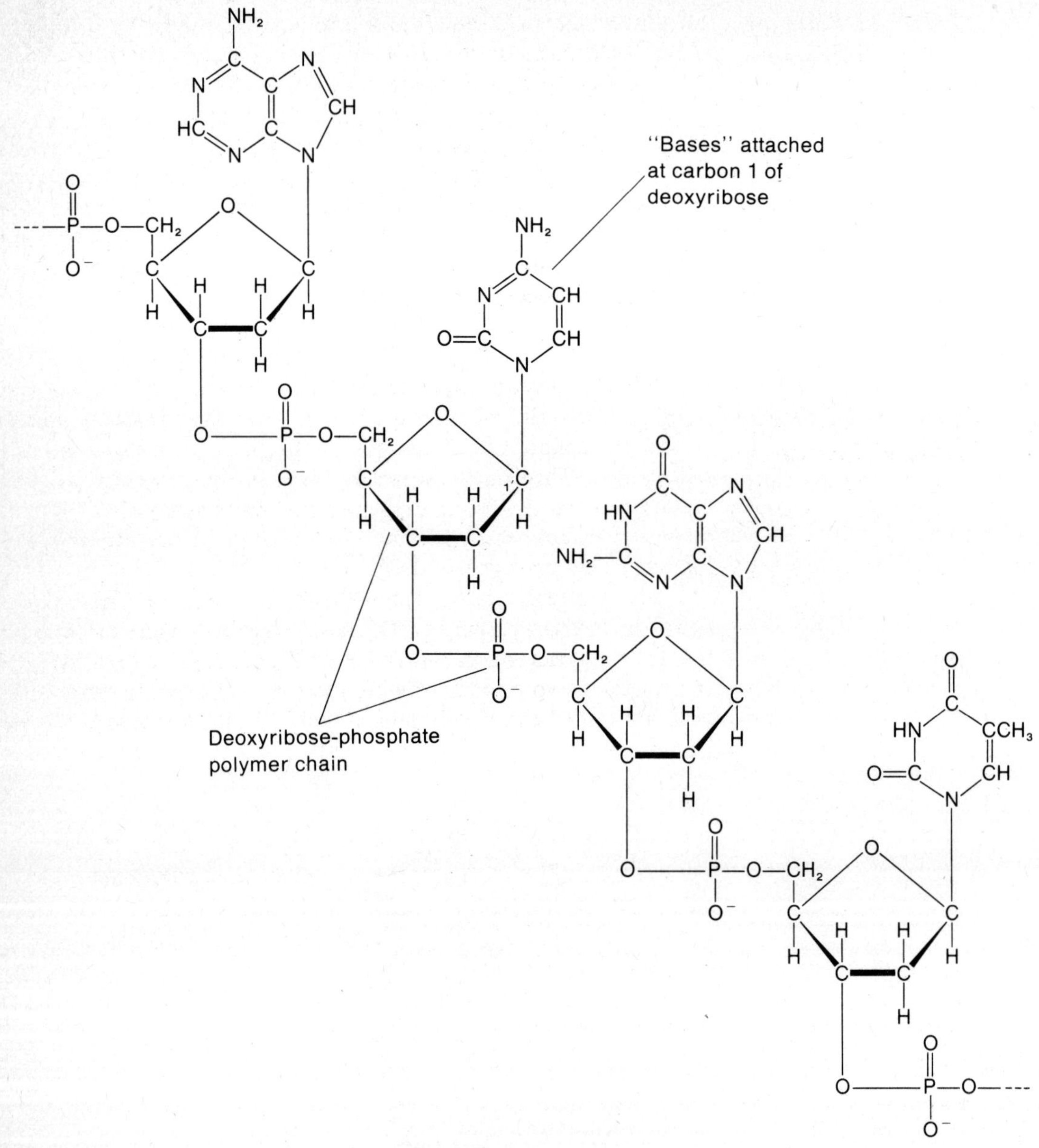

Figure 13.1
Primary structure of DNA

body of the polymer by the 3′ position, and any further chain extension would occur at the 5′ position. This is called the 5′ end. The other end of the polymer is called the 3′ end.

Recently, techniques have been devised to determine the sequence of DNA, and the nucleotide sequence of a virus DNA has been solved. Frederick Sanger, the researcher associated with this milestone in biochemistry, also received the Nobel prize for being the first to determine the amino acid sequence of a protein (Figure 9.2). *The sequence of nucleotide bases on the DNA molecule is the genetic message.* Using DNA as a pattern or template, the genetic information is transferred to RNA, another type of nucleic acid. The RNA enters the cytoplasm and there controls the order in which amino acids are assembled into new protein. The transfer of sequence information from DNA to RNA, and from RNA to protein, was the basic hypothesis or *central dogma* of the researchers who verified the pattern and who are working to understand the details of the system. We shall discuss some of these details as we proceed through the chapter.

13.3 DNA: Secondary, Tertiary Structure

Isn't DNA nucleotide sequence a very fragile way to record a message? Any damage to this polymer—any alteration of the bases—is going to change the amino acid sequence in a protein, that is, cause mutation (Chapter 9). This mutation can be very damaging, as we have seen, and it presumably doesn't happen very often. Apparently DNA must be protected from damage rather well. This protection is provided by the secondary structure of DNA.

Two researchers, James Watson and Francis Crick, did the same sort of X-ray study on DNA that has been done to determine the tertiary structures of protein. They found DNA to be a long, thin rod made of two strands winding around each other in what is now generally called the *double helix* (Figure 13.2). This structure is held

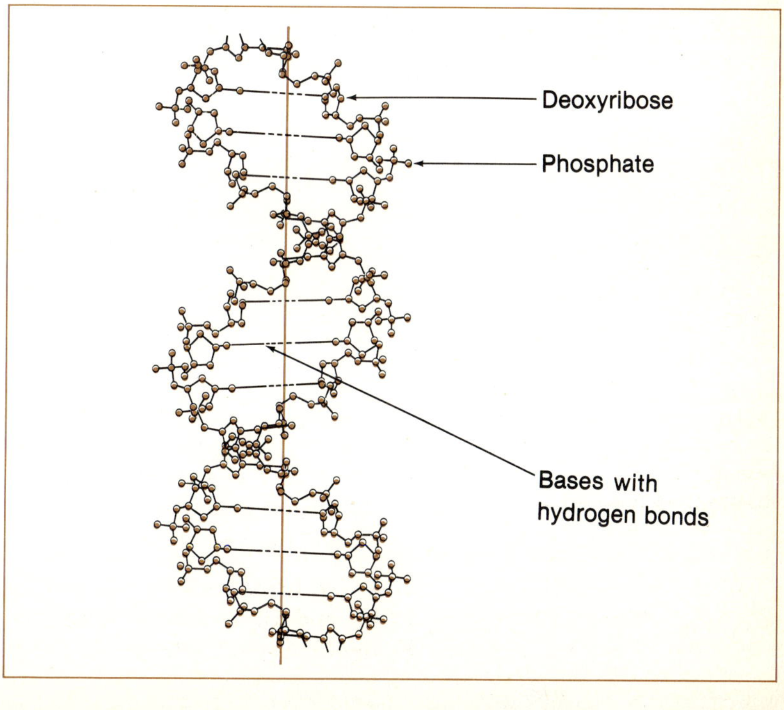

Figure 13.2
The DNA double helix
In this structure, two polynucleotide strands wind around a common axis with the deoxyribose and phosphate groups on the outside and the purine and pyrimidine bases directed toward the center. (From H. R. Mahler and E. H. Cordes, *Basic Biological Chemistry*, 2nd ed., New York, Harper & Row, 1971.)

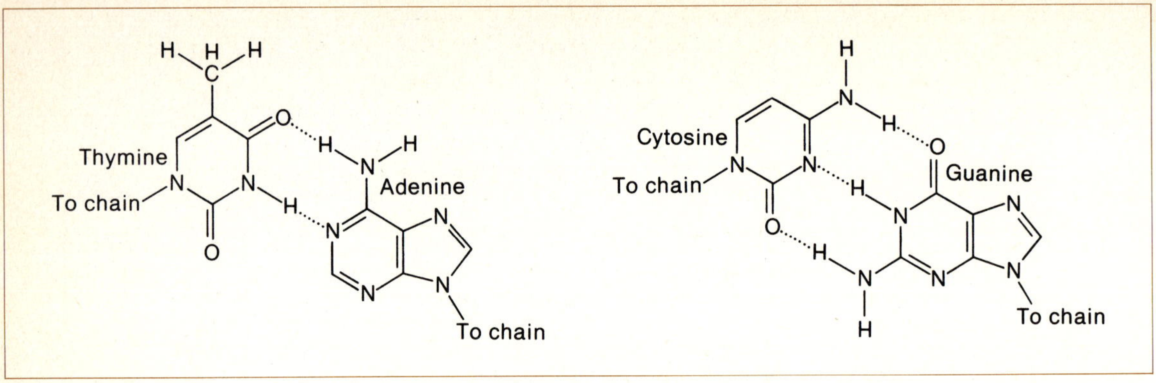

Figure 13.3
Base pairing in DNA
At the center of the DNA double helix the purines and pyrimidines hydrogen bond to one another, adenine uniquely with thymine and quanine uniquely with cytosine.

together in a unique and very important fashion. The sugar phosphate polymer chains are wound around the outside, and the flat bases are parallel to each other and stacked in the center. Notice that this puts the polar parts of the molecule, the sugar and phosphate, on the outside and packs the relatively nonpolar bases in the center. This hydrophobic pattern provides some stabilizing of the DNA helix.

The most interesting stabilizing force is the formation of hydrogen bonds between the bases of one strand and those of the other. There is just the right amount of space in the center of the helix for one purine and one pyrimidine to fit across from each other. Furthermore, when thymine is across from adenine, there are two places where hydrogen bonds can form. If thymine were across from guanine, only one bond would be possible. Therefore, adenine and thymine always pair. In the case of guanine and cytosine, three hydrogen bonds can form. Adenine would be unable to hydrogen bond to cytosine under the same conditions. Guanine and cytosine are thus also uniquely suited to pair (Figure 13.3). These base-to-base hydrogen bonds hold the two strands of the double helix together. But more than that, the sequence of one of the two strands is related to the sequence of the other in a complementary fashion. In other words, the sequence of one strand depends on but is not identical to the other.

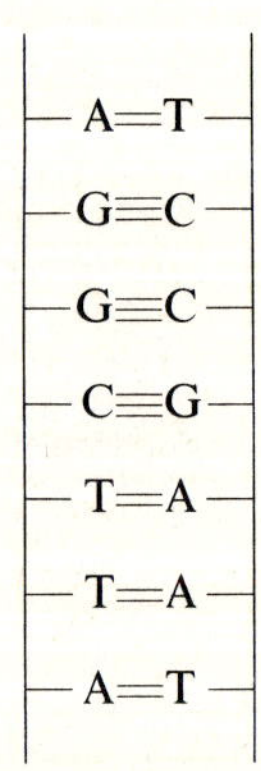

This means that the coded message for protein synthesis is on only one of the two strands. The other strand is the complement of the genetic message and says nothing itself.

Suppose that a base on one strand were chemically altered. The hydrogen bonds holding the strands together are weakened at that point. There are enzymes that repair such damage. How do they know what new base to put in when they snip out the altered one? Easy—they only have to pair with the complementary base on the other strand. The message is not as fragile as it might be, because it comes provided with a reference record, a blueprint for repair.

Exercise 13.1

A segment of DNA has been damaged so that two bases are required to repair it.

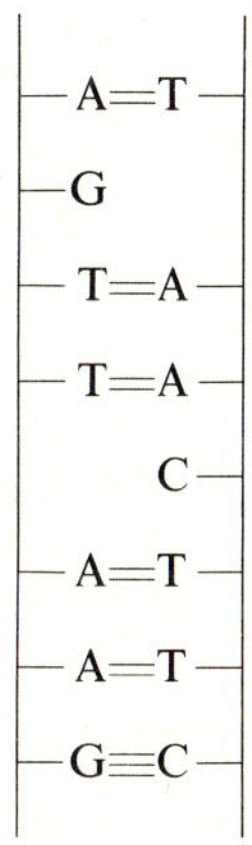

What bases are required?

13.4 Replication

The DNA in any body cell is the same as that in any other body cell. Obviously, in the process of producing all those cells, new copies of the DNA must have been produced. The processes of making new DNA copying the exact sequence of the existing DNA is called *replication.* How is replication accomplished? Once again, the double helix is the key. An enzyme called *DNA polymerase* is responsible for the synthesis of this new DNA. The reaction catalyzed is as follows:

$$\text{dATP} + \text{dGTP} + \text{dCTP} + \text{dTTP}^{1} \xrightarrow[\text{polymerase}]{\text{DNA}} \text{DNA} + \text{pyrophosphate}$$

The reaction does not happen unless there is some DNA present already. It is upon this DNA that the synthesis of new DNA takes place. The existing DNA serves as a template for the new DNA. The process begins with the unwinding of the helix of the existing DNA.

---

[1] Review, if necessary, the system of nucleotide abbreviation (Chapter 10).

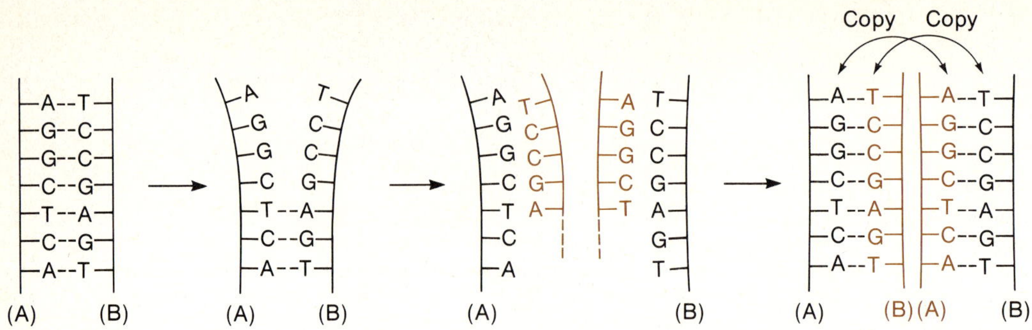

Figure 13.4
DNA replication
During replication, the paired DNA strands separate and a new strand of DNA is synthesized complementary to and attached to each of the separated strands. The end product is two identical sets.

As this happens, the DNA polymerase generates a new DNA chain *complementary* to the newly exposed template DNA chain. Because this is done to both strands of the template DNA, the end product is two complete DNA double helixes where one existed before.[2] Each of these product DNA molecules contains one strand from the template and one newly synthesized strand. If we call the two strands A and B, one of these product DNA molecules has the original A and newly synthesized B, and the other has original B and newly synthesized A. Both the template B and the newly synthesized B are exactly complementary to A, so they are identical to each other (Figure 13.4).

When the duplication of the DNA is complete, the cell can divide, with identical copies of the DNA going to each daughter cell.

13.5 Ribonucleic Acid: RNA

The other nucleic acid of the cell, ribonucleic acid, or RNA, is a partner with DNA in the determination of protein sequence by the genetic message.

The structure of RNA is very similar to that of DNA. In fact, the only differences in the primary structure are the substitutions of ribose for deoxyribose and uracil for thymine. The secondary–tertiary structure is quite different, however. Although RNA can and does form a double helix, the RNA in cells is generally single stranded. Any helix that is found results from the RNA bending backward and winding around itself.

13.6 Transcription

The production of RNA in the cell is called *transcription*. It is catalyzed by the enzyme RNA polymerase. The reaction is virtually identical to that of DNA polymerase:

$$\text{ATP} + \text{CTP} + \text{UTP} + \text{GTP} \xrightarrow[\text{polymerase}]{\text{RNA}} \text{RNA} + \text{pyrophosphate}$$

[2] This simplified description leaves out several components of the system. An excellent discussion with historical background is provided in A. L. Lehninger, *Biochemistry*, 2nd ed., New York, Worth, 1975.

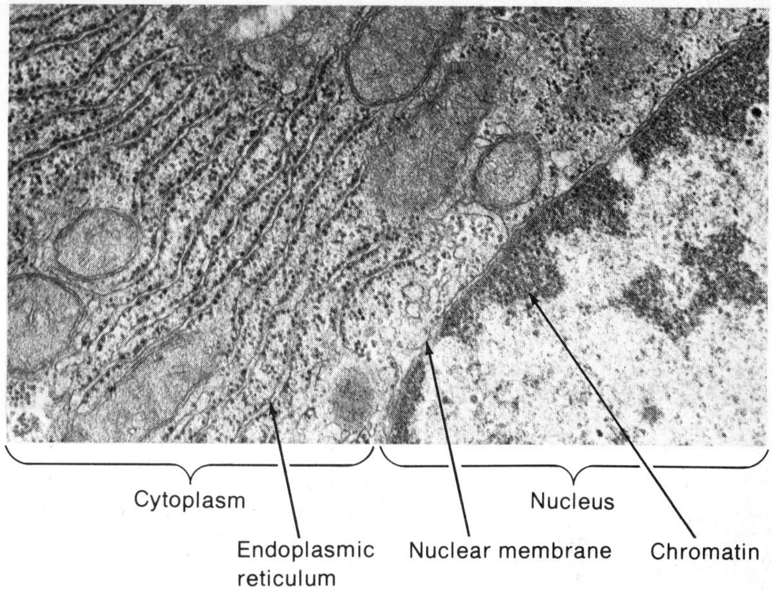

Figure 13.5
Photo of endoplasmic reticulum
This electron micrograph of a portion of a liver cell shows the nucleus with the chromatin (DNA) and cytoplasm with the endoplasmic reticulum. The dark bodies along the endoplasmic reticulum are ribosomes. (From Eastern Michigan University Electron Microscopy Laboratory.)

The reaction requires DNA to serve as a template, and the product has a sequence complementary to that of the DNA template.

Three kinds of RNA are found in the cell. Most of the cell RNA is found in the ribosomes. This is called *ribosomal or rRNA*. Ribosomes are small structures assembled of two subunits, the smaller with one RNA molecule, the larger with two RNA molecules, each with a number of proteins. All the ribosomes of a cell are the same (Figure 13.5). The actual contribution rRNA makes to the structure or function of the ribosome is not well understood.

There is a collection of very small RNA molecules, each with about 75 nucleotides, called *transfer or tRNA*. These are not associated with any cell structure or with protein molecules.

Finally, there is a collection of RNA strands called *messenger or mRNA*. There is a large variety of these but a small number of each kind. These molecules carry the genetic message from the DNA molecule. The sequence of the mRNA determines the protein primary structure.

All three kinds of RNA are produced by transcription of a DNA sequence.

Exercise 13.2

Of the three kinds of RNA in the cell:

a) There are more molecules of __________ than any other kind.
b) The smallest molecules are the ____________________________.
c) The type of RNA of which there is the greatest variety but the least number of molecules of any one variation is the __________

____

Table 13.1
The Genetic Code

UUU	Phe	UCU	Ser	UAU	Tyr	UGU	Cys
UUC	Phe	UCC	Ser	UAC	Tyr	UGC	Cys
UUA	Leu	UCA	Ser	UAA	Stop	UGA	Stop
UUG	Leu	UCG	Ser	UAG	Stop	UGG	Try
CUU	Leu	CCU	Pro	CAU	His	CGU	Arg
CUC	Leu	CCC	Pro	CAC	His	CGC	Arg
CUA	Leu	CCA	Pro	CAA	Gln	CGA	Arg
CUG	Leu	CCG	Pro	CAG	Gln	CGG	Arg
AUU	Ile	ACU	Thr	AAU	Asn	AGU	Ser
AUC	Ile	ACC	Thr	AAC	Asn	AGC	Ser
AUA	Ile	ACA	Thr	AAA	Lys	AGA	Arg
AUG	Met	ACG	Thr	AAG	Lys	AGG	Arg
GUU	Val	GCU	Ala	GAU	Asp	GGU	Gly
GUC	Val	GCC	Ala	GAC	Asp	GGC	Gly
GUA	Val	GCA	Ala	GAA	Glu	GGA	Gly
GUG	Val	GCG	Ala	GAG	Glu	GGG	Gly

Note: The three capital letters (the codon) represent a sequence of nucleotides on mRNA. Each codon designates the amino acid whose abbreviation is to its right.

13.7 The Genetic Code

The order in which the purines and pyrimidines appear in the mRNA is, according to the central dogma, the pattern for the protein primary structure. The nucleotide sequence for a single specific amino acid is called the *codon* for that amino acid. Given that twenty amino acids must be located correctly in the protein sequence, and that there are only four kinds of bases in DNA, it is obvious that each base does not define a specific amino acid. If we predict that a codon is a sequence of two bases, we are still in trouble, because there are only sixteen unique ways that four different bases can pair. A codon must therefore be a sequence of at least three bases. This provides 64 unique codons, more than enough. The results of later research have established that the three-base codon assumption is correct.

All but three of the codons designate an amino acid. Consequently, most amino acids are defined by more than one three-base sequence. Look at the genetic code table (Table 13.1), as it has been solved by researchers. Notice that for several amino acids only the first two bases matter (valine, alanine, or proline). For others the third base need only be a purine or a pyrimidine (lysine, cysteine, or glutamine). This reduced importance of the third base has led scientists to speculate that it is less carefully read in transcription.

Exercise 13.3

You are given this segment of mRNA:

$$_{5'}G\ C\ U\ U\ C\ U\ A\ A\ C\ U\ U\ U\ A\ C\ U_{3'}$$

a) Divide the sequence into 5 three-base segments. What name do we give to such segments?
b) Using the genetic code (Table 13.1), list the amino acids represented by this mRNA sequence.

Does coding in the genes for the protein primary structure really represent the storing of enough information to plan the entire pattern of the organism? Think about the function of the various biochemical compounds discussed so far. Most are like carbohydrates or fats. They are used structurally or represent stored food. Proteins, however, include the enzymes that catalyze every chemical change, the transport molecules, and the antibodies that are selectively produced to combat a specific disease. Proteins also control the passage of chemicals across membranes and constitute most of the hormones that are at the heart of control systems.

Suppose we accept that proteins control the destiny of a cell or organism. The DNA sequence controls only the protein sequence. What about the folding of the protein, which is critical to proper protein function? The sequence of the amino acids itself dictates the folding pattern. As the protein is synthesized, the newly incorporated amino acids change position to establish the hydrogen bonds, electrostatic attractions, covalent interactions, and hydrophobic forces responsible for maintaining the correct tertiary structure in proteins. By the time the last amino acid is in place, the newly synthesized protein is correctly folded. Some proteins respond to these built-in forces for correct folding by spontaneously refolding when they have been denatured. Most proteins, however, do not renature spontaneously. Refolding an entire molecule, with all the attractive forces available at once, is quite different from folding a short segment during synthesis.

13.8 Protein Synthesis

Now we shall deal with the actual machinery by which words written in genetic code on mRNA become a sequence of amino acids in a protein molecule. This process, called *translation*, has four steps:

1. ***Amino acid activation.*** The amino acids are attached to a carrier (tRNA) capable of attachment to the mRNA. The tRNA can be viewed as an adapter to fit between the codon and the amino acid. The combination is an aminoacyl-tRNA.
2. ***Initiation.*** The ribosome, the structure that is the site of synthesis, assembles onto the mRNA and attaches the first aminoacyl-tRNA.
3. ***Elongation.*** The ribosome moves along the mRNA, adding the amino acid of successive aminoacyl-tRNA complexes to the growing protein chain.
4. ***Termination.*** The mRNA signals the end of the protein chain, and the finished protein is released.

Let us consider each step in more detail.

13.9 Amino Acid Activation

For each tRNA molecule there is a correct amino acid. A covalent bond is formed between the amino acid and the tRNA by a correct *activating enzyme*. Energy to form this linkage is provided by ATP. Using alanine as an example, the overall reaction is as follows:

$$NH_2{-}\underset{}{\overset{CH_3}{\overset{|}{C}}}H{-}COOH + ATP + tRNA_{ala} \xrightarrow[\text{synthetase}]{\text{alanyl-tRNA}}$$

$$NH_2{-}\overset{CH_3}{\overset{|}{C}}H{-}\overset{O}{\overset{\|}{C}}{-}tRNA_{ala} + AMP + \text{pyrophosphate}$$

Once the aminoacyl-tRNA combination is formed, the amino acid is ready to be incorporated correctly into new protein.

We know quite a bit about the structure of tRNA. The sequence of nucleotides has been determined for several tRNA molecules. Once scientists saw the sequences, they predicted ways in which the molecule could fold so that the strand would wind back upon itself at places where the sequences were complementary. This would result in a small region of double helix being formed. All the tRNA molecules sequenced were found to be capable of complementary pairing in approximately the same locations. This predicts that tRNA molecules would look much like little clover leaves (Figure 13.6).

The sequence does not indicate what amino acid should bind to that particular tRNA at the point of amino acid attachment. However, let us go to the arm of the tRNA directly across the cloverleaf from the amino acid attachment site. There is a sequence of three bases that is the anticodon, the complement of the mRNA codon. When the tRNA carries the amino acid to the correct place on the mRNA, the anticodon of the tRNA hydrogen bonds to the codon of the mRNA.

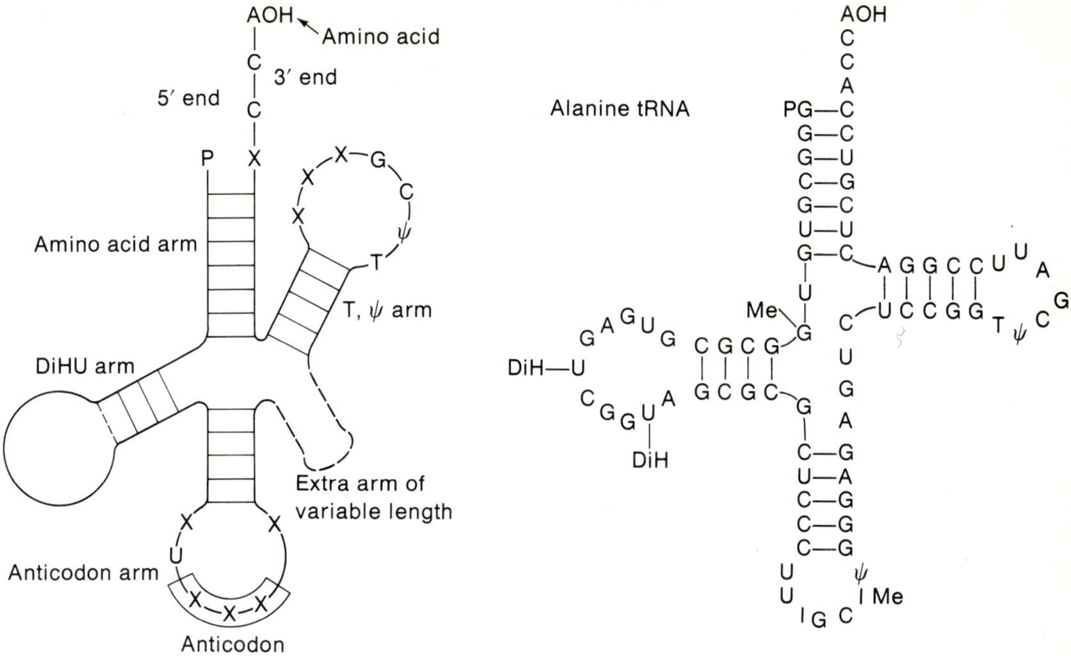

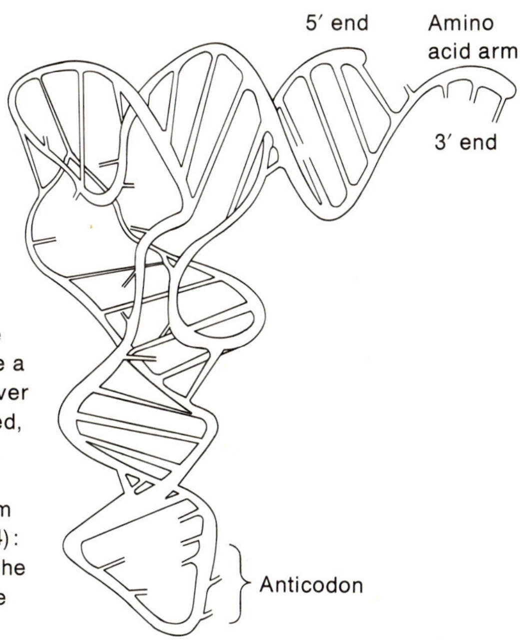

Figure 13.6
The structure of tRNA
Studies have revealed that transfer RNA has four or five arms, giving it an outline like a clover leaf. However, the clover leaf is folded into an L-shaped, three-dimensional structure. There are bases other than A, G, C, and U in tRNA. (From S. M. Kim, *Science 185* (1974): 435–440. Copyright 1974 by the American Association for the Advancement of Science.)

This hydrogen bonding assures that the correct amino acid is ready for addition to the protein chain.

We know the real shape of tRNA. It has been studied by the same X-ray crystallographic techniques that determined the shapes of proteins and of the DNA double helix. It is an L-shaped molecule with the amino acid attached at one end and the anticodon at the other.

13.10 Initiation

The *intiation*, or addition of the first amino acid of the protein chain, is done differently from any later amino acid. The system in the bacterium *E. coli* is best understood. The first amino acid is always methionine. The methionine is attached to a particular tRNA ($tRNA_f$), a different tRNA than is used to insert methionine into the middle of a growing protein chain. After the methionine is attached to the $tRNA_f$, a formyl group is transferred onto the α-amine group of the methionine. The resulting formylmethionyl-$tRNA_f$ is now ready to initiate synthesis of a new protein.

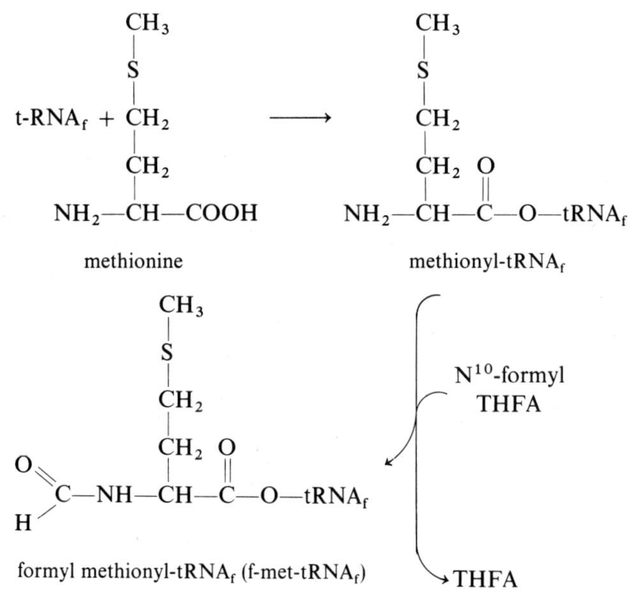

The ribosome is composed of two RNA-protein complex subunits. The smaller of these is called the 30 s subunit and the larger the 50 s subunit.[3] The ribosome breaks into these subunits before

[3] A number followed by s in this context represents a measure of the size of the particle in Svedbergs, an ultracentrifuge molecular weight unit. These sizes (30 s and 50 s) are for bacterial ribosomes. In eukaryotic cells the structures are larger (40 s and 60 s).

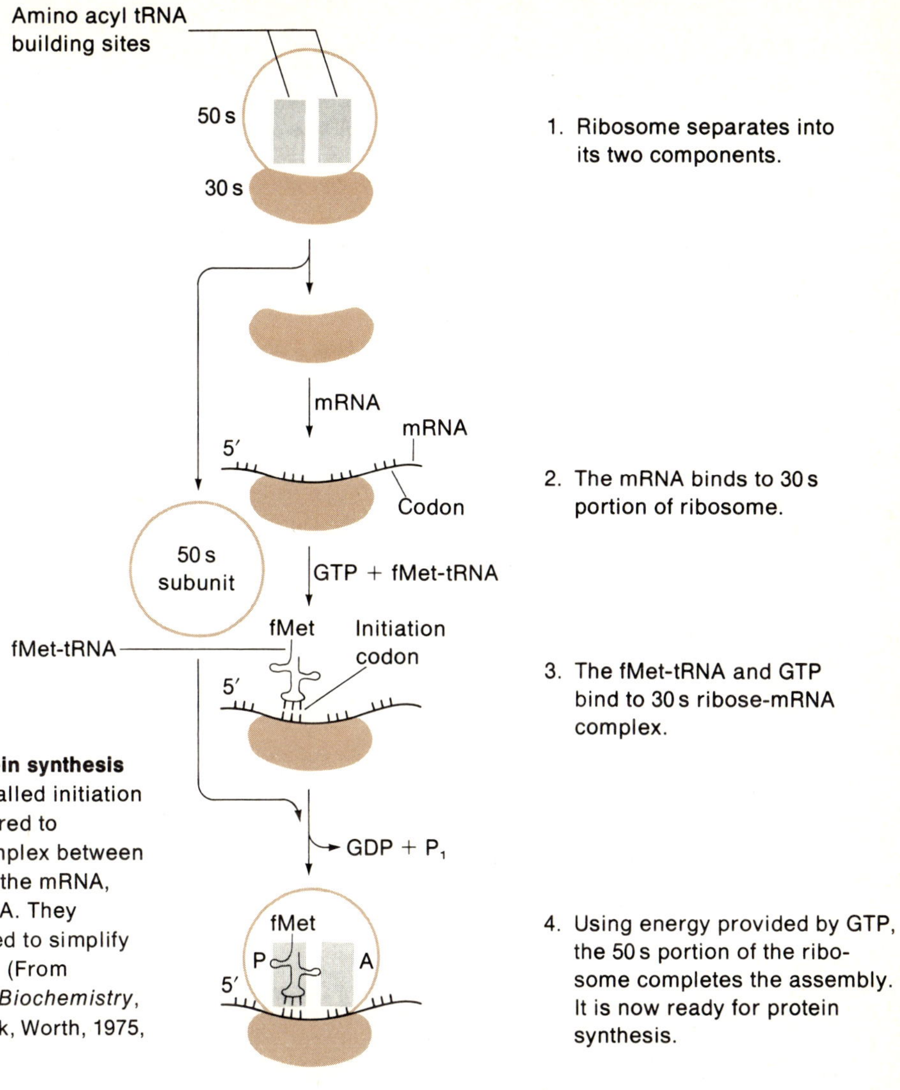

Figure 13.7
Initiation of protein synthesis
Three proteins called initiation factors are required to generate the complex between the 30 s subunit, the mRNA, and the fMet-tRNA. They have been omitted to simplify the presentation. (From A. L. Lehninger, *Biochemistry*, 2nd ed., New York, Worth, 1975, p. 940.)

initiation can occur. The 30 s subunit combines with the molecule of mRNA at an *initiation site*. The f-met-$tRNA_f$ unit binds to the mRNA at the codon complementary to its own anticodon. Some protein *initiation factors* assist the formation of this complex. A molecule of GTP also binds to this complex. In the next step, the 50 s ribosome subunit joins the 30 s subunit to reassemble the ribosome. This requires energy, which is supplied by the GTP. A high-energy bond of GTP is broken, producing GDP + iP (Figure 13.7).

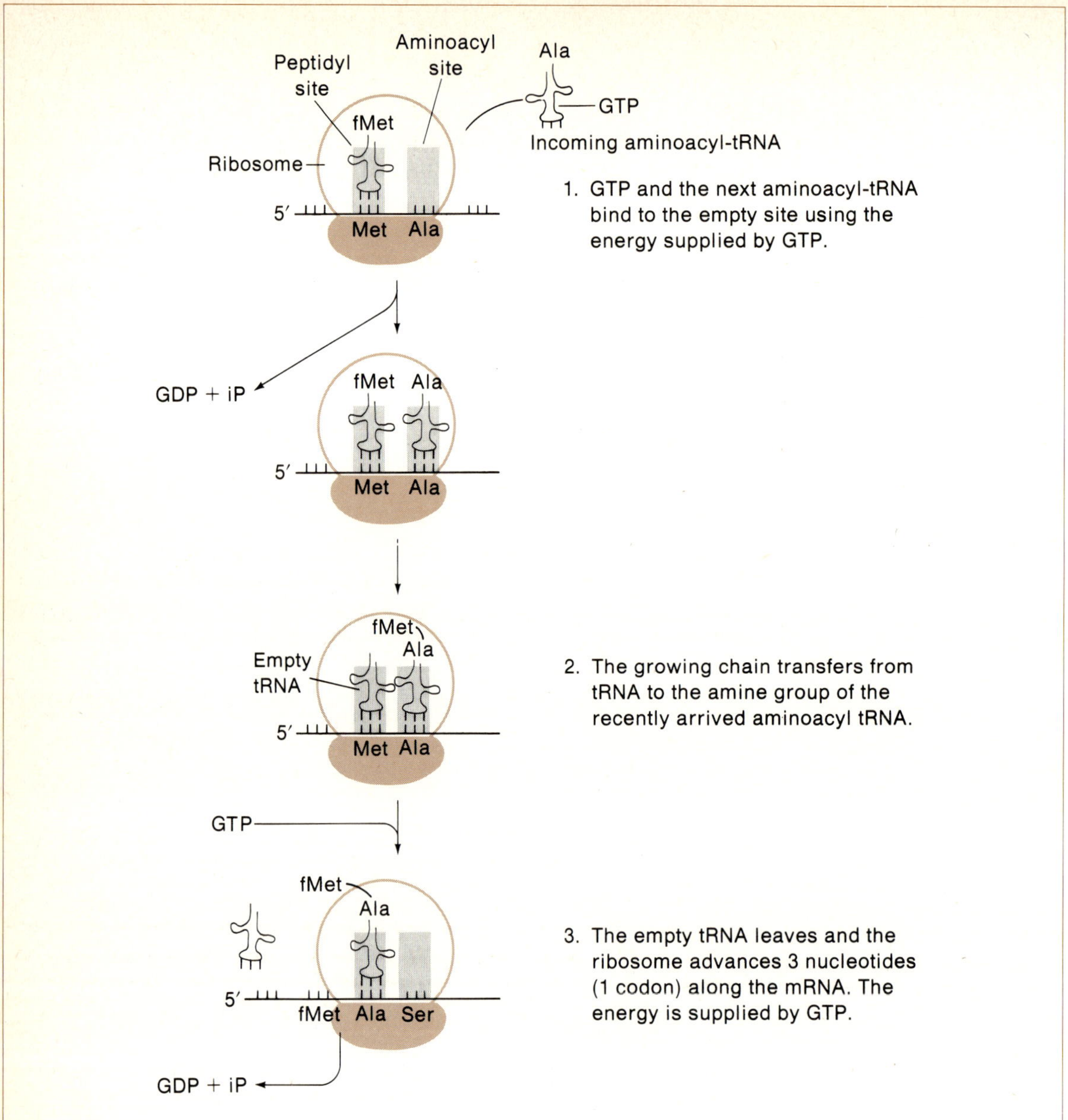

Figure 13.8
Elongation in protein synthesis
Proteins called elongation factors are involved but have been omitted to simplify the presentation. (From A. L. Lehninger, *Biochemistry*, 2nd ed., New York, Worth, 1975, p. 942.)

13.11 Elongation

Now everything is in place. The ribosome is assembled around the mRNA at the beginning of the message. The tRNA carrying the first amino acid of the sequence is also in place.

The ribosome has two places at which tRNA can hydrogen bond to the mRNA. One of these is presently occupied, the other empty. An aminoacyl-tRNA with the correct anticodon to hydrogen bond to the next mRNA codon slips into the empty site. Binding this aminoacyl-tRNA requires energy. The energy is supplied by GTP and an elongation factor.

Now the bond between formyl methionine and the $tRNA_f$ is replaced by a bond between formyl methionine and the amine group of the second amino acid. The reaction is catalyzed by the enzyme peptide transferase. This is the first peptide bond of the new protein. The $tRNA_f$ now has no amino acid. It leaves the ribosome.

The final operation in a step of chain elongation is the shifting of the ribosome. It moves the length of one codon along the RNA. This requires energy, which is supplied by GTP, and another elongation factor. The tRNA with the dipeptide attached was in the site reserved for the incoming aminoacyl-tRNA before the shift. After the shift it is in the site for the previously added aminoacyl-tRNA, and the site for the incoming aminoacyl-tRNA is vacant. The attachment of the correct aminoacyl-tRNA to that now vacant site is the beginning of the next step in chain elongation (Figure 13.8).

13.12 Termination

Chain elongation continues stepwise until all the amino acids have been added. As the chain is lengthening, it is also folding spontaneously into the correct shape. Elongation ends when one of these codons is encountered: UAA, UGA, UAG. These codons do not bind tRNA but bind instead proteins called *release factors*. The release factors cause the peptidyl transferase to hydrolyze the bond between the completed protein and the last tRNA. The ribosome comes apart into its two subunits, releasing the mRNA.

Under normal circumstances a single mRNA has several ribosomes moving along it at the same time, each with a protein under construction. It resembles an automobile assembly line with cars moving along the line coming closer to completion as they go. Such an assembly of ribosomes on mRNA is called a *polyribosome*.

13.13 Antibiotics and Protein Synthesis

Many of the antibiotics function by interfering with the process of reading the genetic information from DNA and creating new protein from that information. Rifamycin and Actinomycin prevent RNA from being synthesized by RNA polymerase.

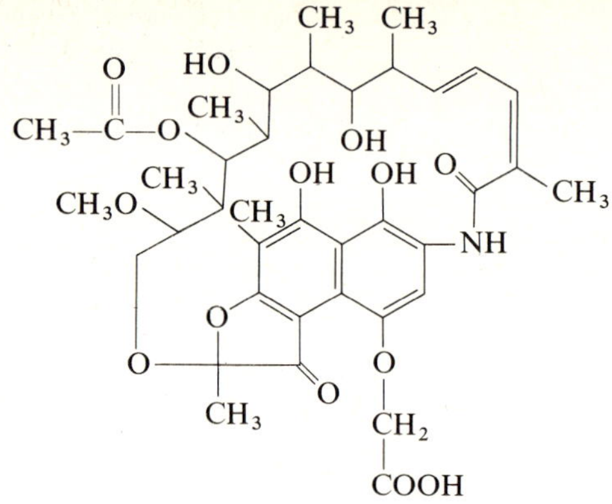

Rifamycin B

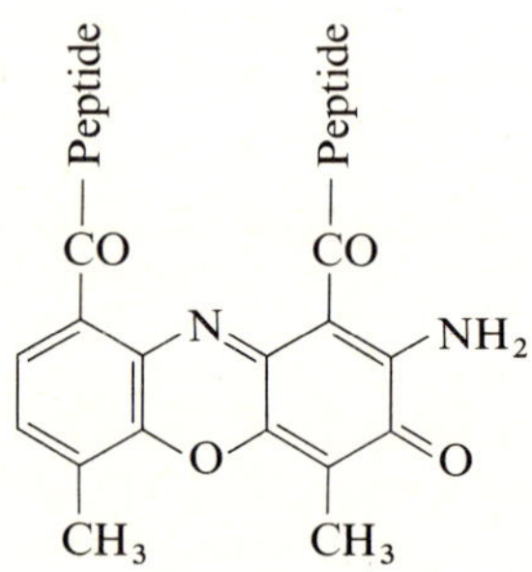

Actinomycin C

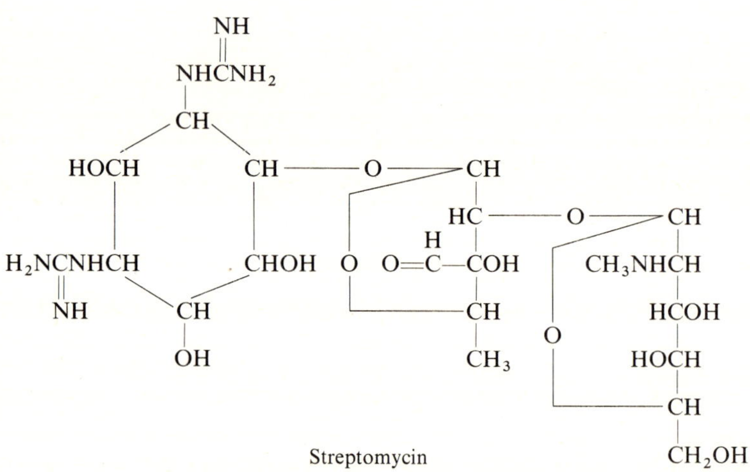

Streptomycin

Streptomycin interferes with the initiation step in protein synthesis. The tetracyclines bind to the ribosome and block aminoacyl-tRNA binding. Chloramphenicol and cycloheximide inhibit peptidyl transferase, chloramphenicol in bacterial cells, and cycloheximide

in higher organisms. Erythromycin blocks translocation-
ment of the ribosome along the mRNA. Puromycin acts
aminoacyl-tRNA, and by binding in place of the correct complex,
it causes the chain to terminate at that point.

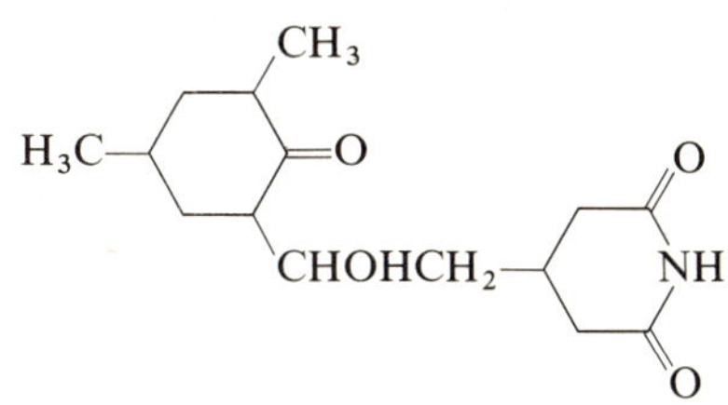

Cycloheximide

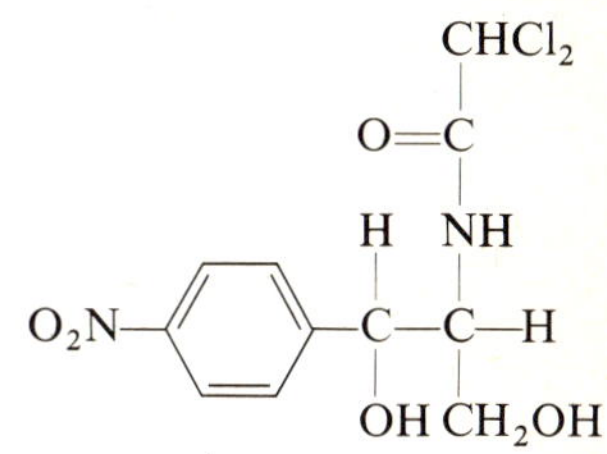

Chloramphenicol

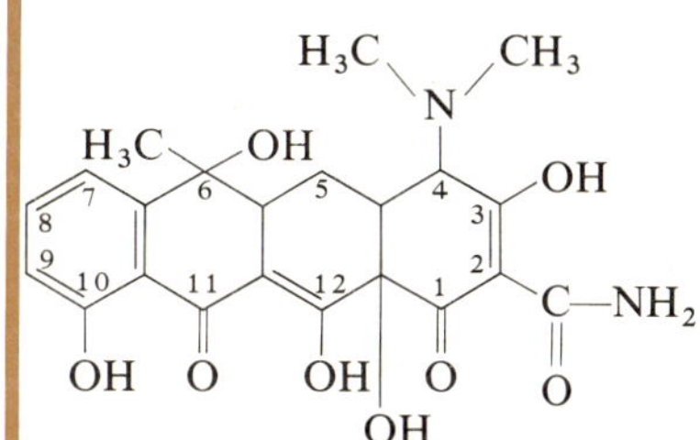

Tetracycline

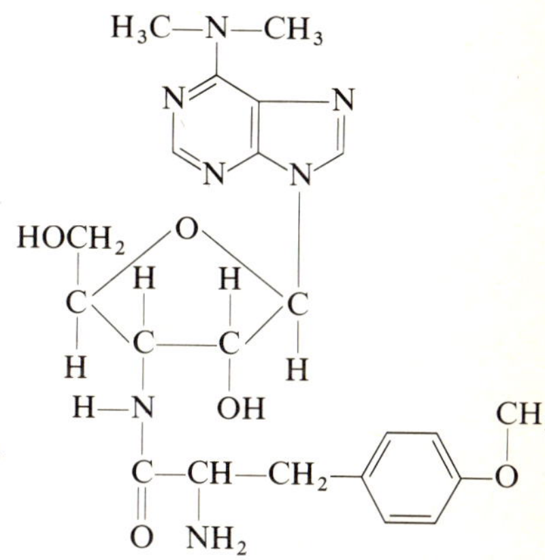

Puromycin

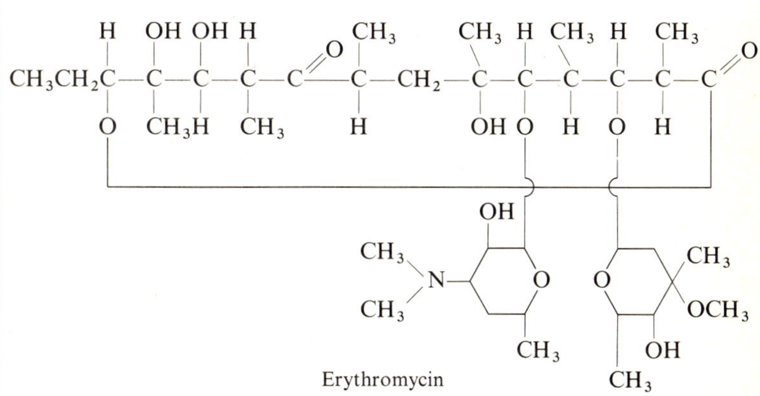

Erythromycin

These antibiotics have been very useful, not only medically, but also in the research laboratory to block processes at a specific point.

13.14 Viruses

Viruses are structures composed of a nucleus of either DNA or RNA and some protein, primarily a coat protein. They adhere to the surface of the cell and insert the nucleic acid into the cell interior. There the nucleic acid utilizes the replication, transcription, and translation machinery of the cell to cause the cell to produce many copies of the virus. Eventually the cell ruptures, releasing the viruses to attack other cells. Plants, animals, and bacteria are all victims of virus attack. In humans, numerous diseases, including colds and flu, are caused by viruses. There is evidence that at least some cancers are caused by viruses.

Summary

DNA is the chemical substance that carries the genetic message. DNA is a polymer with a deoxyribose-phosphate main chain and purine and pyrimidine side groups. Two chains are wound about one another, held together by hydrogen bonding between complementary base pairs. This produces a double helix, generally of very high molecular weight. The genetic message is spelled out in the sequence of the purines and pyrimidines, each three bases denoting an amino acid in a protein sequence. DNA is duplicated or replicated by an enzyme-catalyzed process involving the lining up of complementary bases along unwound stretches of double helix. In a similar fashion, RNA is synthesized (transcription) on the DNA template. Three varieties of RNA are found—rRNA (found in ribosome), tRNA (used to activate amino acids), and mRNA (the carrier of the genetic message for protein synthesis).

Protein synthesis occurs in four stages:

1. Amino acids are activated by attaching them to tRNA.
2. An initiation complex is formed of mRNA, a ribosome, and formyl methionyl-tRNA.
3. The ribosome moves along the mRNA, at each step transferring an amino acid from tRNA to the growing protein chain.
4. At a signal from the mRNA, the protein chain extension ceases.

Several antibiotics function by interfering with one or another of the above processes. Viruses take advantage of this machinery by inserting their DNA or RNA into the cell, causing the cell to reproduce the complete virus particles.

Problems

1. Place the letter of every item in the list on the right required to perform the step indicated in the space provided by each statement.

a) Required to activate the amino acid ____________

b) Required to form the initiation complex ________

c) Required to elongate the chain by one amino acid

____

d) Required to terminate the chain ____________________

A) GTP
B) DNA
C) tRNA
D) ATP
E) mRNA
F) RNA polymerase
G) Amino acid
H) Ribosome
I) Release factors
J) Aminoacyl-tRNA
K) Initiation factors
L) Formyl methionyl-$tRNA_f$
M) Aminoacyl-tRNA synthetase
N) Elongation factors

2. If you trace the events from free amino acid to amino acid residue in a chain, how many molecules of nucleoside triphosphate energy source (ATP or GTP) are required? For what is each used?

3. One of the important clues that led Watson and Crick to propose the double helix was data that reported the relative amounts of each of the purines or pyrimidines in samples of DNA. If that data stated that the proportion of guanine in humans was 20 percent, what would you predict the percentage of cytosine would be? of adenine?

Challenge Problems

4. When a solution of DNA molecules in the form of the double helix is heated, the DNA at some temperature "melts." The melting is actually an unwinding of the helix.

a) The viscosity of the solution (resistance to flow, or "syrupiness," of the solution) changes as it melts. The individual DNA molecules change from rigid rods to crumpled balls as they melt. Would you imagine that the viscosity increases or decreases?

b) The melting point of DNA molecules richer in guanine and cytosine is higher. Higher melting means (higher in molecular weight, harder to unwind, not as soluble). Can you think of an explanation for higher melting point?

5. In 1957 Meselson and Stahl reported some research that was very important to the study of replication. They raised bacteria on a medium with a heavy isotope of nitrogen. The DNA of these bacteria could be separated from that of bacteria raised on a regular nitrogen medium because of its density. If the heavy N bacteria were transferred to a light N medium and allowed to grow for one generation (to replicate the DNA one time), the DNA produced by this replication would have a density representing:

***a*)** All light N.

***b*)** Half light N and half heavy N.

***c*)** All intermediate between heavy and light N. (Intermediate N would represent a DNA double helix one chain of which was heavy N and one light N.)

If the bacteria grew one more generation on the light N medium, the second generation would have a density representing

***a*)** All light N.

***b*)** Half light N and half heavy N.

***c*)** Half light N and half intermediate N.

***d*)** All intermediate N.

14

The Cell, Cellular Metabolism, and Muscle Contraction

Objectives

In this chapter we:

1. *Survey the characteristics of prokaryotic and eukaryotic cells.*
2. *Discuss the problem of cell differentiation.*
3. *View the subject of metabolism in terms of role.*
4. *Learn how energy changes accompany chemical changes, and how organisms manipulate reactions to obtain favorable energy changes.*
5. *Study the structural aspects of striated muscle.*
6. *Relate structure to contraction in muscle.*
7. *Discuss how energy is supplied for muscle contraction, and how contraction is initiated and terminated.*

So far we have spent a great deal of time learning about the molecules of biochemistry—the proteins, carbohydrates, lipids, and so forth. This is a bit like getting to know the actors in a play. We have given these actors a few lines here and there—that is, we have looked at a few reactions. However, we have done nothing to put these actors on a stage; we need to put a complete play, one that tells a cohesive story, in the hands of the actors.

The stage upon which our play will be produced is the cell, and the script is the cell metabolism. All cells have a great deal in common. A full copy of the genetic message is in each cell, for example. The machinery for protein synthesis is in every cell. The enzymes for the cell metabolism are all in the cell. The cell is enclosed in a membrane, a barrier that prevents the cell contents from drifting away and allows only selected substances to enter and leave the cell interior.

14.1 Prokaryotic and Eukaryotic Cells

For all the characteristics cells have in common, there are nonetheless large differences between cells of different organisms or different tissues within the same complex organism. There are two major classes of cells, the prokaryots and the eukaryots.

Bacterial and simple algal cells are prokaryotic (Figure 14.1). They are uncomplicated in organization. The cell is surrounded by a polysaccharide wall. Inside this envelope is the cell membrane, and inside the membrane is the cytoplasm, the semifluid interior. The cytoplasm is an aqueous solution very rich in protein. The genetic material, DNA, is a single tightly coiled molecule in the cytoplasm.

Figure 14.1
Prokaryotic cells
Shown here are cells of blue-green alga (*Plectonema*). Interestingly, the center cell is in the process of dividing. Halfway along the edge of the cell you can see new cell wall forming. The larger lighter regions on either side of this partial division are the masses of DNA that have already separated into two identical sets, one for each daughter cell. The dark streamers in the cell contain the photosynthetic pigments. (Photo by Linda S. Goad, Great Lake Research Division, University of Michigan.)

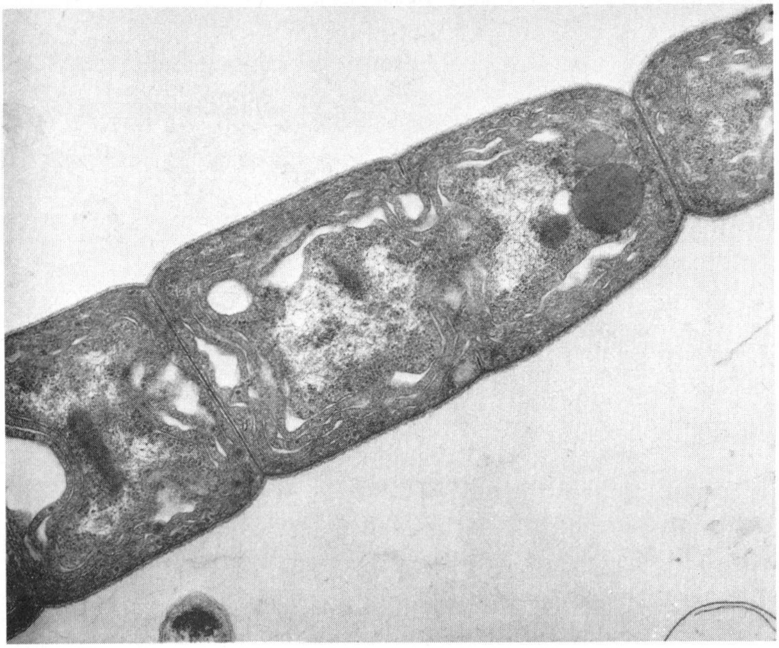

Figure 14.2
Eukaryotic cells
Liver cells are good examples of eukaryotic cells. A number of membrane structures can be seen clearly, including the numerous rounded mitochondria and the long lines of endoplasmic reticulum. The nucleus, which contains the DNA, is clearly defined by the membrane enclosing it. (Photo by Steve Barmatowski, Electron Microscopy Laboratory, Eastern Michigan University.)

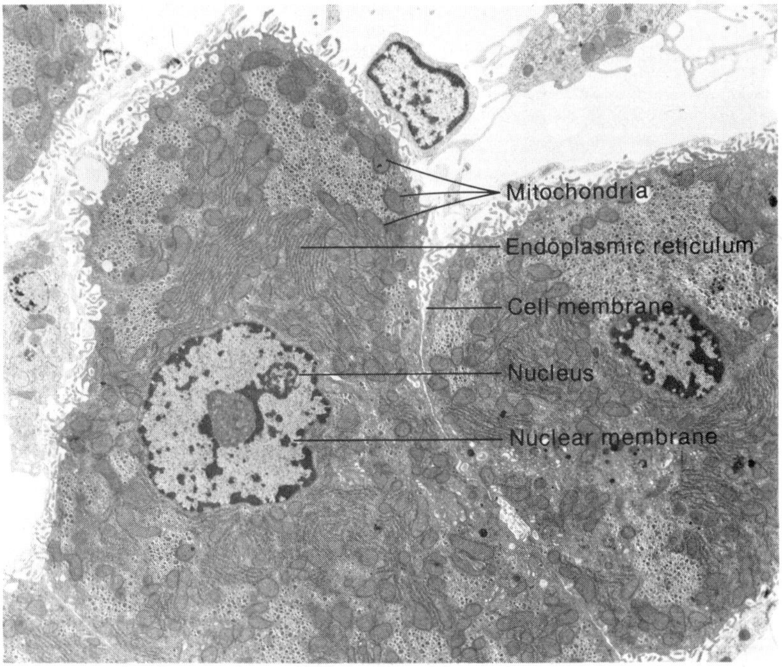

Eukaryotic cells, on the other hand, are very much more complex (Figure 14.2). There is a great deal more membrane. The cytoplasm is enclosed in a membrane, as is true for prokaryots. However, the DNA is contained in a membrane sac called the *nucleus*, rather than simply being coiled in the cytoplasm. In the cytoplasm are a large number of membranous bodies, called *mitochondria*, that specialize in the portion of metabolism that relates to energy production. There is a complex tubular membrane network called the *endoplasmic reticulum*. This may serve as a transport network. The eukaryotic cells we usually deal with are parts of multicellular organisms. A large variety of specialized structures are found on these eukaryotic cells, which have *differentiated*, that is, have taken on a unique function. Obviously, nerve cells, muscle cells, gland cells, liver cells, fat cells, and skin cells all have characteristics that reflect their individual function. Nonetheless all share the basic components described for eukaryotic cells.

While describing the relative complexity and diversity of the eukaryotic cells, consider the genetic message in the DNA of these cells. By comparison with the message in a prokaryotic cell, it must carry enough additional information to describe the components found in a eukaryotic cell but not in a prokaryotic cell. Remember that all the cells of a muticellular organism descended from the original fertilized egg cell. That fertilized egg must have had the message for the special structures in a nerve cell, a muscle cell, a liver cell, a gland cell, and on and on. Eukaryotic cells need to have, and do have, a *great deal* more DNA per cell than do prokaryotic cells. Every cell has the whole genetic message. Why, then, do some of the descendants of the fertilized egg become liver and some muscle? A cell becomes a liver cell when that part of the genetic message not used by a liver cell is repressed. How this comes about is one of the many fascinating, but as yet unanswered, biochemical questions.

14.2 Metabolism

Inside this membrane-wrapped package called the cell many chemical reactions are occurring—the lines read by the actors in our earlier analogy. What is the purpose of running a complicated chemical factory in our insides? Growth and repair is one reason. If we are to grow, we need to add new cells, or if we repair damage, we need to replace cells. Cells are composed of a diverse collection of chemicals—even in a simple bacterial cell we find around 5000 different kinds of compounds. From what? We need raw materials. Those we obtain from our diet. However, our food is not ready-to-use starting material. It has to be chemically converted from whatever diverse assortment we chose to eat into some sort of simple starting unit. But growing isn't all we do. We move, we sense, we think, we do a great many things necessary for survival. These processes are performed by muscle contraction, nerve impulse transmission, and other processes that in the final analysis turn out to be chemical changes. Finally, in all of the above processes we are using energy. Some of our food is going to be burned as a fuel. This is also a chemical process.

Metabolism is the sum of the chemical reactions needed to accomplish all this chemical change. An important generalization is that all this chemistry is not left to chance. It is all under the control and direction of a large number of enzymes, one for each reaction. The study of metabolism is basically the study of the types of enzymes in the cell and the characteristics of those enzymes. More than this, returning to the analogy of the play, we can put the actors on the stage and give each of them lines to read, but utter chaos results rather than a logical story if someone, a director, does not assure that the right lines are read at the right time in the best possible way. In cell metabolism there have to be controls on the enzymes to keep the chemical changes going at the right time and to the correct degree.

14.3 Energy Changes in Cells

Moving to another generalization, chemical changes involve energy changes. Some reactions release energy; others have to have energy added to make them happen. The study of all these energy changes is an important field called *thermodynamics*. The laws of thermodynamics must be obeyed by all systems including living organisms. One of the laws says we can't create energy, only change it from one form to another. If we use energy to create a chemical structure, we get exactly that same amount of energy back when we take the structure apart. Another law says that for any process to occur, there must be an overall transfer of energy from a useful (free-energy) to an unavailable (entropy) form. Some changes that occur may involve an increase in useful energy, but those changes have to be accompanied by another change that decreases a larger amount of useful energy. We will represent this energy transfer by saying more simply that a reaction requires or releases energy. To accomplish chemical changes that require energy, living organisms sometimes resort to coupling these reactions with ones that release energy. For example, binding amino acids to tRNA (Chapter 13) requires energy. The process was coupled to the hydrolysis of a molecule of ATP.

$$\text{amino acid} + \text{tRNA} \longrightarrow \text{aminoacyl-tRNA (reaction requires energy)}$$

$$+ \qquad \text{ATP} + H_2O \longrightarrow \text{AMP} + \text{pyrophosphate (reaction releases energy)}$$

$$H_2O + \text{amino acid} + \text{ATP} + \text{tRNA} \longrightarrow \text{aminoacyl-tRNA} + \text{AMP} + \text{pyrophosphate (coupled reaction releases energy)}$$

In a similar fashion, the binding of the aminoacyl-tRNA to the ribosome-mRNA is a process that requires energy. In each case the energy is supplied by hydrolyzing GTP to GDP and inorganic phosphate. In fact, hydrolysis of ATP or a compound very like it is the most frequent energy releasing reaction to which other reactions are coupled.

The oxidation of compounds releases energy, and the reduction of compounds requires energy. Our source of chemical energy is the oxidation of chemicals in our diet. This is called *catabolism*. When we eat glucose as part of some food, it is transported in the blood to the cells. In the cells it is oxidized, using molecular oxygen that has been transported from the lungs by the carrier hemoglobin. The ultimate end products of this oxidation are carbon dioxide and water. This reaction releases 686 kcal of energy per mole of glucose. One kilocalorie is enough energy to heat a kilogram (therefore a liter) of water 1°C. This means that the energy released by the oxidation of 180 g (1 mole) of glucose to carbon dioxide and water is enough to heat over 9 liters of water from room temperature (~ 25°C) to the boiling point (100°C).

$$C_6H_{12}O_6 + 6O_2 \rightarrow 6CO_2 + 6H_2O + 686\frac{\text{kcal energy}}{\text{mole}}$$

Exercise 14.1

The oxidation of 10 g of glucose releases how much energy?

Exercise 14.2

The oxidation of 1 mole of palmitic acid from fat releases 2340 kcal. If a mole of palmitic acid weighs 256 g, how much energy is released by 10 g of palmitic acid?

Exercise 14.3

Which has more calories (is more fattening), 10 g carbohydrate or 10 g fat?

It is important to realize that the oxidation of glucose does not occur in one step, as the equation shows. Instead there are a very large number of little steps. Why? Here are some reasons:

1. The oxidation is catalyzed by enzymes, and single enzymes don't catalyze such massive changes as the complete oxidation of glucose to CO_2 and H_2O. By their nature they each perform small, simple changes to the substrate.

2. When we oxidize glucose directly to CO_2 and H_2O, as when we burn glucose in a fire, we release the energy in one step as heat. Heat is not the form of energy we want. We want instead to be able to couple the oxidation to reactions that require energy. None of the reactions to which we would like to supply energy requires an amount of energy even remotely close to 686 kcal/mole. If we break the oxidation down into little steps with small releases of energy per step, we will be able to couple these steps to a large number of energy-requiring reactions.

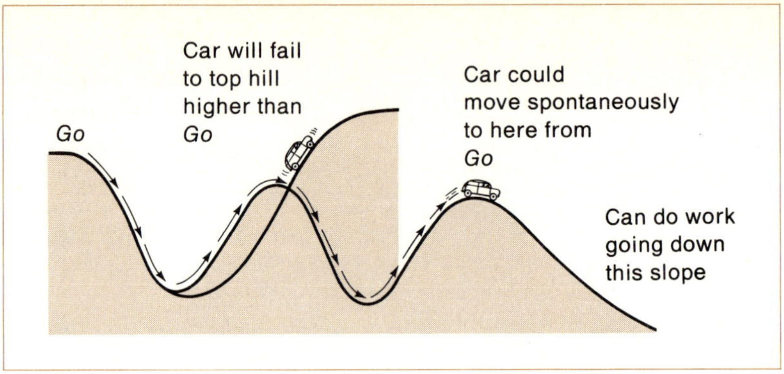

Figure 14.3
Energy
Living systems have to obey the laws of thermodynamics. A system cannot end up with more energy available to do work than it had initially.

14.4 Energy Capture in Metabolism

We are not going to discuss all of metabolism in this text. However, we are going to look at some portions of it to see how it works and how different portions interconnect and relate. Metabolism flows like a river, following its channels, picking up tributaries at points, always going downhill, and capable at points of trapping energy.

How is energy "trapped"? To provide energy we plan to oxidize dietary material. The example with glucose illustrates this. Glucose is oxidized to carbon dioxide and water. This is done in a stepwise fashion employing oxidation operations that consist of removing pairs of hydrogen atoms. As something is oxidized, here the glucose, something else must be reduced. Eventually the hydrogen atoms are transferred onto oxygen atoms to create the water, as indicated in the overall reaction, so the oxygen is what gets reduced. However, the hydrogens are not transferred to oxygen at the time the glucose oxidation intermediates are created. Instead hydrogens are transferred to the enzyme cofactors NAD or FAD. Substrate oxidation is accompanied by cofactor reduction. We find this general pattern with fatty acid oxidation and amino acid oxidation, too.

We have stated that oxidation releases energy. It must therefore be true that reduction adds energy to a compound. As we release energy from a substrate by oxidizing it, we then add energy to the cofactor in the coupled reduction. Overall we release more energy in the oxidation than we trap with the reduction. If this weren't true, the overall process would not occur spontaneously (Figure 14.3). The energy added to the cofactor by reducing it represents the storage of energy eventually to run the biological machinery of the organism. This energy is used to cause muscles to work, to move substances across membranes, and to synthesize new chemicals. The latter, chemical synthesis, is called *anabolism*.

14.5 Cofactor Reoxidation and ATP Formation

As dietary substances are oxidized, the enzyme cofactors NAD and FAD are reduced. Let us summarize where we are and where we want to be:

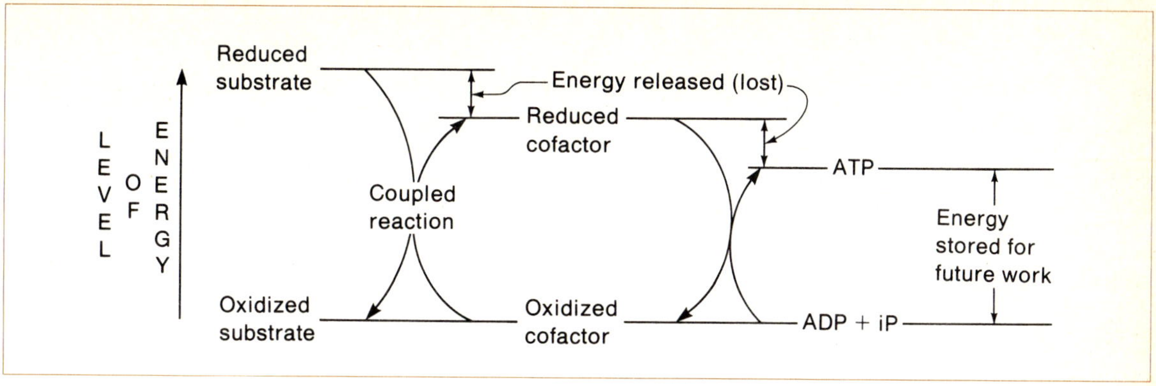

Figure 14.4
Substrate oxidation and ATP formation
In this diagram, vertical distances represent amounts of energy. Three features are significant. First, energy released by a process to the left is used to drive an energy-requiring process to the right. Thus the energy released by oxidation of substrate is used to reduce cofactor. Second, these energy trade-offs must involve (immediately or eventually) a loss of energy. Third, the process depends on coupled reactions, that is, reactions that run concertedly to allow the transfer of the energy.

1. The cofactors are reduced. We don't supply new oxidized cofactor continuously. Instead we reoxidize the reduced cofactor.
2. Energy was required to reduce the cofactors. This energy is released when the cofactor is reoxidized.

The last section stated that energy stored in reduced cofactors is used to run the biological machinery of the cell. This biological machinery includes muscle contraction, movement of substances across membranes, and anabolism, which is the synthesis of new compounds. The energy source that is usually associated with these processes is ATP rather than reduced cofactor. To complete our summary:

3. The reoxidation of cofactor is coupled to the conversion of ADP to ATP (Figure 14.4).
4. The energy stored in ATP is used to run the biological machinery.

More detail about this process is given in Chapter 19.

14.6 Final General Statement About Cell Metabolism

Our discussion of replication, transcription, and translation of the genetic message focused primarily on the process as it occurs in bacteria. This is because bacterial systems were the target of most of the research in that field. They are simpler systems than those found in eukaryotic cells, so our breaking down of the barriers to understanding proceeded more rapidly than they would have had we tackled eukaryotic cells from the outset.

We now must change gears mentally when we discuss the metabolic pathways. The information presented refers to eukaryotic cells. For example, in the chapters to follow, the organization and importance of mitochondria are stressed, and prokaryots don't possess mitochondria.

However, this is a good place to reemphasize the universality of the chemical processes of life. Whether we are focusing on replication or metabolism, cells of all kinds achieve their goals in an amazingly similar fashion. The genetic code is universal, ATP is always the high-energy source, enzymes made of protein catalyze the chemical changes. When we study a process in detail in one organism, we are confident that other organisms use a similar if not identical approach. Specifically, the major metabolic pathways are essentially the same for all cells. The exceptions are of interest and are studied to determine why they are exceptions. Considering the diversity of life forms, isn't this fortunate for the biochemist?

14.7 Muscle Contraction and Biological Energy

Here we shall combine understandings of two biochemical topics to explain muscle function. The topics are protein structure and bioenergetics. Start first with a general description of muscle as revealed by microscopy (Figure 14.5).

Striated muscle is seen to be a bundle of fibers called *myofibrils*. The prefixes myo- and sarco- are common to names of muscle structures. The myofibrils have a variety of bands carrying such names as *A band*, *I band*, *Z line*, and *H zone* (Figure 14.6). The pattern changes as muscle contracts. The *Z* lines move closer together. The A bands are unchanged and the I band shrinks. A cross section of the muscle shows a hexagonal pattern. Cut through the edge of the A band, the hexagonal array shows two types of fibers: *thick* and *thin* (Figure 14.7).

The fibrous material of the myofibril is all protein. Most of this protein is either *myosin* or *actin*. The thick filaments of the muscle cross section are largely myosin. This is a very large protein, with a molecular weight approaching 0.5 million. It has four protein chains in each molecule, two large and two small. The total assembly looks

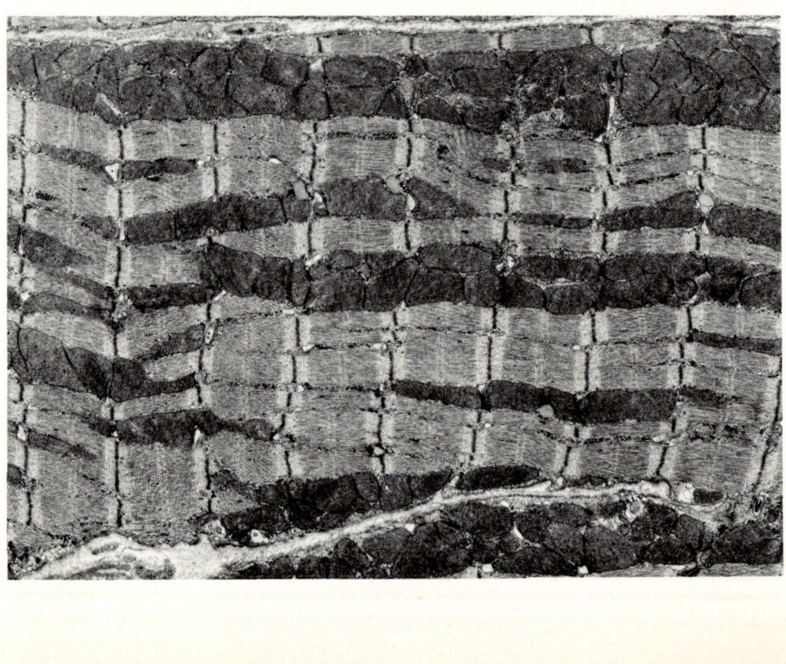

Figure 14.5
A muscle cell
This electron micrograph shows features of muscle cells. The cell boundaries are the white lines running from right to left at the top and bottom. The striped bands are myofibrils, bands of contractile protein. Between the myofibrils are tightly packed mitochondria that supply the ATP to fuel contraction. The tiny dark dots in clusters throughout the cell are glycogen, the substrate that will be oxidized. The structure of liver cells (Figure 14.2) and muscle is a dramatic example of differentiation. (Photo by Glenn K. Walker, Electron Microscopy Laboratory, Eastern Michigan University.)

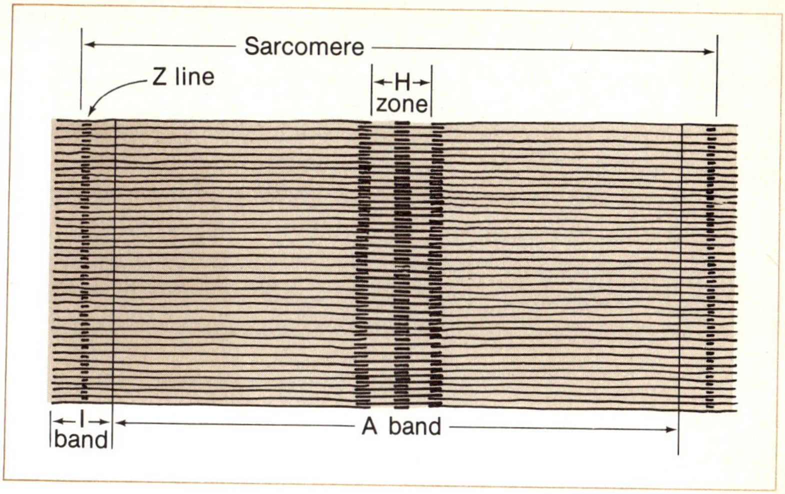

Figure 14.6
Detail of myofibril structure
Locate these structural features on the electron micrograph of a muscle cell (Figure 14.5).

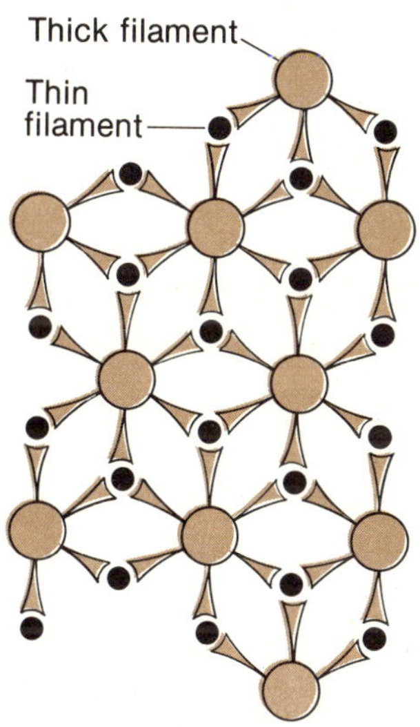

Figure 14.7
Myofibril cross section
A cross section of a myofibril reveals an array of thick and thin filaments. The thin filaments are arranged in hexagons, and each is near three thick filaments.

a bit like a tadpole with a very long tail. The head part has all four protein chains in it. Then the two larger chains extend out of the head, each completely wound into an incredibly long α-helix structure. Why such a long tail? The tails of about 400 myosin molecules wind together to form a rope. This rope is the thick filament. The heads of the myosin are not in the filament but project out sideways in a regular pattern. Looking down the filament you would see them at six positions or angles. There are four binding sites on each myosin head. Two of these sites bind ATP and catalyze its conversion to ADP and inorganic phosphate. The other two sites bind the other major muscle protein actin. Myosin is structurally half fibrous protein, that part building the rope of the thick filament, and half globular, the head with its binding sites.

Actin is a rather spherical molecule of moderate size made of a single protein chain. These individual actin molecules can link together to form a long chain. Then two such long chains wind around one another in a double helix to form a long filament. Actin in this filament form is called *F actin*, or *fibrous actin*. At every seventh actin on this filament we find another muscle protein called *troponin*. Troponin binds Ca^{2+}, the importance of which will soon be discussed. Another muscle protein, tropomyosin, is part of the filament. This long, thin protein fits into the groove formed between the two actin strands as they wind together. Perhaps it is present just to hold the filament together more strongly. This whole assembly is the thin filament of the muscle cross section.

Visualize the muscle structure in a stationary configuration. Long strands of thick filaments are attached by the protruding myosin heads to adjacent thin filaments. Each thick filament has six thin filaments surrounding it. Each thin filament is attached to three different thick filaments. Going back to the microscopic view of the

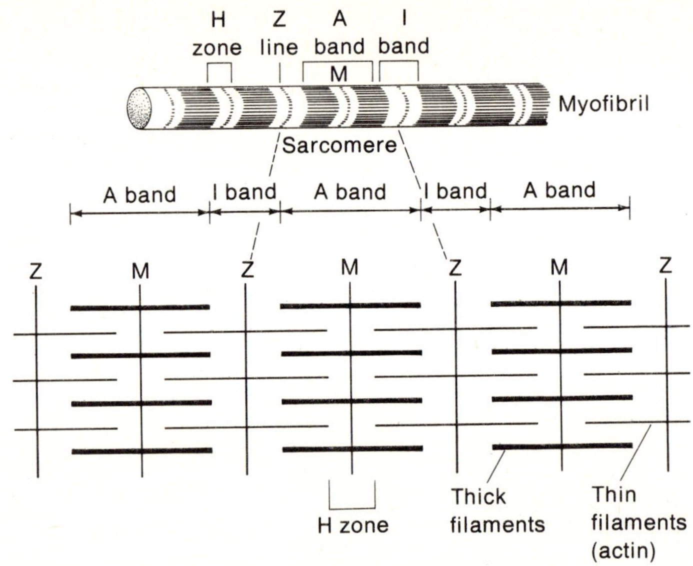

Figure 14.8
The assembly of the myofibril
The thick and thin filaments are proposed to be arranged side by side and to be anchored. This produces the pattern of lines and zones seen under the microscope (Figures 14.5 and 14.6). (From S. C. Keene in W. Bloom and D. W. Fawcett, *Textbook of Histology*, 9th ed., Philadelphia, Saunders, 1968.)

muscle, how do these structures relate to the bands observed on the myofibrils (Figure 14.8)? The A band represents the zone occupied by thick filaments. The I band is the space between the regions containing thick filaments. If we cut through a myofibril at this point and looked at the cross section, we would see no thick filaments, only thin filaments. The H zone is a region with only thick filaments; the thin filaments don't extend that far. The two lines we see in the muscle, the Z line and the M line, are probably anchor points for the filaments. The Z line anchors the thin filaments at their centers and the M line anchors the thick filaments at the center. A protein called the *M-line protein* has been isolated.

14.8 Muscle Contraction

Contraction of muscle is visualized as being the slipping of the thin filaments past the thick filaments, resulting in the distance between Z lines becoming shorter (Figure 14.9). For this to happen, the myosin head had to release one actin and attach to another farther along the thin filament. Such movement requires that the myosin head, while attached to an actin unit, change its shape to shift the two filaments past each other. Then when it releases the actin, it must go back to the original shape, which moves it next to a new actin unit. The shape change uses energy, and the energy is supplied by converting ATP to ADP and inorganic phosphate. This reaction is catalyzed by the myosin head.

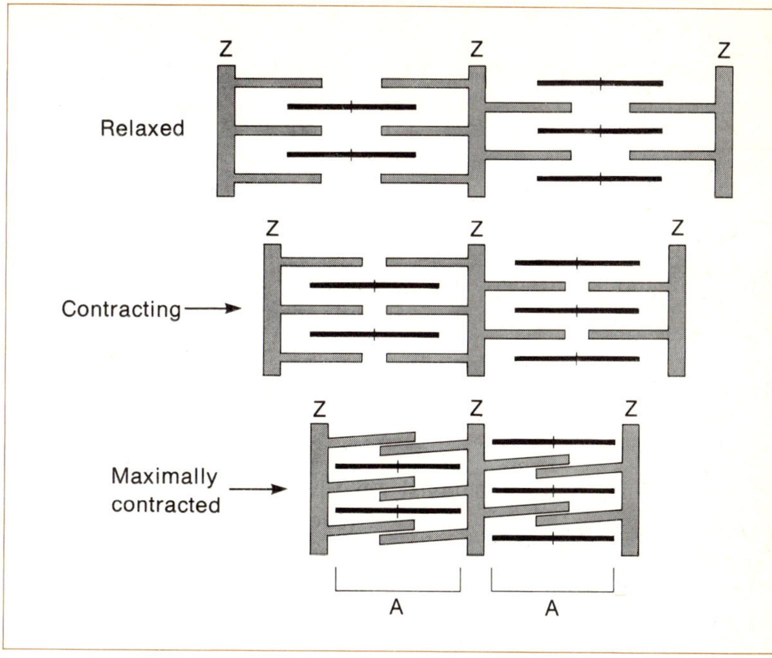

Figure 14.9
Muscle contraction
The distance between the Z lines shortens as the thick and thin filaments move past one another. (From A. L. Lehninger, *Biochemistry*, 2nd ed., New York, Worth, 1975, p. 753.)

14.9 Control of Contraction

The myofibrils are surrounded by ATP all the time. How do they know when to use this ATP to contract and when to leave it alone? A system of tubes comes close to every myofibril. In the tubes the Ca^{2+} concentration is much higher than it is at the myofibril. When the nerve impulse signaling contraction reaches the muscle cell, it causes the walls of these tubes to leak Ca^{2+} into the vicinity of the myofibril. Troponin binds the Ca^{2+}, following which the hydrolysis of ATP by myosin and contraction begins. Proteins that function as pumps in the tube walls pump the Ca^{2+} out into the tube, causing relaxation. This pumping requires energy, which is supplied once again by ATP. This process is covered in more detail in Chapter 21. Thus ATP is used both to contract and to relax the muscle.

The supply of ATP in the muscle cell turns out to be too small to maintain muscle contractions and relaxation for very long. In fact, during hard exercise, vast amounts of ATP are hydrolyzed to provide the energy. Ultimately, the ATP must be supplied by the metabolism of nutrients. How nutrients generate ATP is the topic of the next four chapters.

Summary

The ultimate functional unit in the chemistry of living organisms is the cell. There are two major classes of cells. Bacteria are very simple cells and are termed prokaryots. Prokaryotic cells have a cell wall inside of which is the cell membrane. The membrane contains the cytoplasm, suspended in which is the DNA molecule carrying the genetic information upon which the cell is based. Higher plants and animals have more complex cells and are termed eukaryots. Eukaryotic cells have the same components as do the prokaryotic cells, along with several additional internal membrane compartments and structures. The DNA is in a membrane sac, the nucleus. In the cytoplasm there is a system of tubes called the endoplasmic reticulum and some double membrane-walled bodies called mitochondria.

The sum of the cell chemical reactions is called the cell metabolism. Some of these reactions are directed at breaking down and oxidizing nutrients. These reactions yield energy and are termed catabolism. Other reactions are directed at building up more complex molecules. These changes use up energy and are called anabolism. Most energy is stored and used as phosphate-to-phosphate bonds in the nucleotide ATP. The chemical changes are all performed as numerous simple steps catalyzed by specific enzymes.

An example of a biological use of the energy generated by catabolism is provided by muscle contraction. Muscle is composed of two kinds of filaments interlaced with and attached to one another. The thick filaments are made largely of the protein myosin. The myosin molecules have headpieces with binding sites that attach to the thin filaments, which are made largely of the protein actin. In contraction, the myosin changes shape, causing the filaments to move past one another. It then releases the actin, changes shape back to the original conformation, and attaches a new actin. This shape change uses energy supplied by hydrolyzing ATP. The process is triggered by Ca^{2+} ion entering the cell cytoplasm in response to a nerve impulse. Pumping the Ca^{2+} out again also requires energy.

Problems

1. What would the cross section of muscle show if taken at the following sites:

a) At the I band, not at the Z line.

b) At the H zone, not at the M line.

c) Inside the A band but outside the H zone.

2. In Chapter 13 some reactions were coupled to the hydrolysis of ATP or GTP. One was reviewed in this chapter, the formation of aminoacyl-tRNA.

a) What was the actual reaction that required the energy released by ATP hydrolysis?

b) What reactions required GTP hydrolysis?

3. In the above reactions, two different nucleoside triphosphates (ATP and GTP) served as energy sources. Because they are chemically very similar, it is not surprising that each can serve as a source of energy.

a) Draw the structure of ATP and GTP.

b) Draw an arrow pointing to the bond whose hydrolysis released the energy from each molecule. Is it the corresponding bond in each case?

c) What do the two bonds used for energy share in terms of chemical structure?

15
Carbohydrate Metabolism

Objectives

In this chapter we:

1. *Review the concepts of metabolism and the premetabolism biochemistry of the carbohydrates.*

2. *Trace the conversion of glucose to acetyl CoA step by step, discussing the chemistry of each reaction.*

3. *Look in detail at the energy changes of the first three steps.*

4. *Discuss the controls of this metabolic path.*

5. *Learn the variations of glucose metabolism utilized under anaerobic conditions.*

6. *Study the use of carbohydrates as energy sources for muscle contraction.*

Let us consider a segment of metabolism in detail. Glucose metabolism is selected as the first such topic. Carbohydrates, particularly starch, are the largest source of energy in the human diet. Historically, the paths of carbohydrate metabolism were the first to be deciphered. Pioneer researchers established the occurrence and importance of phosphate esters and detected intermediate three-carbon fragments. Gradually, like assembling a jigsaw puzzle, they uncovered the pieces of information and fitted them into place, until today we have a thorough understanding of the process.

Watch for the following general aspects of metabolism as you study this chapter:

1. Metabolism proceeds by a number of enzyme-catalyzed steps.
2. At each step there is a single product.
3. Each step represents a small change in the substrate, often one of the reactions discussed in early chapters.
4. Enzymes utilize certain cofactors in ways predictable from the chemistry of the reaction.

15.1 Preface to Glucose Metabolism

We have set the stage for our metabolism discussion. We are aware that starch, sucrose, and lactose are the chief dietary carbohydrates (Chapter 6). The digestion of these di- and polysaccharides has been explained (Chapter 12). The monosaccharides, primarily glucose, cross the intestinal mucosa into the blood and are transported by the blood to the individual cells. They are transported across the cell membrane

(Chapter 21) into the cytoplasm of the cell. At this point metabolism begins.

15.2 Glucose Metabolism

In the cell, the first step in glucose utilization is its conversion to a phosphate ester. This is accomplished by the enzyme hexokinase. Here, hexo- is short for hexose. Enzymes that transfer a phosphate from ATP to another molecule are generally called *kinases*. The formation of the ester bond requires energy. The necessary energy is provided by coupling the formation of the ester linkage to the hydrolysis of ATP. In this fashion the ATP serves two purposes: It provides the phosphate and releases the energy to cause the reaction to go readily.

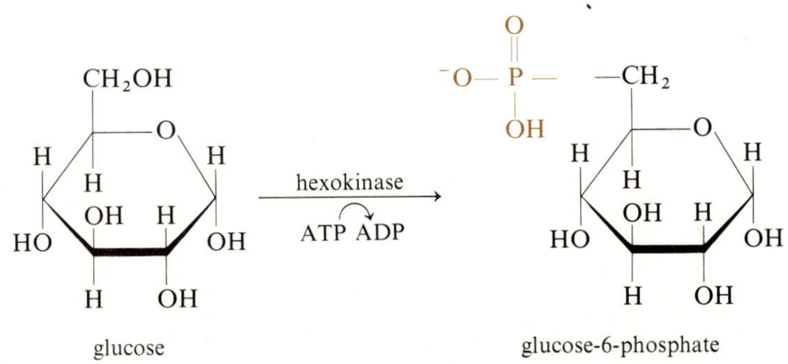

Some researchers believe that the conversion of glucose to the charged phosphate ester helps prevent the glucose from leaking back out of the cell. The nonpolar membrane would be unlikely to allow the charged ester to leak out.

Glucose-6-phosphate is then converted to fructose-6-phosphate. The enzyme, glucose phosphate isomerase, is accomplishing an internal oxidation-reduction. This is perhaps best seen by looking at the open chain form of the carbohydrate.

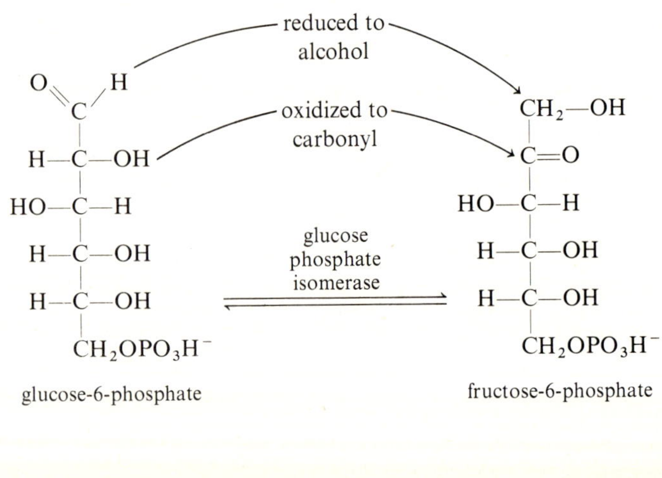

Overall, an alcohol becomes a carbonyl and a carbonyl becomes an alcohol. This sort of trade-off doesn't involve much of an energy change. The enzyme name isomerase is used here because the product has the same atoms as the substrate in a different arrangement.

The fructose-6-phosphate is next converted to a diphosphate. The reaction is much like that catalyzed by hexokinase. The formation of the phosphate ester requires energy, and ATP supplies both the phosphate and the energy. The enzyme is also named as a kinase, here phosphofructokinase.

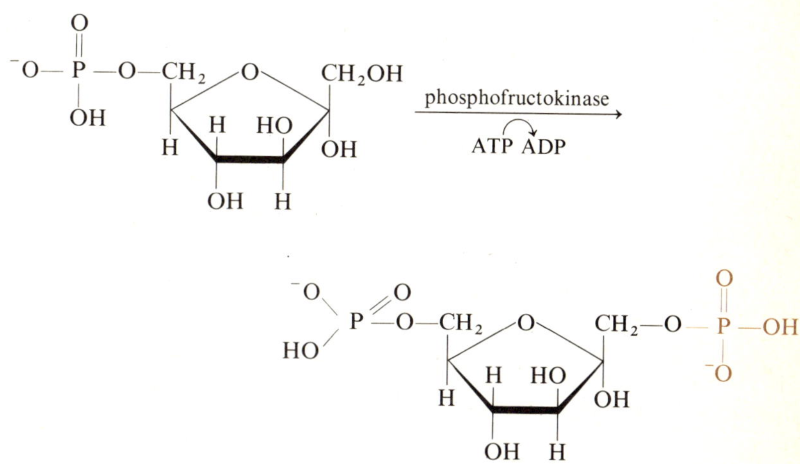

In summary, so far each molecule of glucose has been converted to fructose-1,6-diphosphate at the expense of two molecules of ATP.

$$\text{glucose} + 2\text{ATP} \longrightarrow \text{fructose-1,6-diphosphate} + 2\text{ADP}$$

15.3 Control of Glucose Metabolism

The preceding is the first portion of a pathway for converting glucose to acetyl CoA. However, the pathway must be controlled to prevent over- or underproduction of acetyl CoA.

A major use of acetyl CoA is the production of ATP (Chapters 17 and 18). The level of ATP is therefore a useful measure of the need for glucose metabolism. Phosphofructokinase is an allosteric enzyme inhibited by ATP. If we stop that enzyme, fructose-6-phosphate and glucose-6-phosphate will begin to accumulate. Glucose-6-phosphate inhibits hexokinase. In summary, we have two allosteric enzymes, phosphofructokinase and hexokinase. High levels of ATP inhibit phosphofructokinase, which causes glucose-6-phosphate to accumulate, which inhibits hexokinase. The end result is that glucose ceases to enter the metabolic path.

15.4 Energy Changes

The formation of glucose-6-phosphate from glucose requires about 3300 cal/mole. Hydrolysis of ATP releases about 7300 cal/mole. For the coupled reaction the overall energy picture is therefore

phos. + glucose	⟶	glucose-6-phos.	+3300 cal/mole
H_2O + ATP	⟶	ADP + phos.	−7300 cal/mole
glucose + ATP + H_2O + ~~phos~~.	⟶	glucose-6-phos. + ADP + ~~phos~~.	−4000 cal/mole

Notice that the + sign on the energy means energy must be added to cause the reaction to happen, whereas the − sign means energy is released.

This is an excellent example of coupling two reactions so that the combination is energetically favorable. Conversion of glucose-6-phosphate to fructose-6-phosphate involves a very small change in energy. We must add 400 cal/mole as the reaction runs. However, it is an addition of energy—does this mean the reaction won't work? By itself the energy change means we would have a poor yield of product. However, in series with reactions that release a lot of energy, as we will soon see in the summary of energy changes, the reaction is swept along.

The third step looks very much like the first step in terms of energy:

phos. + fructose-6-phos.	⟶	fructose-1,6-diphos.	+4000 cal/mole
H_2O + ATP	⟶	ADP + phos.	−7300 cal/mole
fructose-6-phos. + ATP + H_2O + ~~phos~~.	⟶	fructose-1,6-diphos. + ADP + ~~phos~~.	−3300 cal/mole

We can add up the total energy change for this series of steps:

glucose	⟶	glucose-6-phosphate	−4000 cal/mole
glucose-6-phosphate	⟶	fructose-6-phosphate	+400 cal/mole
fructose-6-phosphate	⟶	fructose-1,6-diphosphate	−3300 cal/mole
			−6900 cal/mole

Overall the reactions release a large amount of energy and would be expected to go very well.

15.5 The Three-Carbon Compounds

At this point we are ready to break the molecule into two fragments, each with three carbons and one phosphate. Notice that if our objective in metabolizing the glucose is to make ATP, we are doing very badly. We have made no ATP—in fact we've used up two!

The enzyme fructose diphosphate aldolase catalyzes the next step. Fructose-1,6-diphosphate is neatly split into 2 three-carbon fragments.

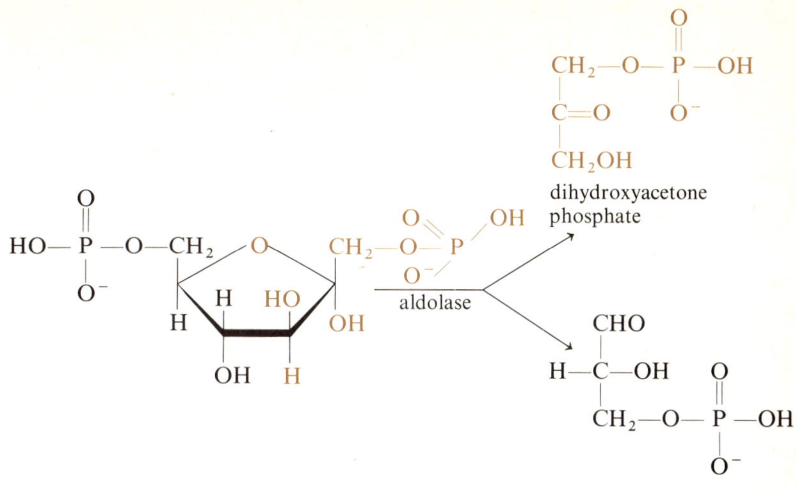

fructose-1,6-diphosphate glyceraldehyde-3-phosphate

These two fragments are very similar. In fact, the next step is to convert the dihydroxyacetone phosphate into glyceraldehyde-3-phosphate.

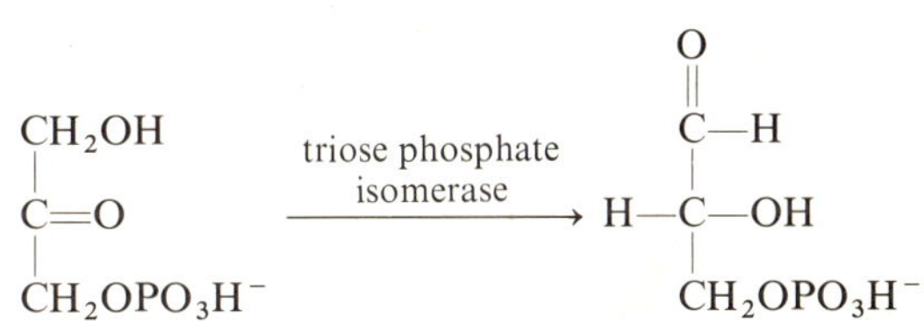

Notice that this reaction, catalyzed by triose phosphate isomerase, is just like the interconversion of glucose-and fructose-6-phosphate catalyzed by glucose phosphate isomerase.

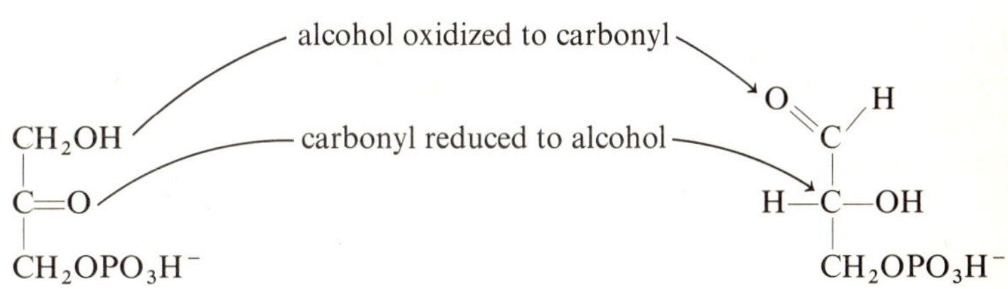

It is another internal oxidation-reduction. Once again this sort of reaction involves very little energy change. The effect of the preceding two steps is that all six carbons are now converted into two glyceraldehyde-3-phosphates.

Let us consider the next two steps at once. In these steps the aldehyde group is converted to a carboxylate group. This oxidation is accompanied by the reduction of the cofactor NAD^+. This reaction is divided into two parts catalyzed by different enzymes. In the first step the oxidation is accomplished, but in place of the expected free carboxylate group the product has the carboxylate bonded to a phosphate in an acid anhydride linkage.

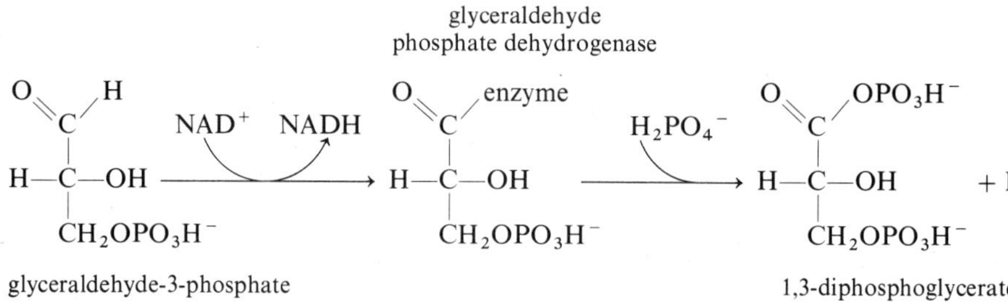

This reaction is not an internal oxidation-reduction of the substrate. The substrate is oxidized and the cofactor is reduced. The enzyme is named not as an isomerase but as a dehydrogenase.

In the second step the phosphate is transferred from the carboxylate group to ADP, the products being 3-phosphoglycerate and ATP.

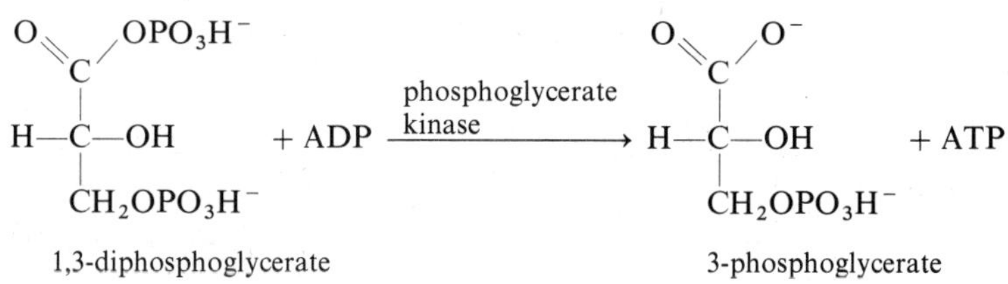

The enzyme, phosphoglycerate kinase, is named for the conversion of 3-phosphoglycerate to 1,3-diphosphoglycerate (the reverse of the reaction as written).

15.6 Energy Consideration in the Aldehyde Oxidation

Energywise, these are very interesting and important steps. The oxidation of the aldehyde to a carboxylic acid releases energy. The coupled reduction of the cofactor NAD^+ is an energy-requiring step that prevents some of the energy of the aldehyde oxidation from being wasted. The formation of the acid anhydride linkage to phosphate is also a step requiring energy. In fact, overall between the reduction of NAD^+ and the formation of the acid anhydride, slightly more energy is used than is released. This is allowed as long as other steps in the sequence release more energy than they use.

In the next reaction the energy used to form the carboxyl-phosphate bond is released by breaking the bond, and part of it is used to bind another phosphate to ADP. At last some ATP has been synthesized! Remember that we are forming an ATP for each three-carbon fragment of glucose. That means that per original glucose two molecules of ATP have been formed. That neatly cancels the ATP debt we incurred informing fructose-1,6-diphosphate from glucose. In addition, per glucose, two molecules of NAD^+ have been reduced. This represents more biological energy captured, as we shall see in Chapter 19.

In summary, the intermediate formation of the carboxyl-phosphate acid anhydride bond can be regarded as a mechanism to store energy by aldehyde oxidation temporarily, then to use it to form ATP.

15.7 Continuing the Pathway

Next we move the phosphate to a different alcohol group, setting the stage for more energy entrapment. Mutase enzymes catalyze such shifts.

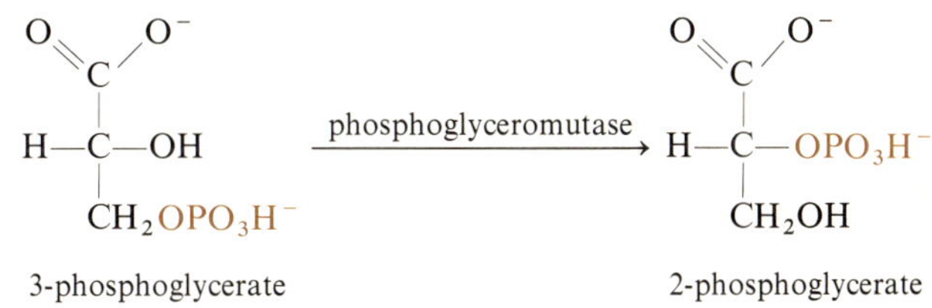

We now create a double bond by the elimination of water from carbons 2 and 3.

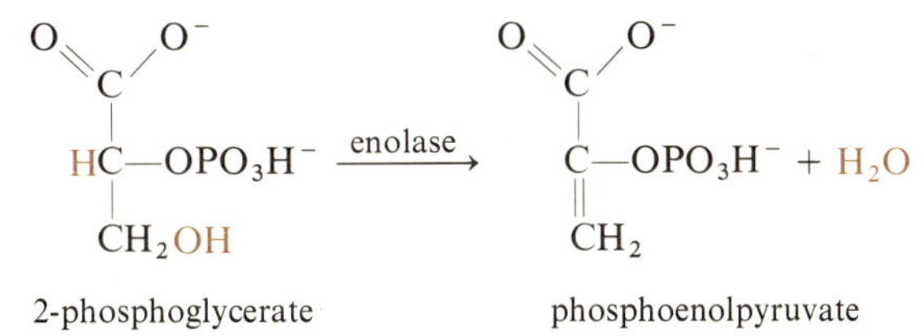

The enzyme is called *enolase*. This is because the product is an unusual structure called an *enol*. An enol has an alcohol group whose carbon is also involved in a double bond. Normally enols convert to ketones with the release of energy.

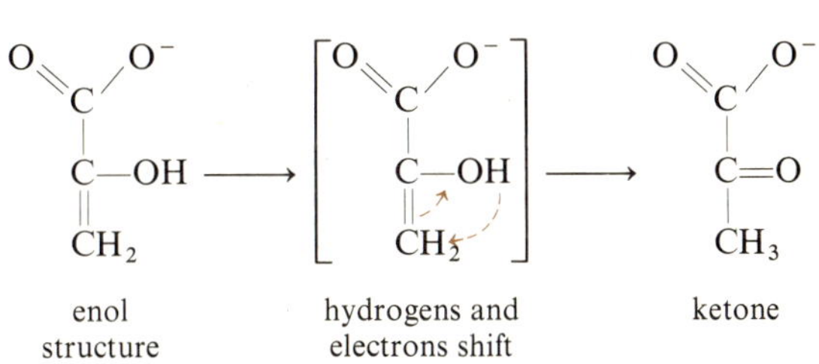

In this case this shift is prevented because the alcohol has been esterified by phosphoric acid. However, if the phosphate were to be removed, the shift would occur and energy would be released. In the next step, that is exactly what happens, and some of the energy released is used to attach the phosphate to ADP.

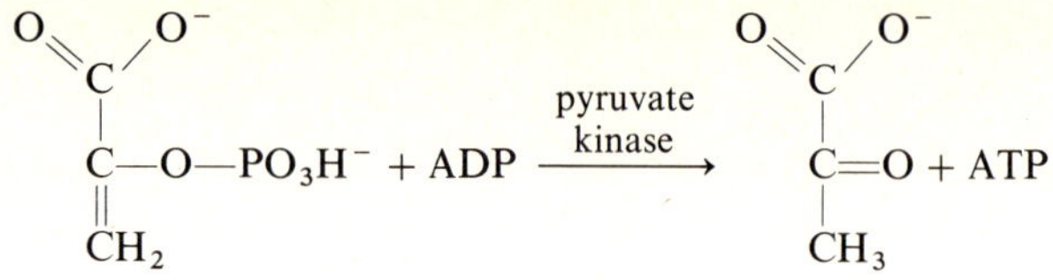

phosphoenolpyruvate pyruvate

Notice once again that the enzyme is named as a kinase, which describes the reverse of the reaction as written above. Pyruvate is a very important metabolic intermediate. It is used for a variety of purposes, some of which we shall see as we proceed.

In order to trace our breakdown of glucose to a logical stopping point, we need to carry the pathway one step further. But what a step! Catalysis is performed by the *multienzyme complex* pyruvate dehydrogenase. Such a complex is an orderly array of enzymes arranged to be very efficient. They transfer the substrate from one enzyme to another without releasing it to wander aimlessly between steps. The overall change catalyzed by pyruvate dehydrogenase is as follows:

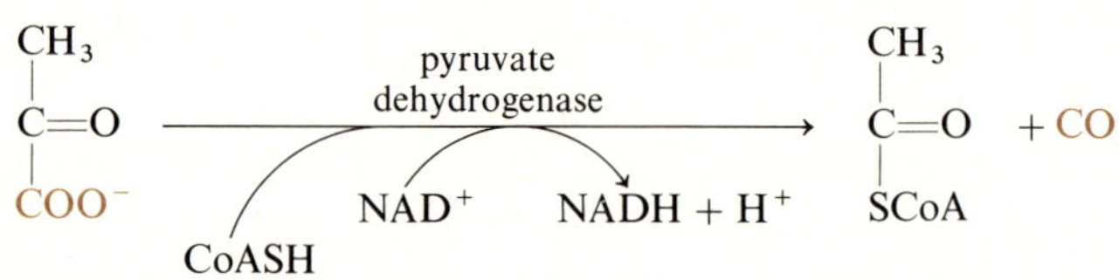

pyruvate acetyl CoA

The reaction has produced CO_2. Of the six original carbons of glucose, two are now CO_2. The CO_2 is the product of oxidation, and this oxidation is accompanied by the reduction of the cofactor NAD^+. Our final product is a structure that is at the crossroads of metabolism—acetyl CoA.

This enzyme system, like phosphofructokinase, is inhibited by high levels of ATP. Pyruvate is useful for a number of reactions in metabolism. Control at this point allows the pyruvate to be supplied to these reactions, but it prevents its conversion to acetyl CoA, which would lead to a further increase in ATP levels.

15.8 Detail of the Pyruvate Dehydrogenase Reaction

In bacteria, pyruvate dehydrogenase involves three enzymes and five cofactors. In mammals, two additional enzymes control the level of activity of the complex. This complex provides examples of several cofactors functioning. Pyruvate first attaches to TPP on enzyme 1. The carboxyl group is released as CO_2 (Figure 15.1). The two-carbon fragment is transferred to enzyme 2, attaching to lipoic acid on the enzyme. Before attachment the lipoic acid is in an oxidized state. After attachment the lipoic acid is reduced, and the two-carbon fragment is oxidized (Figure 15.2).

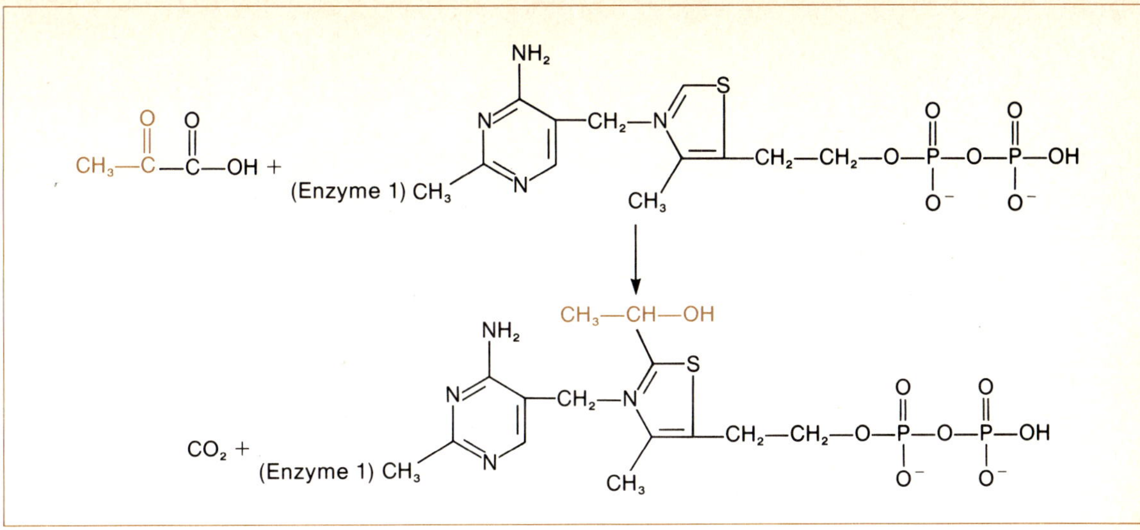

Figure 15.1
The pyruvate dehydrogenase reaction: step 1

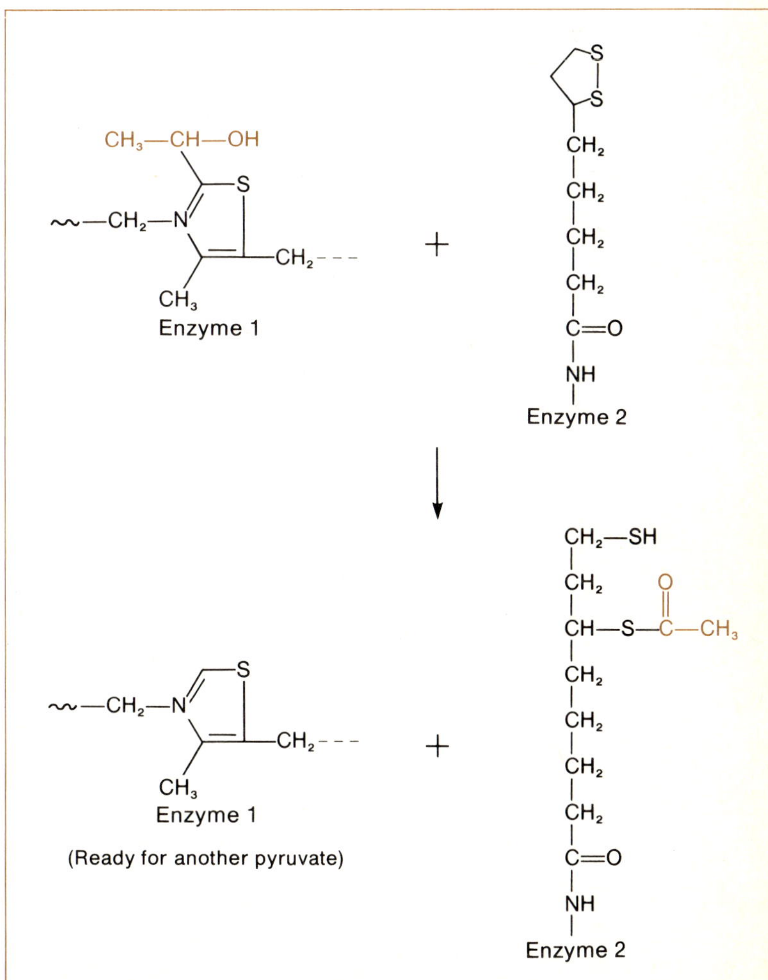

Figure 15.2
The pyruvate dehydrogenase reaction: step 2

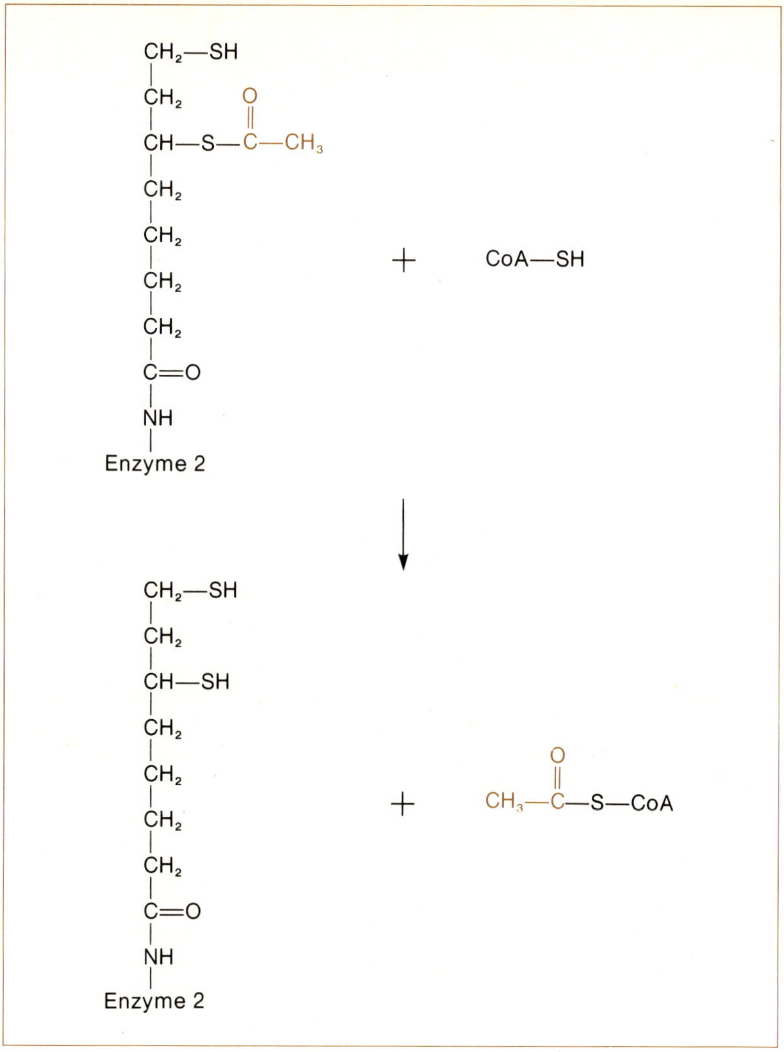

Figure 15.3
The pyruvate dehydrogenase reaction: step 3

Notice that the fragment is now a carboxylic acid attached to the cofactor as a thioester. Lipoic acid has a long flexible attachment to enzyme 2, so it was able to reach right over to enzyme 1 and pick up the substrate. A molecule of CoASH attaches to enzyme 2. The lipoic acid swings over with the acetyl group to a position next to the CoASH. The acetyl group is transferred to the CoASH and leaves as acetyl CoA (Figure 15.3).

We have now generated acetyl CoA. When the reaction sequence began, the lipoic acid was oxidized; now it is reduced. Before the system can process another pyruvate, we need to reoxidize the lipoic acid. Call in enzyme 3 with its flavin cofactor. The lipoic acid swings over next to the flavin and transfers the

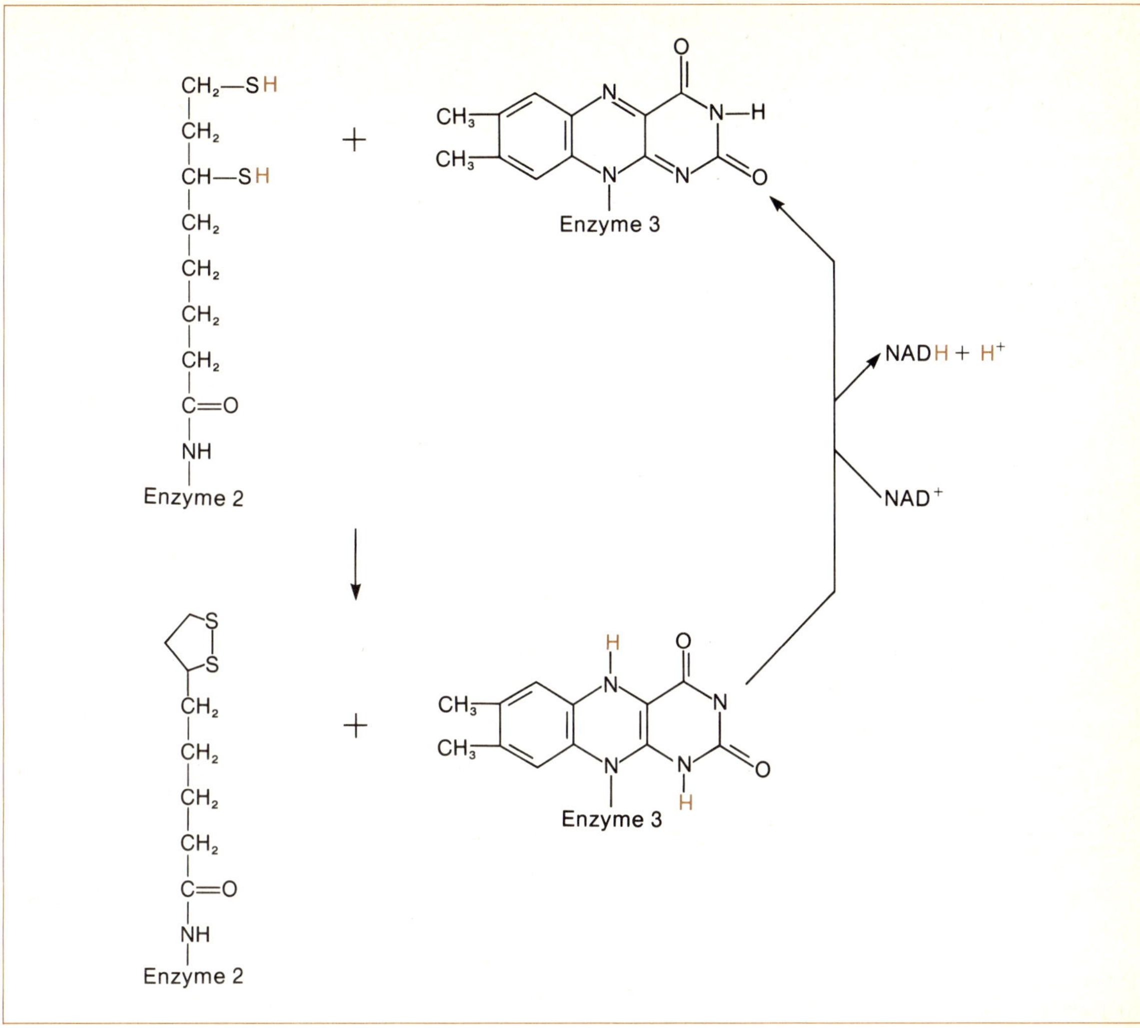

Figure 15.4
The pyruvate dehydrogenase reaction: step 4

two hydrogens to it. An NAD^+ then binds to enzyme 3, the hydrogens are passed on to it, and the reduced NAD leaves (Figure 15.4).

15.9 Anaerobic Carbohydrate Metabolism: Alcohol Fermentation

The above pathways represent the metabolism of glucose in the presence of a supply of oxygen. Ultimately, reduced cofactors are reoxidized by a process requiring oxygen. Oxygen is not always available, however. Some microorganisms, called *anaerobes*, can or perhaps must operate in the absence of oxygen. Metabolism of glucose presents some special problems for anaerobes. An organism's supply of NAD^+ is limited, so it must be recycled. Once it is reduced, as it was in the step catalyzed by glyceraldehyde phosphate dehydrogenase, it has to be reoxidized before it can be used again. With oxygen unavailable, another procedure must be employed. Anaerobic metabolism, also called fermentation, uses the same pathway as far as the formation of pyruvate. At this point yeast cells decarboxylate pyruvate to generate ethanal.

$$\underset{\text{pyruvate}}{CH_3-\overset{\overset{\displaystyle O}{\|}}{C}-COO^-} \xrightarrow[\text{TPP}]{\text{pyruvate decarboxylase}} \underset{\text{ethanal}}{CH_3-\overset{\overset{\displaystyle O}{\|}}{C}-H} + CO_2$$

The enzyme, pyruvate decarboxylase, has a TPP cofactor. The ethanal is then reduced to ethanol, using NADH as the source of hydrogens.

$$NADH + H^+ + \underset{\text{ethanal}}{CH_3-\overset{\overset{\displaystyle O}{\|}}{C}-H} \xrightarrow{\text{alcohol dehydrogenase}} NAD^+ + \underset{\text{ethanol}}{CH_3-CH_2-OH}$$

This serves to reoxidize the NAD reduced earlier in the metabolic scheme, eliminating the need for oxygen. This process is used to ferment starch or other carbohydrates into beer or wine. The decarboxylase step provides the "carbonation" when beer or champagne are fermented in closed vats.

15.10 Lactate Fermentation

Another anaerobic pathway is employed by lactobacilli, the microorganisms that cause milk to sour. The pathway is followed in the usual fashion as far as pyruvate. The pyruvate keto group is then reduced to an alcohol group, using NADH as the source of hydrogens.

$$\underset{\text{pyruvate}}{CH_3-\overset{\overset{\displaystyle O}{\|}}{C}-COO^-} + NADH + H^+ \xrightarrow{\text{lactate dehydrogenase}} \underset{\text{lactate}}{CH_3-\overset{\overset{\displaystyle OH}{|}}{CH}-COO^-}$$

Lactate is the acid that makes the milk sour. The process eliminates the need for oxygen by reoxidizing the NAD that had earlier been reduced.

15.11 Carbohydrate Metabolism and Muscle Energy

Striated or skeletal muscle is capable of operating anaerobically. Consider the value of this ability. In emergency situations, taking direct action, either to fight or to flee, requires maximum exertion of skeletal muscles. Muscles require ATP for contraction (Chapter 14.) The richest source of ATP is the aerobic metabolism of dietary substances such as carbohydrates and fats. However, ATP synthesis is limited by the rate at which oxygen can be supplied. Skeletal muscle does not allow a shortage of oxygen to stop ATP formation. The muscle switches to the lactate fermentation pathway and continues to generate ATP without oxygen. Once too high a concentration of lactate accumulates, this pathway too must be stopped; in other words, the muscle fatigues. As the muscle rests, much of the lactate is converted back to glucose by reversing the pathway for metabolic breakdown. The energy to fuel this resynthesis of glucose is provided by converting some of the lactate through pyruvate to acetyl CoA. In the presence of oxygen, acetyl CoA becomes a rich source of ATP (Chapters 17 and 18). This oxygen is in excess of normal body needs, so during the resting period following vigorous exercise, we continue to breathe hard to supply this extra oxygen.

15.12 Glycogen and Muscle Metabolism

Muscle and liver have reserves of glycogen (Chapter 6), a glucose polymer, which serves as an emergency food reserve. Glycogen is made available by the enzyme glycogen phosphorylase. Instead of adding water to the glycosidic bonds as did the carbohydrate digestive enzymes (Chapter 12), glycogen phosphorylase adds phosphoric acid.

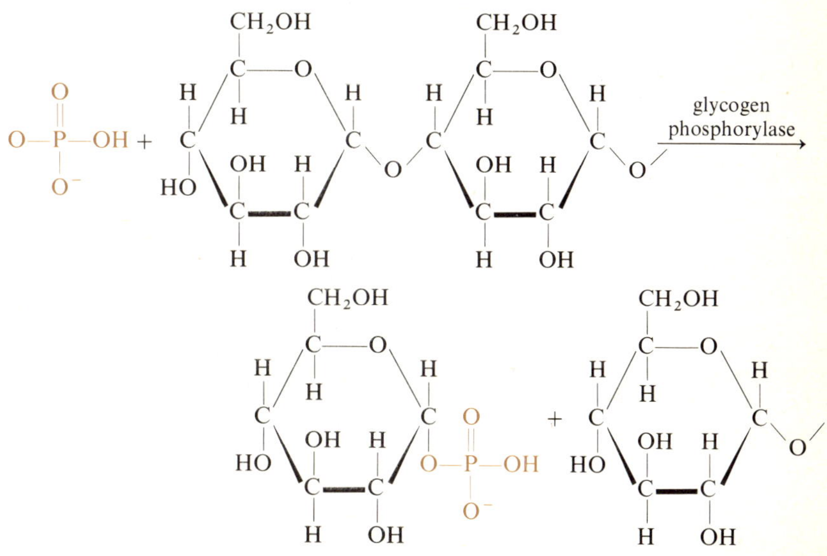

The phosphate is shifted to the 6 position by phosphoglucomutase.

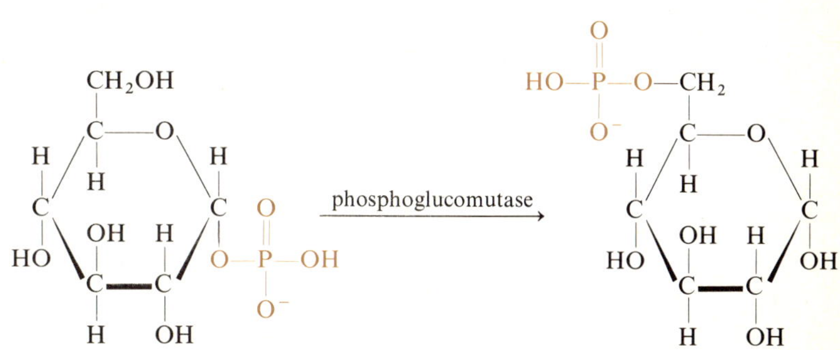

glucose-6-phosphate

Glucose-6-phosphate now enters the regular metabolic scheme.

15.13 Systems for Controlling the Use of Reserves

Because the enzyme glycogen phosphorylase is stored with the glycogen, we require some controls. Otherwise our carbohydrate reserve will be used up and will not be available when we need it. How can the enzyme be designed to respond only when required? Consider what happens when we face an emergency. The adrenals add epinephrine (Chapter 9) to the blood. When epinephrine reaches the muscle, it converts the enzyme adenyl cyclase from the inactive to the active form. This enzyme catalyzes the conversion of ATP to cyclic AMP (Chapter 10) and pyrophosphate.

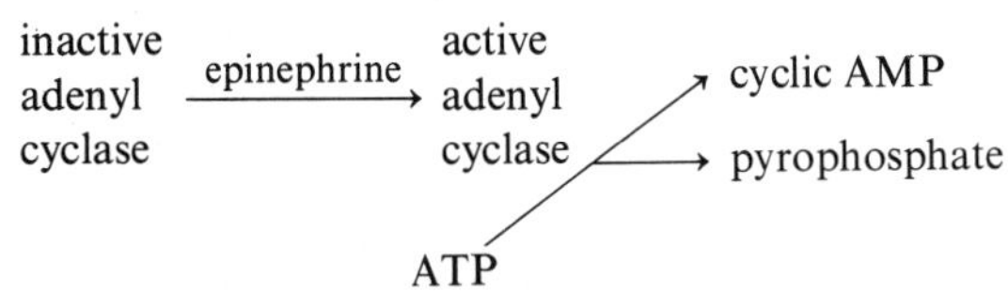

One of the effects of cyclic AMP is to activate the enzyme phosphorylase kinase. Glycogen phosphorylase has been sitting about all this time next to the glycogen but unable to break it down because it has been in the inactive form *phosphorylase b*. Phosphorylase b has two subunits, and in the presence of active phosphorylase kinase and four ATP molecules it joins with another phosphorylase b to become the active phosphorylase a.

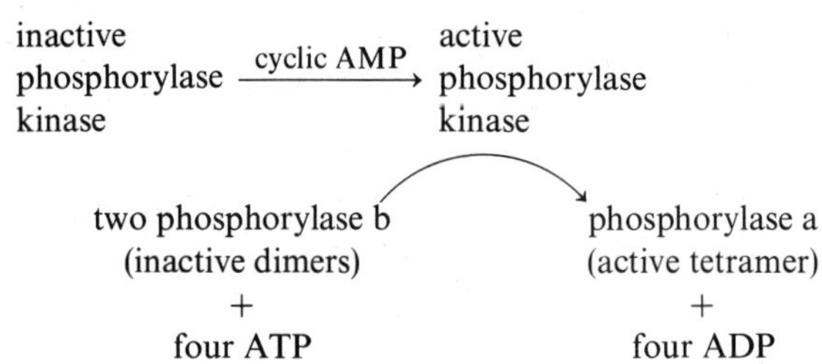

In summary, an emergency triggers the activation of glycogen reserves.

This is not the only way these reserves are activated, however. Suppose you are simply working hard without a traumatic event to stimulate the adrenals to excrete epinephrine. To understand this activation process we must first trace the use of energy reserves available for immediate use in the muscle. The muscle has two high-energy sources, ATP and creatine phosphate. The ATP is used first:

$$\text{ATP} \longrightarrow \text{ADP} + \text{iP}$$

Then creatine phosphate, which also has a high-energy bond, restores the supply of ATP by transferring its own phosphate to the ADP:

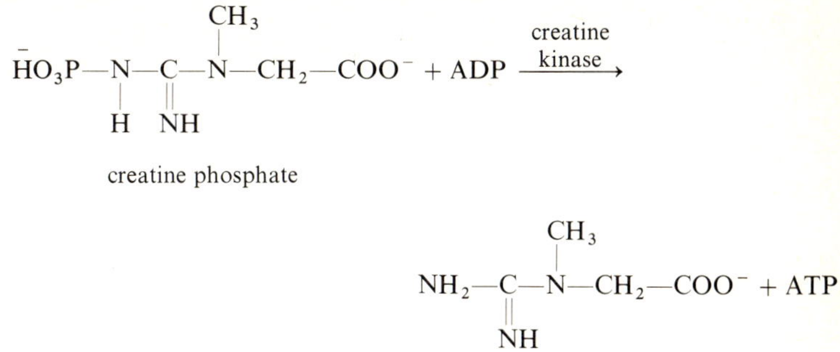

creatine phosphate

creatine

One other reaction increases the supply of immediately available ATP. The enzyme myokinase catalyzes transfer of a phosphate from one ADP to another.

$$\text{ADP} + \text{ADP} \xrightarrow{\text{myokinase}} \text{ATP} + \text{AMP}$$

This allows the utilization of some of the supply of energy represented by the second high-energy bond of ATP. Up to this time, the nucleotides in high concentrations have been ATP and ADP. Now AMP concentrations are increasing. AMP is an allosteric activator of phosphorylase b. This is a different mechanism from the process described earlier, where two phosphorylase b units were joined together to form phosphorylase a. In this activation the phosphorylase b simply binds AMP and becomes active.

$$\underset{\text{(inactive)}}{\text{phosphorylase b} + 2\text{AMP}} \longrightarrow \underset{\text{(active)}}{\text{phosphorylase b} \cdot 2\text{AMP}}$$

In summary, once muscle becomes active and uses the available energy, the level of AMP rises. This AMP specifically activates phosphorylase b, which releases glucose-1-phosphate from glycogen to begin metabolic production of more ATP.

15.14 Control of Glucose Transport

At the beginning of the chapter, we mentioned the link between digestion and metabolism of glucose, transport in the blood. In order to provide a uniform environment for the cells, the concentrations of glucose in the blood are controlled. It is possible to throw this control out of balance by a massive intake of carbohydrates or by fasting, but under normal circumstances glucose levels are very constant.

The blood level of glucose is under the control of hormones. When blood glucose levels are too high, insulin is secreted. Insulin stimulates the transfer of glucose into the tissues, encourages the synthesis of glycogen from glucose, and represses the use of lipids.

When blood glucose is too low, glucagon is excreted, encouraging the breakdown of glycogen and transfer of the glucose so produced into the blood. Insulin and glucagon are peptide hormones produced by different cells in the pancreas. Both exert their effect through increasing or decreasing the cyclic AMP level in the cell (Chapter 9). Epinephrine raises the level of glucose in the blood, as does glucagon, but in response to a stimulus other than existing blood glucose levels.

15.15 Diabetes Mellitus

Diabetes mellitus is a disease resulting from insufficient production of insulin. The problem is very widespread, involving millions of individuals in the United States. Blood glucose levels and consequently urinary glucose levels are elevated. There are likely to be very large volumes of urine. The normal conversion of glucose to body fat is reduced, which, in conjunction with the constant loss of glucose in the urine, leads to weight loss. Tissues turn from metabolizing glucose to metabolizing fat for energy. After a long occurrence of the problem, blood vessel walls degenerate; blindness is a likely result of this degeneration.

Treatment of diabetes involves administration of insulin to establish normal levels and reduction of carbohydrate in the diet. Care must be taken about dosage. Because the body is not controlling the secretion by need, it must accept the dose given. Too much insulin causes a sharp drop in blood sugar, which can lead to coma (insulin shock).

Summary

In this chapter a specific metabolic pathway was introduced in detail (Figure 15.5). The metabolism of glucose was considered from the standpoint first of the sequence of enzyme-catalyzed chemical changes that constitute the pathway, second of the involvement of the cofactors NAD and ATP as they relate to oxygen needs and energy production, third of the mechanisms that control the pathway, and fourth of the energy changes.

Glucose is metabolized by an extended series of steps, beginning with its conversion to a phosphate ester at the expense of ATP. It is converted to fructose and a second phosphate from ATP attached. This diphosphate is split into two triose phosphates. These are oxidized to carboxylic acids, resulting in the reduction of NAD and the formation of ATP. The structure is rearranged and dehydrated to form phosphoenolpyruvate. The phosphate is transferred to ADP, producing ATP and pyruvate, an important intermediate. At this point there is a net gain of two ATP per glucose. In aerobic metabolism the pyruvate is decarboxylated, oxidized, and attached to CoA by a multienzyme complex, producing acetyl CoA. In this pathway the

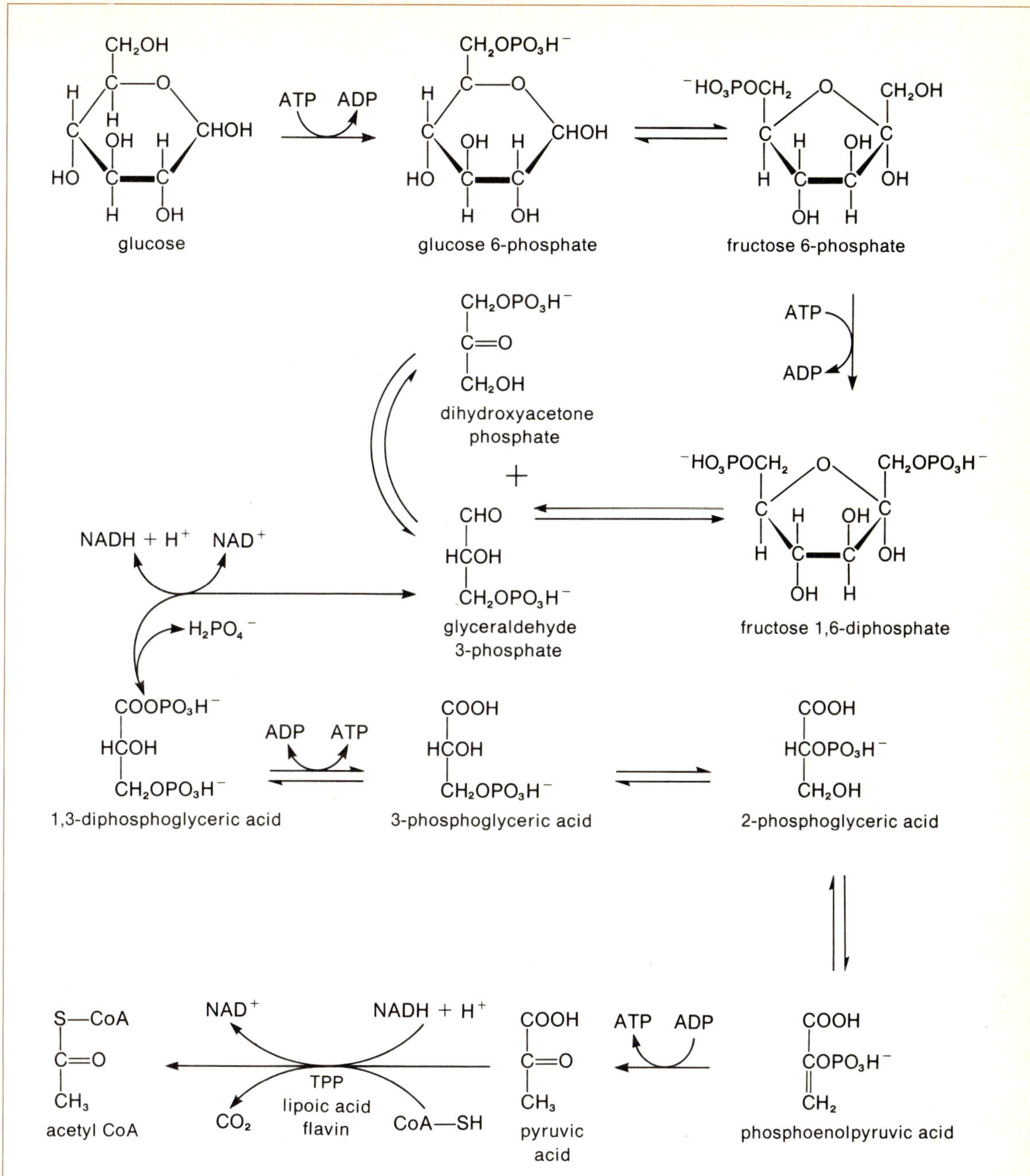

Figure 15.5
The conversion of glucose to acetyl CoA

supply of ATP exerts allosteric control at the formation of fructose diphosphate and acetyl CoA.

Under anaerobic conditions the pyruvate is either reduced to lactate or decarboxylated and reduced to ethanol. In either case the reduction is a mechanism for reoxidizing the NAD reduced earlier. Striated muscle can operate beyond the ability of the organism to deliver oxygen using the anaerobic path to lactate. Glycogen reserves in muscle are mobilized on demand by the enzyme glycogen phosphorylase by one or the other of two pathways. Two molecules of inactive phosphorylase b can be joined together to form active phosphorylase a by a chain of reactions initiated by epinephrine. Inactive phosphorylase b can be converted to active phosphorylase b when the using up of ATP reserves causes the level of AMP to rise. AMP is an allosteric activator of the enzyme.

The concentration of glucose in the blood is under the control of insulin and glucagon. Insufficient insulin production produces a condition called diabetes mellitus.

Problems

1. In each of the following, name the substrate, the product, and the enzyme that catalyzes the reaction. Add any cofactors that have been omitted:

a)

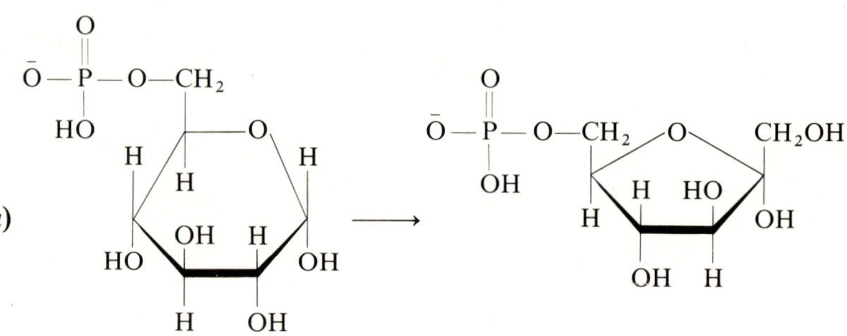

b)

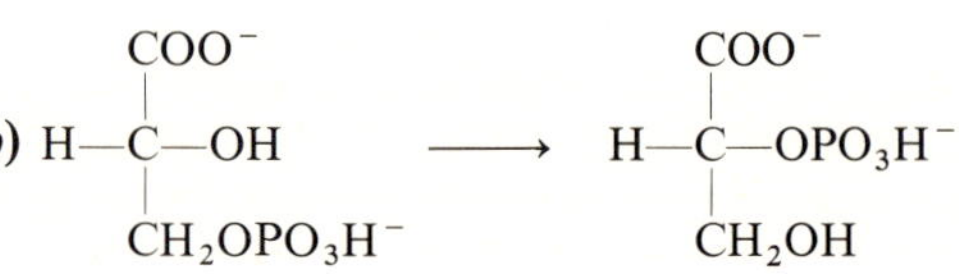

c)

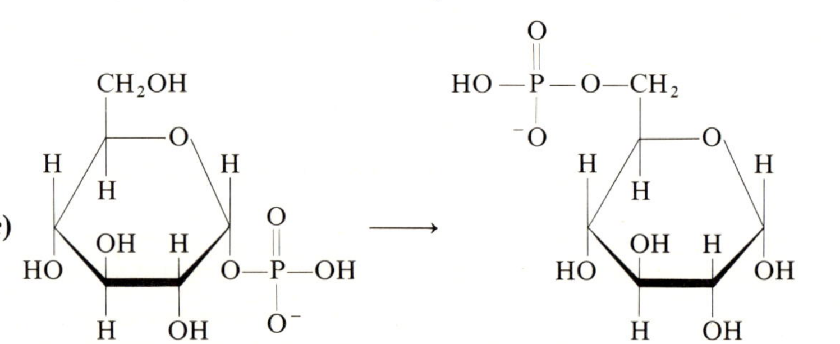

2. Each of the reactions in Problem 1 involves (choose one):

a) Hydrolysis of a glycosidic linkage.

b) Rearrangement of the parts of a structure.

c) Oxidation of the substrate.

d) Addition of phosphate to the structure.

3. You would expect the reactions in Problem 1 to (choose one):

a) Require the input of relatively large amounts of energy.

b) Release relatively large amounts of energy.

c) Not involve a very large energy change.

4. In each of the following, name the substrate, the product, and the enzyme that catalyzes the reaction. Add any cofactors that have been omitted:

a)
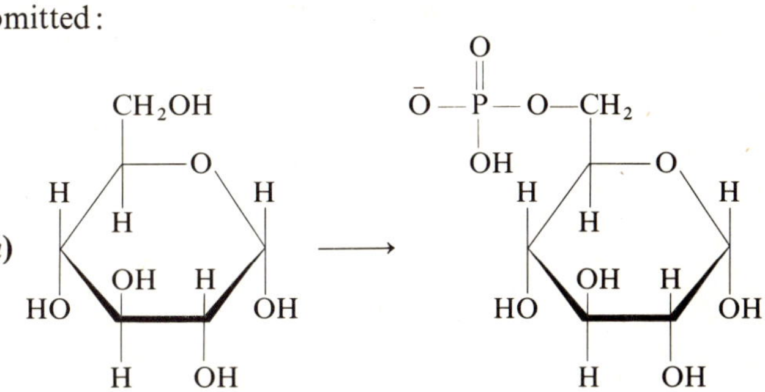

b)
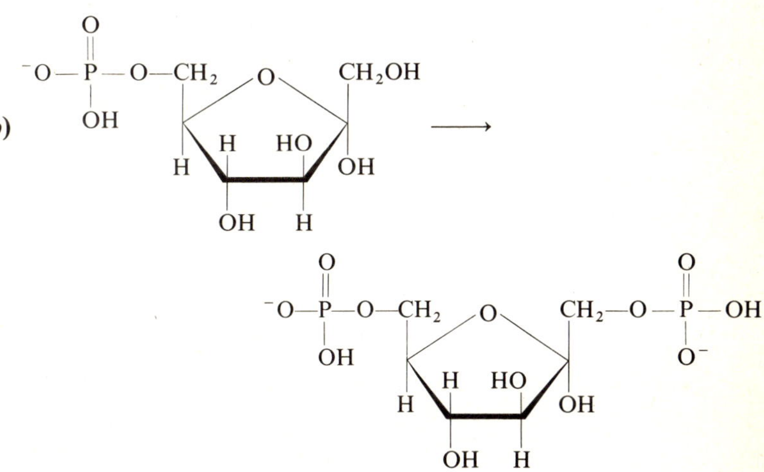

c)
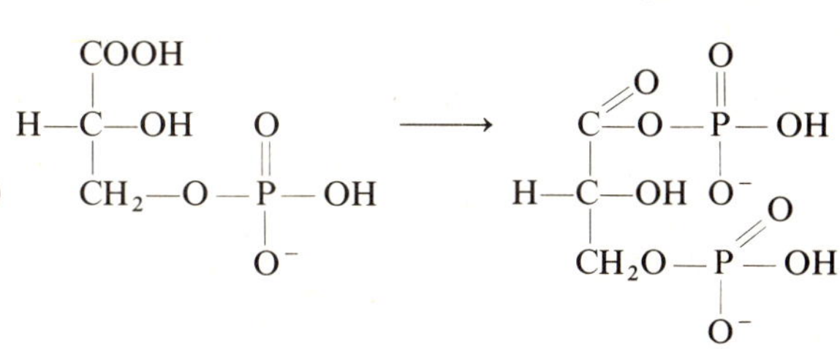

d)
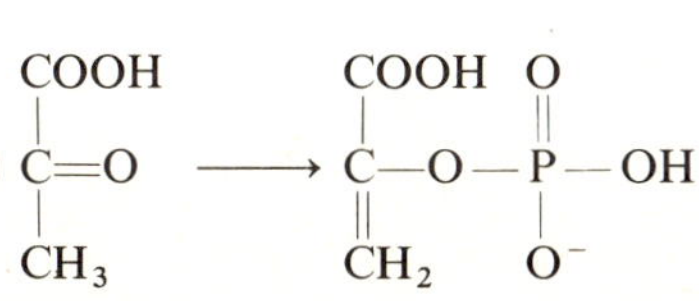

5. Each of the reactions in Problem 4 involves (choose one):

a) Hydrolysis of a glycosidic linkage.

b) Rearrangement of the parts of a structure.

c) Oxidation of the substrate.

d) Addition of phosphate to the structure.

6. Forgetting about changes in cofactors, the changes that are occurring to the substrates of Problem 4 (choose one):

a) Require the input of relatively large amounts of energy.

b) Release relatively large amounts of energy.

c) Do not involve large energy changes.

7. In each of the following, name the substrate, the product, and the enzyme that catalyzes the reaction. Add any cofactors that have been omitted:

a) 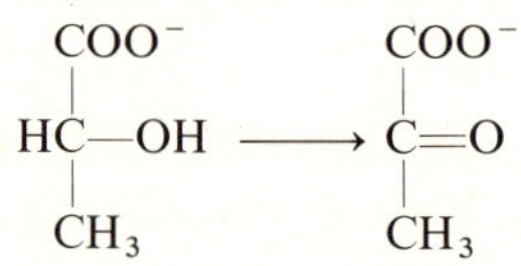

b) 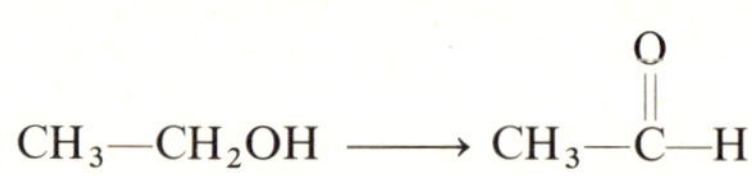

c) 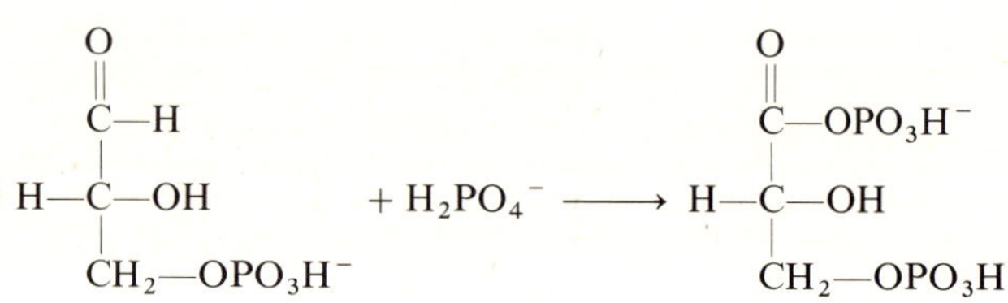

8. Each of the reactions in Problem 7 involves (choose one):

a) Hydrolysis of a glycosidic linkage.

b) Rearrangement of the parts of a structure.

c) Oxidation of the substrate.

d) Addition of phosphate to the structure.

9. The reaction in Problem 7c involved adding phosphate to the structure, as did the reactions in Problem 4.

a) How is 7c different from 4?

b) In each case, what is the source of energy for forming the link to phosphate?

10. It is possible to follow metabolic pathways by use of substrates containing radioactive isotopes at known positions. In fact, this is an important research tool. For example, if a cell were given dihydroxyacetone phosphate containing carbon-14 (labeled) at position 2,

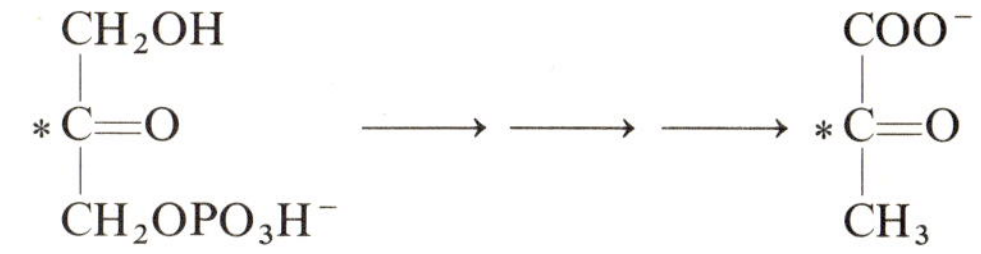

one would expect to see pyruvate also labeled at position 2.

a) In the example, would acetyl CoA be labeled, and if so, where would the carbon-14 be?

b) If glucose labeled at carbon 1 were given to a cell, draw the structure of fructose-1,6-diphosphate with an asterisk at the labeled carbon.

c) Now take the labeled fructose-1,6-diphosphate from part b and show the reaction catalyzed by fructose diphosphate aldolase, indicating where in the products the label would appear.

d) In part c, where would the label appear in pyruvate? Would it appear in acetyl CoA?

e) Repeat parts b–d, starting with glucose labeled at 6, then at 2, then at 4.

11. Indicate the state of glycogen phosphorylase in the muscle of a squirrel caught a long way from its tree by a dog:

a) Before any sign of danger.

b) When the squirrel first sees the dog approaching.

c) After the squirrel has been running for a while.

12. Hexokinase was shown to operate with a sizable release of energy when catalyzing the synthesis of glucose-6-phosphate.

a) What is your prediction for the reaction converting glyceraldehyde to glyceraldehyde-3-phosphate?

b) Would you predict a release of energy for the myokinase reaction?

c) The reaction catalyzed by pyruvate kinase proceeds in the direction of pyruvate and ATP formation with a release of energy. What does this say about the energy released when the phosphate linkage in phosphoenolpyruvate is hydrolyzed, compared with that for the end phosphate in ATP?

13. These are the energy changes associated with a portion of the pathway:

a) glyceraldehyde-3-phosphate + $NAD^+ \rightarrow$ 1,3-diphosphoglycerate + $NADH + H^+$ $\Delta G^0 = +1500$ cal/mole

b) ADP + 1,3-diphosphoglycerate → 3-phosphoglycerate + ATP $\Delta G^0 = -4500$ cal/mole

c) 3-phosphoglycerate → 2-phosphoglycerate $\Delta G^0 = +1060$ cal/mole

d) 2-phosphoglycerate → phosphoenolpyruvate $\Delta G^0 = +440$ cal/mole

e) ADP + phosphoenolpyruvate ATP + pyruvate $\Delta G^0 = -7500$ cal/mole

1. Which of the above reactions would run well as isolated reactions?
2. What is the overall energy change from glyceraldehyde-3-phosphate to pyruvate?

14. What do you anticipate happens to (1) the concentration of glucose in the blood and (2) the amount of glycogen in storage when:

a) The level of insulin increases.

b) The level of glucagon increases.

c) The level of epinephrine increases.

16
Fatty Acid Oxidation

Objectives

In this chapter we:

1. Look at the uptake, transport, and storage of triglycerides.
2. Consider the relationship of structure to function in the cell mitochondria.
3. Learn the pathway by which saturated fatty acids are metabolized to acetyl CoA.
4. Learn the two special pathways by which unsaturated fatty acids are metabolized.
5. Summarize the total changes occurring in these processes.

Triglycerides are an important component of our diets. Because they are not polymeric, they do not require digestion to monomers for absorption as do the proteins and polysaccharides. However, they are not water soluble, so they require detergents to emulsify them in the intestine. Bile salts (Chapter 8) are added to the intestine during digestion. Other surface-active agents are generated in the intestine. For example, phosphoglycerides are enzymatically converted to lysophosphoglycerides and free fatty acids, both of which serve as detergents (Chapter 8).

There are lipases, enzymes that hydrolyze triglycerides into fatty acids and glycerol, in the intestine. However, this hydrolysis is only temporary. Once these products have crossed the intestinal mucosa, they are reassembled into triglycerides. Absorbed triglycerides do not enter the blood directly but are first transported in the lymph, entering the blood finally as chylomicrons (Chapter 7). After a fatty meal, the blood becomes quite milky with these suspended fat droplets. The triglycerides are transported to fat cells in the adipose tissue. On the surface of these cells, lipases convert the triglycerides into fatty acids and glycerol. These enter the cell, are converted back to triglyceride, and are stored in the fat droplet that makes up about 90% of the fat cell. The fat cell also converts glucose from the blood into triglycerides for storage.

In response to the effect of such hormones as insulin and epinephrine on levels of cAMP (Chapter 15), lipases convert triglycerides back to fatty acids and glycerol when metabolic fuel is needed. The fatty acids enter the blood and are transported to tissues attached to serum albumin.

In summary, dietary triglycerides are emulsified in the intestine and transported by the lymphatic system and blood to fat cells. These cells store triglycerides until needed, then release fatty acids in response to hormones to be transported on serum albumins to tissues for metabolism.

Exercise 16.1

Review the history of a triglyceride from the intestine until the time the fatty acid is ready to be metabolized in a heart muscle cell. How many times is the triglyceride hydrolyzed, then reassembled, during that journey?

16.1 Fatty Acid Oxidation

In the previous chapter, the oxidation of carbohydrate was carried as far as the formation of acetyl CoA. In this chapter we do the same for fatty acids. The fate of the long fatty acid molecules is similar to that of a long loaf of French bread: It is shortened a slice at a time until the entire loaf is used. The long fatty acids are oxidized two carbons at a time, each two-carbon unit eventually becoming an acetyl group in acetyl CoA until the entire fatty acid has become acetyl groups. But before this pathway is outlined, it is worthwhile to take a look at the energy factory of the eukaryotic cells in which this process occurs, the mitochondrion.

Figure 16.1
The mitochondrion
Seen at very high magnification, the inner membrane, folding into the interior of the structure, shows clearly. The darker contents are matrix. (Photo by Johanna Huang, Electron Microscopy Laboratory, Eastern Michigan University.)

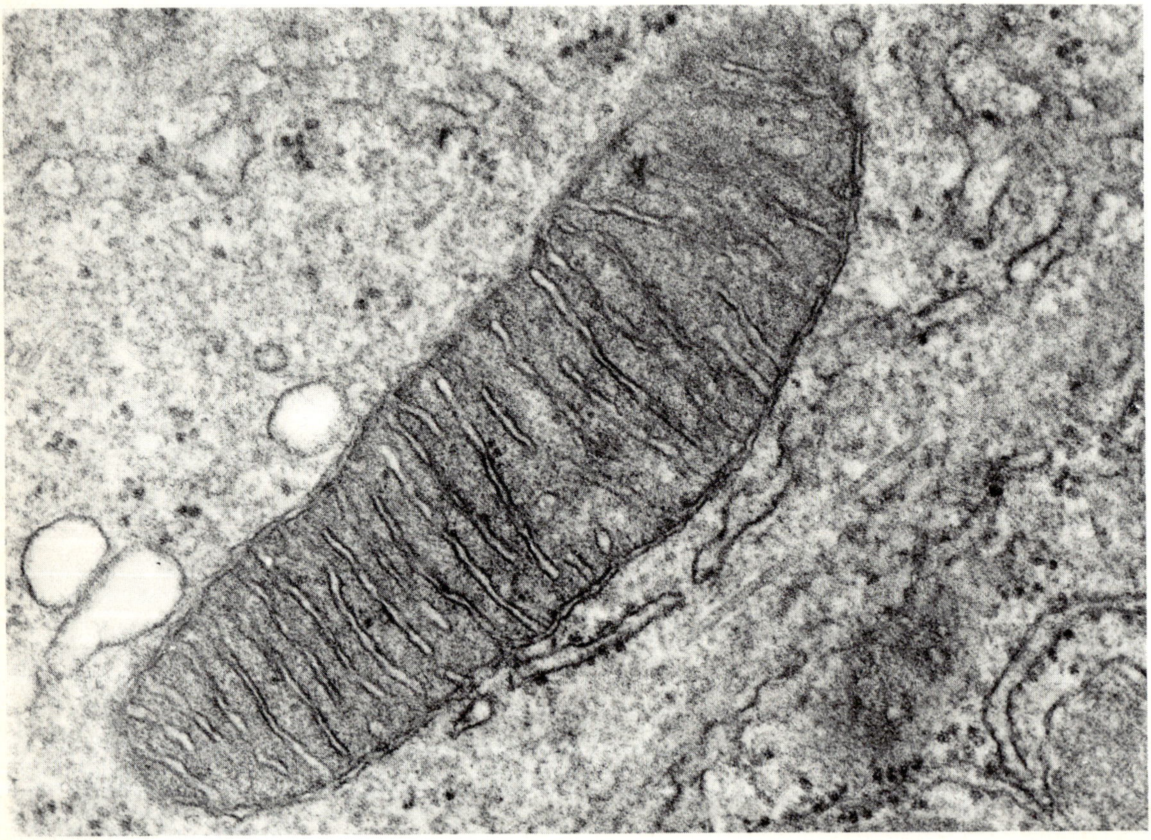

16.2 Mitochondria

Mitochondria are small, double-walled structures varying from spherical to long and thin in shape (Figure 16.1). They are suspended in the cytoplasm of the cell. The more active a cell is in metabolizing food for energy, the more mitochondria it has. Of greatest interest to us is the innermost compartment of the mitochondrion, the matrix, and the inner membrane, which encloses the matrix. Only a few compounds may cross this inner membrane, and those compounds that do enter the matrix must be counterbalanced by compounds moving out. In this fashion the matrix remains quite constant in general composition. For example, whenever a molecule of ATP produced in the matrix leaves to serve the surrounding cell, a molecule of ADP must enter the matrix at the same time.

Chapter 15 is described one reaction that occurs in the mitochondrial matrix. Derivatives of CoA cannot cross the inner membrane, but pyruvate can. The pyruvate produced in carbohydrate metabolism, therefore, enters the matrix before being converted to acetyl CoA by the multienzyme complex pyruvate dehydrogenase. As inferred in the previous chapter, a sizable amount of ATP is generated from acetyl CoA in the matrix.

16.3 Triglyceride Fuel: Entry into the Mitochondrial Matrix

The fatty acids must be "activated" before they can be metabolized. By activated, we mean that they must be converted to thioesters of CoA. Energy for forming the bond between the acid and the CoA is provided by ATP. Here it is not the end phosphate-to-phosphate high-energy bond of ATP that breaks. It is the other phosphate linkage, which results in pyrophosphate being released as a product. This same type of reaction was seen in Chapter 13 when the amino acids were "activated." We will use stearic acid as the example.

$$\underset{\text{fatty acid}}{CH_3(CH_2)_{16}COOH} + CoASH + ATP \longrightarrow$$

$$\underset{\text{fatty acyl CoA}}{CH_3(CH_2)_{16}C(=O)—SCoA} + AMP + \text{pyrophosphate}$$

This occurs in the cytoplasm. There is no transport system into the matrix of the mitochondrion for an acyl CoA derivative. This problem is solved by transferring the fatty acid to an alcohol, carnitine, and transferring the resulting ester into the matrix. In the matrix the carnitine is again replaced by CoA. If sometime you are tempted to

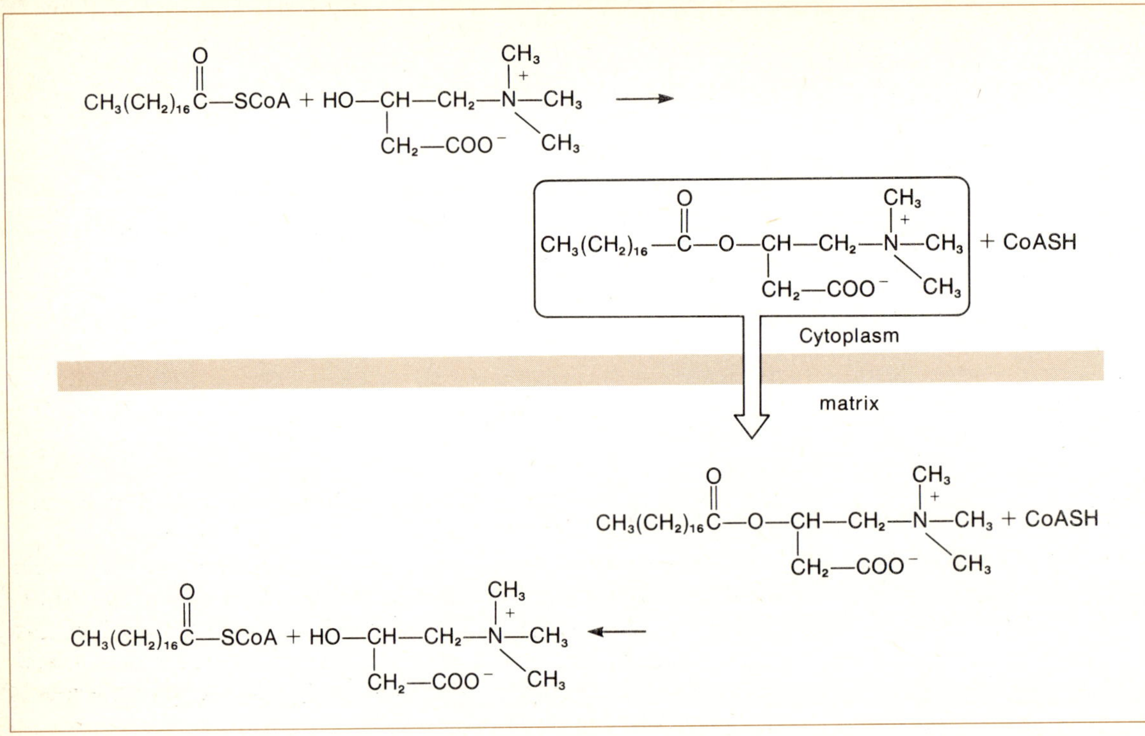

Figure 16.2
Entry of fatty acyl CoA into the mitochondrial matrix

describe the chemistry of life as being brilliantly straightforward and efficient, remember this process (Figure 16.2).

16.4 *In the Matrix: the Cyclic Oxidation Process*

Once in the matrix, a trans double bond is generated in the fatty acid, the hydrogens being transferred to an FAD cofactor bound to an enzyme.

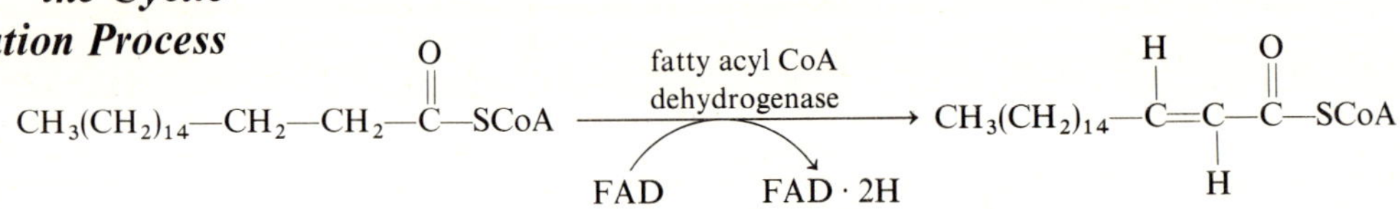

Water is added to the double bond to create the L-β-hydroxy fatty acyl CoA.

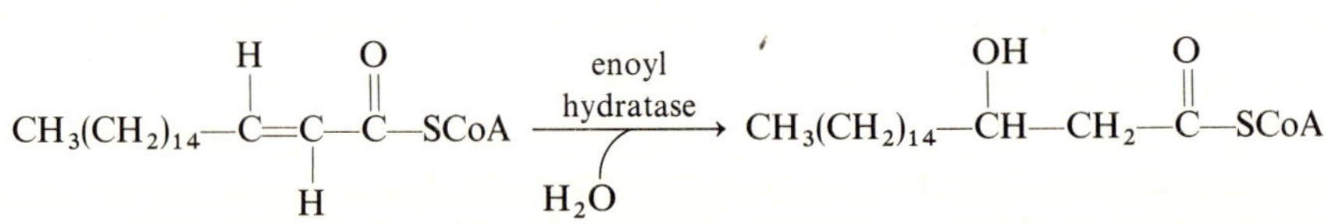

Next the second oxidation step is performed. The alcohol group is converted to a keto group with the hydrogens being transferred to NAD^+. The product is a β-keto fatty acyl CoA.

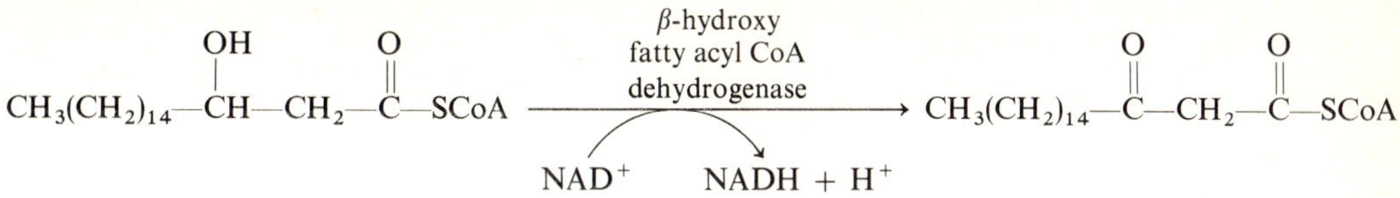

Now we are ready for the final step. The bond between the α and β carbons is broken by adding the hydrogen of CoASH to the α carbon and the rest of the molecule to the β carbon. The product is acetyl CoA and a fatty acyl CoA that is similar to the original fatty acyl CoA except that it is two carbons shorter.

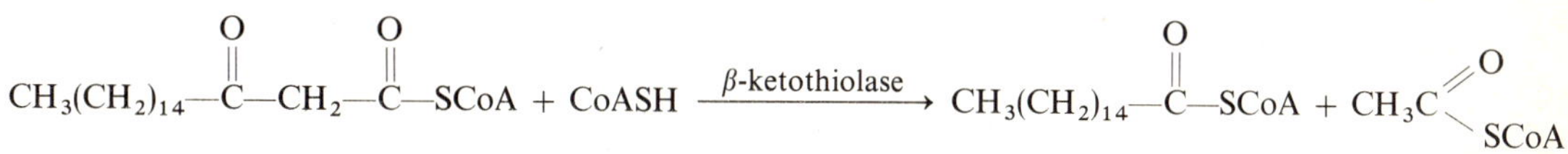

Here we started with stearyl CoA (eighteen carbons) and have now palmatyl CoA (sixteen carbons).

Now the process begins again with the shortened fatty acid. At the end of the next pass through the system, the products will be acetyl CoA and a fatty acyl CoA with two fewer carbons. The final trip through the sequence occurs when the original fatty acid has been reduced to four carbons. The product of that pass is two molecules of acetyl CoA (Figure 16.3).

When unusually large numbers of molecules are being metabolized, the liver sometimes removes the CoASH from the β-keto or β-hydroxy butanoyl CoA intermediates. The free acids are then released into the blood. In peripheral tissues they reenter the mitochondria for energy metabolism. However, some portion of them are removed from

Figure 16.3
Metabolism of butanoyl CoA

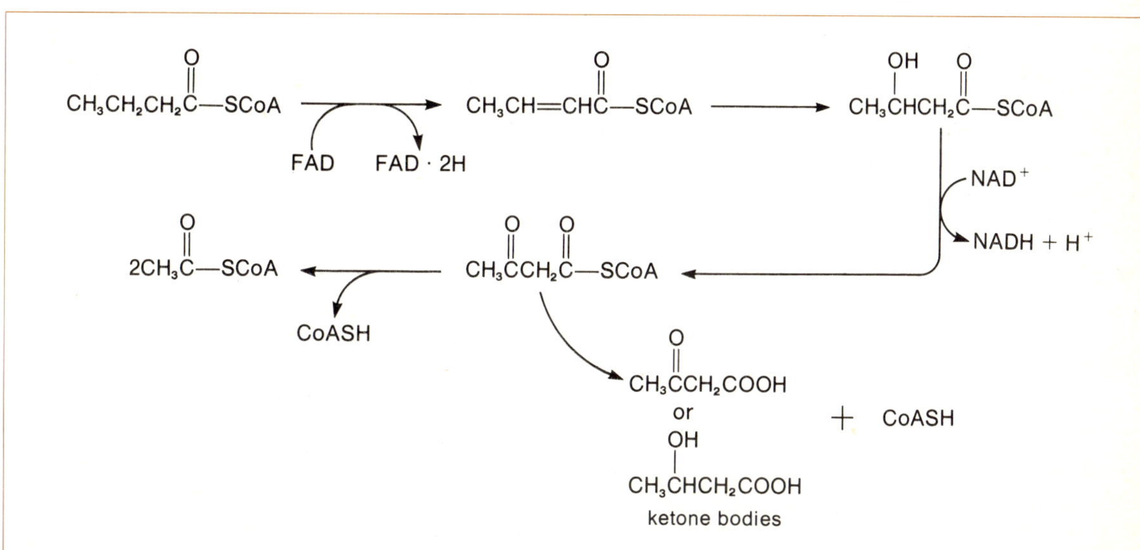

the blood by the kidneys and appear in the urine. Here they are termed *ketone bodies*. Ketone bodies in the urine are characteristic of diabetes patients or individuals who are fasting because of their heightened level of fatty acid metabolism.

16.5 Unsaturated Fatty Acids

The example has been worked through for a saturated fatty acid. Suppose the fatty acid has one or more double bonds. As the chain is metabolized, when the double bond is reached, can we simply add water using enoyl hydratase? We could if the double bond were trans -2,3. However, fatty acid double bonds are usually cis, and the bond could be either -2,3 or -3,4 when it is reached. As a result, we need one mechanism to handle a cis double bond in the correct location and another to handle a cis double bond one position away from the correct location.

Linoleic acid is an example requiring the use of both mechanisms. It is first converted to a CoA thioester. In the matrix the two-carbon segments are removed as acetyl CoA as far as possible (Figure

Figure 16.4
Metabolism of linoleic acid: part 1

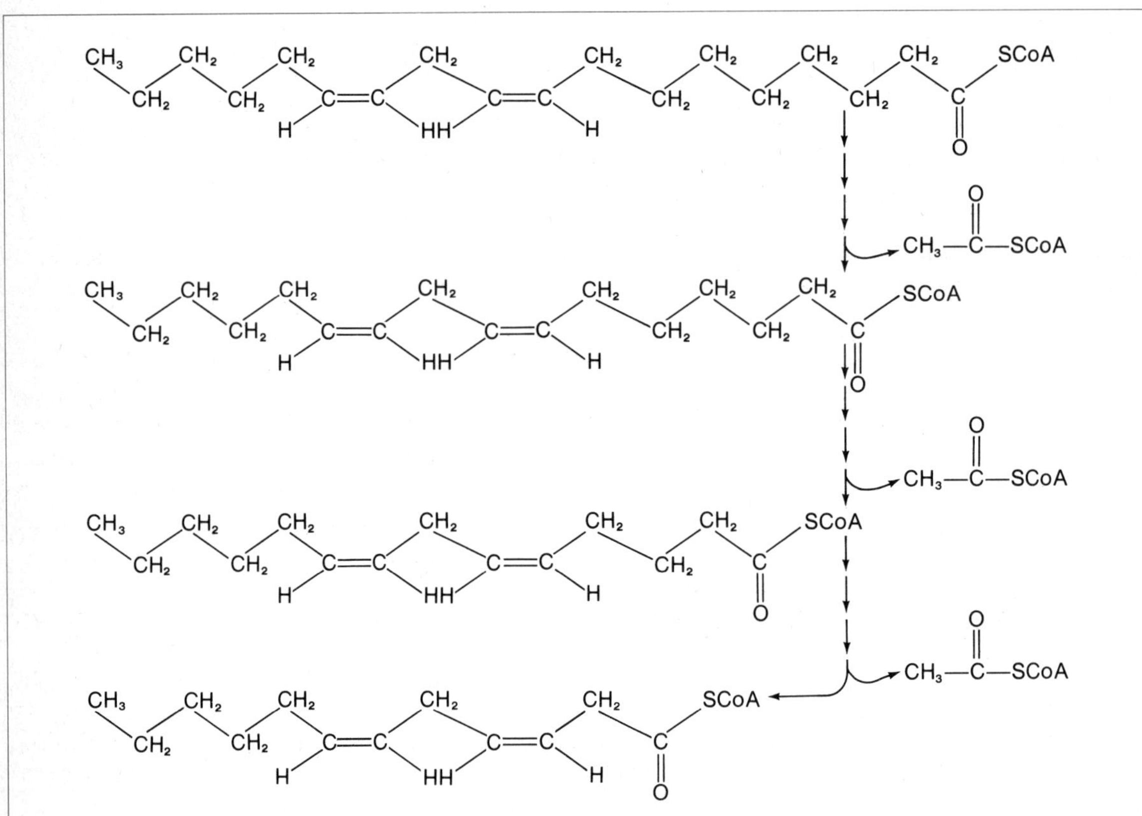

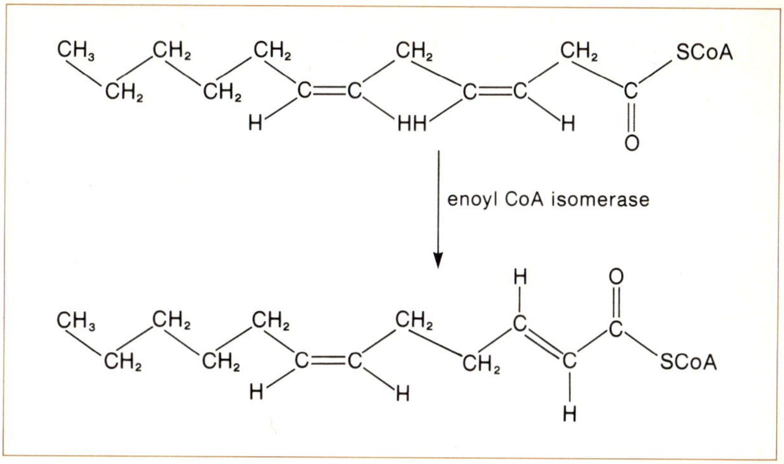

Figure 16.5
Metabolism of linoleic acid: part 2

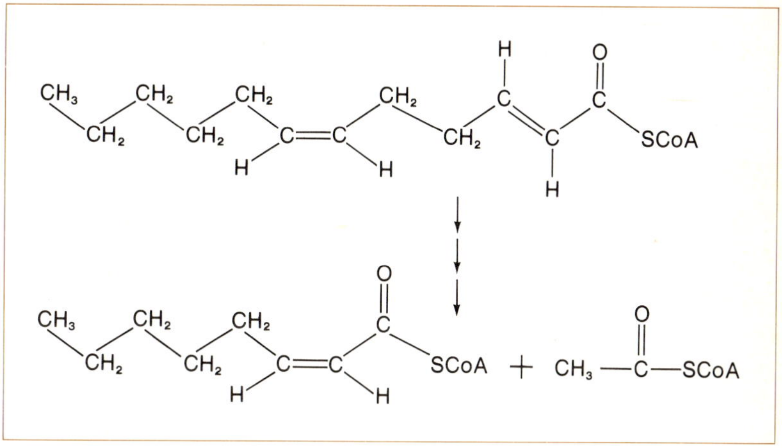

Figure 16.6
Metabolism of linoleic acid: part 3

16.4). Because of the double bond, the next series of reactions encounters difficulty. The double bond is cis -3,4. The double bond normally produced in the pathway is trans -2,3. There is an enzyme, enoyl CoA isomerase, that moves electrons and hydrogens to convert the cis -3,4 bond to trans -2,3 (Figure 16.5). The regular pathway can now handle the molecule to the point of removing one more acetyl CoA (Figure 16.6).

Once again there is an incorrect structure. This time the double bond is cis -2,3. The enoyl CoA isomerase cannot solve this problem. The enoyl hydratase of the regular pathway can add water to this cis double bond, but the product is a D-β-hydroxy fatty acyl CoA rather than the desired L isomer. The β-hydroxy fatty acyl CoA dehydrogenase cannot oxidize the D isomer. However, there is an enzyme, β-hydroxyacyl CoA epimerase, that converts the D isomer to the L isomer. The L product is then metabolized normally (Figure 16.7).

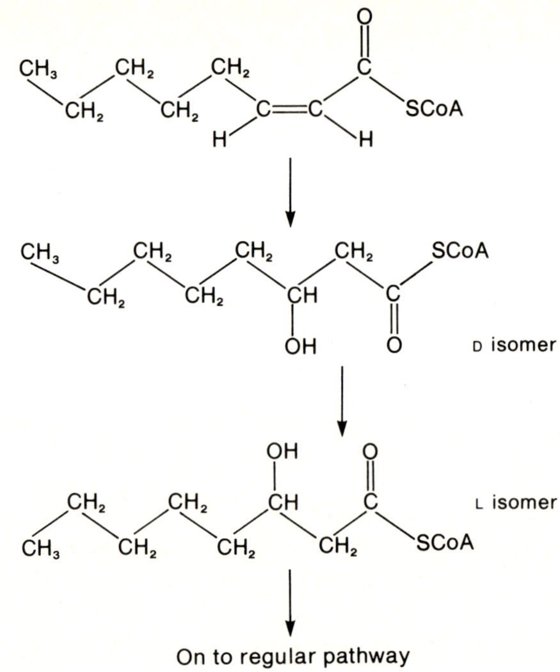

Figure 16.7
Metabolism of linoleic acid: part 4

16.6 Summarizing Cofactor Changes

Go back to the example of stearic acid (C_{18}). What is the result of a molecule of stearic acid going completely through the oxidative pathway?

First an ATP was used to add CoASH to the fatty acid.

$$CH_3(CH_2)_{16}\overset{\overset{\large O}{\|}}{C}—OH + CoASH + ATP \longrightarrow CH_3(CH_2)_{16}\overset{\overset{\large O}{\|}}{C}—SCoA + AMP + \text{pyrophosphate}$$

Then:

$$CH_3(CH_2)_{16}—C\begin{matrix}\nearrow O \\ \searrow SCoA\end{matrix} \longrightarrow 9CH_3C\begin{matrix}\nearrow O \\ \searrow SCoA\end{matrix}$$

This required eight passes through the pathway. (Why not nine? Remember that the last pass produces two acetyl CoA molecules.) Each of these passes causes an FAD and an NAD to be reduced.

$$8FAD\,(+\ 16H) \longrightarrow 8FAD \cdot 2H$$

$$8NAD^+\,(+\ 16H) \longrightarrow 8NADH + 8H^+$$

Adding all these together we have

$$CH_3(CH_2)_{16}\text{—COOH} + \text{ATP} + 9\text{CoASH} + 8\text{FAD} + 8\text{NAD}^+ \longrightarrow 9CH_3\overset{\overset{\displaystyle O}{/\!/}}{C}\text{—SCoA} + \text{AMP} + \text{pyrophosphate} + 8\text{FAD}\cdot 2\text{H} + 8\text{NADH} + 8\text{H}^+$$

Have we generated energy? Not in the form of ATP. However, we have used much of the energy released as the fatty acid was oxidized to reduce the cofactors. These will be used later as a source of energy for making ATP. Also, we are not finished oxidizing the fatty acid. We still have the acetyl CoA in the mitochondrial matrix, and it will be oxidized further.

Summary

Fatty acid metabolism occurs in the inner compartment or matrix of mitochondria. Access to this matrix is restricted to a relatively small number of compounds. Fatty acids are first converted to thioesters of CoASH, ATP energy being used for this synthesis. The CoA is temporarily replaced with carnitine to allow the fatty acid to enter the matrix. The enzymes in the matrix then create a -2,3 double bond in the fatty acid, add water to generate a hydroxy structure, oxidize the hydroxyl group to a keto group, and then add CoASH across the -2,3 carbon-to-carbon bond to produce acetyl CoA and the CoA thioester of a fatty acid two carbons shorter than the original structure. This process is repeated until all the fatty acid carbons are part of acetyl CoA (Figure 16.8). If during this process a cis -3,4 double bond is encountered, it is converted enzymatically to a trans -2,3 bond, and metabolism continues. If a cis -2,3 bond is encountered, water is added, generating the unwanted D isomer of the alcohol. This is converted to the desired L isomer enzymatically, and metabolism continues.

When the liver is metabolizing fatty acids very actively, as occurs in fasting or diabetes, β-hydroxy and β-keto butanoic acids are released into the blood. As these appear in the urine, they are termed ketone bodies.

The end result of fatty acid metabolism is a great deal of acetyl CoA and much reduced flavin and NAD.

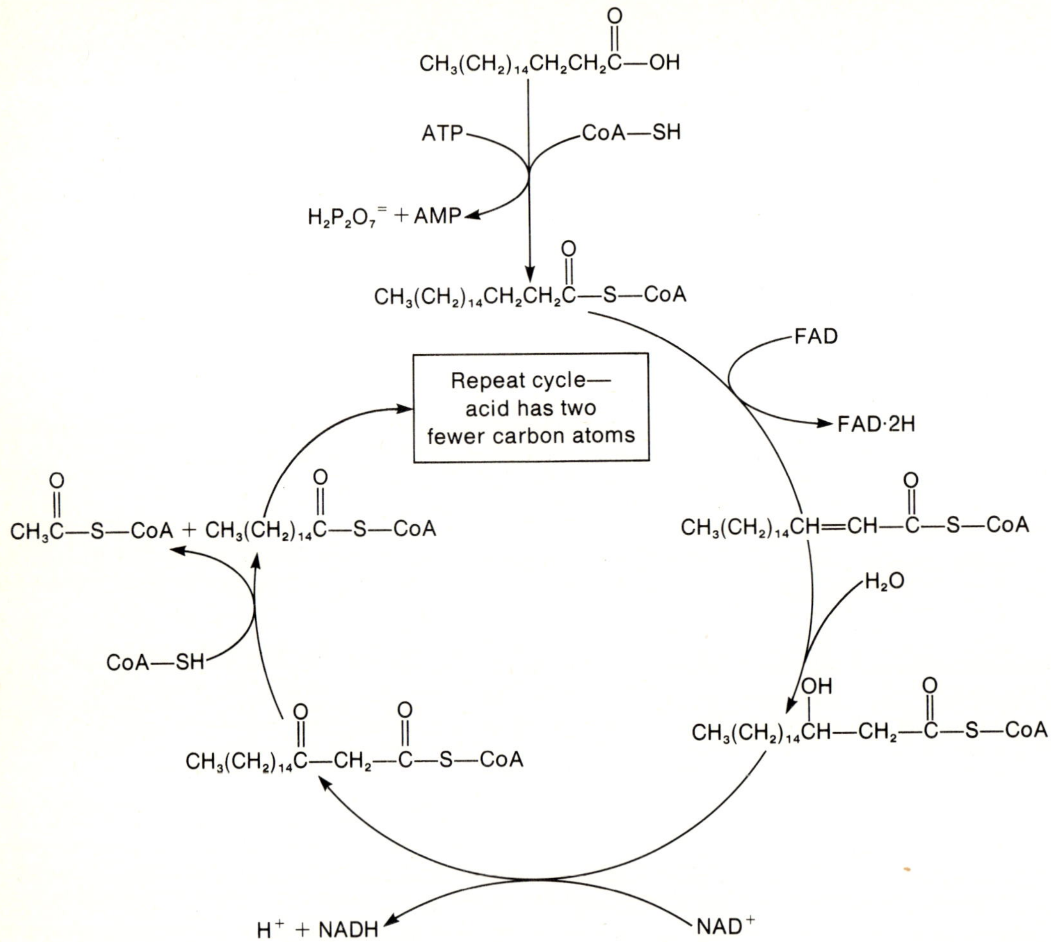

Figure 16.8
Fatty acid metabolism
This summary of the most important reactions in fatty acid metabolism emphasizes the cyclic nature of the process.

Problems

1. Write the reactions for the conversion of butanoyl CoA to β-keto butanoyl CoA. Label each reaction as an addition reaction, an elimination reaction, or a substitution reaction. In which steps is the substrate oxidized?

2. Draw the structure of the product of the logical reaction for which each of these compounds would be the substrate.

a) 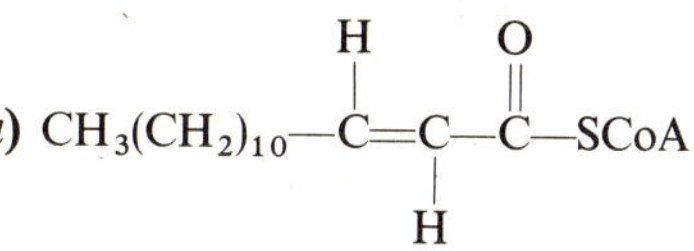

b) 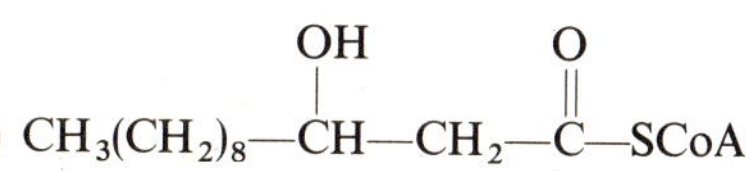

L isomer

c) 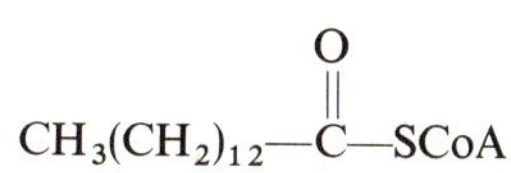

d) 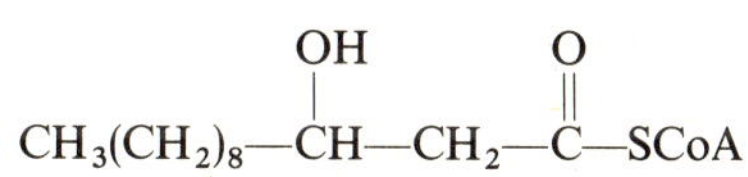

D isomer

e) 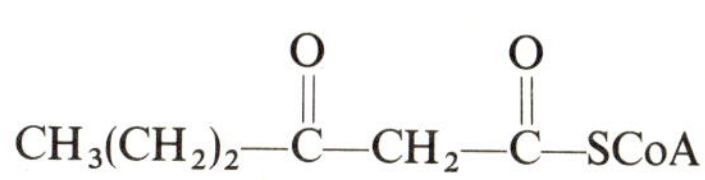

f) 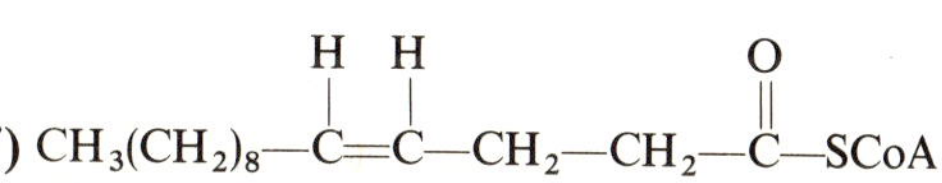

g) 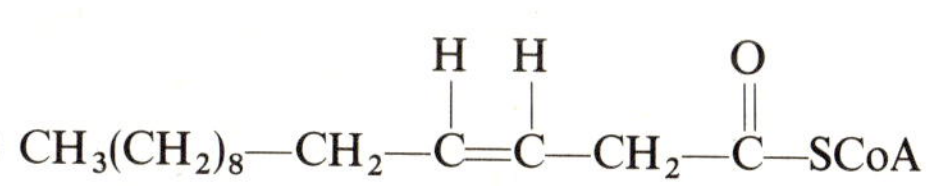

h) 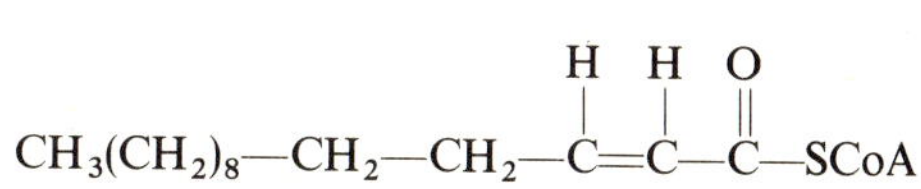

i) 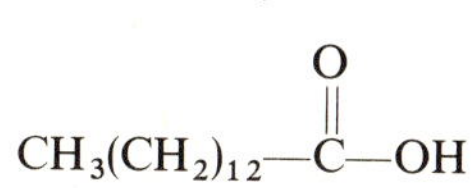

3. Now let us play a game. We will score 3 points every time an NAD is reduced, 2 points every time an FAD is reduced, and 1 point every time a nucleoside triphosphate is formed, and deduct 1 point every time an ATP is used. Form two teams, each with a six-carbon "hand." One hand is a glucose molecule, the other hexanoic acid. Go! After Chapter 18 this game should take on new meaning.

4. In this chapter the production of ketone bodies is described as characteristic of diabetes mellitus. Correlate this with a statement in the discussion of diabetes in Chapter 15.

Challenge Problems

5. Fatty acids are synthesized in the cytoplasm from acetyl CoA. This is why fatty acids normally have an even number of carbons. The pathway involves a completely different set of enzymes from fatty acid catabolism. Before the two-carbon acetyl structure is added to the growing fatty acid, it is first converted to malonyl CoA.

$$CH_3-\overset{\overset{\displaystyle O}{\|}}{C}-SCoA + CO_2 \longrightarrow {}^{-}OOC-CH_2-\overset{\overset{\displaystyle O}{\|}}{C}-SCoA$$

Referring, if necessary, to Chapter 12, decide what cofactor or cofactors would be likely to accompany this reaction.

6. The malonyl group is transferred from coenzyme A to the fatty acid synthesis complex. The attachment is by a thioester linkage, just as it was to coenzyme A, but to an amino acid R group of a carrier protein in the complex. What amino acid is this? Draw the amino acid residue and the malonyl group as they would be joined.

7. Just before the malonyl group is added to the growing fatty acid, both the growing fatty acid and the malonyl group are attached by thioester linkages to the fatty acid synthesis complex, a cluster of enzymes similar to but larger than the pyruvate dehydrogenase complex (Chapter 15). They attach to different proteins in the complex. Perhaps you will be surprised to learn that the growing fatty acid is transferred onto the malonyl group (the mountain coming to Mohammed) rather that the other way around. The fatty acid displaces the malonyl carboxyl group, which leaves as CO_2. Complete the reaction as described:

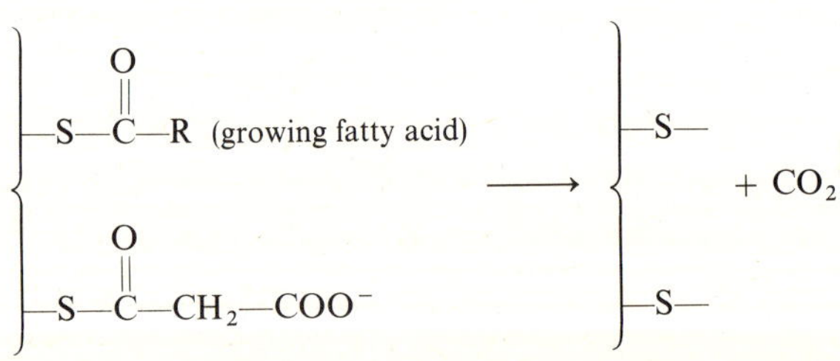

8. At this point we have a β-keto fatty acyl thioester. The reactions to convert the keto group to a methylene group (—CH_2—) are the same as to convert a methylene group to a keto group in fatty acid oxidation. However, they are in reverse order. The flavin and NAD cofactors of fatty acid oxidation are replaced by NADPH. Write the intermediate steps of the sequence:

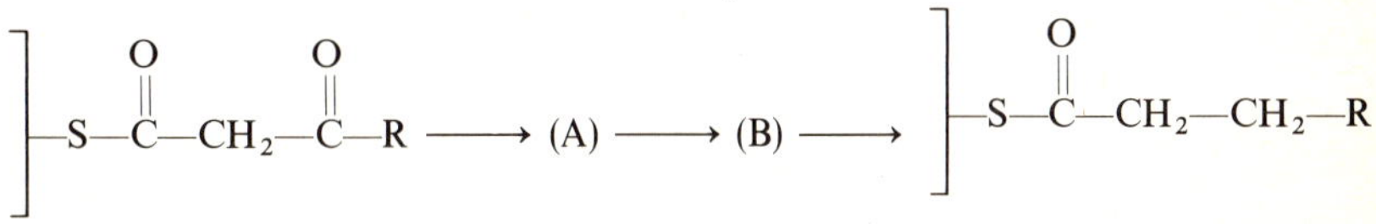

9. In the final step of the sequence of chain extension, the growing fatty acid transfers from one thiol group to the other:

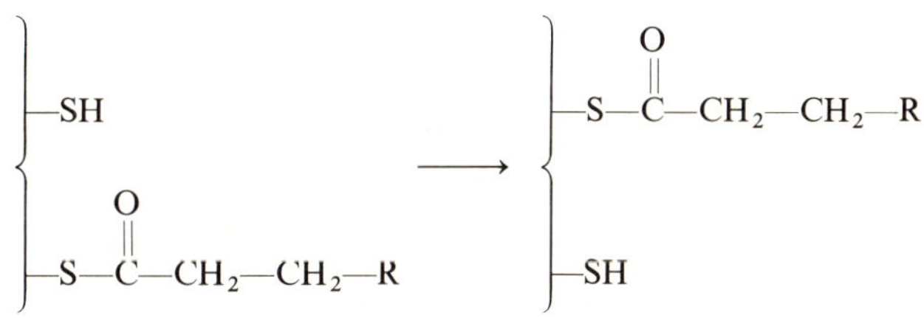

What happens next?

17
The Krebs Cycle

Objectives

In this chapter we:

1. Learn the reactions by which the acetyl group of acetyl CoA is converted to CO_2.
2. Summarize the cofactor changes accompanying the formation of CO_2 from an acetyl group.

We have looked at the metabolism of carbohydrates and fatty acids, in each case carrying the process as far as the creation of acetyl CoA.

What good is acetyl CoA?

We write the overall equations for oxidation of stearic acid or glucose as follows:

$$\underset{\text{stearic acid}}{C_{18}H_{36}O_2} + 26O_2 \longrightarrow 18CO_2 + 18H_2O$$

$$\underset{\text{glucose}}{C_6H_{12}O_6} + 6O_2 \longrightarrow 6CO_2 + 6H_2O$$

How are the CO_2 and H_2O formed?

Of the $6CO_2$ molecules that a glucose should produce, we have made only two. We are worse off with the stearic acid, having made no CO_2 as yet. Oxygen hasn't been involved at all, and we have not generated water. In this chapter we fill in a part of the above equations—the formation of CO_2. The pathway by which this is accomplished is called the *tricarboxylic acid cycle*, or *citric acid cycle*, because it begins with the tricarboxylic acid citrate, or it is called the *Krebs cycle*, in honor of Hans Krebs, who first proposed it.

17.1 A Cyclic Process

If we stand at a distance and get an overview of the Krebs cycle in operation, we see the two-carbon acetyl group being taken into a revolving metabolic wheel. As in a water wheel, the acetyl group enters at one point and is dumped out later in the revolution. Unlike a water wheel, which does not change the water, the acetyl group is chemically altered. When the two carbons are dumped out, they have become two molecules of carbon dioxide. In addition, at several points cofactors are *reduced.* By now, "reducing cofactors" should translate into "structures to which energy is being added." At another point we make GTP, the equivalent of ATP. Glancing about at our surroundings, we see that all this is happening in the matrix of a mitochondrion. Now let us move up close enough to see the process in detail.

17.2 The Reactions of the Krebs Cycle

Fatty acyl CoA was transported into the matrix at the start of fatty acid metabolism, and pyruvate from carbohydrate entered the matrix before it was converted to acetyl CoA. The metabolism of the acetyl CoA thus begins in the matrix. Acetyl CoA is combined with oxaloacetate to form citrate.

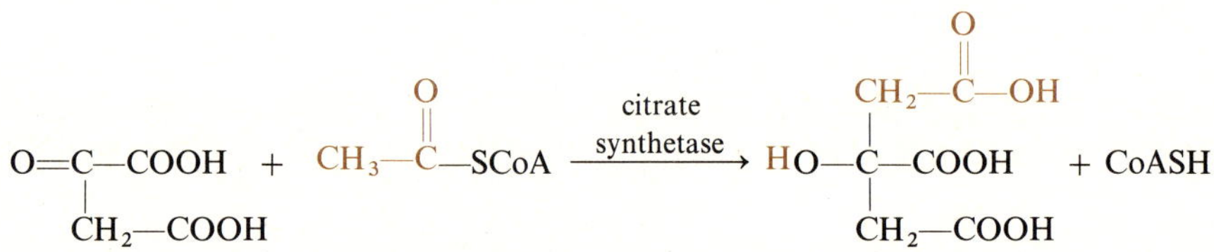

oxaloacetate acetyl CoA citrate

The alcohol group is shifted by the enzyme aconitase, producing isocitrate.

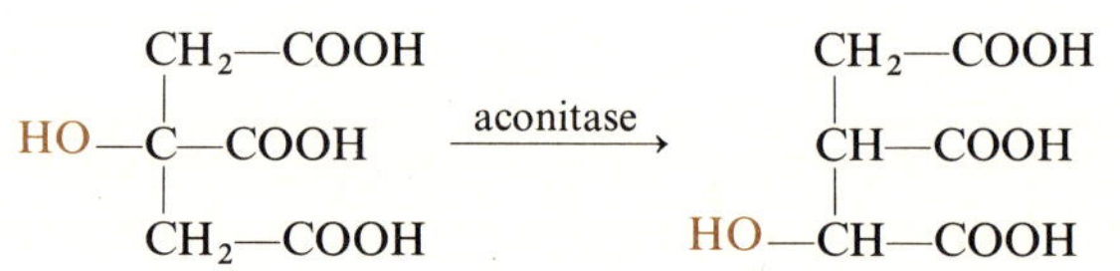

citrate isocitrate

We now release the first of the two carbon dioxides, and at the same time we oxidize the alcohol group to a carbonyl group. As the alcohol is oxidized, NAD^+ is reduced to $NADH + H^+$

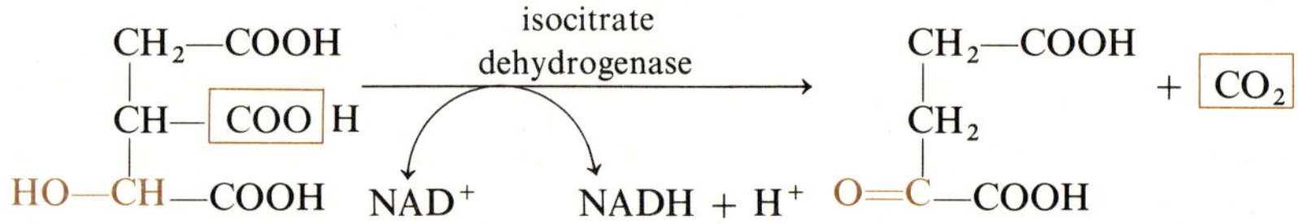

isocitrate α-ketoglutarate

The next step releases the second carbon dioxide. It is a complex process very similar to the step in which pyruvate is converted to acetyl CoA. The same cofactors are involved, and the catalysis is performed by a multienzyme complex.

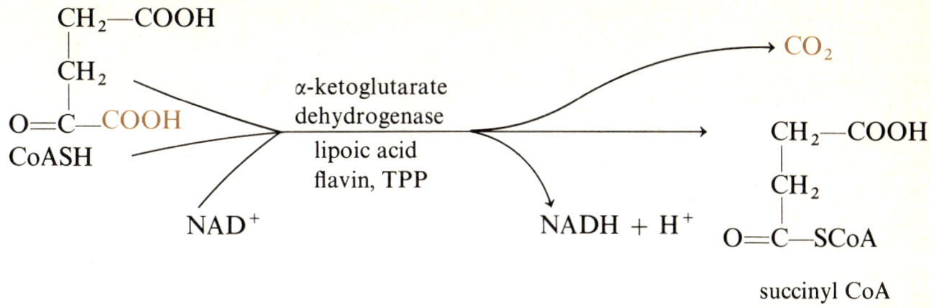

At this point the pathway has released two carbon dioxides, balancing out the two carbons of the acetyl group. In addition, two molecules of NAD^+ have been reduced.

Exercise 17.1

Are the carbons released as CO_2 the same carbons that entered as the acetyl group?

It remains to convert the dicarboxylic acid succinate back to the dicarboxylic acid oxaloacetate with which the cycle began. As a first step in restoring the oxaloacetate, the CoASH is removed from succinyl CoA. The thioester bond is a high-energy bond, which means that energy is released as the bond is hydrolyzed. Rather than waste this energy, part of it is used to add a phosphate to GDP.

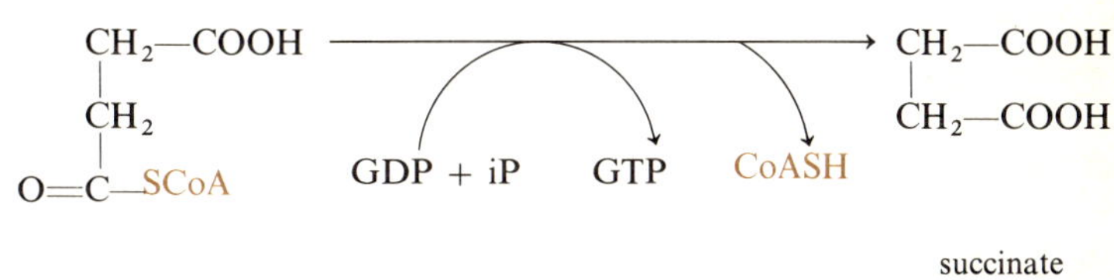

We now begin a series of reactions very similar to those in the oxidation of fatty acids. We are going to create a double bond, add water to it, and oxidize the resulting alcohol to a keto group. First we form the double bond. The hydrogens are transferred to flavin.

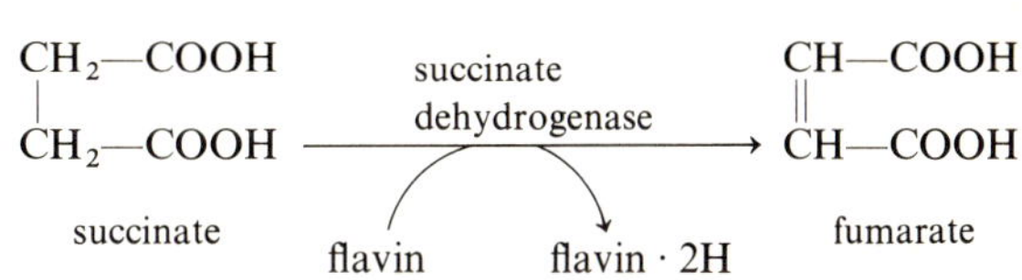

Then the water is added to convert the fumarate to malate.

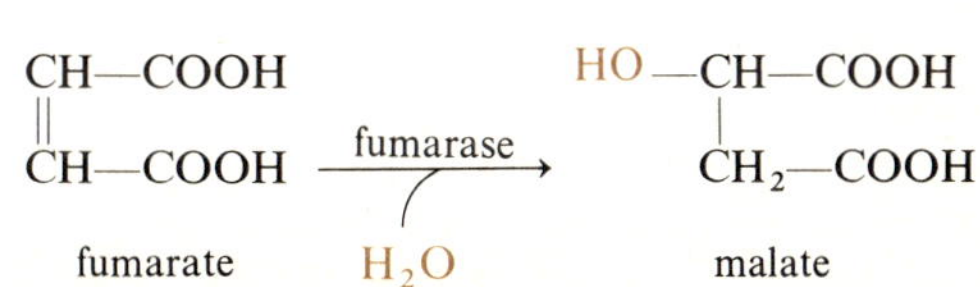

Finally the alcohol is oxidized, the hydrogens being transferred to NAD^+.

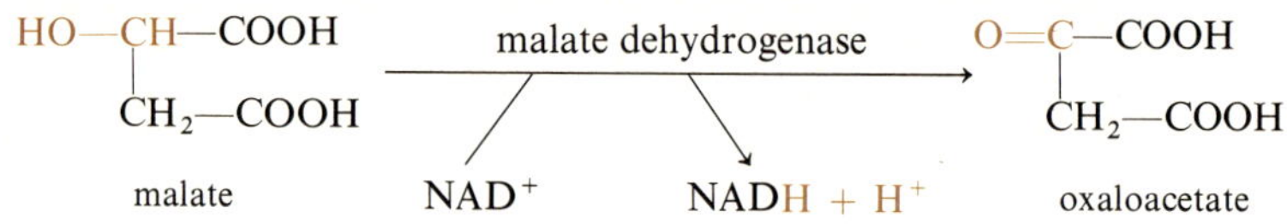

The cycle is complete. We are ready to bring in another acetyl CoA and begin again. In summary:

$$CH_3\overset{O}{\overset{\|}{C}}\text{—SCoA} \longrightarrow 2CO_2 + \text{CoASH}$$

$$3NAD^+ \longrightarrow 3NADH + 3H^+$$

$$\text{flavin} \longrightarrow \text{flavin} \cdot 2H$$

$$\text{GDP} + \text{iP} \longrightarrow \text{GTP}$$

If high-energy compounds such as ATP are our objectives, we have produced one (the GTP), a small yield. We have, however, preserved a fraction of the free energy released on oxidation by the formation of reduced cofactors. We have converted all the glucose or stearate carbons to CO_2.

Summary

The final oxidation of most carbon in the diet to carbon dioxide occurs in the mitochondrial matrix by a cyclic series of reactions usually called either the Krebs cycle or the tricarboxylic acid cycle. The acetyl group of acetyl CoA is condensed with the four-carbon dicarboxylic acid oxaloacetate to form the six-carbon tricarboxylic acid citrate. In the next three steps, cofactor-linked oxidations result in the release of two CO_2 molecules. The rest of the cycle converts succinyl CoA back to oxaloacetate. Overall, each turn of the cycle reduces one flavin and three NAD cofactors. One high-energy bond is formed by direct phosphorylation of GDP (Figure 17.1).

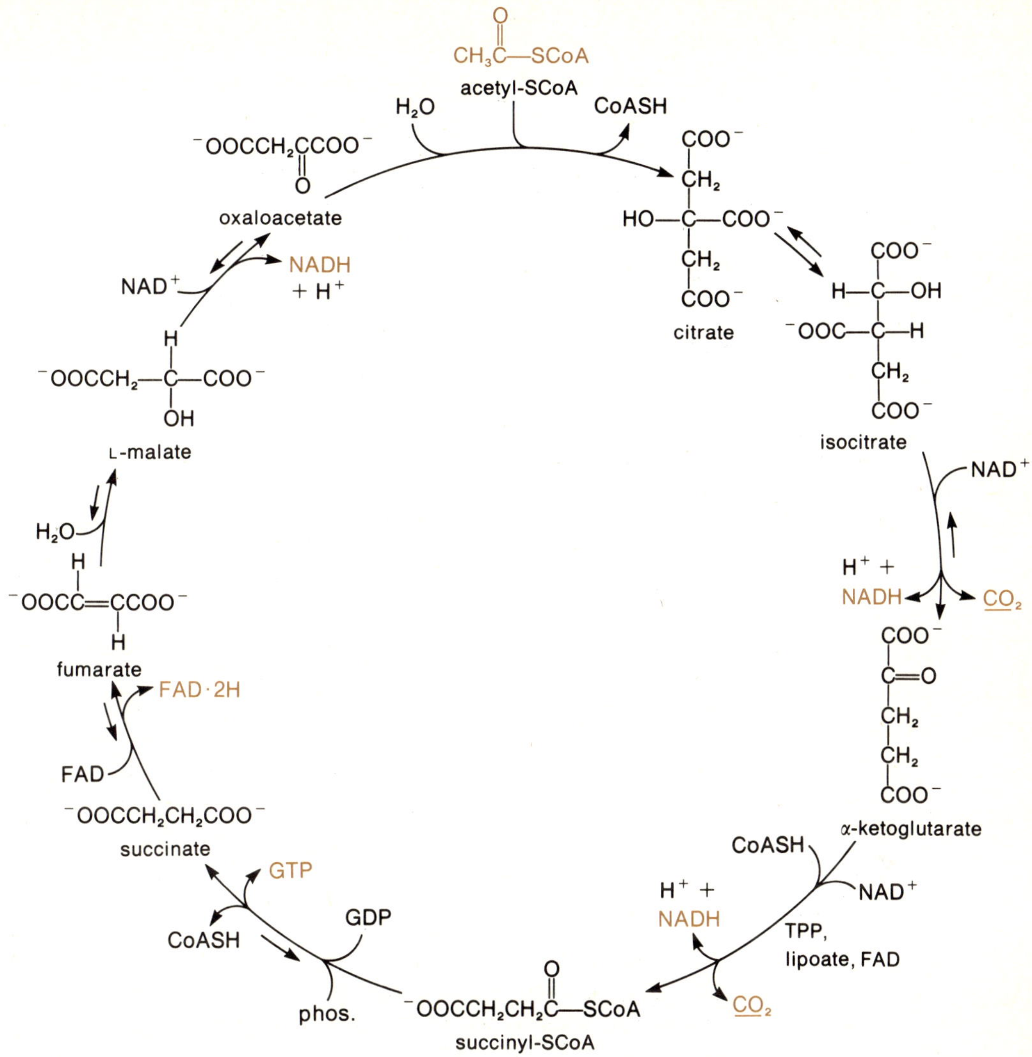

Figure 17.1
The Krebs cycle
This summary of the Krebs cycle focuses attention on the flow of carbon through the cycle and on the cofactor changes that occur.

Problems

1. Complete the following reactions taken from the Krebs cycle, naming the reactant, product, and enzyme. Indicate the cofactors where required:

a) HOOC—CH=CH—COOH ⟶

b) 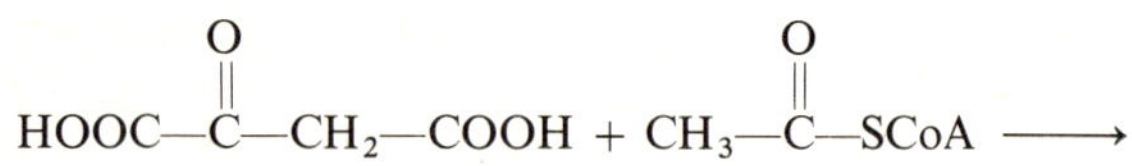

c) 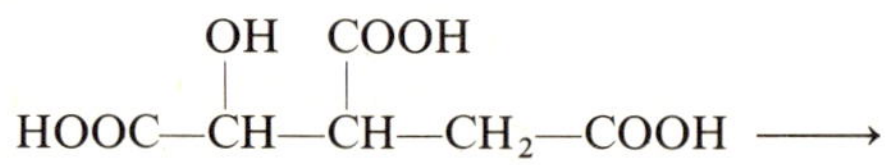

d) 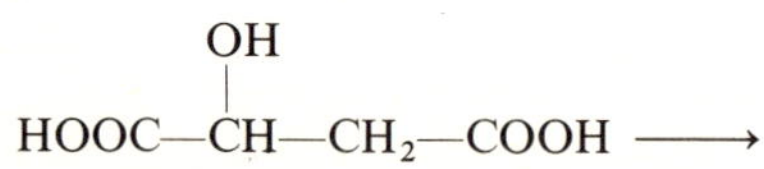

e) 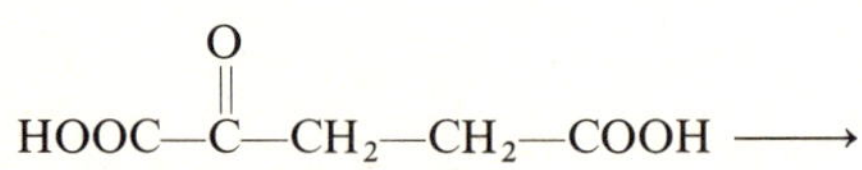

f) 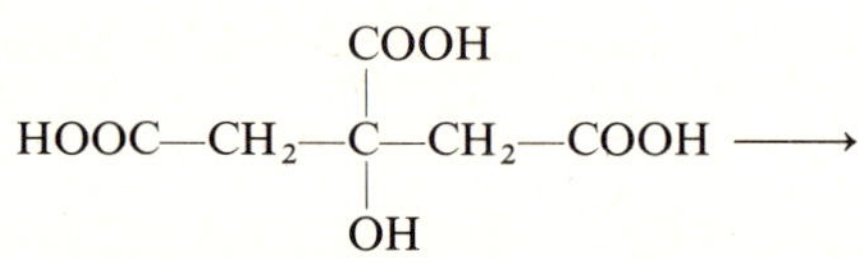

2. For a molecule of stearic acid taken as part of a bite of an ice cream cone and transported eventually to the mitochondrial matrix, indicate all the products we have generated as a result of fatty acid oxidation and the action of the Krebs cycle.

3. Do the same calculation as in Problem 2 for a molecule of glucose found in that same bite of ice cream.

4. Following the rules of the game outlined in Problem 3 of Chapter 16, what scores would glucose and hexanoic acid have accumulated now?

5. The Krebs cycle can be stopped by the presence of malonate, which inhibits the enzyme succinate dehydrogenase.

a) We add malonate to some mitochondria that were busily engaged in metabolizing fatty acids. Predict whether the concentration of succinate would probably increase or decrease. What about fumarate?

b) Suppose to mitochondria poisoned with malonate we add some acetyl CoA. Would you expect all of the acetyl CoA to be converted to CO_2? Would any CO_2 be formed?

c) Suppose to mitochondria poisoned with malonate we add a small amount of acetyl CoA that has carbon-14 (radioactive carbon) in the methyl group position. Would radioactive CO_2 be released?

d) Return to part b. We could have all of the acetyl CoA produce CO_2 if we added something besides malonate to the mitochondria. Which of the following compounds would serve? oxalocetate, fumarate, α-ketoglutarate, citrate, malonate.

6. ***a***) Write the reaction catalyzed by aconitase.

b) Using the reactions listed for alcohol functional groups, can you propose a mechanism for this? (*Hint:* Take the alcohol group completely off the citric acid.)

7. Write all the reactions in which an alcohol group is oxidized to a keto group. Return to Chapter 16, and indicate all the reactions in which an alcohol group is oxidized to a keto group. In each of these reactions, what cofactor is reduced as the substrate is oxidized?

8. Write all the reactions in which a carbon-to-carbon single bond is oxidized to a carbon-to-carbon double bond. Make a similar list for the reactions in Chapter 16. In each of these reactions, what cofactor is reduced as the substrate is oxidized?

18 Oxidative Phosphorylation

Objectives

In this chapter we:

1. Learn the pathway for the transfer of electrons from reduced cofactors to oxygen.
2. Learn the relationship between cofactor reoxidation and ADP phosphorylation.
3. Consider some agents that block this process.

Several times in recent chapters we have mentioned the preserving of energy in the form of reduced cofactors. However, cells cannot endlessly reduce cofactors, as has been the pattern in the last three chapters. Eventually one runs out of oxidized cofactor. All reactions that require oxidized cofactor would then stop, including the metabolism of the food supply. Furthermore, those reduced cofactors are not the most useful form of stored energy to the cell. Any time energy is required, the cell seems to turn to the energy of the phosphate-to-phosphate bonds of ATP. What is needed, then, is a way to transfer the energy stored as reduced cofactor into energy stored in ATP. Such a process is oxidative phosphorylation—"oxidative" for the reoxidation of the cofactor and "phosphorylation" for the adding of a phosphate group to ADP. This process occurs in the matrix of the mitochondria.

The most used oxidation-reduction cofactor in metabolism is NAD. The reoxidation of NAD is clearly our biggest task. Recall that NAD is a cofactor that does not remain bound firmly to the enzyme. Instead it acts like a second substrate. As an enzyme catalyzes the oxidation of a substrate, there is a corresponding reduction of NAD. The reduced NAD then leaves to seek reoxidation elsewhere. The NAD reduction in fatty acid oxidation, in the conversion of pyruvate to acetyl CoA, and in the Krebs cycle occurs in the mitochondrial matrix. In this matrix the machinery for oxidative phosphorylation is right at hand to accomplish reoxidation. Overall this reoxidation involves transferring the hydrogens from the cofactors to oxygen—the same oxygen that was inhaled, bound to hemoglobin, and transported to the tissues. However, there is not a direct transfer from NADH to oxygen. Instead a whole series of transfers occurs—rather like a bucket brigade—between the NADH and oxygen. Let us trace the steps in this transfer chain.

18.1 The First Complex

Three complexes are bound to the membrane surrounding the matrix and are involved in the transfer. Reduced NAD binds to NADH dehydrogenase, the first of these complexes. The complex contains a protein that carries flavin (FMN) and some other proteins that bond iron. The hydrogens are transferred to the FMN, releasing NAD in its oxidized form. Oxidized coenzyme Q then attaches to the complex, and the hydrogens are transferred from the FMN to coenzyme Q. In summary, the process is as follows:

$$\underset{\text{reduced cofactor}}{NADH} + H^+ + \text{enzyme} \cdot FMN \longrightarrow \underset{\text{oxidized cofactor}}{NAD^+} + \text{enzyme} \cdot FMN \cdot 2H$$

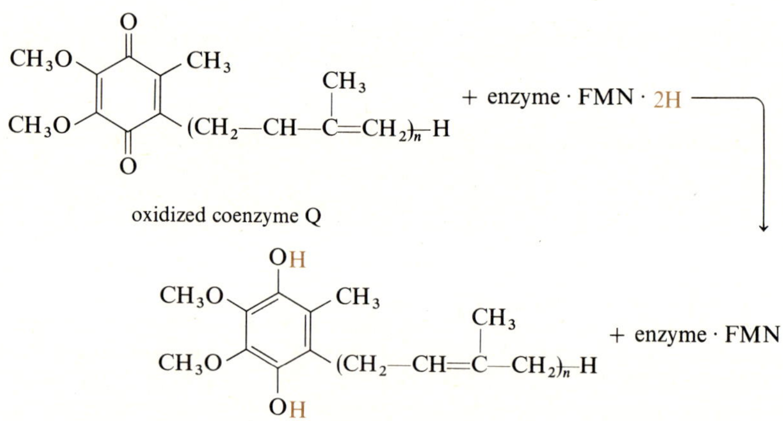

oxidized coenzyme Q

reduced coenzyme Q

Although not shown in the equations, the iron-proteins are thought to be involved as intermediate carriers.

Some enzymes use flavin rather than NAD as the oxidation-reduction cofactor. Succinate dehydrogenase is an example from the Krebs cycle (Chapter 17), and fatty acyl CoA dehydrogenase is important in the metabolism of fatty acids (Chapter 16). The flavin cofactors of these enzymes transfer their hydrogens directly to coenzyme Q in order to become reoxidized.

18.2 The Second Complex

Reduced coenzyme Q then interacts with a second membrane-bound complex. This complex contains two *cytochromes* and more iron-protein. A cytochrome is a protein with a cofactorlike structure attached that is similar to the heme of hemoglobin. The iron in the ring is the key to the function of this structure.[1]

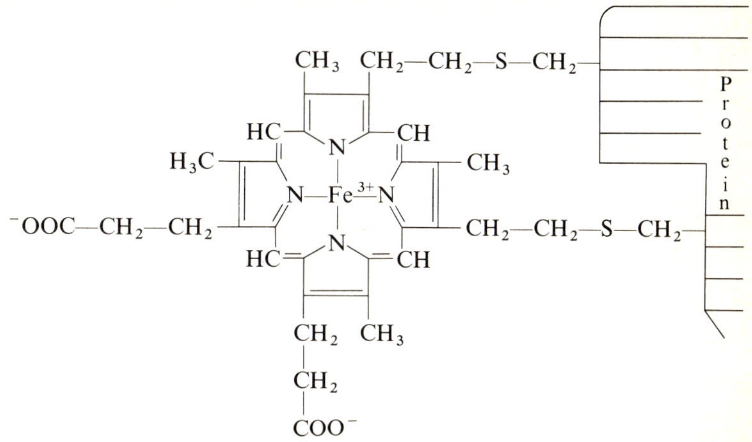

So far hydrogen atoms, each with a proton and an electron, have been transported. At the point where the chain of oxidation-reduction reactions reaches the cytochromes, the protons of the hydrogen atom are released. They interact with oxide ion, produced at the third complex, to form water. Hereafter only the electrons are transported. When a cytochrome accepts an electron, it is carried by the iron. The iron is changed from the +3 to the +2 oxidation state. Up to this point, each cofactor has carried two hydrogens, but as we begin to carry only the electrons, our cytochrome cofactors each carry only one electron.

Two varieties of cytochrome are found in the second complex, b and c_1. The electrons are believed to be transferred first to b, then to c_1. Iron-proteins such as found in the first complex are also involved somehow in the electron transfer. The iron-proteins both here and in the first complex do not have the iron bound in a heme type of structure, and they are often called "non-heme iron."

There is another cytochrome, cytochrome c, that is not membrane-bound. It attaches to the second complex much as coenzyme Q attached to the first complex and accepts an electron after it has passed completely through the second complex. In summary, the reactions at this complex are:

$$O^{2-} + \text{CoQ} \cdot 2\text{H} + 2\ \text{cyto b}(Fe^{3+}) \longrightarrow 2\ \text{cyto b}(Fe^{2+}) + \text{CoQ} + H_2O$$

$$2\ \text{cyto b}(Fe^{2+}) + 2\ \text{cyto c}_1(Fe^{3+}) \longrightarrow 2\ \text{cyto c}_1(Fe^{2+}) + 2\ \text{cyto b}(Fe^{3+})$$

$$2\ \text{cyto c}_1(Fe^{2+}) + 2\ \text{cyto c}(Fe^{3+}) \longrightarrow 2\ \text{cyto c}(Fe^{2+}) + 2\ \text{cyto c}_1(Fe^{3+})$$

[1] See Chapter 23 for an explanation of the bonding of iron in this structure.

18.3 The Third Complex

The cytochrome c carries electrons to the third membrane-bound complex. This complex contains two cytochromes, a and a_3, and copper-protein. Each cytochrome c transfers its electron to this complex. The electrons are transferred through cytochromes a and a_3, finally attaching to the copper. On binding an electron, copper goes from the +2 (cupric) to the +1 (cuprous) state. The final electron acceptor is one of the atoms of molecular oxygen. Upon transfer of the electron to oxygen, the copper returns to the +2 state. The reactions at the third complex are as follows:

$$2\text{ cyto c}(Fe^{2+}) + 2\text{ cyto a}(Fe^{3+}) \longrightarrow 2\text{ cyto a}(Fe^{2+}) + 2\text{ cyto c}(Fe^{3+})$$

$$2\text{ cyto a}(Fe^{2+}) + 2\text{ cyto a}_3(Fe^{3+}) \longrightarrow 2\text{ cyto a}_3(Fe^{2+}) + 2\text{ cyto a}(Fe^{3+})$$

$$2\text{ cyto a}_3(Fe^{3+}) + 2\,Cu^{2+}\text{ protein} \longrightarrow 2\,Cu^{+}\text{ protein} + 2\text{ cyto a}_3(Fe^{3+})$$

$$2\,Cu^{+}\text{ protein} + \tfrac{1}{2}O_2 \longrightarrow 2\,Cu^{2+}\text{ protein} + O^{2-}$$

This oxide ion plus the protons released earlier form water.

18.4 Inhibitors of Electron Transport

Certain compounds are toxic because they block the process at one of the complexes. Amytal (Chapter 9), rotenone, and piericidin block action at the first complex. Rotenone is a plant product first observed being used as a poison by Indians in South America. It has been used as an insecticide by organic gardeners. Piericidin is used as an antibiotic. It is structurally very similar to coenzyme Q and binds at the complex in its place, but it cannot perform the Co Q transport function.

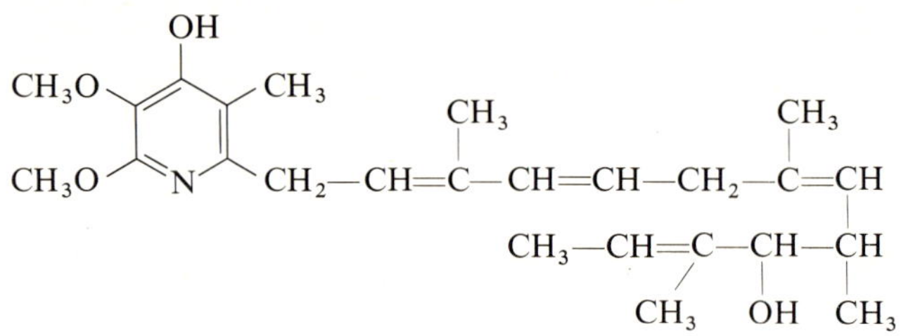

piericidin

The antibiotic antimycin A inhibits the second complex. Well-known toxic substances such as carbon monoxide, cyanide ion, and hydrogen sulfide block the interaction of the third complex with oxygen. Carbon monoxide also binds to hemoglobin, blocking the transport of oxygen to the tissues.

18.5 Summary of Changes

What has been accomplished by electron transport? The reduced NAD and flavin have been reoxidized. Remember our overall equations for oxidation of the sample carbohydrate glucose and of the sample fatty acid stearate.

$$\underset{\text{glucose}}{C_6H_{12}O_6} + 6O_2 \longrightarrow 6CO_2 + 6H_2O$$

$$\underset{\text{stearate}}{C_{18}H_{36}O_2} + 26O_2 \longrightarrow 18CO_2 + 18H_2O$$

Up to this point, we have explained only the formation of CO_2. Now we show the mechanism of use of O_2 and provide the source of H_2O. The equations are now completely explained. However, this process was supposed to generate high-energy phosphate-to-phosphate bonds in ATP. How is that accomplished?

If we measure the amount of energy released as each of the membrane-bound complexes performs its electron-transport function, it is found to be in excess of the amount needed to form a phosphate-to-phosphate bond. By a mechanism under study, the passage of a pair of electrons through each complex can convert an ADP to ATP. As a result, if NAD is reoxidized, three ATP high-energy bonds are formed. Because flavin reoxidation enters the pathway after the first complex, reoxidation of flavin produces only two ATP high-energy bonds.

18.6 Reoxidation of Cytoplasmic NAD

In the oxidation of carbohydrates, an NAD was reduced in the cytoplasm. Only a few compounds can enter the matrix, and NADH is not among them. However, ATP can be formed by reoxidation of this NAD. There is a shuttle system by which molecules that can cross the membrane are reduced in the cytoplasm with the reoxidation of the NAD. The reduced compounds cross the membrane and are reoxidized in the matrix. The reoxidation is accompanied by the reduction of a matrix NAD. In this fashion, the energy represented by the reduced NAD enters the matrix and can be used for ATP high-energy bond production.

18.7 Energy Yields

The total energy in the form of ATP produced by carbohydrate or fatty acid oxidation can now be calculated (Table 18.1).

The energy yield is 148 ATP per stearate, which is an eighteen-carbon structure. If we multiply the glucose yield by three, the total will be 114 ATP per three glucoses. Did you ever wonder why fats are more fattening (have a higher calorie content) than carbohydrates? We see

Table 18.1
Energy Yield from Metabolism

Metabolic Step *Glucose*	***NAD***	***Flavin***	***Direct Phosphorylation***
		Change per Glucose[a]	
glucose → glucose-6-phosphate	—	—	(−1)
fructose-6-phosphate → fructose-1,6-diphosphate	—	—	(−1)
glyceraldehyde-3-phosphate → 1,3-diphosphoglycerate	2	—	—
1,3-diphosphoglycerate → 3-phosphoglycerate	—	—	2
phosphoenol pyruvate → pyruvate	—	—	2
pyruvate → acetyl CoA	2	—	—
isocitrate → α-ketoglutarate	2	—	—
α-ketoglutarate → succinyl CoA	2	—	—
succinyl CoA → succinate	—	—	2
succinate → fumarate	—	2	—
malate → oxaloacetate	2	—	—
Total	10	2	4
	× 3	× 2	× 1
Yield of ATP	30	4	4
		38 ATP/Glucose	
Stearate		*Change per Stearate*	
fatty acyl CoA (saturated) → (double bond)	—	8	—
fatty acyl CoA (ol) → (one)	8	—	—
isocitrate → α-ketoglutarate	9	—	—
α-ketoglutarate → succinyl CoA	9	—	—
succinyl CoA → succinate	—	—	9
succinate → fumarate	—	9	—
malate → oxaloacetate	9	—	—
Total	35	17	9
	× 3	× 2	× 1
Yield of ATP	105	34	9
		148 ATP/stearate	

[a] Remember that each six-carbon glucose forms two–three carbon trioses.

here a measure that they are.[2] This is so because the carbohydrate structure with its abundance of alcohol groups is more oxidized at the start than is the largely hydrocarbon fatty acid.

Summary

Reduced cofactors are reoxidized in the mitochondrial matrix. Initially hydrogen and later just the electrons from the cofactors are transferred through a series of compounds, until finally the electrons are accepted by molecular oxygen. Some of these compounds are found bound to membrane in complexes; others shuttle between these complexes. The addition of phosphate to ADP to form ATP occurs in conjunction with the transfer of electrons. For each NADH that is reoxidized, three ATP molecules are formed, and for each reoxidized flavin, the system forms two ATP molecules. The process can be blocked by the antibiotics piericidin or antimycin A, the barbiturate amytal, or the poisons rotenone, cyanide, carbon monoxide, and hydrogen sulfide.

Problems

1. The *phosphorylation* of oxidative phosphorylation can be summarized by what reaction?

2. In oxidative phosphorylation, the molecules that may be oxidized include: glucose, NADH, stearate, FAD-2H, NAD$^{(+)}$, oxygen. (Indicate all those correct.)

3. Given the following diagram of the structure of the system of oxidative phosphorylation:

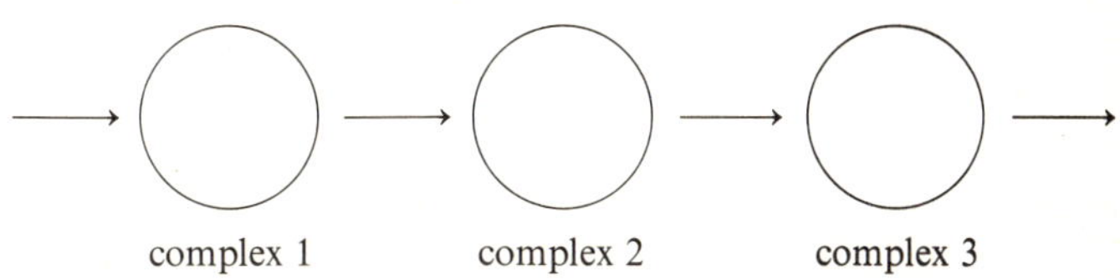

a) Place oxygen at the correct location in the diagram.

b) Here are a list of cytochromes:

$$a, a_3, b, c, c_1$$

Write those that are part of a complex into the correct complex.

---

[2] Now the game problems 16.3 and 17.4 make sense.

c) Write the following into the correct complex or complexes: iron protein, copper protein, flavin.

d) Indicate on the diagram where the system changes from hydrogen transport to electron transport.

e) What compound provides transport of hydrogens or electrons between complex 1 and 2? between 2 and 3?

f) The oxidation-reduction enzymes that are involved in the metabolism of food have either NAD or flavin as the cofactor. Where do each of these cofactors interact with the system as diagrammed?

4. Suppose we have an experimental arrangement that allows us to feed reduced cofactors into a functioning system for oxidative phosphorylation. How many moles of ATP would we produce by adding each of the following?

a) 1 mole NADH

b) 1 mole FAD · 2H

c) 1 mole coenzyme Q · 2H

d) 1 mole cytochrome $c(Fe^{2+})$

e) 1 mole cytochrome $c(Fe^{3+})$

5. Below is a series of metabolic conversions. Determine the energy yield as moles ATP per mole starting material when cofactors are reoxidized by oxidative phosphorylation.

a) ethanol

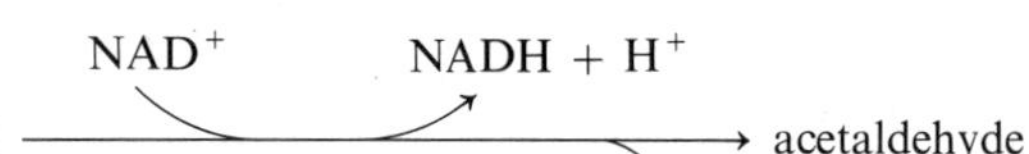

b) citrate

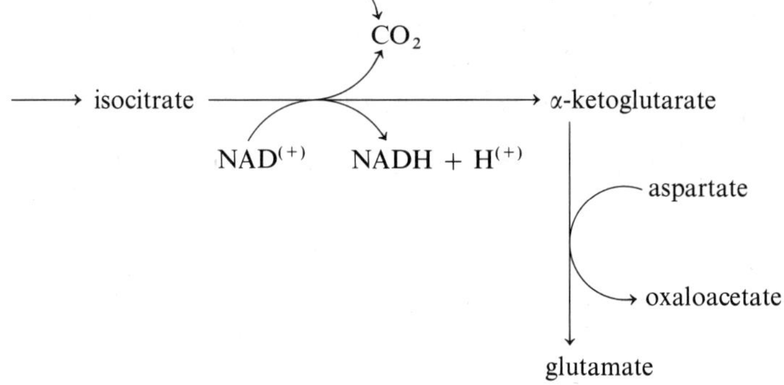

c) succinate

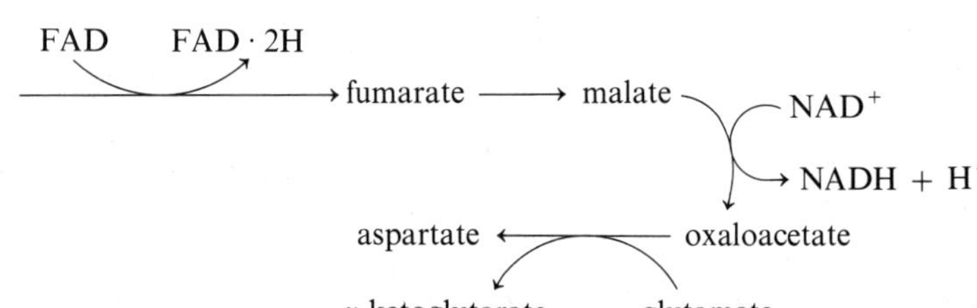

d)

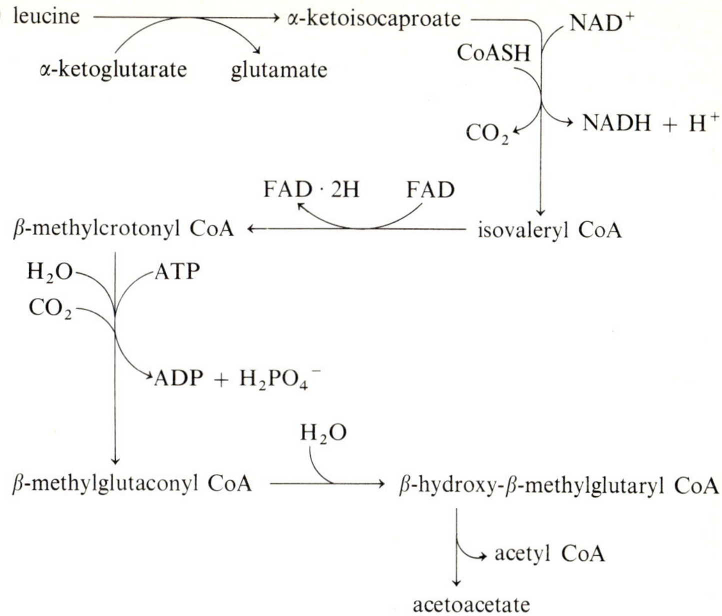

e) How much more ATP would part d generate if the acetyl CoA is converted to $2CO_2$ by the Krebs cycle?

19 Amino Acid Metabolism

Objectives

In this chapter we:

1. *Observe how the amine group of amino acids is converted to urea for elimination principally in the urine.*
2. *Are introduced to pathways for the conversion of selected amino acids.*
3. *Observe examples of genetic diseases based on a lack of certain enzymes of amino acid metabolism.*

The three major nutrients are carbohydrates, fats, and proteins. We saw in Chapter 12 how protein in the diet is enzymatically converted to individual amino acids before it enters the blood. These individual amino acids may be used for protein synthesis (Chapter 13), but some portion will be metabolized to synthesize ATP. Amino acids have more distinctly individual metabolic pathways as contrasted with hexose-type carbohydrates that converge into a common path. As a result, this is a more complex topic than was the metabolism of either carbohydrates or fats. However, all the pathways for amino acid metabolism eventually provide fuel for the Krebs cycle, and the need to remove the amine group is common to the metabolism of all amino acids. Once removed, the amine group must be converted to a compound of low toxicity to be transported and accumulated for elimination. Ammonia itself is relatively toxic and is therefore unsuitable.

19.1 Deamination of Amino Acids

Removal of the amine group of amino acids is accomlished by transamination. In this sort of reaction, the amine group is exchanged for the oxygen of a ketone structure. Specifically, the individual amino

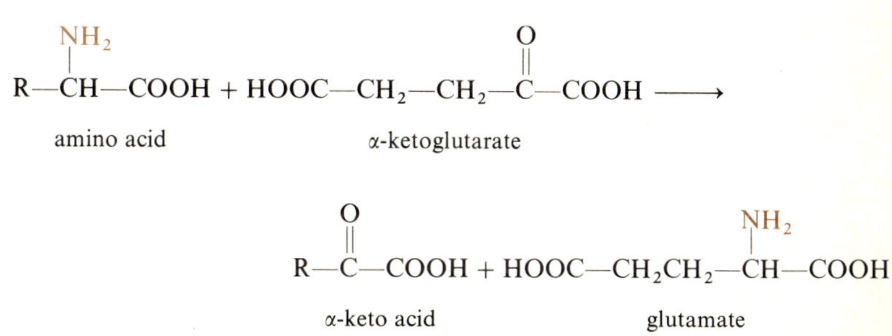

acids and α-ketoglutarate serve as substrates for enzymes called *transaminases*. The amine group of the amino acid and the keto oxygen of α-ketoglutarate trade places. The products of the reaction are glutamate and an α-keto acid corresponding to the amino acid substrate.

Exercise 19.1

Show the substrates and products for these specific transaminase enzymes:

a) $\xrightarrow[\text{transaminase}]{\text{leucine}}$ *b)* $\xrightarrow[\text{transaminase}]{\text{alanine}}$

Transaminases have a pyridoxal cofactor (Chapter 12). These reactions occur in the cytoplasm. Some of the glutamate, which is a product of transamination, enters the mitochondrial matrix. There the amine is removed by the enzyme glutamate dehydrogenase. The products are α-ketoglutarate and ammonia. This process is called *oxidative deamination.*

$$HOOC{-}CH_2{-}CH_2{-}\underset{\displaystyle NH_2}{\underset{|}{CH}}{-}COOH + NAD^+ \xrightarrow[\text{dehydrogenase}]{\text{glutamate}}$$

$$HOOC{-}CH_2{-}CH_2{-}\overset{\displaystyle O}{\overset{\|}{C}}{-}COOH + NH_3 + NADH + H^+$$

This α-ketoglutarate may then return to accept an amine group from another amino acid. The ammonia is, as stated above, an unsuitable product. It will be converted to a less toxic substance, urea, by a pathway presented later in the chapter. First consider the α-keto acids that the amino acids have become.

19.2 Specific Examples of Amino Acid Metabolism

Some amino acids, once they are converted to an α-keto acid, have become an intermediate of another pathway. For example, alanine becomes pyruvate, which is an intermediate of carbohydrate metabolism. Pyruvate is converted to acetyl CoA by pyruvate dehydrogenase. Serine also converts to pyruvate by a pathway involving direct removal of the amine group rather than transamination (Figure 19.1).

Glutamate itself deaminates to become α-ketoglutarate. Some of this α-ketoglutarate is needed for the transamination reaction. However, as an excess of α-ketoglutarate accumulates, it enters the Krebs cycle directly. Some other amino acids are metabolized to form

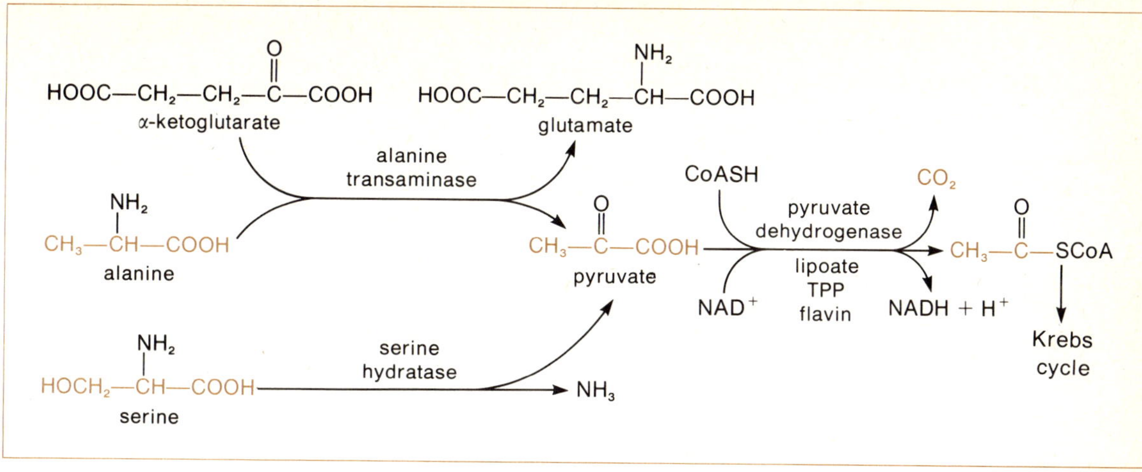

Figure 19.1
Serine and alanine metabolism

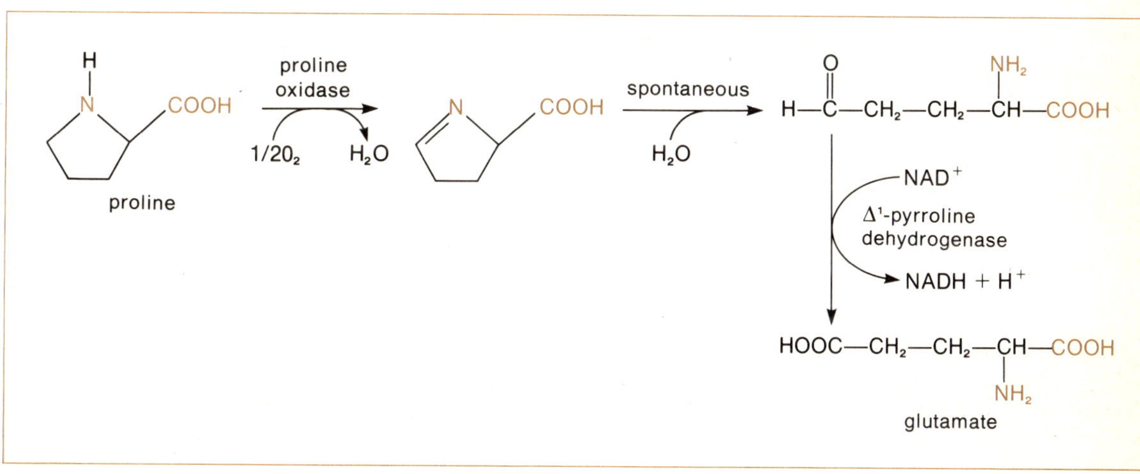

Figure 19.2
Proline metabolism

glutamate and then enter the Krebs cycle by way of α-ketoglutarate. Proline, histidine, and arginine follow this course. The proline pathway is shown in Figure 19.2.

Histidine also has a ring opening reaction (Figure 19.3). Notice several points about this pathway. As with serine, there is a direct deamination rather than a transamination. Notice, too, the use of tetrahydrofolic acid to remove a one-carbon fragment (Chapter 12). The insert is intended to show how the rather magical-looking conversion catalyzed by the second step is really the addition of water

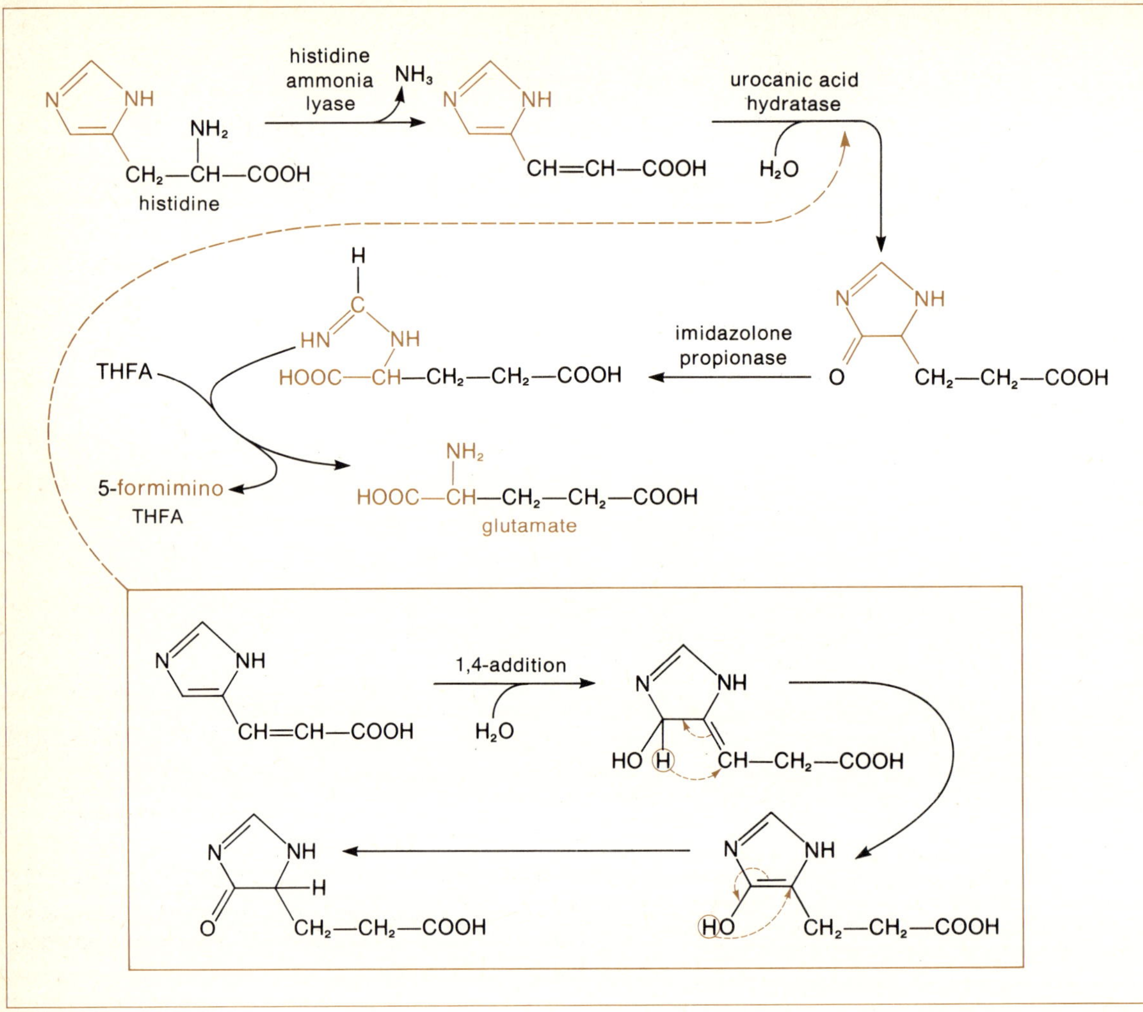

Figure 19.3
Histidine metabolism

to a double bond, the shift of a double bond, and finally an enol-keto shift.

Exercise 19.2

What type of chemical reaction is occurring in the ring opening step?

Phenylalanine and tyrosine metabolism provide another pathway producing intermediates common to other pathways. Phenylalanine is first converted to tyrosine.

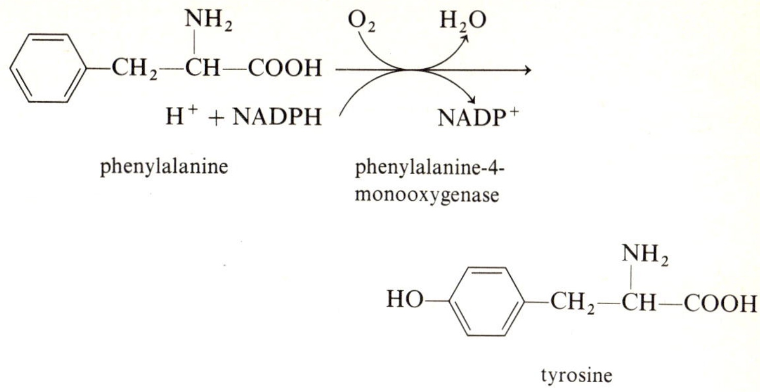

Tyrosine is then converted to an α-keto acid by transamination.

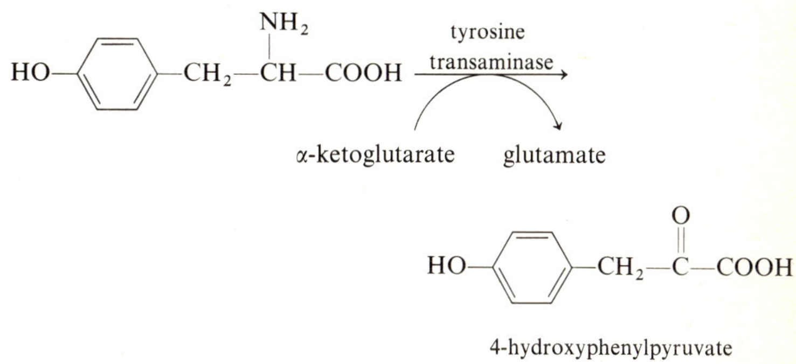

What happens next is rather strange—it does not look like the straightforward reactions to which you have been accustomed. The carbon side chain is actually moved to a different position. The carboxyl is removed as CO_2.

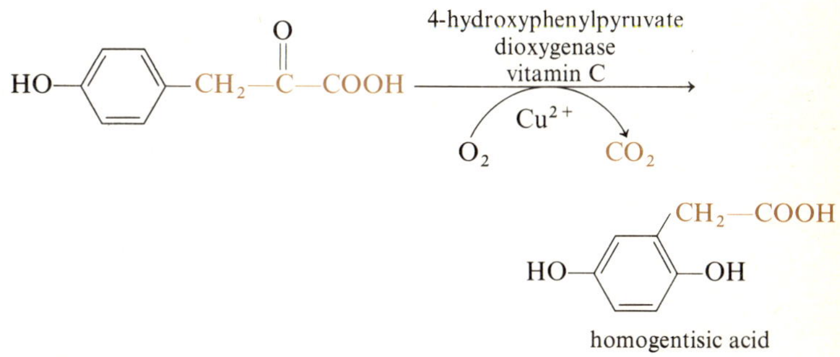

The rest of the pathway consists of the ring opening followed by conversion to fumarate, a Krebs cycle intermediate, and to aceto-acetyl CoA, a fatty acid oxidation intermediate (Figure 19.4).

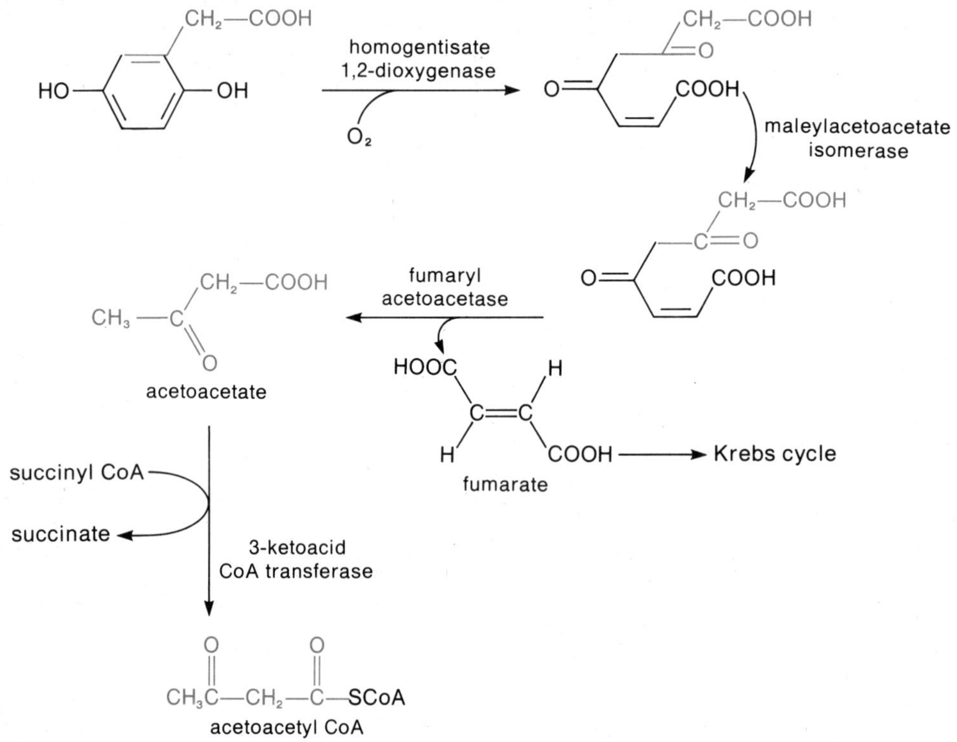

Figure 19.4
Metabolism of homogentisic acid

These reactions represent some of the more complex changes found in metabolic systems.

19.3 Genetic Deficiency Diseases

Obviously, all metabolic reactions are not of uniform importance to the organism. Survival with nonfunctional systems for synthesizing protein or transporting oxygen is out of the question. Survival without the ability to synthesize compounds available in adequate supply from the diet is not only possible, it has become a common pattern for humans. The pathways for amino acid metabolism fall somewhere between the extremes of absolute requirement for the pathway and an indifference to its existence. The loss of nutrition associated with inability to utilize one out of the twenty amino acids is not serious. However, when one enzyme of a pathway has become ineffective, the compound or compounds immediately before that point in the sequence are still synthesized, are not used, and accumulate. This can pose health problems. The difficulty associated with the loss of an enzyme is termed a *genetic deficiency disease*.

One very important example is provided by a genetic deficiency such that phenylalanine-4-monooxygenase is inactive, as occurs in about 0.01 percent of humans. Phenylalanine, then, cannot be converted to tyrosine and is transaminated to form phenylpyruvate. Phenylpyruvate cannot be metabolized further and accumulates. In children, excess phenylpyruvate retards brain development. Because of the high levels of phenylpyruvate detected in the urine, the problem was named *p*henyl*k*eton*u*ria, or PKU. Control of diet to reduce phenylalanine intake is necessary.

A genetic inability to produce active homogentisic acid-1,2-dioxygenase results in an accumulation of homogentisic acid. In the urine, homogentisic acid oxidizes and polymerizes into a black pigment. The problem is termed *alkaptonuria*. Valine and isoleucine metabolism, which have not been outlined, are sometimes disrupted by the absence of an enzyme, α-ketoisovalerate dehydrogenase, which causes α-keto acids to accumulate, resulting in mental retardation. Because these acids give the urine a characteristic odor, it is referred to as *maple sugar urine disease*.

19.4 Formation of Urea: Ammonia Incorporation into Citrulline

Recall that in the mitochondrial matrix, ammonia was generated from glutamate, converting it to α-ketoglutarate. Glutamate in turn had collected the amine groups of other amino acids by transamination during a period when amino acids are being metabolized actively. This could represent a relatively high concentration of ammonia. Ammonia is somewhat toxic. How is this problem handled?

In the matrix, ammonia is combined with CO_2 and phosphate in a reaction utilizing ATP for both an energy source and a phosphate source. The product is carbamoyl phosphate.

$$2ATP + NH_3 + CO_2 + H_2O \xrightarrow[]{\text{carbamoyl phosphate synthetase}} NH_2{-}\overset{\overset{\displaystyle O}{\|}}{C}{-}O{-}\underset{\underset{\displaystyle OH}{|}}{\overset{\overset{\displaystyle O}{\|}}{P}}{-}O^- + 2ADP + H_2PO_4^-$$

carbamoyl phosphate

The enzyme ornithine carbamoyl transferase catalyzes a shift of the $NH_2{-}C(=O){-}$ structure to the R group amine of ornithine, converting it to citrulline.

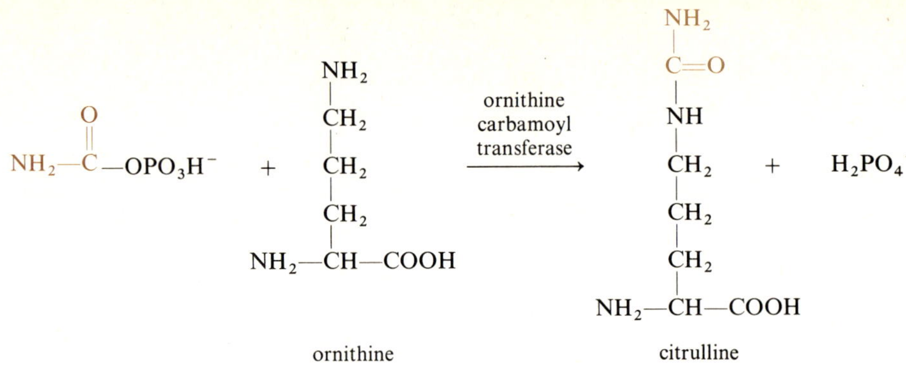

Neither ornithine nor citrulline are amino acids found in protein molecules.

19.5 Conversion of Citrulline to Arginine

Glutamate was the source of the ammonia that has now been incorporated into citrulline. However, the glutamate amine groups enter this system by yet another route. In the cytoplasm, glutamate can transfer its amine group to oxaloacetate, producing aspartate.

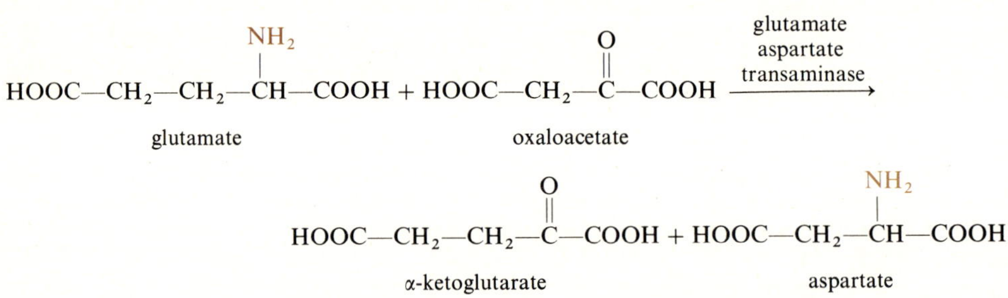

The citrulline leaves the matrix and is joined to this aspartate by the enzyme arginosuccinate synthetase. The energy for this Schiff's base bond formation is supplied by ATP.

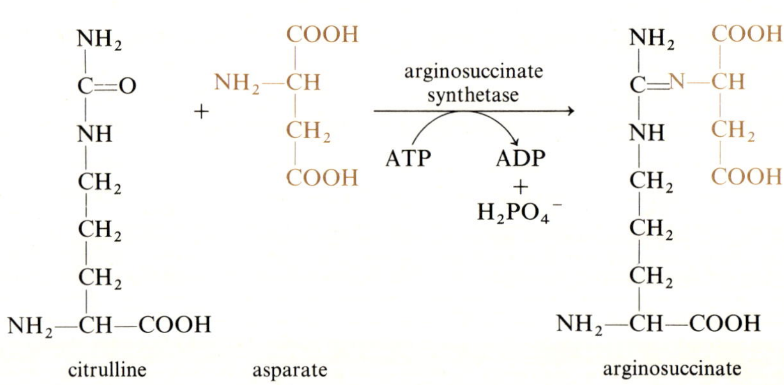

The arginosuccinate is taken apart by arginosuccinate lyase, but the disassembly leaves the amine group on the former citrulline, converting it to arginine.

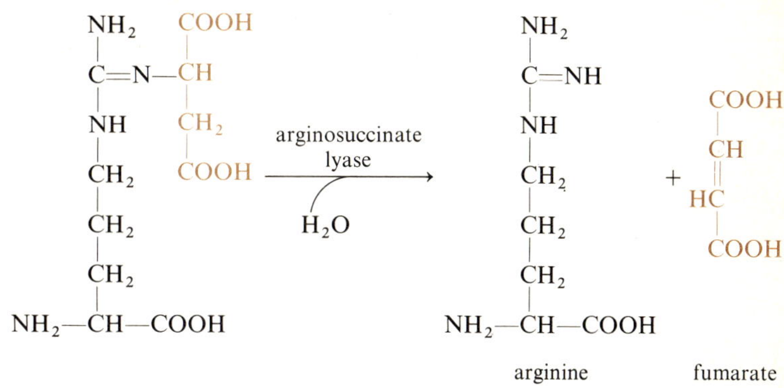

19.6 Completing the Cycle: Hydrolysis of Arginine

The arginine is hydrolyzed to give us back ornithine and a molecule of urea. The ornithine can go back into the matrix for another turn around the cycle. The urea is a detoxified form of the ammonia and is transported by the blood for elimination primarily by way of the urine.

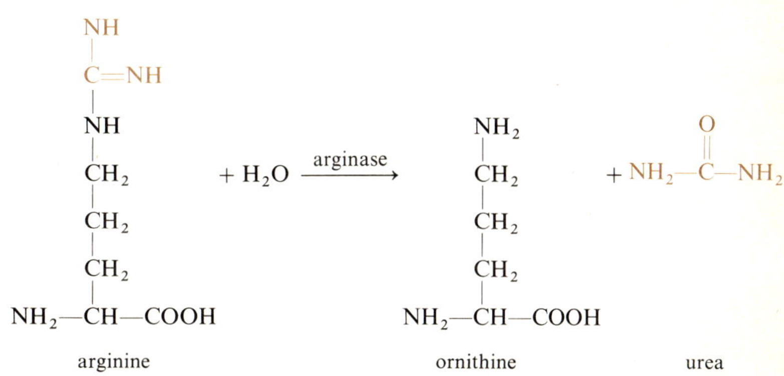

Exercise 19.3

Of the four nitrogens of arginine, mark with one asterisk (*) the nitrogen that came from glutamate by way of oxidative deamination and with two asterisks (**) the one that came by way of transamination with oxaloacetate. Are they the nitrogens of urea?

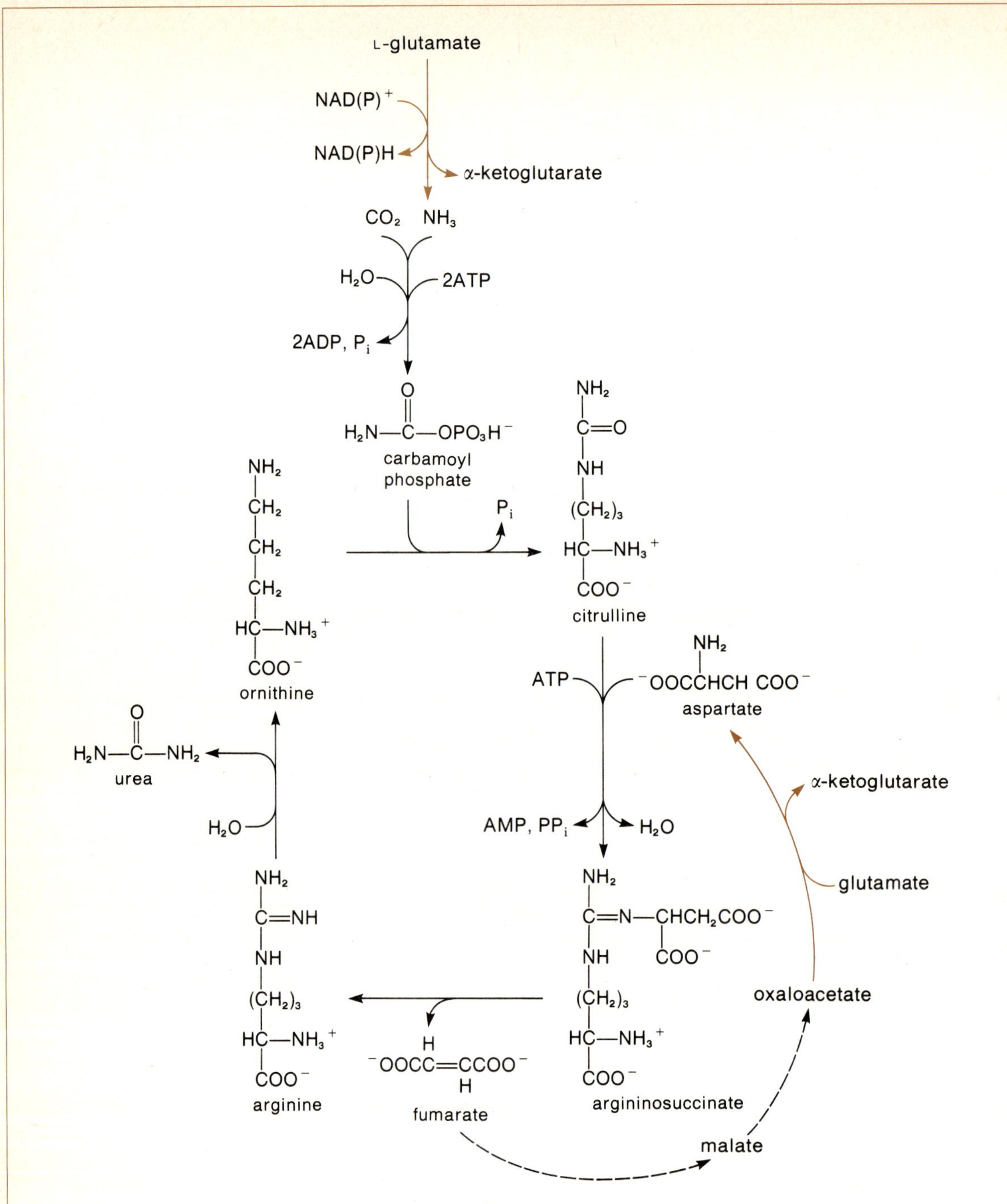

Figure 19.5
The urea cycle
On this summary of the complete pathway, the nitrogen enters as the amine group of glutamate at two points and leaves in the urea molecule.

Summary

Amino acids may be metabolized for energy. The first step is the removal of the amine group, which is often accomplished by transamination to α-ketoglutarate. Although several amino acids share this same first step, the pathways thereafter are individual. The pathways lead to Krebs cycle intermediates, particularly to acetyl CoA, succinyl CoA, α-ketoglutarate, and fumarate. Loss through genetic defect of a specific enzyme in these pathways results in an inability to utilize that amino acid, and in the accumulation in blood and urine of the intermediates preceding the missing enzyme. Such defects can sometimes be survived but produce characteristic, potentially serious disease symptoms. The nitrogen transferred to α-ketoglutarate, producing glutamate, is eliminated as the relatively low toxicity molecule urea. One of the urea nitrogens was removed from glutamate by oxidative deamination to ammonia, incorporation of the ammonia into carbamoyl phosphate, and addition of the $-\overset{O}{\overset{\|}{C}}-NH$ group to ornithine, forming citrulline. The other urea nitrogen left glutamate by transamination to oxaloacetate, then transfer from the resulting aspartate to citrulline, forming arginine. The two nitrogens and carbon thus added to ornithine are hydrolyzed off as urea, which is the chief nitrogen compound eliminated in the urine (Figure 19.5).

Problems

1. Write complete equations for the reactions catalyzed by the following enzymes:

a) cysteine transaminase

b) glycine transaminase

2. Identify the amino acid for whose metabolism the following compounds are intermediates.

a) 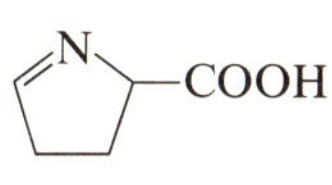

b) 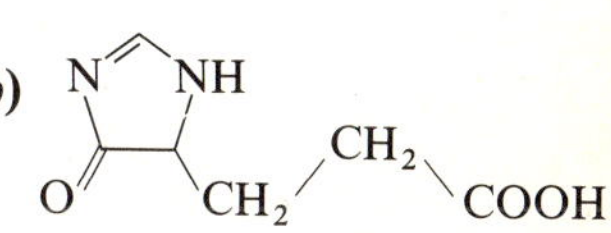

c) 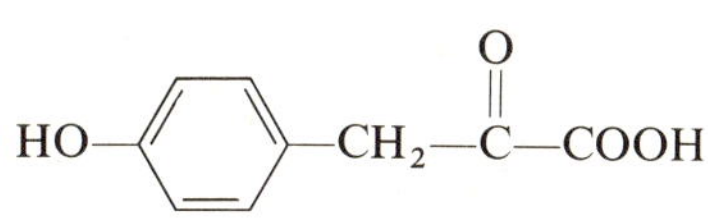

d) $CH_3-\overset{O}{\overset{\|}{C}}-COOH$

e) $H-\overset{O}{\overset{\|}{C}}-CH_2-CH_2-\overset{NH_2}{\overset{|}{CH}}-COOH$

f) $NH{=}CH-NH-\overset{COOH}{\overset{|}{CH}}-CH_2-CH_2-COOH$

g) $HOOC-\overset{H}{\overset{|}{C}}{=}\overset{H}{\overset{|}{C}}-\overset{O}{\overset{\|}{C}}-CH_2-\overset{O}{\overset{\|}{C}}-CH_2-COOH$

3. For each compound in Problem 2 write the reaction for the next step in the particular metabolic pathway.

4. Of all the possible aminotransferase enzymes, which one would you expect to be present at the highest level of activity in a tissue sample? Why?

5. Write the pathway for the conversion of alanine to CO_2, H_2O, and urea using pathways from this and other chapters.

6. ***a)*** Write equations for these three reactions:

   1. The conversion of phenylalanine to tyrosine.
   2. The conversion of proline to the first intermediate in its metabolism.
   3. The conversion of 4-hydroxyphenylpyruvate to homogentisic acid.

   b) What do these reactions have in common?

   c) Have you seen this anywhere else?

   d) The substrates are all (rearranged, oxidized, reduced). (Choose one.)

   e) When substrates have been oxidized in previous chapters, it has involved such cofactors as ______________________________ and ______________________________.

   f) Are such cofactors involved in any of these reactions, and if so, are they used as you would expect them to be?

20 Biosynthesis and Photosynthesis

Objectives

In this chapter we:

1. Recognize that biosynthesis is a reductive process requiring energy.
2. Are introduced to the pathway for biosynthesis of glucose, a pathway largely the reverse of oxidative breakdown.
3. Are introduced to the pathway for biosynthesis of fatty acids, a pathway entirely separate from oxidative breakdown.
4. Learn that reduced carbon for living organisms is created by photosynthesis.
5. Are introduced to noncyclic photophosphorylation, which uses light energy to convert the oxygen of water to atmospheric O_2, phosphorylates ADP to form ATP, and reduces NADP to NADPH.
6. Are introduced to cyclic photophosphorylation, which phosphorylates ADP to form ATP.

Compounds that are taken in as food are catabolized, that is, broken down and oxidized. Cofactors are reduced as food compounds are oxidized. The reoxidation of the cofactors is the source of the high-energy compound ATP. Obviously, for such things as carbohydrates, lipids, and amino acids to be available for metabolic breakdown, they must have been synthesized previously. In fact, if a person consumes more carbohydrate than is needed for energy, the excess carbohydrate is converted to triglycerides for storage in adipose tissue—as some of us understand only too well. We can synthesize some of the amino acids we use to make protein from nonprotein sources. Certainly we synthesize a broad variety of specialized and sometimes complex compounds that serve as hormones, as pigments, and for a number of other special functions.

When we considered the function of metabolism to produce energy, we were oxidizing substrates and producing ATP. As a rule in biosynthesis, we are reducing substrates and using up ATP. In some cases the pathway by which a compound is broken down can simply be reversed to accomplish the biosynthesis. In other cases a completely separate set of enzymes is used. An example of the situation where breakdown is reversed is provided by glucose synthesis, while the other case where a completely separate set of enzymes is used is illustrated by fatty acid synthesis.

20.1 Glucose Synthesis

The pathway for glucose degradation is almost completely reversible. At two steps, however, the energy released by the reaction in the direction of degradation is very large. To reverse those steps, that same amount of energy would have to be added to the reactions. In each case an alternative step that avoids the large addition of energy is provided. The two reactions are shown below, along with the reactions that substitute in the reverse reaction.

Reaction 1

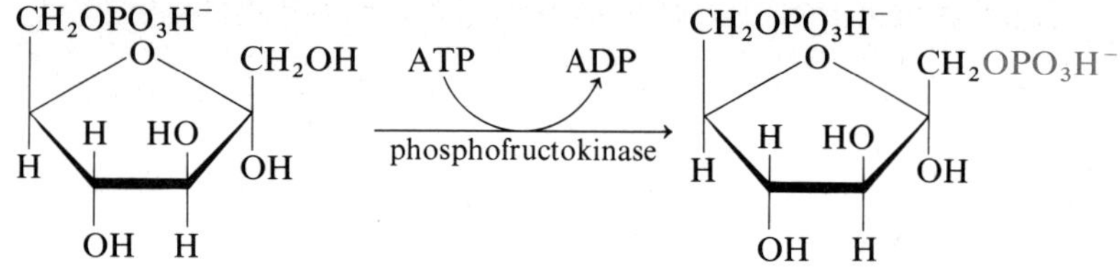

fructose-6-phosphate fructose-1,6-diphosphate

Catabolic reaction

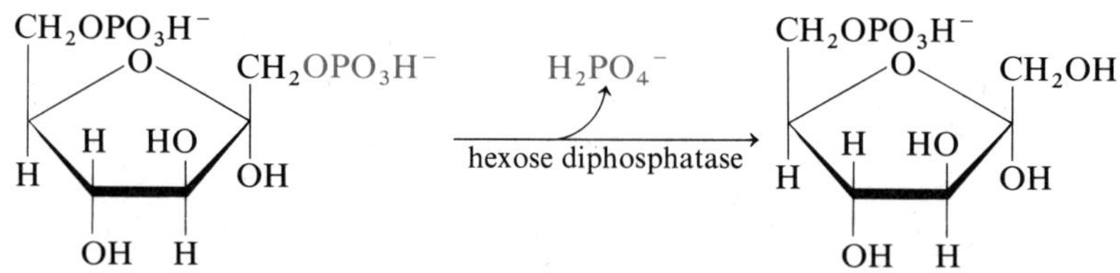

fructose-1,6-diphosphate fructose-6-phosphate

Biosynthetic reaction

Notice that the requirement for forming ATP from ADP, which necessitates a large input of energy, is eliminated in the biosynthetic pathway. The resulting phosphate removal actually releases energy.

Reaction 2

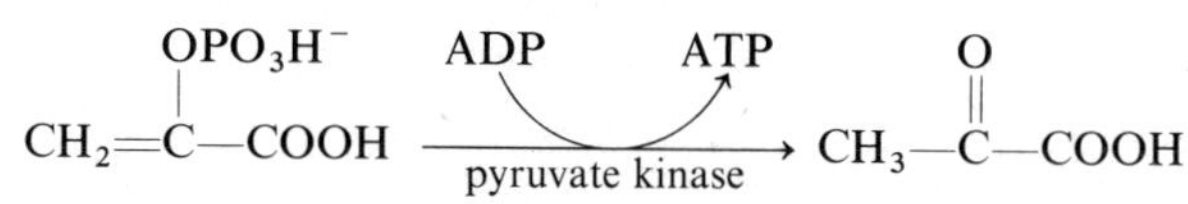

phosphoenolpyruvate pyruvate

Catabolic reaction

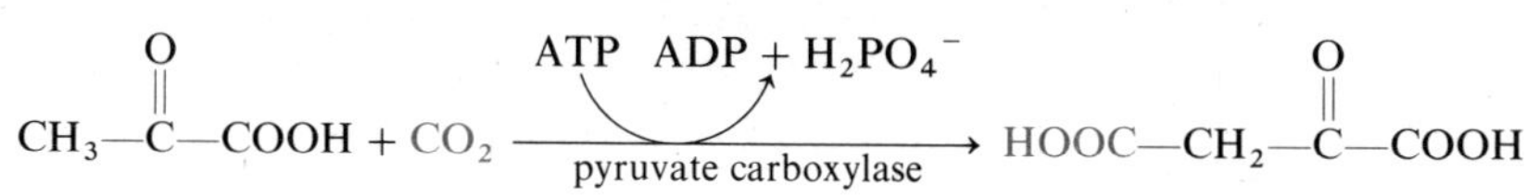

oxaloacetate

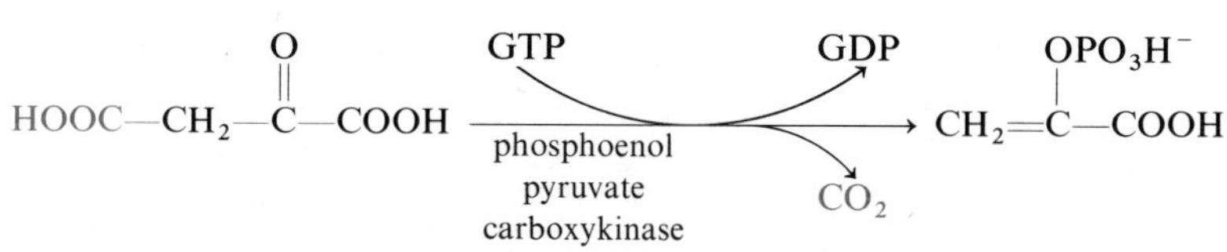

Biosynthetic reactions

Notice that reversal here is accomplished by sacrificing two high-energy phosphate linkages, one from ATP and one from GTP. Reversal of the catabolic reaction would have sacrificed only one such linkage. In summary, glucose is synthesized by reversing all but two of the reactions of glucose catabolism.

20.2 Anaplerotic Reactions

Besides being a step in the synthesis of glucose from pyruvate, the reaction converting pyruvate to oxaloacetate serves a quite different function. If the supply of intermediates in the Krebs cycle drops to too low a level, the cycle is not able to match the pace of reactions that supply acetyl CoA to it. In order for an acetyl group to transfer onto oxaloacetic acid to form citric acid, an oxaloacetic acid must be ready and waiting. If it isn't, the acetyl group must wait until one is available. In such a circumstance, some of the pyruvate from carbohydrate catabolism can be converted to oxaloacetate instead of acetyl CoA until the Krebs cycle is again operating efficiently.

Reactions that build the supply of Krebs cycle intermediates are called *anaplerotic*, or "filling up," reactions. Many of the amino acids are catabolized to α-ketoglutarate, oxaloacetate, succinate, or fumarate, or to pyruvate, which can be converted to oxaloacetate. Fatty acids with even numbers of carbons produce only acetyl CoA, however, and so cannot perform an anaplerotic role.

20.3 Sources for New Glucose Synthesis

The amino acids described above that can supplement the Krebs cycle can also be converted from oxaloacetate to pyruvate, which is the starting material for glucose synthesis. Such amino acids are termed *glucogenic amino acids* (Table 20.1).

Fatty acids and amino acids that catabolize to acetyl CoA cannot be used to increase our level of glucose. This explains why when we are hungry and feel weak and dizzy because of low blood sugar, we can't simply convert some body fat to glucose to handle our problem.

Table 20.1
Glucogenic Amino Acids

Amino Acid	*Intermediate Product*
Glutamate Arginine Histidine Proline Glutamine	α-Ketoglutarate
Isoleucine Valine Methionine	Succinate
Phenylalanine Tyrosine	Fumarate
Aspartate Asparagine	Oxaloacetate

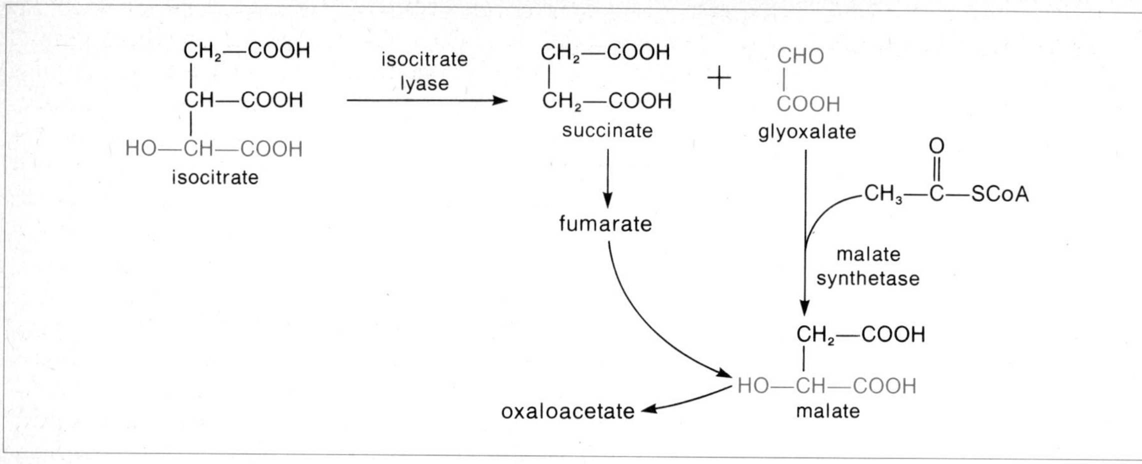

Figure 20.1
The glyoxalate shunt

In plants and microorganisms, however, a modification of the Krebs cycle, called the *glyoxalate shunt*, allows acetyl CoA formed from fatty acids or amino acids to avoid the CO_2-producing steps and generate oxaloacetate, which can then create new glucose (Figure 20.1).

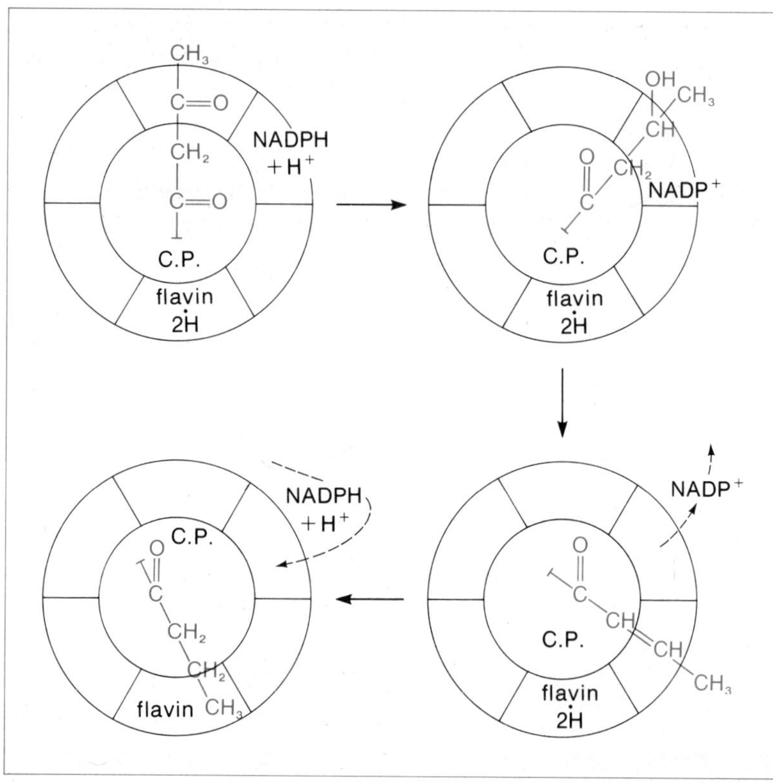

Figure 20.2
Fatty acid synthesis: keto group to saturated chain
In this diagram the central circle represents the carrier protein and the outer segments the six enzymes of the pathway. The group rotates from one enzyme to the next attached to the carrier protein.

20.4 Fatty Acid Synthesis

The catabolism of fatty acids occurs in the mitochondrial matrix. It is a one-way process, incapable of biosynthesis. Biosynthesis of fatty acids occurs in the cytoplasm catalyzed by a multienzyme complex. This remarkable complex consists of seven proteins. Six are enzymes, and the other is a carrier protein (C.P.). The growing fatty acid is attached by a long arm to the C.P. and is moved from one to another of the enzymes. Consider how efficient and well organized such a multienzyme complex is. The product of one enzyme-catalyzed step does not have to drift about looking for the next enzyme in the pathway. Instead the product of one step is already in place for the next step.

Just as fatty acids are metabolized to form acetyl CoA, acetyl CoA provides the carbons for building a new fatty acid. In a simplified view of the process, the joining of two acetyl groups forms an aceto-acetyl group that is attached to the C.P. The C.P. then rotates from enzyme to enzyme. At these enzymes the β-keto is reduced to an alcohol group, that —OH and a hydrogen from the α-carbon are removed to create a double bond, then hydrogens are added to the double bond, producing a saturated acid (Figure 20.2). This series of reactions is simply the reverse of fatty acid catabolism, with a different cofactor.

20.5 Extending the Chain

The complicated part of this process is the mechanism for extending the chain. An interesting concept is involved. Energy is used to add a group temporarily to the acetyl group before its attachment to the complex. The energy used to bind this group is released when the group is removed. However, the energy is not released and discarded. It is used to form a new bond, the bond that lengthens the new fatty acid chain (Figure 20.3). Specifically, the temporary group added is a carboxyl group. The energy for attaching it to the acetyl group is

Figure 20.3
Bond formation
Here the pathway by which energy is supplied to form the chain-extending bond is presented symbolically.

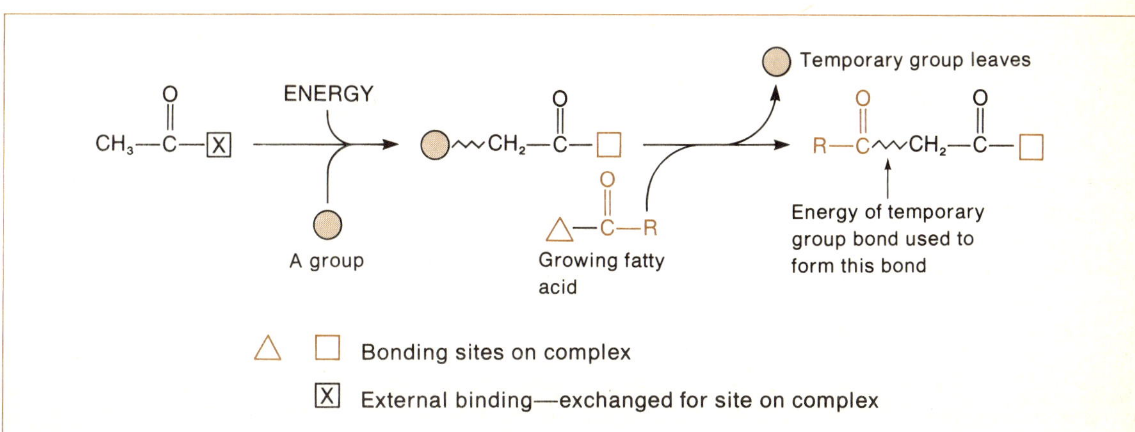

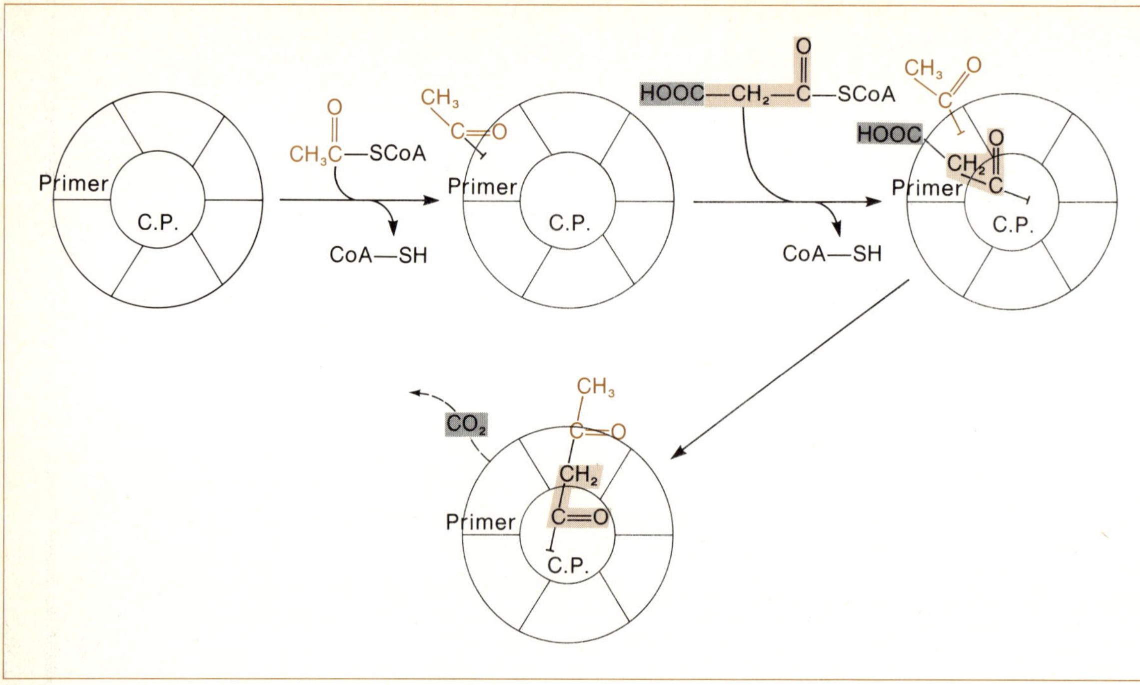

Figure 20.4
Fatty acid synthesis: initiation
A new fatty acid is born by attaching a structure to each of the binding sites—the primer and the carrier protein—then transferring one to the other to form the four-carbon nucleus upon which the new fatty acid will be assembled.

provided by ATP. This is done at another location in the cytoplasm, not at the complex.

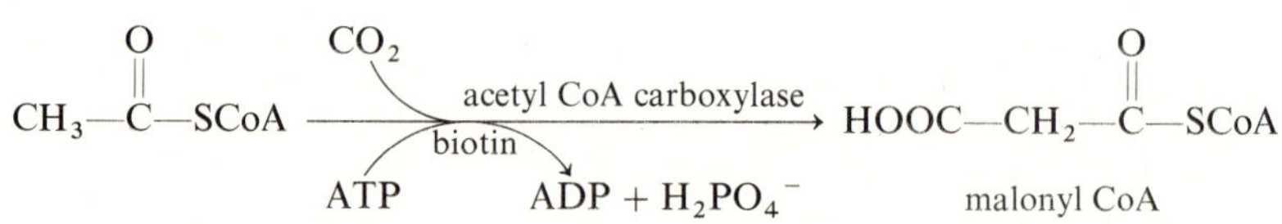

The fatty acid synthesis complex has two binding sites. One is the carrier protein. The other is a primer position and is on one of the other six proteins. At the first turn of the synthesis cycle, an acetyl group is transferred from acetyl CoA, ending up on the primer binding site. Then the malonyl group is transferred from malonyl CoA to the carrier protein. With a group bound to each site we are ready for the chain-extending reaction. The acetyl group leaves the primer site and displaces (replaces) the carboxyl of the malonyl group. The product is acetoacetyl bound to the carrier protein (Figure 20.4).

The acetoacetyl group circulates around the enzymes of the complex, returning to the starting position as a butanoyl group. This is transferred to the primer site. A new malonyl group is added to the carrier protein. Then the entire butanoyl group displaces the malonyl carboxyl group. Next time around the cycle a hexanoyl group is

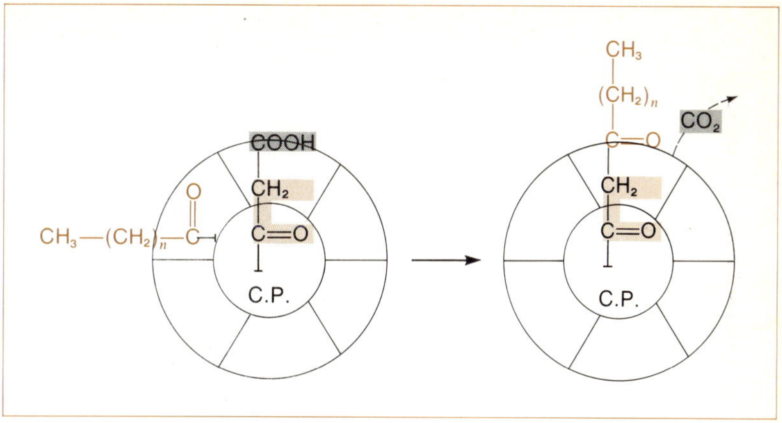

Figure 20.5
Fatty acid synthesis: extending the chain
At each turn of the cycle the chain is extended by two carbons by a sequence very like the initiation step (Figure 20.4).

produced. The process continues until a completed fatty acid is assembled and released (Figure 20.5).

Exercise 20.1

a) In what order do the events of Figures 20.2, 20.4, and 20.5 occur?

b) Where on the finished fatty acid are the two carbons of the first primer acetyl group located?

Choose one of the following:

1. In the middle somewhere.
2. At the methyl end ($CH_3CH_2(CH_2)_n$—COOH).
3. At the carboxy end ($CH_3(CH_2)_n$—CH_2COOH).

20.6 Photosynthesis

In the first part of the chapter we emphasized that biosynthesis is essentially a process in which the carbon compounds substrates are reduced rather than oxidized, as happened in catabolism. Reduction requires that energy be supplied to the reactions. The oxidation of other compounds could provide the energy for this reduction. However, we do not generate energy from oxidation of carbon, then use that energy to reduce carbon with 100% efficiency. Some energy becomes unavailable, in a sense is lost, in chemical reactions. Furthermore, much of the energy generated by carbon oxidation is not used for biosynthesis at all. It is used to contract muscle, to transport substances across membranes, and for other functions. Over a period of time, if we had to support the reducing of one carbon by the oxidation of another, living organisms would utilize all the reduced carbon on earth, thus committing suicide. Fortunately there is a mechanism by which the solar energy we hear so much about today is used to cause

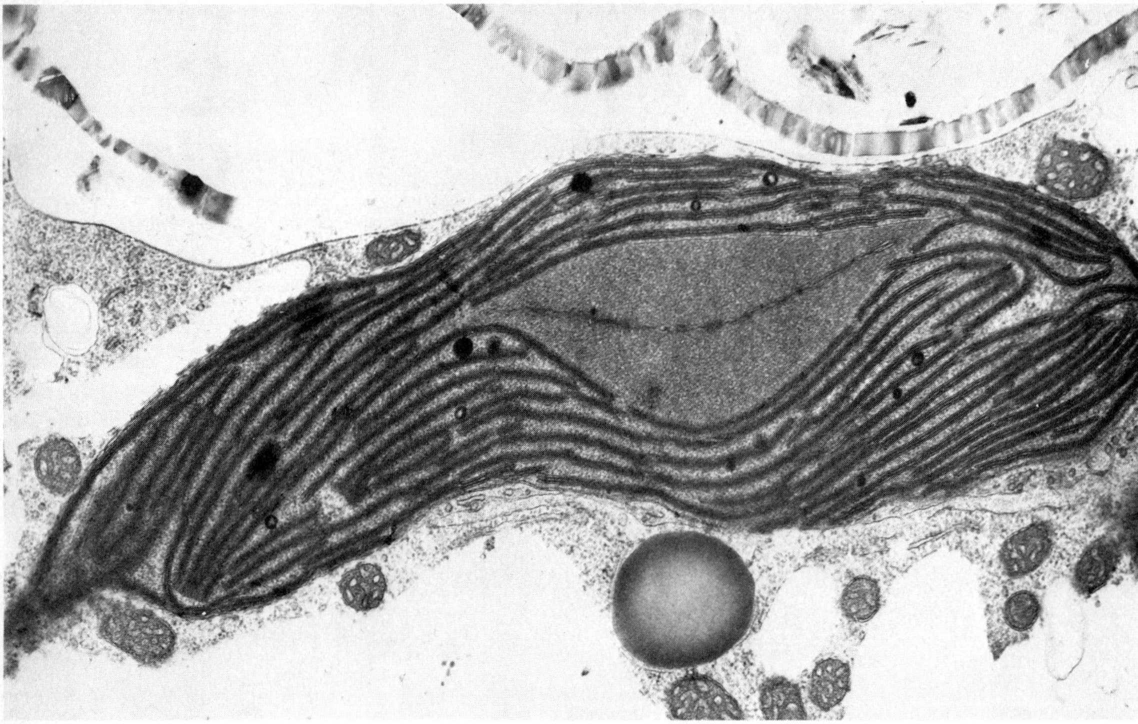

Figure 20.6
A chloroplast
The long bands are grana, structures containing the photosynthetic pigments. The cell wall is the thick band across the top. Much storage space shows, including the large white areas, the dark circle immediately below the chloroplast (starch), and the "smile" in the center of the chloroplast. (Photo by Linda S. Goad, Great Lakes Division, University of Michigan.)

the reduction of carbon. This mechanism is the process of *photosynthesis.* Photosynthesis is, in fact, the source of a high percentage of the energy we use today, because coal and oil are both reduced carbon compounds produced by photosynthetic organisms long ago.

Using solar energy to heat a house requires that we put up a collector to capture the light energy. In plants the collector is in a subcellular particle called the *chloroplast.* (See Figure 20.6.) Chloroplasts are similar to mitochondria. They have a double membrane construction, and the functional region is inside the inner membrane. In this interior are tiny sacs containing pigments that absorb light. You are probably already aware that in plants chlorophyll is one of these pigments. In addition, we find terpenes such as carotenoids and xanthophylls. In algae we find phycobilins, whose structures are based on porphyrin rings that have been broken open.

In plants these pigments are organized into two kinds of assemblies, photosystems 1 and 2. In each of these assemblies are large numbers of molecules of chlorophyll and carotenoids and a single molecule of a key chlorophyll. In photosystem 1 this key chlorophyll is called P700, in photosystem 2, P680. In a photosystem, light energy is absorbed by one of the pigments, converting the structure of the pigment to a high-energy form. This energy is transmitted through the photosystem to the key chlorophyll. The energy is then used to add

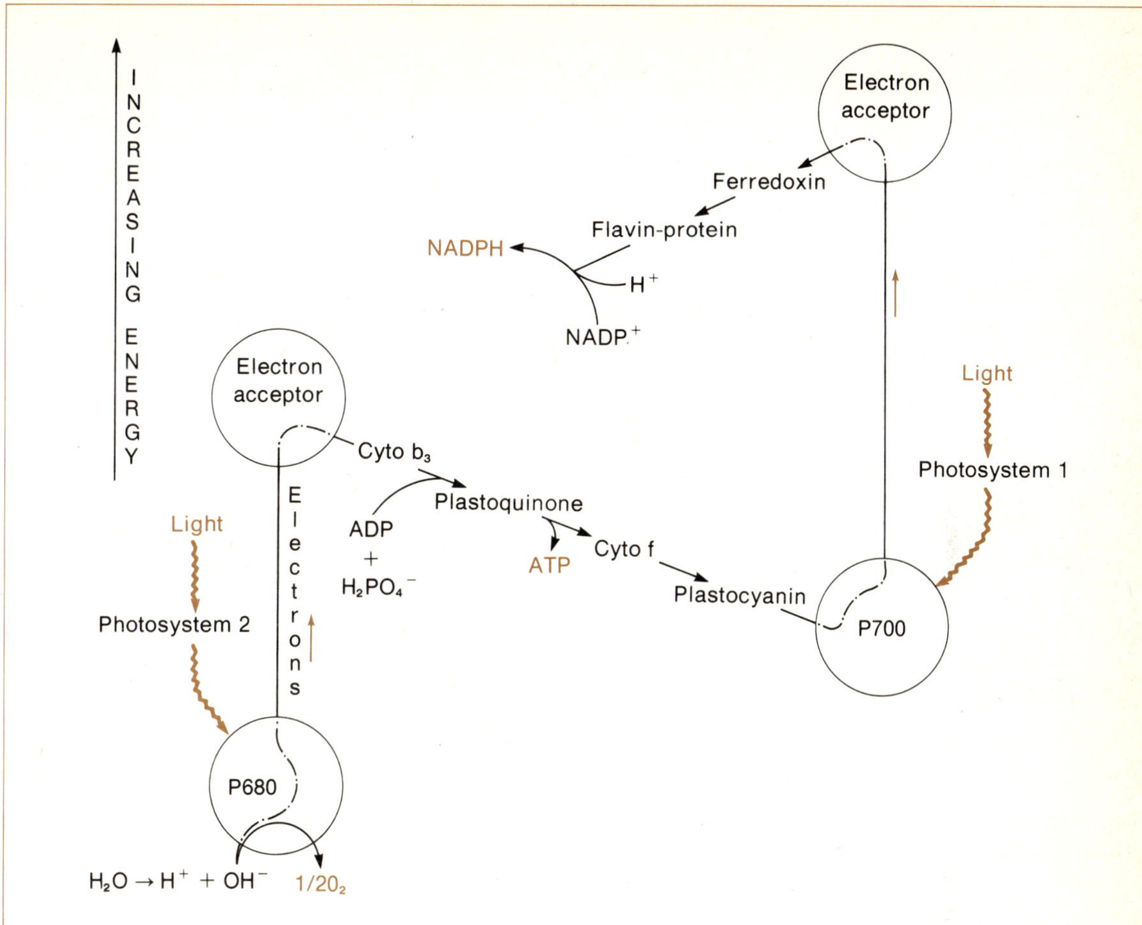

Figure 20.7
Noncyclic photophosphorylation
Noncyclic photophosphorylation generates ATP and oxygen, and reduces NADP.

electrons to a receptor molecule, in other words, to reduce the receptor molecule. A chain of compounds passes these electrons along like a bucket brigade, very much the way electrons were transferred in mitochondria during oxidative phosphorylation. As the electrons are transported, energy is released. Some of this energy is used to add phosphate to ADP, again just as in the mitochondria.

We do not understand the reactions of photosynthesis as well as we do the reactions of oxidative phosphorylation. However, we do recognize that there are two pathways in use, noncyclic (Figure 20.7) and cyclic (Figure 20.8) photophosphorylation. First consider the noncyclic pathway. Light energy absorbed by photosystem 2 (P680) takes electrons from hydroxide ion, generating oxygen. This is the source of oxygen in photosynthesis. These electrons are passed through an acceptor and a series of intermediate carriers, including two cytochromes, plastoquinone and plastocyanin. They complete their journey through this portion of the system by being donated to photosystem 1

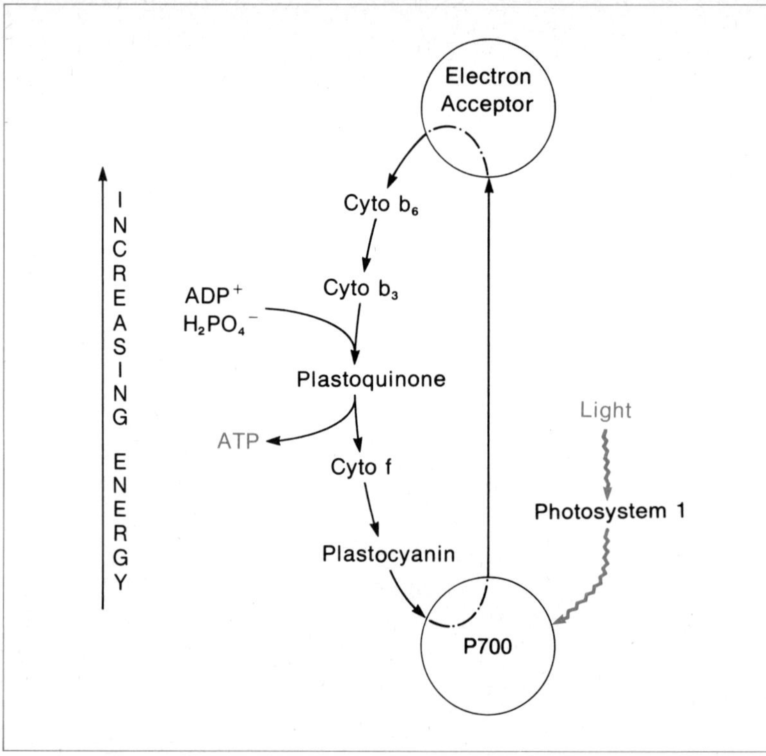

Figure 20.8
Cyclic photophosphorylation
Cyclic photophosphorylation uses light energy to synthesize ATP.

(P700). As the electrons pass along this system, released energy is used to add phosphate to ADP.

Photosystem 1 absorbs light and uses the energy to transfer the newly received electrons to another acceptor. From this acceptor they are transferred to an iron-protein (ferredoxin), then to a flavin-protein. Finally they are used along with hydrogen ions to reduce NADP. In summary, electrons from hydroxide ion with the help of light energy at two light receptors use hydrogen ion to reduce NADP. The oxygen of the hydroxide ion becomes free oxygen. Energy released along the way generates ATP from ADP and phosphate.

Cyclic photophosphorylation is much simpler. Only photosystem 1 is involved. Electrons do not need to be taken from an external source (such as hydroxide ion). Instead they are circulated around and around, driven by light energy absorbed by the photosystem and generating ATP. The pathway uses a part of the noncyclic pathway plus cytochrome b_6. In summary, cyclic photophosphorylation uses light energy to convert ADP to ATP.

Carbon reduction is a separate process, not driven by light. The reactions require the reduced cofactor and ATP. The key reaction is the conversion of a pentose to two trioses by capture of CO_2.

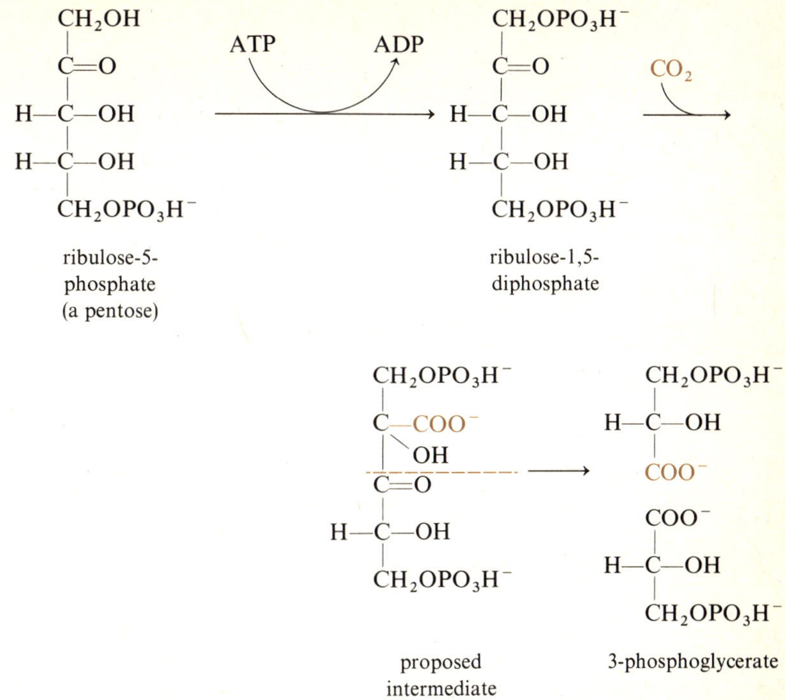

The 3-phosphoglycerate can become glucose by reversal of carbohydrate catabolism. Reduced cofactor is necessary to convert 3-phosphoglycerate to glyceraldehyde-3-phosphate.

Summary

Biosynthesis is a reductive process requiring energy. The pathway may be simply the reverse of catabolism, as with glucose biosynthesis. Only two steps of glucose catabolism are not reversible, and these are bypassed by alternative enzyme catalyzed reactions. Glucose synthesis begins with pyruvate or Krebs cycle intermediates in mammals but cannot begin with acetyl CoA. Therefore the carbons of some amino acids can become part of glucose, but fatty acid carbons cannot. In plants and microorganisms the glyoxalate shunt allows acetyl CoA to generate carbohydrate.

The biosynthesis of fatty acids is an example of a pathway using a completely separate set of enzymes from those used for catabolism. Fatty acid catabolism occurs in the mitochondrial matrix with the fatty acid esterified to CoA. Fatty acid synthesis occurs in the cytoplasm at a very large multienzyme complex. The fatty acid is attached to a carrier protein in the complex during the synthesis.

All the metabolism discussed to this point produces a net oxidation of carbon. Ultimately there must be a way to replenish reduced carbon. Photosynthesis uses the energy of the sun to produce

ATP and reduced NADP. These are used to incorporate CO_2 into carbohydrate. The light-capturing process occurs in subcellular structures much like mitochondria, called chloroplasts. Chloroplasts contain pigments including chlorophyll and various terpenoid structures that absorb light energy and use it to add electrons to a single particular structure, creating a high-energy intermediate. The electrons are transferred from this structure along a series of carriers much like those of electron transport in mitochondria. In noncyclic photophosphorylation the electrons are originally taken from an hydroxide ion of water, generating free oxygen. The energy released as the electrons are passed along is used to generate ATP, and the electrons are finally used to reduce NADP. In cyclic photophosphorylation fewer electron carriers are involved, the same electrons are cycled around repeatedly, and ATP is the only product.

Problems

1. Write the structure and name of the next product in the pathways indicated (include cofactors):

- ***Glucose synthesis***

a)

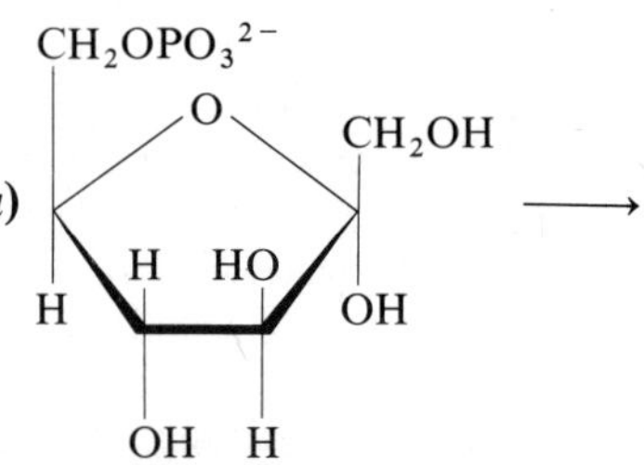

b)

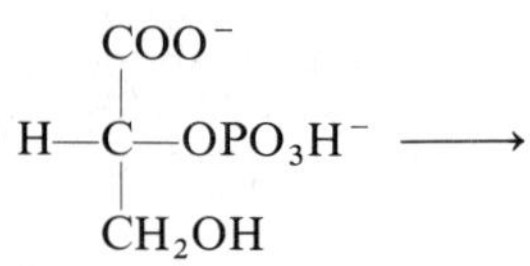

c)

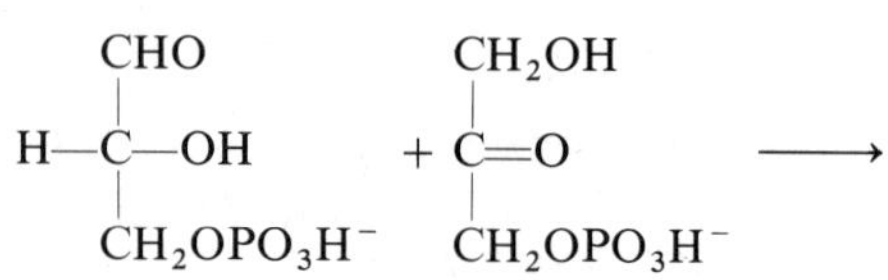

d)

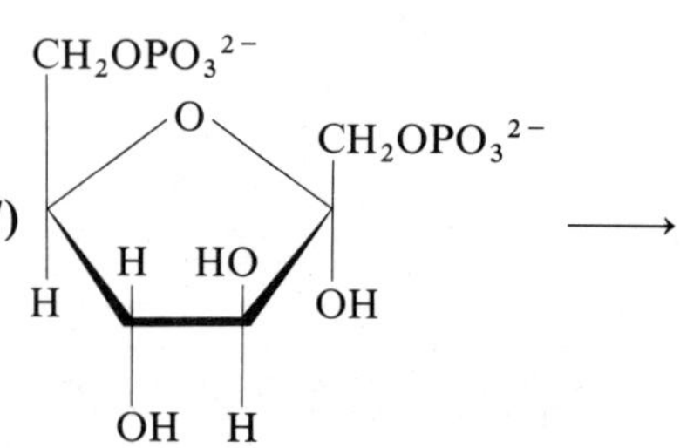

e) $CH_3C(=O){-}COO^- + CO_2 \longrightarrow$

f) $^-OOC{-}CH_2{-}C(=O){-}COO^- \longrightarrow$

- ***Fatty acid synthesis***

a) $CH_3{-}CH_2{-}CH_2{-}C(=O){-}CH_2{-}C(=O){-}S{-}\begin{pmatrix}\text{enzyme}\\ \text{complex}\end{pmatrix} \longrightarrow$

b) $CH_3{-}CH_2{-}CH_2{-}CH{=}CH{-}C(=O){-}S{-}\begin{pmatrix}\text{enzyme}\\ \text{complex}\end{pmatrix} \longrightarrow$

c) $CH_3{-}C(=O){-}SCoA + CO_2 \longrightarrow$

d) $^-OOC{-}CH_2{-}C(=O){-}SCoA + CH_3{-}C(=O){-}S{-}\begin{pmatrix}\text{enzyme}\\ \text{complex}\end{pmatrix} \longrightarrow$

2. Using the glyoxalate shunt, show the reactions by which the carbons of $CH_3{-}C(=O){-}SCoA$ become part of $HC(COO^-)(CH_2OPO_3H^-){-}OH$

Bonus: Pretend that the methyl of the acetyl CoA is radioactive. Is the radioactivity incorporated into the 3-phosphoglycerate, and if so, at what position?

3. Scientists were aware early that plants that were actively engaged in photosynthesis took CO_2 from the atmosphere, returned oxygen to the atmosphere, and produced new carbohydrate. This series of reactions was hypothesized:

$$CO_2 \xrightarrow{\text{light}} \text{reduced carbon} + O_2$$

$$6\ \text{reduced carbon} + 6H_2O \longrightarrow \underset{\text{glucose}}{C_6H_{12}O_6}$$

Does our present understanding of photosynthesis support or disprove this hypothesis?

4. A later hypothesis described the light-catalyzed reaction as being the following:

$$2H_2O + 2A \xrightarrow{\text{light}} 2A \cdot H_2 + O_2$$

where A was proposed as an as yet unknown hydrogen acceptor. Does our present understanding of photosynthesis support or disprove this hypothesis? If it supports the hypothesis, what do we believe A to be?

5. Suppose you run an experiment with an unusual isotope of oxygen (O^{18}). You prepare a sample of CO_2 that contains the O^{18} in place of the usual O^{16}. This is provided to a plant engaged in photosynthesis:

$$6CO_2^{18} + 6H_2O \longrightarrow C_6H_{12}O_6 + 6O_2$$

Does the O^{18} appear in the O_2 produced?

6. In photosynthesis, water serves as an electron donor. There are purple bacteria that carry on photosynthesis in a fashion very similar to the pathway described in the chapter for green plants. A major difference between the pathways is that the bacteria use H_2S rather than H_2O as an electron donor. How will the products of photosynthesis of purple bacteria differ from the products of photosynthesis in green plants, namely $C_6H_{12}O_6$ and O_2?

7. Some compounds to be used by the fatty acid synthesis pathway are prepared with radioactive labels at specific carbons as shown below. Predict where in a molecule of palmitic acid produced by this pathway you would find the labeled carbon.

$$CH_3CH_2CH_2CH_2CH_2CH_2CH_2CH_2CH_2CH_2CH_2CH_2CH_2CH_2CH_2\overset{\overset{O}{\|}}{C}-OH$$

a) Starting with $\overset{*}{C}H_3-\overset{\overset{O}{\|}}{C}-SCoA$.

b) Starting with $^-OOC-CH_2-\overset{\overset{O}{\|}}{C}\overset{*}{-}SCoA$.

c) Starting with $^-OO\overset{*}{C}-CH_2-\overset{\overset{O}{\|}}{C}-SCoA$.

8. Using the list of enzyme names below, deducing the function of the enzyme from the name, label each of the positions of the complex diagram.

a) β-ketoacyl reductase

b) acyltransferase

c) β-ketoacyl synthetase

d) carrier protein

e) enoyl hydratase

f) primer carrier

g) enoyl reductase

9. In the reactions for the synthesis of glucose from pentose and CO_2,

a) Where would the phosphoglycerate be labeled if the system were given carbon-14 labeled CO_2; ribulose-5-phosphate labeled at carbon 5; at carbon 1; at carbon 2?

b) Write the reactions that complete the conversion of 3-phosphoglycerate to glucose.

c) It is readily seen where the ATP is used (as generated in photosynthesis), but where is the reduced cofactor needed?

21
Membranes and Transport

Objectives

In this chapter we:

1. *See the relationship between structure and function in membranes.*
2. *Learn about types of transport across membranes: passive transport, active transport, and cotransport.*
3. *Understand the relationship between ion transport and nerve impulse transmission.*
4. *Learn how nerve impulses are transmitted at the synapse.*
5. *See the relationship between transport and the triggering of muscle contraction.*

I once asked a class what they thought membranes are. A student answered that they were the lines you draw around things you see in the microscope. True enough; as we look at an animal or plant cell through the microscope, much of what we can see and name as "structure" is composed of membrane. Why are membranes so important? Membranes are barriers. The cell membrane functions in part as a bag that prevents the contents of the cell from drifting away. The membranes of the mitochondria hold the machinery of a portion of metabolism in a concentrated and organized package. The nuclear membrane encloses the cell DNA and some of the machinery required to make use of the information coded on the DNA.

By now you recognize that a cell is composed of a large number of different chemicals. These chemicals do not move across membranes at random. The movement of chemicals across a membrane is controlled by the membranes. With membranes we can assure that when one of these chemicals is needed in the cell and is either made or put there, it can be kept in the cell, or perhaps even in a particular part of the cell. Membranes provide an organized functional pattern for the cell.

21.1 Movement of Solutes Across Membranes

Membranes are not barriers in the sense that brick walls are barriers. They are more like brick walls with gates. Molecules and ions can cross these membrane barriers. However, the membrane itself selects what may or may not pass through it. So they are more like brick walls whose gates have guards who ask to see your badge before you can enter or leave. Why is this important? Consider that enzymes are very much affected by pH, concentration, and metal ions. If all these enzymes are to operate properly, we must provide them with a protected environment. When we study isolated enzymes in the laboratory, we are very careful to avoid even traces of contamination and to buffer the solution against changes in pH. Membranes allow the cell to maintain the correct volume, pH, and ion composition—to regulate its internal environment very closely.

The cell can bring necessary metabolites such as glucose or amino acids through the membrane as needed. Interestingly, it can actually concentrate them in the cell interior. It is as if the guards at the gates went out and located everyone who was looking for a job as a gardener and herded them through the gate so they were inside where they were needed. There needs to be an outward flow, too, by which waste products are ejected through gates in the wall.

Finally, membranes can deliberately establish an ionic concentration difference between the interior and exterior of the cell. Such a concentration difference, or *gradient*, gives the membrane an electrical potential that makes it possible for the membrane to respond to stimulation. Such a membrane is called *excitable* and is important in nerve and muscle function, for example.

21.2 Membrane Structure

In Chapter 7, membranes were described as being constructed of a double layer of phospholipid molecules. The polar ends of these molecules form the surface of the membrane, and the nonpolar ends mix with one another in the membrane interior—a good example of hydrophobic bonding (Chapter 11). The lipid bilayer is pictured as having protein molecules loosely attached to the surface, tightly embedded in the surface, or inserted so that they run from one face of the membrane to the other. Some of the proteins attached to or embedded in the membrane surface are enzymes organized in patterns that allow them to pass the metabolic intermediates from one enzyme to the next in a highly efficient fashion.

21.3 Gates or Permeases

Having considered the construction of this barrier, now let us look for a moment at transport, that is, the movement of substances across the membrane. Some of the proteins that are inserted completely through the membrane are the guarded gates in the wall. Proteins associated with the transfer of molecules or ions across the membranes are called *porters*, *carriers*, or *permeases*. Notice that the last term has an -ase ending like an enzyme name. The permeases act very much like enzymes. They bind the compounds to be transported in the same way enzymes bind substrates to their active sites. They are very specific, distinguishing between structurally similar compounds, such as optical isomers, just as enzymes do.

21.4 Passive Transport

In transporting a compound across a membrane, we are moving it from one environment to another. Many of you in a science class have dropped a single crystal of the intensely purple salt potassium permanganate into a cylinder of water. As the crystal dissolves, the purple color at first is localized around the crystal. With time the color moves up the cylinder until eventually it is uniformly distributed throughout the water. A solute spontaneously moves from a region of high concentration (the vicinity of the crystal) to a region of low concentration (the water at the top of the cylinder). This is called *diffusion* and is nothing more than the random distribution of objects that are moving aimlessly. The membrane blocks compounds from simply diffusing into the cell. A compound outside the cell for which there is a permease in the membrane, however, can enter. If the concentration of that particular compound is higher outside the cell than inside, it can diffuse into the cell through the permease gateway. We call this *passive transport.* In passive transport the compound does not become more concentrated inside the cell than it is in the cell surroundings.

21.5 Active Transport

Suppose we want to take molecules from a region of low concentration to a region of high concentration. It can be done, but we will have to expend energy to succeed. Using an absurd example, we can pick up individual molecules in the region of low concentration with super-tiny tweezers and drop them in the region of high concentration. Obviously this involves energy. Remember the discussions of biosynthesis (Chapter 20) or of oxidative phosphorylation (Chapter 18). Reactions that require energy can be made to happen by linking them to other reactions that release a greater amount of energy. Overall, energy is released, allowing the system to work. We can move particular kinds of molecules into a cell even when they already are more concentrated inside the cell, as long as we tie the transport to an energy supply.

Compounds can be transported against a concentration gradient in two ways. What would you guess would be a source of energy for some biochemical process, based on what you know already? Score yourself a point if you named the hydrolysis of ATP. Transport against a concentration gradient at the expense of ATP energy is termed *active transport.*

21.6 Cotransport

To understanding the other mechanism of supplying energy for transport against a gradient, it is necessary to accept and understand this statement: If it requires energy to move a compound into a region of greater concentration, then moving it to a region of lower concentration releases energy. We can move a compound to a more concentrated region by coupling its transport to the movement of another compound from a region of high concentration to a region of lower concentration. This is like the children's seesaw in the park. One child is going up at one end because another is coming down at the other end.

The energy released by one movement is supplied to drive the other transfer. This mechanism is sometimes called *cotransport.*

Exercise 21.1

The transport of three substances (A, B, and C) across a cell membrane is under study. The following observations are made:

a) Compound A can enter the cell from a $0.001M$ solution and eventually reach a concentration of $0.01M$ inside the cell. The transport is accompanied by the hydrolysis of ATP.

b) Compound B can enter the cell from a $0.001M$ solution but never exceeds that concentration inside the cell. There is no apparent link to ATP hydrolysis.

c) Compound C can enter the cell from a $0.001M$ solution, but as C disappears from the external solution, an equivalent amount of A appears. ATP is not being hydrolyzed at that time.

Describe each of these as an example of passive transport, active transport, or cotransport.

21.7 Examples of Transport

One of the most studied membranes is the one surrounding the red blood cells. It is possible to empty these cells of the normal contents, primarily hemoglobin, without breaking the membrane. The intact but empty membrane, called a *cell ghost*, can then be filled with whatever we please. This is obviously an ideal system to study.

We find that red blood cells have permeases for glucose. In fact, it is estimated that there are about 300,000 glucose permeases per cell. Glucose transport is an example of passive transport. It is allowed to enter the cell, but only when the concentration inside the cell is low relative to the extenal concentration (Figure 21.1).

Consider next sodium and potassium ions. Red blood cells, in fact all animal cells, are found normally to have a higher concentration

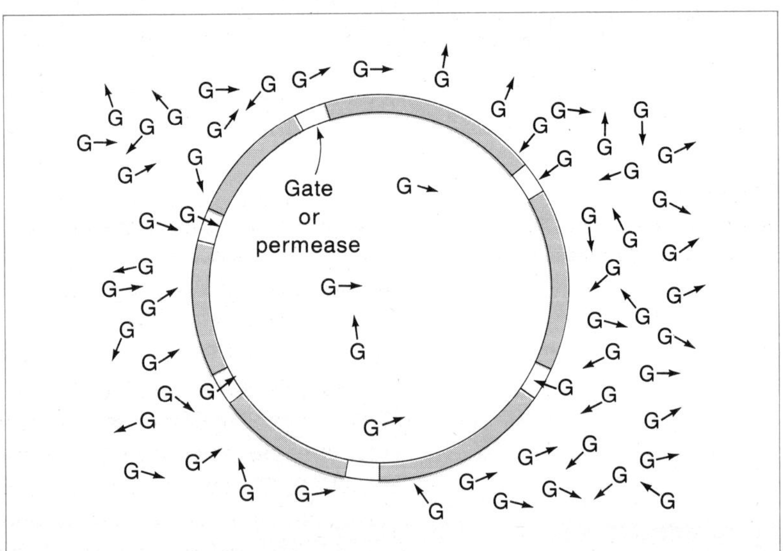

Figure 21.1
Passive transport: glucose entering red blood cells
Glucose molecules move randomly from the region of high concentration by way of the permeases into the lower concentration of the cell interior.

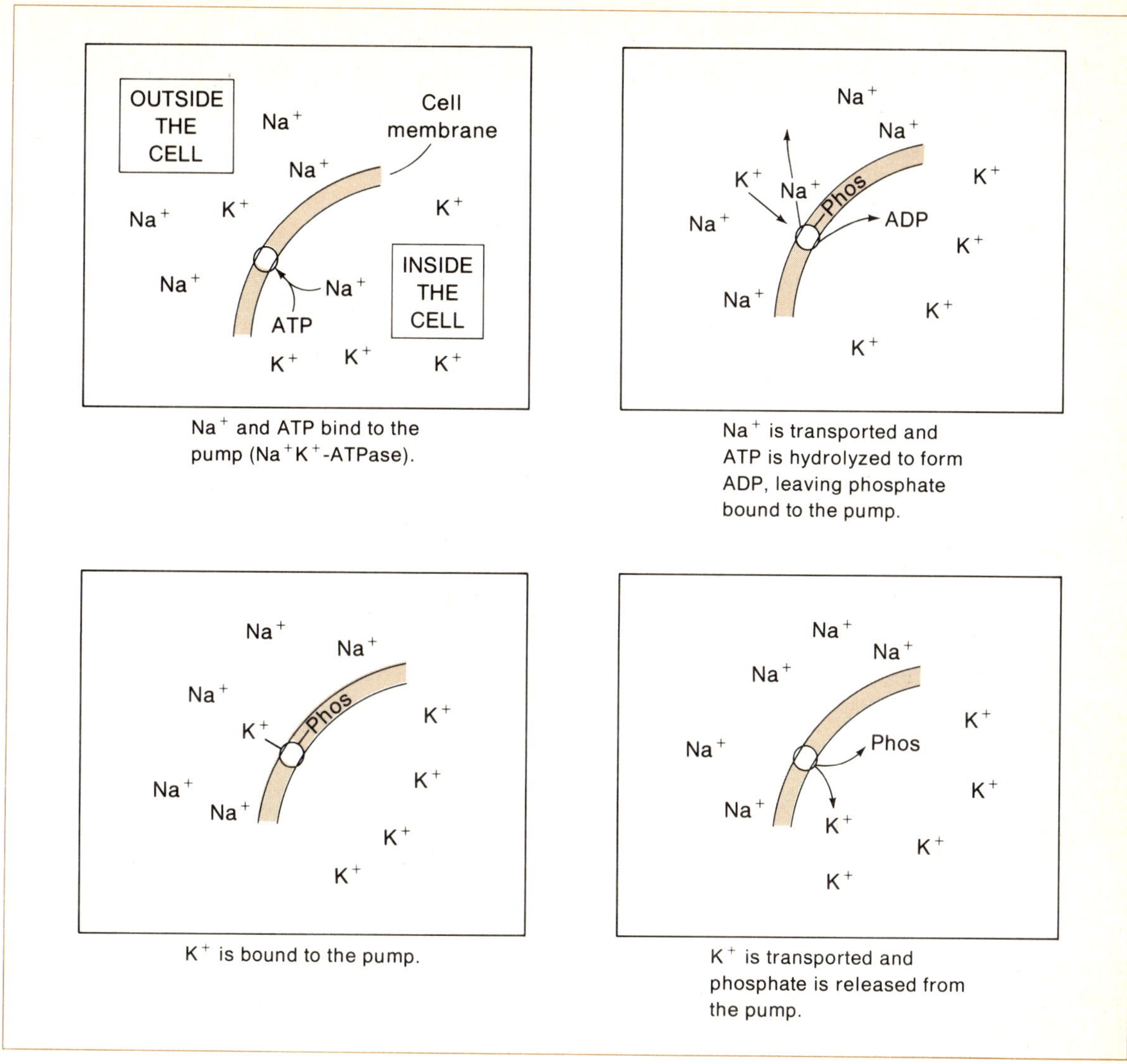

Na^+ and ATP bind to the pump (Na^+K^+-ATPase).

Na^+ is transported and ATP is hydrolyzed to form ADP, leaving phosphate bound to the pump.

K^+ is bound to the pump.

K^+ is transported and phosphate is released from the pump.

Figure 21.2
Active transport: Na^+ and K^+ crossing the cell membrane

of K^+ inside the cell than is present in the fluid surrounding the cell. Typical values would be 4.5 mg/ml potassium in cellular fluids but only 0.16 mg/ml in blood plasma. The opposite is true of Na^+, which is at a concentration of about 3.2 mg/ml in the plasma but less than 0.5 mg/ml in the cellular fluids. How is this maintained? In the cell membrane there is a protein called Na^+K^+-ATPase, or the Na^+K^+ pump. This pump binds sodium ion in the presence of ATP, producing ADP, a phosphorylated pump, and transported sodium ion. In the presence of potassium ion the phosphate is removed from the pump, producing free phosphate and transported potassium ion. The pump operates only to move potassium ion into the cell and sodium ion out of the cell. This is an example of active transport (Figure 21.2).

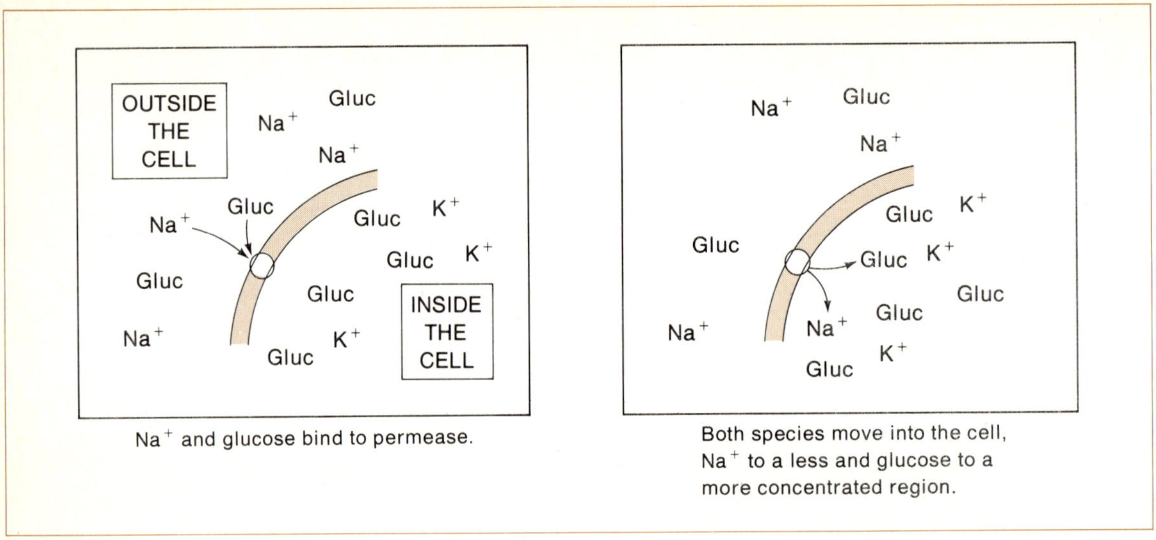

Figure 21.3
Cotransport: glucose into the liver cell

The transport of glucose into cells other than the red blood cell can be performed against a concentration gradient. This is done by transporting a sodium ion at the same time. The sodium ion is moving to a region of much lower concentration, releasing the energy necessary to transport the glucose to the region of greater concentration. This is an example of cotransport. Amino acids are similarly cotransported. Notice that the return of the sodium ion to the outside of the cell later requires the use of the Na^+K^+ pump, which in turn means using ATP. Ultimately both active transport and cotransport depend on ATP energy (Figure 21.3).

21.8 Nerve Action

Nerve cells have long fiber extensions that carry signals or impulses. By these cells sensory messages are carried to the brain and signals are sent from the brain to the muscles. The nerve cell membranes are examples of excitable membranes. Nerve cells have the usual high potassium and low sodium ion concentration in the cytoplasm. As a result, the electrical potential across this membrane is positive on the outside and negative inside the membranes. When a nerve cell is triggered to transmit an impulse, there is an inward rush of sodium ion, which neutralizes and then reverses the potential across the membrane. This is called *depolarization*. A wave of depolarization moves along the nerve fiber. Then, just after the sodium ion has rushed in, potassium ion moves out. The membrane potential is restored, again as a wave along the fiber, by the potassium ion moving out. Once the potential is restored, the abnormal permeability of the membrane to ions ceases.

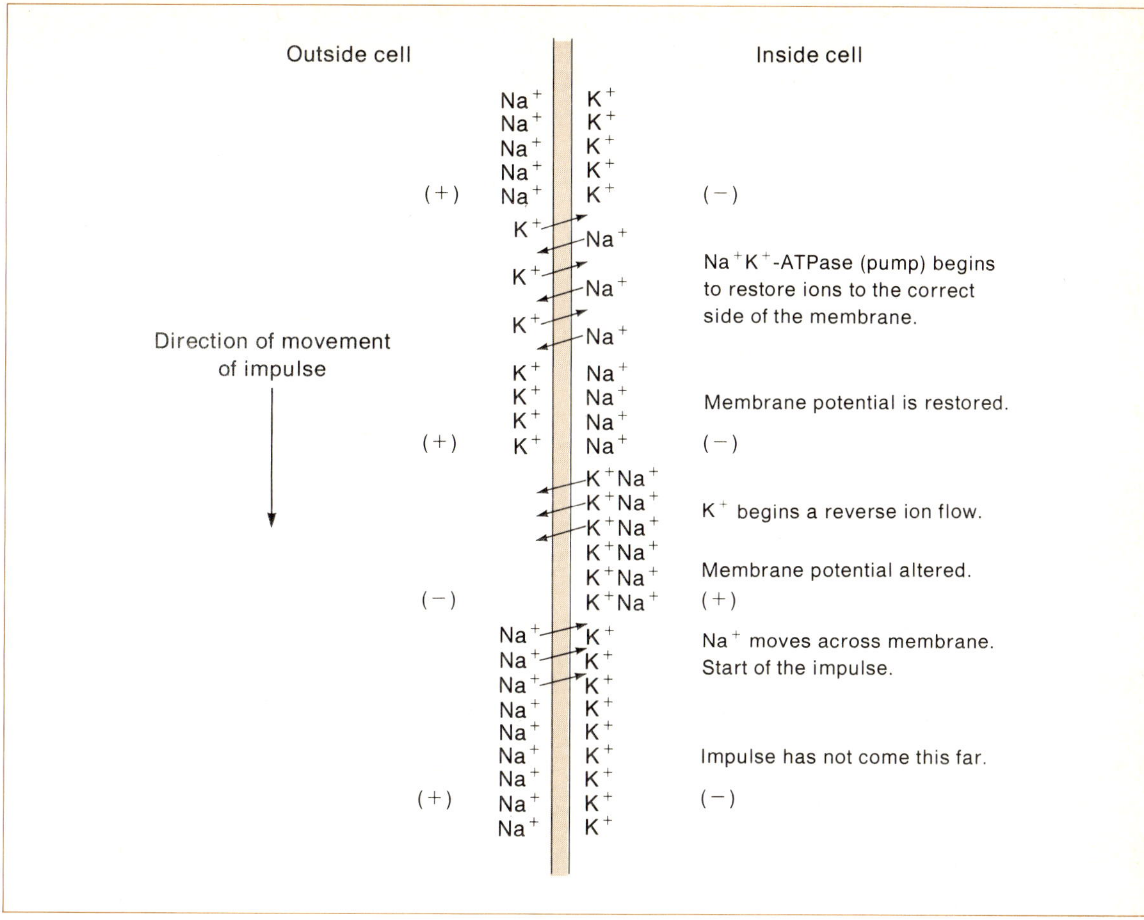

Figure 21.4
Transmission of a nerve impulse

The Na^+K^+ pump must now be employed to put the ions back across the membrane to their normal locations. Because the pump operates on an ATP fuel, nerve impulse transmission requires ATP (Figure 21.4).

21.9 Synapses

Nerve cells are connected either to one another or to other organs, such as muscles, by structures called *synapses*. We shall first look in detail at the synapse between a nerve and a voluntary (striated) muscle. Up to the end of the nerve fiber, the impulse is moving along a continuous membrane structure. However, the membrane is not physically attached to the muscle membrane in a fashion that allows the wave of depolarization to keep on going. There is a gap called the *synaptic cleft*. This gap is bridged by a chemical messenger or transmitter. At the portion of the nerve fiber involved in the synapse (the presynaptic

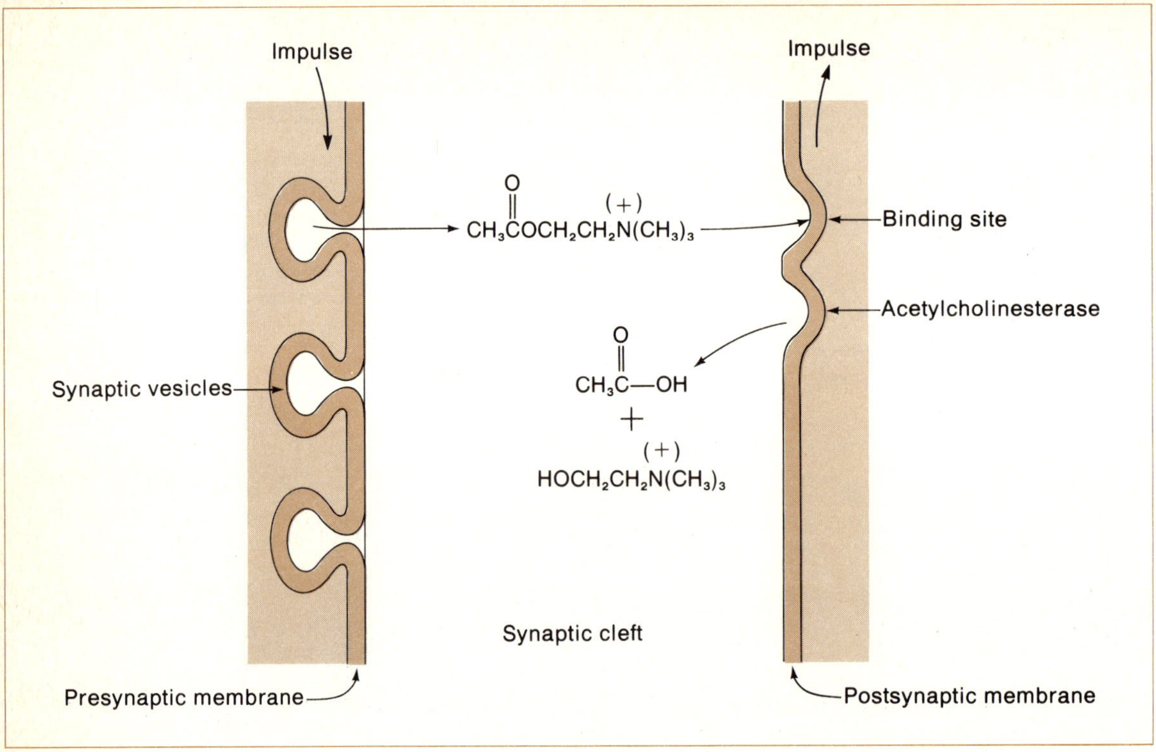

Figure 21.5
The synapse: nerve and striated muscle
The impulse is moving from the nerve ending (left) to the muscle (right).

membrane) there are little sacs called *synaptic vesicles*. The nerve impulses cause the vesicles to empty the chemical messenger, acetyl choline, into the synaptic cleft. At the muscle side of the synaptic cleft (the postsynaptic membrane) a protein binds the acetyl choline. The act of binding stimulates a movement of ions across the muscle membrane. Then at a second and presumably adjacent protein on the postsynaptic membrane, acetyl choline is hydrolyzed into choline and acetic acid. This second protein is the enzyme acetylcholinesterase. This process may sound somewhat awkward and perhaps even a bit slow. However, it isn't at all slow. Several hundred separate impulses can be carried per second (Figure 21.5).

21.10 Stimulation of Muscle Contraction

Membranes and transport are involved in stimulating and ending muscle contraction. A system of membrane-walled tubes runs throughout the muscle. The fluid inside the tubes is rich in calcium ion. When the nerve impulse reaches these membranes, they respond by allowing Ca^{2+} to flow by passive transport into the vicinity of the myofibrils. The Ca^{2+} stimulates hydrolysis of the ATP already present and starts the contraction process. The membrane contains a Ca^{2+}-ATPase molecule, in other words, a calcium ion pump. This pumps the Ca^{2+} back out into the tubes at the expense of ATP and allows the muscle to relax.

21.11 Other Types of Synapse

Other junctions between nerves and organs utilize the same concept for transmission of the nerve impulse but involve different compounds in the process. The nerve-smooth muscle synapse uses norepinephrine as a transmitter. These transmitters are released from presynaptic membrane vesicles, just as acetyl choline was. In the same fashion that acetyl cholinesterase was shown to convert the acetylcholine to a pair of fragments that do not affect transmission of impulses, monoamine oxidases inactivate norepinephrine once it has left the vesicle.

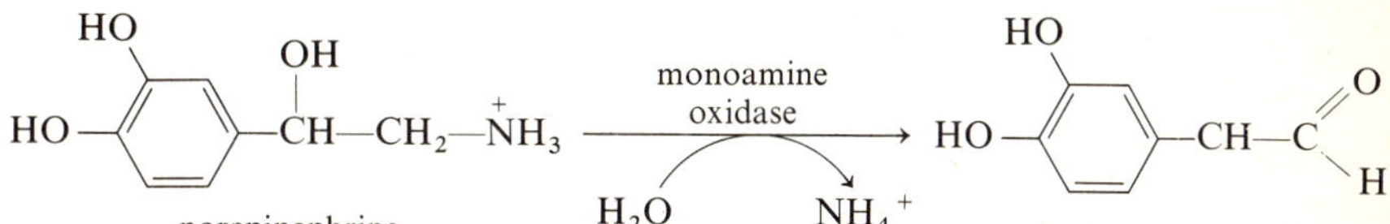

Elsewhere, we encounter epinephrine, dopamine, and serotonin serving as transmitters.

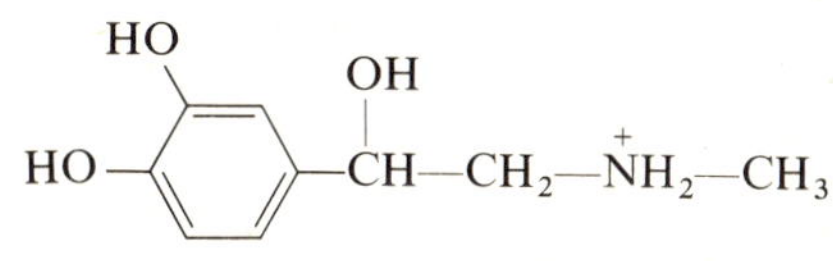

epinephrine

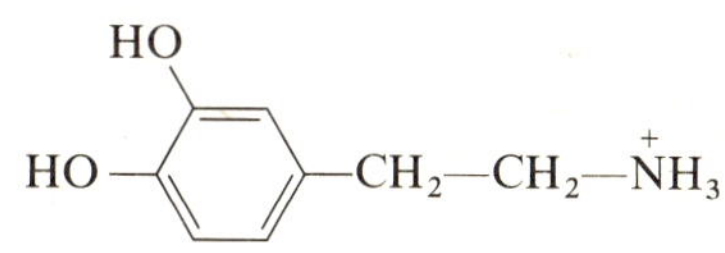

dopamine

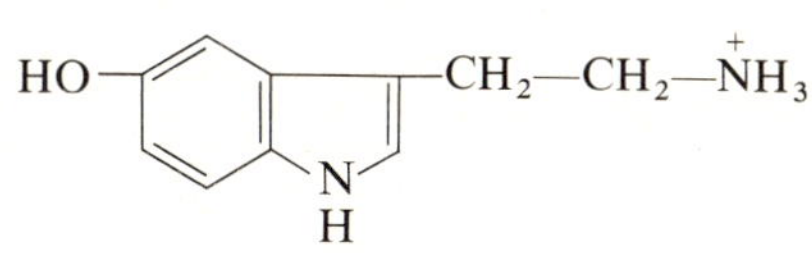

serotonin

21.12 Inhibitors of Synapse Function

Acetylcholinesterase has serine, and therefore an alcohol group, at the active site. One group of compounds that can react with the serine are phosphate derivatives. Of these, diisopropylfluorophosphate is used in research as a reagent to detect and measure serine at the active site of enzymes (Figure 21.6).

Chemical agents that interfere with the synaptic transmission mechanism have a variety of physiological effects. These vary with the type and extent of interference. In the extreme case, total inhibition of the synapse leads to muscle paralysis. The reaction shown earlier with

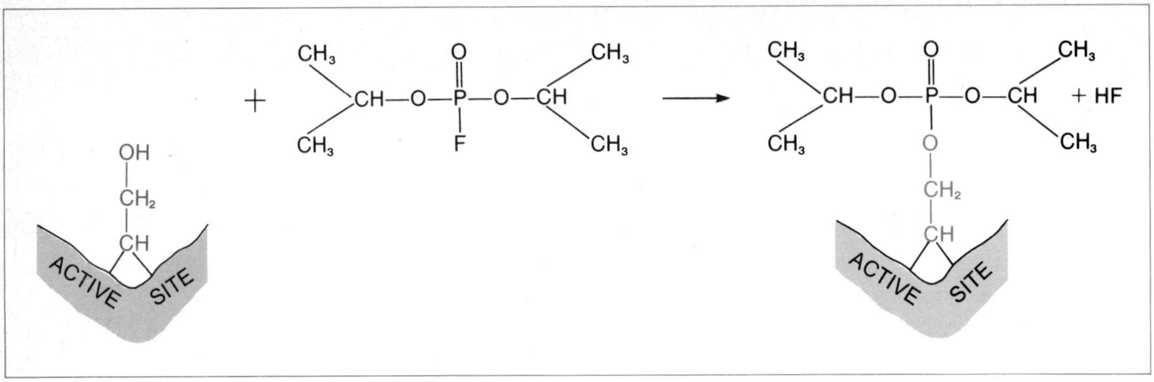

Figure 21.6
The inhibition of acetylcholinesterase at the synapse by diisopropylfluorophosphate

the serine at the active site of acetylcholinesterase destroys the activity of the enzyme. Inhibition of the enzyme results in transmitter not being hydrolyzed. The effect varies from excessive stimulation because of an oversupply of transmitter to paralysis because of a failure to clear transmitter to prepare for the next impulse. Substances classed as nerve gases, such as tabun or sarin, and the insecticides parathion or malathion operate in a similar fashion.[1]

---

[1] Once compounds were created that can inactivate acetylcholinesterase, it became important to generate a nerve gas countermeasure or antidote. Hydroxylamine was found to be an effective chemical group to remove the phosphate esters from the enzymes. However, hydroxylamine is itself poisonous at the dosage that allows it to counteract these agents. Chemists set out to create a new compound that would do the following:

1. Remove phosphate derivatives from acetylcholinesterase.
2. Have lower toxicity than hydroxylamine.
3. Have a structure that would fit the active site of the enzyme. This would allow very low doses to be specifically bound to the enzyme.

Pyridine aldoxime methiodide was created to fit these needs and is effective at very low concentration. Its positive charge binds to the enzyme just as the positive charge of the choline amine group does. The hydroxyl amine group is just the right distance away to reach the phosphate ester.

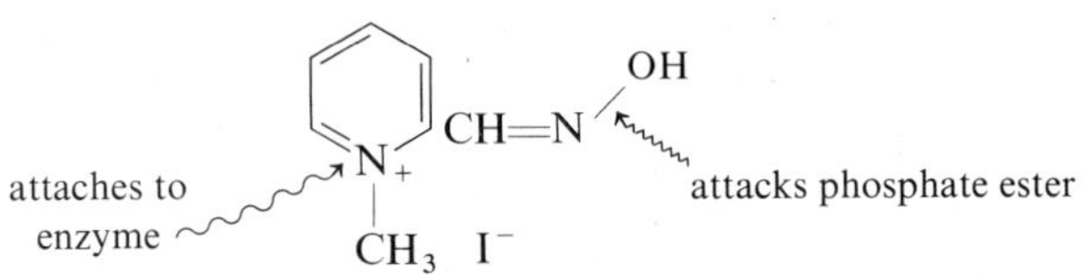

pyridine aldoxime methiodide

Can you see how valuable it was to understand the biochemistry of the enzyme, the chemistry of the phosphate ester, and enough synthetic techniques to produce just exactly the structure that the other studies indicated should work?

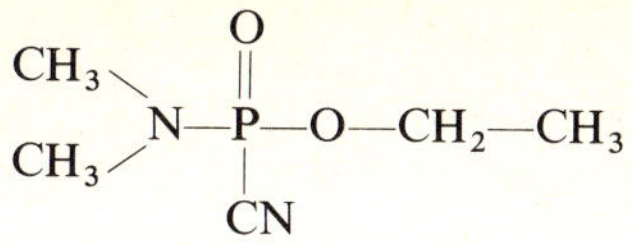

tabun

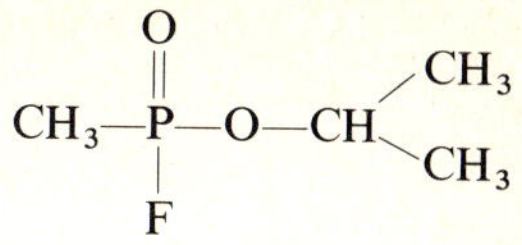

sarin

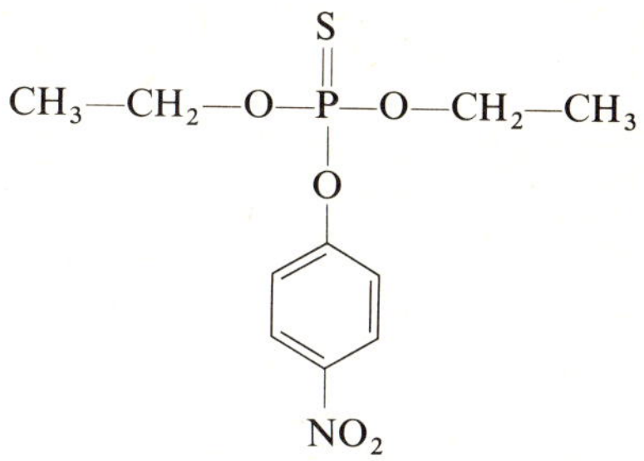

parathion

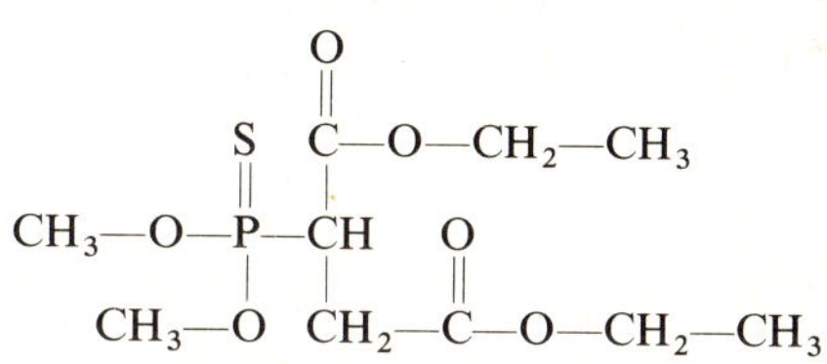

malathion

The insecticides are converted from —P═S to —P═O by the insects themselves, at which time they become very toxic.

Other poisons are known to react with the active site serine. Carbomoyl structures such as physostigmime or neostigmime bromide are effective.

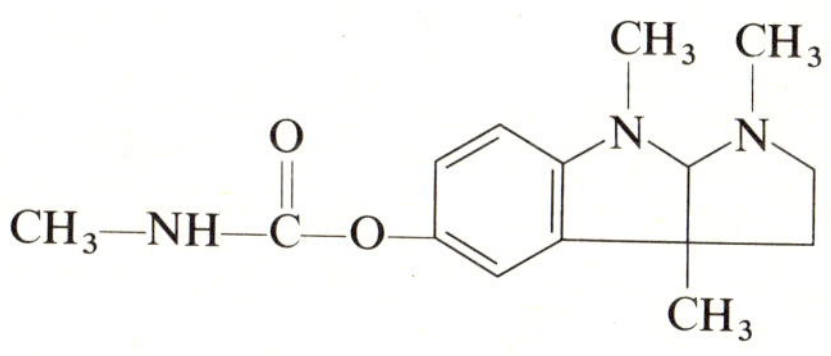

physostigmime

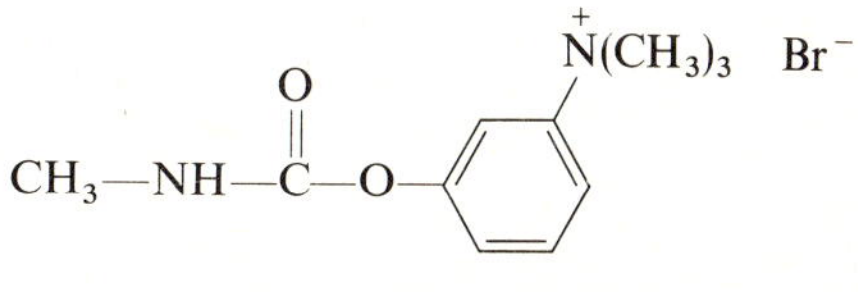

neostigmime bromide

The reaction is as shown in Figure 21.7.

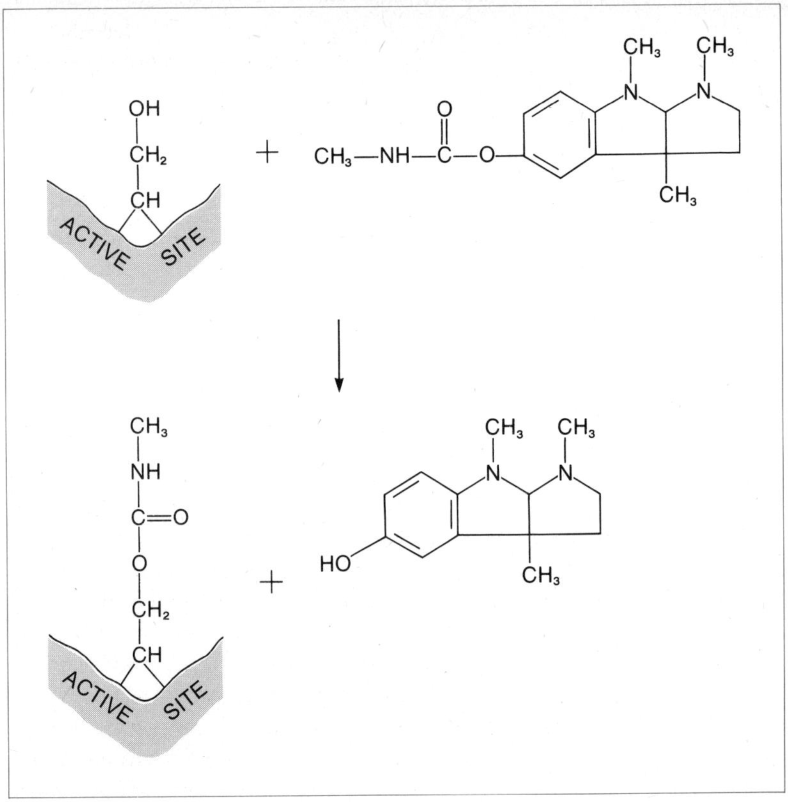

Figure 21.7
The inhibition of acetylcholinesterase at the synapse by physostigmime

Other compounds are active because they resemble acetylcholine structurally. These bind temporarily to the active site of acetylcholinesterase and prevent acetylcholine from being hydrolyzed. Succinyl choline iodide is very similar to acetyl choline and is an effective muscle relaxant in surgery. Normally it is hydrolyzed slowly, and its effects wear off shortly after administration of the drug stops.

$$HOOC{-}CH_2{-}CH_2{-}\overset{\overset{\large O}{\|}}{C}{-}O{-}CH_2{-}CH_2{-}\overset{+}{N}(CH_3)_3 \quad I^-$$

succinyl choline iodide

Exercise 21.2

Locate and label the succinyl group and the choline group. What is the functional group joining them?

The active agent of curare, *d*-tubocurarine chlorine (Figure 10.1), similarly binds with acetylcholinesterase, blocking the enzyme.

Amphetamines may function as stimulants by inhibiting the monoamine oxidases of the synapse—smooth muscle synapse. Enzymes that add a methyl group to the 3-hydroxy of the benzene ring of norepinephrine also destroy the ability to transmit impulses.

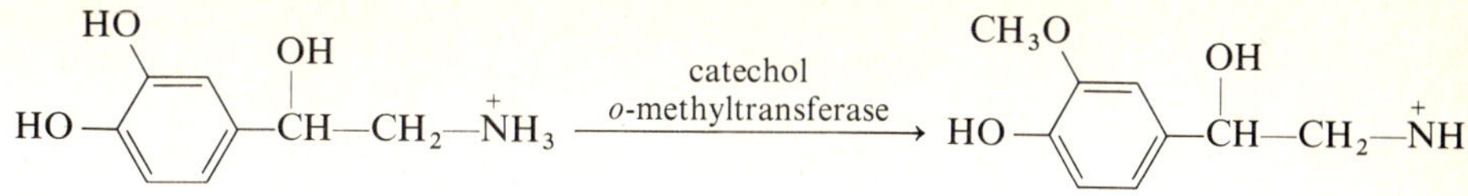

active transmitter inactive

The tranquilizer reserpine may function by blocking return of the transmitter molecules to the vesicles, where they are protected from the monoamine oxidases. The unprotected transmitters are inactivated by the enzyme, leaving the synapse deficient in transmitters and as a result less responsive to stimulation.

Summary

Membranes are lipid barriers with attached or inserted proteins. They separate the cell contents from the environment and subdivide the cell into functional regions. Compounds may cross these barriers if the membrane includes proteins called permeases that are specific for transport of the particular molecule. This transport may be passive; that is, it may resemble diffusion across the membrane. In active transport the molecule may be concentrated by transport across the membrane, the energy for this being supplied by the hydrolysis of ATP. In cotransport, movement of a molecule into a region of greater concentration is driven by the energy released when another molecule moves to a less-concentrated region at the same time.

Concentration of a substance by movement across a membrane is well exemplified by the concentration of potassium ion inside the cell and the concentration of sodium ion in the extracellular fluids. This situation is maintained by an Na^+K^+ pump, a permease performing active transport. This separation of the ions establishes an electrical potential across the membrane. In nerve cells this potential is destroyed by the influx of sodium ion during the transmission of a nerve impulse and is restored by an outflow of potassium ion immediately afterward. At the end of the nerve fiber the message is transferred to an organ through a synapse. In the synapse, sacs of a chemical called the transmitter open as the nerve impulse arrives, emptying the transmitter into the gap between the nerve and the organ, the synaptic cleft. Receptor sites on the other side of the cleft bind the transmitter and stimulate the receptor organ. Enzymes destroy the transmitter, restoring the system to its initial state to await the next impulse. Muscles initiate contraction by stimulation of membranes in the muscle to allow a flow of calcium ion into the region of the contractile protein. This causes contraction to begin. Relaxation occurs when the calcium ions are pumped back out.

Agents that interfere with the function of the synapse cause paralysis, relaxation, stimulation, or tranquilizing effects, depending on the degree of inhibition, the type of inhibition, and the type of synapse affected.

Problems

1. Which of the following statements are true?

a) Any compound that finds its way into the circulation, for example by injection, will enter the cell.

b) Any compound that is inside a plant or animal cell is uniformly distributed throughout the cell.

c) Even if the concentration of an amino acid in a cell is higher than the concentration in the blood plasma, the cell may still transport the amino acid from the blood plasma into the cell.

d) A red blood cell may not concentrate glucose from the plasma.

e) Cotransport can raise the concentration of a compound in the cell to a level higher than it is in the cell environment.

f) Cotransport will cease to operate immediately if the supply of ATP fails.

g) Cotransport will cease to operate eventually if the supply of ATP fails.

2. Ouabain is a steroid obtained from plants that inhibits the Na^+K^+-ATPase of the cell membrane by preventing the removal of phosphate as transport of potassium ion occurs.

a) Will sodium ion continue to be transported actively in a membrane poisoned with ouabain?

b) Will ouabain prevent nerve impulse transmission?

c) Will the cotransport of glucose into cells be inhibited eventually by ouabain?

d) Will the transport of glucose into red blood cells be inhibited eventually by ouabain?

3. Fluoroacetate inhibits the Krebs citric acid cycle. How do you explain the loss of ability of cells poisoned by fluoroacetate to perform active transport?

4. List all the individual events beginning with—(1.) Sodium ion rushes into the nerve cell—that must occur before the cell is ready to accept another nerve impulse. If the nerve is transmitting 500 impulses per second, what is the length of time within which these events must occur?

22
Body Fluids

Objectives

In this chapter we:

1. *Learn about water as a solvent for living systems.*
2. *Encounter the various body fluids.*
3. *Consider the composition and concentration of body fluids, especially blood and urine.*
4. *Consider the need to maintain the correct pH in various fluids and study the mechanism for maintaining that correct pH.*
5. *Learn about the cells found in blood and about blood types.*
6. *Learn about the function of the kidney and the formation of urine.*

Water is a unique substance. I used to be involved in panel discussions open to the public where we discussed such topics as "What is the likelihood of life elsewhere in the universe?" After the astronomer had developed the idea that the number of planets could be incredibly large, I held out for the requirement that an acceptable planet should have a temperature range that would permit liquid water to exist. To be honest, I haven't devoted years of research to the idea, but it is hard to imagine another compound that is as suitable to the existence of life.

Water is a very small molecule. It is easily transported, is fluid so that free movement of dissolved substances is not hindered, and can interact intimately with groups on the surface of large molecules. Yet in spite of its small size, it requires relatively large amounts of energy to convert it to a gas, a very important characteristic for land animals who would otherwise suffer more from loss by evaporation. This high energy requirement exists because water molecules are very strongly attracted to one another. This attraction must be overcome in order for molecules to go from the liquid to the gaseous state. The intermolecular attraction is based both on the polarity of water and on its ability to hydrogen bond to neighboring water molecules. The polarity of water allows other polar, therefore chemically reactive, molecules to be dissolved. The polarity of water permits the charges on ions to be dispersed, thereby assisting in the separation of the ions. Water can serve as a proton donor or acceptor; thus it is the foundation for a system of acid–base chemistry. Yet it is only moderately reactive chemically itself.

Life, we assume, began in the oceans. When animals slowly and painfully crawled out onto dry land, their adaptation to this challenging new environment centered around maintaining and protecting an internal ocean environment around all the cells. Animals could only stray farther away from the ocean as they became adept at preventing body fluid loss and controlling the salt balance internally. Let us look more closely at these body fluids.

22.1 Body Fluids

Water is 45–75 percent of the total body weight. Why such a broad range? Need I say more than to state that muscle tissue is very water rich, whereas adipose tissue is very low in water content. This water is often described as being in three types of locations (Figure 22.1). About half the water is inside cells. Another large fraction of it is generally dispersed throughout the body, either in contact with the outside of the cells or being circulated to a place where it can come into such contact. Here we would include (1) the interstitial fluid, the fluid bathing the cells; (2) the lymph, the interstitial fluid in process of returning to the

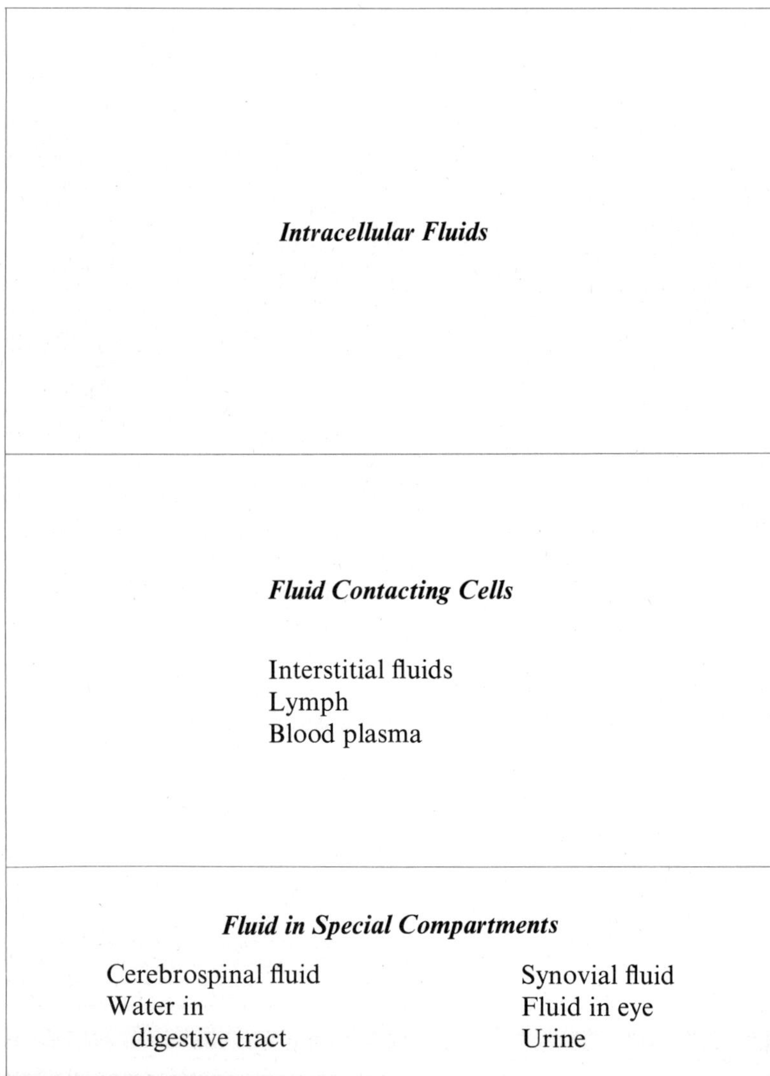

Figure 22.1
Body fluids
The three major components of total body fluids are represented here by areas that approximate the proportion of the whole each comprises.

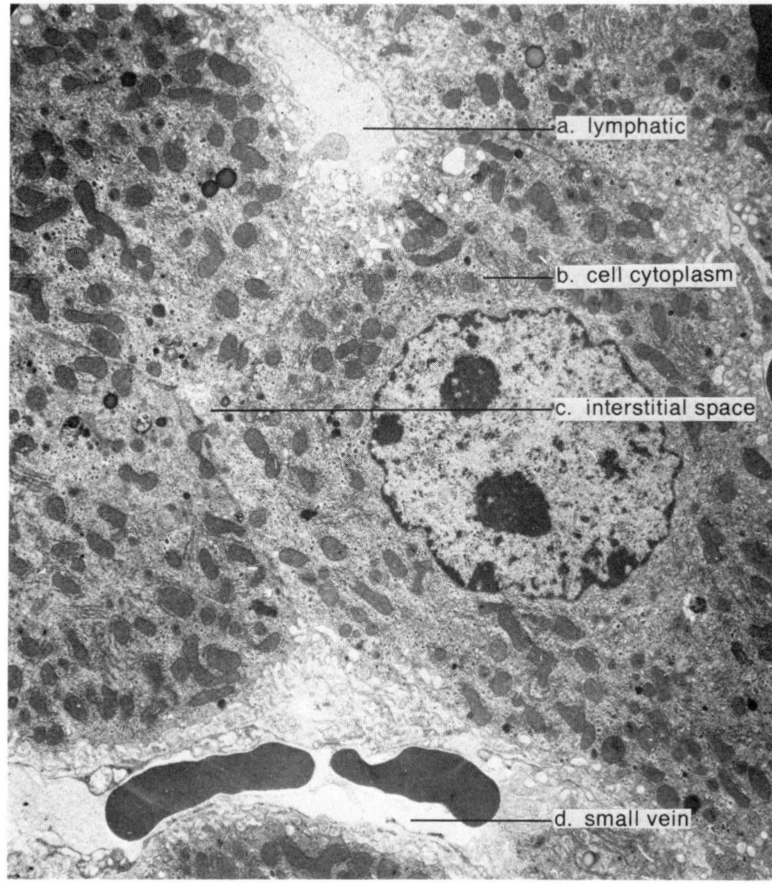

Figure 22.2
Fluids in tissues
The electron micrograph of liver illustrates various body fluid locations. (Photo by Johanna Huang and Cris Edwards, Electron Microscopy Laboratory, Eastern Michigan University.)

blood stream; and (3) the blood plasma (Figure 22.2). The third category of body fluid picks up the special cases, extracellular fluids that are enclosed in special compartments rather than being in or around the general body cell population. Examples include the cerebrospinal fluid (the fluid bathing the brain and spinal cord), the synovial fluid that is the lubricant of the joints, the fluid in the eye, and the urine.

The water intake and loss equal about 1.5 liters daily. Not all of the water is taken in as "beverage" water, that is, as coffee, tea, soft drinks, or water itself. Water also enters body fluids as a product of the metabolism of food.

$$C_6H_{12}O_6 + 6H_2O \longrightarrow 6CO_2 + \underset{\text{metabolic water}}{6H_2O}$$

Metabolic water amounts to about 300 ml per 2000 cal of nourishment, or about 20 percent of the daily intake. Some desert animals minimize their loss of water and utilize this metabolic water for their needs. Water is lost by perspiration, in exhaled air, and as urine. The volume of perspiration varies considerably with the external temperature, body temperature, and muscular activity. When the volume of perspiration is high, the volume of urine will be correspondingly less.

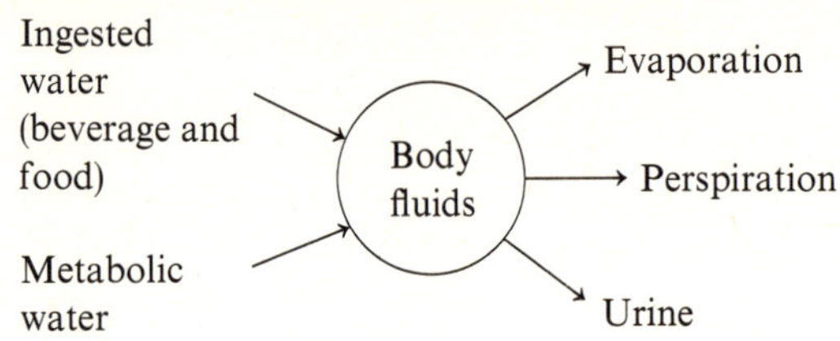

22.2 Fluid Properties: Osmotic Pressure

The fluids in these various compartments differ in composition quite extensively, as seen in Table 22.1. In addition to composition, some general solution properties are important. Concentrations of dissolved materials, sometimes expressed in terms of osmotic pressure, are certainly an important consideration. What is osmotic pressure? Solutions of different concentration separated by a permeable barrier move solvent and solute so as to adjust each side to the same concentration. When two solutions, in this case aqueous solutions, are separated by a barrier that allows only movement of solvent between the compartments, the solvent moves from the more dilute into the more concentrated solution, increasing the volume of the more concentrated solution (and so diluting it). In this case the osmotic pressure is the pressure that can be generated by the movement of solvent to produce uniform concentration. The cellular fluid and the interstitial fluid are two compartments separated by such a barrier, the cell membrane. If for some reason the concentration of the interstitial fluid increases, water moves out of the cells, dehydrating the tissue. If the interstitial fluid becomes dilute, water moves into the cells, causing a swelling of

Table 22.1
Concentrations of Ions in Body Fluids

Ion	*Blood Plasma*	*Interstitial Fluid*	*Intracellular Fluid*
Na^+	142	144	10
K^+	4	4	160
Ca^{2+}	5	2.5	
Mg^{2+}	3	1.5	35
Cl^-	103	114	2
$HCO_3{}^-$	27	30	8
Phosphate$^-$	2	2	140
Other$^-$	22	6	55

the tissues called *edema*. Maintaining a correct balance of concentration between cellular and interstitial fluids is an exceedingly important function of the kidneys.

22.3 Fluid Properties: Acid–Base Balance

Enzymes vary in catalytic effectiveness as pH is changed. Proteins in general can be denatured by a change in pH. Obviously the pH of body fluids is an important property. Let us look first at an overview of the acid–base picture. Most of our food is neutral. However, digestion and metabolism produce a number of acids. Hydrolysis of triglycerides, for example, yields glycerol, which is neutral, and three carboxylic acids. Metabolism produces carbon dioxide, which reacts with water to form carbonic acid. Anaerobic muscle metabolism converts neutral carbohydrate into lactic acid.

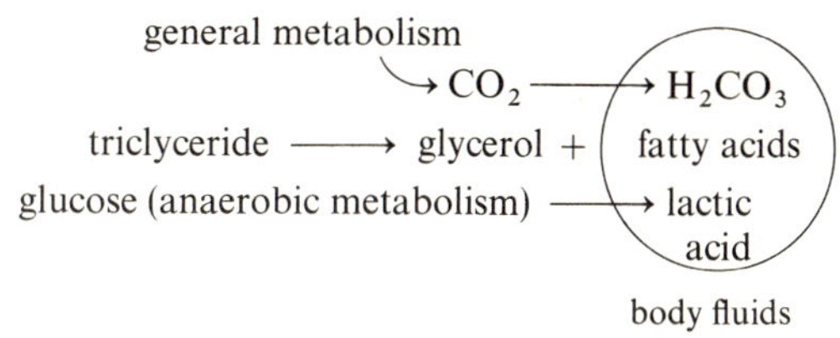

Acids resulting from metabolism are eliminated primarily by two routes. When CO_2, which forms carbonic acid in water, is eliminated through the lungs, acid has been removed from the body. Nonvolatile acid is eliminated by transfer in the kidneys to the urine. For example, in the unusual circumstances of fasting or diabetes mellitus, incomplete metabolism of fatty acids results in β-hydroxybutanoic acid or β-ketobutanoic acid (ketone bodies) appearing in the plasma and being eliminated in the urine.

The pH varies from one body fluid to another, but most are regulated quite closely. Urine is an exception, the pH varying between 4.5 to 8.2 with the diet and the state of health. The blood plasma is the most closely controlled. The normal range of plasma pH is between 7.38 and 7.46. Consider that this means an actual (H_3O^+) range of approximately 4.2×10^{-8} to $3.5 \times 10^{-8} M$, a range of acceptability of only $0.000000007 M$.

22.4 Plasma Buffers

The blood is maintained within this narrow pH range with the help of buffers. The most important buffer is the bicarbonate–carbonic acid system:

$$HCO_3^-/H_2CO_3$$

bicarbonate   carbonic acid

- ***Buffer action***

Added base:

$$OH^- + H_2CO_3 \rightleftharpoons H_2O + HCO_3^-$$

Added acid:

$$H_3O^+ + HCO_3^- \rightleftharpoons H_2O + H_2CO_3$$

This system is particularly effective at buffering acid because the H_2CO_3 formed in removing hydronium ion can further convert to water and dissolved CO_2. This buffer system is present in high concentration ($HCO_3^- = 0.021 - 0.028M$ in plasma) as a direct result of the transport of the waste CO_2 from the cells to the lungs for elimination. This waste CO_2 dissolves in the water of the plasma to form carbonic acid, some of which then reacts with hydroxide ion to form bicarbonate ion.

$$CO_2 + H_2O \rightleftharpoons H_2CO_3 + OH^- \rightleftharpoons HCO_3^- + H_2O$$

waste from cells	part of buffer system	part of buffer system

Proteins account for most of the remaining plasma buffer. The histidine R group is an important buffer group in the neutral range, and each serum albumin molecule has sixteen histidines.

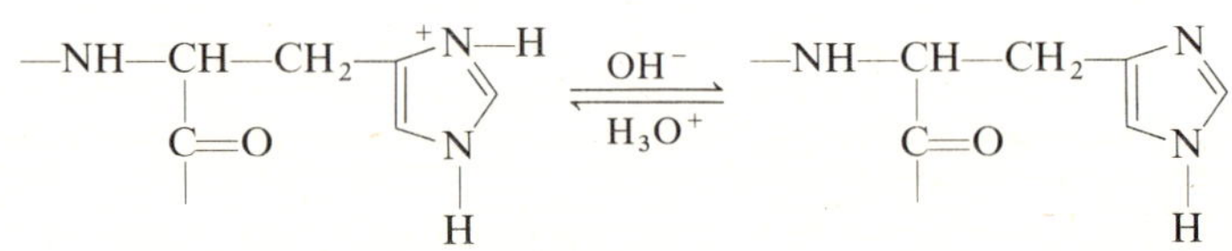

The remaining buffer capacity is associated with hemoglobin and with inorganic or organic phosphate.

22.5 *Acidosis and Alkalosis*

If the blood plasma becomes more acid than the accepted range, the condition is called *acidosis*. The respiratory control center, located in the medulla of the brain, detects the condition and stimulates the depth and rate of breathing. This results in an increase in the loss of CO_2 through the lungs so as to partially restore the pH to an acceptable level. The remainder of the restoration of plasma pH is performed by the kidney. One action of the kidney is the retention of a greater amount of bicarbonate ion. This ion is the basic component of the chief blood buffer. Notice that what is happening is a rebalancing of the ratio of bicarbonate to carbonic acid. In acidosis the ratio becomes too high.

In acidosis, $\frac{H_2CO_3}{HCO_3^-}$ is too high

HCO_3^- is conserved in the kidney increasing → $\frac{H_2CO_3}{HCO_3^-}$ → $H_2O + CO_2$ (exhaled)

The increase in HCO_3^- and decrease in H_2CO_3 restores the balance.

Acidosis results from a variety of conditions. It could originate with too low a respiratory rate, as under the influence of some drugs. A drop in respiratory efficiency can occur, for example, in the case of pneumonia, pulmonary emphysema, or fluid in the lungs (edema) caused by heart failure or some kidney diseases (Bright's disease). The abnormal metabolic production of acids can also cause acidosis. For example, vigorous exercise can cause high levels of lactic acid from the anaerobic metabolism to appear in the blood. Diabetes mellitus or fasting add acids to the blood (Problem 6). Diarrhea causes a loss of bicarbonate ion, resulting in acidosis.

When the blood plasma becomes too basic, the condition is termed *alkalosis*. The respiratory control center slows the breathing depth and rate, preserving the CO_2 to form additional carbonic acid. Among other adjustments, the kidneys allow more HCO_3^- to be lost in the urine. Alkalosis can result from the loss of stomach acid due to vomiting, or by excessive neutralization of stomach acid by antacid preparations. A pituitary tumor (Cushing's syndrome) or administration of steroid hormones that cause loss of K^+ also result in alkalosis.

22.6 Body Fluids: Electrolyte Distribution

The concentration of electrolytes will be such as to maintain correct osmotic pressure, as already indicated. However, the actual ions present vary with the fluids (Table 22.1). As indicated in Chapter 21, the sodium ion is the chief cation of the extracellular fluids, and the potassium ion predominates in the intracellular fluids. Notice also the preferential locations of calcium ion and magnesium ion, and of chloride ion and phosphate ion.

22.7 Blood

We have considered the pH, buffers, and electrolyte composition of blood already. Let us now consider this "circulating tissue" in more detail (Figure 22.3). Blood is the transport system for a variety of substances. Oxygen is carried to the tissues bound to the hemoglobin in the red blood cells. The volatile waste, carbon dioxide, is carried back to the lungs as dissolved CO_2, carbonic acid, and bicarbonate ion. Nutrients such as glucose, fats, and amino acids are transported from the digestive system to the tissues, and urea and some other wastes are transported to the kidney for elimination in the urine. Minerals such

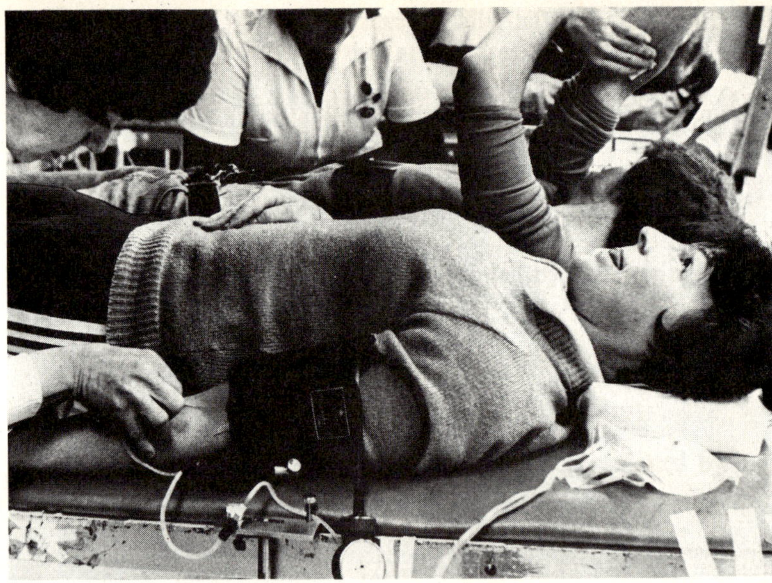

Figure 22.3
Blood donation
Although blood is a complex and vital circulating fluid, the body replaces its accidental or deliberate loss quite rapidly. (Photo supplied by *Eastern Echo* Photo Staff, Eastern Michigan University.)

as manganese and copper required by the tissues are often carried attached to specific proteins.

Blood helps to maintain a uniform environment for the tissues. For example, the body temperature is kept uniform by the circulating fluid. Extra heat generated in one region, say as a result of muscle action, is carried away by the blood. Excess heat is dissipated by increasing the blood flow through capillaries at the surface of the body, thereby allowing the heat to be lost. The careful control of pH in the blood provides for uniform pH in the tissue environment. Hormones that regulate body function are found in blood. The antibody system of defense against infection (Chapter 11) is a blood component.

22.8 Cells in Blood

The blood contains three types of suspended cells. The most numerous are the red blood cells, or erythrocytes. The red blood cells contain primarily the protein hemoglobin and therefore function to transport oxygen. Normal blood contains about 5 million red blood cells per cubic millimeter. If the concentration of red blood cells becomes significantly higher, the condition is termed *polycytemia*. Persons living at high altitudes respond to the lower oxygen pressures by producing additional red blood cells. It is interesting that some olympic competitors have been accused of manipulation of this situation. The charge was that athletes requiring great stamina, such as long-distance runners train at high altitude. Shortly before the games, some of this resultant high red blood cell concentration blood was withdrawn, then returned to the athlete just before the event. The extra red blood cells would then increase the ability of the athlete to transport oxygen to the muscles during competition.

Concentrations of red blood cells significantly lower than normal are termed *anemia*. Anemia can result from decreased production of red blood cells or from increased destruction or loss of the cells. For example, in sickle cell anemia (Chapter 9) the red blood cells have a shorter life expectancy. If the cells are produced at a normal rate and cease to function earlier, there is a reduced cell population in the blood.

The blood groups or types are due to red blood cell characteristics. Their cell membranes may contain blood group substances A and B. These are glycosphingolipids, lipids based on sphingosine and including carbohydrates in the structure. The blood group substances are potential antigens. If the red blood cell membrane contains substance A, the blood is type A. In the plasma there could be antibodies to substance B, but there would be none to substance A. Similarly, if the cell has substance B, the blood could have antibodies to substance A but not substance B. Some red blood cells have both substances, hence type AB blood, and neither antibody. The final group, type O, has neither substance and both antibodies. When typing blood, another substance, the Rh factor, may also be present. If it is, the blood is Rh(+), and if not, it is Rh(−). A serious problem arises when an Rh(−) mother has a baby with Rh(+) blood. The mother may produce antibodies against the infant's red blood cells. In general, it is apparent that blood types must be carefully checked any time blood from one individual is donated to another to avoid disasterous allergic responses.

In addition to red blood cells there are the leukocytes, or white blood cells, and the platelets. There are several varieties of white blood cells. Their function is primarily that of attacking microorganisms to prevent or combat infection. In fact, one response to infection is an increase in the white blood cell concentration above the normal level of 5,000–10,000/mm^3 of blood. The platelets are part of the mechanism for clot formation (Chapter 12).

22.9 Proteins in Blood

The fluid portion of whole blood can be separated from the cells by centrifugation. This must be done soon after the blood is drawn if it is desirable to have the preparation free of hemoglobin from broken cells. Whole blood, when drawn to be used with medical patients, must be kept cool to slow the rate of red blood cell breakage (*lysis*). Even when cool the blood must be dated and discarded after a few weeks because of excessive red blood cell lysis. It is also necessary to add a substance to prevent clotting when blood is drawn. Citrate, which chemically traps the Ca^{2+} needed for clotting, or heparin, a natural blood anticoagulant, are often used.

The array of proteins present in the blood depends on whether or not clotting has occurred. When the cells have been removed without clotting, the fluid is termed *plasma* and contains such proteins of the clotting mechanism as fibrinogen and prothrombin. The fluid of blood that has clotted is called *serum* and lacks those proteins.

Blood proteins are divided into two major groups, the albumins and the globulins. The albumins are small proteins synthesized in the liver. They along with the ions in the blood, are important to maintaining the osmotic pressure balance between blood and tissues. For example, a drop in the albumin concentration causes a movement of water into the tissues (edema). Such reduction in albumin levels can result from a liver disease that interferes with the albumin synthesis, from a type of malnutrition that results in a shortage of amino acids to synthesize the albumins, or from a kidney disease that allows loss of the albumins to the urine. The globulins are divided into subgroups. The α and β globulins, also synthesized in the liver, are associated with transport of lipids and metals. The γ globulins are the immunoglobulins or antibodies (Chapter 11). They are synthesized by plasma cells in the lymph nodes. A number of diseases cause one or another of the protein fractions to change in concentration. For example, in cases of kidney damage, the albumins escape into the urine. The γ globulins increase during an infection or chronic liver disease. In multiple myeloma, a malignant proliferation of the lymph node plasma cells, a dramatic increase in a specific γ globulin occurs. Tissue damage as in injury or myocardial infarction produces a distinct pattern called the *acute phase reaction.* Electrophoresis (Chapter 9) of plasma proteins produces patterns that are valuable diagnostically.

22.10 Kidney Function

The kidneys serve to clear the blood of unwanted dissolved material. An adult human has about 6 liters of blood in circulation. More than 1 liter per minute passes through the kidneys! Clearly all the blood is likely to be exposed to the purifiying efforts of the kidneys very frequently.

In the kidneys, blood passes through a bundle of fine capillaries called the *glomerulus.* These capillaries are in close contact with a system of fine tubules in a structure called the *nephron.* Water and small dissolved substances from the plasma filter through the capillary and tubule walls, finally ending up inside the nephron tubules. This fluid is the first step in urine formation and is called the *glomerular filtrate.* Of the liter of blood passing through the kidneys each minute, about 125 ml becomes glomerular filtrate. Under normal circumstances the glomerular filtrate is free of protein because these molecules are too large to filter across the capillary wall. The filtrate flows along a tube that has a blood capillary nearby. As it flows, such substances as glucose, electrolytes, amino acids, and water are transported back into the blood plasma. Most of the Na^+, Cl^-, and water are reabsorbed. The sodium ions are shifted back into the plasma by an active transport process (Chapter 21). As the positively charged ion is transported, the negative ions (primarily Cl^-) follow in response to the need to balance the charge. The movement of solute back into the plasma results in the plasma's becoming more concentrated than the filtrate. Water from the filtrate then spontaneously moves back into the plasma. This water shift is simply a movement from a region of high water concentration (dilute solution) into a region of low water concentration (concentrated solution). This is therefore an example of passive transport.

Consider how important this process is to land animals. A major problem to these animals is that of maintaining an internal "oceanic" environment, that is, a salt-water environment similar to the ocean in which life evolved. It is not inaccurate to view land animals with their surface coatings of waxes or oils designed to minimize evaporative loss of water as invaders of an alien environment wearing protective space suits. All this would be a futile effort if the creatures then excreted a very dilute urine, throwing away great quantities of precious water.

Grazing animals face an additional problem. Plants do not contain quantities of sodium ion; however, sodium ion is required by animals (Chapter 23). The mechanism of sodium ion retention in the kidney allows these animals to survive with only infrequent sodium ion intake, for example by deliberate intake of sodium-containing minerals at "salt licks."

In addition to the machinery for Na^+ retention, there are systems for the retention of glucose and of the various amino acids. In summary, in the kidney everything is initially discarded, then those things we wish to retain are specifically transported back into the blood.

Because many of the kidney transport processes used to recover desirable solutes from the filtrate are active transport mechanisms, a great deal of ATP must be generated in the kidney. Kidney tissue has a high rate of metabolism.

There are some hormonal controls of kidney function. The steroid aldosterone stimulates Na^+ retention. In Addison's disease the adrenal cortex produces insufficient corticosteroids, one of which is aldosterone. Some of the symptoms of this disease result from excessive loss of sodium ion in the urine. Vasopressin, a hormone of the posterior lobe of the pituitary, stimulates water retention in the kidney. There is a complex relationship controlling blood pressure involving a peptide, angiotensin, and the hormones vasopressin and aldosterone. Secretion of aldosterone and vasopressin, both of which encourage increasing the total volume of body fluids, is stimulated by the active form of angiotensin. Increasing the volume of the plasma fluids by restricting water loss in the kidneys raises the blood pressure. Understanding these relationships thoroughly can provide tools for the treatment of hypertension (excessively high blood pressure). Hypertension is an important factor leading to heart problems.

22.11 Urine

Unlike the blood plasma, the volume and composition of urine is extremely variable. In the long term, an output of too little urine (oligouria), too much urine (polyuria), or no urine at all (anuria) can

Table 22.2
Approximate Composition of Normal Urine

	Solute	Range (g/24 h)
Positive ions	Sodium	2–4
	Potassium	1.5–2
	Magnesium	0.1–0.2
	Calcium	0.1–0.3
Negative ions	Sulfate	0.2–0.6
	Phosphate	2.2–5
	Chloride	3.5–8.8
	Bicarbonate	0–3
Nitrogen compounds	Ammonia	0.5–1.2
	Uric Acid	0.24–0.6
	Urea	13–40
	Creatinine	0.8–2.1
	Amino acids and peptides	3–6.5

indicate a serious health problem. In the short run, variation in urine production is likely only to reflect variation in water and salt intake, the presence of urine-stimulating (diuretic) substances in the diet, or the relative loss of water and salts through the alternatives of perspiration or moisture in exhaled air. The kidneys, in conjunction with aldosterone and vasopressin, respond to such variations in conditions to maintain or restore the plasma to an optimum volume and composition.

For example, if you perspire heavily due to vigorous exercise and you do not drink a compensatory amount of water, the urine volume will drop. If the salt lost in perspiration is not counterbalanced in the diet, the salt content of the urine drops. However, diabetes mellitus, the disease associated with insufficient insulin production, leads to a persistent polyuria. A more extreme example, diabetes insipidus or vasopressin insufficiency, leads to the production of perhaps 50 liters of urine daily.

The pH of urine varies between such extremes as 4.7 and 8.0. A vegetable diet results in a more alkaline urine, and protein leads to a more acid urine. However, the loss of HCl by vomiting causes an alkaline shift also, and fever results in an acid shift. In fact, any adjustment required in plasma pH (see Section 22.5) produces a corresponding alteration of the pH of the urine. The dissolved solids in urine vary with diet but generally are within the ranges listed in Table 22.2.

A urine analysis is a routine procedure in a physical examination. The presence or concentrations of several compounds is of clinical significance.

1. ***Plasma albumins (proteinuria).*** The smallest of the plasma proteins may appear in the urine after severe exercise, but their continued presence is a sign of kidney disease.
2. ***Glucose (glucosuria).*** Unless blood levels are very high, glucose will be completely reabsorbed in the kidney. Excessive carbohydrate intake does cause temporary glucosuria. However, an insulin deficiency (diabetes mellitus) is suspected when glucose is found continuously.
3. ***Ketone bodies (ketonuria).*** Ketone bodies are incompletely oxidized fatty acid fragments (Chapter 16) and are in the blood and urine when fatty acid oxidation is occurring at a very high level. Fasting, forcing use of body fat, is one cause. Low carbohydrate utilization as in diabetes mellitus is another.
4. ***Blood.*** Stones or other damage to kidneys, urinary tract infections, and other conditions lead to blood in the urine.
5. ***Bile.*** Normally the liver removes bile from the blood adequately. Blockage in the bile duct preventing the passage of bile into the intestine (obstructive jaundice), or failure of the liver to remove bile from the blood (hepatitis), results in its accumulation in blood and subsequently in the urine.

Clearly, tests that analyze for levels of these compounds in blood and urine reveal a great deal of information about the state of health of an individual. A major part of a medical technologist's responsibility is to perform these tests. Furthermore, the tests must be done with great accuracy and with careful controls, because an incorrect result could be extremely misleading diagnostically.

Summary

Water is the solvent for living systems and is uniquely well adapted to the role because of its small size and high intermolecular attractive forces. It represents 45–75 percent of the body weight and is found inside the cells, in the fluids surrounding the cells, in blood and lymph, in a few special compartments such as the eyes and central nervous system cavity, and in the urine. Important solution properties of these body fluids include osmotic pressure, which relates to the concentration of dissolved compounds, pH, and the distribution of dissolved ions, particularly sodium and potassium ion.

Blood contains a number of compounds that are being transported, such as oxygen, nutrients, wastes, and hormones. Some of the transport and functions, such as the combating of disease by antibodies, is performed by the proteins in the blood. Blood also contains red blood cells, white blood cells, and platelets. Blood types (A, B, AB, and

O) depend on the presence or absence of glycosphingolipids A and B on the red blood cell membranes.

In the kidneys the plasma filters into a tubular collecting system. Cells and proteins do not enter the filtrate, but other dissolved compounds do. Specific desired solutes, such as sodium ion, bicarbonate, glucose, amino acids, and water, are returned to the plasma. The remaining fluid is the urine. Urine varies considerably in composition with water and salt intake, diet, and state of health. Analysis of urine and blood provides important information about an individual's state of health.

Problems

1. List the types of fluids found in a human being.

2. As a person diets and loses weight, what happens to the percentage of the remaining body weight that is water?

3. On the basis of what has been discussed in this chapter, explain these observations: A leaf of lettuce is placed in distilled (pure) water and becomes very firm and crisp. Another leaf is placed into salt water and becomes limp and wilted. What is the name of the solution property involved here?

4. If the kidney retains extra HCO_3^-, what happens to the pH of the blood plasma?

5. When the blood becomes more acid, the respiratory control center changes the rate of respiration. Does it increase or decrease the rate, and why does this change the blood pH?

6. Which of the following would tend to decrease the blood pH, which would increase the blood pH, and which would do neither?

a) You are diving in deep water and stay under too long.

b) You have stomach flu and are vomiting.

c) You have indigestion and take "bicarbonate" to ease the discomfort.

d) You run the mile in the shortest time you have ever achieved.

e) You are desperate to lose weight and fast for several days.

7. Which of the cells in blood:

a) Is present in greatest numbers?

b) Is involved in blood clotting?

c) Devours bacterial cells in the blood?

8. What is the difference between blood plasma and blood serum?

9. What happens to the volume of urine if:

a) A person drinks a large amount of water?

b) A person artificially raises the level in the blood of vasopressin by injection?

c) A person takes a large amount of a diuretic?

10. Very weight-conscious individuals often notice that reduction of the level of salt consumed causes an immediate weight loss. Reduction of salt in the diet is used also to control blood pressure. Can you explain these two comments in terms of the material in this chapter?

23 Bioinorganic Chemistry

Objectives

In this chapter we:

1. Learn about coordination compounds.
2. Consider the role of metals in enzyme function.
3. Learn about calcium in bones and teeth.
4. Specifically study the role of the elements iron, zinc, copper, cobalt, molybdenum, manganese, chromium, boron, selenium, arsenic, and iodine in living organisms.
5. Learn about the bacterial process of nitrogen fixation.
6. Consider the problems created by the release of cadmium, mercury, and lead into the environment.
7. Briefly consider the possibility of a relationship between cancer and some trace elements.

We have focused our attention primarily on the organic components of living organisms. The organic molecules were composed largely of carbon, hydrogen, oxygen, nitrogen, and some sulfur and phosphorus. In Chapters 21 and 22 we discussed the importance of metal ions, particularly sodium ions and potassium ions, in the body fluids. Now we consider the functions of other elements. Some of the elements are present in easily detected and measured amounts. Certainly a human being has sizable amounts of calcium, magnesium, and chlorine, along with smaller quantities of iron and zinc, and a long list of elements present only in traces.

23.1 Coordinate Covalent Bonding

Before proceeding further, it is important to review the chemistry of metal complexes (coordination compounds). For many biologically significant metals the function involves this type of bonding.

Chemical bonding (Chapter 1) occurs when atoms stabilize by eliminating or filling incomplete electron shells. Metals have only a few electrons in the outer shell and stabilize by losing those electrons and becoming metal ions. Metals also stabilize by covalent bonding. However, they do not have electrons to share as do the carbons, oxygens, and nitrogens found in organic structures. When a metal bonds covalently, the other atom or group of atoms must supply the electrons to be shared. A covalent bond wherein the electrons come from one of the two atoms only is called a *coordinate covalent bond.* The group contributing the electrons is called the *ligand*, and the product is a *complex* or *coordination compound.*

The pattern of the complex is that of a simple three-dimensional geometric shape. The number of ligands required depends on the number of pairs of electrons the metal requires to form a more stable structure. In most biochemically significant complexes the metal is either hexavalent (six ligands) or tetravalent (four ligands).

Try making a complex yourself. Add concentrated ammonia to a solution of copper sulfate. The deep blue color resulting is due to the complex ion.

$$Cu^{2+} + 4(:NH_3) \longrightarrow \begin{matrix} & NH_3 & \\ H_3N: & \underset{\cdot\cdot}{\ddot{Cu}} & :NH_3 \\ & NH_3 & \end{matrix}^{2+}$$

complex ion
(deep blue)

Notice that the ligands are uncharged, so the complex ion product still has the +2 charge of the cupric ion.

The chlorophyll molecule provides an excellent example of a metal complex (Chapter 15). In this structure, four pairs of electrons are provided by the prophyrin ring system.

23.2 Metals and Enzymes

Some enzymes require metal ions in order to be active. Table 23.1 gives a number of examples of enzymes that require metal ions. It is thought that the metal ions interact with the enzyme and/or the substrate by way of the formation of coordinate covalent bonds. A number of the amino acid functional groups have electrons that can be donated to the metal, for example, alcohols, amines, and carboxylic acids. In a hypothetical example, a hexavalent +2 metal ion could be bound to the active site by some coordinate covalent bonds, the rest of the positions being occupied by water molecules. The presence of the metal ion on the active site might contribute to holding the enzyme molecule in the correct shape. As the substrate approaches, the water molecules are displaced from the metal ion and bonds are formed with the substrate instead. Now the metal ion is helping to hold the substrate onto the active site. Finally the charge of the metal ion could contribute to the catalysis of the reaction.

23.3 Calcium and Magnesium in Body Fluids

Blood contains about 0.1 g of calcium per liter in blood serum. About 40 percent of the blood serum calcium is bound to the serum proteins, the rest existing as the free ion in solution. The small amount of magnesium in the blood is about 35 percent protein bound. The parathyroid glands are of prime importance in the maintenance of correct blood calcium levels. Serum calcium levels drop when there is a reduced level of parathyroid hormones. When there is insufficient serum calcium, the person so afflicted develops muscle tremors, disorientation, and convulsions. Too-high levels of calcium ion leads to depositions of calcium carbonates in soft tissues.

The degree of absorption of calcium from the digestive tract depends largely on the calcium salt present. Thus in the presence of

Table 23.1
Metal Ion Requirements of Representative Enzymes

Metal Ion	*Enzyme Name*
Mg^{2+}	Enolase
	Propionate thiokinase
	Propionyl CoA carboxylase
	α-Ketoglutarate dehydrogenase
	Succinyl CoA synthetase
	Acetyl CoA carboxylase
	Aminoacyl tRNA synthetase
	DNA polymerase
	Glutamine synthetase
	Hexokinase
	Phosphoenolpyruvate carboxylase
	Phosphoglucomutase
	Phosphoglycerate kinase
	Pyruvate carboxylase
	RNA polymerase
	Pyruvate decarboxylase
	Deoxyribonuclease
	Glycerokinase
Ca^{2+}	Lipase
	Phospholipases A, C, and D
	Thrombin (blood-clotting component)
	Muscle ATPase activity
Fe^{2+}, Fe^{3+}	Aconitase
	Succinate dehydrogenase
	Homogentisate oxygenase
	Glycerol phosphatase
Zn^{2+}	Alcohol dehydrogenase
	Lactate dehydrogenase
	Triose phosphate dehydrogenase
	Glutamate dehydrogenase
	Carboxypeptidase
	Carbonic anhydrase
Mn^{2+}	Malate dehydrogenase
	Isocitrate dehydrogenase
Cu^{2+}	Monoamine oxidase
	Cytochrome oxidase
Co^{2+}	Glycyl-glycine dipeptidase

oxalate ion, which forms a very insoluble calcium salt, the rate of calcium uptake is sharply reduced. Vitamin D increases the degree of calcium ion absorption. A deficiency of this vitamin in childhood leads to improper bone formation, a problem called *rickets* (Chapter 8).

Calcium and magnesium are distributed unevenly between the cell interior and extracellular fluids much as sodium and potassium are. Calcium is found primarily in the extracellular fluid, while magnesium is accumulated by the cells. Both these ions serve as enzyme activators (Table 23.1), and in view of their nonuniform distribution it is not surprising that calcium is primarily an activator of extracellular enzymes while magnesium is required by intracellular enzymes.

We take advantage of the normal scarcity of Ca^{2+} inside the muscle cell ($10^{-7}M$) by including a protein that binds Ca^{2+} and triggers muscle contraction when we allow a little Ca^{2+} to leak into the cell (Chapters 14 and 21).

23.4 Calcium in Bones and Teeth

In adult humans there is ordinarily more than 1 kg of calcium, almost all of this in the bones and teeth. The hydroxy and carbonate salts of the apatite ion are chief forms of calcium found in these structural tissues. The apatite ion in turn is a complex structure of calcium and phosphate.

The calcium found in bones and teeth is not in a fixed, unchanging state but is constantly exchanged with blood serum calcium. This exchange is surprisingly rapid considering the insolubility of the apatite salts. On reflection, this sort of mobility is essential if the bones and teeth are to be capable of growth. Pregnancy and the period of nursing are critical times as extra calcium is needed for the embryonic bone and for the milk. In the absence of an adequate dietary calcium supply these needs are filled at the expense of the mother's bones and teeth. Because children are born with bones incompletely calcified, an adequate supply of calcium in their diet is essential.

Organisms other than vertebrates utilize calcium structurally. The hard shell (exoskeleton) of arthopods includes calcium carbonate along with chitin in its construction. The shells of mollusks are elaborate calcium carbonate structures.

23.5 Inorganic Compounds Present in Lower Concentration

How many times have you heard someone talking about achieving a state of extreme health, strength, mental alertness, or general well-being through careful control of the diet or by taking pills that boost the supply of some trace nutrient? Is this nonsense or not? First recognize that humans are creatures who have presumably always had a varied diet. Whatever was ripe or "catchable" became dinner. Humans therefore can adapt to quite a variation in diet. However, we do require small amounts of certain nutrients. We discussed vitamins and essential fatty acids. Some inorganic substances are also required in small amounts in the diet. Let us now review these mineral components of the body that are present in smaller amounts.

23.6 Iron

Iron represents only about 0.004 percent of the weight of an adult human. About 70 percent of this iron is in the hemoglobin structure, and most of the rest is in storage as ferratin, an iron protein found in liver, bone marrow, and spleen. Ferratin has variable iron content. The protein portion alone is called *apoferratin*, and the iron is attached as ferric hydroxide in amounts as high as 23 percent by weight.

Iron generally is conserved well by the body so that an intake of 5–20 mg per day is sufficient to meet the usual needs. However, menstruation involves a loss of about 80 mg per period. Pregnancy, too, creates extra demands for iron. Iron in the ferrous form is absorbed primarily from the small intestine. It is stored as ferratin and is transported in the blood attached to transferrin, a plasma protein belonging to the β globulins.

Iron is often found complexed by the nitrogens of a prophyrin ring system, as is the case for hemoglobin (Chapter 11) and the cytochromes (Chapter 18). Two iron porphyrin enzymes, peroxidase and catalase, are also known. Both have H_2O_2 as substrate.

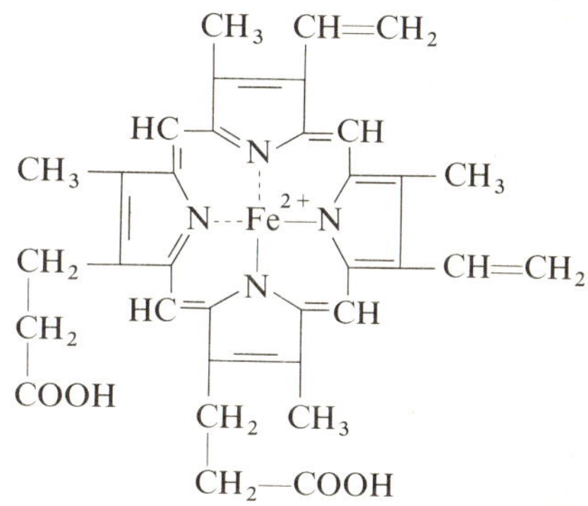

heme

Notice that four pairs of electrons are supplied by the nitrogens of the porphyrin ring. Iron, however, is hexavalent. The other two ligands attach above and below the porphyrin ring. These ligands are functional groups of the associated protein.

In recent years, nonporphyrin iron protein structures have also been recognized. For example, each photosynthetic subunit has ferredoxin, an oxidation-reduction intermediate. An iron protein is associated with nitrogen fixation by soil bacteria. Nonporphyrin iron proteins are now recognized as part of the sequence in mitochondrial oxidative phosphorylation (Chapter 18). Iron also activates some enzymes (Table 23.1).

23.7 Zinc

Zinc comprises about 0.003 percent of the weight of an adult. Although zinc is present in relatively sizable amounts, and although the results of zinc deficiency are known, why the deficiency symptoms are produced is not well understood. Zinc is transported in the blood bound to proteins or in solution inside the red blood cells.

A shortage of zinc in the diet retards normal growth, and if it persists, it causes dwarfism. If zinc is restored to the diet of these dwarfs, their growth rate increases. Other symptoms of zinc deficiency include anemia, enlarged spleen and liver, and atrophy of the testicles. Zinc has been established as an activator for some enzymes (Table 23.1).

23.8 Copper

Copper is an essential element for humans; the daily requirement is about 2.5 mg. A copper deficiency inhibits growth, causes metabolic disruption, and inhibits red blood cell formation. Copper is required for the incorporation of iron into hemoglobin. Several enzymes have copper activators (Table 23.1). Copper is transported in blood either bound to an α globulin called ceruloplasmin or loosely bound to plasma albumin. It is stored on several proteins, hepatocuprein in the liver, erythrocuprein in red blood cells, and cerebrocuprein in the brain. Milk is deficient in copper, so newborn mammals must utilize stored copper until they are weaned.

In plants the photosynthetic subunit contains copper, and in crustacea the blood oxygen carrier is a copper protein called *hemocyanin*.

23.9 Cobalt

Small amounts of cobalt are needed in the diet, presumably to be used primarily by intestinal bacteria for the synthesis of vitamin B_{12}, a porphyrin structure complexing Co^{3+} (Chapter 12). At present only a few reactions are recognized as requiring B_{12} as a cofactor.

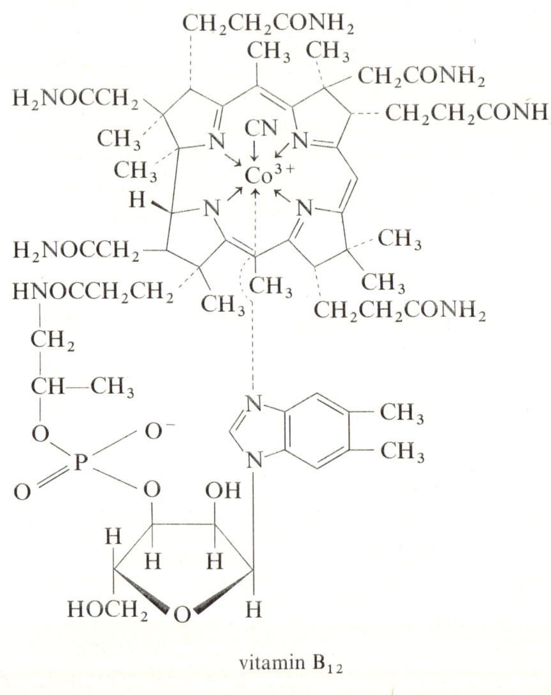

vitamin B_{12}

There also is an unresolved cobalt requirement for the incorporation of iron into heme.

In Australia the Coast disease of sheep has been traced to a deficiency of cobalt in the grazing lands.

23.10 Molybdenum

A few flavoproteins requiring iron are also found to require molybdenum for activity. These include xanthine oxidase and aldehyde oxidase (both 2Mo/8Fe/2FAD). Xanthine oxidase catalyzes a step in the formation of uric acid, the form for excretion of purines. A disorder is described, xanthinuria, where xanthine rather than uric acid appears in the urine. This is due to the congenital lack of xanthine oxidase. Gout, a disease characterized by deposits of uric acid in joints, has been combated by inhibitors of xanthine oxidase, thus preventing the uric acid from being formed.

23.11 Nitrogen Fixation

Nitrogen fixation is a process requiring molybdenum. In the roots of a few types of plants are nodules of bacteria that "fix" nitrogen. The bacteria that fix nitrogen use an enzyme system, nitrogenase, that converts atmospheric nitrogen into ammonia. This enzyme is a three-protein complex, two iron proteins and an iron–molybdenum protein. There are 2 molybdenums and 32 irons in the complex. Energy from possibly twelve molecules of ATP is required to cause this reaction to run. This system is of enormous importance as a source of nitrogen in a form usable by agricultural plants. A crop such as soybeans provides its own nitrogen-fixing bacteria on its roots. Research is under way to extend this natural process to other crops.

Enormous amounts of energy are required to produce the nitrogen fertilizers. However, these fertilizers have been the key to increased food output in recent years. This energy could be saved without a loss in food production if the bacterial nitrogen fixation process were associated with more crops.

23.12 Utilization of Nitrates by Plants

When plants use nitrates from the soil, they must convert them to ammonia. This is a two-step process:

$$NO_3^- \xrightarrow{\text{enzyme}} NO_2^-$$

$$NO_2^- \xrightarrow{\text{enzyme}} NH_3$$

The first of the two enzymes is an iron–molybdenum enzyme.

23.13 Manganese

Manganese is transported in the blood attached to the protein transmanganin as Mn^{3+}. It is the activator for several enzymes (Table 23.1). In mice, manganese is necessary in the diet to prevent a specific congenital defect, the mechanism of which is unknown.

23.14 Chromium

Studies with rats have established that a chromium-deficient diet decreases the effectiveness of insulin. Although not a replacement for insulin, it seems to make the insulin already present function more effectively. In humans, tissue chromium levels decrease with age. Rates of glucose utilization, normally decreased in older people, can be improved by dietary chromium supplements.

23.15 Boron

Boron is essential to higher plants. There is no recognized requirement in animals, however.

23.16 Selenium

A small amount of selenium in the diet (below three parts per million) may not be essential to animals, but it is useful in promoting the growth rate of sheep and other livestock. At higher levels (above 10–15 ppm) this element has a toxic effect, as demonstrated by the problems of cattle in selenium-rich pastureland. The toxic properties of selenium are somewhat counteracted by arsenic.

23.17 Arsenic

As with selenium, small amounts of arsenic seem to be beneficial to livestock. Arsenic is transported in the erythrocytes and concentrates in skin, hair, and nails. Seafood, especially shellfish and crustacea, contain high levels (up to 100 ppm) of arsenic.

23.18 Iodine

Almost all the iodine in humans is associated with thyroxin or its simple derivatives. These are found in blood either free or bound to the protein thyroglobulin.

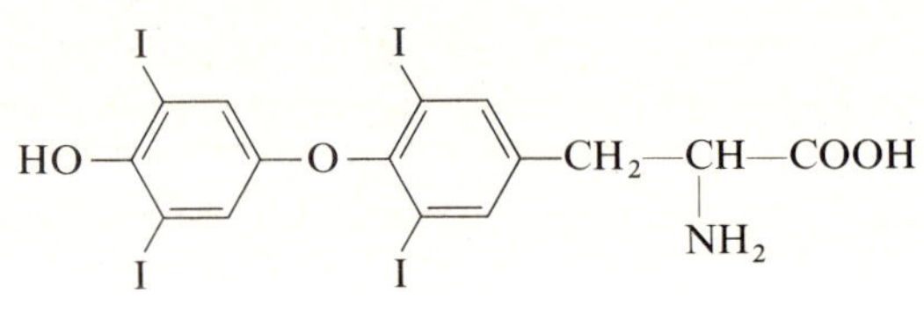

thyroxin

Exercise 23.1

From what amino acid would you guess that thyroxin is derived?

Goiter is a problem wherein the thyroid gland becomes oversized but not due to the growth of a tumor. Where the soils are deficient in iodine, such that crops will not provide an adequate supply of iodine for thyroxin production by the thyroid, the gland enlarges in an attempt to raise thyroxin production. This has been a problem in the Midwest. It has been alleviated by adding iodide salts to the table salt (iodized salt).

23.19 Elements Without Known Function

Because of their natural abundance, or at least partly because the industrialization of our society has made them unnaturally abundant, a number of elements found in the body may serve no function. For example, rubidium, which can substitute to a degree for potassium, constitutes almost 0.002 percent of body weight. It has no apparent function.

All the group IIA elements (Be, Ba, Sr, Ra) can incorporate into bone in place of calcium. In the case of strontium, this came to be of concern when it was recognized that the level of radioactive ^{90}Sr was increasing in the atmosphere due to atomic explosions and was appearing in milk because of its similarity to calcium.

Vanadium associates with lipid material. Nickel can bind to RNA and may have a role in pigmentation. Titanium and aluminum accumulate in the lung with age but have low toxicity. Tin, largely from cans, is retained in the intestine and heart but seems to have very low toxicity.

Other elements that have been detected in humans include the following: bromine, germanium, antimony, lanthanium, niobium, ruthenium, thallium, zirconium, beryllium, bismuth, silver, gold, gallium, cesium, tellurim, and lithium. Little is known of the effects of of most of these.

23.20 The Unholy Three

Chiefly we obtain metals through our diet, the water we drink, and the air we breathe. This includes both useful and harmful metals. Of these sources, water and food are not surprising. We are made abundantly aware of the potential hazards to the health resulting from water pollution. However, the idea of heavy metals carried to us through the air may seem strange at first consideration. Actually, airborne metals can be more threatening than waterborne metals. The metals are put into the air by smelting operations, by combustion of wastes containing metals, or by burning coal or petroleum, both of which contain metals. These metals can become part of fine particles suspended in the atmosphere. These particles when inhaled can lodge in the lung. From there the metal can be absorbed more effectively than from the stomach or intestine. For example, about 5 percent of the total cadmium in food is actually absorbed and enters the blood as compared with about 40 percent of the cadmium trapped in the lungs.

Three metals have come to be of great concern because of their toxicity and the unnatural levels of them in the environment as a result of human activities. These are cadmium, lead, and mercury.

23.21 Cadmium

In recent years, world output of cadmium has been around 30 million lb, about half produced in the United States. Cadmium is usually obtained as a by-product of zinc smelting, because zinc ore often also contains cadmium. Cadmium is used primarily as a corrosion-resistant coating, in pigments, in plastics stabilizers, and in alloys. Most of the uses of cadmium are nonrecyclable. In other words, our production of 30 million lb/yr is going to be scrapped (coated metals) or burned (plastics). To this cadmium entering the environment we add a quantity released by the burning of coal or oil.

We sometimes measure the tenacity with which something is retained in the body as the half-life of the substance. For example, if you took a 1.0-mg dose of something, one half-life would be the length of time required to excrete 0.5 mg. For cadmium the half-life is 10–25 years. With such a long half-life our tendency is to continuously increase the amount of cadmium in our bodies. A child at birth may have only 1 μg (0.001 mg), but a typical 50-year-old person has 30 mg, a 30,000-fold increase in total body cadmium.

Poisoning by airborne cadmium produces emphysema (lung damage) and a disposition toward heart attack. The maximum recommended exposure to airborne cadmium is 0.1 g/m^3 air (10^{-5} ppm). Cigarette smoke has been reported to have 10,000 times this concentration of cadmium (1 ppm). The most dramatic instance of cadmium poisoning, however, was in Japan and involved cadmium in river water. An industrial plant discharged cadmium into water used to grow crops. People in that area were dwarfed and had serious bone embrittlement.

23.22 *Mercury*

World production of mercury is about 20 million lb/yr. It is used for plastics, paints, electrical products, and agricultural or pharmaceutical chemicals. Once again many of the products are discarded or burned without any attempt at recycling. Also as with cadmium, the burning of coal or oil releases mercury into the atmosphere.

Our difficulty with mercury compounds is the vast difference in toxicity displayed by different chemical forms. Some of the more dramatic problems with mercury have involved metallic mercury entering water as waste where paper is processed or at chlor-alkali plants (plants converting salt into chlorine gas and sodium hydroxide). The metallic mercury was not considered seriously toxic. In the water, however, mercury is converted by bacteria into highly toxic alkyl mercury compounds, which are then taken up by fish. The fish may then be eaten by wild predators or humans.

$$Hg^0 \longrightarrow Hg^{2+} \longrightarrow \underbrace{CH_3Hg^+ \longrightarrow (CH_3)_2Hg}_{\text{alkyl mercury compounds}}$$

Such problems have been encountered in Japan, Sweden, and the United States. Another problem arose when seeds intended for planting only were treated with a fungicide containing mercury to prevent loss of the seeds on storage. These seeds were eaten by mistake, causing widespread poisoning in Iraq and a few cases in the United States. Mercury poisoning causes brain and nerve damage with symptoms similar to encephalitis. Once again the half-life of mercury is long, approximately one year.

23.23 Lead

World consumption of lead runs over 3 million tons annually, of which the United States uses between a third and a half. Lead is used in storage batteries, solder, type metal, bearing alloys, paint pigments, glass products, and as an antiknock additive in gasoline. The picture on recycling lead is much better than for cadmium or mercury. Lead storage cells are largely recycled. Of the three metals under consideration here, lead presents perhaps the greatest threat to health. In 1971, the National Research Council reported that urban Americans had blood lead levels 50 percent of the amount at which biochemical effects appear and 25 percent of the amount at which symptoms of lead poisoning occur. Clinical disorders appear when the intake exceeds 1 mg per day.

How does the lead get into our bodies? Evidence indicates that airbone lead is an important source. Coal, for example, contains relatively high levels of lead (over 50 ppm). Much of the gasoline burned in automobiles has tetraethyl lead or tetramethyl lead added as an "antiknock" compound.[1] The exhaust of the car has a variety of lead compounds present. Newer cars with catalytic mufflers as part of the air pollution control system must use lead-free gas, because the lead fouls the catalyst. As a result, more cars are using lead-free gas every year, reducing this source of airbone lead.

Another serious health problem originates with the use of lead pigments in house paints. The lead washes into the soil, of course, but more importantly, especially in the case of interior paintwork, young children pick up the paint fragments from an old painted surface and swallow them. Regular testing for blood lead levels has become an important public health service, particularly in older sections of large cities.

Like mercury, lead causes brain and nerve damage, producing symptoms similar to those of encephalitis.

23.24 Trace Elements and Cancer

Recent research has indicated a possible relationship between trace elements in the diet and the occurrence of cancer. Low levels of sodium selenite reduce the frequency of certain tumors in laboratory mice. A relationship between selenium in the diet and cancer in the human population has been inferred by a broad-based statistical study. However, the National Research Council's Committee on Nutritional Misinformation has stated that the United States diet probably contains sufficient selenium. Zinc appears to counteract the protective effects of selenium. Further, laboratory animals fed a zinc-deficient diet also showed a reduction in incidence of tumors. Similarly, lowered levels of iron or copper reduced the frequency of some types of cancers.

---

[1] The octane number of a fuel represents its predisposition to preignite or knock. Preignition is harmful to the engine. The lead compounds raise the octane number. It can be done in other ways, enriching the fuel in branched hydrocarbons, but those processes are more expensive.

In some tests, nickel increased the frequency of cancer but was counteracted by the presence of manganese. These results are preliminary and are not completely understood, but they do point the way to additional research.

Summary

Metals can form coordinate covalent bonds, covalent bonds wherein the shared electrons are donated by the partner in the bond, the ligand. Such structures explain the function of metals in enzyme catalysis. Sodium and calcium ions are found in extracellular fluids, while potassium and magnesium ions are found in cells. Calcium blood levels are under hormonal control. Most calcium is in the bones and teeth. Iron is largely in hemoglobin but is also required in enzymes and electron transport systems. Zinc is a common enzyme activator. Copper is required for a few important systems, as are cobalt, manganese, molybdenum, and iodine. Iron and molybdenum are necessary for nitrogen fixation by soil organisms and for utilization of nitrate by plants. Chromium increases the effectiveness of insulin. Small amounts of selenium and arsenic are useful, but larger amounts are poisonous.

Of the harmful metals, three—cadmium, mercury, and lead—are the most threatening. They may enter the body with food, water, or as particles in the air. They all are retained by the body for long periods of time, and they can damage the brain and nervous system.

Problems

1. Show the formation of a coordinate covalent bond between BF_3 and NH_3.

2. Write the formula and indicate the final charge of the following complexes:

a) A hexavalent complex between Fe^{2+} and NH_3.

b) A tetravalent complex between Cu^{2+} and Cl^-.

c) A hexavalent complex between Mg^{2+} and H_2O.

3. The common blue crystals of copper sulfate are found to have this composition: $CuSO_4 \cdot 5H_2O$. Four of the waters are bound in the crystal differently from the other water. Can you propose an explanation?

4. Insufficient calcium ion uptake causes muscle contraction problems (tremors and convulsions). Why should the supply of calcium affect muscle contraction?

5. Anemia could be due to insufficient iron in the diet. It could also be due to a shortage of what other metals in the diet?

Appendix 1 Chemicals: How Many Are There?

Like it or not, the world around us is filled with chemicals. The clothes we wear, the foods we eat, the magazines we read, and virtually all of the other things that nurture our civilization are made possible by the use of chemicals. Recognition of these facts, and of some of the potential dangers of chemicals, often prompts the question How many chemicals are there? A definitive answer to that question has proved elusive, but one measure of the answer is provided by the American Chemical Society's Chemical Abstracts Service (CAS). As of November 1977, CAS's unique computer registry of chemicals contained 4,039,907 distinct entities. The number of chemicals in the register, moreover, has been growing at an average rate of about 6000 per week.

CAS began the computer registry in 1965 to assist in the indexing of chemical substances reported in the scientific literature. The system identifies chemicals on the basis of an unambiguous computer-language description of their composition and molecular structure and automatically assigns a permanent identifying number to each unique substance. The registry contains all compounds that have been mentioned in the literature since 1965.

The system is not set up to provide a detailed breakdown of different classes of chemicals. Some broad generalizations can be made, though. About 96 percent of the chemicals, for example, contain carbon. The average compound in the registry, if it is possible to define such a thing as an average compound, contains 43 atoms, 22 of which

---

are hydrogen. The fictitious average compound also contains one and a half ring systems with eight atoms per ring.

About 3.4 million of the chemicals are organic or inorganic chemicals whose structures are fully defined. Some 3 million of these contain at least one ring system. Another 258,000 entities are coordination compounds, which require a somewhat different registration procedure. About 59,000 are organic compounds whose structures are not completely defined; these generally are compounds in which the exact location of a substituent or a double bond or the site of an esterification is uncertain. About 120,000 entries are listed only by name or molecular formula; these are specifically identified substances for which not enough is known or has been disclosed about structure to permit machine structure registration. The registry also lists 72,000 alloys, 120,000 polymers, and 10,000 mixtures with specific names.

The list contains some apparent duplication in that stereoisomers are listed individually. There are, for example, four listings for aspartic acid: D-aspartic acid, L-aspartic acid, DL-aspartic acid, and aspartic acid of unspecified stereochemistry. Since each of these has its own biological characteristics, however, they can justifiably be considered distinct entities.

The vast majority of chemicals in the registry are esoteric materials that have been isolated from natural products or synthesized for research purposes. A more interesting problem, then, might be to define the number of chemicals that are in everyday use. That problem is much more difficult, but at least a partial answer may soon be available. As part of the Toxic Substances Control Act, the Environmental Protection Agency (EPA) has been charged with maintaining an inventory of chemical substances manufactured, imported, or processed in the United States for commercial purposes. Because there are so many different names for many of the chemicals used in industry, EPA has contracted with CAS to process reports submitted by manufacturers, determine the precise identity of the reported chemicals, and create and maintain a computer file on the chemicals and their manufacturers.

To initiate the project, CAS has submitted to EPA a preliminary list of some 33,000 chemicals that are thought to be in common use. The complexity of the registration problem is illustrated by the fact that CAS has already found in its files more than 183,000 different names for those chemicals. Current estimates from EPA, moreover, indicate that there may be as many as 50,000 chemicals in everyday use, not including pesticides, pharmaceuticals, and food additives. EPA estimates that there may be as many as 1500 different active ingredients in pesticides. The Food and Drug Administration estimates that there are about 4000 active ingredients in drugs and about 2000 other compounds used as excipients to promote stability, cut down on growth of bacteria, and so forth. FDA also estimates that there are about 2500 additivies used for nutritional value and flavoring and

3000 chemicals used to promote product life. The best estimate thus is that there are about 63,000 chemicals in common use. Small wonder then that determination of the safety of all commonly used chemicals is a massive project that may never be completely finished.

Thomas H. Maugh II

Appendix 2 Plastics in Cars

To many people "plastic" is synonymous with "cheap" and "easily broken." Visions of prizes in cereal boxes come to mind. However, properly designed structures made of plastic can prove as serviceable as metal structures without weighing as much.

This one word weight underlies the attractiveness of plastics to the auto manufacturer. The federal government has imposed fuel mileage regulations on the automobile industry. Fleet mileage, the average mileage for the mix of cars sold, must be 18 miles per gallon (mpg) in 1978, 22 mpg in 1981, and 27.5 mpg in 1985. The mileage an automobile delivers directly relates to the weight of the car—what mass must be accelerated each time the car starts up. As a rough estimate, saving 400 lb of weight raises the gas mileage by 1 mpg.

Obviously, the easiest way to make a car lighter is to make it smaller. To date, about two-thirds of the weight reduction in cars has been achieved by "downsizing." This can only be pursued to the point where the car ceases to be practical. Improved steels and aluminium have been used to cut weight. However, plastics offer the best

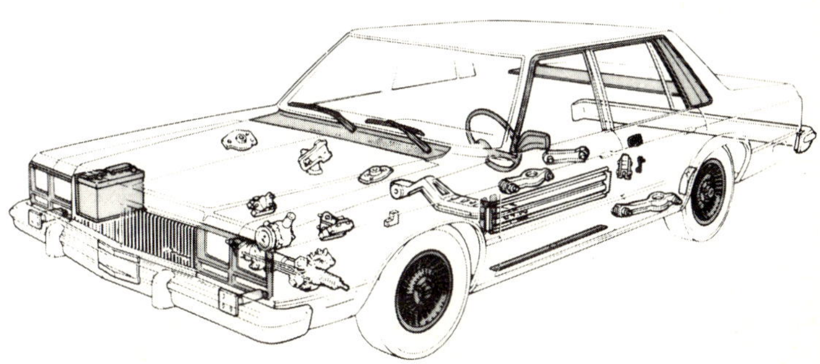

A 1978 car
The materials made of plastic to reduce weight are shaded. By 1985, the amount of shading on a comparable diagram will be much greater. (Drawing courtesy of Ford Motor Company, Dearborn, Michigan.)

solution to the problem. The 1979 autombiles include over 200 lb of plastics each. By 1985 it is estimated there will be 350 lb of plastics per car, representing an increase from 4.5 percent to 12 percent of the total car weight. We are going to see everything from major body panels to supports, gas tanks, and even drive shafts made of plastic.

Plastics have advantages in addition to low weight. They are corrosion resistant, require less expensive machinery for manufacture than some alternative materials, and require less energy during manufacture. Just between 1977 and 1979 alone, the estimated total use of plastics by auto companies is increasing from 800,000 to 974,000 metric tons. Such a large market is obviously very important to the chemical industry.

Appendix 3 Glossary of Terms

acetal (*full acetal*) (*6*)* — The product of a substitution reaction between a hemiacetal and an alcohol.

acid anhydride (*7*) — The product of the condensation of two carboxylic acids with the loss of water.

acid chloride (*7*) — A carboxylic acid in which a chlorine is substituted for the hydroxyl group.

acidosis (*22*) — The condition when blood plasma becomes more acid than normal.

active transport (*21*) — The movement of a compound across a membrane into a region of higher concentration using energy provided by the hydrolysis of ATP.

Addison's disease (*8, 22*) — A disease in which the adrenal cortex produces insufficient corticosteroids.

addition polymer (*3*) — Polymers formed from alkene monomers that extend the polymer chain by adding one monomer to the double bond of another.

addition reaction (*3*) — Some chemical agents composed of two parts joined by a single bond will separate, one part joining to one carbon of a double or triple bond by the electrons once holding the parts together while the other part joins to the other carbon by a pair of pi electrons from the bond. In the final product the alkene or alkyne and the agent will be joined into one molecule, and one bond of the former double or triple bond will be gone.

adipose tissue (*1* — Body tissue composed of fat cells and serving to store triglycerides.

---

* The number in parentheses indicates the chapter(s) in which the term is used.

adrenal cortex (*8*) The outer portion of the adrenal gland, a tissue that synthesizes steroid hormones.

alcohol (*5*) An organic molecule containing an —OH functional group.

aldose (*6*) An aldehyde carbohydrate.

aliphatic compounds (*4*) Organic compounds lacking the benzene-type ring system.

alkaloid (*10*) A compound produced by plants usually containing a nitrogen heterocyclic ring.

alkalosis (*22*) The condition when the blood plasma becomes more basic than normal.

alkane (*2*) A compound composed of carbon and hydrogen atoms joined by covalent single bonds.

alkene (*3*) A hydrocarbon containing a carbon-to-carbon double bond.

alkyl group (*2*) A side group that is hydrocarbon in chemical character.

alkyne (*3*) A hydrocarbon containing a carbon-to-carbon triple bond.

allergy (*11*) The response to the release of histamine as a result of antibodies produced to environmental antigens such as pollens, wool, animal fur, and the like.

allosteric effectors (*12*) Molecules whose binding to an enzyme elsewhere than the active site controls the catalytic effectiveness of the enzyme.

allosteric enzymes (*12*) Enzymes whose rate of catalysis is governed by the binding of a control molecule at a specific (allosteric) site.

α helix (*11*) A type of secondary structure in which the protein chain forms coils that are hydrogen bonded between one turn of the coil and the next.

amide (*7*) The product of a condensation reaction between a carboxylic acid and an amine.

amine (*9*) The result of substituting an alkyl group for one or more of the hydrogens of ammonia. These are classed as follows:

primary amine has one alkyl group.

secondary amine has two alkyl groups.

tertiary amine has three alkyl groups.

amino acid (*9*)	A carboxylic acid with an amino group on carbon 2.
amino acid activation (*13*)	The attachment of an amino acid to the correct tRNA.
amino group (*9*)	An $—NH_2$ substituted on a structure.
amino sugar (*6*)	A carbohydrate in which a hydroxy group has been replaced by an amino group.
anabolism (*14*)	Biological synthesis of new compounds.
anaerobic metabolism (*15*)	Metabolism carried on in the absence of oxygen; also called fermentation.
analgesic (*7*)	A drug that relieves pain.
anaplerotic reactions (*20*)	Reactions that produce additional Krebs cycle intermediates.
androgens (*8*)	Male sex hormones.
anemia (*22*)	The condition of abnormally low concentrations of red blood cells in the blood.
antibodies (*11*)	Protein structures with pairs of identical bonding sites shaped to attach to specific foreign objects in the blood.
anticholinergic (*10*)	A drug antagonistic to the transmission of nerve impulses.
anticodon (*13*)	The three consecutive nucleotides on tRNA that are complementary to and can hydrogen bond to the codon on mRNA.
antigens (*11*)	Proteins or polysaccharides that stimulate the formation of antibodies.
antipyretic (*7*)	A drug that reduces fever.
aromatic compounds (*4*)	Benzene, benzene derivatives, and other ring compounds with conjugated double bonds and benzenelike reactivity.
asymmetric carbon (*5*)	A carbon with four different groups attached.

β structure (*11*)	A type of secondary structure in which protein segments line up side by side with hydrogen bonds between the adjacent segments.
bile (*8*)	A fluid rich in bile salts that is stored in the gall bladder awaiting the entry of food into the intestine.

bile acids (*8*) Steroid alcohols containing side-chain carboxyl groups and found in the bile.

bile salts (*8*) Bile acids whose carboxyl groups have been converted to amides by condensation with the amine group of either glycine or taurine.

biosynthesis (*20*) The generation by a cell of new molecules from simpler starting materials.

blood group substances (*22*) Glycosphingolipids formed on red blood cell membranes that are associated with the blood type.

blood plasma (*22*) The fluid portion of the blood including the blood-clotting proteins.

blood serum (*22*) The fluid portion of the blood excluding the blood-clotting proteins.

bond angles (*1*) Two atoms joined to a third, thus forming an angle with the third atom as the apex and the other two as the extensions.

---

carbohydrate (*6*) A polyhydroxy aldehyde or ketone.

carbonium ion (*3*) A positively charged carbon structure formed during the course of a reaction.

carbonyl group (*6*) A carbon connected to an oxygen by a double bond.

carboxylate ion (*7*) The resulting negative ion when a carboxylic acid ionizes (COO^-).

carboxylic acid (*7*) An organic structure containing a —COOH functional group.

carcinogen (*4*) A compound that increases the likelihood of occurrence of cancer.

carrier (*21*) A permease.

catabolism (*14*) The biological degradation of dietary or stored food.

cerebrospinal fluid (*22*) The fluid bathing the brain and spinal cord.

chiral carbon (*5*) *See* asymmetric carbon.

chloroplast (*20*) A subcellular structure in plants that contains the light-absorbing pigments and electron transport compounds for photosynthesis.

chromosome (*13*) One of the molecules of DNA found in plants or animals.

circular DNA (*13*) A DNA molecule wherein the polymer chain is in the form of a closed loop.

codon (*13*) Three consecutive nucleotides on mRNA that constitute a genetic code word.

cofactor (*12*) A nonprotein factor necessary for enzyme activity.

collagen structure (*11*) A secondary structure found in the structural protein collagen in which three protein chains wind around one another to form a helix stabilized by hydrogen bonds between the adjacent chains.

combustion (*2*) The reaction of compounds with oxygen accompanied by the release of heat energy.

complementarity (*5*) The requirement for the correct functioning of many biomolecules that they fit together exactly.

complementary base pairs (*13*) Unique patterns of hydrogen bonding selectively joining guanine to cytosine and adenine to either uracil or thymine.

complex lipids (*7*) Lipids that have components in addition to glycerol and carboxylic acids.

condensation polymer (*7*) A polymer formed by condensation reactions.

condensation reaction (*7*) The reaction that joins two molecules together with the loss of a small molecule, usually water.

congenital defect (*9*) A malfunction or malformation in an organism arising before birth due either to mutation (genetic disease) or to a flaw in development.

conjugated double bonds (*4*) A pair of double bonds separated by one single bond.

coordinate covalent bond (*23*) A covalent bond in which the electrons are supplied by one of the partners in the bond.

coordination compounds (*23*) Compounds formed using coordinate covalent bonds.

corticosteroids (*8*) A family of steroid hormones produced by the adrenal cortex and regulating salt and water retention, carbohydrate utilization, and lipid utilization.

cotransport (*21*) The movement of a compound across a membrane into a region of higher concentration using energy provided by concerted movement of another compound into a region of lower concentration.

coupled reactions (*14*) Reactions that are not favored energetically that can occur when run simultaneously with reactions that are favored energetically.

covalent compounds (*1*) Compounds whose atoms are joined by the sharing of electrons.

covalent single bond a bond formed by the sharing of two electrons.

covalent double bond a bond formed by the sharing of four electrons.

covalent triple bond a bond formed by the sharing of six electrons.

Cushing's syndrome (*8*) A condition caused by an adrenal tumor and characterized by high levels of corticosteroids.

---

dehydrogenase (*6*) An oxidoreductase.

denaturation (*11*) The unfolding of the tertiary structure or quaternary structure of a protein, especially to change properties or function.

deoxyribonucleic acid (DNA) (*13*) A polynucleotide with a deoxyribose phosphate polymer chain and the bases thymine, adenine, guanine, and cytosine as side groups.

depolarization (*21*) The change in potential across a nerve cell membrane that results when the membrane rapidly becomes more permeable to sodium ion.

detergent (*7*) A surface-active agent.

diabetes mellitus (*22*) A condition in which insufficient insulin is produced, resulting in lowered utilization of carbohydrates by the cells.

diastereoisomers (*5*) Isomers of a compound with more than one asymmetric carbon that are not mirror images.

differentiated cells (*14*) Cells in a multicellular organism that have modified structures related to specialized functions.

diglyceride (*7*) A glyceride in which two carboxylic acids are esterified to hydroxy groups of glycerol.

disaccharide (*6*) Two monosaccharides joined by a glycosidic bond.

disulfide (*5*) An organic molecule with two sulfur atoms joined by a single covalent bond, each having an alkyl group attached.

diuretic (*22*) A compound that stimulates an increase in urine volume.

DNA *See* deoxyribonucleic acid.

double helix (*13*) A helical structure composed of two polymer strands twisted about one another.

---

edema (*22*) The accumulation of fluid in the tissues.

electronegative atoms (*1*) Atoms at the upper right corner of the periodic chart that are able to attract electrons more strongly.

electrophoresis (*9*) A technique for separating proteins and other charged molecules by causing them to migrate toward an electrode in response to the molecular charge at a particular pH.

elimination reaction (*5*) A reaction in which atoms are removed from adjacent carbons to convert a single bond to a double (or even triple) bond.

elongation (*13*) The addition of successive amino acids to the growing protein chain.

emphysema (*23*) Damage to the lungs resulting from any of a number of external factors.

emulsion (*7*) Polar and nonpolar liquids mixed stably by use of a surface-active agent.

enantiomers (*5*) Optical isomers that are mirror images of one another.

encephalitis (*23*) A term describing the symptoms of inflammation of the brain as the result of a wide variety of disorders.

endopeptidase (*12*) An enzyme that hydrolyzes specific peptide bonds in the middle of a protein chain.

enzyme active site (*12*) The region of the enzyme surface at which the substrate is bound.

essential amino acid (*9*) An amino acid that is required for protein synthesis but cannot be synthesized by the body.

ester (*7*) The product of a condensation reaction between a carboxylic acid or mineral acid and an alcohol.

estrogens (*8*) Female sex hormones.

ether (*5*) An organic molecule in which two alkyl groups are attached to an oxygen.

excitable membrane (*21*) A membrane with an electrical potential caused by the difference in ion concentrations on either side of the membrane that can leak ions in response to a stimulus and so change the potential.

exopeptidase (*12*) An enzyme that hydrolyzes peptide bonds at the end of a protein chain.

---

fatty acid (*7*) A carboxylic acid that has the structural characteristics of those formed in glycerides.

feedback control system (glandular) (*8*) A system of regulation wherein hormone production by a gland is inhibited by a product of the target tissue.

feedback inhibition (*12*) When the first enzyme in a metabolic sequence is allosteric and the effector is the final product of the sequence.

fermentation (*6*) Metabolism in the absence of oxygen.

fibrous protein (*11*) Protein whose function is structural and whose shape is elongated.

formyl group (*6*) An aldehyde group named as a side group.

fractionating tower (*2*) A tall tower in a refinery that sorts out the hydrocarbons of petroleum into fractions with similar boiling point.

free radicals (*2*) Unstable species produced when a covalent bond is broken such that one electron stays with each atom.

free rotation (*3*) The situation when groups joined by a single bond and not otherwise blocked can rotate around the bond with respect to one another.

freons (*2*) A family of hydrocarbons wherein the hydrogens have been replaced by chlorine and fluorine.

functional groups (*1*) Covalently joined groups of atoms that are reactive and whose reactions are predictable.

fungicide (*4*) A compound that kills or restricts the growth of fungus.

γ ***globulins*** (*11*) A blood plasma protein fraction containing the antibodies.

gas chromatography (*7*) A technique for separating volatile compounds by passing them as gases through a column packed with a substance that retards some compounds more than others.

genetic code (*13*)	The identity of the specific amino acid or stop signal designated by each possible combination of three nucleotides on mRNA.
genetic deficiency disease (*9, 19*)	A disorder resulting from the inability to produce one of the enzymes of a metabolic pathway in an active form.
geometric isomers (*3*)	Compounds containing a double bond in which the groups attached to each of the double bond carbons are nonidentical; such groups can exist in two arrangements, termed the cis and the trans isomer.
gland (*8*)	Tissue producing hormones.
globular proteins (*11*)	Proteins folded into a somewhat spherical shape.
glomerular filtrate (*22*)	The initial fluid filtered from the blood in the process of forming urine.
glomerulus (*22*)	The capillaries in the kidney adjacent to the tubules of the nephron through which the plasma filtrate passes in the first step of urine formation.
glucogenic amino acids (*20*)	Amino acids that are metabolized into Krebs cycle intermediates, permitting them to be used for glucose synthesis.
glucosuria (*22*)	The appearance of glucose in the urine.
glycerides (*7*)	Lipids based on the esterification of glycerol.
glycosidic bond (*6*)	The acetal formed between the hemiacetal of one carbohydrate and an alcohol of another or between the two hemiacetals.
goiter (*23*)	The overgrowth of the thryoid gland due to an inadequate iodine supply.
gout (*23*)	A disease characterized by uric acid accumulations at joints.
gum (*6*)	A plant polysaccharide often extracted for commercial use.

---

hemiacetal (*6*)	The product of the addition reaction between an aldehyde and an alcohol.
hemicellulose (*6*)	A structural polysaccharide other than cellulose found in plants.
hepatitis (*22*)	A disease of the liver.
heterocycle (*10*)	A ring compound in which carbon atoms and other atoms are included in the ring structure.

hexavalent metal (*23*) — A metal requiring six ligands to form a complex.

hexose (*6*) — A carbohydrate with six carbons.

hormone (*8*) — A chemical produced in glands, transported in the circulatory system, and creating a regulatory effect on a target tissue.

hydrogen bond (*1*) — A weak bond resulting from the attraction of a hydrogen attached to an electronegative atom by another electronegative atom.

hydrolysis (*7*) — Breaking a bond by adding water to the bond.

hydrophobic bonding (*11*) — The organization and stabilization of structures like proteins that occurs because water preferentially associates with itself or with polar groups leaving nonpolar structures to associate with themselves.

hypertension (*22*) — A state of health characterized by excessively high blood pressure.

hypothalamus (*8*) — A gland near the brain that produces hormones to regulate the pituitary.

---

initiation (*13*) — The formation of a ribosome —mRNA-formylmethionyl $tRNA_f$ complex.

interstitial fluid (*22*) — The fluid bathing the cells in higher animals.

intracellular fluids (*22*) — The fluids inside the cell.

ionic compounds (*1*) — Compounds formed when electrons are totally donated from one atom to another.

ionization (*of acids*) (*7*) — The loss of hydrogen ion by an acid, usually to create a negative ion.

isomers (*3*) — Molecules that have the same atoms differently arranged.

isomerase (*15*) — A type of enzyme whose catalyzed reaction involves the exchange of position of the groups already present on the substrate.

isoprenoid compounds (*3*) — *See* terpenes.

isozymes (*12*) — Enzymes that catalyze the same reaction but are found in more than one structural form.

IUPAC (*2*) — Acronym of the *I*nternational *U*nion of *P*ure and *A*pplied *C*hemistry, which decides questions of nomenclature and values of constants to assure good communication among scientists.

ketone bodies (*16, 22*)	Acetoacetate or β-hydroxybutanoate, which appear in the blood and urine during periods of very high use of fatty acids for energy production.
ketose (*6*)	A ketone carbohydrate.
kinase (*15*)	A type of enzyme whose catalyzed reaction involves the transfer of a phosphate from ATP to a substrate.
lactone (*10*)	A cyclic ester.
Lewis structures (*1*)	Drawings of covalent molecules where outer shell electrons are represented as dots around the symbols of the atoms, and covalent bonds are represented by placing the dots between symbols.
ligand (*23*)	A group contributing electrons to a metal atom or ion in a metal complex.
lipids (*7*)	A group of biochemically significant compounds characterized by solubility in organic solvents.
lymph (*22*)	The interstitial fluid in the process of returning to the blood serum.
lymphocyte (*11*)	A white blood cell.
main chain (*2*)	The longest linear chain of carbons in an organic molecule.
matrix of the mitochondrion (*16*)	The compartment inside the inner membrane of the mitochondrion.
mercaptan (*5*)	A thiol.
meta (m-) substitution (*4*)	Groups separated by one position on the benzene ring.
metabolic water (*22*)	Water that arises as the product of metabolic oxidation reactions.
metabolism (*14*)	The sum of all the cell chemistry.
metal complexes (*23*)	Coordination compounds composed of metal atoms or ions surrounded by a series of electron-donating groups.
methoxy group (*5*)	A methyl ether group.
micelle (*7*)	The droplet of nonpolar liquid held in water suspension by a layer of surface-active agent.

monoglyceride (*7*)	A glyceride in which one carboxylic acid is esterified to a hydroxy group of glycerol.
monomer (*3*)	The small units linked together to form polymers.
monosaccharide (*6*)	A single polyhydroxyaldehyde or ketone molecule.
mucopolysaccharides (*6*)	Animal polysaccharides of two types of monomers, often a uronic acid and an amino sugar.
multienzyme complex (*15*)	A quaternary protein structure, usually very large, containing more than one enzyme of a metabolic sequence and arranged to facilitate the transfer of metabolite from enzyme to enzyme.
mutase (*15*)	A type of enzyme that catalyzes the shift of a group, particularly a phosphate group, to a different position on the substrate.
mutation (*9*)	The altering of genetic information in such a fashion that the message for sequence of amino acids of a specific protein is altered.

nerve gas (*21*)	A compound that destroys the ability of nerve cells to stimulate striated muscle by inhibiting the enzyme (acetylcholinesterase) that hydrolyzes the transmitter (acetylcholine).
nitrogen fixation (*23*)	The conversion of atmospheric nitrogen to ammonia by bacteria found associated with the roots of some plants.
nonpolar covalent bond (*1*)	A covalent bond consisting of equally shared electrons.
nucleoside (*10*)	A structure composed of either a purine or a pyrimidine heterocycle connected to the carbohydrate D-ribose.
nucleotide (*10*)	A nucleoside with phosphate esterified to a ribose alcohol group.

obstructive jaundice (*22*)	A blockage in the bile duct preventing the passage of bile into the intestine.
octane number (*23*)	A number measuring the relative resistance of a motor fuel to preignite or knock.
optical isomers (*5*)	Isomers of a structure with one or more asymmetric carbons.
optical rotation (*5*)	The twisting of the plane of polarized light that occurs when it is passed through a solution containing one isomer of a structure that has an asymmetric carbon.

ortho* (*o*-) *substitution (*4*) Two groups on adjacent carbons of a benzene ring.

osmotic pressure (*22*) The pressure that can be generated by the movement of water across a barrier from a region of low concentration to a region of high concentration.

oxidative phosphorylation (*18*) The process by which the cofactors reduced in metabolism are reoxidized, accompanied by the conversion of ADP to ATP.

oxidoreductase (*6*) An enzyme that catalyzes an oxidation or reduction of the substrate.

---

para*(*p*-) *substitution (*4*) Two groups on carbons diametrically across a benzene ring.

passive transport (*21*) The movement of a compound across a membrane such that the compound is always moving into a lower concentration solution.

pentose (*6*) A carbohydrate with five carbons.

peptide (*9*) A short polymer chain of amino acids.

peptide bond (*9*) The amide linkage joining amino acids together.

permease (*21*) A protein in a membrane that is specific for the transport of a particular compound or small group of compounds across the membrane.

***pH* (*of a solution*)** (*7*) A measure of a solution's relative acidity or basicity, specifically the negative log of the H_3O^+ concentration.

phenol (*5*) An organic molecule with an —OH functional group attached to a benzene ring.

phenol coefficient (*5*) Used as a measure of the antibacterial properties of a compound, it is the ratio of a concentration of phenol to the concentration of the compound that is equally effective (abbreviated P.I.).

pheromones (*3*) Compounds produced by animals to communicate.

phosphoglycerides (*7*) Glycerides in which one ester linkage includes phosphoric acid instead of a carboxylic acid.

photophosphorylation (*20*) The portion of the pathway in photosynthesis that includes the light-absorbing pigments and the electron transport system.

photosynthesis (*20*) The use by plants of light energy to reduce carbon dioxide.

pi bond or pi electrons (*3*) Pairs of electrons present in addition to those of the sigma bond to form a double or triple bond.

pituitary (*8*)	A gland near the brain that produces hormones to regulate other glands (also called the hypophysis).
pK (*of an acid*) (*7*)	A measure of acid strength, specifically the negative log of the acid ionization constant.
pleated sheet (*11*)	A ridged surface formed by several rows of protein segments arranged in the β-structure pattern.
polar covalent bond (*1*)	A covalent bond consisting of unequally shared electrons.
polarized light (*5*)	Light that has passed through a polarizing filter.
polyamide (*7*)	A condensation polymer wherein the monomers are joined by amide linkages.
polycytemia (*22*)	The presence of abnormally high concentrations of red blood cells in the blood.
polyester (*7*)	A condensation polymer wherein the monomers are joined by ester linkages.
polymer (*3*)	Molecules of very high molecular weight composed of numerous small units joined together.
polyribosome (*13*)	A complex of mRNA and several ribosomes each engaged in protein synthesis.
polysaccharide (*6*)	Many monosaccharides joined by glycosidic bonds.
polyuronide (*6*)	A polysaccharide composed of uronic acid monomer units.
porter (*21*)	A permease.
postsynaptic membrane (*21*)	The synapse portion of the cell membrane of the organ connected to the nerve cell.
precipitin reaction (*11*)	The appearance of a flocullant precipitate due to the formation of the antigen–antibody complex.
presynaptic membrane (*21*)	The nerve cell membrane at the synapse.
primary structure of a protein (*9*)	The sequence of the amino acids of the protein.
prostaglandins (*7*)	A group of regulatory compounds based on the fatty acid derivative prostanoic acid.

protein (*9*)	A long polymer chain of amino acids.
proteinurea (*22*)	The appearance of protein in the urine.
proton acceptor (*9*)	A chemical structure capable of receiving a hydrogen ion.
proton donor (*9*)	A chemical structure capable of giving up a hydrogen ion.

---

quaternary structure (*11*)	Two or more folded protein chains assembled to create the functional form of the protein.

---

racemic mixture (*5*)	A sample containing equal amounts of each of a pair of enantiomers.
rancidity (*7*)	The spoiling of glycerides by hydrolysis of esters or oxidative cleavage of the carboxylic acid chain.
reducing sugar test (*6*)	A color reaction used to detect hemiacetal carbohydrates.
replication (*13*)	The synthesis of new DNA on the template of existing DNA.
Rh factor (*22*)	A substance whose absence or presence is recorded as part of the blood type.
ribonucleic acid (*13*)	A polynucleotide with a ribose phosphate polymer chain and the bases uracil, adenine, guanine, and cytosine as side groups.
rickets (8, *23*)	The situation of improper bone formation resulting from a childhood deficiency of vitamin D.
ring compound (*2*)	A structure in which one end of a chain of atoms reacts with the other to form a closed loop.
RNA	*See* ribonucleic acid.

---

saturated structure (*3*)	An organic compound lacking double bonds.
secondary structure (*11*)	Regions of local orderliness stabilized by hydrogen bonding between peptide bonds within the total folding or tertiary structure of a protein.
side group (*2*)	Anything other than hydrogen attached to the main chain.
sigma bond or sigma electrons (*3*)	The pair of electrons located between any two atoms that are covalently joined.

Simmond's disease *(8)* — A condition caused by failure of the system regulating the adrenal cortex and characterized by low levels of corticosteroids.

soap *(7)* — A surface-active agent that is chemically a fatty acid salt.

sphingolipids *(7)* — Any lipid containing the sphingosine structure.

steroid *(8)* — An isoprenoid family of compounds sharing a system of four connected rings.

strained rings *(2)* — Ring compounds with three or four atoms in the ring having bond angles less than 109.5°.

substrate *(12)* — The molecule whose chemical change is catalyzed by an enzyme.

surface-active agent *(7)* — A molecule that has a polar or charged end and a nonpolar end so that it will dissolve at the interface of a polar–nonpolar liquid mixture.

synapse *(21)* — The connection between a nerve cell and a cell of another organ.

synaptic cleft *(21)* — The gap in the synapse between the presynaptic membrane and the postsynaptic membrane.

synovial fluid *(22)* — The fluid that serves as a lubricant at the movable joints.

---

termination *(13)* — The stopping of protein synthesis with the release of product.

terpenes *(3)* — Naturally occurring compounds assembled from the branched five-carbon unit isoprene.

tertiary structure of protein *(11)* — The total folding pattern that a protein undergoes to achieve its final shape.

tetravalent metal *(23)* — A metal requiring four ligands to form a complex.

thermodynamics *(14)* — The study of energy changes.

thermoplastic *(3)* — Polymers that soften and melt on heating.

thioester *(5)* — An ester in which one of the ester oxygens has been replaced by a sulfur.

thioether *(5)* — An organic molecule that has two alkyl groups attached to a sulfur.

thiol *(5)* — An organic compound with an —SH group.

transcription (*13*) The synthesis of new RNA on the template of existing DNA.

translation (*13*) The synthesis of new protein following the genetic message as transmitted by mRNA.

transmitter (*21*) The compound that crosses the synaptic cleft and binds to the postsynaptic membrane, thereby stimulating it.

triglyceride (*7*) A glyceride in which carboxylic acids are esterified to all three hydroxy groups of glycerol.

unsaturated structure (*3*) An organic compound containing carbon-to-carbon double or triple bonds.

uronic acid (*6*) A carbohydrate in which the highest numbered alcohol group has been oxidized to a carboxylic acid.

van der Waals forces (*1*) Very weak attractive forces that exist between all atoms that are very close.

vitamin (*12*) An organic molecule that must be supplied in small quantities in the diet.

volatile compound (*5*) A liquid compound with a low boiling point, inferring its tendency to evaporate rapidly.

wax (*7*) The ester of a fatty acid and a long chain alcohol.

xanthinuria (*23*) A disease wherein xanthine appears in urine in place of uric acid.

zymogens (*12*) Inactive precursors of enzymes.

Appendix 4 Glossary of Reactions by Chapter

Chapter 2

36 $CH_4 + 2O_2 \longrightarrow CO_2 + 2H_2O$ (+ heat)

(combustion)

38 $Cl_2 \xrightarrow{\text{light}} 2\ :\!\ddot{\underset{\cdot\cdot}{Cl}}\cdot$

$RH + :\!\ddot{\underset{\cdot\cdot}{Cl}}\cdot \longrightarrow R\cdot + HCl$

$R\cdot + Cl_2 \longrightarrow RCl + :\!\ddot{\underset{\cdot\cdot}{Cl}}\cdot$

(free radical substitution)

Chapter 3

53 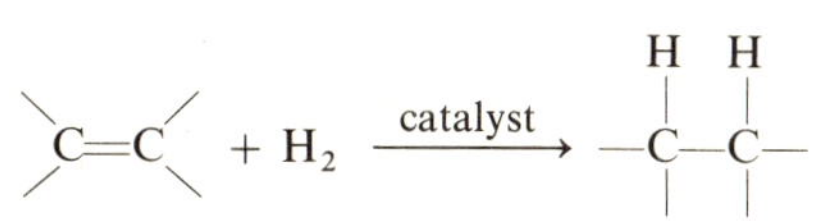

53 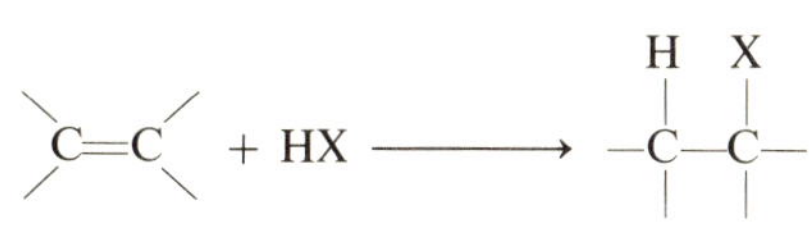

(X=Cl, Br, I)

53 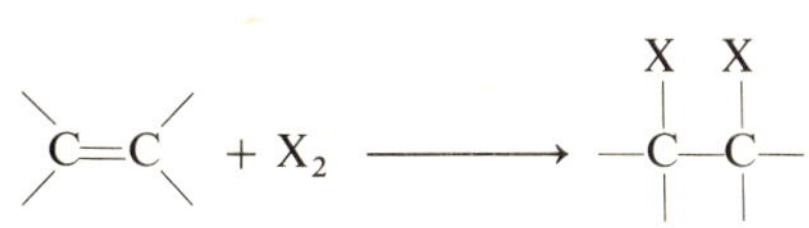

53 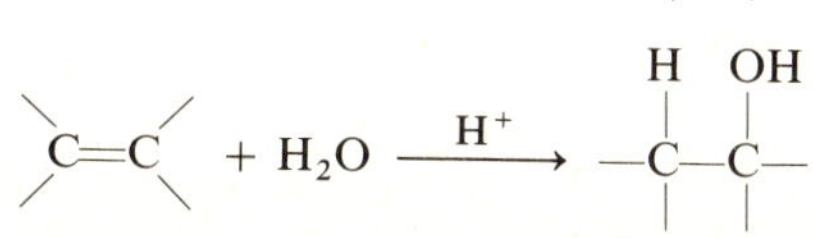

57 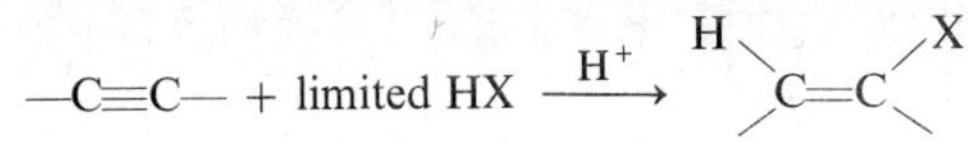

57 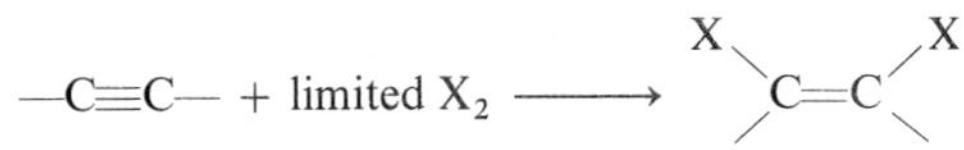

60 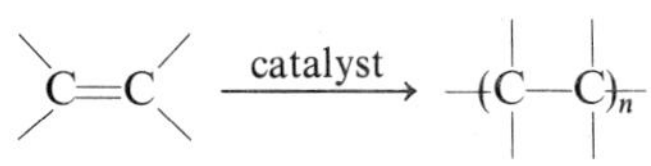

(addition polymerization)

Chapter 4

70 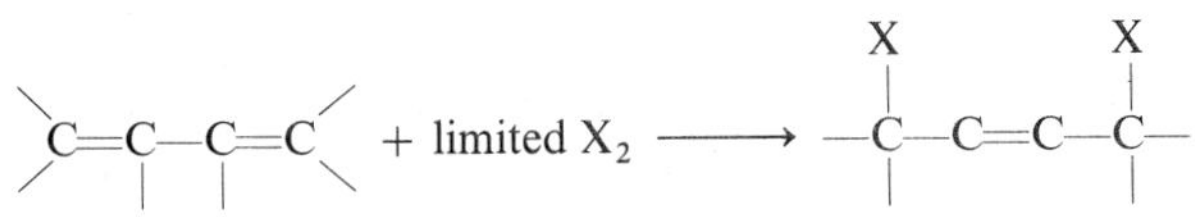

71 

72 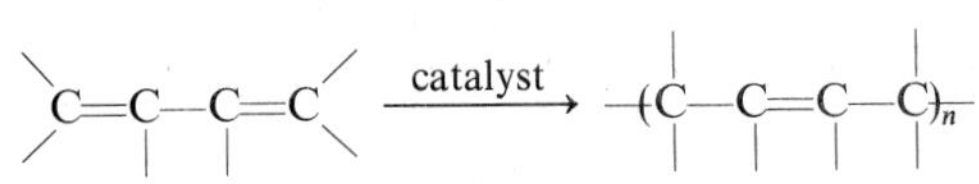

73 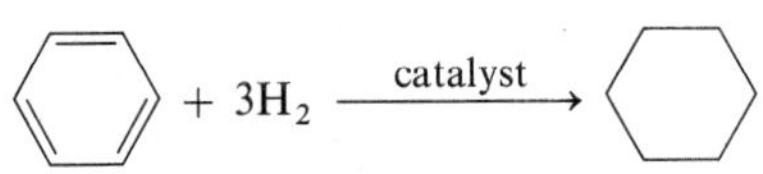

77 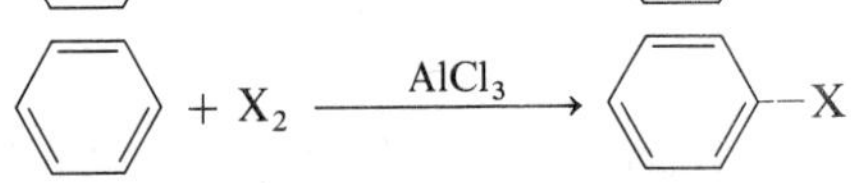

77 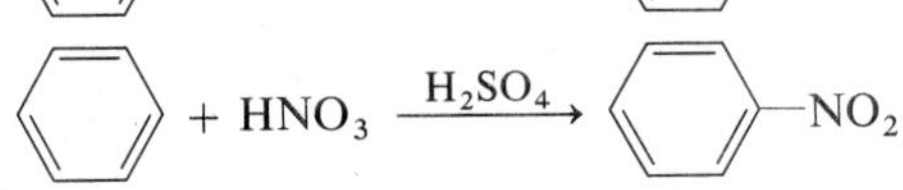

77 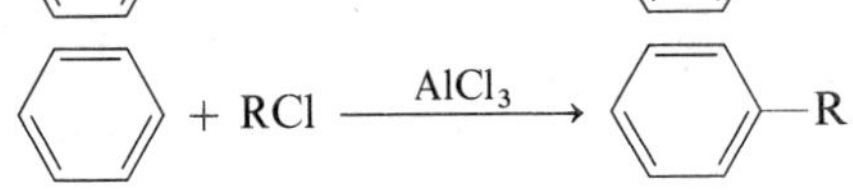

Chapter 5

91 $ROH \xrightarrow{PCl_3} RCl$

91 $ROH \xrightarrow{PBr_3} PBr$

91 $ROH \xrightarrow{SOCl_2} RCl$

91 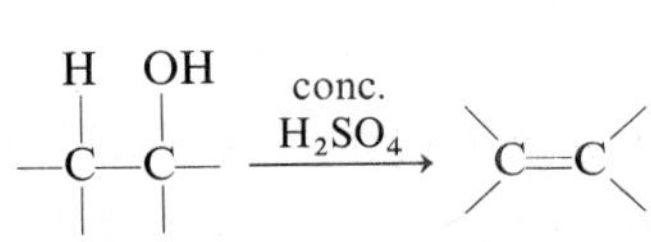

92 $$R{-}CH_2OH + (O) \xrightarrow[\text{agent}]{\text{oxidizing}} R{-}\overset{O}{\overset{\|}{C}}{-}H + H_2O$$

92 $$R{-}\underset{}{\overset{OH}{\overset{|}{C}H}}{-}R + (O) \xrightarrow[\text{agent}]{\text{oxidizing}} R{-}\overset{O}{\overset{\|}{C}}{-}R + H_2O$$

94 $$R{-}Cl \xrightarrow[\text{NaOH}]{\text{aqueous}} ROH$$

94 $$R{-}\overset{O}{\overset{\|}{C}}{-}H + H_2 \xrightarrow{\text{catalyst}} R{-}CH_2OH$$

94 $$R{-}\overset{O}{\overset{\|}{C}}{-}R + H_2 \xrightarrow{\text{catalyst}} R{-}\overset{OH}{\overset{|}{C}H}{-}R$$

99 $$R{-}SH + HgCl_2 \longrightarrow R{-}S{-}HgCl + HCl$$

100 $$2R{-}SH + \tfrac{1}{2}O_2 \longrightarrow R{-}S{-}S{-}R + H_2O$$

111 $$-\underset{|}{\overset{H}{\overset{|}{C}}}-\underset{|}{\overset{Cl}{\overset{|}{C}}}- \xrightarrow[\text{alcohol}]{\text{KOH in}} \;\rangle C{=}C\langle$$

Chapter 6

125 $$R{-}\overset{O}{\overset{\|}{C}}{-}H + (O) \xrightarrow[\text{agent}]{\text{oxidizing}} R{-}\overset{O}{\overset{\|}{C}}{-}OH$$

128 $$R{-}\overset{O}{\overset{\|}{C}}{-}H + R'{-}OH \xrightarrow[\text{HCl}]{\text{dry}} R{-}\underset{OR'}{\underset{|}{\overset{OH}{\overset{|}{C}}}}{-}H$$

128 $$R{-}\overset{O}{\overset{\|}{C}}{-}R + R'{-}OH \xrightarrow[\text{HCl}]{\text{dry}} R{-}\underset{OR'}{\underset{|}{\overset{OH}{\overset{|}{C}}}}{-}R$$

130 $$R{-}\underset{OR'}{\underset{|}{\overset{OH}{\overset{|}{C}}}}{-}H + R''{-}OH \xrightarrow{H^+} R{-}\underset{OR'}{\underset{|}{\overset{OR''}{\overset{|}{C}}}}{-}H$$

130 $$R{-}\underset{OR'}{\underset{|}{\overset{OH}{\overset{|}{C}}}}{-}R + R''{-}OH \xrightarrow{H^+} R{-}\underset{OR'}{\underset{|}{\overset{OR''}{\overset{|}{C}}}}{-}R$$

Chapter 7

155 $$\mathrm{R{-}\overset{O}{\overset{\Vert}{C}}{-}OH + H_2O \rightleftharpoons R{-}\overset{O}{\overset{\Vert}{C}}{-}O^- + H_3O^+}$$

158 $$\mathrm{R{-}\overset{O}{\overset{\Vert}{C}}{-}OH + NaOH \longrightarrow R{-}\overset{O}{\overset{\Vert}{C}}{-}O^- + Na^+ + H_2O}$$

161 $$\mathrm{R{-}\overset{O}{\overset{\Vert}{C}}{-}OH + R'{-}OH \xrightarrow{H^+} R{-}\overset{O}{\overset{\Vert}{C}}{-}O{-}R' + H_2O}$$

161 $$\mathrm{R{-}\overset{O}{\overset{\Vert}{C}}{-}O{-}R' + \underset{(KOH)}{NaOH} \longrightarrow R{-}\overset{O}{\overset{\Vert}{C}}{-}O^- + \underset{(K^+)}{Na^+} + R'{-}OH}$$

167 $$\mathrm{R{-}\overset{O}{\overset{\Vert}{C}}{-}OH + NH_3 \xrightarrow{\Delta} R{-}\overset{O}{\overset{\Vert}{C}}{-}NH_2 + H_2O}$$

172 $$\mathrm{R{-}\overset{O}{\overset{\Vert}{C}}{-}OH \xrightarrow{PCl_3} R{-}\overset{O}{\overset{\Vert}{C}}{-}Cl}$$

172 $$\mathrm{R{-}\overset{O}{\overset{\Vert}{C}}{-}OH \xrightarrow{SOCl_2} R{-}\overset{O}{\overset{\Vert}{C}}{-}Cl}$$

172 $$\mathrm{R{-}\overset{O}{\overset{\Vert}{C}}{-}OH \xrightarrow{PCl_5} R{-}\overset{O}{\overset{\Vert}{C}}{-}Cl}$$

172 $$\mathrm{R{-}\overset{O}{\overset{\Vert}{C}}{-}Cl + R'{-}OH \longrightarrow R{-}\overset{O}{\overset{\Vert}{C}}{-}O{-}R' + HCl}$$

173 $$\mathrm{R{-}\overset{O}{\overset{\Vert}{C}}{-}Cl + NH_3 \longrightarrow R{-}\overset{O}{\overset{\Vert}{C}}{-}NH_2 + HCl}$$

173 $$\mathrm{R{-}\overset{O}{\overset{\Vert}{C}}{-}Cl + R'{-}NH_2 \longrightarrow R{-}\overset{O}{\overset{\Vert}{C}}{-}NH{-}R' + HCl}$$

Chapter 9

225 $$\mathrm{R{-}\overset{O}{\overset{\Vert}{C}}{-}R + R'{-}NH_2 \longrightarrow R{-}\overset{N{-}R'}{\overset{\Vert}{C}}{-}R{-} + H_2O}$$

227 $$\mathrm{R{-}NH_2 + H^+ \rightleftharpoons R{-}NH_3{}^+}$$

Appendix 5 Answers to Selected Exercises and Problems

Chapter 1

Exercise 1.1. ***a)*** 38 kcal (exactly the same amount of energy is required to break the bond as was released during its formation).

b) N_2 (since the greatest amount of energy was released during its formation).

c) O_2 weakest (least energy required to break it), N_2 strongest (most energy required to break it).

Exercise 1.2. Nitrogen, in group VA, has five electrons in its outer shell and therefore requires three electrons to establish an octet (8).

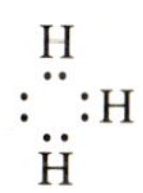

Exercise 1.3.

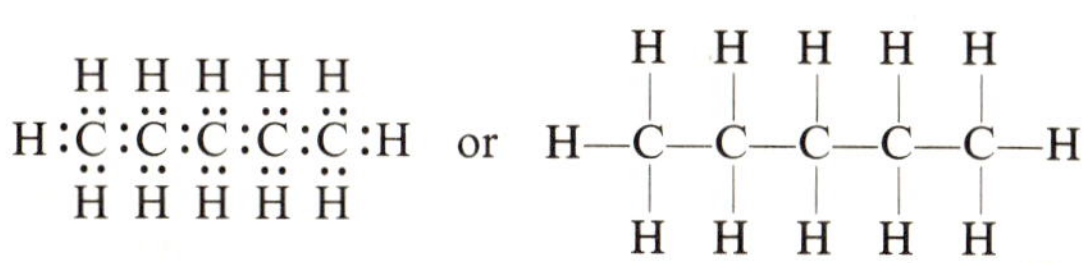

Exercise 1.4.

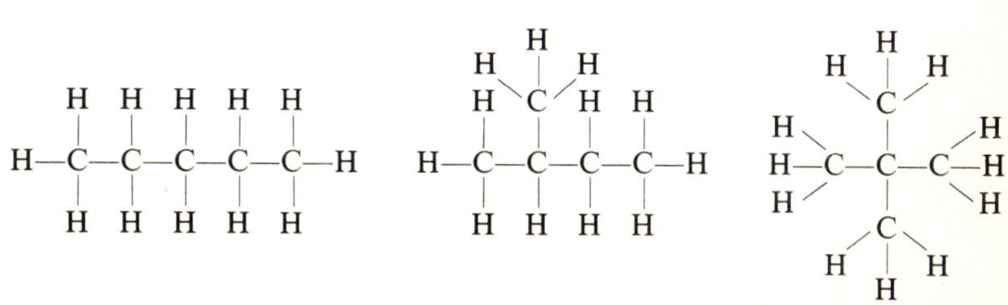

Exercise 1.5. No. This would require hydrogen to provide two electrons, and it has only one to contribute. Furthermore, hydrogen would end up sharing four electrons, and only two are required to satisfy it.

Exercise 1.6.

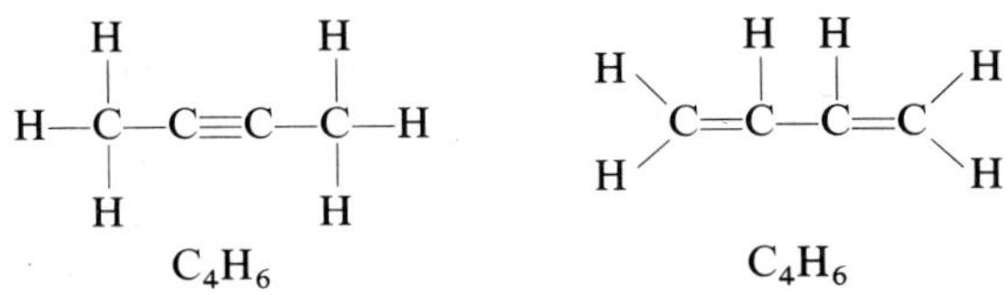

Carbon always has four lines leading to it, counting a double bond as two lines and a triple bond as three lines. Because each line represents a pair of electrons, four bonds are eight electrons—an octet.

Exercise 1.7. F, F, C, respectively

Exercise 1.8. $H{:}\ddot{\underset{..}{Cl}}{:}$ There are three pairs of unshared electrons on chlorine.

Problem 1. Chlorine, in group VIIA, has seven electrons in its outer shell. It requires one more electron for an octet.

a) Ionic ***b)*** Covalent ***c)*** Covalent

Problem 3. ***a)*** 180° ***b)*** 109.5° ***c)*** 109.5° ***d)*** 120° ***e)*** 120°

Problem 5. F

Problem 7.

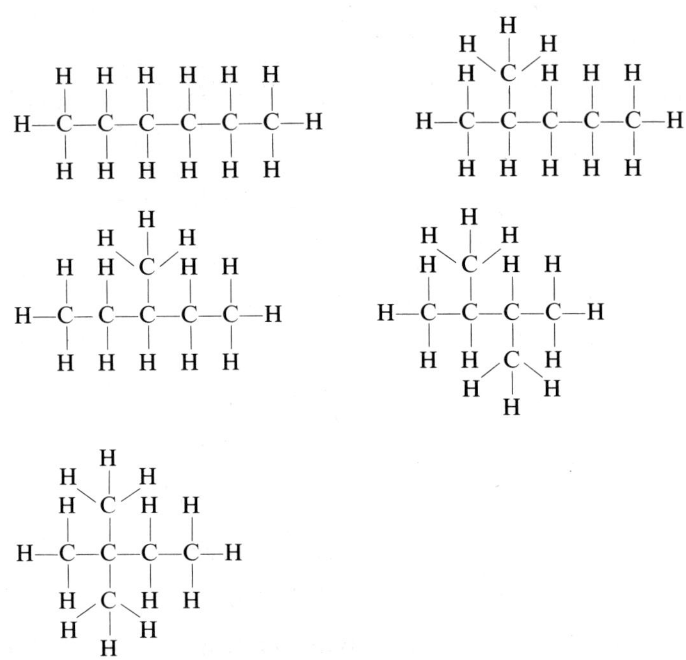

Problem 9. ***a)*** alkene ***b)*** alcohol ***c)*** ketone ***d)*** ether
e) carboxylic acid ***f)*** alkene, aldehyde
g) amine, thiol (mercaptan) ***h)*** alkyne, ester ***i)*** amide

Chapter 2

Exercise 2.1 ***a)***

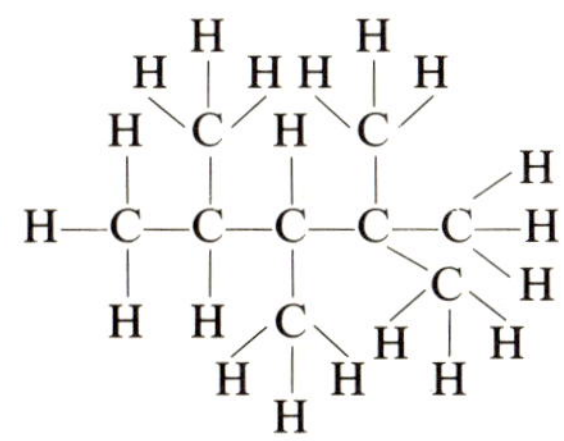

b) $CH_3—CH_2—C(CH_3)_2—CH_2—CH(CH_3)—CH_3$

Exercise 2.2 ***a)***

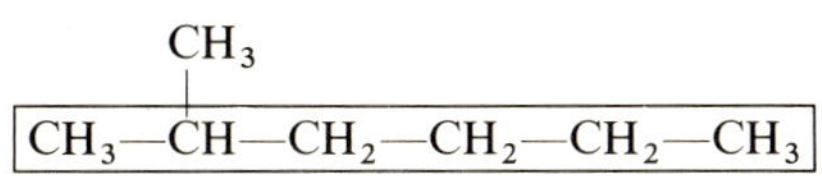

or

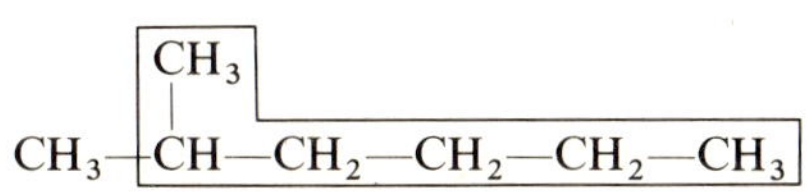

b)

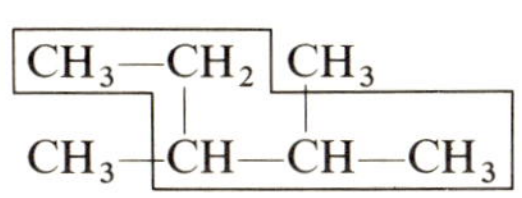

or

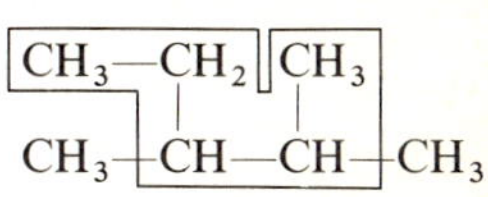

c)

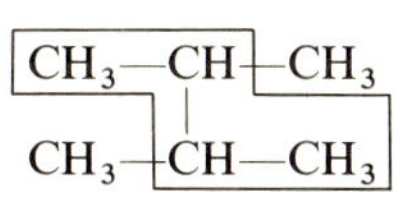

or

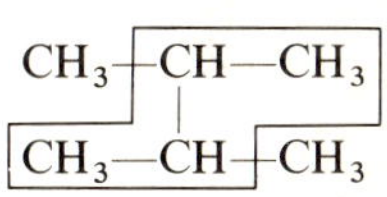

d)

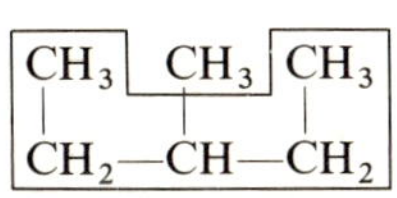

Exercise 2.3. ***a)*** 3-methylhexane

b) 2,3-dimethylbutane

c) 3-ethylpentane

d) 3-isopropylhexane *or* 2-methyl-3-ethylhexane

Exercise 2.4. ***a)*** 2-chloropropane ***b)*** 1-chloropropane

c) 1-chloropropane ***d)*** 1,3-dichloropropane

e) 2-iodopropane ***f)*** 1-nitropropane

g) 1-bromopropane ***h)*** 1-fluoropropane

i) 1,3-dibromopropane *j*) 1-chloro-2-iodopropane

k) 1-nitropropane *l*) 1,2,3-trichloropropane

Exercise 2.5. 1,1-dichlorocyclohexane, $C_6H_{10}Cl_2$ (10)

1-bromo-2-methylcyclopentane, $C_6H_{11}Br$ (11)

Exercise 2.6. Freon 11, trichlorofluoromethane

Freon 12, dichlorodifluoromethane

Freon 13, chlorotrifluoromethane

1,1,2,2-tetrachloro-1,2-difluoroethane

1,2-dichloro-1,1,2,2-tetrafluoroethane

Freon C-318, octafluorocyclobutane

Problem 1. *a*) butane *b*) cycloheptane

c) 2-chloropentane *d*) 2-methylpentane

e) 2,3-dimethylbutane *f*) 1,1,2-trichloroethane

g) trifluoromethane *h*) 2-chloro-2-methylpropane

i) 4-bromo-1,2-dimethylcyclohexane

j) 1,1,3,3-tetrachlorocyclopentane

k) 1,2,4,5-tetrachloro-3-isopropylpentane

Problem 3. *a*) 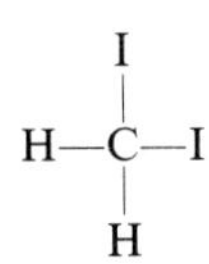

b) 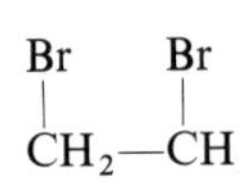

c) 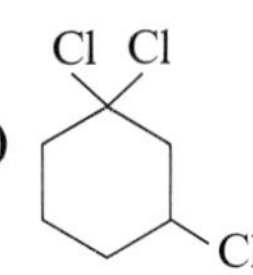

d) 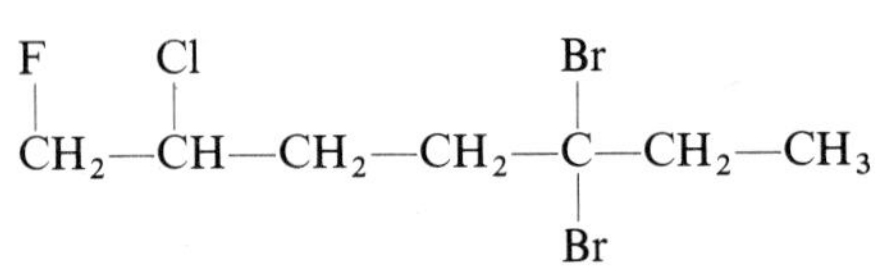

e) 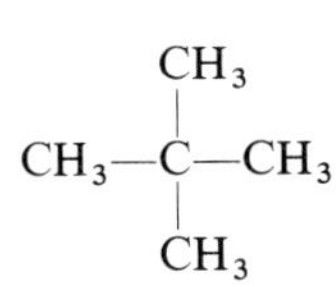

f) 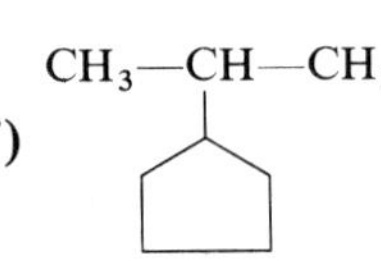

g) 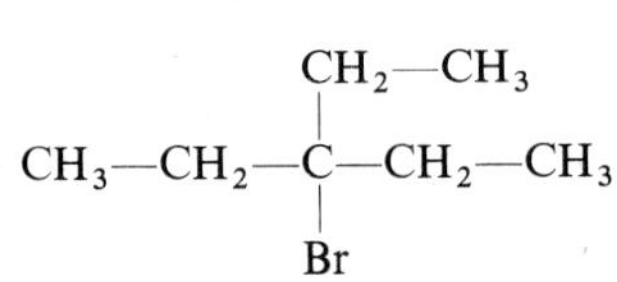

h) 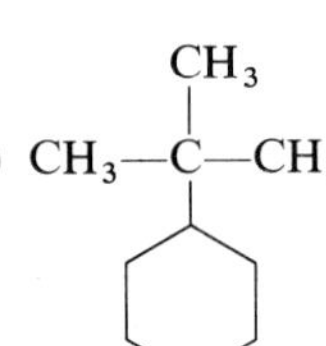

Problem 5. $2C_6H_{14} + 19O_2 \longrightarrow 12CO_2 + 14H_2O$

Problem 7. $C(Cl)_3—CFCl—CF_3$
$C(Cl)_3—CF_2—CClF_2$
$C(Cl)_2F—C(Cl)_2—CF_3$
$C(Cl)_2F—CClF—CClF_2$
$C(Cl)_2F—CF_2—C(Cl)_2F$
$CClF_2—C(Cl)_2—CClF_2$

Problem 9. Carbon monoxide results from the combustion of hydrocarbons with insufficient oxygen. The catalytic muffler becomes hot when converting unburned hydrocarbons and carbon monoxide to carbon dioxide because the reactions are exothermic.

Chapter 3

Exercise 3.1. Most per molecule: d, b, a, c

Most per gram: b, d, a, c

Exercise 3.2. ***a***) 2-pentene ***b***) 1-bromo-1-propene

c) 2-bromo-1-propene ***d***) 3-methyl-1-pentene

e) 2,3-dichloro-2-butene ***f***) 4-methyl-2-hexene

Exercise 3.3. c

Exercise 3.4. 1-chloro-1-propene only

Exercise 3.5.

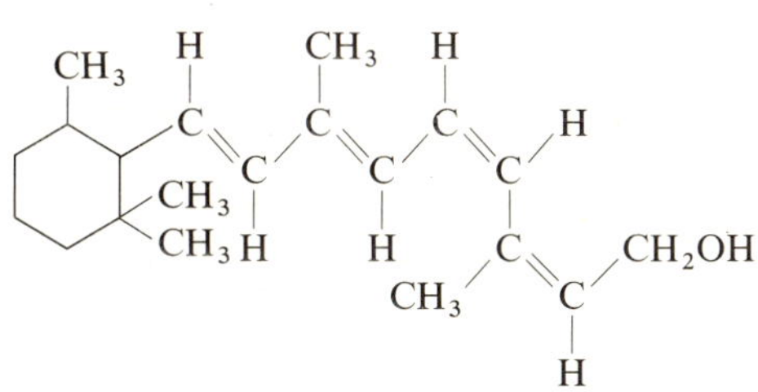

Exercise 3.6. ***a***)
Br
$CH_3—CH—CH_2—CH_3$

b) 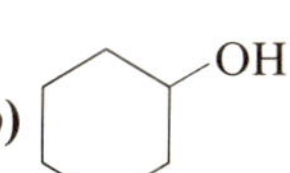

c) $CH_3—CH_2—CH_3$

d)
Cl Cl
$CH_2—CH—CH_2—CH_3$

Exercise 3.7. ***a)*** 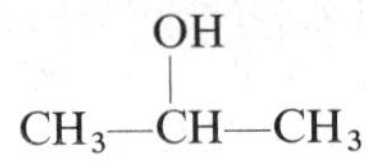 ***b)*** 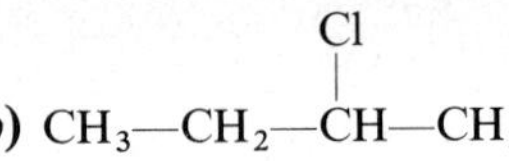

c) 

Exercise 3.8. ***a)*** $CHBr{=}CHBr$ ***b)*** $CH(Br)_2{-}CH(Br)_2$

c) $CH_2{=}CHCl$ ***d)*** $CH_3{-}CH(Cl)_2$

e) $CH_3{-}CH_2{-}C(Br)_2{-}CH_3$

Exercise 3.9. Markovnikov's rule is followed.

Problem 2. ***a)*** 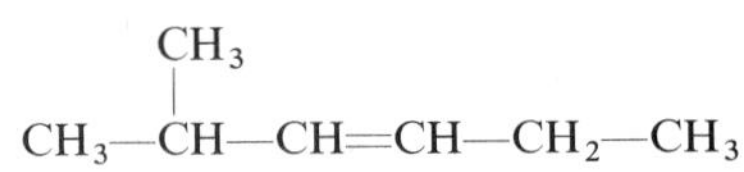 ***b)*** 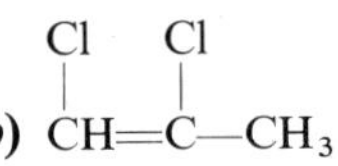

c) $CH_2I{-}C{\equiv}C{-}CH_3$ ***d)*** 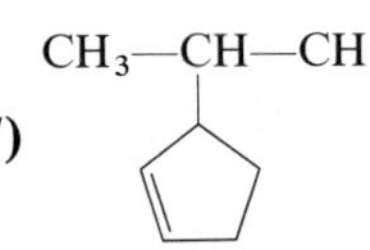

e) 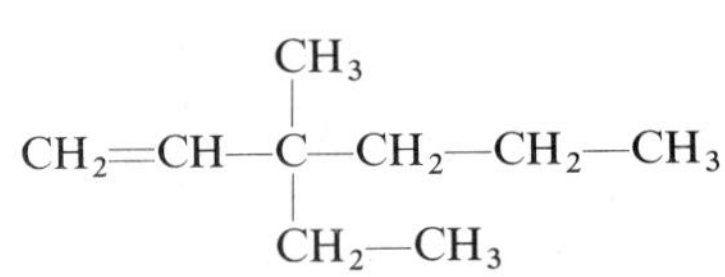 ***f)*** 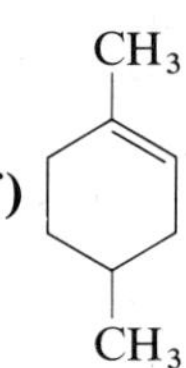

g) 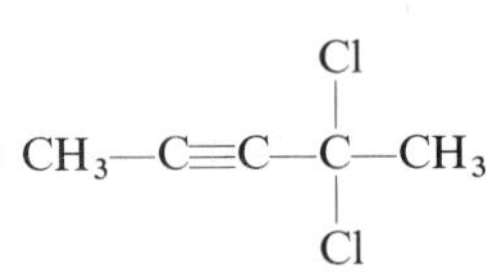 ***h)*** 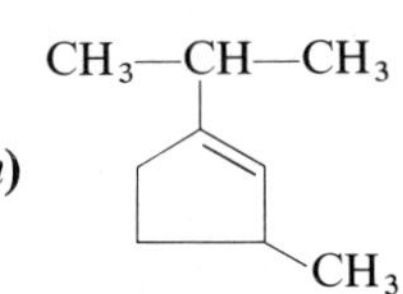

Problem 4. b only

Problem 6. ***a)***
1. alkene, alcohol
2. alkene, aldehyde
3. alkene, aldehyde
4. alkene, aldehyde
5. alkene, alcohol
6. alkene, alcohol
7. alkene, ketone
8. alkene, alcohol
9. alkene, alcohol
10. alkene, alcohol
11. ester
12. carboxylic acid
13. alkene, aldehyde
14. aldehyde
15. alkene, aldehyde
16. alkene, ester
17. alkene, alcohol
18. alkene, alcohol
19. alkene
20. ester
21. alkene, alcohol
22. alkene, alcohol
23. alkene, ketone, carboxylic acid
24. alkene, ester

***b*)** There would be geometric isomers of 1–5, 13, 15–19, 21–24.

***c*)** 1. trans 2. cis 3. cis 4. trans 5. cis 13. cis
15. trans 16. cis 17. cis 18. trans, trans 19. cis
21. trans 22. trans, trans 23. trans 24. cis
Both isomers are drawn in the case of 3 and 4.

***d*)** The isoprenoid structures are 1–6, 8–10, 21, 22.

Problem 8.

***a*)** $CH_3{-}CH_2{-}CHBr{-}CH_3$

***b*)** 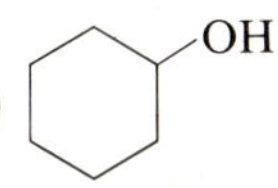

***c*)** $CH_2Cl{-}CHCl{-}CH_3$

***d*)**
$$\begin{array}{ccccccc} & & & & CH_3 & & \\ & & & & | & & \\ CH_3 & - & CH & - & CH & - & CH_3 \\ & & | & & & & \\ & & CH_3 & & & & \end{array}$$

***e*)** $CH_3{-}CBr{=}CHBr$

***f*)** $CH_3{-}CH_2{-}CCl_2{-}CHCl_2$

***g*)** $CH_3{-}CH_2{-}CH_2{-}CH{=}CH_2$

***h*)** $CH_3{-}CH_2{-}CHCl{-}CH_3$

***i*)**
$$\begin{array}{ccccccc} & & & & CH_3 & & \\ & & & & | & & \\ CH_3 & - & CH_2 & - & C & - & CH_3 \\ & & & & | & & \\ & & & & OH & & \end{array}$$

***j*)** 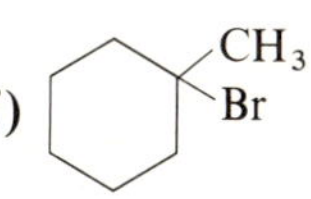

***k*)** $CH_2{=}CHCl$

***l*)** $CH_3{-}CHCl_2$

***m*)**
$$\begin{array}{ccccccccc} & & Br & & & & & & \\ & & | & & & & & & \\ CH_3 & - & C & - & CH_2 & - & C & = & CH_2 \\ & & | & & & & | & & \\ & & CH_3 & & & & CH_3 & & \end{array}$$

Problem 10. $CH_2{=}CCl_2$

Problem 12. Softness relates to melting point, the lower melting being the softer. The melting point relates to the number of double bonds, the larger number of double bonds per molecule being the lower melting. Margarine producers convert liquid triglycerides to solids by reacting them with hydrogen, eliminating double bonds. Reacting the liquid triglyceride with less hydrogen leaves more double bonds per molecule, resulting in a lower melting point, producing a softer margarine.

Chapter 4

Exercise 4.1. a, b, d, e, g

Exercise 4.2. $CH_2{=}CH{-}CH{=}CH_2$ + limited HCl $\longrightarrow$ $CH_3{-}CH{=}CH{-}CH_2Cl$

Exercise 4.3. ***a)*** fluorobenzene ***b)*** ethylbenzene ***c)*** bromobenzene

Exercise 4.4. ***a)*** 1,3-dibromobenzene or *m*-dibromobenzene

b) 4-chloroiodobenzene or *p*-chloroiodobenzene

c) 2-chlorofluorobenzene or *o*-chlorofluorobenzene

d) 1,4-dinitrobenzene or *p*-dinitrobenzene

Exercise 4.5. ***a)*** Correct ***b)*** Correct ***c)*** Very unlikely

d) Very unlikely ***e)*** Correct

Problem 2. ***a)*** 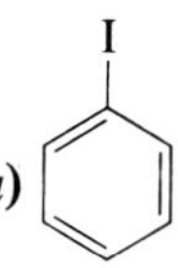

b) $CH_2{=}CH{-}CH_2{-}CH_2{-}CH{=}CH_2$

c) 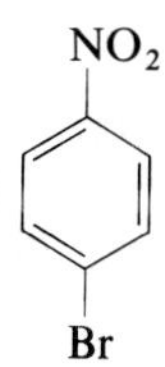

d) $CH_2{=}C(CH_3){-}CH(CH_3){-}CH{=}CH_2$

e) 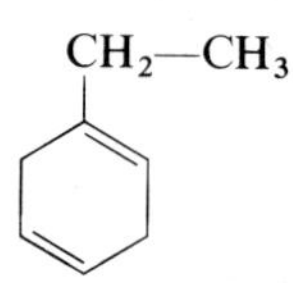

f) 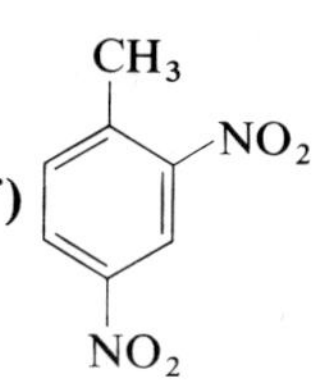

Problem 4. Problem 1: b, c, f, g, h Problem 2: none

Problem 6. Two are possible: 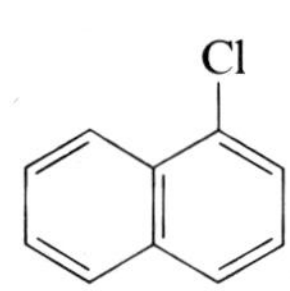 and 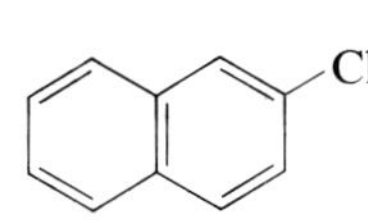

Problem 8. $-CH_2{-}CH{=}CH{-}CH_2{-}(CH_2{-}CH{=}CH{-}CH_2){-}CH_2^-$
The repeating unit is the result of 1,4-addition of the monomer: $CH_2{=}CH{-}CH{=}CH_2$.

Yes, it does exhibit cis-trans isomerism.

Problem 10. ***a)*** $:Cl:Cl: \longrightarrow :Cl^+ + :Cl:^-$ ***b)*** $:Cl:^-$ ***c)*** 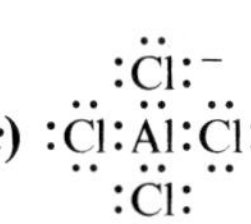

d) OH^-; H^+ reacts with OH^- to produce H_2O. Removing the OH^- as it forms speeds the reaction to form NO_2^+.

e) $H_2SO_4 + OH^- \longrightarrow H_2O + HSO_4^-$

Chapter 5

Exercise 5.1.

a) 1-butanol

b) 2-butanol

c) 3-pentanol

d) 2,4-pentanediol

e) 2-hydroxypropanoic acid

f) 2-methylcyclopentanol

g) *p*-nitrophenol; 4-nitrophenol

h) 2-methylphenol; *o*-cresol

i) 2,4,6-trichlorophenol

Exercise 5.2.

a) $CH_3—CHCl—CH_3$

b) $CH_3—CH_2Cl$

c) CH_3Br

d) $CH_2{=}CH_2$

e) 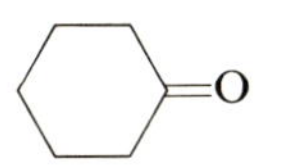

f) $CH_3—\overset{\overset{\displaystyle O}{\|}}{C}—CH_2—CH_3$

g) $CH_3—CH_2—CH_2—\overset{\overset{\displaystyle O}{\|}}{C}—H$

h) No reaction

i) $SOCl_2$ or PCl_3; (0), oxidizing agent; conc. H_2SO_4

Exercise 5.3.

a) $CH_3—CH_2—OH$

b) 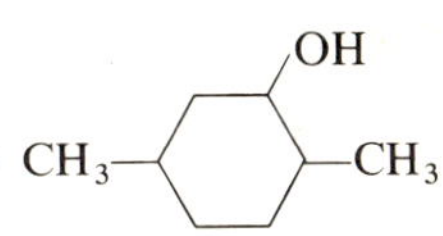

c) $CH_3—\underset{\underset{\displaystyle CH_3}{|}}{CH}—\overset{\overset{\displaystyle OH}{|}}{CH}—CH_3$

d) 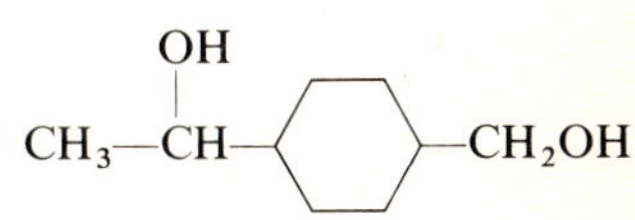

Exercise 5.4.

a) diisopropyl ether

b) cyclopentyl *n*-propyl ether

c) *p*-chloroanisole or 4-chloroanisole

d) 1,3-dimethoxycyclopentane

Exercise 5.5. a, b, c, e

Exercise 5.6.

a) There are two asymmetric centers, and therefore four isomers.

b) D
L

Exercise 5.7.

D
L
D

Exercise 5.8.

D	D	D	D	D	D	D	D
D	D	D	L	D	L	L	L
D	D	L	D	L	D	L	L
D	L	D	D	L	L	D	L
L	L	L	L	L	L	L	L
L	L	L	D	L	D	D	D
L	L	D	L	D	L	D	D
L	D	L	L	D	D	L	D

Exercise 5.9.

***a*)** One center, two isomers
***b*)** No centers
***c*)** Two centers, four isomers
***d*)** Two centers, four isomers
***e*)** One center, two isomers
***f*)** Three centers, eight isomers
***g*)** Two centers, four isomers
***h*)** Three centers, eight isomers

Problem 2.

***a*)** 2-pentanol
***b*)** resorcinol
***c*)** 1,3-propanediol
***d*)** 2-methyl-2-propanol
***e*)** *p*-bromophenol or 4-bromophenol
***f*)** 1,2-dichloro-1-propanol
***g*)** 2,4,6-trimethylphenol
***h*)** 2-isopropylcyclopentanol
***i*)** 2-cyclopentyl-2-propanol
***j*)** *m*-cresol or 3-methylphenol
***k*)** isopropyl *n*-propyl ether
***l*)** *p*-methoxyphenol
***m*)** ethane thiol or ethyl mercaptan
***n*)** diisopropyl sulfide

Problem 4.

***a*)** $CH_2Br—CHOH—CH_3$

***b*)**
$$CH_3—CH_2—\underset{\underset{CH_3—CH—CH_3}{|}}{\overset{\overset{OH}{|}}{C}}—CH_2—CH_2—CH_3$$

***c*)**
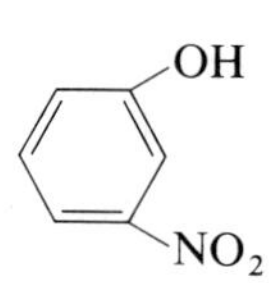

***d*)** $CH_3—CH_2—CH_2—SH$

***e*)**
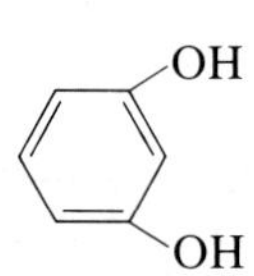

***f*)**
$$CH_3—O—\overset{\overset{CH_3}{|}}{CH}—CH_3$$

g) 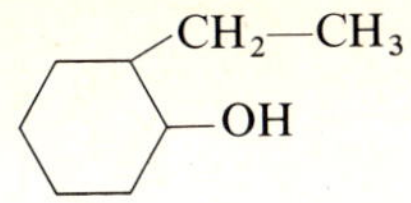

h) $CH_2OH—CH_2—CHOH—CH_3$

i) 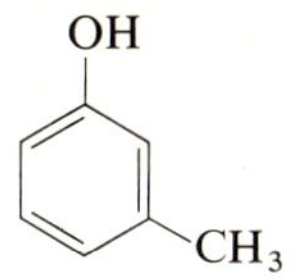

j) 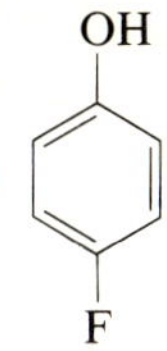

k) 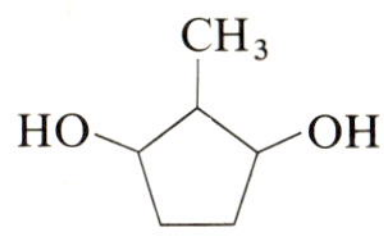

l) 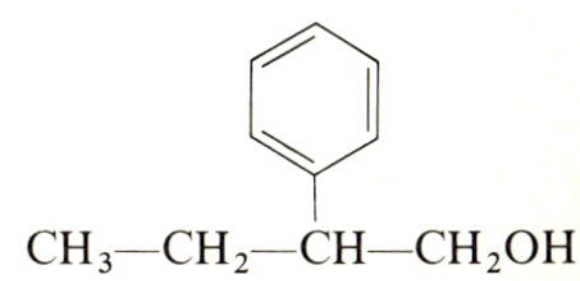

m) 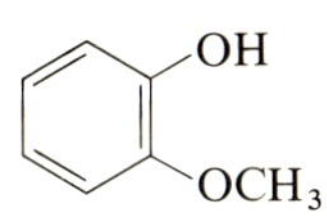

n) 3-ethyl-2-methyl-3-hexanol

Problem 6.

a) (0), oxidizing agent

b) $SOCl_2$ or PCl_3

c) conc. H_2SO_4

d) conc. H_2SO_4

e) H^+, H_2O

f) $SOCl_2$ or PCl_3

g) aqueous NaOH

h) (O), oxidizing agent

i) H_2, catalyst

Problem 8.

a) $CH_3—CH{=}CH_2$

b) 

c) $CH_3—\underset{\displaystyle CH_3}{\underset{|}{C}}{=}CH_2$

d) no reaction

e) $CH_2{=}CH—CH{=}CH_2$

f) $HC{\equiv}CH$

g) $HC{\equiv}CH$

Problem 10.

a) $CH_2{=}CH_2 \xrightarrow[H^+]{H_2O} CH_3—CH_2OH$

$$CH_2{=}CH_2 \xrightarrow{HCl} CH_3—CH_2Cl \xrightarrow[NaOH]{aqueous} CH_3—CH_2OH$$

b) $$CH_2{=}CH_2 \xrightarrow[H^+]{H_2O} CH_3{-}CH_2OH \xrightarrow[\text{oxidizing agent}]{(O)} CH_3{-}\overset{\displaystyle O}{\overset{\|}{C}}{-}H$$

c) $$CH_2{=}CH_2 \xrightarrow{HCl} CH_3{-}CH_2{-}Cl$$

$$CH_2{=}CH_2 \xrightarrow[H^+]{H_2O} CH_3{-}CH_2OH \xrightarrow{SOCl_2} CH_3{-}CH_2Cl$$

Problem 12.

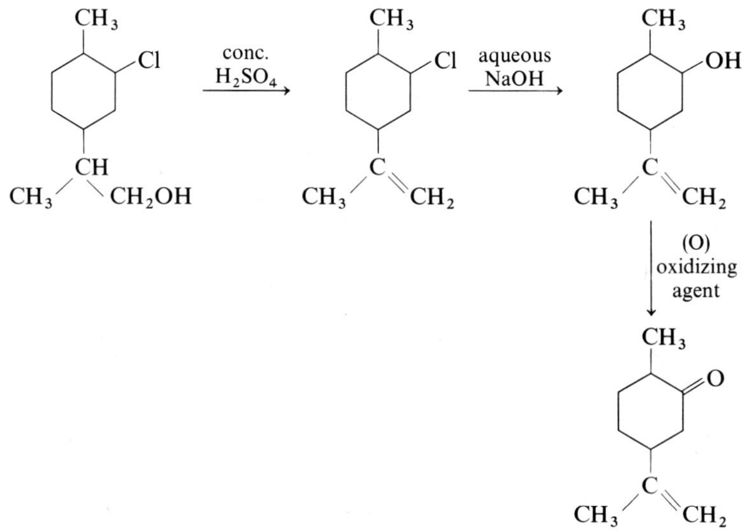

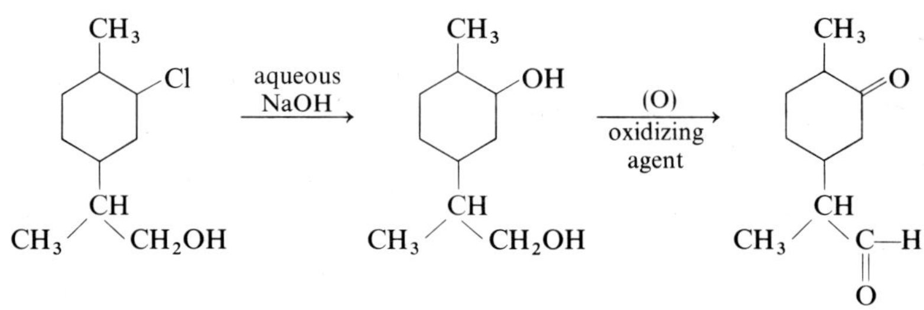

Problem 14. ***a)***

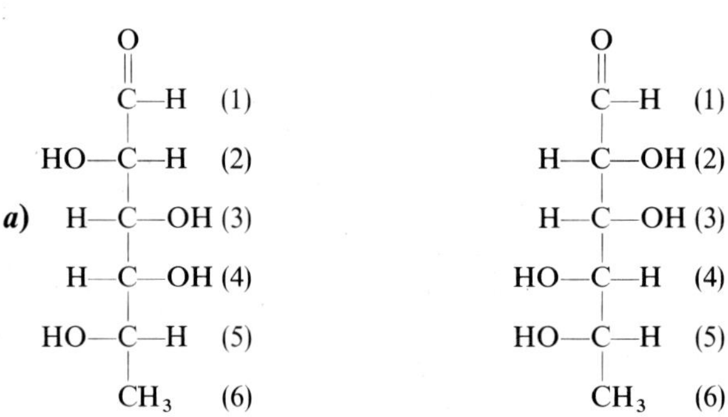

b) 2, 3, 4, 5 on each structure

c) D-fucose: 3, 4   D-rhamnose: 2, 3

d) L-fucose: 2, 5   L-rhamnose: 4, 5

e) diastereoisomers

f) 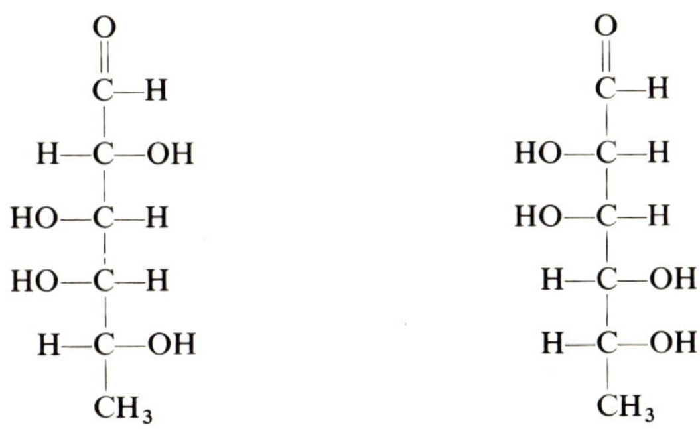

g) 16

h) Any structure with the same set of covalent bonds except the structures used to answer part f, for example,

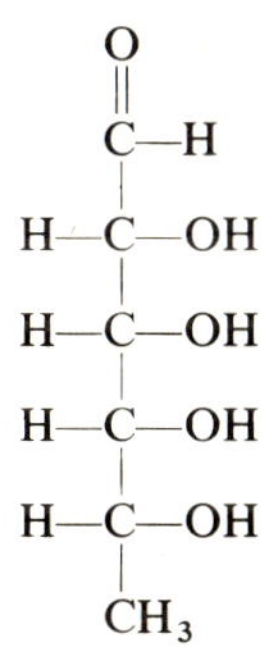

Problem 16. ***a)*** 2, 3-butanediol

b) 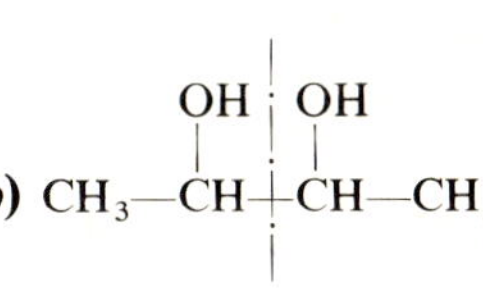

c) A. 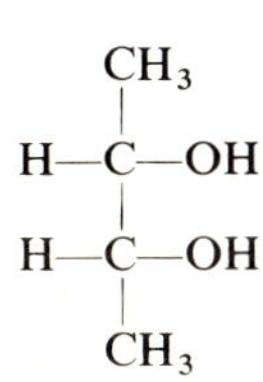

B. 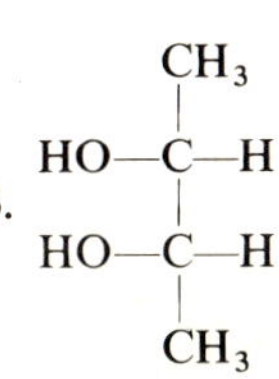

C. 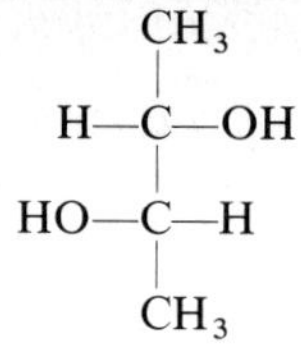

D. 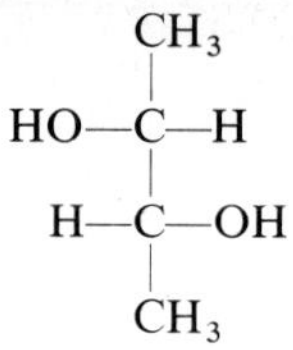

d) Enantiomers

e) Diastereoisomers

f) No

g) 28° to the right

h) No rotation

i) Only 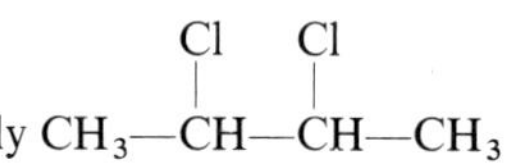

Chapter 6

Exercise 6.1. *a*) 2-pentanone *b*) 3-hexanone

c) 2,4-dimethyl-3-pentanone *d*) propanone; no

e) cyclohexanone, no *f*) 1,5,6-trichloro-3-hexanone

Exercise 6.2. propanone

Exercise 6.3. *a*) pentanal *b*) 4-methylpentanal *c*) 3,3-dichlorobutanal

Exercise 6.4. methanal (formaldehyde)

Exercise 6.5. *a*) ketohexose *b*) aldopentose *c*) aldohexose

Exercise 6.6. 9; diastereoisomer

Exercise 6.7. *a*) $CH_3—CH_2—CHOH—CH_2—CH_3$ *b*) 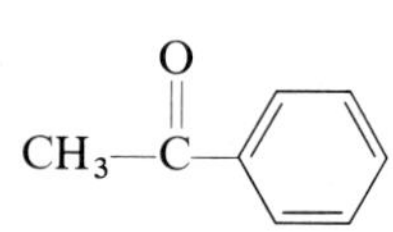

c) 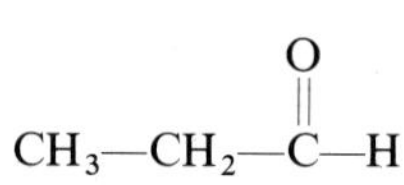

d) 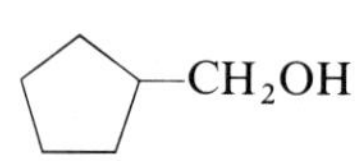

Exercise 6.8. Oxidation; b, c.

Exercise 6.9. Such an oxidation requires breaking a carbon-to-carbon bond.

Exercise 6.10. **a)** 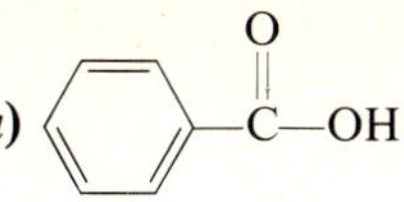 **b)** 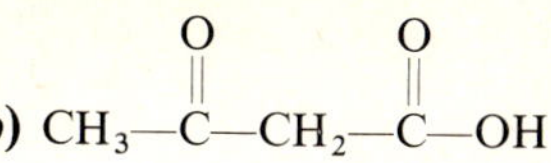

Exercise 6.11. **a)** 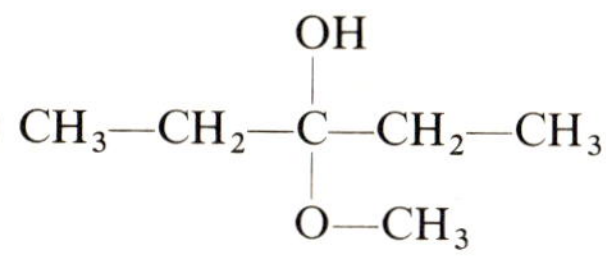 **b)** 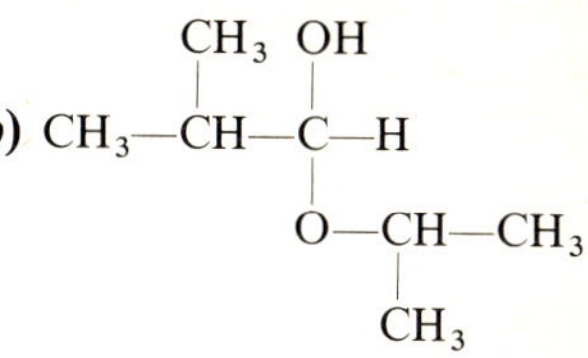

Exercise 6.12. Diastereoisomers

Exercise 6.13. **a)** 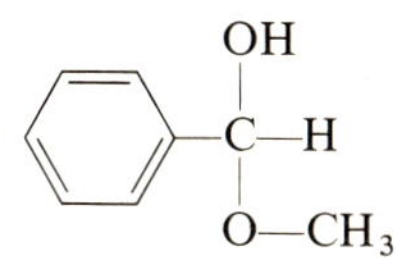 **b)** 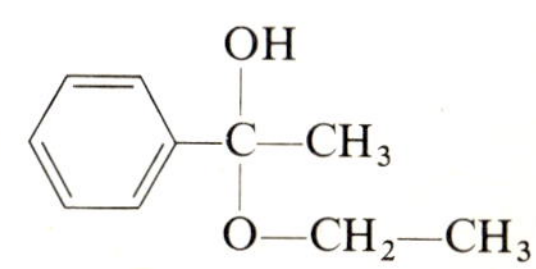

c) 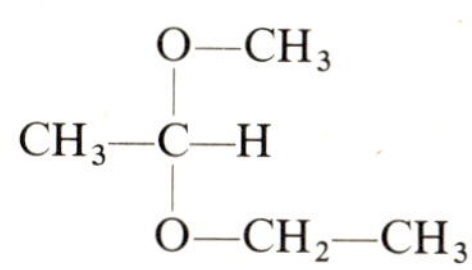 **d)** 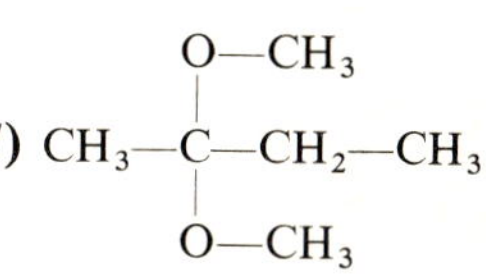

e) 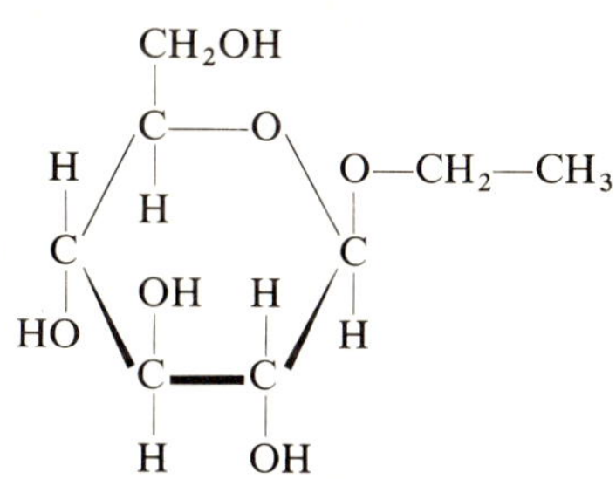

f) 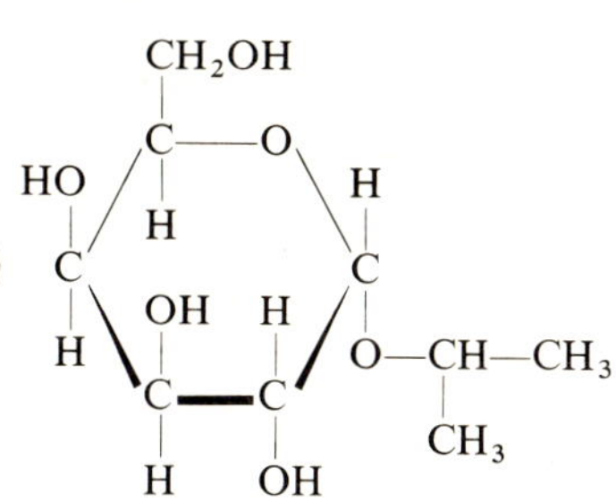 **g)** 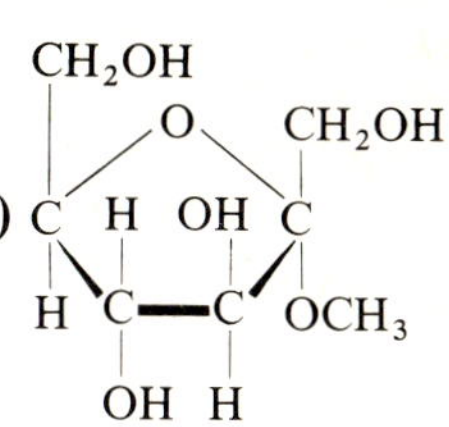

Exercise 6.14. Diastereoisomers

Exercise 6.15. glucose α, fructose β

Exercise 6.16.

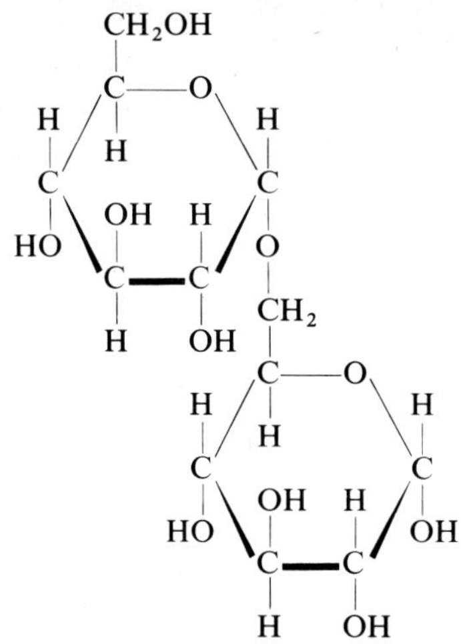

Problem 1.

a) 3-chlorobutanal

b) 1-phenyl-2-propanone

c) 1,3-dihydroxypropanone

d) 2,6-dimethylcyclohexanone

e) o-ethylbenzaldehyde

f) p-nitroacetophenone

g) 4,4-dibromo-3-methylpentanal

h) 2-formylcyclopentanone

i) 2,4-pentanedione

Problem 3.

a)

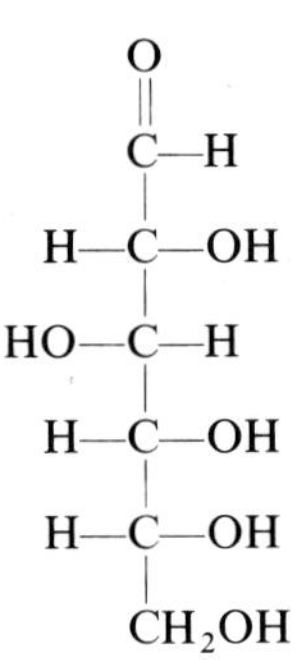

D-glucose

b)

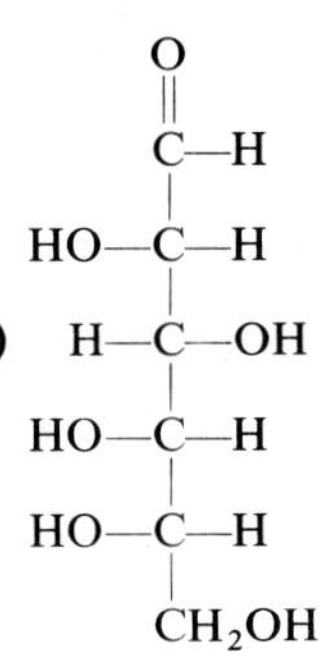

L-glucose

c)

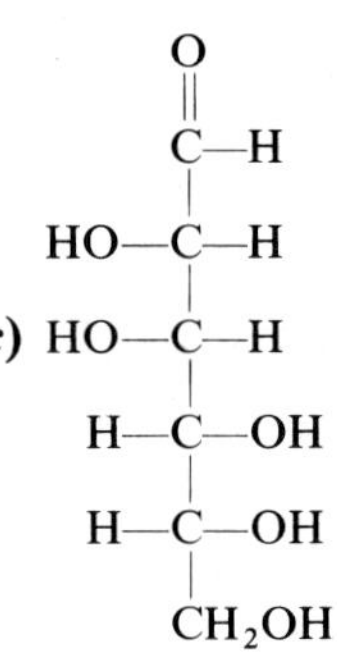

D-mannose

d)

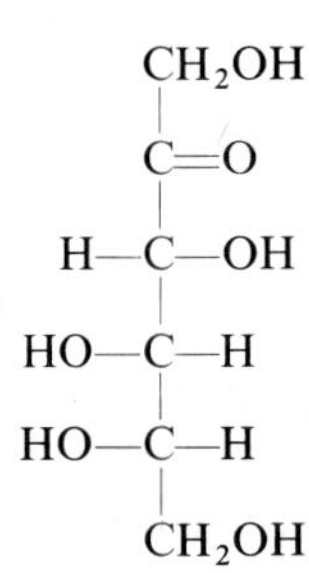

L-fructose

Problem 5.

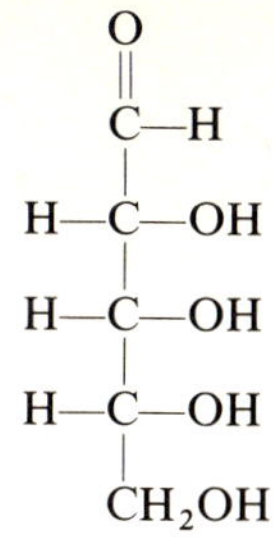

D-ribose

Problem 7.

a) Hemiacetal ***b)*** Full acetal

c) Neither ***d)*** Hemiacetal

e) Full acetal

f) Full acetal (between the glucoses) and hemiacetal (right-hand glucose)

Problem 9.

a)

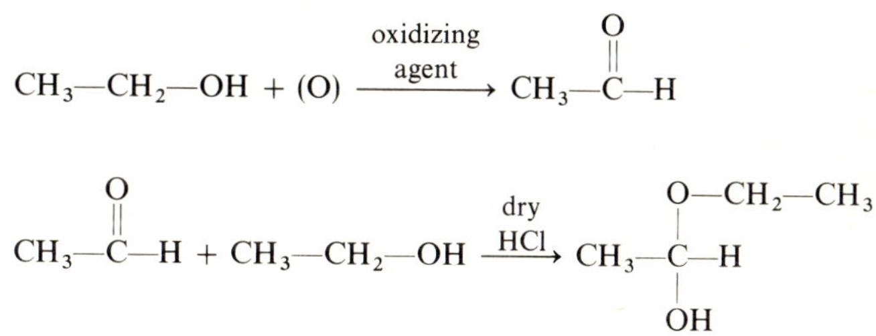

b)

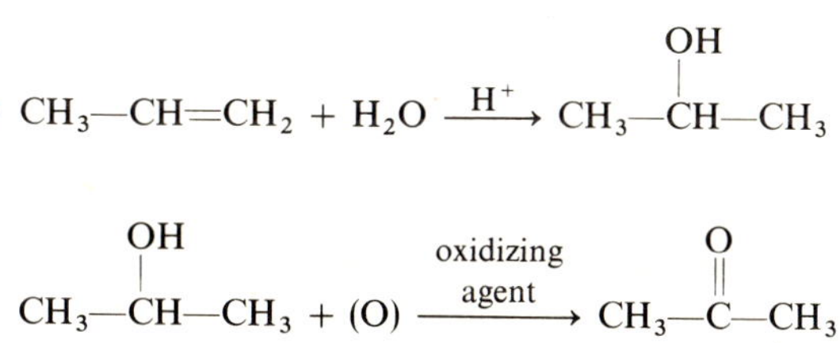

c)

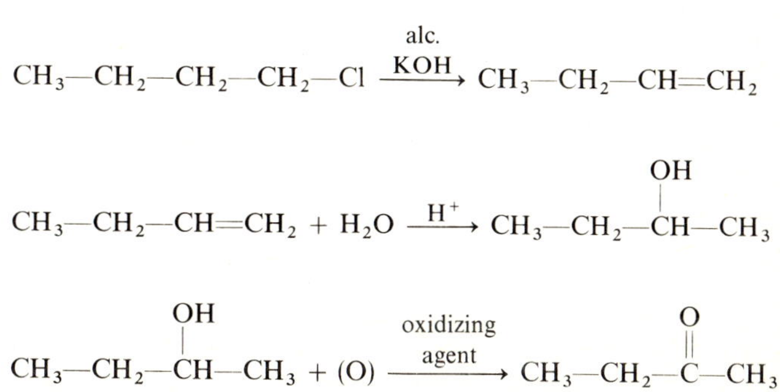

d)

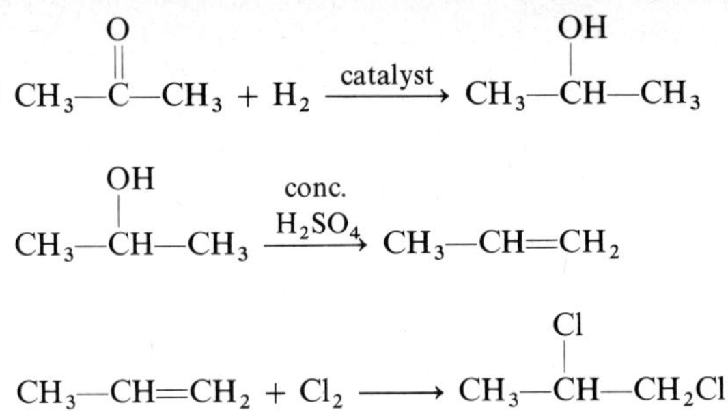

Problem 11. *a)* From left to right they are galactose, glucose, and fructose.

b) α-1,6

c) Break the bond between galactose and glucose. The disaccharide fragment is sucrose.

Problem 13. Chondroitin sulfate and heparin differ in the location of the sulfate ester linkage to carbohydrate alcohol groups. Also, in heparin the sugar amine is bonded to sulfuric acid, while in chondroitin sulfate it is attached to acetic acid. Notice that the starred positions of the galactosamine in chondroitin sulfate are at carbons 4 and 6, these being the numbers used to differentiate the two forms of the polysaccharide.

Chapter 7

Exercise 7.1. *a)* ethanoic acid (commonly called acetic acid)

b) 3,3-difluoropentanoic acid *c)* *m*-chlorobenzoic acid

d) 3-methylbutanoic acid *e)* 3,5-dinitrobenzoic acid

f) trichloroethanoic acid (trichloroacetic acid)

Exercise 7.2. *a)* sodium propanoate *b)* potassium salicylate

Exercise 7.3. 2.1; 7.0

Exercise 7.4. Ten times as great

Exercise 7.5. b

Exercise 7.6. *a)* ethyl ethanoate (ethyl acetate)

b) isopropyl butanoate (isopropyl butyrate)

c) methyl propanoate (methyl propionate)

d) ethyl benzoate *e)* methyl salicylate

***f*)** butyl ethanoate (butyl acetate)

Exercise 7.7.

***a*)** $CH_3-CH_2-C(=O)-Cl$ ***b*)** $CH_3-CH_2-C(=O)-O-CH_3$

***c*)** $CH_3-CH_2-C(=O)-NH_2$ ***d*)** $CH_3-CH_2-C(=O)-O-CH(CH_3)-CH_3$

***e*)** 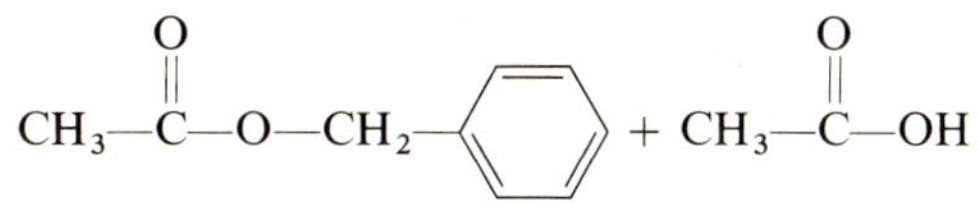

***f*)** $CH_3-C(=O)-OH + CH_3-CH_2-OH$

***g*)** $CH_3-CH_2-C(=O)-NH-CH_2-CH_3$

***h*)** $CH_3-CH_2-C(=O)-O^{-}-Na^{+} + CH_3-CH_2-OH$

Exercise 7.8. No

Problem 1.

***a*)** methanoic acid ***b*)** 2-methylpropanoic acid

***c*)** 3,3-dichlorobutanoic

***d*)** trifluoroethanoic acid (trifluoroacetic acid)

***e*)** *o*-nitrobenzoic acid ***f*)** 2-phenylpropanoic acid

***g*)** anthranilic acid ***h*)** salicylic acid

***i*)** phthalic acid ***j*)** sodium pentanoate

***k*)** sodium *p*-chorobenzoate ***l*)** benzamide

***m*)** propanoic anhydride

***n*)** isopropyl ethanoate (isopropyl acetate)

Problem 3.

***a*)** 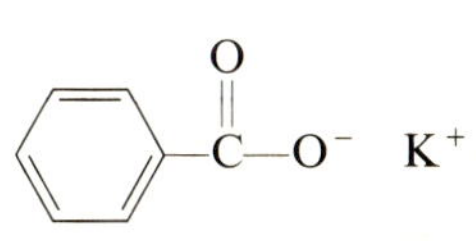

***b*)** $CH_3-CH_2-C(=O)-O-CH_2-CH_3$

c) 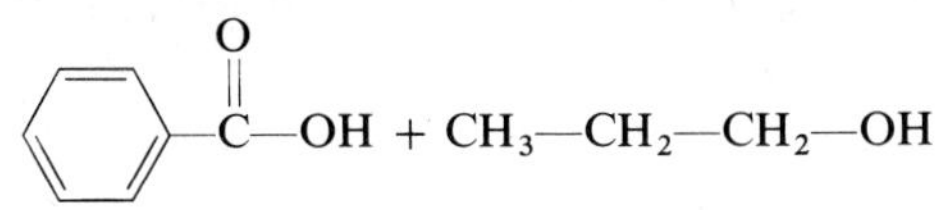

d) $CH_3—CH_2—C(=O)—NH_2$

e) 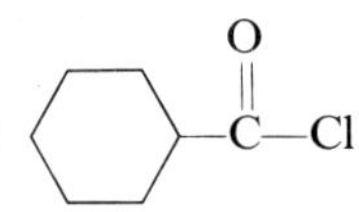

f) $CH_3—C(=O)—O—CH(CH_3)—CH_3$

g) $CH_3—CH_2—C(=O)—O—CH_3 + CH_3—CH_2—C(=O)—OH$

h) 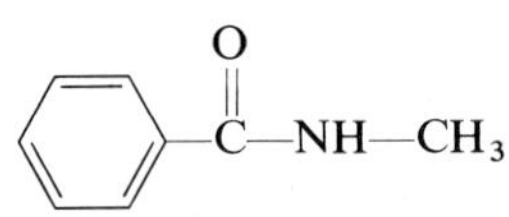

i) $\text{—(}C(=O)—CH_2—CH_2—C(=O)—O—CH_2—CH_2—O\text{)}_n\text{—}$

j) $H—C(=O)—O^-Na^+ + CH_3—CH_2—OH$

k) 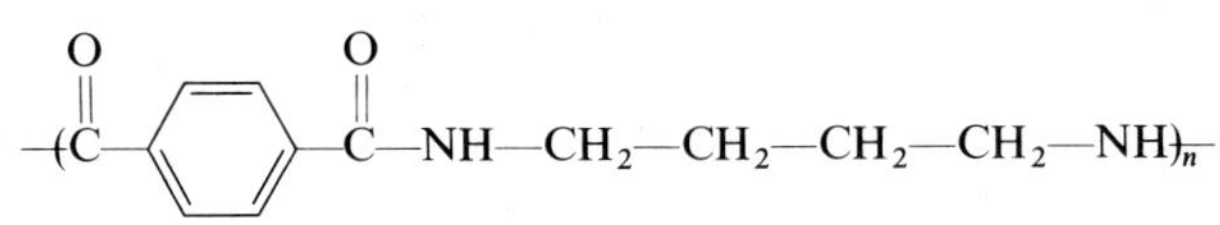

Problem 5. a., K; b., I; c., G; d., D; e., C; f., H; g., A; h., E; i., B; j., J; k., F

Problem 7. Use the acid chloride instead of the acid as shown.

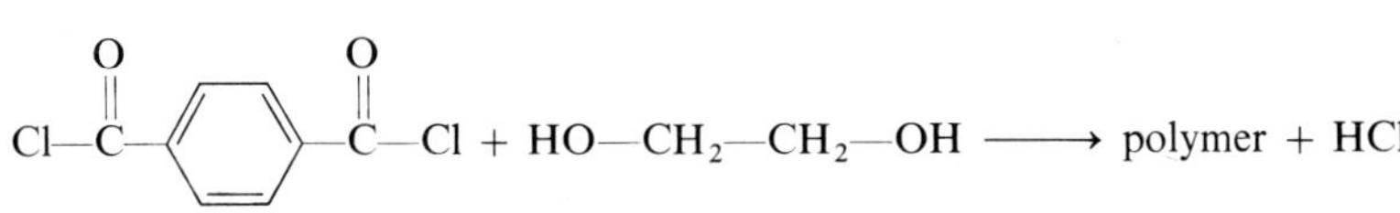

The hydrogen chloride, which is produced instead of water, is a gas and leaves the polymer preparation rapidly. There is no product comparable to the H_2O or HCl in the addition polymerization reactions.

Problem 9. lowest melting: d, b, c, a

Problem 11. Use either the acid chloride or the acid anhydride form of the desired carboxylic acid. The reaction will then proceed rapidly at a much lower temperature.

Problem 13. Most soluble: a, c, b

Problem 15. ***a***) 1. 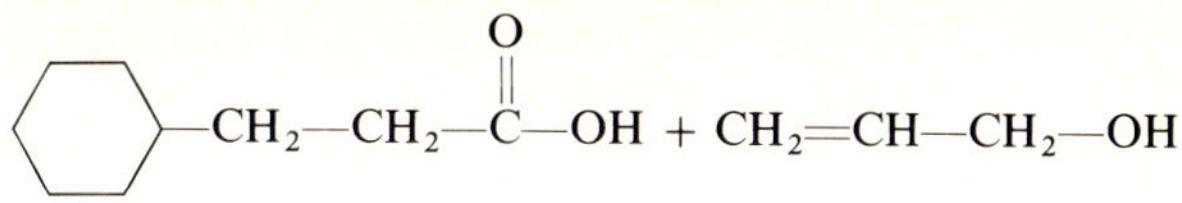

2. 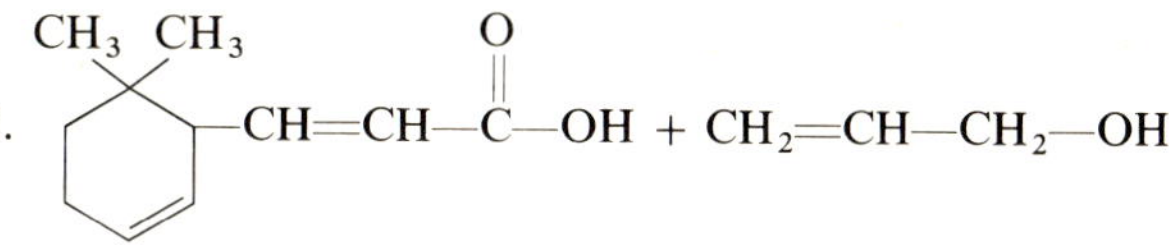

3. $CH_3—CH_2—CH_2—\overset{O}{\overset{\|}{C}}—OH + CH_3—CH_2—CH_2—CH_2—OH$

4. 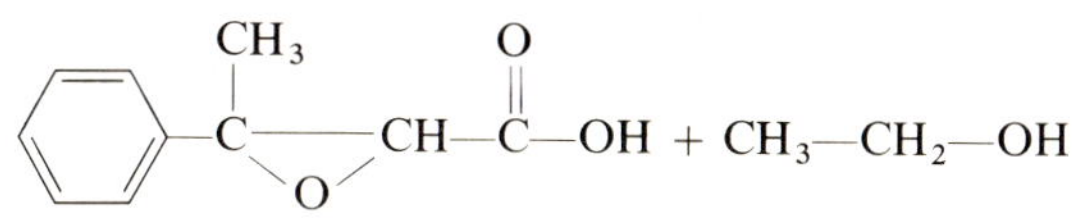

b) The ester is internal; that is, the carboxylic acid has reacted to form an ester with an alcohol on the same molecule. The positions on the carbon chain of the carboxylic acid are designated as follows:

$$\overset{\varepsilon}{C}—\overset{\delta}{C}—\overset{\gamma}{C}—\overset{\beta}{C}—\overset{\alpha}{C}—\overset{O}{\overset{\|}{C}}—OH$$

In a Δ (δ or delta) lactone the alcohol group with which the carboxylic acid reacted was attached to the δ carbon, whereas in the γ lactone it was attached to the γ carbon. An ε lactone would be follows:

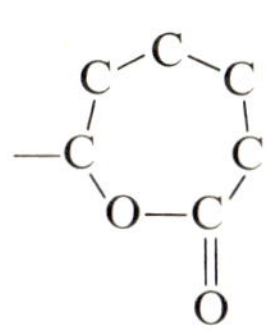

c) The three-membered ring is a cyclic ether.

d) 1. 3-cyclohexylpropanoic acid
3. butanoic acid
10. propanoic acid
11. 5-hydroxydecanoic acid
16. 3-methylbutanoic acid
17. 4-methylpentanoic acid
21. 5-methyl-4-hexenoic acid
24. *p*-methoxybenzoic acid
27. hexanoic acid
29. nonanoic acid
32. 2-methylpropanoic acid
33. phenylethanoic acid

e) The ester linkage is not polar and is incapable of forming hydrogen bonds the way carboxylic acids and alcohols do. The forces between molecules in solution are therefore smaller, less energy is required to separate them, and as a result the boiling point is lower. Because esters are lower boiling (more volatile), they are present in relatively high concentration in the atmosphere near a liquid sample. Detection by smelling requires this volatility, and smell makes an important contribution to flavor.

Chapter 8

Exercise 8.1.

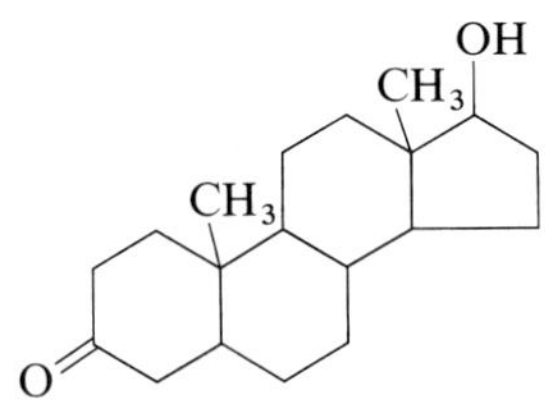

Problem 2. 30; 6; tri-

Problem 4. ***a)*** Compare your answer with the pattern on page 201.

b)

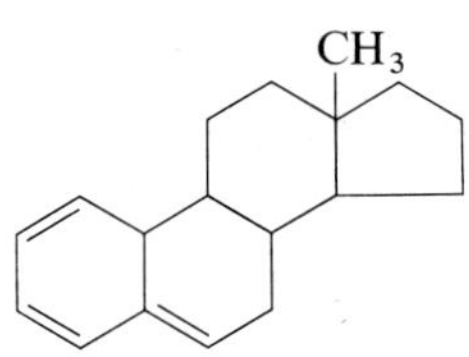

estra-1,3,5-triene

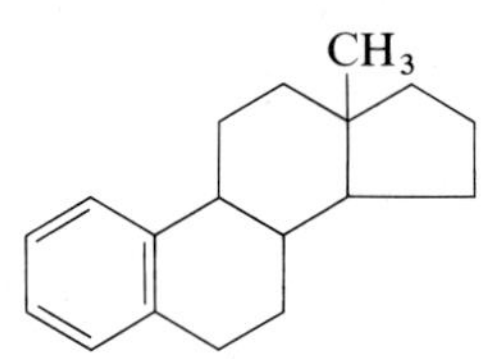

estra-1,3,5(10)-triene

In both cases the bonds are conjugated, but the structure on the right is aromatic.

c) A group attached to an aromatic ring is in the same plane as the ring, rather than being above or below the plane of the ring. There is therefore only one orientation in space. The designations α and β were used in the same sense when describing the orientation of the carbohydrate hemiacetal or acetal groups.

d) B and C

e)

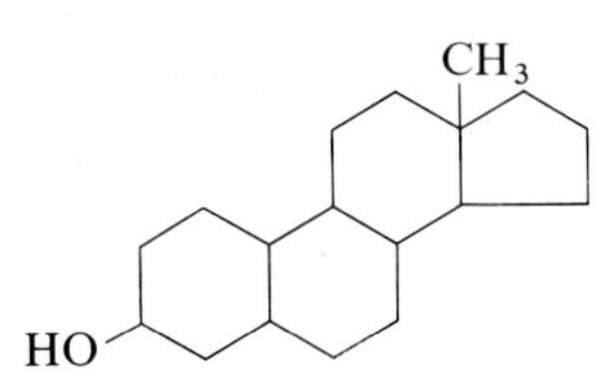

f) A. 17β-hydroxyestrane
B. 14α-hydroxyestrane
C. $3\alpha,7\alpha$-dihydroxyestrane
D. 3-hydroxyestra-1,3,5(10)-triene

g) A.

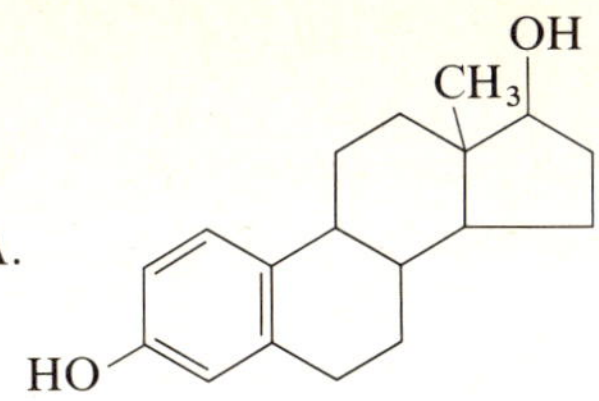

estradiol

B.

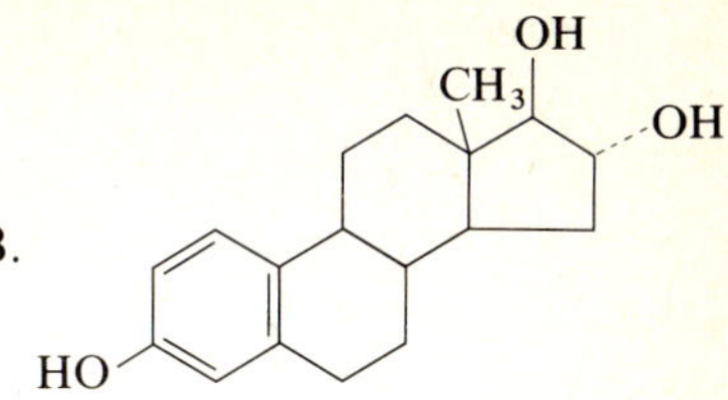

estriol

C.

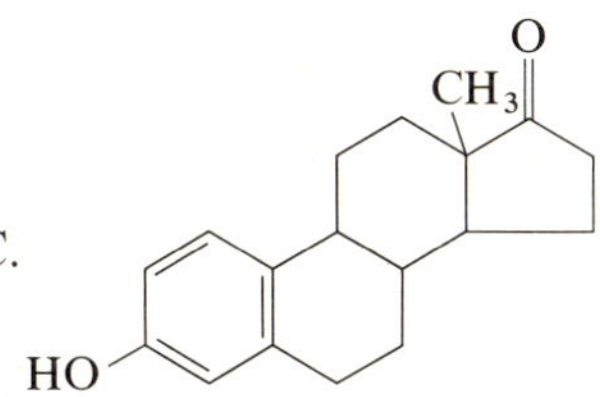

estrone

Problem 6. *a*) androstane

b)–*e*)

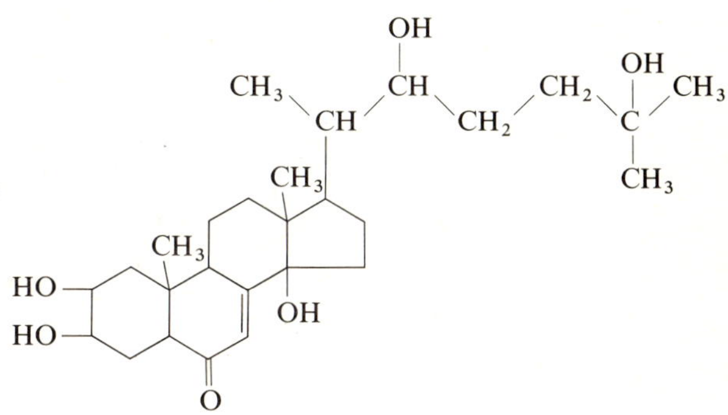

ecdysone

Chapter 9

Exercise 9.1.

a) ethyl amine

b) isopropyl amine

c) diethyl amine

d) diphenyl amine

e) diethyl methyl amine

f) *m*-chloroaniline

g) butyl methyl amine

h) ethyl dimethyl amine

i) anthranylic acid or *o*-aminobenzoic acid

j) 2-aminobutanoic acid

k) *m*-aminobenzaldehyde

Exercise 9.2.

a)
$$CH_3-CH_2-\overset{\overset{\displaystyle O}{\|}}{C}-CH_2-CH_3 + CH_3-CH_2-\overset{\overset{\displaystyle NH_2}{\|}}{CH}-CH_2-CH_3 \longrightarrow$$

$$\begin{array}{c} CH_3-CH_2-CH-CH_2-CH_3 \\ | \\ N \\ \| \\ CH_3-CH_2-C-CH_2-CH_3 \end{array}$$

b) 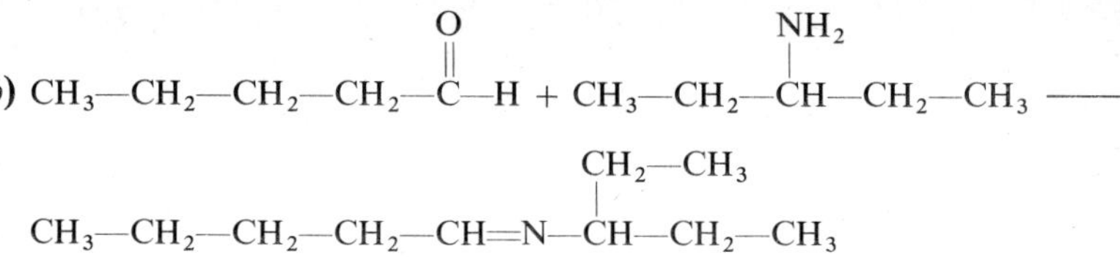

Exercise 9.3 no; 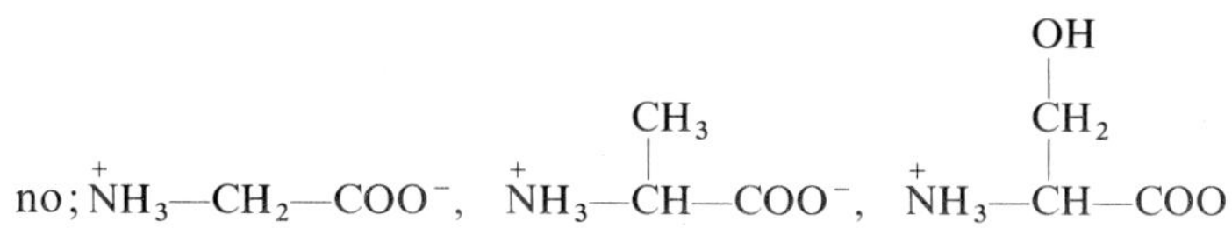

Exercise 9.4. 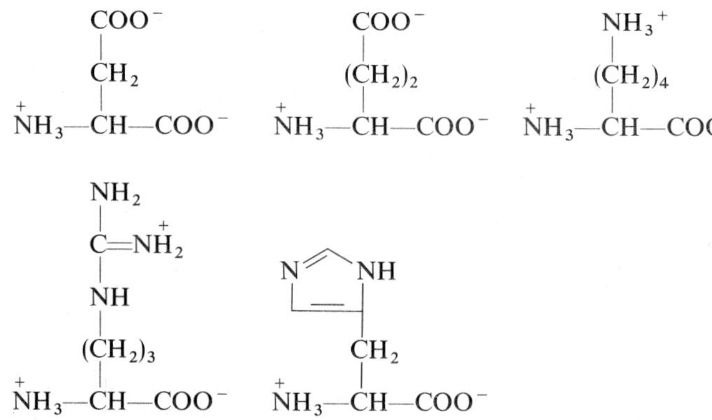

Exercise 9.5. *a*) 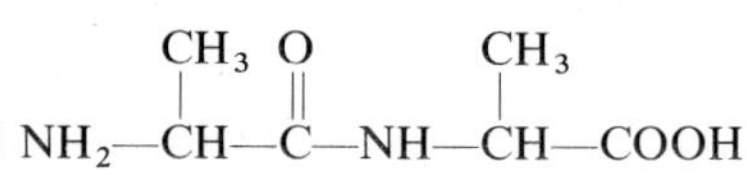

b) Two: alanyl glycine and glycyl alanine

Exercise 9.6. In hemoglobin Koln the normal and replacement amino acids both have uncharged R groups. Electrophoresis would not be likely to separate these two proteins. In all others a change in charge takes place between normal and mutant sequences.

Exercise 9.7. *a*) butyl ethyl dimethyl ammonium bromide

b) dicyclohexyl isopropyl methyl ammonium hydroxide

Problem 2. *a*) $CH_3—CH_2—NH—CH_2—CH_3$ *b*) 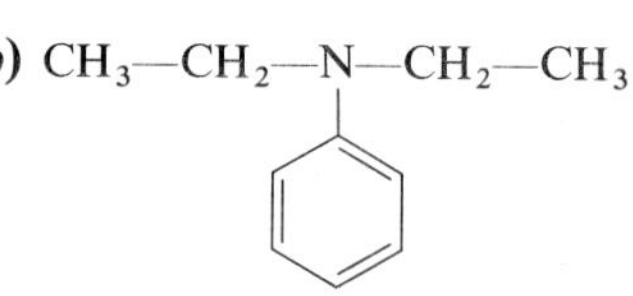

c) 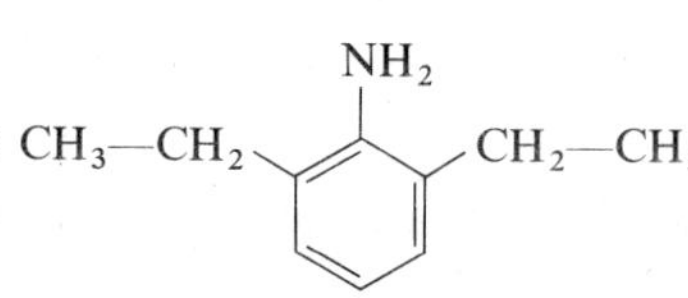

d) 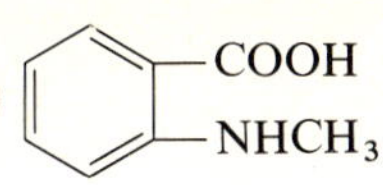

e) 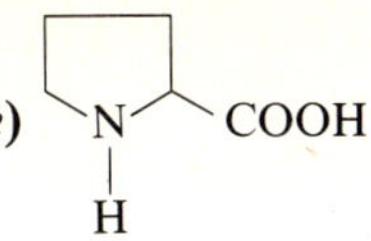

f) $(CH_3—CH_2)_4N^+Br^-$

Problem 4. *a)* 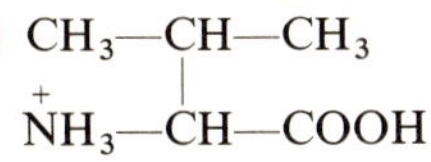

pH 1.0

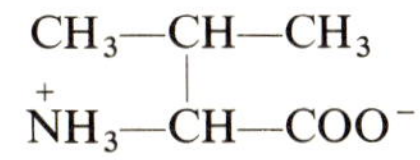

pH 6.0

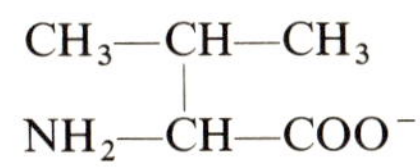

pH 11.0

b) 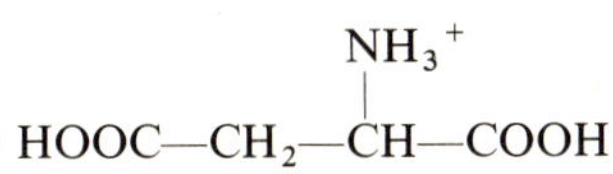

pH 1.0

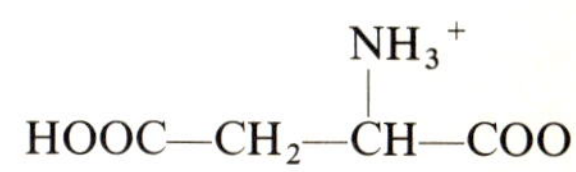

pH 3.0

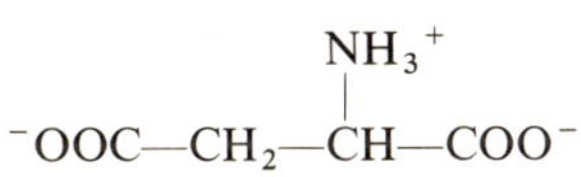

pH 6.0

At pH 7.0 aspartic acid would migrate toward the positive pole in an electrophoresis apparatus.

c) 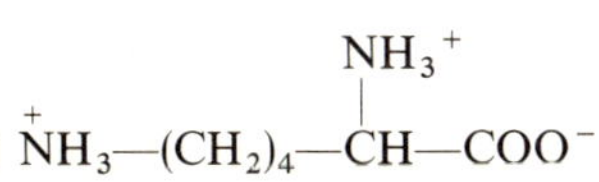

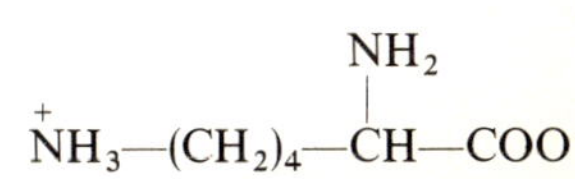

At pH 9.0 a solution would contain about 50 percent of each of these forms.

d) 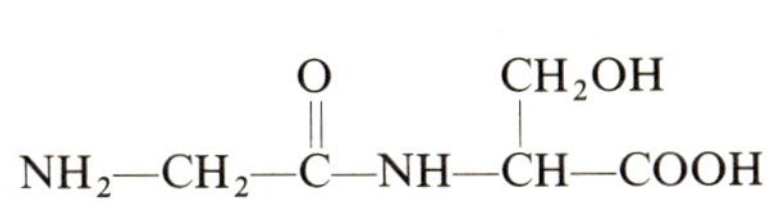

e) 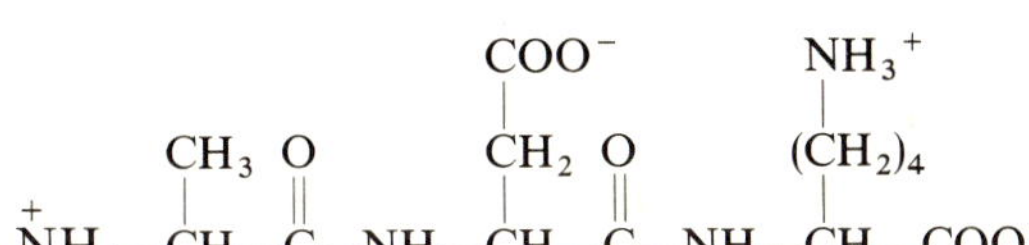

In an electrophoresis apparatus at pH 7.0 it would not move very rapidly.

Problem 6. ***A***) lysozyme ***B***) hemoglobin ***C***) pepsin

Problem 8. ***a***) proline ***b***) isoleucine, threonine

c) tyrosine; *p* ***d***) asparagine, glutamine

e) methionine

f) phenylalanine, tryptophan, serine, aspartic acid, asparagine, cysteine, tyrosine, histidine

g) glycine

Problem 10. It would seem very unlikely that a single amino acid change in a single protein would accomplish "monster synthesis," particularly since the most likely effect of such a modification is either to destroy the ability of the protein to function properly or to leave the protein functionally unaltered.

Chapter 10

Exercise 10.1. ***a***) purine ***b***) indole

c) pyridine ***d***) This is not a heterocycle.

Exercise 10.2. ***a***) lactone ***b***) ether

c) lactone ***d***) ether

e) anhydride

Exercise 10.3. ***a***) amine ***b***) amine

c) amide ***d***) amine

Exercise 10.4. 4-formyl-3-hydroxy-5-hydroxymethyl-2-methylpyridine

Exercise 10.5. ***a***)

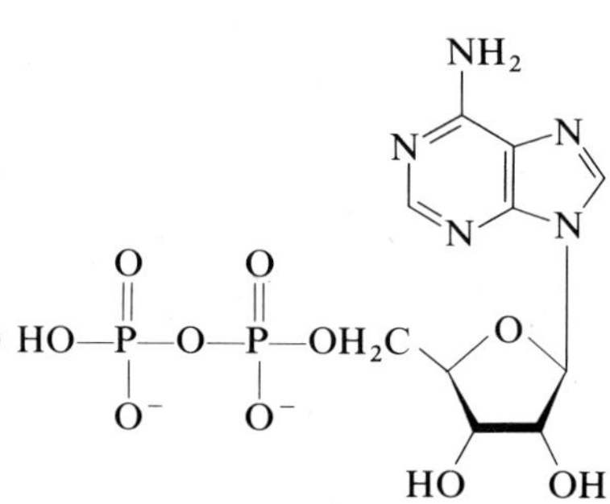

b)

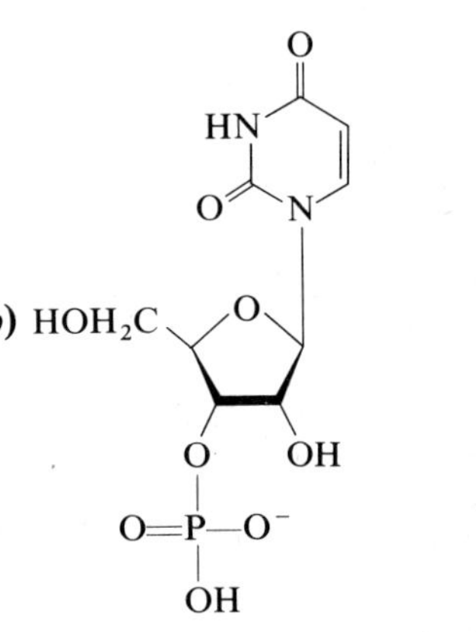

c) 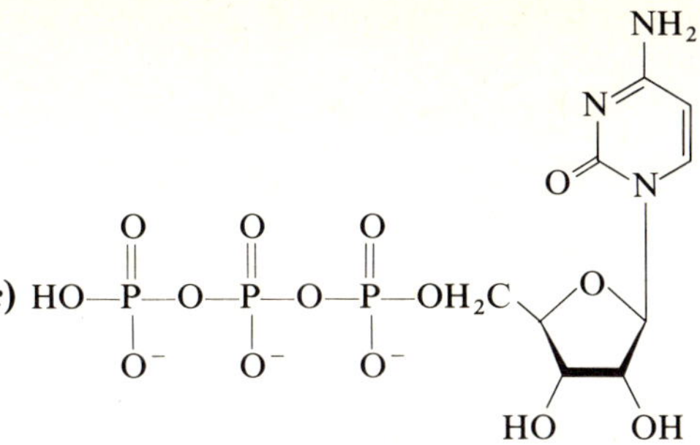

d) 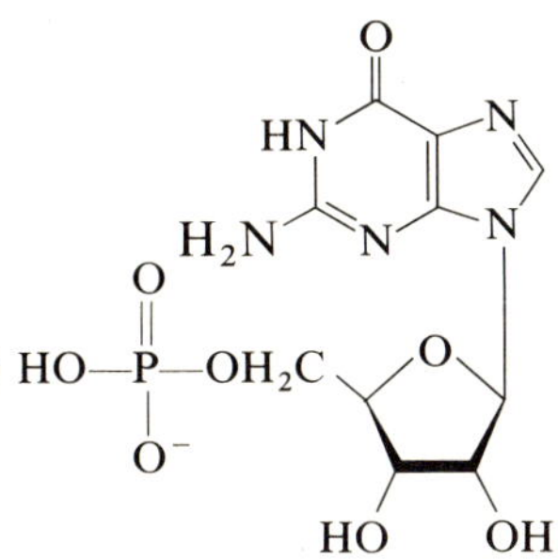

Problem 2. *a*) pyran, 3; pyridine, 4; piperadine, 4; pyrazine, 2; pyrimidine, 4; piperazine, 2; furan, 2; tetrahydrofuran, 2; pyrrole, 3; pyrrolidine, 3; imidazole, 3; quinoline, 8; indole, 7; purine, 7

Problem 4. *a*) 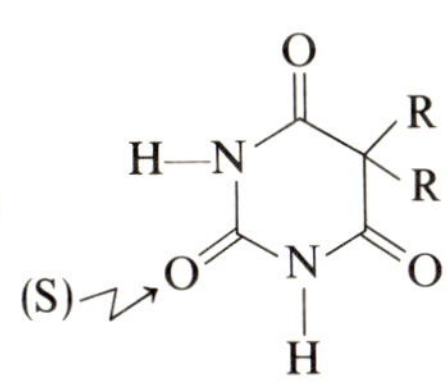

b) 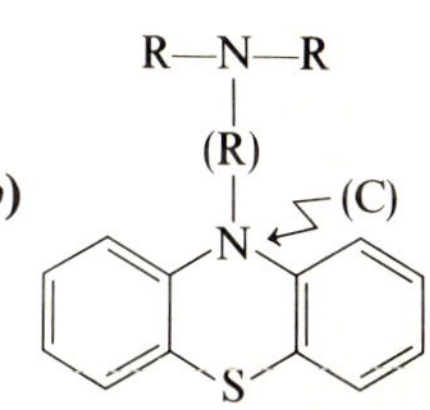

c) The molecules have similar shapes and locate functional groups and hydrophobic regions in the same place. Such similar structures all can attach to the same binding site, for example to an enzyme-active site. This binding may inhibit or stimulate the activity of the molecule to which they bind.

Chapter 11

Problem 1.

a) E	*b*) A	*c*) D	*d*) A
e) C	*f*) B	*g*) A	*h*) B, C, D

Problem 3. *a*) B *b*) C *c*) D *d*) A

e) C *f*) B *g*) A

Problem 5. *a*) β structure *b*) collagen structure

c) α helix

Problem 7. *a*) pH 7.2, 25%; pH 7.4, 32%; pH 7.6, 45%

b) $H_2O + CO_2 \rightleftharpoons H_2CO_3$. Blood is more acidic in the distant capillary.

c) At a low pH, blood is less saturated with oxygen and hence is releasing it more readily. In the tissues where the oxygen is needed, the oxygen is more readily released.

Chapter 12

Problem 1. *b*) The substrate must be bounded to the enzyme strongly enough and therefore long enough to allow time for conversion of the substrate to product. However, if substrate were held very firmly, the active site would clear very slowly for the next conversion of substrate to product and the enzyme would be a less efficient catalyst.

Problem 3. *a*) A, B, C *b*) B, C *c*) B, (D)

d) B, D *e*) A, B, D, E *f*) A, B, C

g) B, D, F *h*) A, B, C *i*) B, D, F

Problem 5. Because blood clots by a cascade mechanism, a small amount of triggering material produces a large amount of clot formation in a brief time interval.

Problem 7. Dicumarol and warfarin and effective against unwanted clot formation because they resemble but do not function as vitamin K. Vitamin K does not have a direct role in clot formation. Instead it is needed for the formation of structures that are involved in clot formation. In freshly drawn blood these structures are already present.

Problem 9. *a*) The rate of reaction is higher than the base (no addition) reaction for any given substrate concentration when ATP is added; therefore it is a positive effector. The rates for addition of CTP are lower than the base rate; hence CTP is a negative effector.

b) A product should slow its own rate of formation in a feedback inhibition pattern, and CTP inhibits the rate of catalysis by the enzyme. Therefore it is logical to suspect that the pathway is the first step in CTP synthesis.

Problem 11. ***a)***

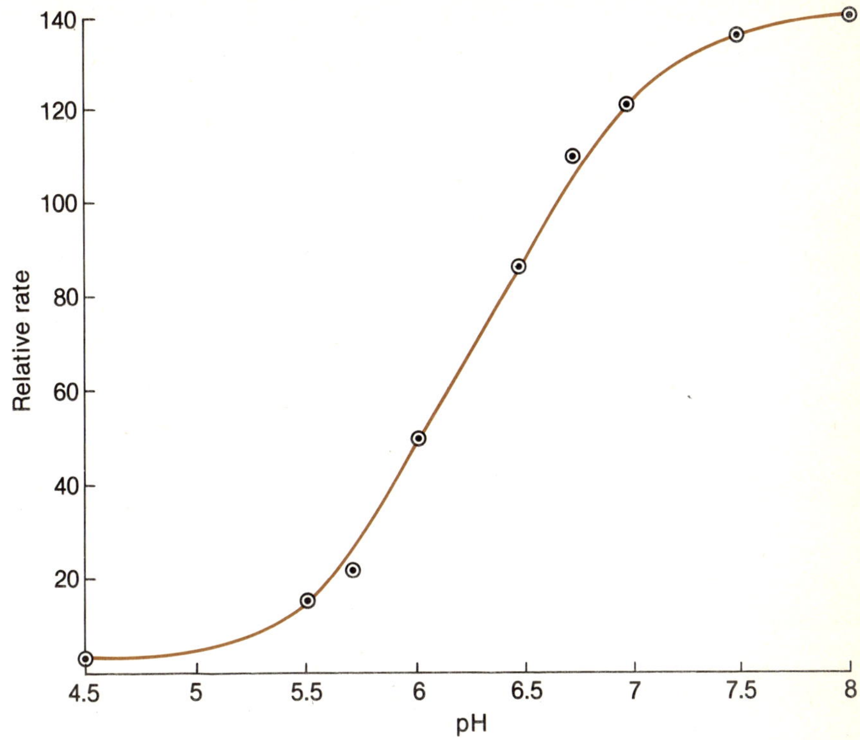

b) pK = pH 6.3

c) histidine (pK = 5.5 – 7)

d) trypsin: pH optimum 8
pepsin: pH optimum 2
trypsin: pK's = 6.5 and 9.2, indicating histidine and either lysine or the N-terminal amine pepsin: pK = 3, indicating aspartate, glutamate, or a C-terminal carboxyl

Problem 13. ***a)*** (N-terminal) serine, lysine, phenylalanine, alanine, tyrosine, arginine, valine, methionine (C-terminal)

b) seryl lysine, phenylalanyl alanyl tyrosyl arginine, and valyl methionine

c) seryl lysyl phenylalanine, alanyl tyrosine, and arginyl valyl methionine

d) Trypsin hydrolyzes any peptide bond that includes the carboxyl group of either lysine or arginine. As a result, all fragments obtained by hydrolysis with trypsin except the C-terminal fragment should have either lysine or arginine as the C-terminal amino acid. This fragment logically would have to be glycyl lysine.

e) One peptide has a C-terminal lysine, the second a C-terminal arginine, and the third a C-terminal methionine. The first two, therefore, ended at places where the trypsin hydrolyzed a peptide bond, and the third with neither lysine nor arginine is the C-terminal portion of the original structure.

f) Chymotrypsin hydrolyzes the peptide bond whose carboxyl group belongs to tryptophan, phenylalanine, and tyrosine (Table 12.1). Therefore, arginyl valyl methionine, the only peptide fragment not ending with one of these three amino acids, must be the C-terminal fragment.

g) The extra amino acid in the longer fragment must be the N-terminal amino acid (A) because either methionine or valine must be C-terminal to the entire original peptide.

h) Because hydrolysis with trypsin establishes lysine as C-terminal, the peptide must be seryl lysine.

i) Because hydrolysis with chymotrypsin establishes phenylalanine as C-terminal, and because the internal dipeptide is established (h) to be seryl lysine, the complete tripeptide must be seryl lysyl phenylalanine.

j) Tyrosine must be C-terminal, therefore the sequence must be alanyl tyrosine.

k) Because trypsin was used for hydrolyzing the original peptide, arginine must be C-terminal (position D).

Of the eight amino acids in the original peptide, we know that alanine and methionine are beyond the C-terminal and of this sequence. The other two amino acids must be serine and lysine, which are part of the tripeptide seryl lysyl phenylalanine whose sequence was solved in i. The phenylalanine must therefore be N-terminal in this tetrapeptide (position A).

Alanine and tyrosine must be positions B and C. Because in j the dipeptide was determined to be alanyl tyrosine, then position B must be alanine and position C must be tyrosine.

l) methionine

m) In the original peptide, serine would have been labeled with DNFB.

In the trypsin fragments, we would label serine in seryl lysine, phenylalanine in phenylalanyl alanyltyrosyl arginine, and valine in valyl methionine using DNFB.

In the chymotrypsin fragments, we would label serine, alanine, and arginine in the three fragments with DNFB.

Problem 15. ***a)*** Four isozymes: AAA, AAB, ABB, BBB

b) Four subunits: AAAA, AAAB, AABB, ABBB, BBBB

c) Proteins that are similar but separate well by electrophoresis must have at least one amino acid difference that involves a change in charge.

d) The detection method for locating the bands of lactate dehydrogenase on the electrophoresis strip utilizes the reaction catalyzed by the enzyme. Only that one enzyme is detected as a result.

Chapter 13

Exercise 13.1. C pairing with G and G pairing with C

Exercise 13.2. *a*) ribosomal RNA *b*) transfer RNA *c*) messenger RNA

Exercise 13.3. *a*) GCU, UCU, AAC, UUU, ACU; codons

b) alanine, serine, asparagine, phenylalanine, threonine

Problem 2. *a*) ATP: amino acid activation

b) GTP: binding amino acyl-tRNA to codon on mRNA

c) GTP: shifting the ribosome one codon along the mRNA

Problem 4. *a*) Decreases: Imagine pouring wire through a funnel first as little straight rods, second with the rods rolled into little balls.

b) Harder to unwind: Unwinding requires breaking the base-to-base hydrogen bonds; G and C have three such bonds each, and A and T have only two.

Chapter 14

Exercise 14.1. One mole of glucose releases 686 kcal:

$$\frac{10\text{ g}}{180\text{ g/mole}} \times \frac{686\text{ kcal}}{\text{mole}} = 38.1\text{ kcal}$$

Exercise 14.2.

$$\frac{10\text{ g}}{256\text{ g/mole}} \times \frac{2340\text{ kcal}}{\text{mole}} = 91.4\text{ kcal}$$

Exercise 14.3. 10 g of fat 91.4/38.1, or 2.4, times as fattening as 10 g of carbohydrate.

Problem 2. *a*) Forming a bond between tRNA and the amino acid carboxyl group.

b) Joining the two halves of the ribosome together, binding the amino acyl tRNA to mRNA, and moving the ribosome along the mRNA.

Chapter 15

Problem 1.

a) Substrate: glucose-6-phosphate
Product: fructose-6-phosphate
Enzyme: glucose phosphate isomerase

b) Substrate: 3-phosphoglycerate
Product: 2-phosphoglycerate
Enzyme: phosphoglyceromutase

c) Substrate: glucose-1-phosphate
Product: glucose-6-phosphate
Enzyme: phosphoglucomutase

Problem 3. ***c)***

Problem 5. ***d)***

Problem 7.

a) Substrate: lactate
Product: pyruvate
Enzyme: lactate dehydrogenase
Cofactor: NAD^+

b) Substrate: ethanol
Product: ethanal
Enzyme: alcohol dehydrogenase
Cofactor: NAD^+

c) Substrate: glyceraldehyde-3-phosphate
Product: 1,3-diphosphoglycerate
Enzyme: glyceraldehyde phosphate dehydrogenase
Cofactor: NAD^+

Problem 9.

a) In 4 the cofactor ATP was required in each case. In 7c no ATP was involved.

b) In 4 the high-energy phosphate-to-phosphate bond was broken, providing the necessary energy to form the new bond between substrate and phosphate. In 7c energy was supplied by the oxidation of the aldehyde group to a carboxylate group.

Problem 11.

a) The glycogen phosphorylase would be inactive—in the phosphorylase b form.

b) The squirrel would release adrenalin, which would trigger the conversion of inactive phosphorylase b into the active phosphorylase a.

c) A quantity of the muscle ATP would have become AMP. The inactive phosphorylase b is allosterically activated by AMP, becoming active phosphorylase b.

Problem 13. 1. b, e 2. −9000 cal/mole

Chapter 16

Exercise 16.1. It is hydrolyzed in the intestine, reassembled after crossing the mucosa; hydrolyzed at surface of fat cell, reassembled inside the cell; hydrolyzed inside fat cell, transported to tissues for metabolism.

Problem 2.

a)
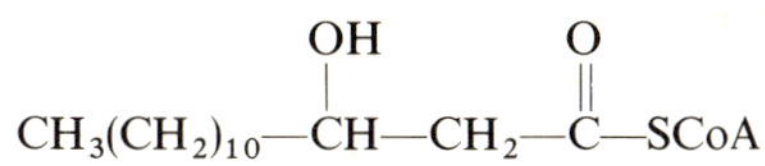

b)
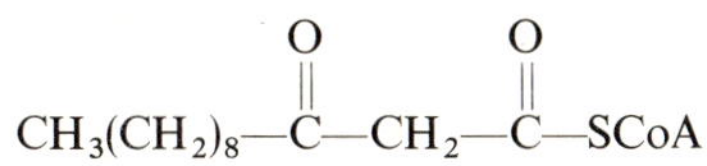

c)
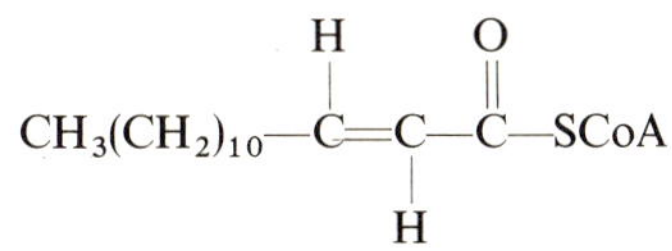

d)
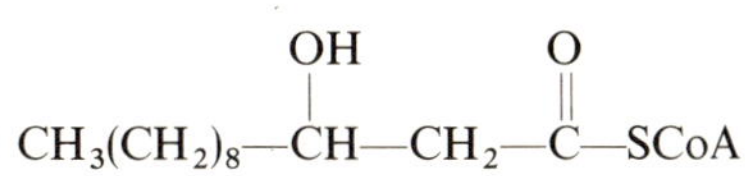

L isomer

e)
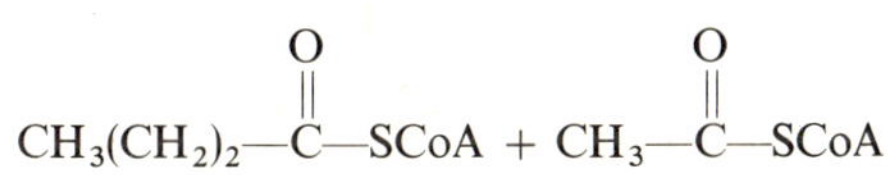

f)
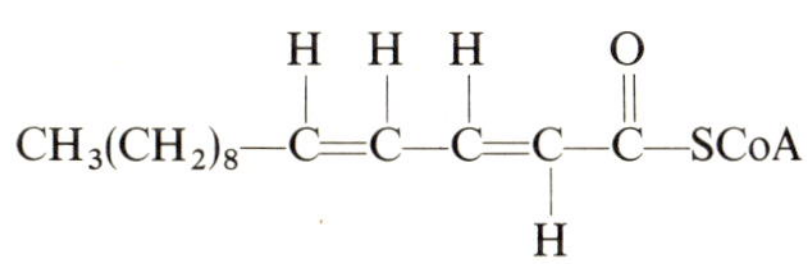

g)
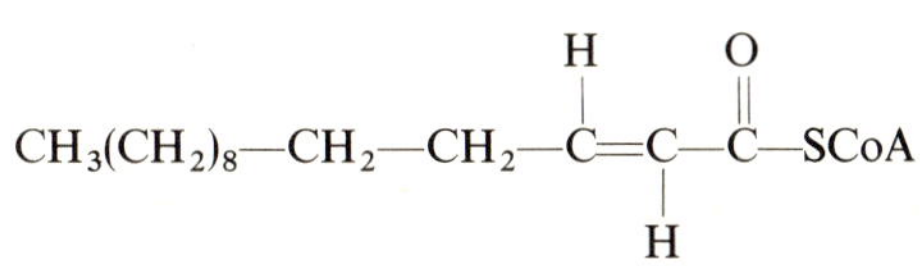

h)
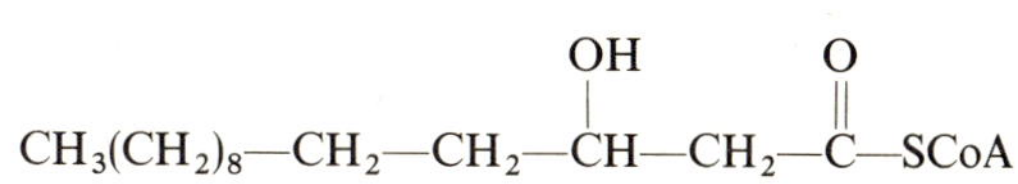

D isomer

i)
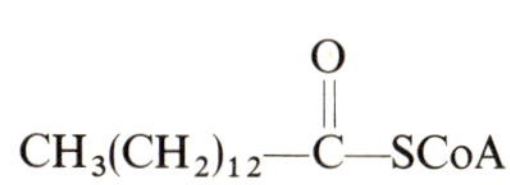

Problem 4. In diabetes mellitus an insufficiency of insulin reduces the rate of transfer of glucose into the tissues. Metabolic needs are then met by metabolizing fatty acids in place of glucose. The high rate of fatty acid metabolism results in ketone bodies entering the blood.

Problem 6. The amino acid is cysteine:

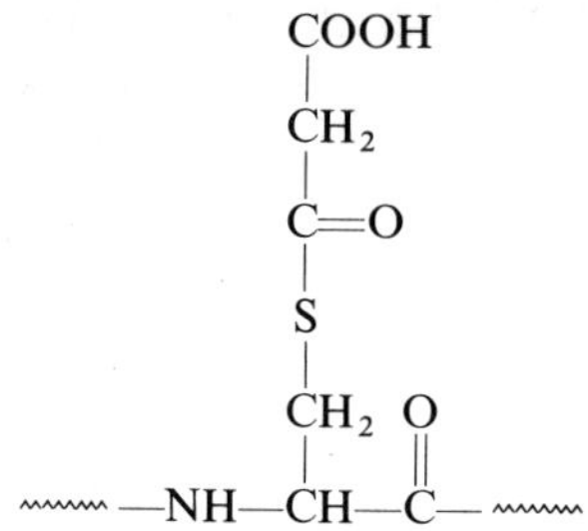

Problem 8. ***a)***

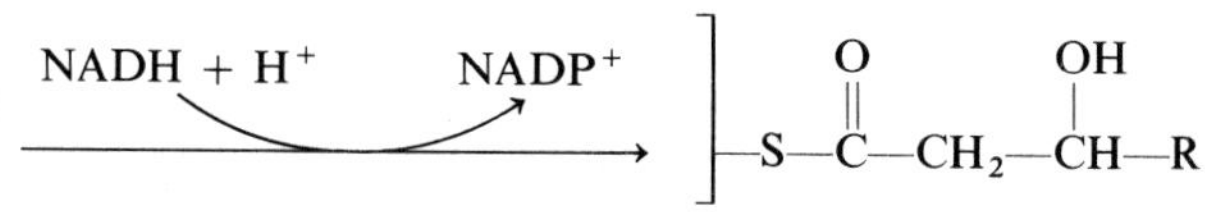

b)

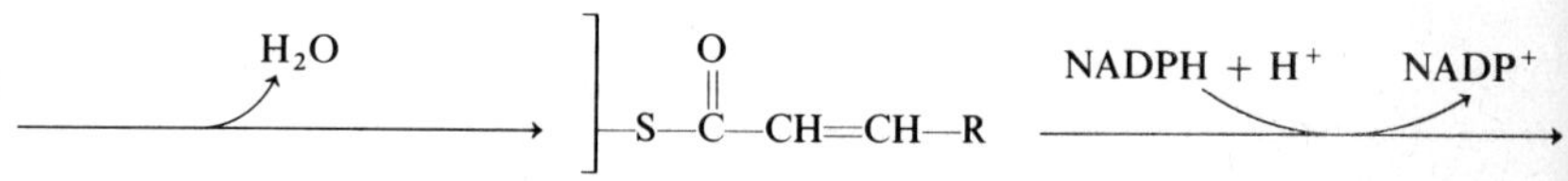

Chapter 17

Exercise 17.1. No

Problem 2. Stearic acid: $CH_3(CH_2)_{16}COOH$

First convert to CoA thioester:	−ATP
Then eight turns of fatty acid oxidation:	+8 NADH
	+8 $FADH_2$
Finally nine turns of the Krebs cycle:	+9 GTP
	+9 $FADH_2$
	+27 NADH
	+18 CO_2

Total: 18 CO_2, 35 NADH, 17 $FADH_2$, 8 GTP

Problem 4. Glucose:

4 ATP	=	4
2 $FADH_2$	=	4
10 NADH	=	30
		38

Hexanoic acid:

2 ATP	=	2
5 $FADH_2$	=	10
11 NADH	=	33
		45

Problem 6. ***a)***

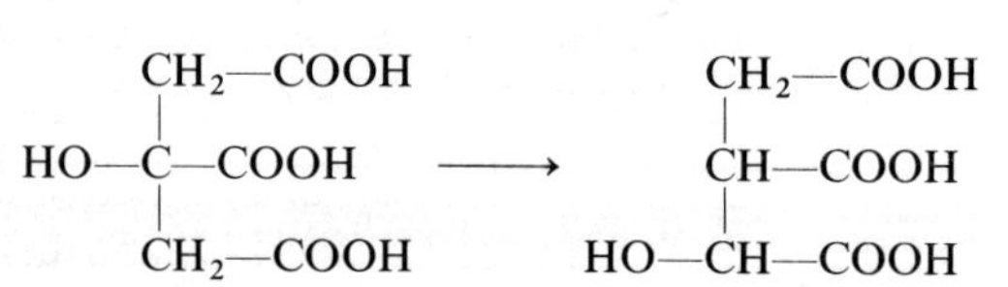

b) 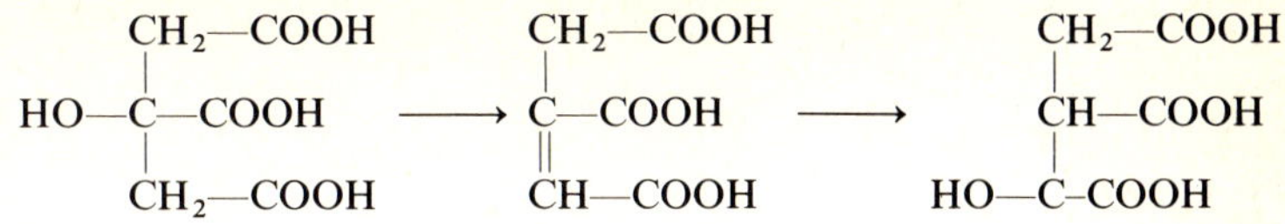

Problem 8. ***a***) 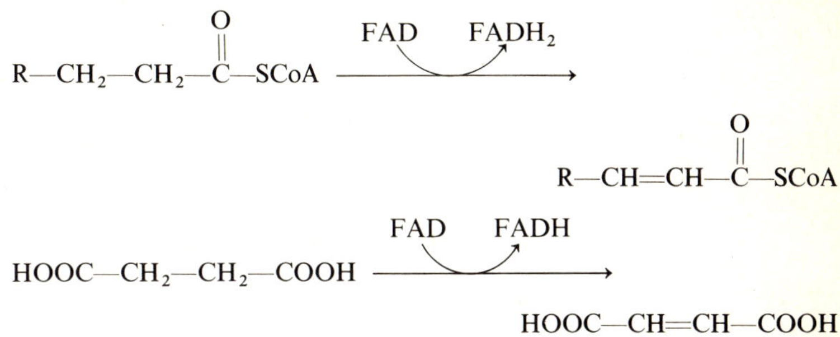

b) Flavin

Chapter 18

Problem 1. ADP + $H_2PO_4^-$ $\longrightarrow$ ATP + H_2O

Problem 3. ***a–d***) 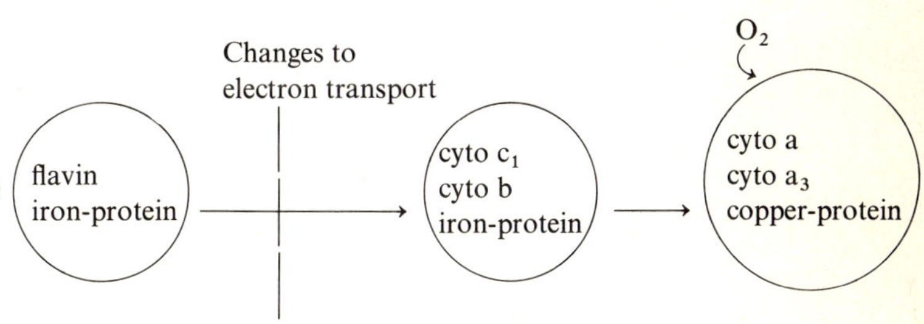

e) Complex 1 → complex 2, coenzyme Q
Complex 2 → complex 3, cyto c

f) NAD interacts with complex 1; flavin interacts with coenzyme Q

Problem 5. ***a***) Three ATP/ethanol ***b***) Three ATP/citrate

c) Five ATP/succinate ***d***) Four ATP/leucine

e) Twelve ATP more/leucine

Chapter 19

Exercise 19.1. ***a***) Substrates: leucine, α-ketoglutarate
Products: 2-keto-4-methylpentanoate, glutamate

b) Substrates: alanine, α-ketoglutarate
Products: 2-ketopropanoate (pyruvate), glutamate

Exercise 19.2. Hydrolysis of a cyclic amide

Exercise 19.3.

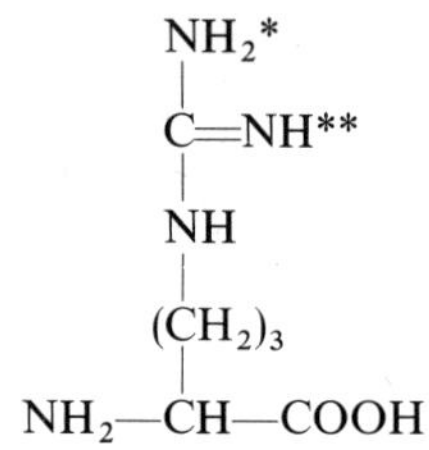

They are the two nitrogens of urea.

Problem 2.

a) proline

b) histidine

c) phenylalanine, tyrosine

d) alanine, serine

e) proline

f) histidine

g) phenylalanine, tyrosine

Problem 4. Aspartate transaminase. One of the nitrogens of every urea must be transferred from glutamate to oxaloacetate to form the aspartate that combines with citrulline.

Problem 6. ***a***) 1.

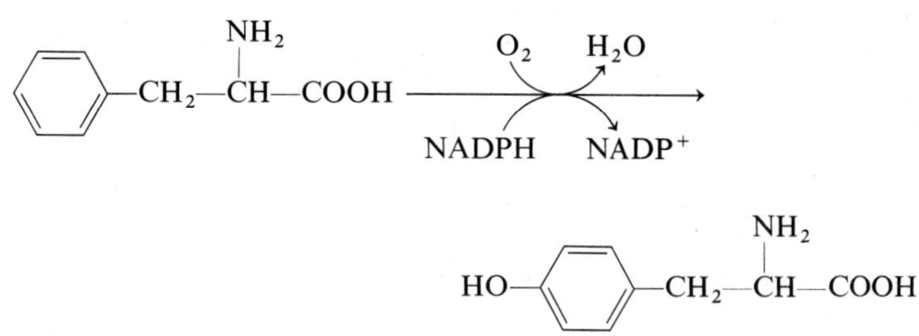

2.

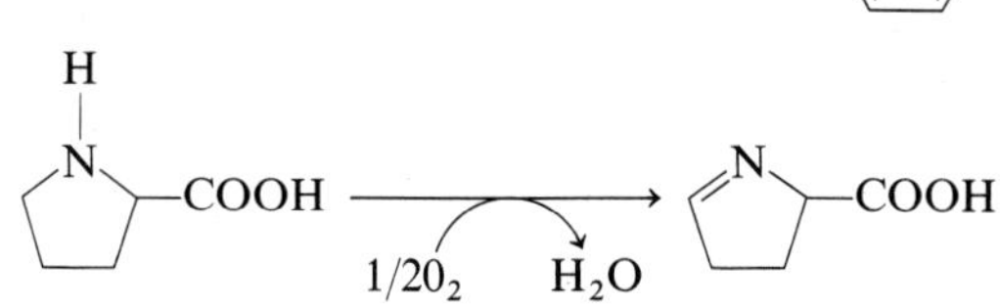

3.

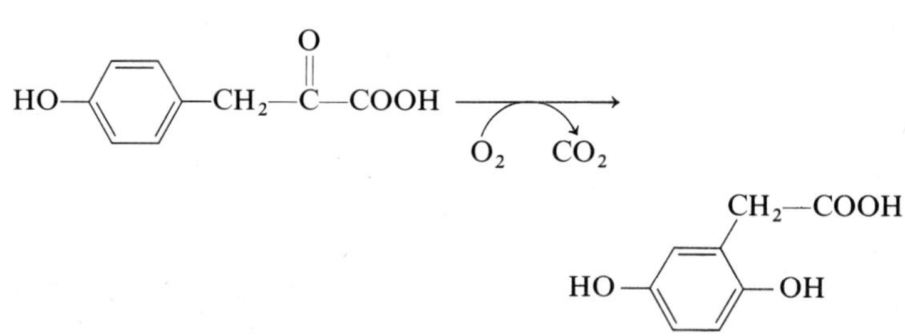

b) Oxygen is a substrate ***d***) Oxidized ***e***) NAD, flavin

f) NADPH in 1 is oxidized. Previously the oxidation of a substrate was accompanied by the reduction of such a cofactor.

Chapter 20

Exercise 20.1. ***a)*** 20.4, 20.2, 20.5

b) The acetyl methyl group becomes the terminal methyl (highest-numbered carbon) in the fatty acid. The acetyl carbonyl becomes the methylene adjacent to it.

Problem 2.

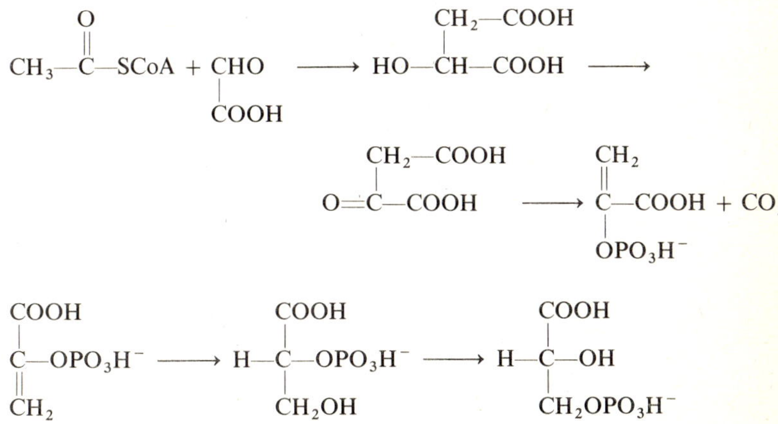

Problem 4. The hypothesis is supported, and A represents $NADP^+$.

Problem 6. Instead of oxygen, free sulfur is formed.

Problem 8.

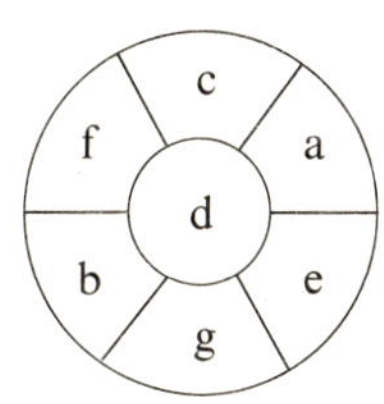

Chapter 21

Exercise 21.1. ***a)*** Active transport ***b)*** Passive transport ***c)*** Cotransport

Exercise 21.2.

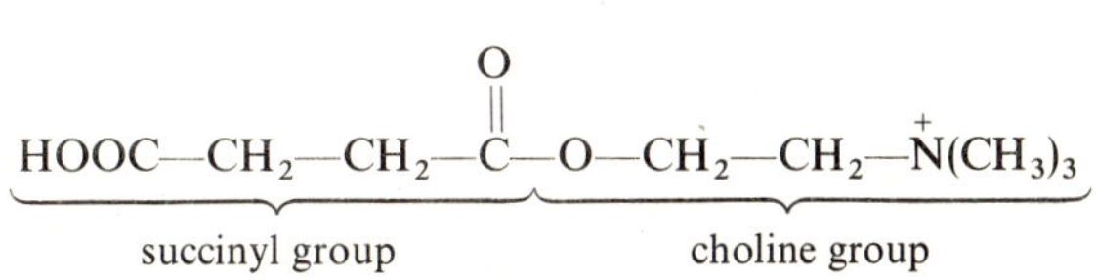

They are joined by an ester linkage.

Problem 1. ***a)*** False; it must cross the cell membrane.

b) False; there are membrane-bound compartments inside the cell.

c) True; amino acids enter the cell by active transport or cotransport.

***d*)** True; red cells take up glucose by passive transport.

***e*)** True.

***f*)** False; as long as the transport partner is available, cotransport can continue.

***g*)** True; the transport partner must be actively transported back following cotransport.

Problem 3. Once the Krebs cycle is blocked, cofactor reduction is blocked. This stops oxidative phosphorylation, the source of ATP. In the absence of ATP, active transport halts.

Chapter 22

Problem 1. Intercellular fluids, interstitial fluids, lymph, blood plasma, urine, cerebrospinal fluid, synovial fluid, fluid in the eye, water in the digestive tract.

Problem 3. When the leaf is in distilled water, water enters the relatively more concentrated intracellular fluid. When the leaf is in salt water, water leaves the cell. Cells packed tightly with water produce a firm, crisp leaf. The property is osmotic pressure.

Problem 5. The respiratory control center increases the rate of respiration, which speeds the removal of CO_2 from the blood. CO_2 in the blood reacts with water to form H_2CO_3, which ionizes to produce H_3O^+. Removing CO_2 removes a source of H_3O^+.

Problem 7. ***a*)** Red blood cells ***b*)** Platelets ***c*)** Leukocytes

Problem 9. ***a*)** Urine volume increases

***b*)** Urine volume decreases

***c*)** Urine volume increases

Chapter 23

Exercise 23.1 Tyrosine

Problem 1.

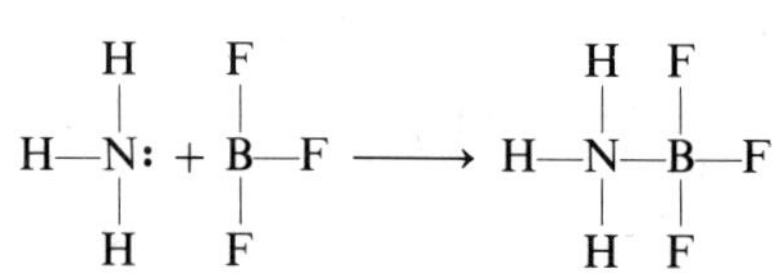

Problem 3. The waters are ligands contributing electrons to the Cu^{2+}.

Problem 5. Zinc, copper, cobalt

Index

80 81 82 83 6 5 4 3 2 1